Papermaking Science and Technology

a series of 19 books
covering the latest
technology and
future trends

Book 7

Recycled Fiber and Deinking

Series editors

Johan Gullichsen, Helsinki University of Technology
Hannu Paulapuro, Helsinki University of Technology

Book editor

Lothar Göttsching, Darmstadt University of Technology
Heikki Pakarinen, UPM-Kymmene Oy

Series reviewer

Brian Attwood, St. Anne's Paper and Paperboard Co. Ltd.

Book reviewer

Peter Seifert, Thermo Black Clawson Inc.

Published in cooperation with the Finnish Paper Engineers' Association and TAPPI

PI

TAPPI PRESS

Cover photo: Suomen Kuvapalvelu, John Lythgoe

ISBN 952-5216-00-4 (the series)
ISBN 952-5216-07-1 (book 7)

Published by Fapet Oy
(Fapet Oy, PO BOX 146, FIN-00171 HELSINKI, FINLAND)

Printed by Gummerus Printing, Jyväskylä, Finland 2000

Printed on LumiMatt 100 g/m^2, Stora Enso Fine Paper, Imatra Mills

Foreword

Johan Gullichsen and Hannu Paulapuro

PAPERMAKING SCIENCE AND TECHNOLOGY

Papermaking is a vast, multidisciplinary technology that has expanded tremendously in recent years. Significant advances have been made in all areas of papermaking, including raw materials, production technology, process control and end products. The complexity of the processes, the scale of operation and production speeds leave little room for error or malfunction. Modern papermaking would not be possible without a proper command of a great variety of technologies, in particular advanced process control and diagnostic methods. Not only has the technology progressed and new technology emerged, but our understanding of the fundamentals of unit processes, raw materials and product properties has also deepened considerably. The variations in the industry's heterogeneous raw materials, and the sophistication of pulping and papermaking processes require a profound understanding of the mechanisms involved. Paper and board products are complex in structure and contain many different components. The requirements placed on the way these products perform are wide, varied and often conflicting. Those involved in product development will continue to need a profound understanding of the chemistry and physics of both raw materials and product structures.

Paper has played a vital role in the cultural development of mankind. It still has a key role in communication and is needed in many other areas of our society. There is no doubt that it will continue to have an important place in the future. Paper must, however, maintain its competitiveness through continuous product development in order to meet the ever-increasing demands on its performance. It must also be produced economically by environment-friendly processes with the minimum use of resources. To meet these challenges, everyone working in this field must seek solutions by applying the basic sciences of engineering and economics in an integrated, multidisciplinary way.

The Finnish Paper Engineers' Association has previously published textbooks and handbooks on pulping and papermaking. The last edition appeared in the early 80's. There is now a clear need for a new series of books. It was felt that the new series should provide more comprehensive coverage of all aspects of papermaking science and technology. Also, that it should meet the need for an academic-level textbook and at the same time serve as a handbook for production and management people working in this field. The result is this series of 19 volumes, which is also available as a CD-ROM.

When the decision was made to publish the series in English, it was natural to seek the assistance of an international organization in this field. TAPPI was the obvious partner as it is very active in publishing books and other educational material on pulping and papermaking. TAPPI immediately understood the significance of the suggested new series, and readily agreed to assist. As most of the contributors to the series are Finnish, TAPPI provided North American reviewers for each volume in the series. Mr. Brian Attwood was appointed overall reviewer for the series as a whole. His input is gratefully acknowledged. We thank TAPPI and its representatives for their valuable contribution throughout the project. Thanks are also due to all TAPPI-appointed reviewers, whose work has been invaluable in finalizing the text and in maintaining a high standard throughout the series.

A project like this could never have succeeded without contributors of the very highest standard. Their motivation, enthusiasm and the ability to produce the necessary material in a reasonable time has made our work both easy and enjoyable. We have also learnt a lot in our "own field" by reading the excellent manuscripts for these books.

We also wish to thank FAPET (Finnish American Paper Engineers' Textbook), which is handling the entire project. We are especially obliged to Ms. Mari Barck, the project coordinator. Her devotion, patience and hard work have been instrumental in getting the project completed on schedule.

Finally, we wish to thank the following companies for their financial support:

A. Ahlstrom Corporation
Stora Enso Oyj
Kemira Oy
Metsä-Serla Corporation
Metso Oyj
Raisio Chemicals Ltd
Tamfelt Corporation
UPM-Kymmene Corporation

We are confident that this series of books will find its way into the hands of numerous students, paper engineers, production and mill managers and even professors. For those who prefer the use of electronic media, the CD-ROM form will provide all that is contained in the printed version. We anticipate they will soon make paper copies of most of the material.

List of Contributors

Ackermann, Christiane, D. Sc. (Tech.), Senior Research Scientist, IfP*, Paper Technology

Göttsching, Lothar, D. Sc. (Tech.) Dr. h.c., Professor, IfP*

Hamm, Udo, D. Sc. (Tech.), Research Manager, IfP*, Environmental Protection Management

Holik, Herbert, M. Sc. (Tech.), Head of Research and Development Department, Voith Sulzer Paper Technology, Ravensburg/Germany

Lassus, Anders, Lic. Tech., Development Manager Recycling Chemicals, Raisio Chemicals, Raisio/Finland

Pakarinen, Heikki, M. Sc. (Tech.), Research Manager, UPM-Kymmene Corp., Research Department (Paper Divisions), Valkeaskoski/Finland

Putz, Hans-Joachim, D. Sc. (Tech.), Research Manager, IfP*, Paper Technology

Renner, Katharina, M. Sc. (Tech.), Senior Research Scientist, IfP*, Paper Technology

Schwarz, Michael, M. Sc. (Tech.), D. Sc. (Tech.), Manager Process Technology, Voith Sulzer Paper Technology, Ravensburg/Germany

*IfP = Department of Paper Science and Technology
Darmstadt University of Technology/Germany

Preface

Lothar Göttsching and Heikki Pakarinen

Development of the use of recycled fibers in the paper industry is a success story of the second half of the 20th century. It is a development that was unseen in the 1950s. At the threshold of the next century, recycled fibers are an indispensible raw material in the global paper industry to satisfy the need for fibers in industrialized and developing countries.

Recovered paper as the base material for recycled fibers does not come from forests. It is collected from paper converters, printers, and distributors. It also comes from used paper products from consumers in private homes, offices, and institutional settings. Using a sophisticated infrastructure, the collection of used paper is a successful business venture especially in densely populated countries with a high paper consumption per capita.

This book covers the entire path taken by recovered paper from its collection and sorting through processing in the paper industry to make new paper. The book also includes the recycling of process residues, their combustion for energy recovery, or their disposal in landfill sites. Current legislative actions especially for Western Europe, Japan, and the United States are reviewed, but these are obviously subject to rapid change due to new laws, ordinances, and regulations. This book containing 13 chapters is the result of a Finnish and German project.

Chapters 1–4 describe the statistical background, legal issues, and procedures for the collection and sorting of recovered paper. Questions concerning the quality control of recovered paper receive special consideration. The presentation of data relating to physical characteristics uses the significant recovered paper grades in Germany.

Chapter 5 concentrates on the unit operations and equipment in recycled fiber processing. It discusses fundamentals and applications of the different process steps such as slushing, screening, cleaning, fractionation, deinking, dispersing, and beating. Processes and equipment from various suppliers provide examples. These companies constantly develop and offer on the world market improved technologies and tools to upgrade the increasing amounts of recycled fibers.

Chapter 6 focuses on stock preparation lines using newsprint, supercalendered paper, lightweight coated paper, fluting, liner, and folding boxboard as examples. Closed water loops and an environmentally sound waste management receive special attention as do the possibilities of energy efficient recycled fiber processing. Highlighted in Chapter 7 is deinking for the increasingly important production of deinked pulps (DIP) as substitutes for virgin mechanical and chemical fibers. The chapter concentrates on the chemistry of flotation and wash deinking. For the chemicals used such as peroxide, caustic soda, water glass, and chelating agents, the focus is on natural and synthetic surfactants with their individual benefits and disadvantages. The text also includes enzymatic deinking although it is still under development.

The deinkability of the various conventional and nonimpact inks is the subject of Chapter 8. It includes the parameters of printing substrate (coated and uncoated wood-containing and wood-free papers), printing inks, effects of aging, and methods to evaluate deinkability. These topics have interest to the manufacturers of binders, pigments, and printing inks within the paper chain for recycling of printed products.

For hygienic and graphic paper products, deinked pulp requires bleaching. Chapter 9 therefore covers the application of different oxidizing and reducing bleaching agents and the process engineering and equipment for bleaching of deinked pulp. The chlorine-free bleaching process represents the technical state-of-the-art. The search continues for biologically degradable chelating agents.

Using an intensive evaluation of technical literature, Chapter 10 describes the papermaking potential of recycled fibers. It begins with the chemistry and morphology of fibers and continues through the effects of recycling on sheet properties and the influence of residual contaminants and inks on final quality. Using several different models, the chapter presents the calculation of the age distribution of recycled fibers in a national recycling system. Of special importance are the observations concerning upgrading of recycled fibers by mechanical treatment and the characteristics of deinked pulp fibers regarding runnability.

A major challenge in the processing and use of recycled fiber pulps is the presence of various tacky substances or stickies. They occur in small concentrations. Chapter 11 describes the sources of these substances, their development during recycled fiber processing, and their gradual elimination by screening, cleaning, flotation, and passivation measures. This comprehensive chapter includes a description of analytical methods for quantification of stickies using different principles such as screening or adsorption.

The processing of recovered paper also produces solid waste in the form of rejects, deinking sludges, sludges from effluent treatment, and inorganic residues or ashes from mill waste incineration. Chapter 12 starts with the description of the compositions and characteristics of various components of solid waste. It then covers their handling in a mill by dewatering and drying and their final fates. Also presented are the wide range of different combustion technologies, the resulting emissions, and the relevant legal threshold values. Disposal by composting, in agriculture and the use of solid waste in other parts of industry is described. The chapter also focuses on the use and final disposal of nonrecycled or nonrecoverable used paper products by landfilling, composting, and incineration with energy recovery. New developments such as hydrolysis or pyrolysis are also included.

Chapter 13 concerns environmental aspects such as ecological labeling in Germany, Northern Europe, and the European Union; the influence of paper recycling on the global CO_2 balance, and the consequences of the greenhouse effect. Of particular current interest are the conclusions concerning harmful substances in recovered paper. This focuses on heavy metals and chloro-organic compounds such as dioxins, furans, pentachlorophenol, and polychlorinated biphenyls. Carefully compiled findings show that these chloro-organic compounds do not constitute any danger for the consumer of paper containing recycled fibers. This is due to chlorine-free pulp bleaching or the prohibition in force in many countries of additives containing pentachlorophenol. Recovered paper, new paper made from it, and process residues are very benign.

As editors of this book, we thank all the authors for their valuable contributions. They toiled endlessly to gather and record the latest findings and experiences to provide experts in the industry and research organizations and nontechnicians, students, and representatives of nongovernmental organizations with a comprehensive, current insight into this rapidly developing field. Some texts were approved by outside experts. We therefore thank the following reviewers:

Bernd Böcking, Lawyer, GesPaRec/Bonn (legal issues)

Dr. Klaus-Dieter Kibat, VDP/Bonn (legal issues)

Stephan Meißner, Lawyer, Papermakers' Center/Gernsbach (legal issues)

Dr. Erich Frank, BASF Drucksysteme/Stuttgart (printing inks)

Dr. Hans Ulrich Süß, Degussa/Hanau (bleaching)

Dr. Hermann Onusseit, Henkel/Düsseldorf (adhesives)

Dr. Christian Möbius, CM Consult/Augsburg (solid waste).

In conclusion, we express our very special thanks to the Finnish Paper Engineers' Association for taking the initiative for this series of publications and for the invitation to contribute to Book 7. Special mention is also due to the valuable work provided by Peter Seifert of Thermo Black Clawson, Middletown, Ohio, as reviewer and to TAPPI, Atlanta, Georgia, for the organizational and linguistic services.

Darmstadt, November 1999
Lothar Göttsching

Valkeakoski, November 1999
Heikki Pakarinen

Table of Contents

CHAPTER 1

General aspects and basic statistics

Lothar Göttsching

General aspects and basic statistics

1 General aspects

The paper industry is the exclusive user of recovered paper as a secondary raw material. Various mechanical and chemical processes generate recycled fibers for the production of paper and board. These processes use different recovered paper grades that contain either chemical or mechanical fibers or mainly an undefined mixture of both. Some paper and board grades produced can use recycled fibers exclusively. This would include corrugating medium and test liner, for example, in Central, Western and Southern Europe or newsprint in Germany. Other grades use blends of recycled and virgin fibers. Examples are newsprint and liner produced in North America or Northern Europe. The proportion of recycled fibers in the furnish can vary from about 5% for fine papers to 100% depending on the paper or board grade or geographic region.

At 128 million tons in 1997, the volume of recovered paper used globally in the paper industry is nearly identical with the volume of chemical pulp at about 140 million tons. The volume of recovered paper is almost four times that of mechanical pulp [1].

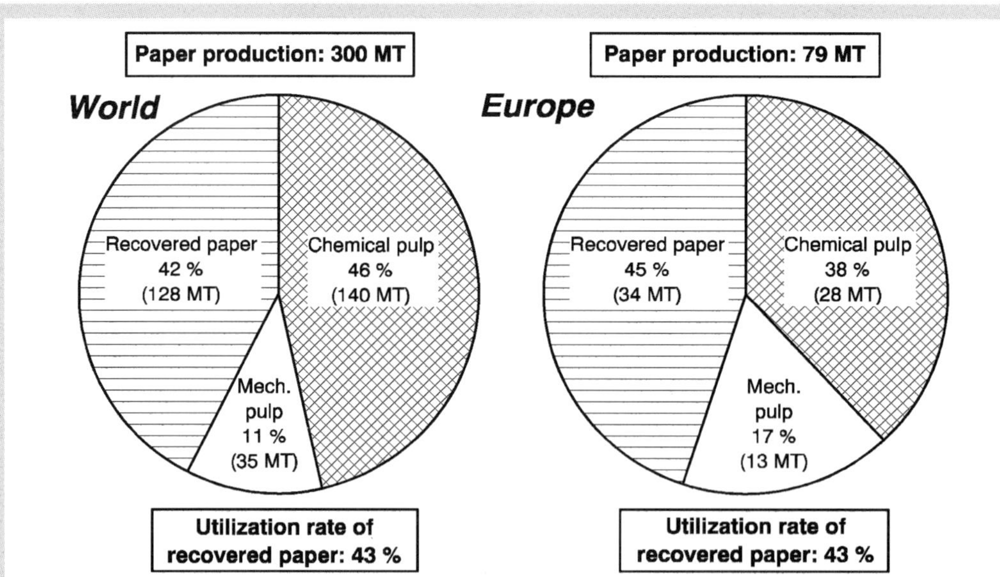

Figure 1. Fiber material consumption for paper and board production in Europe and the world for 1997[1]. (MT: million tons)

Recycled fibers therefore play a very important role in the paper industry today as a substitute for virgin pulps. Figure 1 shows global and European usage values. In many countries of the world, the paper industry could not exist as it does today without recycled fibers. In Central, Southern, or Western Europe, this is mainly due to economic reasons. In places such as Japan, Korea, Taiwan, or Mexico, the additional reason is the scarcity of native pulpwood resources.

In the era before the introduction of industrialized paper production 200 years ago, the most common fiber furnish was secondary fibers recovered from used textiles. These were rags based on hemp, linen, and cotton. Only after the invention of mechanical wood pulp in 1843 and chemical wood pulp in the second half of the 19th century was paper production no longer as reliant on recycled material as in the previous 2000 years.

Before industrialized paper production and the invention of the paper machine in 1799, stationery or writing paper made from rags was recycled to produce low-grade board. As early as 1774, Claproth in Göttingen, Germany, improved the processing of used, hand-made stationery. His process removed optically disturbing inks or printing ink. Today, we call this method "deinking".

Compared with the situation today, recycling of used paper products manufactured from wood pulp fibers had little importance in the first half of this century. Only in the 1950s and 1960s was increasing use made of recycled fiber furnish especially for the production of packaging paper and board. The 1970s saw increasing use of recycled fibers in graphic and hygienic papers. Use of recovered paper accelerated in the 1980s. Between 1980 and 1996, the global consumption of recovered paper increased approximately 6% annually. Figure 2 shows that paper production grew only about 3% annually during the same period. In contrast, the annual growth in chemical and mechanical pulp consumption was about 2%. During this period, the consumption index of recovered paper therefore increased by 140 points compared with 65 points for

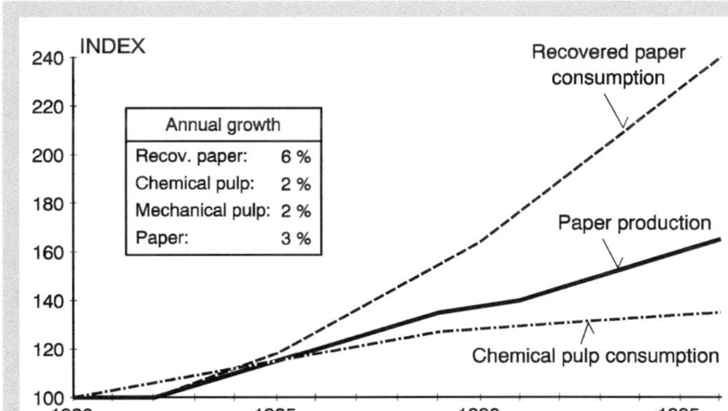

Figure 2. Global development of paper production, chemical pulp, and recovered paper consumption.

paper production. Increasing application of improved deinking processes also influenced the growth in use of recovered paper in the 1980s and 1990s. This allowed use of recycled fibers in higher grade hygienic papers, newsprint, and recycled graphic papers such as copying and writing papers.

Recovered paper use is most attractive in densely populated regions with a high paper consumption per capita. The region must also have a paper industry with sufficient technology and a long tradition in recycled fiber processing. The most prominent regions are Japan and Europe outside Scandinavia. Recovered paper is not the only material increasingly collected and re-used in developed countries. National recycling management enforced by legal measures encourages the population to collect metal, glass, and plastics also. Consumers in many countries participate in recycling. They are involved as consumers of commodities while producing secondary raw materials for re-use.

Cost competition and the legal requirements in many countries primarily drive the use of recovered paper. The impact of environmentalists through "green" movements and the level of acceptance in the market of paper made from recycled fibers are additional driving forces that vary by country. Recovered paper use is an environmentally-friendly issue according to the recycled fiber processing paper industry, environmentalists, governmental authorities, and often even the marketplace. It is accepted that recycling preserves forest resources and energy used for production of mechanical pulps for paper manufacturing.

Processing of recycled products requires relatively little fresh water per ton of paper produced. The solid waste rejects and sludge such as deinking sludge from recovered paper processing mills typically present a problem. The proportion of such residues is 5%–40% depending on the recovered paper grade processed and the paper grade produced. The average rate of rejects and sludge totals about 15%, calculated on the recovered paper input on an air-dried basis. Because landfilling of organic matter has no future in many countries, most organic waste requires burning to reduce its volume. This uses its heat value and contributes to self-supporting incineration. Effective, clean incineration technologies are available that control flue gas emissions. The final waste (ashes) can be discarded or used as raw materials in other industries. An increasing volume of rejects and sludges can be used in brick works, cement industry, and for other purposes (See Chapter 12).

2 Basic statistics

Consideration of recovered paper use and recovery of used paper products requires definition of two statistical parameters:

- Recovered paper utilization rate, in percent, is the amount of recovered paper used as raw material in the paper industry, in tons, divided by paper production, in tons, on an annual basis, multiplied by 100.
- Recovery rate, in percent, is the amount of recovered paper for material recycling, in tons, divided by paper consumption, in tons, on an annual basis, multiplied by 100.

The draft of an environmental agreement on the recovery of paper and board products between the European Union and the European paper chain gives a more comprehensive definition of recovery rate in percentage as 100 times the recovered paper for material recycling plus export, composting, and energy recovery, in tons, divided by paper consumption, in tons, on an annual basis.

The above mentioned parameters can be calculated globally or for a country or region such as the European Union. Calculating these factors for certain paper grades or categories of paper grades such as newsprint or graphic papers globally or for a country or region is often even more relevant. Naturally, this calculation presupposes that the data for recovered paper used or collected are recorded and documented using the same criteria in each country.

Calculating the recovered paper utilization rate gives no indication of the proportion of recycled fibers (with or without minerals) actually ending up in specific paper grades. This is because the stock yield in recovered paper processing differs by paper grade produced and is usually not statistically documented but only estimated. This factor is often of more interest for technical, quality, ecological, and economic reasons than the recovered paper utilization rate. Many years of industrial experience have provided the following ranges of pulp yields for recovered paper processing for different production segments:

- Packaging papers and board 90%–95%
- Graphic papers 65%–85%
- Hygienic papers 60%–75%
- Specialty papers 70%–95%
- Market DIP (wood-free) 60%–70%
- Market DIP (wood-containing) 80%–85%.

The yields of recycled fiber pulps for packaging papers and board are high because these are mostly "brown" recovered papers processed without deinking. The yield loss results primarily from the removal of nonpaper components such as metal, glass, plastics, wood, and textiles as rejects from slushing, cleaning, and screening.

Besides a careful coarse and fine cleaning and screening in several stages, a flotation deinking process is necessary to improve the optical characteristics for graphic papers. The low yield of about 65% applies to the preparation of wood-containing deinked pulp (DIP) for higher grade printing papers such as lightweight coated (LWC) paper with ink removal and partial deashing. The higher yields relate to deinked pulp for the production of newsprint. Depending on the system concept, this includes a one- or two-stage flotation with yields of 80%–90%.

Processing of deinked pulp for hygienic paper is a special case. The multistage deinking process including flotation and washing must ensure removal not only of printing ink and other impurities but also almost all fillers and coating pigments. The required residual ash content of less than 5% and the removal of fiber fines cause the low yield of 60%.

Specialty papers cover a wide spectrum with greatly varying quality requirements. The high yield of about 95% relates to processing of ordinary recovered paper grades for production of automobile board and core board. Wallpaper is an example of a low yield material.

Especially exacting demands for optical cleanliness, brightness, or luminance and freedom from ash, stickies, and other impurities are necessary for market DIP produced from wood-free mixed office waste (MOW) to permit its use as a pulp substitute in wood-free printing paper and stationery. For this reason, low yields of 60%–70% result when processing MOW depending on the ash content of the recovered paper grades processed. Wood-containing market DIP for newsprint is less demanding which results in a higher yield ranging between 80% and 85%.

When estimating the required proportion of recycled pulp in furnishes containing recycled fibers and virgin fibers, the recovered paper utilization rate itself is not as important as the proportion of recycled fibers processed considering the process-specific yield in recovered paper processing. The following discussion refers to the recovered paper utilization rate defined above according to common convention.

2.1 Utilization rate of recovered paper

Recovered paper utilization rates globally and nationally can provide comparisons. Such statistics can give a false impression of the recycling activities in different countries as the following discussion explains.

The range of national utilization rates extends globally from 5% to more than 100%. Figure 3 shows the twelve largest papermaking countries accounting for 80% of

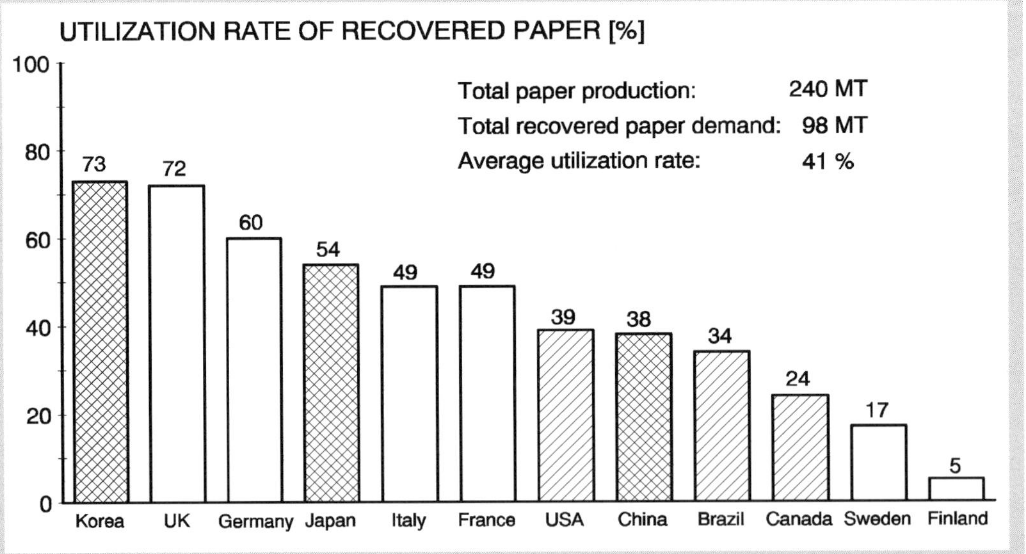

Figure 3. Utilization rate of recovered paper for the twelve largest papermaking countries in the world in 1997[1].

the global paper production in 1997 with 240 million tons. Simultaneously, they consumed about 80% of the world's recovered paper volume. Heading the list are Korea and the United Kingdom. Despite its scarcity of domestic wood resources, Japan has a utilization rate less than 55%. The lowest utilization rates are for Finland and Sweden. This is due to a paper production of which at least 85% is for export, the low population figures and related low level of national paper consumption and recovery relative to paper production.

Among the leading industrialized nations, United Kingdom, the Netherlands, and Spain have raised their recovered paper utilization rate to a high level above 70% followed by Germany and Switzerland with about 60%. Some countries such as Denmark or Ireland record levels of more than 100% due to the structure and size of their paper industry. A utilization rate in excess of 100% occurs because the figures include the loss of yield in recovered paper processing as rejects and sludge. In addition, the share of packaging paper and board in the production of these countries is exceptionally high.

Figure 4 shows the utilization rate for the three largest paper production regions in the world where the paper industry uses 65% of the global recovered paper volume: the United States, Europe, and Japan. Japan continues to lead the list today although the utilization rate stagnated in the 1990s. The United States and Europe have both shown a continuous increase in the rate to 39% and 43% in 1997, respectively. The gradient is steeper for the United States than for Europe.

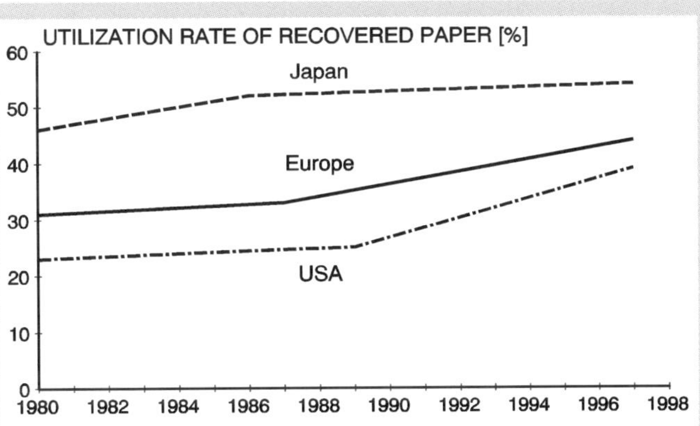

Figure 4. Utilization rate of recovered paper for Japan, United States, and Europe[1].

Utilization rates for different countries should not be compared without further comment. Knowing the structure of the production program of the different national paper industries in the main product categories of packaging papers and board, graphic papers, hygienic papers, and specialty papers is important because the utilization rates for these product segments can differ significantly as Fig. 5 shows for Germany.

In Germany, Europe and globally, packaging papers and board have the highest recovered paper utilization rate. Demanding quality specifications have to be fulfilled by these recycled fiber-based papers and board to ensure trouble-free converting to corrugated board containers or folding boxes and adequate functional characteristics of the finished products.

Occupying second place in many countries is the utilization rate of hygienic papers. Due to high yield losses in flotation and washing steps, the actual proportion of recycled fibers in hygienic paper produced in Germany is not higher than 45% on average. The figure of 45% deinked pulp comes from a 65% yield of a 70% utilization rate.

Specialty papers now achieve a 48% utilization rate due to inclusion of core board made from more than 100% recovered paper. Many grades of specialty paper such as cigarette paper, filter paper, or banknote paper can never use recycled fibers. This causes stagnation of the average utilization rate in this product segment.

Finally, the level of the utilization rate for graphic papers has increased to 37% in Germany. Due to the wide spectrum of the produced paper grades, a distinction is necessary for this product segment because newsprint has reached a level of about 115%. The utilization rate of the other graphic papers averages 17%. Included among these other graphic papers are wood-containing and wood-free papers that are coated or uncoated. Some use 100% recycled fiber furnish such as copy paper (so-called recycled paper). The latest development in the use of recycled fibers is occurring with supercalendered (SC) and lightweight coated (LWC) papers. These currently do not contain more than about 20% deinked pulp in individual grades.

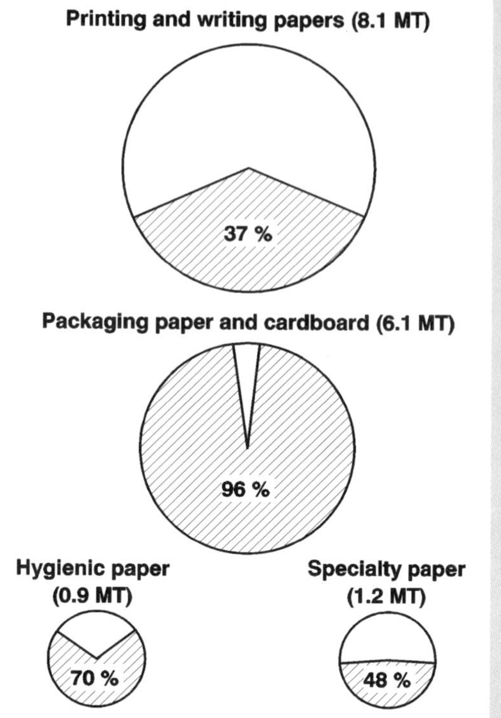

Figure 5. Recovered paper utilization rates for different paper grades produced in Germany in 1998[1].

The proportions of the paper production segments vary from country to country resulting in different national recovered paper utilization rate levels. A higher utilization rate means the fibers undergo the following steps of the complete recycling loop more frequently:

- Paper production
- Paper converting and printing
- Consumption of paper products
- Collection of used paper products
- Sorting of recovered paper
- Processing of recovered paper.

Chapter 10 discusses how to estimate the number of recycling loops of recycled fibers in accordance with the recovered paper utilization rate and other factors.

2.2 Recovery rate of used paper and paper products

As far as national recycling activities are concerned, an assessment of the paper recovery rates of different countries is more relevant than a comparison between national recovered paper utilization rates. Figure 6 shows the development of recovery rates (recovered paper for material recycling) since 1991 for Europe and nine selected European countries. Due to intensified collecting in several countries, the European average rose from 40% to 50% in 1997 corresponding to a total volume of recovered paper of 35 million tons in that year. During the same period, the recovery rate stagnated in Japan at 51%–53%. The rate for the United States was 45% for 1997 and should increase to 50% by 2000. This will also improve recovered paper export.

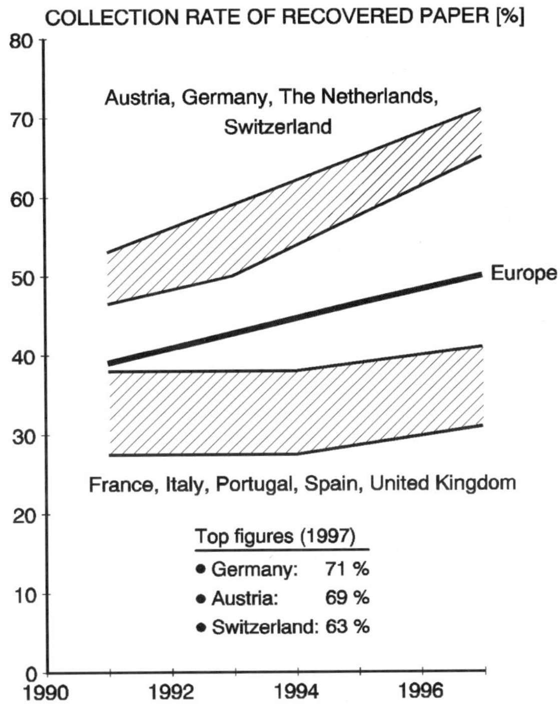

Figure 6. Ranges of paper recovery rate for different European countries[2].

Europe requires a distinction between two categories of countries. Some have a high recovery rate, others have a lower one. Due to comprehensive, country-wide collection measures, Austria, Germany, the Netherlands, and Switzerland have succeeded in achieving increasing recovery rates of 63%–71% in only a few years. In France, Italy, Portugal, Spain, and the United Kingdom, no corresponding development occurred between 1991 and 1997. The recovery rate therefore stagnated between 31% for Italy and about 40% for France, Portugal, Spain, and the United Kingdom. As a result of this stagnation, growth in paper production during this period depended on increasing imports of recovered paper from countries with higher recovery rates. This satisfied the industry demand for recovered paper and maintained the utilization rate at a high level.

Germany is the largest exporter of recovered paper in Europe with 3.0 million tons in 1998. Most goes to destinations within Europe. Small volumes of recovered paper go to Indonesia and Taiwan. Because of the large demand for fiber material in Asia, Europe could easily activate its still dormant recovered paper potentials. The United States did this by exporting about 8 million tons in 1997 primarily to Canada, Mexico, and the Pacific Rim.

The ceiling for the paper recovery rate in an industrialized country with a well-developed infrastructure should be about 75%. The theoretical limit of the recovery rate is about 80% since approximately 20% of the used paper products is not collectible. This include hygienic papers, specialty paper products, long-life paper products, papers used in industrial applications, and papers used for other purposes in private homes such as fuel or compost material. Considering the theoretical recovery rate of 80%, the recovery rate of 71% already achieved in 1997 in Germany is the equivalent of a paper collection efficiency of about 90%. Collecting additional used paper is probably not economically feasible. Additionally collected material would probably not satisfy the quality requirements of the paper industry.

Production residues from paper converting such as the corrugated board industry or printing houses and the used paper products from packaging in product distribution such as supermarkets are totally collected and recycled in industrialized countries. A high collection rate also applies to private homes provided corresponding collection systems are available in sufficient numbers with convenient pick-up and drop-off systems (See Chapter 3). Some potential exists to increase the level of paper collection in offices for the production of higher grade recovered paper. These would be largely wood-free paper grades that are scarce due to increasing needs of the paper industry.

2.3 Deinked pulp (DIP) capacities

The first deinking system working according to the flotation principle was installed in the late 1950s in the United States. The first European operation started production at a Dutch paper mill in 1959. This unit had a capacity of 10 t/d for production of hygienic papers. Figure 7 shows flotation capacities for different continents. In 1997, this capacity totaled 25 million tons, corresponding to 20% of global recovered paper use. For 1975–1995, the annual growth of the flotation deinking capacity in Europe was 15%, clearly more than the annual increase in paper production of 4.5%.

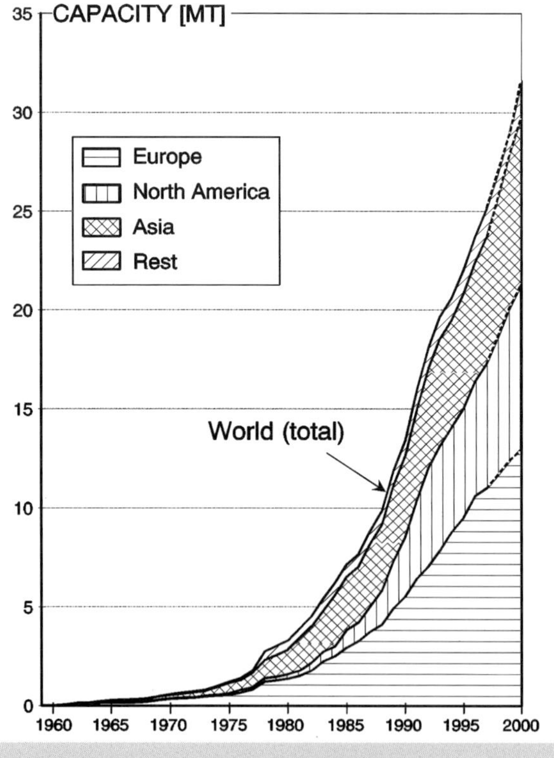

Figure 7. Global flotation deinking capacities.

The largest share of the global deinking capacities is in Europe. North America and Asia account for a share of 25% each. The deinking capacities in Latin America, Africa, and Oceania are negligible as Fig. 8 shows.

Figure 9 shows use of DIP by paper grades. Globally and in Europe, the use of DIP in newsprint dominates followed by hygienic papers and printing and writing papers. Included under the heading of other grades is a wide spectrum of paper grades such as magazine papers, white lined chipboard (top layer), wallpapers, and molded products such as egg cartons. One can assume that the share of deinked fibers in wood-containing printing papers such as newsprint, SC, and LWC papers will increase in the future.

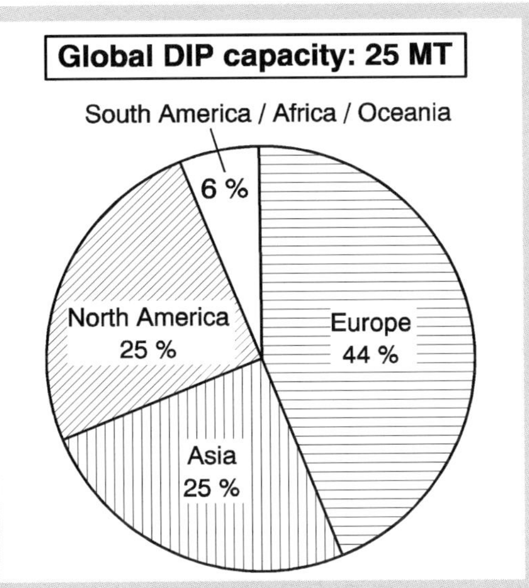

Figure 8. Production capacities of DIP using flotation by continents for 1997.

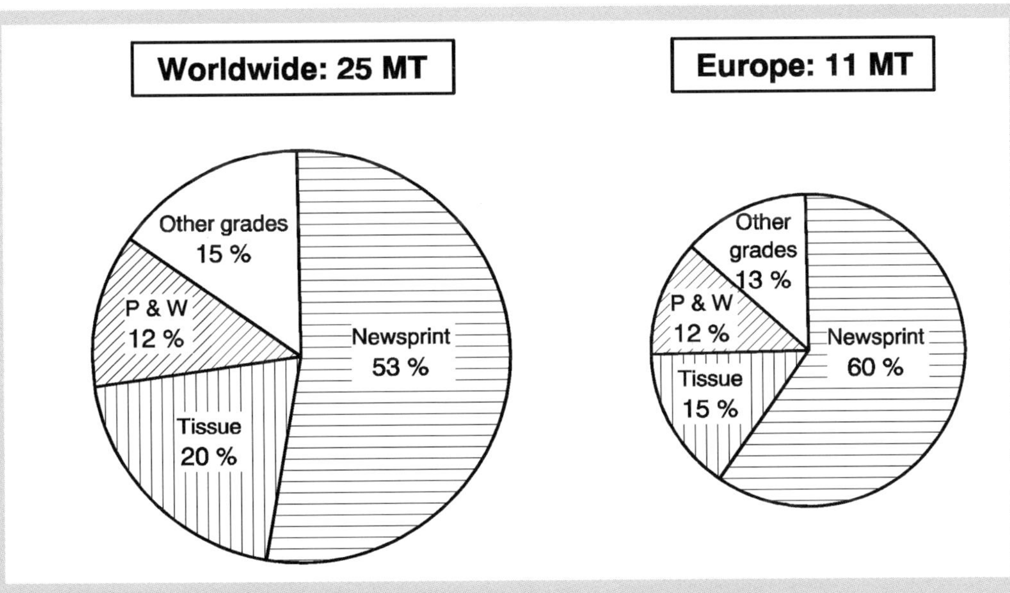

Figure 9. Use of DIP by paper grades in 1997. (P&W: printing and writing papers).

2.4 Future development of recovered paper utilization

The global recovered paper utilization rate approached 43% in 1997. This corresponds to a smaller share of recycled fibers used for paper and board manufacture due to rejects and sludge generated by recovered paper processing. Therefore, the average recycled fiber content (including minerals) amounts to 36.5% using an average yield in recovered paper processing of 85% (Fig. 10).

Looking at the next decade, this utilization rate will further increase globally but less dramatically than in the past. By approximately 2010, a balanced utilization rate of about 50% will be reached. Using the same basis of an average of 15% rejects and sludges for all processing technologies for packaging papers and board, graphic papers, and hygienic papers, the average content of recycled fibers in paper might be about 50% minus 7.5% or 42.5%. Since papers and board might then contain an average of 15% fillers and coating pigments, about 42.5% virgin pulps as mechanical, semichemical, and chemical fibers would be necessary for the global production of paper and board at the end of this decade.

Due to decreasing availability of suitable recovered paper grades and because of demanding quality requirements of paper and board produced, it is hardly to be expected that the recovered paper utilization rate will exceed the 50% level in the long-term.

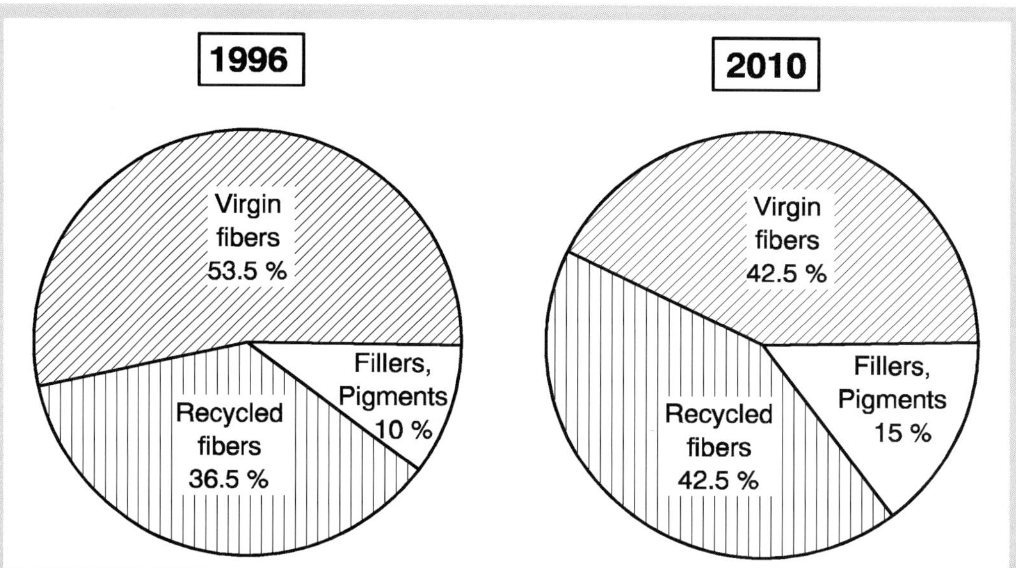

Figure 10. Relative global raw material demand today and tomorrow.

References

1. *Anon., "Papier '99 – Ein Leistungsbericht" (VDP, Ed.), Bonn, Germany, 1999, 81 pp.*
2. *Kibat, K.-D., Wochenbl. Papierfabr. 126(16):742(1998).*

CHAPTER 2

Legislation for use of recycled fibers

Lothar Göttsching and Hans-Joachim Putz

Legislation for use of recycled fibers

Limited capacities of landfill sites and (municipal) incineration plants, increasing waste disposal costs, and environmental awareness of the general public are the most important reasons for the efforts by all governments in industrialized countries to limit or even reduce the quantities of waste from private homes, businesses, and industrial operations. For example, the annual population growth of the United States from 1960–1993 was 1.1%. The generation of municipal solid waste (MSW) grew 2.7% per year[1]. Prevention of waste and material recycling are the most important means of limiting the generation and growth of municipal and industrial waste.

A wide range of legislation in various countries tries to promote material recycling and reduce further the generation of waste that requires disposal in appropriate facilities. Waste- and recycling-related legislation can be categorized into general regulations for different material categories, special regulations for paper and board products, and regulations relating to private households for reduction of municipal waste. This area is subject to rapid change involving testing of many different political approaches. Some approaches oriented toward paper products are the following:

- Directives and ordinances
- Procurement policy guidelines
- Voluntary agreements.

Other regulations that cover packaging materials or graphic paper products are as follows:

- Packaging ordinance for packaging material with set recovery and recycling rates (e.g. in Austria or Germany)
- Voluntary agreement of the graphic paper chain with set recovery and recycling rates (e.g. in Germany)
- Stipulation of minimum contents of recycled fibers for example in newsprint (in the USA)
- Executive orders for purchasing graphic papers with a minimum content of recycled fibers by public funds (e.g. in the USA).

These regulations set responsibilities for taking back used paper products as packaging material independent of the public disposal system and recycling them. Alternatively, they may specify the content of secondary fibers used to manufacture a certain paper grade. Other efforts that describe recycling objectives in general or stipulate the installation of collection systems aim at household and municipal waste[2].

The legal conditions relating to recycling objectives and waste disposal measures globally are changing rapidly. This may mean that the laws, guidelines, and voluntary agreements described in the following text may already be out-of-date by the time this book reaches print. Simultaneously, internationally no uniform objectives exist regarding recycling of recovered paper. The survey in this book must therefore cover examples from a limited number of countries. For this reason, the following statements concentrate less on statistical objectives and more on the basic intentions of waste prevention strategies and recycling philosophies.

1 Europe

At least in Europe, Germany was the pioneer for the following three-pronged waste strategy with priority in the order listed:

- Prevention of waste
- Re-use or recycling of used paper products
- Disposal or incineration of waste.

The European Union (EU) has subsequently adopted this concept with its waste management directives in which the European Waste Guideline 91/156 EWG of 1991 and the EU Packaging Directive 94/62 EG of 1994 have special importance. The latter also places top priority on preventing packaging waste and covers packaging made from paper and board, glass, metal, plastic, and composite material. Other main principles are re-use of packaging material, recycling of packaging material and other uses of packaging waste, and a consequent reduction of the amount of waste for final disposal.

The EU member states must establish systems for take-back, collection, and utilization of used products as secondary raw material. All parties involved in the production, converting, import, and distribution of packaging and packaged products must become more aware of the degree to which packaging becomes waste. They must also take responsibility for this waste in accordance with the "polluter pays" principle. Final consumers play a decisive part in the prevention and use of packaging and packaging waste. They therefore require education. The efficient separation of waste at its source such as in private households has major importance for a high recycling level.

Independent of the packaging material, a distinction exists among the following three categories of packaging:

- Sales packaging or first packaging. This is packaging offered to the final user or consumer at the sales outlet as a sales unit such as a bottle or can of beer, a box of washing powder, or a cigarette box.

- Re-packaging or second packaging. This is packaging containing a certain number of sales units given to the final user or consumer at the sales outlet or that serves only to stock the shelves. Removing this packaging does not influence the characteristics of the goods. Examples are a case, tray, or pack of bottles or cans. (This category of packaging is of minor importance).
- Transport packaging or third packaging. This is packaging that facilitates the handling and transport of several sales units in such a way that direct contact with them and transport damage are avoided. Examples are corrugated containers for carrying one or many purchased items, e.g., a new TV set or a dozen jars of jam. Containers for road, rail, ship, or air transportation are not part of the category of transport packaging.

According to the EU Packaging Directive, the term "utilization" means the following for various packaging materials:

- Material recycling: The reprocessing of packaging materials in a production process for the original purpose or for other purposes including organic use but not combustion for generation of energy
- Energy recovery: The use of combustible packaging materials for energy production purposes by direct combustion with or without other types of waste but with heat recovery
- Organic utilization: The aerobic treatment (biological use) or anaerobic treatment (biogas production) with microorganisms in a controlled process of the biologically degradable components of the packaging materials to produce stabilized organic residue or methane. Landfill disposal is not a form of organic utilization.

The EU Packaging Directive has the following objectives:

- By five years after the directive became effective (2001 at the latest), national legislation must guarantee that at least 50% and at most 65% by weight of used packaging material must be used according to the aforementioned aspects.
- Within this utilization objective and in the same period, at least 25% and at most 45% by weight of the total packaging material contained in the packaging waste and at least 15% by weight of each individual packaging material must be processed by material recycling.

All EU member states had to implement this packaging directive into national law by June 1996. All countries except United Kingdom have set very high objectives for the collection of used packaging averaging 70%–85% of the total material by the year 2001 as Table 1 shows. Table 2 shows that the material recycling objectives in the individual EU member states actually exceed the levels of the EU directive[3].

Table 1. Objectives for paper and board packaging recovery rates in European countries.

	Schedule	Goals	Remarks
EU directive*	from 2001	50% – 65% of total packaging material	European law
Austria	from 2001	Not more than 99 000 t to be deposited per year	Amendment 1996 to packaging ordinance
Belgium	since 1996 from 2001	30% 80%	Interregional decree
Finland	from 2001	75%	Council of State Decision
France	since 1994 from 2002	100% for industrial packaging material 75% for sales packaging	Lalonde Decree No. 92–377 of 1992
Germany	since 1995 since 1996 from 1999	100% of transport packaging 64% of sales packaging 70% of sales packaging	Packaging ordinance 1st amending law of packaging ordinance (1998)
Sweden	since 1997 from 2001	30% of cardboard and board packaging and 65% of corrugated board 70% of cardboard and board packaging and 65% of corrugated board	Intermediate goal by law until EU directive implemented
The Netherlands	from 2001	85%	Voluntary agreement
U.K.	since 1998 from 2001	40% 50%	Industry objective

The EU members Greece, Ireland, and Portugal can prolong the time period until the end of the year 2005

Table 2. Objectives for paper and board packaging recycling rates in European countries.

	Schedule	Goals	Remarks
EU directive*	from 2001	25% – 45% of total packaging material 15% of each individual material	European law
Austria		No recycling target figures	Amendment 1996 to packaging ordinance
Belgium	since 1996 from 2001	25% 50%	Interregional decree
Finland	from 2001	53%	Council of State Decision
France			No fixed recycling rate by law
Germany	since 1995 since 1996 from 1999	100% of transport packaging 64% of sales packaging 70% of sales packaging	Packaging ordinance 1st amending law of packaging ordinance (1998)

Table 2. Objectives for paper and board packaging recycling rates in European countries. (continued)

	Schedule	Goals	Remarks
Sweden	since 1997 from 2001	30% cardboard and board packaging and 65% of corrugated board 70% of cardboard and board packaging and 65% of corrugated board	Intermediate goal by law until EU directive implemented
The Netherlands	from 2001	85%	Voluntary agreement
U.K.	since 1998 from 2001	8% 15% of each individual material	Industry objective

**The EU members Greece, Ireland and Portugal can prolong the time period until the end of the year 2005*

All regulations are based on the principle of producer responsibility. Producers and retailers bear the specific responsibility for the entire life of a product from its production until its final use. The producer is the most important player since he sets to a high degree the conditions for waste disposal. In the opinion of the EU Commission, the producer makes the decisions that determine how the product will be handled as waste. This covers the concept, use of specific materials, composition, and marketing. With careful use of natural resources, renewable raw materials, or harmless substances, the producer can help avoid waste and design products so they will be suitable for utilization. The EU Commission has adopted a spirit of cooperation emphasizing the active participation of all parties involved to meet the objectives of the waste policy. The objectives will be reached only with the active cooperation of authorities, private and public organizations, environmental organizations, and especially the general public as final consumers[4]. The revision of the Packaging Directive is already prepared by the EU and should be issued in 2001.

Since Germany pioneered in these environmental objectives, the following text will concentrate primarily on the German legal framework and voluntary agreements in industry regarding collection and utilization of used paper products. Then the situation in selected European countries will be discussed briefly.

1.1 Germany

Figure 1 shows the previous and current legislative background in Germany regarding the following:

- Recovery and recycling of used paper products
- Municipal waste management
- Industrial waste management.

Waste Management Act (1986)

- Prevention of waste
- Recycling of waste
- Environmentally-friendly disposal of waste

Packaging Ordinance (1991)

- Sets recovery targets for recycling (e.g. 70 % recovery of paper and board sales packaging by 1999)

Proposed **Waste Paper Ordinance** was replaced by

Voluntary Agreement of the Graphic Paper Chain (1994)

- Sets recovery targets for recycling (60 % recovery of printed material & office paper by 2000)

Recycling and Waste Management Act (1996)

- Replaced Waste Management Act of 1986 and aims at comprehensive product responsibility

Municipal Waste Management Provision (1993)

- By 2005 no solids must be landfilled that contain more than 5 % organic matter

Figure 1. Legislation on recovery and recycling of recovered paper in Germany.

The requirements of the EU Packaging Directive are already part of national laws, ordinances, or regulations. This results in increased efforts for paper recovery and recycling. The Waste Management Act in force since 1986 covers waste management including recovered paper issues and focuses on the following targets:

- Prevention of waste
- Recycling of used products as secondary raw material
- Environmentally-friendly final disposal of residual waste.

In 1972, the Waste Disposal Act was enacted that was the precursor of the Waste Management Act. In 1986, the above targets of prevention of waste, material recycling, and final disposal of waste were incorporated into the original act, as its fourth amending law. Thanks to these modifications, the principle of prevention received first priority. The principle of recycling requires separate collection of recyclable components of household waste such as paper, plastic, glass, metal, and composite material. Separation into recyclable components is also required for any industrial waste including waste generated by processing of recovered paper, paper manufacturing, and paper converting.

In the framework of the Waste Management Act, the federal government can issue ordinances such as the Packaging Ordinance of 1991. It requires collection, recovery, and recycling of transport packaging, re-packaging, and sales packaging. For paper- and board-based transport packaging, a 100% recovery and recycling rate had already been established. In the Packaging Ordinance for paper products, a recovery rate and a utilization rate for sales packaging of 64% beginning in 1996 was necessary. According to the first amendment, this rate must increase to 70% of the total tonnage of paper- and board-based sales packaging by 1999 at the latest.

Responsibility for collection of other paper products such as printed material (e.g., old news and magazines) from private households and small commercial enterprises remains with local authorities. In contrast, the collection of used packaging is not part of the responsibility of communities. Collection of sales packaging including paper and board, composite material, plastic, glass, and metal is organized and financed through a private organization named Duales System Deutschland GmbH (DSD).

The industry, wholesale, and retail trade organizations established DSD in 1990 for that purpose. DSD signed contracts with different organizations who run the business of collection of sales packaging material carrying the "Green Dot." Licensing fees are based on the type and size of packaging material (glass, plastic, aluminium, tinplate, paper and board, and composite material) and are paid by the producer of the packaging or by the company filling the packages to finance the costs for collection and sorting of sales packaging with Green Dots. Since the fees become a component of the sales price of the goods, the final consumer actually pays the fees (Note that the Green Dot does not serve as an eco label.).

In 1994, the German federal government issued the Act for Promoting Closed Substance Cycle Waste Management and Ensuring Environmentally Compatible Waste Disposal commonly abbreviated as Recycling and Waste Management Act. This act is more comprehensive in its regulatory scope than the Waste Management Act replaced in 1996. Using a basis of comprehensive product responsibility, the object of the new act is to implement improved recycling management in the ecological and social market economy of Germany. It does this by ensuring environmentally compatible waste disposal. The product responsibility of manufacturers of packaging material and distributors of packed consumer goods should guarantee the development and production of environmentally sound products.

In the early 1990s, the German federal government and its Ministry of Environment contemplated an exclusively paper-related ordinance setting goals for recovery and recycling of graphic paper products. This intensively discussed Waste Paper Ordinance was not enacted because the associations of German paper manufacturers, paper importers and wholesalers, printers, publishers, and others succeeded in negotiations with the German government in producing a voluntary agreement. It encompasses recovery and recycling of used graphic paper products from printed material (news, magazines, directories, catalogs, brochures, inserts, direct mail, etc.) and office papers including copy papers and computer printouts.

Recovery rates of used graphic paper products
(related to total consumption of graphic paper in Germany)

• By Dec. 1994	53 %
• Jan. 95 - Dec. 96	55 %
• Jan. 97 - Dec. 99	58 %
• From Jan. 2000	60 %

- Recovered printed material and office papers can be used in the German paper industry and in foreign paper industries as well
- Recovered printed material and office papers can be used in manufacturing of any paper grade (Printing and writing, packaging paper and board, hygienic and specialty paper) and in other industries as well

Figure 2. Voluntary agreement on recovery and recycling of recovered graphic papers.

This agreement enacted in 1994 sets targets for material recycling as a percentage of the annual consumption of graphic papers per Fig. 2. The recovery rate of printed material and office papers was to increase from 55% in 1995 to 60% in 2000. In 1998 the recycling rate of the total used graphic papers including office papers already achieved about 80%. Reasons for this unexpected development are the strong increase of recycling capacities in the German paper industry, improved recovery efficiency, and increased material recycling abroad. The federal government and its Ministry of Environment will probably review the voluntarily agreed recovery rates by the year 2000.

The collected products consisting of used graphic paper may be re-used for paper manufacturing in Germany or abroad. In addition, recycling may occur in any paper and board grade or in other materials such as molded products or particle board. Considering these goals, the approach of the Voluntary Agreement is liberal because it does not force the paper industry to produce paper and board with set recovered paper utilization rates of individual paper and board grades. Apart from the primary aim of that agreement, other goals include the following items:

- Manufacture of graphic papers and printed products should use fibers, additives, fillers, coating pigments, printing inks, and adhesives that do not adversely influence recycling.
- Research and development should contribute to intensifying recycling of used graphic paper.
- Importers of paper and paper products should encourage their suppliers to use recycled fibers and conduct research and development to increase recycling of recovered graphic paper.

Between 1995 and 1999, the Alliance for Graphic Paper (AGRAPA) in Bonn was the group responsible for implementation of the Voluntary Agreement. It arranged collection trials in cooperation with five local communities in different parts of Germany to obtain detailed information on costs of collection, effectiveness of different collection systems, and quality of recovered paper considering its composition as original paper grades, amount of impurities, and chemical and physical characteristics.

According to the German Pulp and Paper Association (VDP), the establishment of the Voluntary Agreement is a political and economical success especially compared with the Packaging Ordinance. The Voluntary Agreement should improve the utilization of recovered paper from households and commercial enterprises in three specific areas: economy, quality, and material recycling.

As mentioned above, the Packaging Ordinance resulted in the establishment of a private but monopoly-like and much criticized organization: DSD. This organization is responsible for collection of packaging material using paper and board, plastic, composite material, glass, and metal but only for sales packaging with Green Dots. For the collection and recovery of transport packaging such as old corrugated containers or used paper sacks, other organizations have responsibility. They charge their clients who are often the manufacturers of corrugated containers or paper sacks a licensing fee.

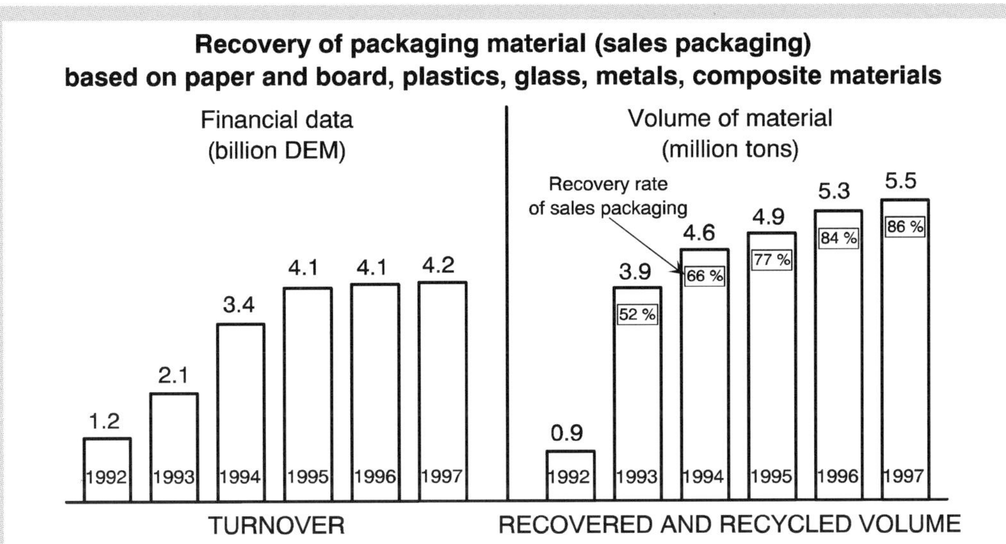

Figure 3. Recovery of sales packaging by Duales System Deutschland (DSD) and its budget.

In 1993, the DSD budget of 2.1 billion DEM financed by the Green Dot licensing fees as Fig. 3 shows did not cover the collection costs of 2.9 billion DEM. This is because many manufacturers and commercial customers of sales packaging did not contribute to the budget of DSD although their packaging material entered the recovery system managed and financed by DSD. The situation improved in 1994 when the budget amounted to 3.4 billion DEM because 3 800 additional clients or companies paid for the licensing fees. In 1996, the budget of DSD balanced for the first time. With a budget of 4.1 billion DEM, 5.3 million tons of sales packaging were recovered and mainly reused by recycling processes. For the packaging material recovered on behalf of DSD, material recycling is guaranteed for paper, glass, and metallic products but not for the total volume of composite materials and plastics. These are difficult to recycle. They therefore have some use as a carbon source such as in blast furnaces as replacement for coal. This replacement is not regarded as energy recovery.

Energy recovery from sales packaging does not agree with the philosophy of the Packaging Ordinance since it requires recycling of used material for manufacturing new products. The recovery of sales packaging material for recycling still increased continuously from 52% in 1993 to 86% in 1997 when it reached a total of 5.5 million tons. One-third of the recovered material comprises paper, board, and composite material (1.4 million tons + 0.4 million tons).

Contractors of the DSD are traditional recovered paper merchants and newly established waste management companies. Locally, almost no competition exists between such contractors. In the field of recovery of pre-consumer waste from industrial and commercial outlets – outside the responsibility of DSD – keen competition did exist. It led to a restructuring between groups of traditional dealers of recovered paper and waste management companies as efficient newcomers in this field.

Today in Germany, the paper collection business is not driven by market forces, but purely by supply due to legislation. The key object is reduction of the volume of solid waste for disposal by landfilling or incineration in municipal plants. In addition, dumping of organic waste by landfilling has no future in Germany because of ecological issues (emission of methane-containing biogas and polluted leaching water). To control the environmental problems of landfills including their effect on landscape and public acceptance, an additional regulation is the so-called Technical Guideline: Municipal Waste Management Provision issued in 1993.

The Municipal Waste Management Provision for disposal of municipal and industrial waste foresees that by the year 2005 no solids that contain more than 5% organic matter can be landfilled. The highly organic sludge and other residues of the paper industry will therefore require incineration before dumping as ash. Of course, re-use of sludge and other residues in other industries such as the cement industry and brick works or in compost is strongly desired. The recovery of energy from nonrecyclable paper waste and other lignocellulosic residues is permitted. Incineration guarantees the minimization of hazardous substances although the amounts present in paper and biomass waste are usually insignificant and noncritical. Although biomass combustion does generate carbon dioxide emissions, this type of reduction of waste volume is permitted (See Chapter 13).

According to the most current and comprehensive legislation in Germany, the aforementioned Recycling and Waste Management Act, the production of waste must first be prevented as far as practical. Waste must either be re-used, recycled, or used for energy recovery. The type of use of waste that is most compatible with the environment should have priority. The federal government can issue ordinances giving priority to material recycling or to energy recovery for certain types of waste. If a regulation in force or about to be issued does not define priority, energy recovery is only permissible when several conditions are fulfilled:

- The heat content of the waste in question is at least 11 MJ/kg (11 GJ/t)
- A combustion efficiency of at least 75% is achieved
- The resulting energy is either used by the recovering entity or supplied to a third party.
- The waste generated by energy recovery, i.e., the ash can be disposed of with no further treatment.

Waste from renewable raw materials such as paper mill rejects and sludges or excess waste paper has use for energy recovery if it meets the prerequisites of the last three items. This means that energy recovery must include co-generation, the traditional technology of power plants in the paper industry. The paper industry has agreed to the principle that nonrecyclable paper products may be used for energy recovery[5].

1.2 Other European countries

1.2.1 Austria

The legal framework conditions for the collection, recycling, or final disposal of packaging material are defined in Austria by the Waste Management Act of 1990. A Packaging Ordinance and a Recovered Packaging Material Objective as defined in 1993 with amendment in 1996 are based on this law. The objective defines the maximum residual amounts of packaging material from paper, cardboard, board, and corrugated board that may still be deposited in landfill sites by the year 1998.

To ensure achievement of these objectives, manufacturers and distributors of packaging materials (sales packaging, re-packaging, transport packaging) are obliged to accept their returned packaging materials after they have been used without charge. All used packages can be taken back by a previous member of the supplier chain or must be re-used or recycled in the sense of the ordinance. If this obligation is not transferred to a third party, manufacturers and distributors must collect 80% of all transport and sales packaging material from paper, cardboard, board, and corrugated board as of 1999. After removal of nonpaper components and paper not subject to this ordinance 90% of the collected packaging materials must be recycled.

The possibility to fulfill the recycling requirements individually or with a third party forms the basis of the so-called ARA system (Altstoff-Recycling Austria = secondary material recycling in Austria) established by industry in Austria in 1993. As a national collection and recycling system, this organization relieves packaging operations in Austria of their recycling obligations. Within the ARA organization, the ARO (Altpapier-Recycling Organisationsgesellschaft = recovered paper recycling organization) is the industrial recycling operation of the Austrian paper industry. It is responsible for the collection and recycling of paper and board packaging. To finance its activities, ARA transfers some revenues obtained from licensing fees to ARO.

Since the amendment to the Packaging Ordinance in 1996, the percentages for the collection of transport and sales packages have been omitted. In addition, the stipulated goal for recovered packages states that after 1998 not more than 140 000 t annually of packaging material from paper, cardboard, board, and corrugated board may be deposited in landfills. After 2001, this quantity falls to 99 000 t per year.

Presently, no legal recovery and recycling decrees exist for graphic papers or graphic paper products respectively.

1.2.2 Belgium

In the beginning of the 1990s, the establishment of an ecological tax for paper and board products was discussed in Belgium. The tax rates should be calculated on the content of recycled fibers contained in different paper grades. Besides paper and cardboard packaging, graphic papers and hygienic papers also were intended to become subject to the tax. Originally scheduled for implementation in 1994, the law has been postponed several times. Since no method is available for determining the content of recycled fibers in a paper product, the law will not be enacted.

According to the interregional decree, 80% paper and cardboard packaging material should be collected by 2001. Of this, 50% should be recycled; at least 15% material recycling for each category of material.

1.2.3 Finland

The Finnish Waste Act from 1994 was totally rewritten according to the corresponding EU Waste Directive. The law essentially adopts the principle of producers' responsibility ranging from the obligation to provide information to setting recycling goals. Municipalities are responsible for collection of recovered material from private households. They may commission other organizations or private businesses to carry out these activities on their behalf. The municipality still remains liable for fulfillment of its legal obligations.

In Finland, a Council of State Decision is in force since 1997 for packaging and packaging waste. It requires all domestic packers and importers of ready-packed products to take responsibility for the recovery of the packaging material they have put in the domestic market. The targets by June 2001 are at least 61% recovery and 42% recycling for the total packaging waste (comprising different materials) with at least 15% by weight of each used packaging material to be recycled. For paper and board packaging waste, the rates are 75% for recovery and 53% for recycling. Recycling includes the reprocessing of the material in a production process or its use as a raw material for its original or other purposes including organic recycling (composting) but excluding energy recovery. Energy recovery is part of the targets of recovery that comprise the separation of the packaging material and further use of the material or energy in the packaging waste.

The Council of State Decision on collection and recycling of selected graphic paper products is in force since January 1999. The paper product categories include news, magazines, advertising papers, catalogs, telephone directories, envelopes, office papers, and similar paper products sold and used on the domestic market. 70% of the described graphic paper products (imported paper products are also included) should be collected and recycled by 2000. The first priority is recycling into new paper or board, but energy recovery is also permissible. This rate will increase to 75% by 2005.

1.2.4 France

In France, various ordinances for different products stipulate the principle of producer responsibility. Special efforts have occurred in the packaging sector where the establishment of Eco-Emballages marked the formation of a collection system comparable with ARA in Austria or DSD in Germany.

Complete collection is already stipulated for industrial paper and board packaging. By 2002, the aim is collection of 75% of paper and board (sales) packaging material originating from private households. In France, re-use comprises recycling into new paper or energy recovery, but legislation does not presently define specific recycling rates.

No specific recovery or recycling rates are currently under consideration for used graphic paper products.

1.2.5 Sweden

Swedish legislation regulates producer responsibility for various packaging materials. Phased regulations cover different paper products that have been in force since 1994 for packaging, news, and magazines. For packaging materials, the Ordinance on Producer's Responsibility for Packaging (No. 1235: 1994) defines a recovery and a recycling rate of 30% for paper, cardboard and board packaging material. For corrugated board packaging, a recovery and recycling rate of 65% including re-use or material recycling is to be realized. Starting in July 2001, the rates for cardboard will rise to 70% recovery. Corrugated material will experience no change.

For office and administration paper, a voluntary agreement exists between the Ministry of Environment and the recovered paper merchants dating from 1997 to recover 50% after 2000. According to the Ordinance on Producer's Responsibility for Recovered Paper (No. 1205: 1994), paper manufacturers and importers must guarantee that 75% of all news, magazines, catalogs, and direct mailings used in Sweden are collected by 2000. A system for separated collection of paper in private households has been in use for many decades.

1.2.6 The Netherlands

In the Paper Fiber Covenant that is a sub-covenant of the Packaging Covenant, the goal for the collection of paper and board packaging material until 2001 is 85%. In the "Agreement on a take back levy", municipalities are guaranteed that all household collected paper and board will be taken by recovered paper merchants at a minimum price of zero. This means that the collection and the collection costs of recovered paper from households are the responsibility of the municipalities.

The Paper Fiber Covenant was signed by the Minister of Environment on behalf of the Ministers of Economic Affairs and Agriculture, the Association of Municipalities, and Paper Recycling Netherlands (PRN). The associations of printers, publishers, carton makers, corrugated box makers, recovered paper merchants, liquid board packaging makers, and the paper and board industry participate in PRN. The "Agreement on a take back levy" commits the PRN parties to exercise the take back guarantee with a minimum price of zero. A fund will operate when recovered paper prices are too low to cover the costs of the recovered paper merchants.

For excess recovered paper, a private company, the Surplus Management Ltd., founded by the paper and board industry will handle the surplus quantity of paper and board.

1.2.7 United Kingdom

Legislation in the United Kingdom presently concentrates on taking back and recycling packaging material in accordance with the EU Packaging Directive. A goal is 50% returns by 2001. In the interim, further extensions are under discussion within the Producer Responsibility Group (PRG). From 1998–2000, the aim is to achieve recovery rates for paper and board packaging materials of 40% and recycling rates of 8% for all packaging materials. By 2001, the corresponding rates will be 50% for recovery and 15% for recycling.

No recovery or recycling rates are presently defined for used graphic paper products.

2 United States

Many individual states have issued laws and promulgated rules that promote recycling, reduce the amount of waste material, or prohibit individual objects from being disposed in landfills. About 40 states have passed recycling laws or implemented voluntary goals and directives. Most states have determined a recycling goal between 20% (Maryland) and 70% (Rhode Island) for used paper products. With its Act 335, Wisconsin issued the most comprehensive and far-reaching recycling and waste material prevention law as early as 1989. Besides numerous promotional and educational programs, the law established Responsible Units (RUs) for local recycling activities. RUs receive grants to administer these activities[6,7].

In contrast to European directives that define specific recovery rates for paper and board packaging and partly for printed material, a recycled fiber content for specific new paper grades is effective in the United States. This began in 1976 with the Resource Conservation and Recovery Act that led to the introduction by the U.S. Environmental Protection Agency (EPA) of a set of Procurement Guidelines for governmental agencies in 1991.

Dissatisfaction with these guidelines motivated the U.S. Recycling Advisory Council (RAC) to promulgate a minimum recycled fiber content (including post-consumer recovered paper) for the paper procurements of the federal authorities. President Clinton renewed the EPA acquisition guidelines by the Executive Order called Greening the Government by Waste Prevention, Recycling, and Federal Acquisition in 1998. According to this order, all writing, printing, and office papers purchased after 1998 with public funds must contain at least 30% post-consumer recovered paper or not less than 50% recovered material including pre-consumer recovered paper. All price increases must be offset by a reduction in the amount of waste material produced and by lower consumption. This constitutes a rare example of an officially decreed consumption limitation.

In 1995, the EPA issued a Guidance Document on the Acquisition of Environmental Preferable Products and Services. This was the first comprehensive articulation of its "green products" policy. The document contains seven principles to ensure that environmentally-friendly products receive greater consideration for orders placed by state procurement agents. This considered quantity and quality. The first principle of this environmental preference did not consider whether one "green product" can substitute for another. Instead, it questioned whether a function must be improved and realized this while minimizing negative effects to the environment. The intention was that federal agencies should acquire more environmentally-friendly products and services and initiate voluntary pilot projects.

For paper and paper products, the EPA Office for Solid Waste has published a Paper Products Recovered Materials Advisory Notice. This guidance note promotes paper recycling by centralized acquisition and expands the markets for papers containing recycled fibers. Two important agencies for paper acquisition in the United States are the General Services Administration (GSA) and the Government Printing Office. The EPA guidance notice has special importance in that it extends beyond governmental agencies and is also applicable on a wide scale for purchases in the private sector. Such purchases represent 95% or more of total paper requirements.

EPA has defined conditions for the minimum recycled fiber content in different paper grades. A two-tier level exists consisting of post-consumer fiber content and total recycled fiber content. According to the American Forest & Paper Association (AF&PA), almost all pre-consumer recovered paper is collected and recycled. Therefore, EPA believes that the two-tier approach leads to greater use of post-consumer recovered paper. To date, the EPA has not defined environmentally preferable paper in an even stricter sense. For example, it has not taken into account major environmental influences on paper production such as the type of forests from which the virgin fiber furnish comes.

Table 3 shows the minimum content of recycled fibers for different paper grades. It includes a comparison of the recycled fiber content proposed by RAC for government acquisitions since 1992 with the EPA guidelines. At first glance, the EPA conditions appear to be stricter. The statistical values must be interpreted very carefully since definitions differ somewhat[2].

Table 3. RAC and EPA guidelines for government paper products[2].

Product Line	RAC 1992		EPA 1995	
	Total recycled fiber content	Post-consumer fiber content	Total recycled fiber content	Post-consumer fiber content
Newsprint	40%	40%	40%–100%	40%–85%
Printing & writing	40%–50%	10%–15%	10%–50%	10%–20%
Tissue	60%–80%	50%–70%	20%–100%	20%–60%
Construction paper	80%	65%	n/a	n/a
Packaging paper	40%	20%	5%–40%	5%–20%
Uncoated paper: corrugated	40%	35%	30%–50%	30%–50%
Uncoated paper: other	100%	60%	n/a	n/a
Coated paperboard	90%	45%	100%	45%–100%

n/a: not announced

Besides these general regulations, ordinances, and guidelines, many federal regulations exist in the United States to avoid waste material. Typically, the laws cover the fields of separation of discarded paper at the source, procurement preference, and recycled fiber content in newsprint. These regulations achieve the aim of the EPA to reduce the amount of waste material for disposal in landfills or in incineration plants. These regulations include the following:

- Mandated separation of newspapers by households
- Federal, state, and municipal procurement policies favoring recycled materials
- Guidelines requiring consumers of newsprint to use recycled fiber-containing newsprint for a certain portion of their total requirements (Table 4).

Table 4. U.S. guidelines (state regulations and voluntary commitments) for recycled fiber content in newsprint[8].

State	Penalties	Goal	Timetable	Voluntary (V) Mandatory (M)
Arizona	Not to exceed $1 000/violation	40% 50%	1998 2000	M
California	Subject to civil penalties	40% 50%	1998 2000	M
Colorado	n/a	30%	1998	V
Connecticut	$5 per ton shortfall, (minimum $2 500, maximum $100 000)	45% 50%	1998 1999	M
Florida	$5 per ton shortfall	50%	1995	M
Hawaii	n/a	n/a	n/a	V
Illinois	Attorney General offices for prosecution	28%	1993	M
Indiana	n/a	23%	1996	V
Iowa	n/a	30% 40%	1998 2000	V
Kentucky	n/a	n/a	yearly	M
Louisiana	n/a	30% 40%	1997 2000	V
Maine	n/a	40%	1995	V
Maryland	$10 per ton shortfall	40%	1998	M
Massachusetts	n/a	31% 40%	1997 2000	V
Michigan	n/a	24%	1995	V
Missouri	Civil penalty of not more than $100 per day of violation	40% 50%	1996 2000	M
New Hampshire	n/a	31% 40%	1997 2000	V
New York	n/a	31% 40%	1998 2000	V
North Carolina	$15 per ton shortfall	40%	1998	M
Ohio	n/a	31% 40%	1998 2000	V
Oregon	n/a	7.5%	1995	M
Pennsylvania	n/a	50%	1995	V
Rhode Island	Administrative penalties	31% 40%	1998 2000	M

Table 4. U.S. guidelines (state regulations and voluntary commitments) for recycled fiber content in newsprint[8]. (continued)

State	Penalties	Goal	Timetable	Voluntary (V) Mandatory (M)
South Dakota	n/a	n/a	n/a	V
Texas	Mandatory enforcement measures	20% 30%	1997 2000	M
Vermont	n/a	31% 40%	1997 2000	V
Virginia	n/a	25% 30%	1997 2000	V
Washington, DC	$10 per ton shortfall	20%	1994	M
Wisconsin	1% of total annual newsprint cost multiplied by recycling status	40% 45%	1998 2000	M

n/a: not announced

The table shows that the state regulations are mandatory or voluntary. Goals of recycled fiber content in newsprint differ significantly by state. The various states have no general definition either for the penalties or for the basis of the recycling goals. In some states, the percentage figures relate to the annual consumption of the following:

- Recycled fiber content in newsprint
- Recycled fiber content in newsprint in relation to total newsprint consumption where recycled fiber-containing newsprint consists of not less than 40% post-consumer recovered paper
- Recovered paper utilization in relation to newsprint consumption.

Such regulations therefore force mills in the USA or Canada to invest in paper recycling plants to produce papers with the stipulated proportion of recycled fibers. Customers of paper are limited in their choice of supplier. A lack of methods also exists to monitor the recycled fiber content in paper and board grades produced with certain proportions of recovered paper.

Against this background, guidelines for a national recovered paper collection rate, recycling rate, or both make more sense because they give the paper industry considerably greater freedom in its efforts to achieve the targets particularly concerning technical and economical matters. Depending on the availability of recovered paper grades and technical possibilities of recovered paper processing, corporate decisions could ensure the recycled fiber content of future paper production on a national level. This also allows newsprint mills that operate near major cities to make use predominantly of recovered paper as fiber material. Mills located in rural, thinly populated areas can continue to produce newsprint from virgin fibers. Both paper grades can be marketed everywhere and compete directly with one another. For this reason, AF&PA has specified a recovered paper collection rate of 50% of paper consumption by 2000 as a goal for the American paper industry. In 1997, the rate was 45%.

3 Japan

Japan is the world's second-largest consumer of recovered paper at 16.5 million tons in 1997. Only the United States has greater volume. Considering the structure of the paper grades produced, negligible paper exports and imports, and the fact that Japan in the middle1980s had already achieved high utilization rates, a consideration of the political goals for recovered paper recycling is useful. The Waste Disposal Law enacted in 1970 was amended in 1991 when the Resource Recycling Facilitation Law called the Recycling Law was added. Article 1 describes the purpose of this law: "Considering that Japan relies on importing many important resources, but a large part of these resources are now being discarded without being used, the purpose of this law is to provide the basic mechanism for promoting the use of recyclable resources, and thereby promote the healthy development of the nation's economy."

According to Ministerial Ordinance No. 53 of the Ministry of International Trade and Industry (MITI), a further increase in the utilization rate of recovered paper was expected. Article 1 of this ordinance states: "Taking into consideration that the low ratio of recovered paper used in printing paper, information-related paper, and wrapping paper, and the different uses of recovered paper depending on the kind of paper, the industry shall, in cooperation with consumers as well as central and local governments, strive to increase to 55% the amount of recovered paper that is used in domestic paper production by the end of the Fiscal Year 1994 (April 1994 to March 1995)." Paper producers must also do the following:

- Install processing systems for the utilization of recovered paper and provide the necessary storage space for recovered paper
- Improve recovered paper processing technology in cooperation with machine manufacturers and chemical suppliers
- Develop a recovered paper utilization plan for each fiscal year
- Publicize statistical data on the percentage use of recovered paper.

The MITI contribution to increase the utilization rate of recovered paper consists of regulations that create financial and tax incentives on the recovered paper market. These include reduced corporate taxes, special rate of depreciation allowance, low credit rates for purchasing machines for pre-treating recovered paper such as bale presses and wiring machines, and a model for collection systems. Paper producers were offered reduced tax rates and special depreciation rates for machines for processing recovered paper and favorable loans for installation of deinking systems.

MITI itself launched several advertising campaigns for use of recycled fiber products. Two private associations also promoted use of recycled paper among consumers. The Paper Recycling Promotion Center introduced "Green Mark Labels" for magazines, toilet paper, and books produced from recycled fibers. The Japan Environment Association developed the Eco-Mark project for products manufactured with recycled fibers. Unfortunately, due to the unclear definition of the recycled fiber contents in paper and board products, both programs had only limited influence on the consumers' decisions.

The objective of a 55% utilization rate for recovered paper in 1994 was determined in 1989 when the rate was 50%. This rate was accepted without reservation by the industry since it represented a realistic goal. Nevertheless, a utilization rate for recovered paper of only 53% occurred in 1994 although the market had an oversupply of recovered paper. Some factors that contributed to this failure to reach the projected utilization rate were the following:

- Reduced production share of those grades of packaging papers and board that contain large proportions of recycled fibers
- Increased demand for high-quality paper products
- Rising costs for processing recovered paper due to environmental legislation
- Efforts by industry to increase profits
- Inadequate legislation.

This failure by the paper industry to achieve the goal of recovered paper use of 55% motivated the Ministry of Health and Welfare (MHW) to accuse MITI of taking an excessively lenient approach toward paper producers. Economic studies concerning the use of recycled papers at private businesses showed that the consumers of recycled papers either demanded identical quality as with papers manufactured from virgin fibers or much lower prices for them. The studies also revealed that a precondition for raising the level of use of recovered paper is improvement of public knowledge and understanding of recycled paper products.

To overcome these problems, MITI established a Study Committee on Basic Issues of the Japanese Paper and Pulp Industry. This committee comprised representatives of newspaper, banking, and trading companies and the pulp and paper manufacturing, paper distribution, paper converting, printing, and recovered paper industries. MITI supported the research and consulting activities of this committee, used it to establish a new political orientation, and helped companies develop new strategies. In 1994, the Study Committee presented a new program called "Recycle 56" aimed at achieving a recovered paper utilization rate of 56% by 2000. In 1997, a utilization rate of 54% occurred. The new target for recycling of recovered paper will only be reached if all society actively supports efforts to reduce waste and improve recycling. The participation of the recovered paper trading companies and paper mills alone is not sufficient. The industry calls for government support for the following:

- Establish a policy of improving the separation of recovered paper at the source
- Expand collection of office papers by establishing a support system
- Expand consumer demand for recycled paper
- Negotiate an agreement between industry and consumers on standard qualities for printing and writing paper with a clear statement on the content of recycled fibers

- Promote increasing transparency on the domestic and export market for recycled paper products
- Subsidize improved recovered paper processing methods
- Re-evaluate the tax system to promote increased use of recycled products
- Establish a policy to reduce transport costs of recovered paper
- Develop alternative recovered paper uses other than recycling into paper and board products
- Increase recycling for energy recovery. This means that low grade recovered paper can also have use for energy recovery.
- Increase international recycling with increased export of recovered paper.

For the paper sector, the political guidelines proposed by MITI and the Study Committee consider a lower economic efficiency, the development of the paper industry in East Asian countries, the increasing demand for a better market access, and the growing concern for environmental legislation. The future structure of the paper industry calls for the following:

- Promotion of internationalization including the internationalization of enterprises and markets
- Promotion of environmental stewardship that should set new targets for the utilization rate of recovered paper and develop new uses for recovered paper.

MITI and the sector representatives are working to reach consensus on legislation that is enforceable. Recycling and recovered paper issues are a matter of principle, but the best methods of implementing these principles are not clear. An assumption exists that the industry should be recovering and using more recovered paper in ever-increasing production, but very little discussion of ways to reduce consumption of paper has occurred[9].

References

1 *Anon., "Characterization of Municipal Solid Waste in the United States", 1994 Update, EPA530-R-94-042, U.S. EPA, Washington, D.C.,1994.*

2 *Anon., "Towards a Sustainable Paper Cycle", International Institute for Environment and Development (IIED), London, 1996, 258 pp.*

3 *Kibat, K.-D., Wochenbl. Papierfabr. 124(14/15):656(1996).*

4 *Kibat, K.-D., Wochenbl. Papierfabr. 126(16):742(1998).*

5 *Göttsching, L., Hamm, U., Plätzer, E., et al., "Analysis of Waste Paper Recycling and Disposal Options in Germany", Sub-Study No. 15 of IIED Study: Towards a Sustainable Paper Cycle, IIED, London, 1996, 185 pp.*

6 *Anon., "Waste Reduction and Recycling Initiative", PUBL-IE-041, REV 2/93, Wisconsin Department of Natural Resources, Madison, USA, 1993.*

7 *Bilitewski, B., Härdtle, G., Marek, K., et al., "Waste Management", Springer-Verlag, Berlin, 1994, 699 pp.*

8 *Anon., Information from Internet Web-Site of American Forest & Paper Ass. available [online]http: www.afandpa.org, 1998.*

9. *Carbonnier, S., "Paper Recycling and the Waste Paper Business in Japan", Sub-Study No. 16 of IIED Study: Towards a Sustainable Paper Cycle, IIED, London, 1996, 113 pp.*

CHAPTER 3

Collection systems, sources, and sorting of recovered paper

Hans-Joachim Putz

Collection systems, sources, and sorting of recovered paper

In general, the collection of recovered paper is accomplished by:

- Collection in private households and small commercial enterprises
- Collection from industrial and business operations, sites where unwrapping is carried out (e.g., supermarkets), offices, authorities and administration, including returns of recovered paper from converting facilities (e.g., printing houses and corrugated board plants) as well as over-issues of news and magazines.

Depending on the origin of the collected recovered paper, a clear distinction exists between pre- and post-consumer recovered paper. The U.S. EPA defines these two types of recovered paper as follows[1]:

- Post-consumer recovered paper comprises paper, paperboard, and fibrous material recovered from retail stores, offices, and homes after these products have served their end-uses as consumer items and papers separated from municipal solid waste
- Pre-consumer recovered paper is paper and paperboard material from manufacturing and converting and finished paper from obsolete inventories.

1 Collection systems and equipment

For paper recovery, a distinction is made by collection systems and equipment according to their mode of operation. In the “drop-off” system, generators of the used paper product as end-users carry the material to a centralized collection point. Alternatively, when the material is collected at each individual point of generation, e.g., a household or an office, we speak of a “pick-up” system.

The two systems operate somewhat differently depending on whether pre- or post-consumer recovered paper is handled as discussed first.

1.1 Pre-consumer recovered paper

The collection of pre-consumer recovered paper in the form of over-issues and converting residues such as shavings and cuttings uses pick-up systems traditionally organized via the recovered paper dealers using containers at individual points of generation. Pick-up systems collect recovered paper from industrial and business operations. Containers are exchanged or emptied at regular intervals either with or without the benefit of stationary compressing compactors. Collection vehicles, that compress refuse are seldom used.

The traditional recovered paper dealers are facing increasing competition from waste management companies. They tend to offer their services to industrial and business operations and to the retail trade not only for paper but also for the collection of other materials such as glass, plastics, or wood and thus provide greater convenience and flexibility.

Most pre-consumer collections are therefore handled in the pick-up mode.

1.2 Post-consumer recovered paper

Recovered paper from operations where unwrapping occurs (e.g., supermarkets) belongs to post-consumer recovered paper. In these facilities, the same collection systems are installed as for collection of pre-consumer recovered paper, i.e., in the pick-up mode. In contrast, different collection systems are available for recovered paper collection from private households. The choices made are based on population structure (population density and rural or urban regions), housing structure (inner city, high-rise developments, single-family housing) and the customary use for the collected recovered paper. A distinction exists here between pick-up and drop-off systems both of which have been in long-term practice for example in Germany.

1.2.1 Pick-up systems

Container collection

Paper pick-up collection systems placed in households primarily use containers with a volume of 120 or 240 L as mono-bins or larger multicompartment containers. The latter permit separate collection of different secondary materials such as paper, metal, or glass. Special vehicles are necessary for the collection of multicompartment containers to ensure that the different materials remain separated. Larger mono-bin containers of 1.1m^3 volume frequently serve large apartment houses solely for recovered paper collection.

With the most common collection of used paper products from private households using mono-bins, printed graphic paper products are mixed with paper and board packaging material. Avoiding this requires two paper containers or a multicompartment container for the separate disposal and collection of used graphic paper products and paper and board packaging. Use of the recovered paper is then possible in different recycling areas without the labor-intensive manual sorting that is necessary with mixed paper collection when a graphic paper grade for deinking has to be generated.

Curbside collection

Besides container collections, curbside collections can occur at certain intervals by charitable or private organizations. This usually involves collection of bundles of printed graphic paper products. The population collects and bundles used paper products. Bundles of newspapers and/or magazines are put at the curbside at specific times. The collecting organizations pick them up and usually deliver them to nearby mills and are paid by them for the service.

Unfortunately, strongly fluctuating prices for recovered paper have almost completely discouraged curbside collections in Germany. In Switzerland such collections of recovered paper continue to have major importance for the national paper industry. The difference may be due to the fact that the Swiss paper industry has supported and financed the curbside collection all the time even when it was more economic to buy the recovered paper exclusively from the recovered paper dealers.

1.2.2 Drop-off systems

Container collection

Larger containers above 5 m^3 volume have frequent use in drop-off systems. These are at central, publicly-accessible locations and are emptied at regular intervals. They can be single-compartment containers intended only for collecting paper or multicompartment containers for collecting different materials such as paper, glass, and metal that remain separate within the container. Besides the greater space that is necessary for multicompartment containers, the decision on whether to install single-compartment or multicompartment containers depends on the availability of suitable vehicles and the occurrence of the materials concerned at the installation sites.

Containers can simply be emptied into a suitable refuse collection vehicle for single-compartment containers or exchanged. For multicompartment containers, the refuse collection vehicle must suit the multicompartment system of the container to avoid mixing the individual materials. With multicompartment containers, exchanging or emptying the entire container when collecting the materials is necessary despite the fact that individual compartments may not be totally full. This means the collection frequency must consider the material whose compartment fills first.

Avoiding of overflowing containers or compartments is essential. Otherwise, the tendency to throw materials in still empty containers or compartments for other materials increases, or the sites can turn into rubbish dumps. Single-compartment containers offer the basic advantage that the collection intervals of single containers can be linked to the occurrence of material. Containers that fill more quickly can be emptied more frequently than ones that fill more slowly. Larger containers can also be selected. In principle, separate collection of used graphic paper products and packaging material is possible with two different containers or a multicompartment container.

Recycling centers

Some regions in Germany collect materials at so-called drop-off recycling centers where municipal employees accept a range of different materials. Here, the collection has supervision that is not available with container collection in public places. A disadvantage of the drop-off recycling centers is that disposal is available only at certain times. The ability to dispose of difficult materials such as paints, varnishes, used batteries, or motor oils can compensate for this disadvantage[2].

1.3 Efficiency of different collection systems

For the different systems, the specific amount of paper collected per inhabitant depends on many different marginal conditions such as collection frequency, population density, and sizes of the containers used. Analyses of this in 1994 in Germany for different cities, communities, and regions revealed the following average specific recovered paper collection figures[3]:

- Drop-off container system: 32 kg per capita and year
- Curbside collection: 43 kg per capita and year
- Recycling center: 45 kg per capita and year
- Pick-up container system: 53 kg per capita and year.

Densely populated regions should therefore use a pick-up container system although the space required for an additional paper container at home requires consideration. In less populated areas, drop-off container systems are useful. Finding the most suitable locations for the containers (clean and attractive, accessible by automobile, and close to home) is especially important to ensure that the collectable volume and the standard of purity of the recovered paper satisfy the requirements. Bundled paper collections in densely populated areas are always suitable if containers cannot be installed in sufficient numbers.

A few regions of Germany collect used papers from private households as one component of the so-called multicomponent recyclable container collection. Here, one container is used in private households for collecting several materials such as printed products, packaging materials including liquid packaging, plastic, metal, and textiles. Glass is usually collected separately due to the danger of breaking and subsequent occurrence of glass splinters in the paper and board fraction even after manual sorting.

Although paper recovery by this multicomponent container collection is possible, it requires especially intensive and consequently expensive manual sorting. Generally, contamination of the moisture-absorbing recovered paper with liquid and foodstuff residues from liquid packages, plastic bottles, and tin cans can also occur. The paper industry therefore has major reservations about the use of recovered paper from this collection system. In accordance with European Standard EN 643 (See Chapter 4 on "Recovered paper grades"), recovered paper from this collection system requires express identification as such.

Recovered paper from multicomponent collection systems has not been used for materials and articles of paper and board for contact with foodstuffs that must comply with international or national hygiene regulations. It is not permissible to mix recovered paper from multicomponent collection systems with other recovered paper grades without special identification.

According to Table 1, post-consumer paper collection by the different systems in Germany has changed considerably in only two years in favor of a more comprehensive recovered paper collection documented by an increase in the recovery rate from 50% to 59% during this period. In 1997, Germany had a recovery rate of 71%.

Table 1. Collection of post-consumer paper in Germany by different recovery systems.

Collection system	1992[2]	1994[4]
Pick-up container	44%	37%
Drop-off container	39%	53%
Curbside collection	14%	8%
Other systems	3%	2%

1.4 Municipal solid waste

Despite all efforts to recover paper as a raw material for the paper industry, domestic waste also includes paper and board although in increasingly smaller quantities. In the United States according to the EPA definition, municipal solid waste (MSW) consists of durable goods, containers and packaging, food scraps, yard trimmings, and miscellaneous wastes from residential, commercial, and industrial resources. In 1996 in the United States, MSW contained about 38% paper and paperboard of which about half was recycled. Finally, about 19% paper content remained in MSW and went to landfill or to incineration plants[5].

In Germany, according to the most recent National Household Waste Analysis (NHWA) from 1985, the share of paper, paperboard, and composite paper contained in household waste was only about 18%. This result was achieved much sooner than in the United States due to the totally higher level of separate paper recovery in Germany and also due to different categories of waste counted as domestic waste.

The most recent results from individual German cities reveal even lower figures[2]. In principle, the paper industry takes a negative stance regarding the use of recovered paper that is sorted from general domestic waste collections. This is also reflected in EN 643 by the specification that "paper recovered from general refuse is not suitable for use in the paper industry".

2 Sources of recovered paper

2.1 Germany

At the international level, no data are available for the different sources of collection systems of recovered paper by tonnage. In Germany, VDP (Pulp and Paper Industry Association) has conducted recovered paper surveys for several years. In 1997, the following three sources of recovered paper are indicated:

- Recovered paper trade, municipal authorities, communal organizations, private collectors (87% by recovered paper tonnage)
- Internal plant containers, collections, and participants in recovered paper trading companies (10%)
- In-house paper converting plants (3%).

A comparison of the latest German data with surveys from the year 1993 shows that the procurement of recovered paper via third parties (especially the recovered paper traders and municipal authorities) has developed in favor of the paper industry's own activities within only a few years. This trend is a consequence of the intensified recovered paper collection due to the increased demand for such paper and the legal situation.

2.2 Western Europe

Figure 1 provides a total view of paper recovery in Western Europe in 1996. The figure includes the tonnages of paper at all production and consumption stages. It considers the following four product groups:

- Packaging papers and board
- Graphic papers
- Hygienic papers
- Specialty papers.

Almost 73 million tons of paper and board were produced in Western Europe. For this, 32 million tons (44% utilization rate) of recovered paper and 50 million tons of other raw materials were used. The European market demanded 66.1 million tons of paper and board. The export totaled 6.8 million tons.

Considering the residues from converting and printing such as cuttings and shavings, a market supply of 60.8 million tons of converted paper and board products was realized. After recovery of 1.5 million tons of over-issues, a product use of 59.3 million tons was established. By subtracting the volumes of nonrecoverable paper and board products, and the unusable recovered paper, the theoretical recovered paper potential was 50.7 million tons. This is the maximum tonnage of paper products that could be recovered from industrial operations, business, and final consumers including households.

26.2 million tons of used paper products were recovered through household and industrial collection. If one considers recovered paper as material from converting, printing, and over-issues, 33 million tons of recovered paper were collected in Western Europe. After subtracting the net export of 0.4 million tons and the use of recovered paper outside paper production of 0.6 million tons, 32 million tons of recovered paper remain for use in paper production. The 24.5 million tons of paper not recovered from the end-users (including the 8.6 million tons of unusable recovered paper) are disposed of by landfilling, incineration, and composting.

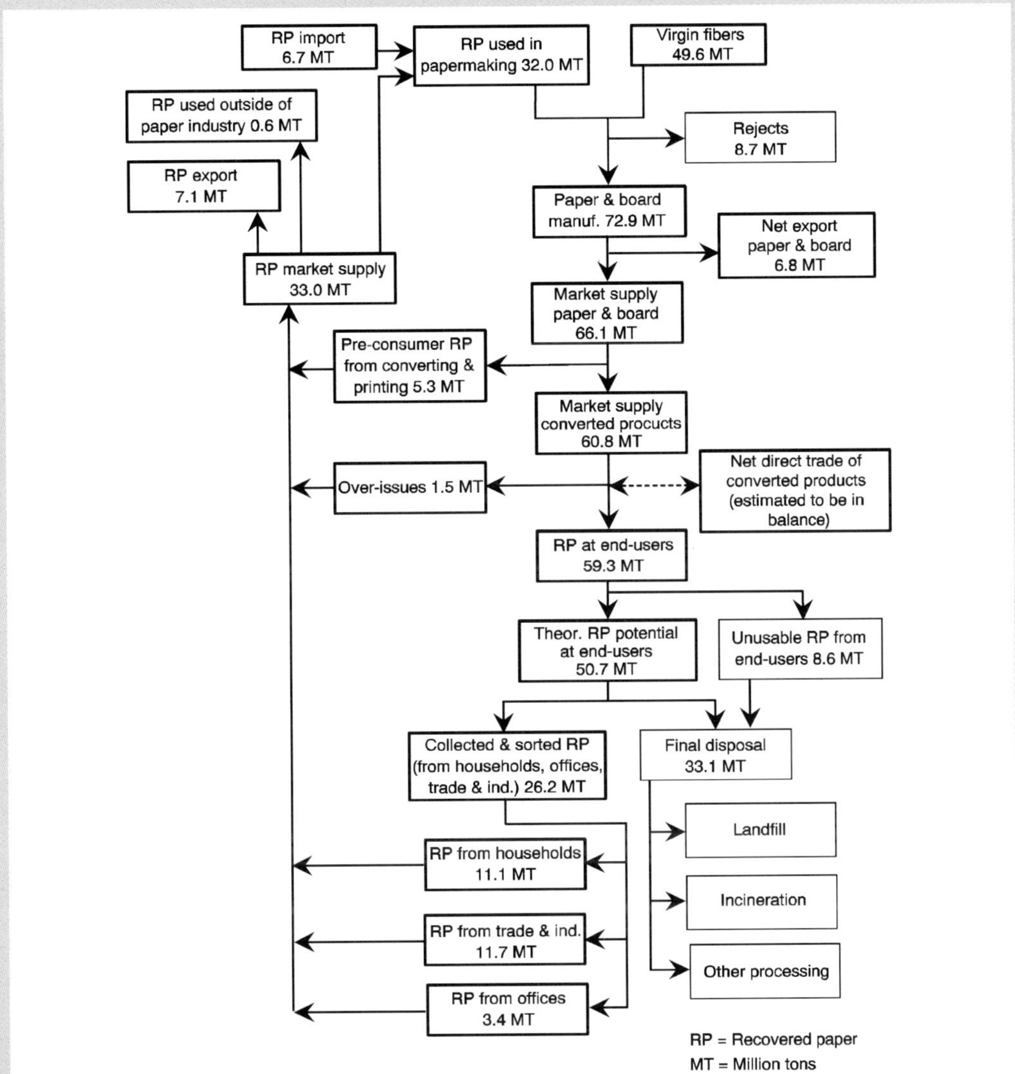

Figure 1. Diagram of paper, paper products, and recovered paper in Western Europe in 1996[6]. (MT: million tons)

3 Sorting, handling, and storage of recovered paper

3.1 Sorting

Despite many technical developments, the sorting of recovered paper following its collection continues to be predominantly a manual, labor-intensive activity that is increasing as the recovered paper becomes more and more heterogeneous and contaminated with nonpaper components. If recovered paper is a component of a multicomponent collection of different materials for example, its sorting will be especially troublesome and will require machines such as grid drums to remove small components, magnetic separators to separate metals, and blowers to remove especially light weight components from the material stream. Nevertheless, the received paper fraction requires manual sorting.

Recovered paper intended for use as a raw material for the paper industry is usually collected separately from other materials so that an exclusively manual sorting is sufficient. Besides an inclined conveyor and a speed-adjustable sorting belt that is usually installed at a higher level, no further mechanical auxiliary equipment is necessary. Several persons work at this sorting belt to remove unusable materials manually. This includes nonpaper components such as wood, metal, glass, plastic, and paper and board that are detrimental to production such as liquid packaging. The workers throw these items into containers provided for this purpose below the sorting belt.

The amount of work involved in the manual sorting depends primarily on the origin of the recovered paper. Pre-consumer recovered paper such as cuttings or shavings from paper converting requires only a very low level of sorting. It can often go to recovered paper processing after inspection on the spot. In contrast, post-consumer recovered paper such as from household collections always requires sorting to satisfy the quality standards specified for the different recovered paper grades. This is especially true for recovered graphic paper for deinking.

A distinction is made between positive and negative sorting:

- With positive sorting, the material to be recovered is removed from the incoming stream. This sorting technology produces a relatively clean material. A disadvantage is the low specific sorting output per person and unit of time.

- Because separately collected recovered paper consists primarily of paper and board with relatively small amounts of unusable materials, the negative sorting method occurs almost without exception to obtain a greater sorting productivity. In this case, only the unusable materials are removed from the incoming material stream. The sorted recovered paper can be marketed in accordance with its composition such as mixed recovered paper or origin such as recovered paper from department stores.

Negative sorting usually results in a lower quality level since unusable materials can be overlooked and may remain in the sorted product. Excessively high speeds of the sorting belt or insufficient personnel can cause a high proportion of unusable materials even in the sorted, recovered paper.

The growing demand for recovered paper for deinking has caused a situation in which recovered paper from post-consumer collections that consists primarily of a higher proportion of graphic papers and a lower proportion of packaging materials undergoes a modified negative sorting. Besides the unusable materials, packaging materials are also separated into different containers. The remaining recovered paper should then be free of unusable and packaging materials. This allows offering it as recovered paper for deinking. The removed packaging paper and board fraction is marketed as mixed paper or department store waste depending on its composition. With this type of sorting, small-component packaging or torn material such as corrugated board containers cannot be completely sorted by a justifiable effort. This results in a certain amount of packaging material remaining in the recovered paper for deinking.

Accordingly, paper mills with deinking operations demand a separate collection of (white) graphic paper and (brown) packaging materials for post-consumer recovered paper. From the start, this ensures that the proportion of packaging materials contained in the collected graphic paper is considerably smaller. This improves the quality of the sorted recovered paper for deinking.

The sorting productivity depends highly on the recovered paper collection system.

- For the sorting of mixed papers and boards collected in containers, sorting productivity of 0.7–1.2 t per person and hour (t/p·h) is possible.
- Sorting productivity decreases to 0.7–1.0 t/p·h when generating recovered paper for deinking from mixed collected papers.
- The sorting productivity for recovered graphic papers collected in bundles is 1.0–1.5 t/p·h.
- If recovered paper is collected as a component in a multicomponent material collection system, the sorting productivity drops independently of the additional mechanization of the sorting system to 0.4–0.7 t/p·h.

A dramatic reduction in the sorting productivity occurs if the recovered paper undergoes compression before sorting such as in compactor-type refuse vehicles to increase the load. Recovered paper treated in this way can usually only be marketed as mixed recovered paper[2].

3.2 Handling

The additional handling of sorted recovered paper depends on transport conditions of the recovered paper operation, the storage conditions at this operation and at the mills, and the additional agreements between the dealer of secondary raw material and the mill. It may, for example, include shredding.

3.2.1 Loose recovered paper

The simplest approach is to load the sorted recovered paper as loose material on large trucks and transport it directly to the paper mill. Due to the large space requirements resulting from the low settled bulk of such paper, it is stored in large quantities only in the paper mill where appropriate unloading equipment and storage bunkers are installed in accordance with safety regulations. These storage facilities must always be covered by roofs.

The trade and supply of loose recovered paper involves almost exclusively recovered paper for deinking. Settled apparent density is approximately 330 kg/m^3. In Germany in 1997, about 75% of this recovered paper grade was supplied as loose material. The other recovered paper grades are primarily delivered in bales.

3.2.2 Baled recovered paper

To produce bales, the sorted recovered paper enters the hopper of a stationary baling press where it is compressed hydraulically at pressures of 8 MPa. This produces compact bales having a typical size of 1.2 m x 1.0 m x 0.8 m weighing 500–600 kg. These are then tied with wire within the baling press. With larger baling presses and higher pressures, bales of larger size and weight can also be produced.

The individual bales can be simply transported and stacked by fork-lift trucks equipped with bale clamps. Bale density ranges from 300 to 900 kg/m^3 depending on the recovered paper grade. Additional transport can use almost all modes such as truck, railcar, and ship. For this reason, the trade, transport, and storage of recovered paper in bales is a global business[7].

3.2.3 Shredding of files

For confidential material, the recovered paper trader agrees to shred products such as office papers from the information source. After sorting but before baling, the recovered paper undergoes shredding into tiny particles by cutters, shredders, cutting mills, punches, or other suitable machines. Independent of the type of machine used, this treatment is generally called shredding. According to the German standard DIN 32757-1, Table 2 shows the five different security levels for the shredded recovered paper depending on the size of the particles produced.

Table 2. Threshold values for particle sizes of shredded paper for different security levels.

Level	Reproduction of information	Suitable for	Particle form (threshold values)		
			area [mm^2]	width [mm]	length [mm]
S1	Possible without special aids	Making general material illegible	≤ 2 000 (ø 50 mm)	≤12	–
S2	Possible with aids and time	Making internal material illegible	≤ 800 (ø 32 mm)	≤6	–
S3	Possible with considerable effort	Confidential material	≤ 594 ≤ 320 (ø 32 mm)	≤2 ≤4	– ≤80
S4	Possible with special devices	Confidential material	≤ 30 (ø 6 mm)	≤2	≤15
S5	At present not possible	Highly confidential material	≤ 10 (ø 3.6 mm)	≤0.8	≤13

(. . . .) = The standard does not state the diameter; this serves solely for information purposes.

Selecting the security levels should consider that recovered paper shredded by machines to the S4 and S5 security levels are suitable for use in fiber recycling only to a limited degree. To satisfy the requirements for fiber recycling, the lowest possible security level should be selected. If one can meet the demands of data protection by mixing the recovered paper, compressing it, or doing both operations, shredding may not be necessary at all thus saving additional costs.

3.3 Storage

Recovered paper especially in bales is stored in the recovered paper trade and the paper industry. Loose recovered paper is stored in special bunkers or enclosed facilities with roofs. Bales are stored under a roof or in the open air. Mixed recovered paper that is traded at a low price is often stored outdoors. Inside storage is used more with higher quality and therefore more expensive grades. Recovered paper stored outdoors is exposed to weather influences such as rain, snow, frost, or sunshine.

Knowing the extent of physical changes that occur in the recovered paper depending on the duration of storage is interesting. One study examined individual bales containing different recovered paper grades stored for 20 months in enclosed areas and outdoors with examination at intervals of about 2 months[8]. The following figures show examples of several characteristics of samples taken from bales stored outdoors. Figure 2 shows that the long fiber content that is so important for several strength characteristics gradually decreases with the length of storage time. When stored under roof, the long fiber content remains constant at the original level. The decrease of

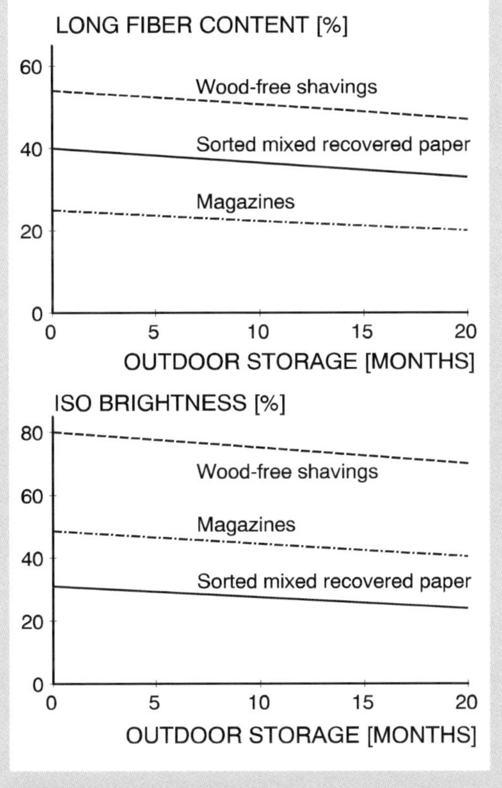

Figure 2. Effect of outdoor storage on long fiber content and brightness of various recovered paper grades.

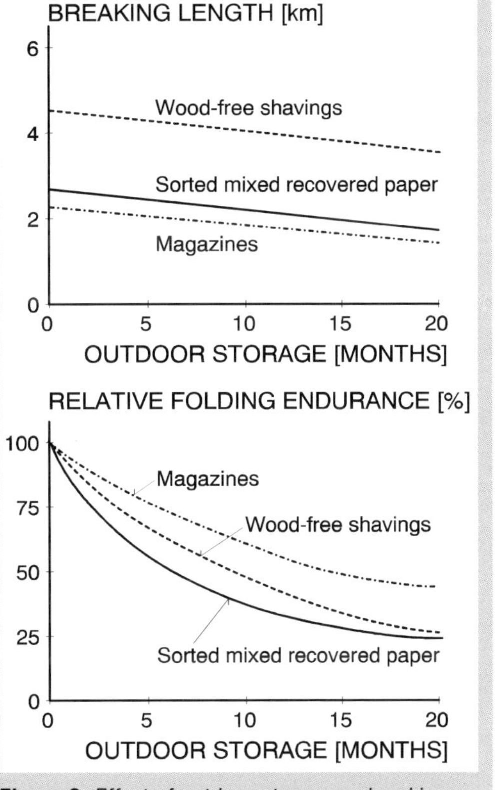

Figure 3. Effect of outdoor storage on breaking length and folding endurance of various recovered paper grades.

strength properties of bales stored outdoors is due to increasing moisture and microbiological degradation processes. The reduction in the long fiber content is reflected in the deterioration of breaking length and folding endurance as Fig. 3 shows. Folding endurance suffers significantly with increasing outside storage periods.

For brightness, a linear drop occurs with increasing outdoor storage time as Fig. 2 shows. No changes occur with storage under cover. The same trend occurs with all other optical characteristics and the strength characteristics[8]. This project did not consider the deinkability of recovered graphic paper grades. With increasing age, offset printed news and magazines become more difficult to deink due to poorer ink detachment as a result of oxidation processes in the binders (See Chapter 8). For this reason, storage periods for recovered graphic paper grades for deinking should be as short as possible.

If the average storage time of recovered paper in the trade and at a paper mill does not exceed a few months, the losses of characteristics that occur during outdoor storage will not be too serious. If the storage extends over a longer period, it should preferably be dry and under a roof to protect the material from the weather. Baling of the recovered paper in an air-dried state should also be assured. If recovered moist paper is stored too long, a rotting process will commence. This will cause an accelerated microbiological decomposition of the paper inside bales and even in loose paper piles.

References

1. *Mulligan, D.B., in Secondary Fiber Recycling (R.J. Spangenberg, Ed.), TAPPI PRESS, Atlanta, 1993, p. 75.*
2. *Bilitewski, B., Härdtle, G., Marek, K., et al., "Waste Management", Springer-Verlag, Berlin, 1994, 699 pp.*
3. *Bilitewski, B., Heilmann, A., Apitz, B., et al., "Wissenschaftliche Untersuchung und Begleitung von Modellversuchen zur getrennten Erfassung graphischer Papiere. Phase I: Erfassung und Bewertung vorhandener Sammelsysteme", INTECUS, Dresden, Germany, 1994, 47 pp.*
4. *Renner, K., Putz, H.-J., and Göttsching, L., Wochenbl. Papierfabr. 124(14/15):662(1996).*
5. *Andersen, S.L., in Recycling (S. Abubakr, Ed.), TAPPI PRESS, Atlanta, 1997, p. 3.*
6. *Böcking, B., private communication, data based on evaluations of GesPaRec/Bonn, VDP/Bonn and INTECUS/Dresden, Germany, 1998.*
7. *Miller, B., in Secondary Fiber Recycling (R.J. Spangenberg, Ed.), TAPPI PRESS, Atlanta, 1993, p. 87.*
8. *Göttsching, L., Das Papier 38(10A):V95(1984).*

CHAPTER 4

Recovered paper grades, quality control, and recyclability

Hans-Joachim Putz

Recovered paper grades, quality control, and recyclability

1 Recovered paper grades

1.1 Europe

In Europe, recovered paper trades according to the specification of different recovered paper grade lists. Besides national lists[1], the following European grade lists also exist and are issued by

- Confederation of European Paper Industries (CEPI)[2]
- Bureau International de la Récupération (B.I.R.)[3]
- European Organization on Standardization (CEN), European Standard EN 643[4].

These three European lists came from different organizations and have different histories. The paper industry compiled the CEPI list, but the recovered paper trading companies published the B.I.R. list. The European Union established the European Standard EN 643 as a standard list for recovered paper grades.

In 1999, a combined CEPI and B.I.R. working group agreed upon a common "European List of Standard Grades of Recovered Paper and Board" (See Annex)[5]. Following approval, a working group of the European Organization on Standardization (CEN) will revise the European Standard EN 643. Subsequently, national recovered paper grade lists will then be withdrawn so that only one list will be valid throughout Europe. Because the paper industry is not yet familiar with the numbering system of this new list, the following text mainly references the system of the still valid version of EN 643.

The above grade lists are quite similar, but they do differ in the descriptions of individual grades. The numbers of listed grades also differ. Individual recovered paper grades that are less relevant due to the small quantities in which they occur are contained only in some lists and not in others.

All the lists contain the most important grades of recovered paper that occur in the largest quantities. These have the following four groups:

- Ordinary qualities
- Medium qualities

- High qualities
- Kraft qualities.

The German and the B.I.R. recovered paper grade lists and the common European list of B.I.R. and CEPI contain a fifth group called “Special qualities” that is not part of the European Standard EN 643. In the new European list, the five groups are named “grades” whereas in the older lists the groups are called “qualities”.

Recovered paper grades are generally not specified with quantified quality characteristics. Only the proportion of unusable materials in the recovered paper is limited in some lists. Unusable materials are defined according to EN 643 as materials that are unusable in production. They consist of nonpaper components and paper and board detrimental to production. Nonpaper components are any foreign matter in the recovered paper that can cause damage to machines, interruption to production, or reduce the value of the finished product. Such items include the following:

- Metal
- Plastic
- String
- Glass
- Textiles
- Wood
- Sand and building materials
- Synthetic materials
- “Synthetic papers.”

According to EN 643, paper and board detrimental to production are those components treated such that for a basic or standard level of equipment they are unsuitable as raw material for the manufacture of paper and board, actually damaging, or whose presence makes the whole consignment of recovered paper unusable. Only the B.I.R. grade list gives a more precise description of unusable paper and board components. It includes those paper and board grades that become unsuitable for use as raw material for paper production or are even detrimental to the production process due to the uniform treatment applied to all components including the following:

- Bituminized board
- Carbon paper
- Parchment and grease proof paper
- Wet strength paper
- Waxed paper.

Especially in the (white) recovered paper grades intended for deinking, (brown) packaging materials and colored papers are also considered as unusable components.

According to EN 643 and the CEPI list, recovered paper must be supplied free of nonpaper components and unusable paper and board. Certain amounts of unusable materials may be included only if this has been agreed individually between the paper mill and the supplier. In addition, the German list limits the proportion of unusable materials in recovered paper for a small number of (ordinary) grades to a maximum of 1% by weight. The B.I.R. list notes that the purity of recovered paper as a secondary raw material resource depends on the collection method and manual sorting. Homogeneity of pulp as is common with virgin fibers is therefore not possible. The permissible share of unusable materials in the B.I.R. grade list is therefore between 1% and 5% by weight. For the grade "Unsorted mixed recovered paper", none of the above grade lists specify limits for the proportions of unusable materials.

Unless otherwise stated, the names and identifying numbers of the recovered paper grades used below agree with the still valid EN 643. The naming of recovered paper in all lists uses one of the following criteria:

- Places of occurrence, such as supermarket waste (A4) or shredded office waste paper (unsorted) (A12)
- Former paper grades contained in the recovered paper such as mixed news and pamphlets (A9) or white wood-free continuous stationery free from colored carbonless copy papers (C9)
- Mixed category such as mixed waste paper (unsorted) (A0) or mixed papers and boards (sorted) (A2).

The descriptions of individual recovered paper grades are as follows:

- Primarily solely descriptive without further specifications, e.g.:
  - Corrugated container waste (A5)
    Used cases, sheets, or cuttings of corrugated board
  - Printed white multi-ply board (C10)
    New cuttings of lightly printed white multi-ply board without grey plies
- One specification stating which grades should not be part of the recovered paper grade or limited to a certain percentage, e.g.:
  - Colored letters (B7)
    Correspondence, in mixed colors with and without print of printing or writing paper, free from carbon paper and hard covers
  - White newsprint (C12)
    Shavings and sheets of white unprinted newsprint, free from magazine paper
- Minimum or maximum demand for a certain paper grade as components of the recovered paper grade, e.g.:

- Mixed papers and boards (sorted) (A2)
  A mixture of various qualities of paper and board, containing less than 40% news and magazines
- Supermarket waste (A4)
  Used paper and board packaging, containing at least 70% of corrugated board with the rest being solid boards and wrapping papers.

A satisfactory description of a wood-containing recovered paper mix with commonly used for deinking called "Sorted graphic paper for deinking" originating primarily from household collections is only part of the latest German recovered paper grade list and the common European list of B.I.R. and CEPI. The deinking material described in the German list under the identifying number D39 has the following characterization:

- Sorted graphic papers from post-consumer collection including a minimum proportion of 40% each of news and magazines, a total maximum proportion of nonpaper components, and paper and board unsuitable for deinking of 1%.

In this special case, the limit to a maximum 1% of paper and board unsuitable for deinking includes packaging papers, corrugated board containers, and board boxes as well as colored papers. In the new European list the identifying number is 1.11. The description is almost identical, but the maximum figure agreed for nondeinkable paper and board should be reduced over time to a maximum of 1.5%.

The special grades on the German list comprise only wet labels (X01) and unsorted recovered paper from multicomponent collection (X09). In contrast, the B.I.R. listing contains nine special qualities including three wet strength grades, white waxed shavings, two aluminum foil or polyethylene coated board grades, impregnated petroleum kraft paper, textile boxes, and siliconized paper.

The common European list of B.I.R. and CEPI comprises seven special grades such as unsorted mixed recovered paper and board, mixed packaging, liquid board packaging, wrapper kraft, wet labels, and unprinted as well as printed white wet strength wood-free papers.

The examples of these grade names show that the specifications of recovered paper in accordance with the official grade lists do not define physically verifiable quality criteria such as moisture content or optical and mechanical characteristics. In contrast, such definitions are available for chemical pulp as virgin fibers using optical characteristics or minimum strength parameters depending on the freeness.

Figure 1 shows that the use of recovered paper in paper production in the CEPI countries (European Union plus Norway and Switzerland) varies greatly by grade of recovered paper. In 1996, the consumption of recovered paper of all CEPI countries was as follows:

- 40% OCC grades
- 27% deinking grades
- 23% mixed grades
- 10% other grades including high grades.

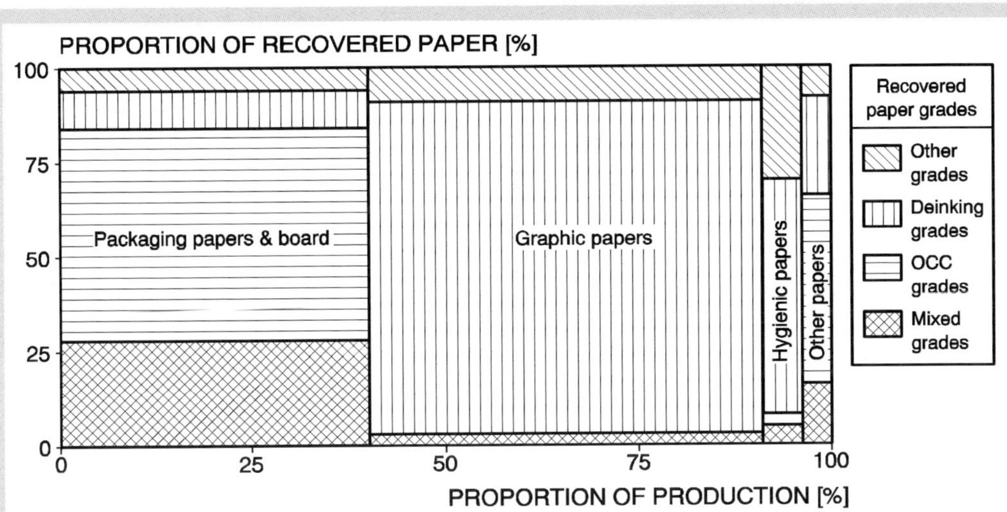

Figure 1. Recovered paper utilization by grades in the CEPI countries in 1996.

Whereas mixed grades and OCC grades (84% by weight) primarily find use in the production of packaging papers and board, almost all deinking grades (88% by weight) go to graphic paper production. The latter recovered paper grades are also used for hygienic papers with an addition of nearly one-third other higher grades. The production of other papers requires almost equal proportions of all four recovered paper grade groups.

1.2 United States and Japan

Neither the United States nor the Japanese grade lists for recovered paper use the European classification of four or five different groups. Instead, they define 26 grades in Japan and 51 grades in the United States of which five are currently not in use. The names of these grades are based on the original paper or packaging materials. Some grade descriptions generally refer to the paper and board products as printed or unprinted and wood-containing or wood-free papers according to what is assumed to be contained in the recovered paper grade concerned.

1.2.1 United States

The United States grade list does not permit the presence of individual paper or board grades in several recovered paper grades including coated, printed, or waxed papers[6]. In some recovered paper grades, the contents of mechanical fibers or of colored papers are limited. Unusable materials in the recovered paper are divided into outthrows and prohibitive materials. Outthrows are defined as all papers that are manufactured or treated or in such a form that renders them unsuitable for consumption in the recovered paper grade specified. Prohibitive materials have the following definition:

- Any materials that by their presence in the packaging of paper stock in excess of the amount allowed will make the packaging unusable as the grade specified
- Any materials that may be damaging to equipment.

For both definitions, a material can be an outthrow in one recovered paper grade and a prohibitive material in another grade. For example, carbon paper is unsuitable in mixed paper (#2) and is therefore classified as an outthrow, whereas it is unusable in sorted white ledger (#40). The classification in this case is prohibitive material.

Prohibitive materials are allowable in 15 of the 46 recovered paper grades but not in the remaining 31. The total including both outthrows and prohibitive materials ranges from 0% to 10% by weight.

The American grade list also includes 33 specialty grades called 1-S to 33-S that are not part of the regular grades. Generally, they require a special processing system. The recovered specialty paper grades do not have more definite specifications. The grade name reflects their origin. The majority of these paper and board grades relate to the following:

- Waxed cup cuttings, waxed boxboard cuttings, and waxed corrugated cuttings
- Wet strength corrugated cuttings, white or brown, printed and/or colored wet strength waste
- Polycoated milk carton, diaper stock, bleached kraft, or boxboard cuttings.

Other special grades include asphalt laminated corrugated cuttings, carbonless treated ledger, or textile boxes. The term “soft” describes the American grade list for short fiber stock consisting of predominantly soda pulp, hardwood fibers, or both. “Hard” refers to long fiber stock predominantly sulfite or sulfate made of softwood fibers.

The most important grades considering consumption in the United States are the OCC grades (49%) followed by deinking grades (29%) which include old news and high quality recovered graphic paper grades. The remaining 22% are mixed grades and pulp substitutes (wood-free recovered papers).

1.2.2 Japan

Two-thirds of the 26 grades on the Japanese list are pre-consumer recovered paper. They refer to grades that occur at printing plants, bookbinders, corrugated board plant operations, or in board or paper converting plants[7]. Remarkably, the Japanese grade list does not recognize any mixed recovered paper grades. In nearly half of all grades, no objectionable materials at all are permitted. In the remaining grades, limits of objectionable materials range from 1% to 3% by weight.

In 1997, old corrugated containers (#22) and kraft browns (#18 to #21) had a share of 48% of the total recovered paper tonnage followed by old news (#16) with 22% and the remaining grades (#1 to #15, #17, #23 to #26) with 30%. The symbol “#” is used to indicate the identifying number of the recovered paper grade.

1.3 Recovered paper grades in different countries

Although an international comparison of recovered paper consumption by grades is difficult due to the different grade definitions on different continents, Table 1 shows that deinking grades comprise the largest category (nearly 40%) in Germany. In the CEPI countries, United States, and Japan, the largest category comprises the OCC grades (40% to 50%). The proportion of mixed grades used in paper production in Germany and the CEPI countries is 30% and 20%, respectively. This share in the United States is 14%. Due to the absence of collection and trade of these recovered paper grades in Japan, no mixed grades are used at all. The category "Other grades" in Germany, the CEPI countries, and Japan primarily comprises higher grades, whereas this category in the USA covers pulp substitutes and the higher grades for deinking. The latter were assigned to the deinking grades in Table 1.

Table 1. Recovered paper consumption by grades in different countries for 1995. (MT: million tons.)

Country	Mixed grades	Deinking grades	OCC grades	Other grades	Total [MT]
Germany	29%	38%	29%	4%	8.6
CEPI countries	20%	31%	40%	9%	30.5
USA	14%	29%	49%	8%	30.3
Japan	–	39%	51%	10%	15.7

2 Quality control of recovered paper

Because of the heterogeneous composition of a recovered paper delivery whether supplied in bales or loose, the quality control of recovered paper is often limited to recording the delivered quantity, determining the moisture content, and includes a visual inspection. Naturally, the visual control can only cover the surface of a loose paper delivery or the outside surfaces of paper bales. Still based on this judgement, most recovered paper is accepted, rejected, or complaints are made.

2.1 Equipment for sampling

Representative sampling has major importance for objective evaluation of a recovered paper delivery. Only then can reliable results be obtained. Mostly samples are taken manually from broken bales, paper piles, or from conveyor belts yielding less reliable results.

2.1.1 Sampling of baled recovered paper

For baled recovered paper, core drilling systems can provide a sampling from inside the bales. At the end of the 1970s, such a core drilling system became available for use in quality testing of recovered paper at the Department of Paper Science and Technology (IfP) in Darmstadt, Germany (Fig. 2). The unit has a cutting head and an internal plastic sampling tube with 50-mm diameter and 750-mm length that places the sample in a plastic bag. A conical lid transmits the torque from the motor to an external metallic tube[8]. By placing two drillings diagonally opposite one another in the direction of compression of the bale, samples can be removed from across the full length of a bale. The sample is taken objectively and thus suitable for gravimetric determination of the moisture content or for more extensive physical and chemical tests.

After manual sorting of the drill core, proportions of nonpaper components and the composition of the recovered paper sample by paper and board grades can be determined. Pulp and sheet characteristics can be measured after repulping and processing of the sample. Depending on the recovered paper grade and type of compression (hard or soft), the core drilling system permits 8 to 15 drillings per hour resulting in sample weights between 300 and 700 g per drilling. Accordingly, samples can be taken from 4–7 bales per hour. Depending on the subsequent manual sorting and additionally desired sample processing to determine the recycled pulp or handsheet characteristics, considerable additional time is necessary. Then sufficiently detailed and statistically defined data for the supplied recovered paper can be obtained[8,9].

Figure 2. IfP core drilling device.

The Centre Technique du Papier (CTP) in Grenoble has continued development of this core drilling technique, and the unit is commercially available. The core drill sleeve has been lengthened to 1.5 m and its diameter reduced to 40 mm. One drilling therefore covers the entire length of a bale. Moreover, the system is designed in such a way that physical strength is no longer required to advance the drill (Fig. 3). A sampling process takes 5 min maximum. This allows sampling of about 12 bales per hour. The weight of the samples ranges from 700 to 1 000 g per core[10].

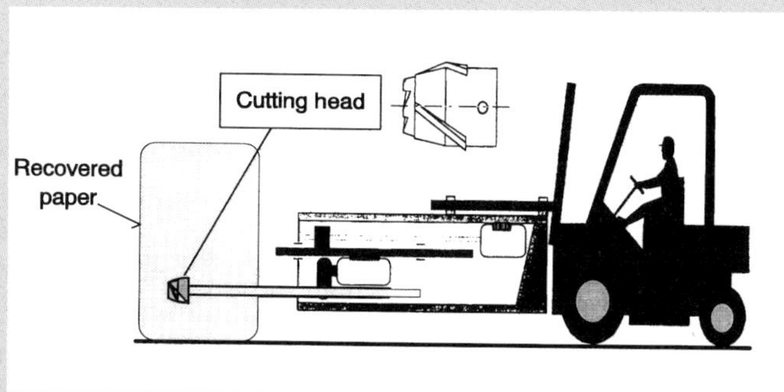

Figure 3. Commercial core drilling device[10].

2.1.2 Sampling of loose recovered paper

For statistical reasons, sampling of loose recovered paper is more difficult than sampling of bales. Theoretically, there can be no argument against an arbitrary and random sampling from a recovered paper pile or stream. The person taking the sample already subjectively influences his choice of samples so that it is not a random one. For example, if he thinks the recovered paper batch contains a high proportion of magazines, he will tend to gather more magazines when removing the samples. Also, one "grab" of the loose paper may contain more of one kind of paper than another, e.g., a magazine is always complete because it is stapled or glued. A newspaper might be not complete because it is only folded and parts can be lost before or during grabbing. Subjective and nonrandom sampling can clearly influence all quality data obtained from the sample.

No sampling device currently exists for loose recovered paper. A research project of the International Research Association for Deinking Technology (INGEDE) in Munich to analyze the composition of wood-containing loose deinking material has started with a development that permits representative sampling[11]. By positioning a bin of 60 cm diameter below the ejection point of a sorting belt, sampling was carried out on about 30 kg. This sampling process for loose recovered paper was compared with that for baled recovered paper at a recovered paper dealer, where two parallel sorting belts were run. One had baled material, and the other had loose material of the same grade. The analysis of loose paper sample from the described bins began with the determination of its composition by manual sorting followed by homogenization in a pilot plant pulper and final disintegration in a laboratory pulper. Pulp and handsheet properties measured were compared with those of the samples obtained from drill cores (about 1 kg recovered paper per bale) from the parallel produced bales using standard test procedures. Both sampling methods gave the same results in terms of paper composition and physical pulp and sheet characteristics.

The above described sampling possibilities from a sorting or conveyor belt are usually not available at a paper mill. A different approach is necessary to perform quality control on loose recovered paper. Therefore, INGEDE has developed a process that is based on visual inspection of recovered paper on reception (INGEDE Method 07)[12]. The inspecting personnel look for any indications of the presence of unusable components during unloading in the mill. Subsequently, the unloaded delivery is checked for typical smells or rotting. For the estimation of the recovered paper composition, the individual components on a defined observation surface (minimum 30 m^2) are counted or estimated and multiplied by previously determined weight factors depending on individual paper grades.

A special aspect of this control process is that the weighting factors for the different paper products (e.g. news and magazines) contained in the recovered paper deliveries and the control personnel as well are calibrated repeatedly. For this purpose, a larger sample is removed from the recovered paper batch using a front-end loader and emptied via a suitable sample receptacle. The same procedure is used in cases of noticeable moisture in the recovered paper by removing a larger sample for gravimetric analysis described below. The recovered paper that falls by chance into this receptacle (minimum 30 kg) is taken for objective quality evaluation. It is therefore much less influenced by subjective factors than would be the case if the samples were taken arbitrarily. The content of the receptacle is then sorted into its different components (e.g. unusable materials such as board and corrugated board packaging, colored papers, plastics, and the composition of the usable papers such as news and magazines), weighed, and compared with the results of the previously established visual control results. In this way, the control personnel can be trained to improve the reliability of the visual recovered paper control at reception.

2.2 Quality characteristics

2.2.1 Moisture content

The moisture content of recovered paper is especially important for economical reasons at paper mills because water can represent an expensive cost factor. Moist bales or moist zones in loose recovered paper can also cause rotting due to biological degradation of the organic material. Because of the conditions of collection and storage that occur partially outdoors, one cannot expect that the material should have the same moisture content as the originally produced papers. The level of the moisture content will instead depend on several factors for different recovered paper grades:

- Material composition such as type of fiber furnish, recycled fiber content, and ash content
- Climatic conditions during collection, transport, baling, and storage of the recovered paper
- Storage conditions of the recovered paper (outdoors, under cover, or in enclosed facilities).

Independent of whether the recovered paper is baled or loose, an exact gravimetric analysis of the moisture content must always be performed by determining the weight of a recovered paper sample before and after drying. Circulating air ovens at 105°C have use for drying according to the method of determination for the moisture content of paper and board (ISO 287-1985 or TAPPI 550 om/93). The drying process must proceed to the point where the weight remains constant. This requires several hours of drying. Typical moisture contents for most common recovered paper grades is between 6% and 13% in Central Europe[9].

To shorten the drying time, microwave devices and so-called high-speed dryers provide a faster determination of the moisture content. No examinations of the comparability of these measured values with the classical gravimetric method are published. Therefore, the accuracy of specific microwave drying methods should be verified individually.

The challenge of an adequate moisture content determination always exists in the representative sampling of the recovered paper. With loose recovered paper, several printed products such as news and magazines are usually taken at random. The quantity taken as a sample must be sufficiently representative and not so large to exceed the capacities of the available drying facilities.

For recovered paper supplied in bales, the bales cannot be opened to inspect their content or to remove samples to determine the moisture content since stacking the bales will no longer be possible. For this reason, bales can usually only be sampled from the outside unless a core drilling system is available for taking the samples. In the case of manual sampling, samples of the recovered paper should not be taken from the outside of the bales since recovered paper is frequently stored outdoors. This causes the moisture content of surface samples to be different from that inside.

Using moisture checking devices equipped with a sword sensor pushed into the bales to determine the moisture content quickly is also possible. Sufficient dimensions and stability of the sword sensor are preconditions for reliability. Depending on the device manufacturer, the measurement uses the principle of conductivity, electrical resistance, or the dielectric characteristics of paper. These indirect measurements of the moisture can occur very quickly to record several measured values from different bales to calculate an average for a recovered paper delivery. Such devices do not provide absolute values. The coefficients of variation of these measurements compared with gravimetic moisture content determination are considerably greater[9].

With indications of excessive moisture contents determined using indirect measuring systems, bales should be opened to take samples from the inside of the bales for gravimetric verification. The quick measurement of the recovered paper moisture using indirect measuring processes is useful since orienting values result that are sufficiently accurate to indicate which recovered paper batch requires closer analysis for its moisture content.

2.2.2 Composition and other characteristics

Besides the moisture content of recovered paper deliveries, almost no other quality characteristics are quantitatively recorded. If any other parameters are considered, they relate to the aforementioned estimations regarding the material composition of unusable materials in the recovered paper. One reason for this is the highly labor-intensive aspect of carefully performed quantitative analyses in contrast to the relatively low price of most recovered paper grades. Because recovered paper prices are subject to major fluctuations, higher prices are paid sometimes. This would then make more exact monitoring of recovered paper upon reception worthwhile. Scarcities of recovered paper always accompany such periods so complaints by a paper mill during a shortage of raw materials could endanger its supply.

Brown grades

Purchased recovered paper should undergo quantitative controls upon reception despite the preceding remarks. More attention is necessary especially to the unusable materials in all recovered paper grades. These contain nonpaper components that not only raise the overall price but also require removal during processing at a high cost. In the absence of published international information, the following figures refer to analyses of German recovered paper grades that comprise almost three-quarters of the total volume used including brown and white grades.

Nonpaper components

In the grade "Sorted mixed papers and boards" (A2), the proportion of nonpaper components between 1984 and 1994 increased from an average of 1.6% to 4.5%[13]. Figure 4 shows that in 1984 for about two-thirds of all examined mixed recovered paper bales the proportion of nonpaper components was still below the threshold value of 1% as defined by the German recovered paper grade list for the identical grade, B12. This proportion decreased to less than 20% in 1994 especially due to changes in the collection structure in Germany indicated by the increase in the recovered paper collection rate from 38% (1984) to 60% (1994).

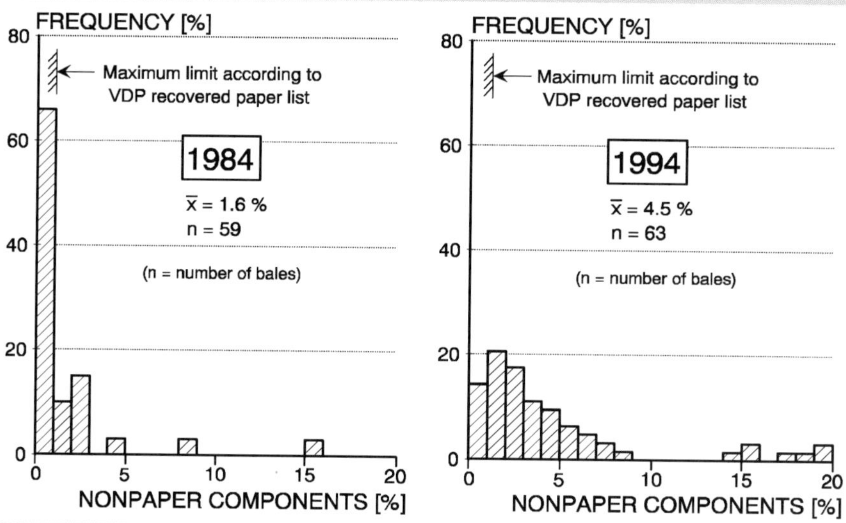

Figure 4. Amount of nonpaper components of German sorted mixed papers and boards (A2) in 1984 and 1994.

Composition

Besides the content of unusable materials in mixed (brown) recovered paper grades, the composition of the recovered paper by paper and board grades is also important. The composition of "Sorted mixed papers and boards" (A2) is limited by grade definition to a maximum of 40% old news and magazines. In "Supermarket waste" (A4), a minimum content of 70% corrugated board is stipulated. The remainder should consist of solid board and wrapping papers.

According to tests, the proportions stipulated in EN 643 are far from reality as Fig. 5 shows. In sorted mixed papers and boards (A2), the stipulated proportion of news and magazines called graphic papers is exceeded, whereas the amount of corrugated board, solid board, and wrapping papers usually falls short of the target for supermarket waste (A4). In both cases, recyclers of these recovered paper grades (the manufacturers of test liner, corrugating medium, and board) must tolerate increasing ash contents in the products with accompanying negative consequences for mechanical characteristics such as CMT (Concora Medium Test), RCT (Ring Crush Test), or bending stiffness.

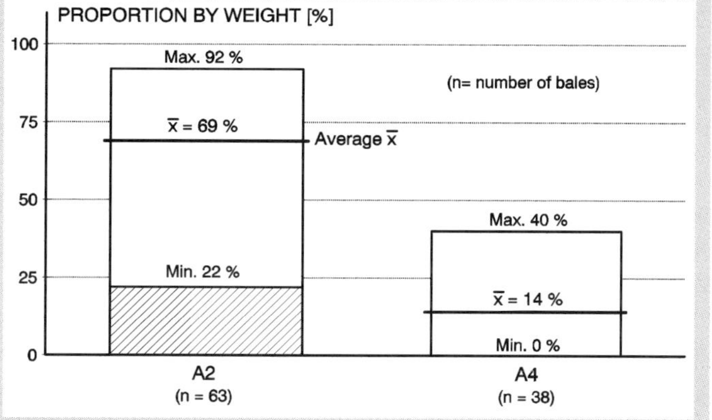

Figure 5. Ranges and averages of the proportion of graphic papers in German recovered paper grades A2 and A4 for 1994.

Pulp and handsheet characteristics

Despite fairly clear increases in the use of recovered papers during the last decade in all main segments of paper production in Germany, analyses of recovered paper bales of the A2 and A4 grades have not revealed any significant change in the fiber quality to date that would indicate an imminent "recycling collapse" as Fig. 6 shows. A trend has existed since 1984 for a slight increase in ash content by about 3%-points to 17% for grade A2 and 11% for grade A4 due to a generally more intensive use of minerals and increased recycling in the graphic sector particularly in the case of newsprint.

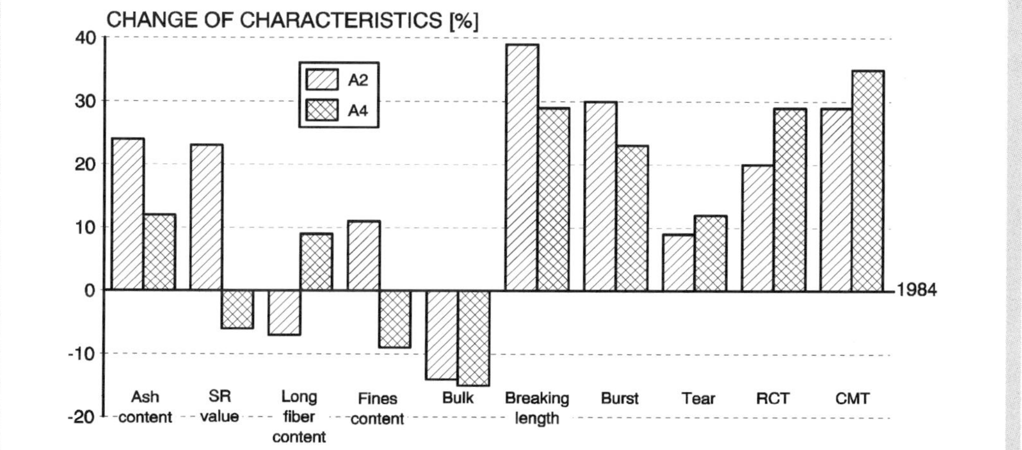

Figure 6. Change of characteristics between 1984 and 1994 of the German recovered paper grades A2 and A4.

The drainage resistance of the repulped, nonrefined samples has developed differently for both recovered paper grades according to the changes in fiber materials. It has increased by 8 SR units in the A2 grade and decreased by 2 SR points for the A4 grade. The increase in the Schopper-Riegler (SR) value is accompanied by an increase in the fiber fines content and a simultaneous reduction in the long fiber content. Reduced drainage resistance of grade A4 proceeds in the opposite direction. Since ISO brightness also has a 6% increase, it becomes evident that the recovered paper grade A2 of 1994 compared with 1984 contains a larger proportion of graphic papers. Similarly, the graphic paper proportion in grade A4 has dropped indicated by a loss of ISO brightness of 2%.

Noticeable for both recovered paper grades is a reduced bulk that results from the higher total ash content, the reduced content of mechanical fibers, and the greater number of recycling loops with repeated mechanical stress of the fibers. For all tested mechanical parameters such as breaking length, burst, tear, RCT, CMT, an increase of at least 10% has nevertheless occurred for both recovered paper grades.

While the increase in the long fiber content in grade A4 compared with 1984 may be the reason for the total enhanced mechanical characteristics, this is especially due to the higher Schopper-Riegler value for grade A2. The probable reasons that most mechanical characteristics of the tested recovered paper grades have actually increased despite the clearly intensified practice of recycling are as follows:

- Improved recovered paper processing technology in the previous paper manufacturing
- Decreased consumption of virgin mechanical pulp compared with virgin chemical pulp in the German paper industry[13].

White grades

Wood-containing recovered paper for deinking

For wood-containing recovered paper grades destined for deinking, attention is especially necessary for packaging materials as unusable papers and board besides the nonpaper components that may be in the delivered sorted material. Unusable paper and board components can cause quality losses, particularly in terms of optical properties. One must for example know that the presence of only 1%

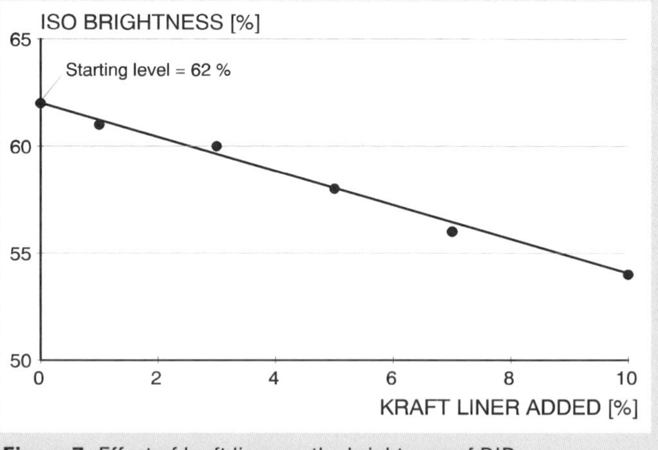

Figure 7. Effect of kraft liner on the brightness of DIP.

kraft liner in recovered paper for deinking can lower the brightness of the processed deinked pulp by about 1% ISO as Fig. 7 shows.

In 1995, analyses of recovered paper batches provided an insight into the quality of recovered paper for deinking (D39 on the German recovered paper grade list) when considering the collection system[14]. Besides the type of collection system, results showed that the quality of sorting by the recovered paper dealer and the time of year for sampling influenced the sorting result. The content of catalogs in the material for deinking increases significantly independent of the collection system with publication of new, seasonal catalogs.

For the amounts of unusable papers and board in the recovered paper for deinking, clear differences existed depending on the investigated collection systems. Figure 8 shows that the lowest content of unusable papers and board at 2.6% is in the bundle collection. The drop-off containers and pick-up mono-bins produced 2–3 times higher proportions. Surprisingly, the tested multicomponent container collection system showed a proportion of unusable papers and board that is between bundle collection and drop-off containers. This is where good manual sorting by a dealer has best results. For this reason, no generalization is possible regarding quality for all the recovered papers collected by a multicomponent container collection system[14].

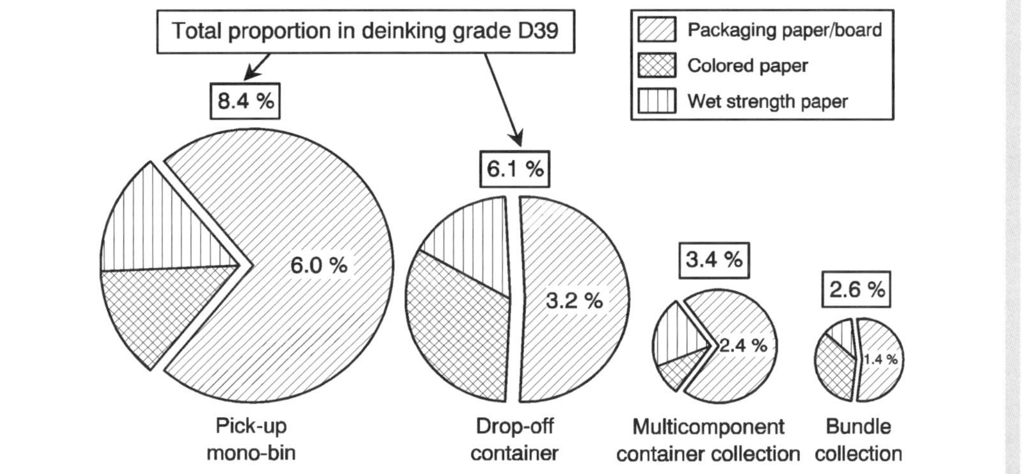

Figure 8. Proportion of paper and board detrimental to production in sorted recovered paper for deinking (D39) influenced by German collection systems (1995).

For all collection systems, an examination of the unusable papers and board shows that packaging papers and board always predominate in this category. The highest amount of brown papers detected in the material for deinking from the pick-up mono-bin is 6%. Insufficient sorting of the mixed recovered papers collected from households, disposal of small sales packages, and tearing of larger transport packages before disposal are causes for the high content of packaging material in the (sorted) recovered graphic paper from the pick-up mono-bin.

The bundle collection not only supplies the lowest content of unusable papers and board but also the lowest amount of nonpaper components at 0.5% as Fig. 9 shows. The largest amount of nonpaper components is in the sorted recovered paper from the multicomponent container collection system. Drop-off containers reveal nonpaper components of about 1%.

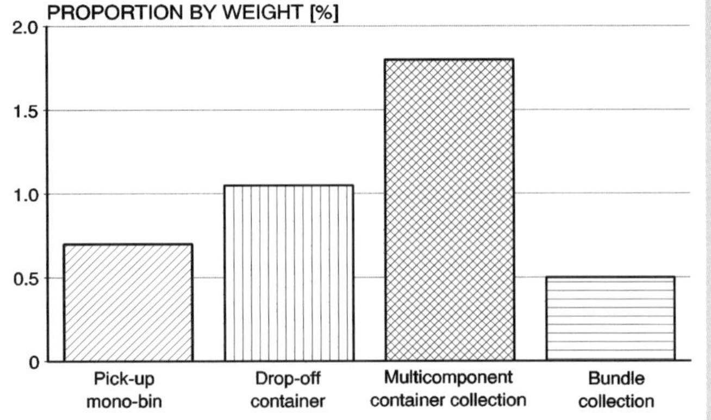

Figure 9. Nonpaper components in sorted recovered paper for deinking (D39) influenced by German collection systems in 1995.

In Germany, the share of the recovered paper collected by different post-consumer collection systems varies considerably. In 1994, 53% of post-consumer recovered paper was collected via drop-off containers, 37% via pick-up mono-bins, 8% via bundle collection, and 2% via multicomponent container collection systems. Considering this collection structure, the German deinking material (D39) obtained from post-consumer collection agreed with Fig. 10 in 1995 on average. The proportions of news and magazines in the material for deinking are about 40% each as it should be. Wood-free papers are 9%. Other bright papers account for 4% (mainly recycled graphic papers). At an average of 6%, the amount of unusable papers and board in the sorted wood-containing material for deinking is unacceptably high. The target value according to the German recovered paper grade list is 1% and 1.5% according to the new B.I.R. and CEPI lists, including nonpaper components.

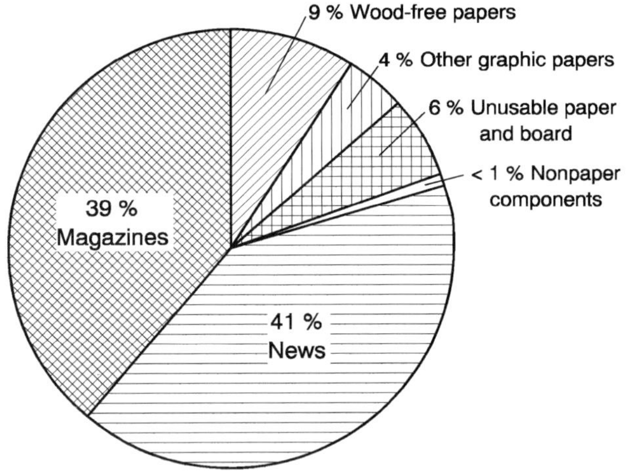

Figure 10. Average composition of sorted recovered paper for deinking (D39) derived from German households in 1995.

An examination of the age structure of newspapers and magazines collected by different methods revealed a circulation time between printing and supply of the used printed products to the paper mills of only one month for more than three-quarters of the old newspaper (Fig. 11). More than 90% total of all newspapers arrive at the paper mills within three months after printing. Magazines are usually older when they reach processing. About two-thirds of all magazines are a maximum of three months old. These results indicate a fairly young material for deinking assuring good deinkability. Note that the data stem from 1995, a year of major demand for this recovered paper grade. In other years, the printed material may be somewhat older[14].

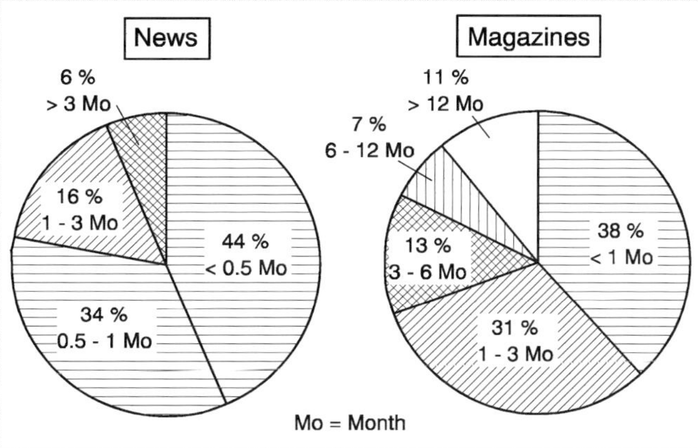

Figure 11. Age distribution of newspapers and magazines in sorted recovered paper for deinking (D39) derived from German households in 1995.

Wood-free recovered paper for deinking

Another important recovered paper grade especially for the production of hygienic papers is colored letters (B7) that are called colored ledger in the United States. According to the grade description in EN 643, these should consist of office papers, printing or writing paper, and mixed colors with and without print and should be free from carbon papers and hard covers. This recovered paper grade collected in Germany and also imported showed a highly heterogeneous composition according to a test in 1995.

Paper mills account for the diverse composition of colored letters by making internal visual distinctions between different sub-grades. According to Fig. 12, the average composition of the colored letters was 57% wood-free and 19% wood-con-

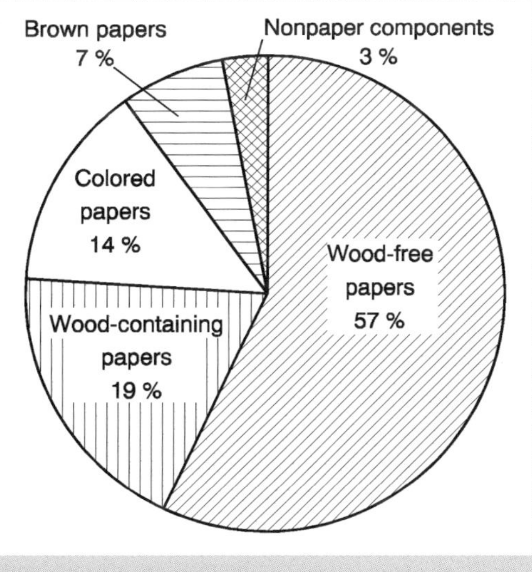

Figure 12. Average composition of colored letters (B7) used in Germany in 1995.

taining papers. The proportion of colored papers was 14%. The proportions of packaging papers and board (7%) and nonpaper components (3%) are unacceptably high.

A comparison of the two grades for deinking (D39 on the German recovered paper grades list) and colored letters (B7 according to EN 643) shows that the variations of all characteristics for the D39 samples are about half as great as those for the colored letters as Fig. 13 shows. Likewise, the share of nonpaper components is considerably greater in colored letters. The two recovered paper grades were comparable regarding the average values of strength and optical characteristics of the samples prepared where the deinking process applied for colored letters was washing and for the wood-containing material D39 was the flotation process. The resulting brightness increase is greater in the wood-containing DIP, but the final brightness is identical for both recovered paper grades[14].

Characteristics		Colored letters (B7)	Recovered paper for deinking (D39)
COMPOSITION			
• Nonpaper components	[%]	0.0 - (3.0) - 9.0 [9]	0.1 - (0.8) - 3.5 [3.4]
• Dry substance	[%]	76 - (92) - 98 [22]	83 - (90) - 93 [10]
PULP PROPERTIES			
• Schopper-Riegler value	[SR]	29 - (40) - 55 [26]	52 - (58) - 63 [11]
• Fiber fraction R30 (McNett)	[%]	11 - (21) - 32 [21]	21 - (24) - 27 [6]
• Ash content (575 °C)	[%]	9 - (18) - 29 [20]	20 - (23) - 27 [7]
HANDSHEET PROPERTIES			
• Breaking length	[km]	2.6 - (3.6) - 5.2 [2.6]	3.4 - (3.6) - 3.8 [0.4]
• Tear	[mNm/m]	915 - (1165) - 1835 [920]	1015 - (1090) - 1205 [190]
FILTER PAD PROPERTIES			
• R_{457} before deinking	[% ISO]	37 - (57) - 83 [46]	41 - (44) - 49 [8]
• After deinking		Washing	Flotation
R_{457}	[% ISO]	34 - (62) - 94 [60]	56 - (60) - 65 [9]
a*	[-]	-8.4 - (1.4) - 8.1 [16.5]	-1.2 - (-0.9) - 0.1 [1.3]
b*	[-]	-11.0 - (3.7) - 41.0 [52]	4.8 - (6.0) - 7.2 [2.4]

Legend: Min - $(\bar{x})$ - max, [range]

Figure 13. Ranges and averages of characteristics of colored letters (B7) and recovered paper for deinking (D39) used in Germany.

A French pilot test for collecting mixed office paper showed that after implementation of a suitable collection system in individual offices, municipal authorities, and universities the quality of recovered office paper improved considerably. While the proportion of colored and brown papers, old newspapers and magazines, as well as contaminants in the tested recovered office paper was totally 12%–32% before the pilot test, reducing this share to an average of 1% was possible. This was the result of early segregation of white office paper into separate collecting receptacles and other office papers in a second container.

Experience shows that informing and instructing office personnel are necessary to obtain good separation. The collected office papers underwent extensive processing tests. For separately collected white office paper after a simple flotation, more than 80% ISO brightness with a speck area of 400 mm^2/m^2 was achieved. In contrast, mixed office paper achieved despite double flotation and two-stage bleaching a brightness of only 70%. The speck area in the deinked pulp was almost twice as high[15].

3 Recyclability of paper products

The evaluation of recyclability of paper and board products such as folding boxes, polyethylene coated liquid containers, or envelopes with pressure sensitive adhesives and transparent windows has decisive importance for improving recovered paper use. In this context, good recyclability refers to the ease of processing of recovered paper into a new product. Materials added to papers such as printing ink or adhesives should be benign during recovered paper processing and not affect water loops, paper production, and paper converting. The processing residues (rejects and sludges) enriched with nonfibrous components should not be harmful to the environment and offer possibilities for suitable use in other industrial processes, energy recovery, or final disposal by landfilling.

Guidelines and test methods for evaluating technological aspects of recyclability of finished products are necessary to develop recycling-friendly paper and board products. This is the only way that new products can satisfy the requirements for a more completely closed paper cycle.

The test methods described below are also useful to determine whether a packaging material belongs to a group of paper or board packaging or to a category based on composite materials. In Germany, this classification of sales packaging material has direct economic consequences for the license fees to be paid for the collection of used sales packaging by DSD (Duales System Deutschland). This institution is responsible for packaging made of glass, plastics, metal, and paper, board, cardboard, and composite materials (See Chapter 2). Pure paper and board products are charged with a lower license fee than composite materials such as polyethylene coated liquid containers.

Alternatively, the testing of the recyclability of packaging materials and printed products (graphic products) can evaluate the recycling-friendly characteristics of these products, i.e., how they behave in a simulated lab processing procedure discussed below. After disintegration of the packaging material, the reject mass is recorded. Dirt content, transparent spots, and stickies are evaluated using the screened accept. With graphic paper products such as old newspapers or magazines, the deinking behavior is also evaluated.

3.1 Determination of nonpaper components

The main factors hindering the recyclability of paper and board products are the amounts and nature of nonpaper components. This includes substances added subsequent to the time of conversion such as adhesives, plastic foils, and aluminum foil. In a draft of the Papiertechnische Stiftung (PTS) on the determination of the content of nondefiberable components from composite materials and packaging made of paper, board, and cardboard, the content of nonpaper components is determined after pulping in a laboratory disintegrator followed by screening through 0.7 mm diameter holes[16]:

- When the dry reject mass is less than 5%, the packaging is categorized "paper product", for the purposes of the (lower) DSD license fees.
- With a reject mass more than 5%, the residue must first be investigated for remaining fiber content possibly using a microscope followed by careful removal of the fibers. If the remaining residue of nonfibrous material is still more than 5%, the packaging belongs to the category of "composite materials" for the (higher) DSD license fees.

This test is not applicable to wet strength paper and board packaging materials.

Naturally, the proportion of nonfibrous components does not provide a sufficient indication of the recyclability of a paper or board product. For example, a foil coating of more than 5% by weight is undesirable due to the resulting higher amount of retained material. If the foil is sufficiently thick corresponding to a basis weight of more than 12 g/m^2, it will resist disintegration into small particles. The result is that the coarsely fragmented foil can be separated efficiently during industrial recovered paper processing. In contrast, a redispersible coating or adhesive does not contribute to the nonfibrous components in the residue but may still have a strongly negative effect on the recycling of a packaging product due to a possibly tacky character.

3.2 PTS test method on recyclability

Testing recyclability by the PTS method RH 021/95 evaluates the recyclability of paper or board packaging materials or of printed products on a laboratory scale[17]. Note that the tests exclusively deal with an individual packaging material or printed product as (potential) components of recovered paper and not with generally supplied recovered paper grades that may possibly contain these paper products in different quantities. Because the objective of paper recycling is not to recycle nonpaper materials, the materials examined must only satisfy the specification "suitable for recycling." For this purpose, the nonpaper components are not permitted to disturb recovered paper processing and paper production, place an unjustifiable stress on process water and effluent, or negatively influence the quality of the new paper produced from the recycled fiber pulp. A nonpaper material is suitable for recycling if it can be sufficiently separated from the repulped recovered paper by recognized technical processes and recycled or disposed without damage to the environment.

The first requirement is to subdivide the packaging material or printed product intended for recycling (referred to as paper product in the following) into the two categories of I or II in Fig. 14. This subdivision does not constitute any quality evaluation. It simply considers the fact that the intended use of the new paper and board product containing recycled fibers makes different demands on the paper products used. Depending on the categorization of a paper product, the content of residues also has importance for the recyclability test by the PTS method[18]:

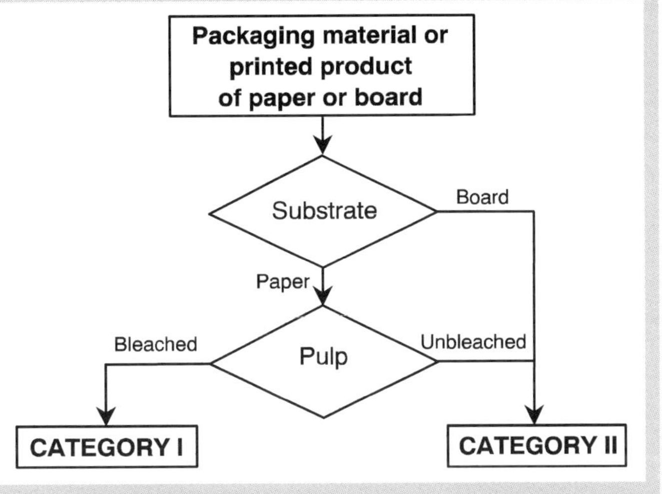

Figure 14. Differentiation of paper products according to their utilization possibilities[18].

- Category I includes paper products from white or bleached fiber furnish that can be processed to deinked pulp due to their composition and character. They have use primarily for production of graphic paper grades and hygienic paper.
- Category II comprises all other products especially board and cardboard products including unbleached and brown papers that need not satisfy optical requirements such as brightness. They are not suitable to produce new white products but are suitable to make brown grades of board, test liner, or corrugating medium.

The test for category I includes first the same steps as in category II. That is why the latter is discussed first below.

Category II

With category II paper and board products that are usually collected in mixed recovered paper grades, the first step is to determine the residual mass after screening the repulped sample on a perforated plate with holes of 0.7-mm diameter as Fig. 15 shows:

- With a residue mass of more than 50%, the paper or board product is classified as “not recyclable.”
- For 20%–50%, the material is “recyclable but requires improvement in the product design” provided that the other test criteria are satisfied.
- If the residue mass is less than 20% and all other test criteria are satisfied, the product is categorically classified as “recyclable.”

When the rejected mass is below 50%, the test also includes screening using a 0.15 mm slotted plate and the production of a handsheet from the screen accepts. The test sheet should not adhere to the blotter board or the cover sheet used in the Rapid-Köthen handsheet mold according to the German standard. The handsheets produced may contain only a small number of transparent spots or optical inhomogeneities (dirt specks). If adhesion does occur, the evaluation is "not recyclable due to tacky impurities."

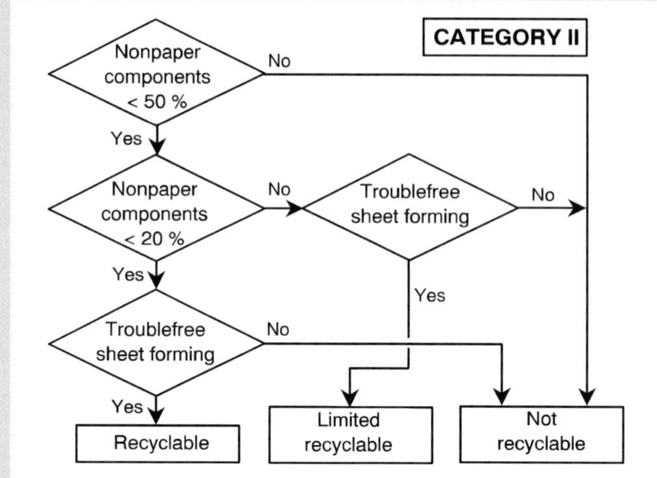

Figure 15. Evaluation scheme for paper and board products of category II (packaging material)[18].

When transparent spots or optical inhomogeneities are observed but no adhesion occurs, the product is classified as "recyclable with reservation due to optical inhomogeneities in the processed pulp" although the number and area of these defects are not numerically defined. Losses of quality caused by the dirt count can possibly be reduced by an effective dispersion in industrial processing.

Category I

For paper products belonging to category I, a flotation deinking process is also part of the test procedures as Fig. 16 shows. Besides having a screening residue mass of less than 5%, the fiber accepts of the slotted screen undergo flotation. The resulting final accept is also used for sheet formation where the same criteria as for category II paper products apply. In addition, an evaluation occurs using the luminosity values of the flotation-treated printed and unprinted products according to the standard deinking process in the PTS laboratory flotation cell. This calculates the ratio of the luminosity values (luminosity after flotation of the printed paper divided by luminosity after flotation of the unprinted paper):

- With a ratio of more than 0.8, a printed product is "recyclable."
- With a ratio of 0.7– 0.8, the printed product is "recyclable with reservation due to unsatisfactory deinkability."
- With ratios less than 0.7, the evaluation is "not recyclable as a component of material for deinking due to poor deinkability."

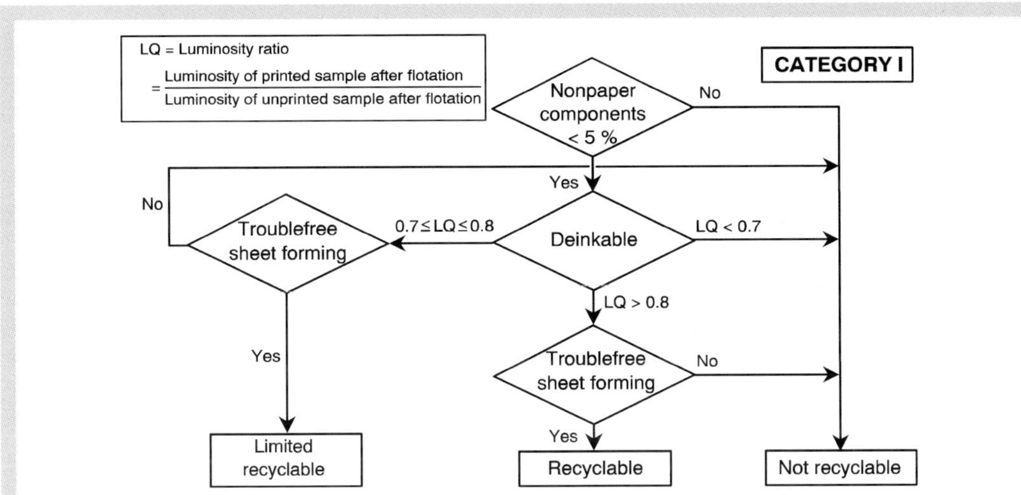

Figure 16. Evaluation scheme for paper products of category I (printed products).

The same criteria as for category II paper products apply regarding tacky impurities, transparent spots, and other optical inhomogeneities[17].

3.3 FINAT test method on recyclability

The FINAT (Fédération Internationale des Fabricants et Transformateurs d'Adhésives et Thermocollants sur Papiers et Autres Supports) test method FTM 19, "Assessing the recycling friendliness of self-adhesive labels," describes a similar procedure[19]. The test is especially applicable for self-adhesive labels.

Although the test process is largely comparable with the PTS RH 021/95 method, individual details do not completely agree with the PTS method. This method also combines screening with a slotted plate and flotation. The evaluation is restricted exclusively to the occurrence of tacky impurities, dirt specks, and transparent spots in the handsheet produced from screened and deinked fibers. No evaluation of the optical characteristics occurs.

3.4 INGEDE test method on recyclability

The INGEDE method on recyclability is related to printed products. The method recognizes that the recyclability of a printed product includes both deinkability and the potential of stickies formation that requires separate assessment as Fig. 17 shows. The latter is not yet finalized while the test method to determine the deinkability is already established.

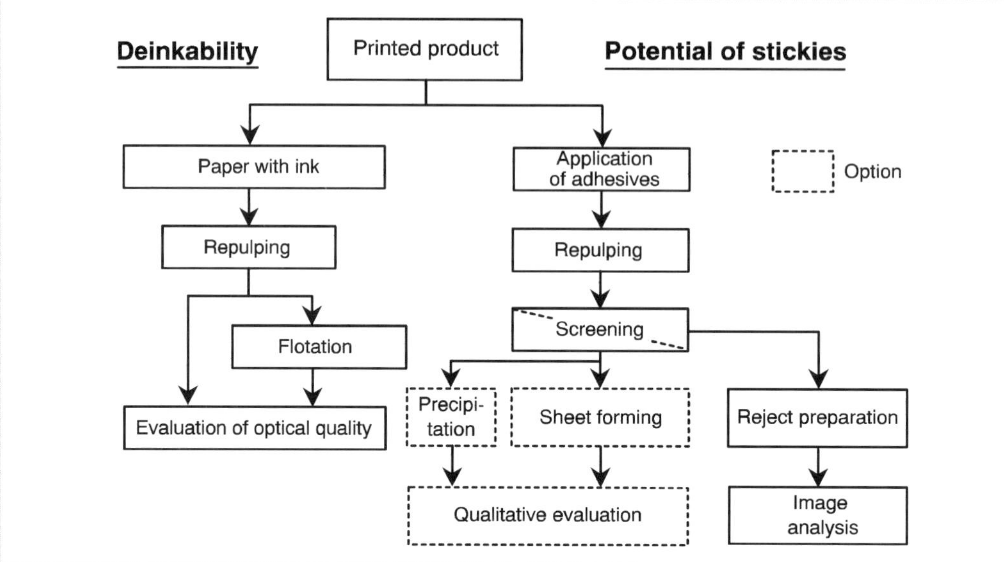

Figure 17. Scheme for testing the recyclability of printed products according to INGEDE method.

Before testing a printed product, it is classified by the printing process applied (offset or rotogravure) and the paper grade used (in terms of coated or uncoated paper). In the standard procedure (described in the INGEDE Method 11[20]) the printed product is slushed with a deinking chemical formulation and deinked in the PTS laboratory flotation cell (Fig. 18). After flotation the following parameters are determined for the following reasons:

- Luminance (Y) of the deinked pulp to determine if it reaches a sufficient level of a white appearance
- Light absorption coefficient k measured at 700 nm wavelength to calculate the "Ink Elimination (IE)" by the flotation process[21]
- Color value a^* to determine a red or green shaded discoloration of the DIP
- Total dirt speck area by image analysis to assess the cleanliness of the DIP
- Luminance difference ($\Delta Y_{Filtrate}$) measured on a membrane filter (diameter: 45 mm, pore size: 0.45 μm) between a blind (tap water) sample and the filtrate of the filter pad preparation to assess discoloration of the process water by soluble and finely dispersed ink components, e.g., flexographic inks.

By examination of all five parameters, it is possible to make a comprehensive assessment of the deinking behavior of a printed product. To classify a printed product as "recyclable", all five parameters must fulfill minimum requirements that differ to some extent both according to the category of printing process used and according to the type of paper. The minimum requirements of the various categories of printed products have been agreed upon by the relevant partners in the paper chain: papermakers, publishers, printers, and ink manufacturers[22].

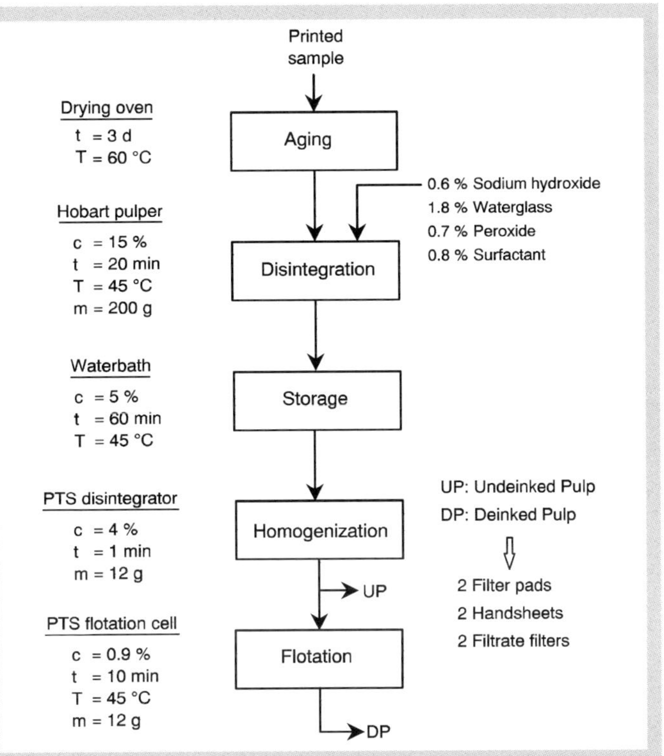

Figure 18. Deinkability test method for printed products according to INGEDE method.

3.5 Effect on process water

The PTS and FINAT test methods have the disadvantage that dissolved or colloidal substances contained in the filtrate of the dewatered recycled fiber pulp are not assessed for their effects on a recycling process. Nonfibrous components that are soluble in the neutral or alkaline aqueous environment of the deinking process such as redispersible or soluble additives and that do not adversely influence sheet formation are therefore classified as recyclable. No account is taken of possible enrichments in closed water loop systems of a paper machine or the stock preparation line. The possible presence of stickies resulting from interaction with other substances or the additives used in paper production such as retention agents or coating binders also do not receive consideration.

In the existing test methods for determining the recyclability of paper and board products, no threshold values are defined for an increasing load of the process water as COD or total organic carbon (TOC) due to the presence of nonfibrous components in a product. The complex interaction of soluble or redispersible nonfibrous components of a paper or board product is difficult to simulate and therefore predict in a closed process water system.

The basic demand of recycling-friendly paper and board products is therefore that nonfibrous components should have the maximum possible wet disintegration resistance to ensure they can be removed as early as possible in the recovered paper processing operation. The development of adhesives that can be screened after pulping and do not disintegrate or dissolve too strongly is especially desirable for the recovered paper processing industry. It is an indispensable precondition for the design of recycling-friendly paper or board products.

References

1. Anon., "Altpapierliste der Deutschen Standardsorten und ihre Qualitäten", BDE/Köln, bvse/Bonn, VDP/Bonn, Germany, 1997, 2 pp.
2. Anon., "CEPI List of European Standard Qualities of Recovered Paper", CEPI, Brussels, 1995, 8 pp.
3. Anon., "General List of Standard Qualities of Recovered Paper", B.I.R., Bureau International de la Récupération, Barcelona, Spain, 1996, 22 pp.
4. Anon., DIN EN 643, "List of European Standard Qualities of Waste Paper", Beuth Verlag, Berlin, 1994, 8 pp.
5. Anon., "European List of Standard Grades of Recovered Paper and Board", B.I.R./Brussels, CEPI/Brussels, 1999, 11 pp.
6. Anon., "Recovered Specifications Circular 1998, Guidelines for Paper Stock: PS-98 Domestic Transactions", Institute of Recovered Recycling Industries (ISRI), Washington, D.C., 1998, p. 27.
7. Anon., "1997 Statistics of Recovered Paper in Japan", Paper Recycling Promotion Center, Tokyo, 1998, 24 pp.
8. Phan-Tri, D. and Göttsching, L., Wochenbl. Papierfabr. 112(6):167(1984).
9. Phan-Tri, D. and Göttsching, L., Wochenbl. Papierfabr. 113(10):343(1985).
10. Galland, G., Das Papier 51(6A):V173(1997).
11. Renner, K., Putz, H.-J., and Göttsching, L., "Erfassungssysteme für Altpapier – Bestandsaufnahme und Bewertung neuer Erfassungsstrategien zur Qualitätsverbesserung von Altpapier", INGEDE Report 4094 IfP, Darmstadt, Germany, 1997, 73 pp.
12. INGEDE-Methode 07, "Eingangskontrolle von Altpapierlieferungen der Sorte Deinkingware (D39) in loser Lieferform", INGEDE, Munich, 1998, 6 pp.
13. Putz, H.-J., Wochenbl. Papierfabr. 124(3):74(1996).
14. Renner, K., Putz, H.-J., and Göttsching, L., Wochenbl. Papierfabr. 124(14/15):662(1996).
15. Gobbey, A., "DIP quality versus office wastepaper collection", 3rd Advanced Training Course on Deinking Technology, CTP, Grenoble, 1997, Chap. 25.

16. *PTS-Methode (draft), "Bestimmung papierfremder Bestandteile (Gehalt an nicht zerfaserbaren Bestandteilen) in Verpackungen aus Papier, Karton und Pappe", Papiertechnische Stiftung, Munich, 1997, 2 pp.*

17. *PTS-Methode RH 021/95, "Kennzeichnung der Rezyklierbarkeit von Packmitteln aus Papier, Karton und Pappe sowie von grafischen Druckerzeugnissen", Papiertechnische Stiftung, Munich, 1996, 6 pp.*

18. *Großmann, H., Das Papier 48(10A):V63(1994).*

19. *Anon., FINAT Test Method FTM 19, "Assessing the Recycling Friendliness of Self-Adhesive Labels", FINAT, The Hague, The Netherlands, 1997, 6 pp.*

20. *INGEDE-Methode 11, "Bewertung der Rezyklierbarkeit von Druckerzeugnissen – Prüfung der Deinkbarkeit", INGEDE, Munich, 1999, 7 pp.*

21. *INGEDE-Methode 10, "Quantitative Ermittlung des Druckfarbenaustrags beim Deinken von Altpapierstoffen (Ink Elimination IE)", INGEDE, Munich, 1999, 3 pp.*

22. *Ackermann, C. and Putz, H.-J., "Kriterien für die Rezyklierbarkeit von graphischen Druckerzeugnissen", 1999 INGEDE Seminar "Verbesserung der Recyclierbarkeit graphischer Druckerzeugnisse", Munich, p. 73.*

CHAPTER 5

Unit operations and equipment in recycled fiber processing

CHAPTER 5

CHAPTER 5

Herbert Holik

Unit operations and equipment in recycled fiber processing

1 General overview

Similar to virgin fibers, recycled fiber stock must meet defined quality criteria for a given furnish. In this sense, quality refers to the fiber characteristics such as fiber type and condition and to additives and contaminants. Recycled fiber processing systems are significantly more complex than systems for virgin fibers for two reasons. First, the recovered paper usually is a mixture of various fiber types or paper grades. More importantly, it also contains a certain proportion of contaminants and other detrimental substances. These have the following classification according to their origin:

- Additives used at the previous paper production process such as fillers and dyes, coating components, and functional and process chemicals
- Substances added according to application such as printing inks, coatings, foil laminations, and adhesives
- Materials mixed with the paper during its life cycle and subsequent collection including wires and strings, sand and stones, or paper clips and folders.

Within wide regional variations, recycled fibers partly, largely, or completely replace the various kinds of virgin fibers in many paper and board grades today. For this reason, fiber quality and cleanliness have comparable demands. For example, fiber quality defines mechanical paper characteristics and printability. Cleanliness covers optical, chemical, colloidal, microbiological, and processing aspects such as sand and content of stickies.

Selection and monitoring of recovered paper furnish play very important roles in cost-effective recycled fiber processing and maintaining adequate quality of the finished stock. The primary tasks of processing are removal of contaminants, elimination of their effects, or both as much as necessary to meet quality requirements. Figure 1 shows an overview of some substances encountered in recycled fiber pulps with their approximate particle size ranges. The extent to which they require removal depends on how much they interfere with paper manufacturing or product quality requirements.

Separation processes vary according to the type and quantities of contaminants for removal. As separation criteria, they use certain characteristics of these materials that require clear discernment from fiber characteristics. These include particle size, shape, and deformability, density, and surface properties concerning flotation and sedimentation. Table 1 summarizes characteristic particle size and specific gravity of some typical contaminants.

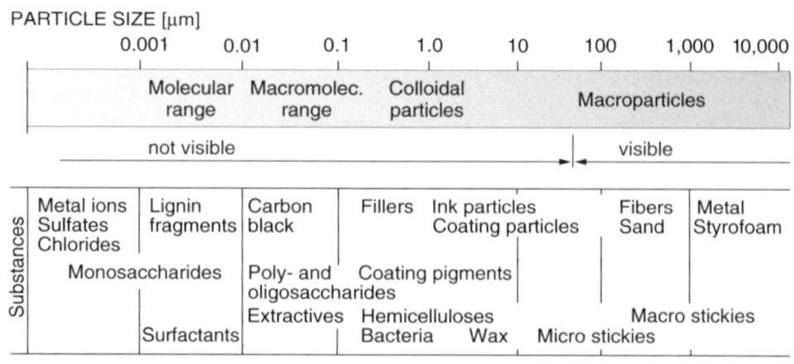

Figure 1. Substances in recovered paper processing.

Table 1. Specific gravity and particle size of contaminants in recovered paper processing.

Type of contaminant	Specific gravity [g/cm^3]	Particle size [µm]				
		<1	<10	<100	<1 000	>1 000
Metal	2.7–9					
Sand	1.8–2.2					
Fillers/coating particles	1.8–2.6					
Ink particles	1.2–1.6					
Stickies	0.9–1.1					
Wax	0.9–1.0					
Styrofoam	0.3–0.5					
Plastics	0.9–1.1					

Separation processes have the following classification according to particle characteristics:

- Screening: particle size, shape, and deformability
- Washing: particle size and shape
- Centrifugal cleaning: specific gravity, size, and shape of particles
- Flotation: surface properties and size of particles.

Separation processes and their efficiencies depend on suspension properties such as consistency and flake and debris content. The same processing stages can have use several times in a stock preparation line depending on the type and quantity of contaminants to provide better effectiveness and increased reliability of processing machinery (See Chapter 6).

Figure 2 shows efficiency scatter of various separation processes vs. debris particle size. The separation efficiency of cleaners is excellent at large particle sizes if their density sufficiently differs from that of water and wetted fibers. Otherwise, the separation effect can decrease almost to zero.

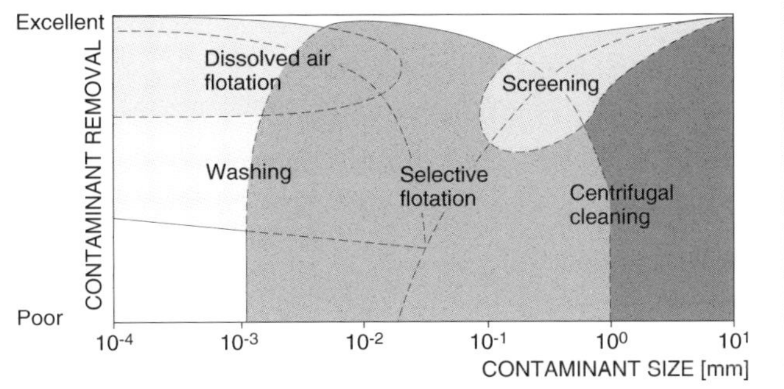

Figure 2. Efficiency ranges of unit processes for contaminant removal in recovered paper processing.

Large cubic particles separate very well during screening, but smaller, flat, or deformable particles hardly separate in this process step. Flotation removes only hydrophobic particles but with great efficiency and over a wide size range. Hydrophilic particles remain in the suspension. Washing removes small particles. The efficiency depends on the amount of extracted water. Dissolved air flotation (DAF) or microflotation separates small particles up to the 10 μm range very efficiently[1].

Besides separation, the following processes also play an important role in recycled fiber preparation. Machinery requirements again vary according to stock consistency and contamination. The main unit processes and their tasks are as follows:

- Slushing and deflaking: Breaks down the recovered paper first into a pumpable condition enabling coarse separation that is normally included in this stage and then by deflaking if necessary
- Fractionation (a special kind of screening): Separates a fiber mixture according to defined criteria such as size or deformability
- Dispersing and kneading: Makes residual dirt specks and stickies smaller or floatable, detaches ink still adhering to the fibers, mixes in bleaching agents, and conditions the fibers technically
- Refining: Modifies the fiber morphology and surface characteristics to ensure the required bonding properties and paper quality
- Dewatering (thickening and pressing): Removes water from the suspension making certain processes possible such as dispersing or more cost-effective such as bleaching or storage (Drainage also separates the water loops for ionic trash and temperature control).
- Bleaching: Endows yellowed or colored (brown) fibers with the required brightness or luminance
- Storing and mixing: Prevents the suspension components from demixing or mixes them together.

Important criteria for the evaluation of efficiency and cost-effectiveness of each processing stage are the specific energy demand and stock consistency range. As Table 2 shows, such variations can be wide. This can be due to the widely differing technology available. The primary reason is that energy demand depends considerably on how much work has occurred in the previous stages concerning deflaking and contaminant removal and on the requirements placed on the pulp slurry from the machinery side in the downstream stages. The figures do not include the energy demand for pumping the pulp slurry from one process step to the other.

Table 2. Specific energy demand and operating consistency of unit processes in recovered paper processing.

Unit process	Specific energy demand [kWh/t]	Operating consistency [%]
Slushing pulper	10–40	3–18
Drum pulper (screening)	15–20/40	3.5 – 20/(3.5–6)
Deflaking	20–60	3–6
Screening	5–20	0.5–4
Tailing screen	20–40	1–4
Washing	1.5–20	0.7–1.5 ⇒ 5–12
Dissolved air flotation (DAF)	10–20	< 0.3 ⇒ 0.01
Flotation (selective)	20–50	1–1.3
Centrifugal cleaning	4–8	< 0.5–4.5 (< 6)
Thickening	1–10	0.5 ⇒ 5 (10)
Dewatering-Screw press	10–15	4–10 ⇒ 25–40
Dewatering-Double wire press	2–4	3–10 ⇒ 25–50
Dispersing	30–150	22–32
Refining LC (per SR unit)	3–25	3–5.5
Refining HC (per SR unit)	10–60	approx. 30
Storing	0.02–0.1	3.0–5.5 (12)
Mixing	0.2–0.5	3.5–4.5

For stock consistency, no clear distinction exists in the industry between low consistency (LC), medium consistency (MC), and high consistency (HC). The definitions of the actual range depend on the kind of processing stage as Table 3 shows.

Table 3. Consistency ranges and their different naming in various process stages.

Unit operation / Naming	Pulping	Screening	Centrifugal cleaning	Bleaching	Refining
LC (low consistency)	< 6%	<1.5%	< 1.5%	–	3%–6%
MC (medium consistency)	<12%	< 4.5%	< 2.5%	10%–15%	10%–13%
HC (high consistency)	<19%	–	2.5%–6%	25%–35%	28%–35%

Stock preparation is essentially the three-level process of Fig. 3[2,3]. The first process is the production level. This is the direct conversion of recovered paper into finished recycled fiber stock in the machine chest. This level includes the following functions:

- Break down into individual fibers (slushing and deflaking) and retention in suspension
- Separation of fibers and contaminants (slushing, screening, flotation, washing, and centrifugal cleaning)
- Dewatering of fibers and rejects
- Fiber treatment (refining and dispersing)
- Residual contaminants treatment (dispersing).

The second level is the recovery of fibers, other solids, and water from the reject flows from the production level. Separation processes are again necessary here for sorting fibers and contaminants and removing solids from water. The third level is the discharge where the rejects, sludges, and process water undergo processing. This requires separation of the solids and water.

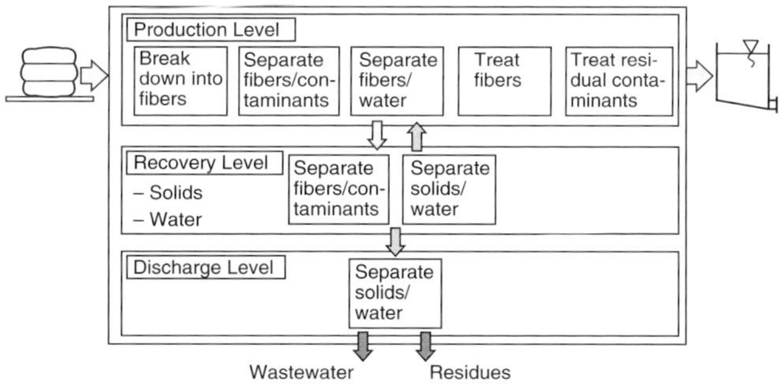

Figure 3. Three process levels in stock preparation with their subsystems for recovered paper processing[3].

A detailed description of the individual processes involved in recycled fiber processing occurs below. These require appropriate combination in the various stock preparation systems to meet requirements set by the raw material and product quality. Chapter 6 describes some typical layouts of recovered paper processing systems.

2 Slushing and pulping

2.1 Objectives and system integration

The purpose of slushing or pulping (repulping) in recovered paper processing is to break down the raw material into individual fibers as much as possible to form a suspension that can at least be pumped. The operation primarily comprises the following processes:

- Feed the system with a pre-determined rate
- Break down of the raw material into individual fibers
- Removal of solid contaminants such as foils, stickies, and printing ink from the fibers

- Removal of solid contaminants at an early stage that might otherwise break down into excessively small particles
- Mixing of process chemicals into the suspension such as deinking chemicals or bleaching agents.

To save energy, the first step is often limited to coarse slushing and heavy particle separation. Fine disintegration with a disk screen or deflaker usually follows.

For recovered paper with very high wet strength, a disperger is necessary to break down all the flakes into individual fibers. Using thermal, chemical, or both types of treatment during the pulping process is often more cost-effective.

Residual flake content therefore depends highly on the slushing method, the chemicals used, and the coarse separation system. These also influence total system design requirements and the final product characteristics.

2.2 Principles

At slushing, the mechanical strength of the raw material must be reduced by wetting. In this process, the fiber-to-fiber hydrogen bonds break. Figure 4 shows how much wetting reduces the strength of some secondary raw materials[4]. The tensile strength of nonwet strength grades decreases by 86%–98% and by 60%–80% for wet strength grades. As Fig. 5 shows, strength reduction occurs very rapidly (to the extent of 95% within one minute for low basis weight grades). Rapid and complete wetting of the incoming recovered paper is therefore extremely important for efficient slushing.

For grades that are difficult to pulp, tensile strength may be reduced further by heating to more than 75°C. If necessary, chemicals are also added. These are acidic or alkaline depending on the wet strength agent[5].

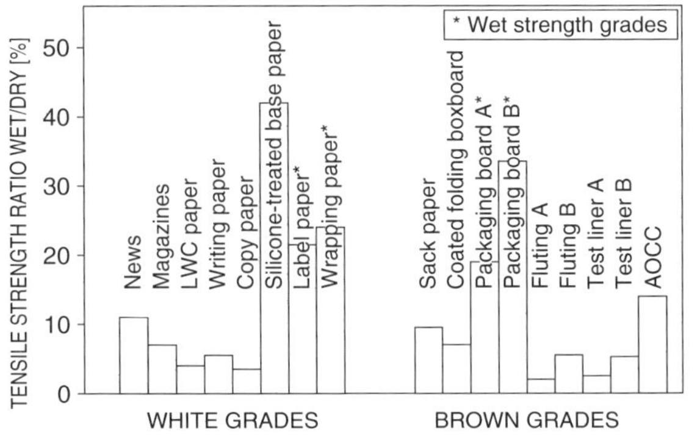

Figure 4. Tensile strength reduction of various recovered paper grades due to wetting[4].

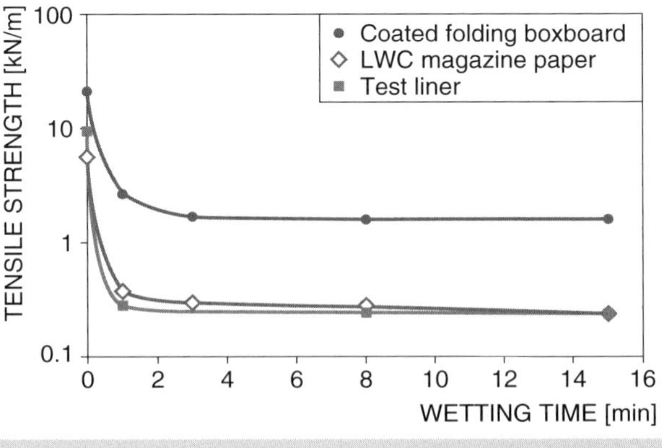

Figure 5. Tensile strength of three recovered paper grades vs. time after wetting[6].

The disintegrating forces applied by a suitable method must be greater than the strength of the raw material or the adhesion force of contaminants to the fibers. Pulping forces should not be excessive or contaminants will break excessively.

Very little detailed knowledge currently exists about slushing mechanisms. A good assumption is that acceleration, viscosity, and clinging such as on the rotor generate deflaking forces[6]. Sudden motion and resultant inertia apply acceleration forces to the flakes. To apply viscosity forces, the suspension of solids in water must have a velocity difference with the flakes. Viscosity forces are responsible for the fiber-to-fiber friction that causes flake disintegration due to shear stress. High shear forces exist at the interfaces between slow moving stock in the pulper and the individual pulp jets that emanate from the rotor at high velocity. The flakes wrapping themselves over the rotor, screen plate holes, or impact bars cause clinging forces. To obtain a defibering effect, these forces must act in pairs and in opposite directions.

2.3 Machinery and effects

Slushing uses pulpers or drum pulpers. Table 4 gives an overview of applications with characteristic features and operating conditions of pulpers and drum pulpers. Pulper feed systems[7] vary considerably depending on whether the raw material comes in bales or as loose, unbaled recovered paper.

2.3.1 Pulpers

Pulpers normally comprise a cylindrical stainless steel vat and a concentric or sometimes eccentric impeller with blades for clearing any extraction screen plate that may be installed. Since the impeller is subject to extreme wear, its wearing parts must be easily replaceable or repairable by resurfacing. For better flow circulation to the pulper center, the vat walls usually have vertical elements, guide elements, or both. In some cases, impact bars are fixed on the vat floor immediately adjacent to the impeller periphery to increase the pulping energy transfer. Figure 6 shows the basic layout of an LC pulper for stock consistencies up to about 6%. Pulpers of this type now have a volume as high as 160 m^3.

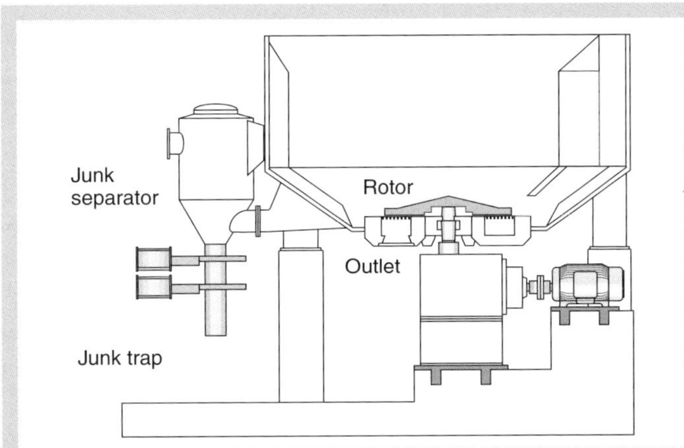

Figure 6. LC pulper.

The impeller that rotates at a peripheral speed of 12–20 m/s causes the pulping forces. Since acceleration forces influence pulping considerably, the wet breaking length of the unprocessed material is a critical parameter for slushing.

Table 4. Overview of pulper and drum pulper applications, design features, and operating modes.

Type of pulper	Feed		Operating conditions					Machine characteristics				
	Furnish	Condition	Consistency [%]	Capacity min [t/d]	Capacity max [t/d]	Slushing time (total) [min]	Operation	Axis	Effective element	Circumf. speed [m/s]	Screening sieve	Pulper diameter [mm]
HC pulper	News Magazines	Loose Dewired bales	<19	30	400	15–25	Batch	Vertical	Helical rotor	12–16	No (yes)	< 7 100
MC pulper	News Magazines	Loose Dewired bales	<12	140	500	20–30	Batch	Vertical	Helical rotor	13–17	No (yes)	< 6 000
LC pulper	Fluting Liner High wet strength grades	Loose Opened bales	<6	200	1 600	5–40	Contin. (ragger, Junk separator)	Vertical Horizontal Inclined	Flat rotor with vanes	15–20	Yes	< 8 000
Drum pulper	News Magazines Fluting Liner	Loose	<20 (<5)	100	1 600	20–40	Contin.	Horizontal	Drum	1.5–2	Yes (drum)	< 4 250

Wet breaking length is defined in the same way as dry breaking length. The only difference is that tensile strength and basis weight values are in the wet condition for calculating wet breaking length. Figure 7 shows the ratio of wet to dry breaking length for the same material used in Fig. 4. Due to a different water absorption potential for various paper grades relative to their basis weight, the ratios of wet breaking length in Fig. 7 are not proportional to those of tensile strength. Clearly, the breaking length decreases more than tensile strength due to wetting. This can be 90%–98% for nonwet strength grades.

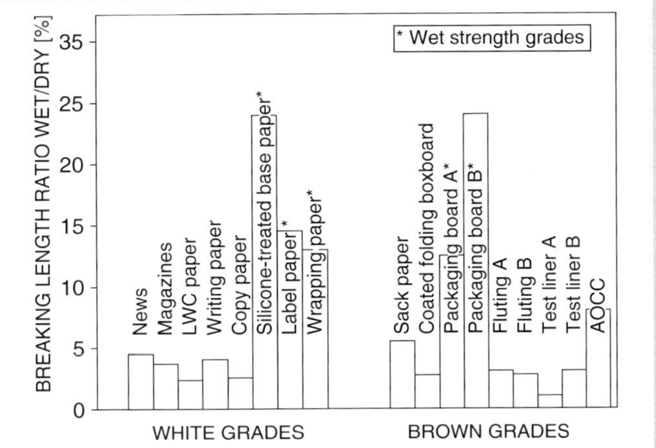

Figure 7. Reduction in breaking length of various paper grades due to wetting[4].

For good slushing results, the pulper vat geometry, impeller geometry, and peripheral speed require proper coordination. This ensures not only rapid feeding of the raw material into the slurry for wetting but also cost-effective pulping[8].

Pulpers operate at high consistencies as HC pulpers (up to about 19% stock consistency), at medium consistencies as MC pulpers (up to about 12% stock consistency), and at low consistencies as LC pulpers (up to about 6% stock consistency) depending on raw material. HC and MC pulpers generally operate intermittently. LC pulpers usually operate continuously. Pulper impellers vary according to stock consistency range as does the pulper cleaning system.

Slushing curves for the various recovered paper grades can assess pulper operating characteristics. With batch operation, these indicate the flake content vs. time and energy input. Flakes are small pieces of paper or board not disintegrated into individual fibers. Figure 8 shows two slushing curves with different impellers and different stock consistencies using magazines as raw material. Clearly, the steepest reduction curve of flake content occurs in an HC pulper by comparison with other types for this kind of stock.

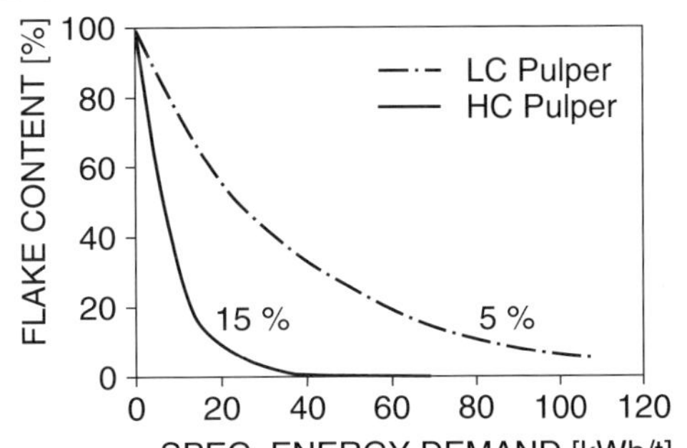

Figure 8. Deflaking characteristics of pulpers at different stock consistencies vs. specific energy demand[9].

Slushing characteristics depend on the wet strength of the raw material. During batch pulping of mixed recovered paper with different wet strengths by individual paper grades, the flake content of higher wet strength components therefore decreases more slowly compared with the components with low wet strength. This is because materials difficult to disintegrate need a longer pulping time.

Figure 9 explains this principle using three stocks (A, B, and C) with different slushing characteristics and a mixing ratio of 1:1:1. The resultant composition (A, B, C) of residual flake content varies considerably according to pulping time, i.e., specific energy input. For the same reason, the pulper content composition in continuous operation under steady state conditions is not the same as the input composition.

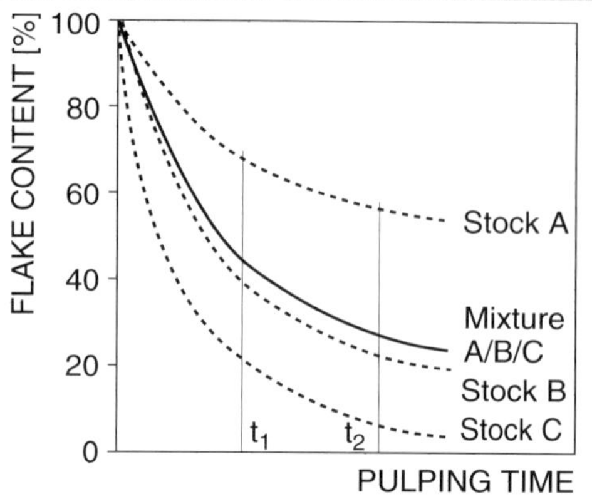

Figure 9. Pulping characteristic for a recovered paper mixture of components A, B, and C for feed component mix 1:1:1 vs. time[6].

HC and MC pulpers

HC and MC pulpers usually operate in batch mode, i.e., with separate phases for water and recovered paper input, slushing, dilution, dumping, and separate flushing of residues if necessary. This means that only about two-thirds of the total operating time is available for actual pulping. To shorten the total time necessary, all emptying and filling aggregates are dimensioned as generously as possible. This increased outlay is a disadvantage of the HC pulper, but the main advantage is economy of pulping energy as Fig. 8 shows and gentle slushing with low breaking down of contaminants[9].

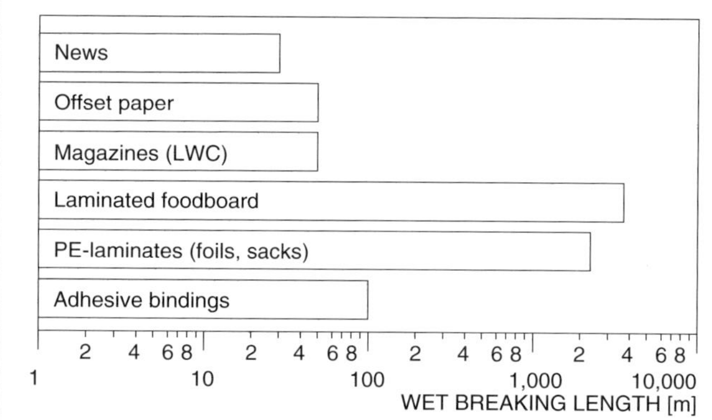

Figure 10. Wet breaking length of recovered paper grades compared with the wet breaking length of polyethylene laminates[10].

To avoid excessive breaking down, the pulper operating forces should be adequate for good disintegration but have the least possible peaks. Figure 10 shows typical wet breaking length values for various recovered paper grades and some common contaminants. The strength values of these substances vary significantly with the wet breaking length of adhesive bindings (stickies) being closest to that of the fibers. Chemicals and elevated temperature reduce the strength of stickies more.

Figure 11 shows pulping test results on foil coated food packaging[9]. The forces applied were obviously sufficiently high to achieve good slushing results in the case of the board component. They were also sufficiently small not to break down the plastic laminate by increasing energy demand as the diagram shows.

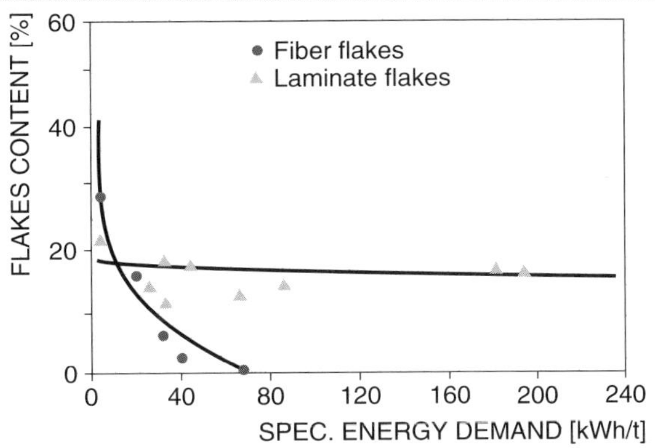

Figure 11. Flake reduction during pulping of foil coated food packaging material[9].

Another slushing task usually performed most effectively in the HC range is ink detachment in the deinking process. The forces required for this depend on the strength of the ink film and its attachment to the fibers. For example, Fig. 12 shows how quickly the number and size of ink particles still adhering to the fibers decrease with increasing specific energy application[10].

Figure 13 shows a typical HC pulper with characteristic spiral rotor. Besides slushing, the individual zones of this rotor also have a transport function. The upper part transports the pulp slurry downward, and the bottom part primarily has a recirculation function. Figure 14 shows two types of spiral rotor. One is for HC pulpers, and the other is for MC pulpers.

The input to HC and MC pulpers can comprise loose material or bales. Bales can have manual or automatic dewiring. Breaking down the bales after dewiring facilitates wetting in the pulper to shorten the slushing time. Typical raw materials for these types of pulpers are deinking materials (old newspapers and magazines) or coated grades

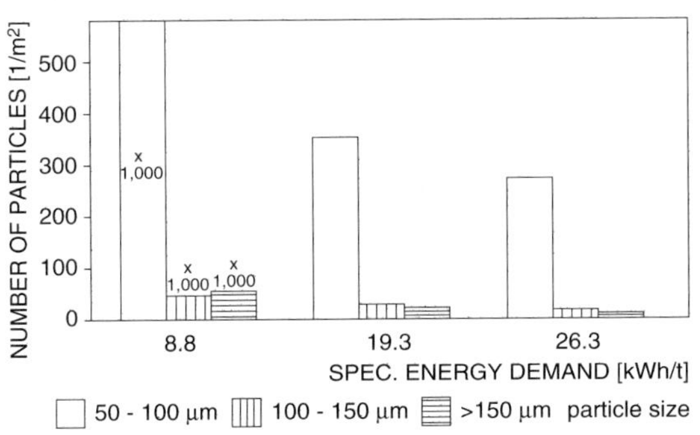

Figure 12. Number of ink particles still adhering to fibers at different levels of specific defibering energy demand[10].

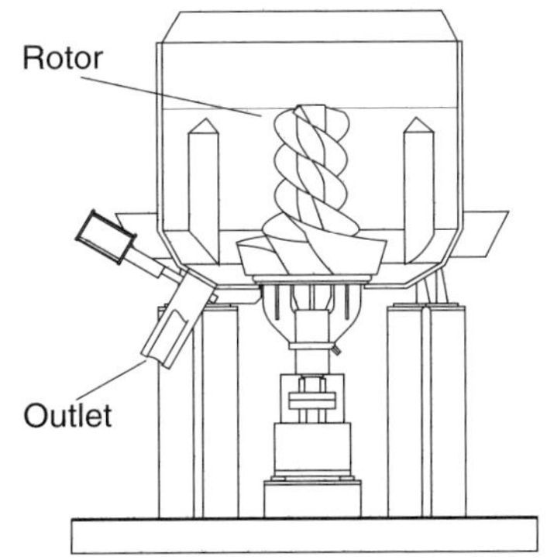

Figure 13. Schematic arrangement of an HC pulper.

Figure 14. Spiral rotors of HC (left) and MC (right) pulpers.

although wet strength furnish can also be slushed cost-effectively with chemicals at temperatures of about 75°C or more.

In batch pulpers, drive torque changes with increasing slushing time with the highest peaks up to 200% of rated load at the beginning of slushing. As soon as the recovered paper has been pulped to the required residual flake content, dilution water is added and the pulp slurry is pumped out. Depending on the system type, dumping may be through a screen plate in the pulper to retain coarse trash that remains in the pulper and then must be removed in an additional disposal stage.

Normal practice is to empty the pulper completely in a single operation. This includes coarse trash separation and further slushing in a downstream machine. Figure 15 shows this kind of dumping system with an HC pulper and downstream disk screen for coarse screening and further deflaking. The accepts from this machine are stored in the dumping chest, and the rejects are pumped via a buffer storage tank into a drum screen. Rejects from this drum screen are disposed after dewatering, and the accepts are recirculated upstream of the disk screen.

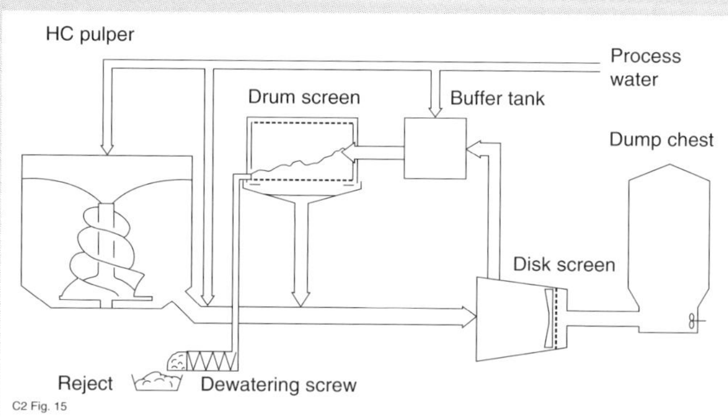

Figure 15. HC pulping subsystem for recovered paper: pulper, dumping system, and dilution water feed system.

LC pulpers

Recovered paper input to LC pulpers is in loose form or bales that require opening but not necessarily dewiring. LC pulpers have most frequent use in recovered paper processing for manufacturing packaging paper and board. The photograph of an LC pulper in Fig. 16 shows the impeller and screen plate.

Figure 16. Interior of an LC pulper with impeller and extraction screen plate.

The rotor shaft can be vertical, horizontal, or oblique. In most cases, LC pulpers operate in the continuous mode with constant removal of the almost completely slushed stock through the screen plate. Impeller blades keep this free. The extraction screen plate hole diameter can vary from 6 to 20 mm depending on system concept and recovered paper grade. Smaller holes result in smaller flakes in the accepts but require longer pulping times. With larger holes, the pulping time is shorter but flakes in the accepts are larger by size and amount. The hole diameter can only be optimized by considering the downstream screening system that must not be overloaded by excessive flake size or quantities. Easily pulpable stocks are often reduced to a flake content of 15%–20%. With more difficult stocks, these figures approach 20%–40%.

To prevent excessive trash concentration in continuously operating pulpers, their ongoing removal is necessary. Figure 17 shows a typical pulper discharge system for this purpose. Part of the suspension is extracted from the pulper and pumped via a junk separator to a disk screen that also acts as a deflaker. The accepts are pumped forward while the rejects are sorted in a drum screen. Accepts from this drum screen are recirculated to the pulper while the rejects are disposed.

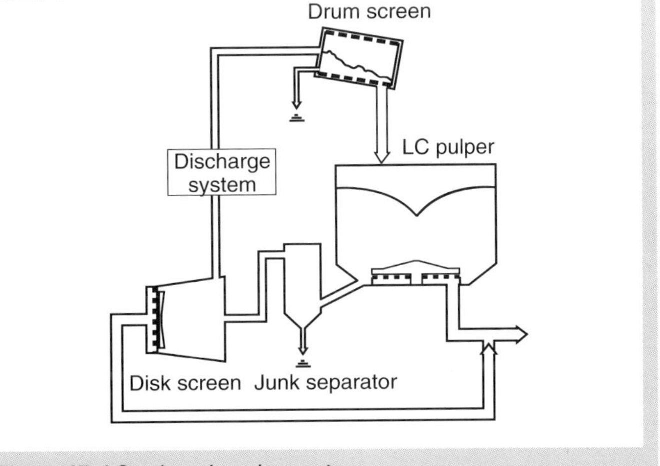

Figure 17. LC pulper dumping system.

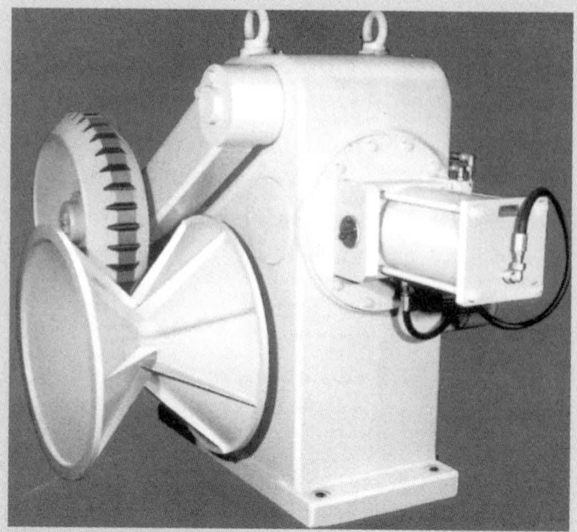

Figure 18. Ragger.

Bale wires, strings, plastics, foils, and textiles are often removed from the pulper by a ragging device consisting of a winch as Fig. 18 shows followed by a tail cutter. This system operates on the principle that this type of junk forms itself into a tail under certain conditions.

To start ragging, a length of barbed wire is placed into the tub. Due to the intensive rotary flow in the neighborhood of the impeller, spinning materials become entangled. The tail accumulates and becomes increasingly longer. It is continuously pulled from the pulper by a winch at a speed that must not exceed the tail growth rate. If extracted too quickly, the tail becomes too thin and may fracture. If the extraction rate is too slow, the tail becomes too large, and the impeller may destroy it. If the tail fractures, the loose parts will continue spinning in the pulper and may settle on the bottom. This will bring the rotor movement to a complete halt. For this reason, the tail extraction rate can vary over a wide range from zero to 100 m/h. Conditions for efficient ragging operation are adequate rotary flow in the pulper with stock consistency about 3.5%–5.5%, adequate pulper size, and appropriate ragger positioning in relation to the pulper feed point.

LC pulpers are particularly suitable for slushing wet strength grades in batch operation. For more intensive slushing on the bottom of the pulper vat, stationary bars are fitted around the rotor periphery. The recovered paper can be further treated by heating to 75°C or more and adding chemicals although this is more cost-effective with HC pulpers.

To assess the need for steam and chemicals and the resulting higher outlay in operating costs, wet breaking length can be a criterion. According to long-standing experience, a value of about 600 m (about 25 kWh/t specific pulping energy demand) seems to be the limit for cost-effective slushing without thermal or chemical measures. These are necessary for raw materials with breaking length values much higher than this value.

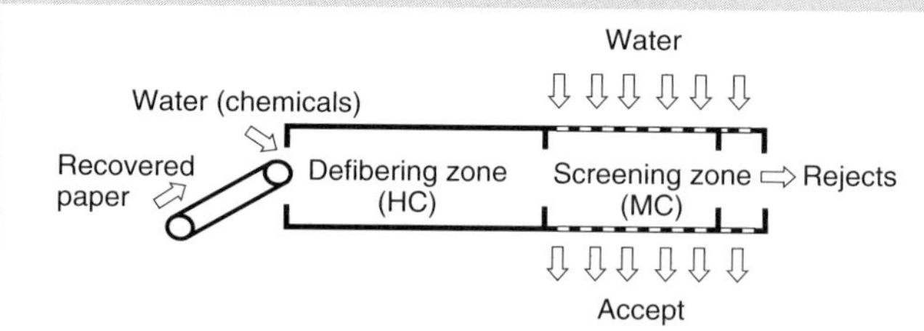

Figure 19. Schematic arrangement of a drum pulper.

2.3.2 Drum pulpers

The drum pulpers of Figs. 19 and 20 have use for recovered paper grades with low wet strength. Their most common applications are for deinking stock based on old newspapers and magazines. They also are finding increasing use for brown grades with low wet strength.

Figure 20. Drum pulper.

Drum pulpers rotate at circumferential speeds of 100–120 m/min with diameters of 2.5–4 m and lengths up to 30 m. They obviously require considerable space. They are driven on the periphery by friction using truck tires or by gear ring and pinion. They gently incline toward the rejects end.

The drum has two zones. One is for slushing, and the other is for screening. These either form a rigid unit or are separated and rotate in opposite directions. The slushing zone requires about two-thirds of the drum length. The screening zone uses the other third. The loose recovered paper or broken down dewired bales are fed into the slushing zone with water and any chemicals required such as for deinking. Inside the slushing zone are carrier claws for the stock. Drums operate at a consistency of about 14%–20% in the slushing area.

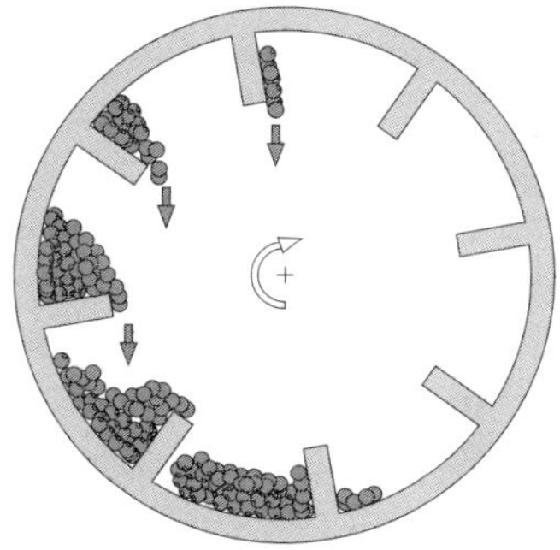

Figure 21. Stock motion in a drum pulper.

As the drum rotates, the stock moves upward as Fig. 21 shows. Depending on the filling level, a flowing or rolling motion occurs that exerts shear forces for efficient slushing. The stock that falls freely from the top of the drum undergoes the greatest slushing forces. Since these forces are generally lower than in pulpers, the contaminant particle size remains sufficiently large for good screening. The wet strength components of recovered paper mixtures cannot be slushed and therefore end up in the rejects[11].

In the screening zone of the drum, the stock consistency is reduced with dilution water to about 3%–5% in the accepts. The hole diameter is about 6–9 mm. The coarse contaminants and flakes usually comprising wet strength materials reach the end of the drum as low-fiber rejects. The accepts are deflaked and coarse screened.

3 Deflaking

3.1 Objectives and system integration

Deflaking breaks down small pieces as flakes of undisintegrated paper into individual fibers. In recovered paper processing, the flakes to be broken down comprise sized, coated, and wet strength grades. In broke processing, the broke of wet strength papers requires deflaking. Although deflaking machines are normally used for this task, disk screens also have a certain deflaking effect but to a smaller extent. Deflakers and disk screens can both handle stock consistencies of 3%–6%.

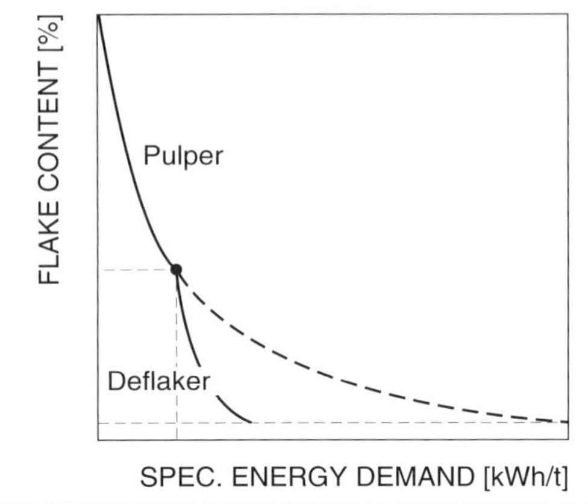

Figure 22. Flake reduction vs. specific energy demand during pulping followed by fine slushing in a deflaker.

For recovered paper that is more difficult to break down, discontinuing slushing at a high flake content and then using a deflaker is more cost-effective. Figure 22 clearly shows the reasons for this. With decreased flake content, the energy demand for further deflaking in the pulper increases rapidly as the flattening of the curve shows. By interrupting slushing at a suitable point and continuing with a deflaker, significant energy savings are possible. The logical outcome of this is a processing sequence comprising coarse deflaking in a pulper followed by fine deflaking with a deflaker.

Another way to save energy with a clean suspension is by screening with subsequent feed forward of the heavily flaked screen reject via a deflaking stage. Extremely difficult stocks to break down are usually treated with chemicals at elevated temperatures to reduce fiber bonding for easier deflaking. A disperger is also useful for efficient flake reduction.

Deflaker efficiency depends greatly on contamination of the recycled fiber pulp. Since deflakers have fine-grade fillings for more intensive breakdown of flakes, coarse trash easily blocks them. For this reason, they require protection by efficient upstream cleaning and screening. To save technical outlay and costs, disk screens are often useful instead.

3.2 Principles

In a similar manner to slushing in pulpers, deflakers break down flakes principally by forces induced by wrapping, viscosity, acceleration, or a combination of these forces. The forces are much higher in deflakers than in pulpers. The probability is also much higher for flakes to break down because they are forced through shear and impact zones in the deflaker fillings. Nevertheless, the deflaking forces are limited. For grades with excessive wet strength, higher dispersing forces are necessary. Reducing flake strength with high temperature treatment and chemicals is often more cost-effective. Figure 23 shows that this provides deflaking in the pulper and deflaker. The chemicals used for this are alkaline or acidic depending on the type of wet strength agent.

Disk screens are similar to pulpers not only geometrically but also regarding flow characteristics. Since the screen disk has finer holes and energy intensity is high in these relatively small machines, deflaking is more effective and energy more efficient than in pulpers. Disk screens do not approach the effectiveness of deflakers primarily because not all flakes actually pass through an area of high deflaking forces.

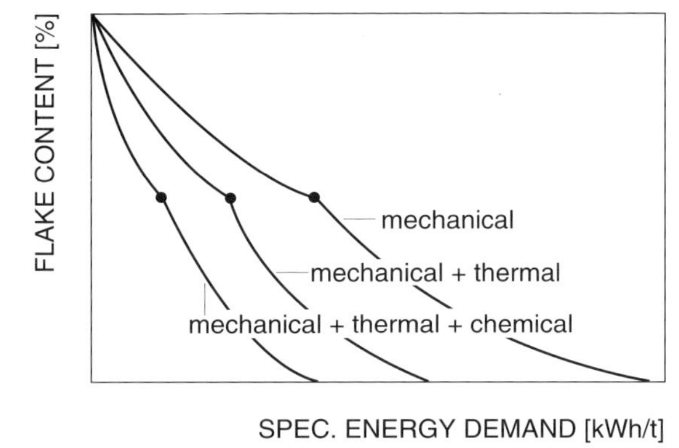

Figure 23. Flake reduction of wet strength grades by mechanical, mechanical-thermal, and combined mechanical-thermal-chemical treatment.

3.3 Machinery

3.3.1 Deflakers

Figure 24 shows a schematic diagram of the deflaker photographed in Fig. 25. This unit has two stator disks and one rotor disk each, having three slotted intermeshing rings operating at a peripheral speed of about 25–40 m/s. The suspension containing the flakes must pass radially through the rotor and stator slots. Acceleration and shear forces are induced in the flakes with wrapping over the impact edges to assist their break down[12].

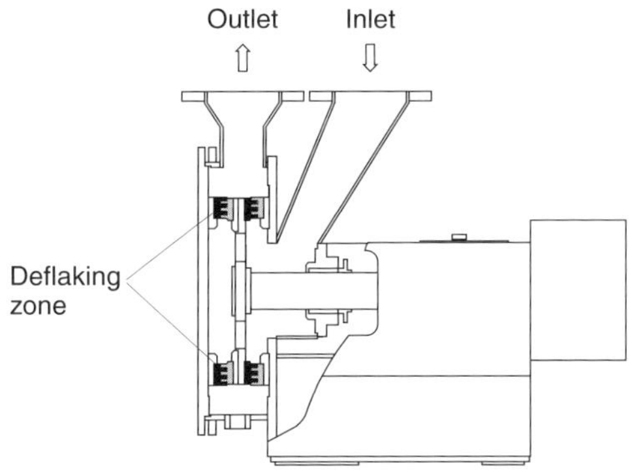

Figure 24. Schematic cross section through a deflaker.

Deflaker fillings can vary between fine and coarse according to flake size, flake content, and trash content in the pulp. Ideally, the degree of disintegration after deflaking should be at least 95%, i.e., a residual flake content below 5%.

Figure 25. Final assembly of a deflaker with cover opened.

Other types of deflakers may include conical working elements with pockets to force the suspension to cross a high shear working zone several times. This type of deflaker also provides for the separation of very coarse contaminants.

For stocks with low deflaking resistance, the specific energy demand is about 20–60 kWh/t with deflaking often occurring in a single pass. For stocks that are more difficult to deflake, two or more passes through the deflaker may be necessary. Several passes are generally more cost-effective than one for a given amount of deflaking energy. This means connecting deflakers in series rather than in parallel is better. This ensures a high throughput per machine. Deflaker system design must consider that pressure increases or releases depending on filling type, throughput, and stock consistency.

Other types of deflakers have holes and disk type fillings between which the pulp slurry must flow in a radial direction. During the transition from inside to outside, the flakes tear.

3.3.2 Disk screens for deflaking

Screens and especially disk screens have a certain deflaking effect. Those with additional impact bars on the disk near the rotor blade ends have the greatest deflaking effect. Figure 26 shows a disk screen opened to reveal the rotor, screening disk, and impact bars.

Disk screens have primary use for coarse screening of suspensions with a high flake and debris content. Their advantage over deflakers for this purpose is that they are insensitive to debris that they separate simultaneously. Their disadvantage is a limited deflaking effect. This is adequate for mixed recovered paper processing in the production of packaging paper and board. It reduces reject rates in the secondary stages of coarse screening. The following section covers disk screens in greater detail.

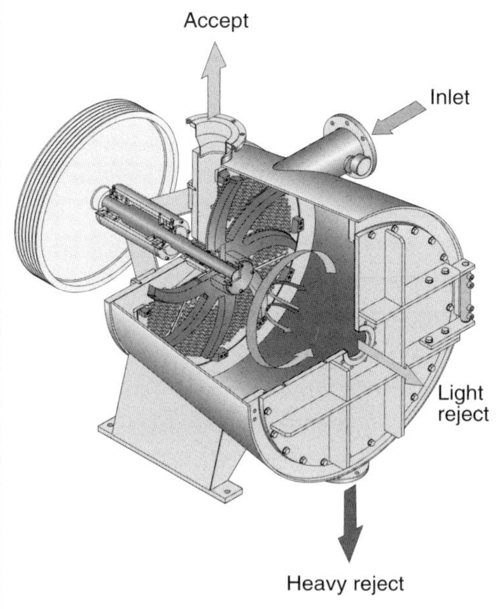

Figure 26. Typical disk screen with cover opened.

4 Screening and fractionation

4.1 Objectives and system integration

The purpose of screening is to remove debris and solid contaminants from the recycled fiber pulp. This uses appropriate screen configurations according to particle size, shape, and deformability of the contaminants. Separating fibers according to characteristics such as their length or flexibility is also necessary sometimes. This kind of separation is called fractionation[13,14]. Strict separation of long and short fibers is not yet possible. Only enrichment of long fiber concentration in one line is possible.

Fractionation usually uses normal screening machinery. Only slight but effective differences exist in the screen equipment and their operating conditions when arranged for a strong fractionation effect with mass flow split according to requirements.

The pulp slurry is screened through holes or slotted openings that are smaller than many of the particles to be separated but much larger than fibers. The openings are also larger than many debris particles that can still be separated to some extent as explained later. Clearing devices that often help transport the retained particles to the reject outlet prevent screen plugging.

Screening is the primary separation process in processing recovered paper whose high debris content places special demands on the screening process. Attempting to remove a large quantity of these particles in one operation can hinder screening and cause rapid abrasion of critical machine components including the screen and rotor. In addition, flat particles such as plastic foil or pieces of undisintegrated paper (flakes) interfere with screening and require removal. Particularly demanding is the screening of deformable particles and especially stickies that are partially broken down or reshaped under high loading.

Complete screening of recycled fiber pulp in a single stage is normally not possible. Depending on the composition of the pulp slurry, different kinds of machines are necessary with different types of rotors and screens and different openings. The main principle is to remove debris particles as soon as possible while simultaneously preventing them from breaking down excessively for fine screening. Appropriate screening also enhances machine efficiency in the subsequent processing stages.

Trash separation starts as early as in the slushing stage where coarse heavy particles, plastic foils, strings, etc., undergo removal. The first step in screening is often reduction of trash content to the point where finer screens have use in the next stage. The object may also be to enable deflaking to occur after which finer and more effective screening is possible.

Avoiding fiber loss during screening is impossible. Rescreening the first stage screen rejects in a second, third, or even fourth stage can reduce such losses significantly. The last stage (tailing screen) determines the fiber losses of the screening system.

The cleanliness efficiency of a screening system increases with higher reject mass flow. Cleanliness efficiency defines the effectiveness of a screen or screening system to remove detrimental particles from the suspension. A higher reject flow gives greater fiber losses. Since cleanliness efficiency and fiber losses also depend on the number of stages, screen selection and system operation are always compromises between maximum cleanliness efficiency, minimal fiber losses, and investment costs.

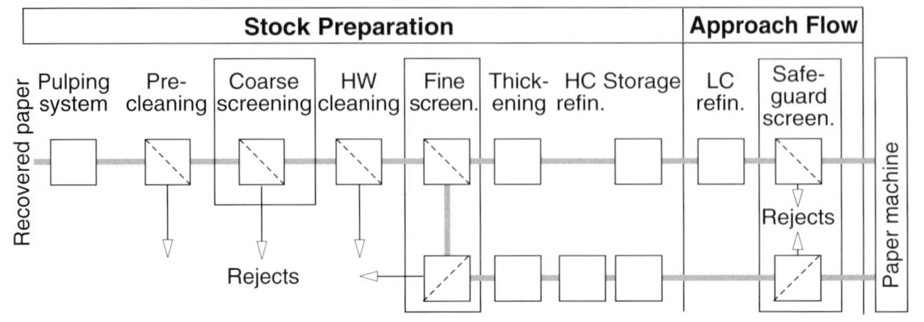

Figure 27. Screening integration in a stock preparation system for test liner using department store waste and mixed recovered paper.

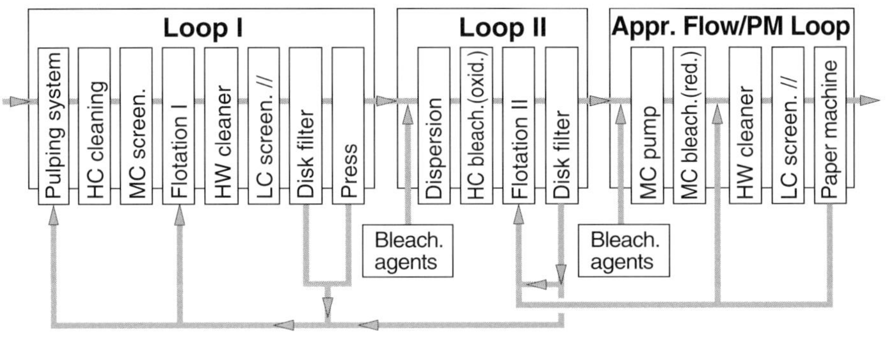

Figure 28. Screening integration in a deinking stock preparation system for improved newsprint using old newspapers and magazines.

Figure 27 shows the system integration of various screening stages with coarse screening in the MC range (disk screen with holes) and fine screening in the LC range (cylindrical screen with slots). The figure shows a stock preparation system for test liner using department store waste and mixed recovered paper. Figure 28 shows a deinking stock preparation system using old newspapers and magazines for improved newsprint with MC cylindrical screen (holes) and LC cylindrical screen (slots). From these two systems, one can derive the screen operating ranges for stock consistency and fineness for a given screen using system positioning. Screening is primarily in the MC range of 3%–4% and the LC range at about 1%.

4.2 Principles

4.2.1 Separation probability

Particles are rejected on the screen according to their size, shape, and deformability.

Nondeformable particles whose dimensions exceed the screen opening size are retained. The removal of particles with one or two dimensions smaller than the screen opening is a matter of probability[15,16].

As Fig. 29 shows, the probability of rejecting particles with one, two, or all three dimensions smaller than the screen openings is greater with the least possible number of contacts between the screen surface and the particle. Preventing a retained particle from contacting the screen again requires rejection as quickly as possible from the machine (short dwell time). This usually means high reject flows. Concentration of debris particles near the screen must also be avoided requiring good mixing in the inlet side of the screen. The quantity of fibers being removed with the reject will also increase due to this mixing.

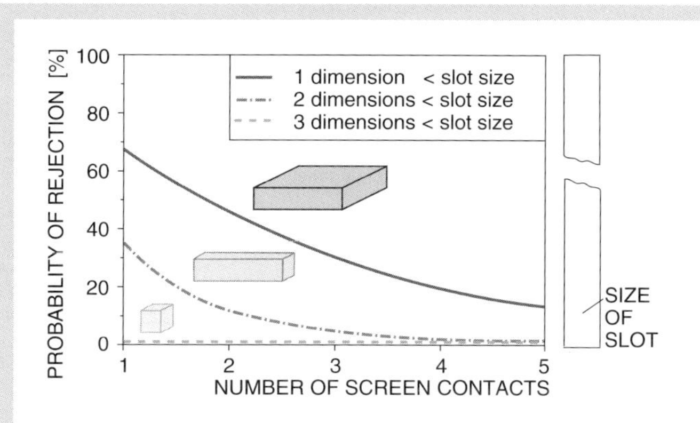

Figure 29. Probability of rejecting particles whose dimensions in one, two, or all three directions are smaller than the screen perforations vs. screen contact frequency[15].

4.2.2 Influence of inflow conditions on cleanliness efficiency

Cleanliness efficiency also depends on factors other than screen perforation size. For particles with only one or two dimensions smaller than the screen perforation, cleanliness efficiency also depends on their orientation due to flow patterns near the inlet to a perforation.

The suspension's approach flow to the very proximity of the screen openings strongly depends on its surface contours[16–18]. In bar-type screen baskets for example, the bar profile angle relative to the tangent influences flow conditions[15,16]. Figure 30 shows these flow conditions for a basket screen with large bar profile angle. Clearly visible are the large vortices and the flow separation point downstream from the bar tip.

Figure 30. Flow pattern at the inlet of a profiled bar-type screen basket.

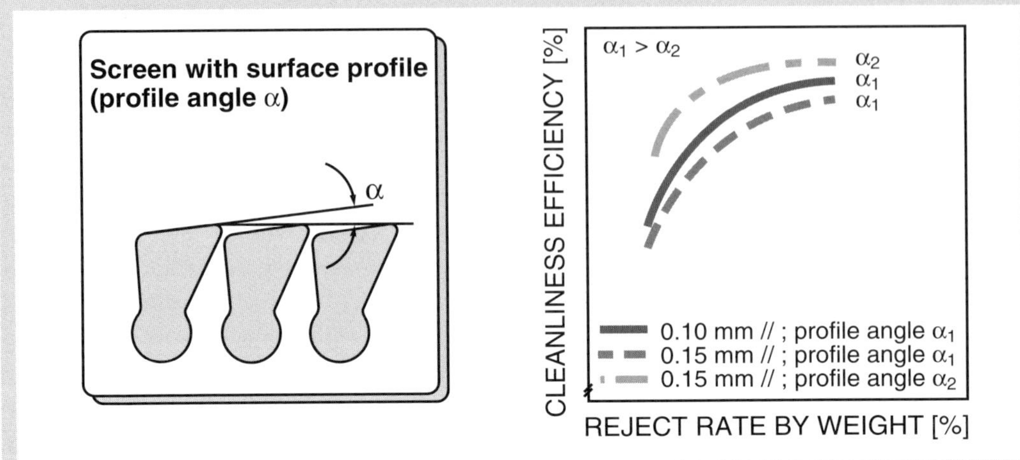

Figure 31. Effect of profile angle and slot width on cleanliness efficiency[15].

Figure 31 shows the cleanliness efficiencies of slotted cylindrical screens with different profile angles, α_1 and α_2. This illustrates the common fact that for the same profile angle, α_1, the cleanliness efficiency at a slot width of 0.1 mm is greater than at 0.15 mm. At a slot width of 0.15 mm and a smaller profile angle, α_2, the cleanliness efficiency is better than with a slot width of 0.1 mm and larger angle $\alpha_1 > \alpha_2$. This is due to the different flow patterns and vortex formation at the basket screen slot intake zone as Fig. 30 shows. Different contaminants and different screen operating conditions can lead to somewhat different results.

4.2.3 Particle deformability and shear forces in the screen

Deformable debris particles of low strength such as stickies can undergo crushing or deforming in the screen by shear and pressure forces and pass through the openings more easily. The shear forces are due to a velocity differential between the rotor and screen surface. At high stock consistencies, the shear forces are much larger for the same velocity differential. Rotor geometry and speed also influence the forces[15,16].

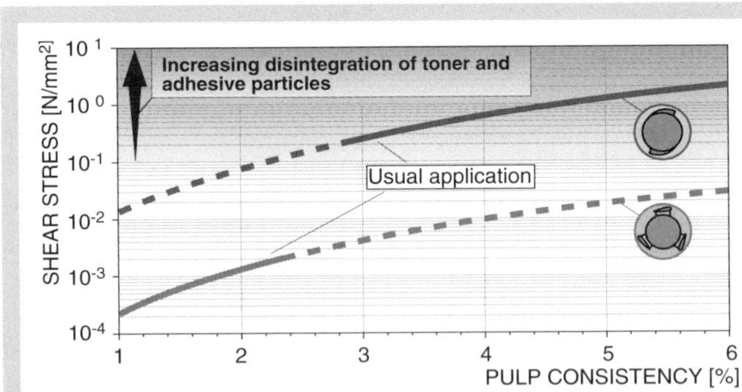

Figure 32. Shear forces generated by two typical rotor types at different stock consistencies and shear strength of stickies in the suspension[15].

Figure 32 summarizes the shear stresses occurring in screens. The two rotor types shown for MC and LC screening generate different shear stresses during operation. Comparing these with the shear strength of stickies (See section on "Dispersion" also) clearly shows that some stickies are deformed or broken down with a typical screening geometry due to the higher shear forces in the MC range but not in LC screening. These high stock consistencies demand higher shear forces from the rotor for efficient fluidization of the suspension to ensure adequate mobility of the fibers and debris particles.

Pressure forces primarily apply to the screen itself. Deformable particles are pressed through the openings due to the pressure differential between screen inlet and outlet[19]. The two curves in Fig. 33 show how the cleanliness efficiency for deformable stickies (soft stickies) is much lower than for solid particles. For a given screen basket, this efficiency decreases even more at higher specific throughput, i.e., higher mean velocity through the slots (the through-velocity) and pressure differential[20].

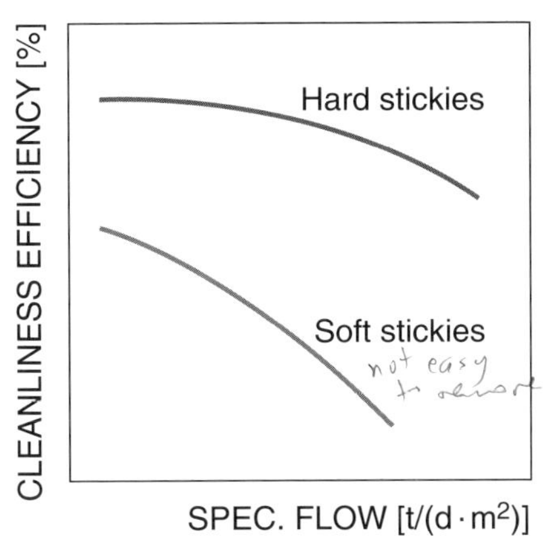

Figure 33. Different cleanliness efficiencies of hard and soft stickies depending on specific flow through screen[15].

4.2.4 Flow through screen openings

Rotors clean the screens primarily by suction or pressure pulsation. The pulse mode and amplitude depend on the rotor geometry, spacing between rotor and screen, and peripheral speed. The pressure pulses cause a local or periodical change in the suspension through-velocity with peaks reaching a multiple of the average[15,16].

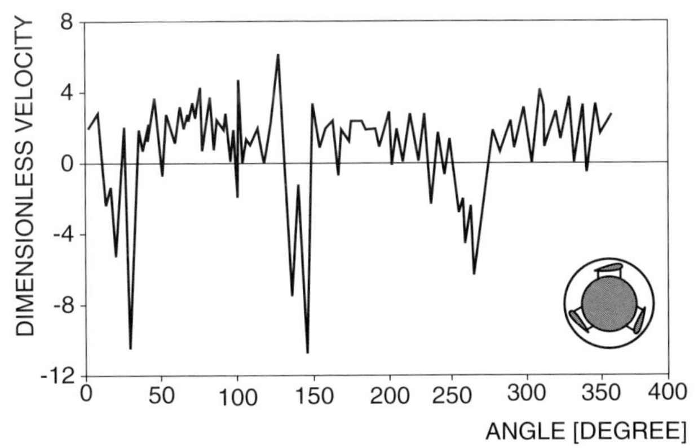

Figure 34. Screen through-velocity at the screen opening outlet per rotation of a foil rotor[15].

Figures 34 and 35 show actual screen through-velocities of a pressurized cylindrical screen for two different rotor types plotted over a single rotation. These diagrams represent local velocities at the screen opening outflow into the accepts zone. With the foil rotor, through-flow in the production phases (positive flow) is relatively uniform as Fig. 34 shows. The three short flushing phases (negative flow) can also be seen.

With the step rotor type, the production and flushing phases are approximately equal. Effective velocity in the production phase is therefore much higher for the same specific throughput so that the pressure differential is also much higher.

Not only the flow characteristics vs. time are important but also their velocity amplitudes. Maximum through-velocity with the foil rotor is about 5–10 times mean velocity. With a step rotor, it can be as much as 20 times.

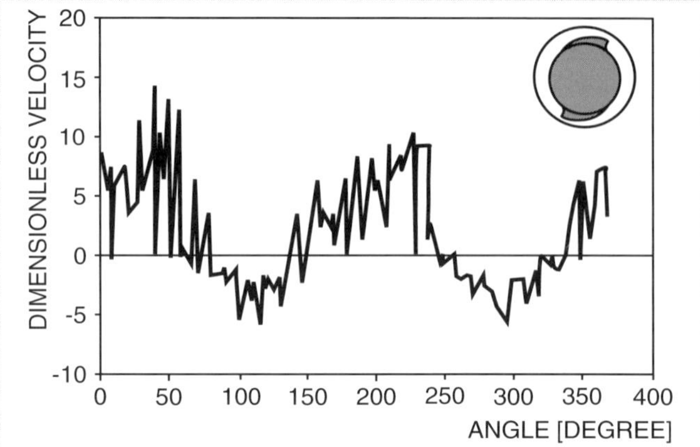

Figure 35. Screen through-velocity at the screen opening outlet per rotation of a step rotor[15].

Mean screen through-velocity is a theoretical value calculated from flow volume divided by total screen open area. As mentioned above, a higher effective through-velocity, i.e., a higher differential pressure, means a greater number of deformable particles such as stickies can be forced through the openings. The mean through-velocity of different screening machine types is therefore an unreliable indication of actual through-flow conditions.

4.2.5 Uniform flow through cylindrical screens

Uniform flow over the entire perforated area of cylindrical screens brings the advantage of highest possible specific throughput while still avoiding plugging. This is optimum screening for a given throughput. Most types of cylindrical screens have considerable differences in local throughput. The following text explains the primary reason for this.

Assuming that the input pressure is about the same at all points on the basket, the accept flow velocity will increase along the basket toward the outlet.

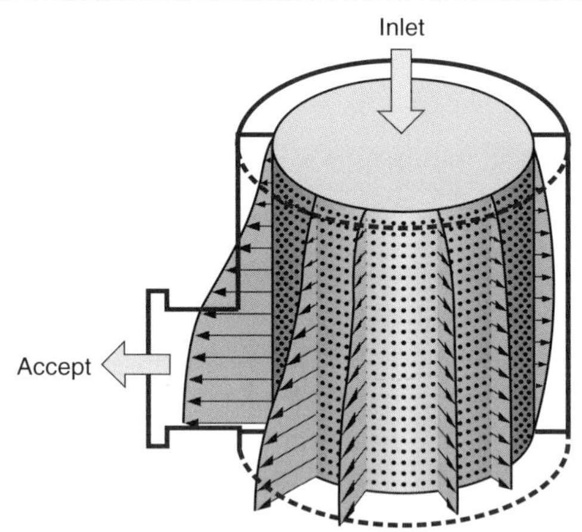

Figure 36. Calculated local flow through a screen basket with concentric and cylindrical outer accept chamber based on pressure differential across screen basket due to flow velocity increase toward the screen outlet[16].

This is because the flow rate and the suspension velocity both increase with constant cross section of the accept zone. Using the Bernoulli energy theorem, this causes a static pressure reduction on the accept side that decreases toward the outlet. This is therefore the point of greatest pressure differential across the screen basket and consequently the point of greatest throughput as Fig. 36 shows[21].

Figure 37. Pressure screens with conical housing and fish-mouth outlet for equalizing through-velocity over the entire basket surface area [16].

In screens with a large accept chamber, this effect is smaller. Flow uniformity through the screen basket and therefore performance are better in screens with conical housing as Fig. 37 shows. Uniformity also improves with the fish-mouth shape and low position of the accept outlet below the basket[16,22].

4.2.6 Assessment of separation processes

Criteria such as separation ratio, enrichment, and cleanliness efficiency define and assess separation processes including screening. These criteria not only apply to individual screens but also to entire separation process steps such as screening systems. For greater simplicity, the definition of suspension in the following explanation is a mixture of debris particles, fibers, and water.

In screening, the debris particles separated from the inlet flow are enriched in the inlet chamber of the screen and removed with the reject. Figure 38 defines the total flow balance and the relative reject mass flow for an individual machine or complete system.

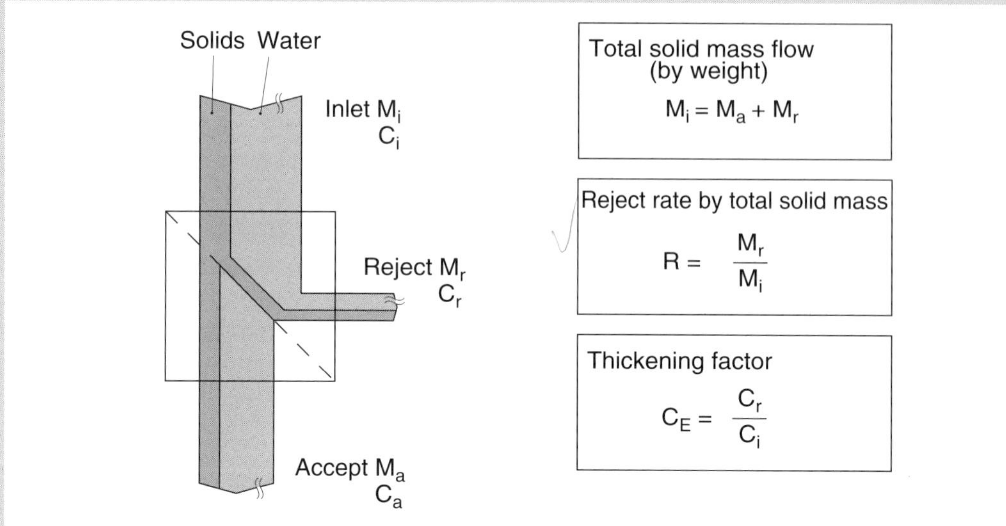

Figure 38. Mass balance and definitions in a separation system.

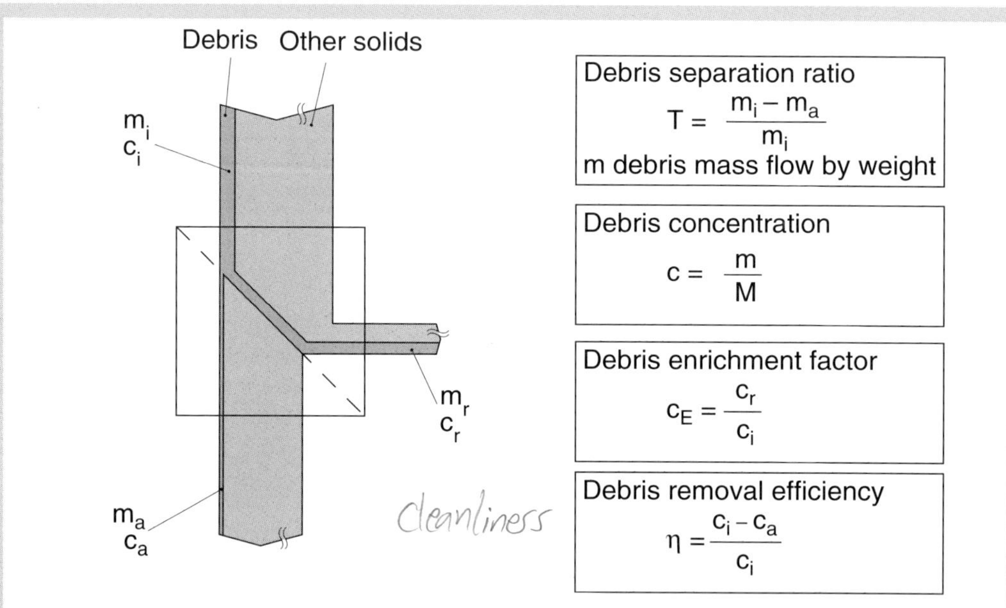

Figure 39. Definitions in a separation process.

Figure 39 shows other balancing definitions for debris particle content. These are the separation ratio, *T*, concentration, *c*, enrichment factor, c_E, and cleanliness efficiency, η, defined as follows:

- The separation ratio, *T*, or debris reject efficiency[23] indicates the change in debris particle mass flow between inlet, m_i, and accept, m_a. *T* relates to the inlet debris particle mass flow, m_i.
- The debris particle concentration, *c*, is the ratio of debris particle mass flow, *m*, to total solids flow, *M*, fibers including contaminants.
- The debris enrichment factor, c_E, is the ratio of reject debris particle concentration, c_r, to inlet debris particle concentration, c_i.
- The debris removal efficiency, defined as cleanliness efficiency, η, is the difference between debris particle concentration in the inlet, c_i, and accept, c_a, divided by the inlet debris particle concentration, c_i.

In many cases, the difference between separation ratio, *T*, calculated from the mass flow balance and the cleanliness efficiency, η, calculated from debris concentration is not sufficiently clear despite the very different magnitudes[23,24]. If the debris particle concentrations in the accept and reject do not change compared with the inlet for example, the separation ratio, *T*, would not be zero but would correspond to the value, *R*, of the relative rejects flow. This can be 0%–100%. The cleanliness efficiency, η, would be 0% for all values of relative reject flow. Only at very high efficiencies do the numerical values of cleanliness efficiency, η, and separation ratio, *T*, approach each other.

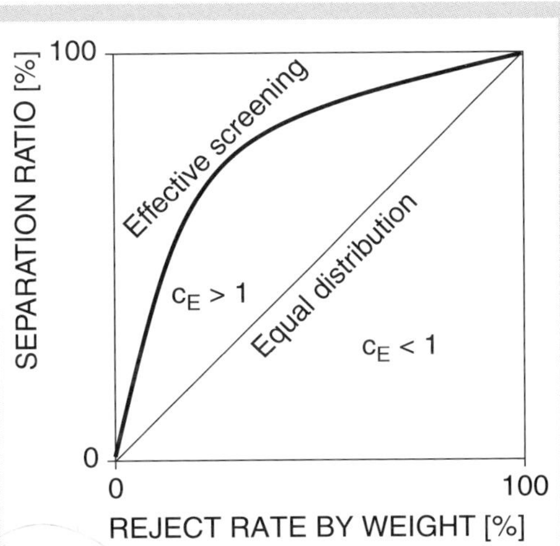

Figure 40. Separation diagram.

A separation diagram best illustrates these relations. Figure 40 shows screen separation ratio, *T*, as a function of relative reject flow, *R*. At all points above the 45° line, i.e., on every point of the screening curve shown, the debris particle concentration in the reject is higher than in the inlet.

All operating points on the 45° line show a separation ratio greater than zero. At this point, the inlet mass flow is only split into accept and reject according to the relative reject flow and concentrations remain unchanged. Operating points below the 45° line would signify a reduction of debris particle concentration in the reject.

The cleanliness efficiency, η, in Fig. 41 is represented by the ratio, x/y, of effective separation $(x = T_x - R_x)$ to the maximum separation theoretically possible $(y = 100 - R_x)$. This relationship derives from mass balances for the efficiency definition in Fig. 39. The value of η_x for a given separation ratio, T_x, can also be derived geometrically from this diagram where the dashed straight line through the points (100%/100%) and (R_x/T_x) intersects the ordinate.

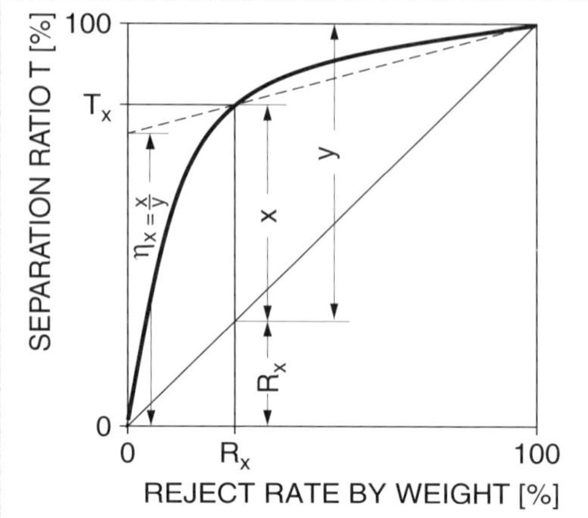

Figure 41. Relationship between separation ratio, T, and screening efficiency, η.

4.2.7 Difference between screening and fractionation

The relations given above apply not only to debris particle separation but also to fiber fractionation, i.e., to the separation of specific fiber fractions from the suspension. If separation of the long fiber fraction as completely as possible from a suspension containing both long and short fiber fractions is desirable for instance, the result of fractionation can be assessed using a curve in the separation diagram.

In addition to the debris particle separation curve, Fig. 42 also includes a long fiber fractionation curve. This clearly shows that far from complete separation only a certain enrichment of the long fiber component in the rejects occurs. The efficiency of long fiber enrichment is defined in the same way as debris separation efficiency. Since the screening curve for short fiber components lies below the 45° degree line, short fiber concentration in the reject is lower than in the inlet.

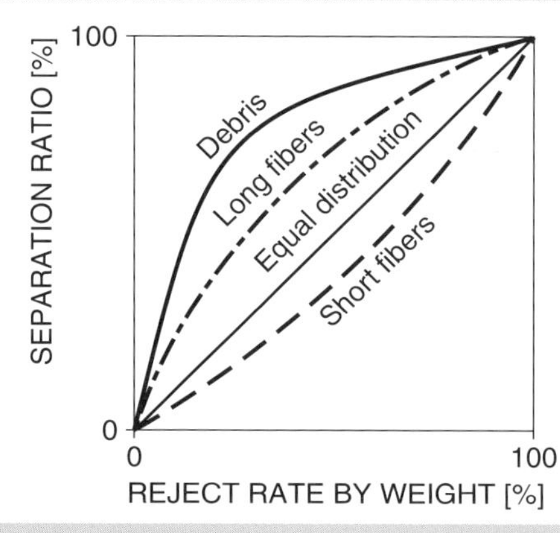

Figure 42. Separation diagram for debris, long fibers, and short fibers.

Both separation diagrams in Fig. 43 show the different goals of screening and fractionation. The primary object of screening is to remove only the solid contaminants with the same distribution of fiber fractions in the rejects and accepts as in the inlet. The

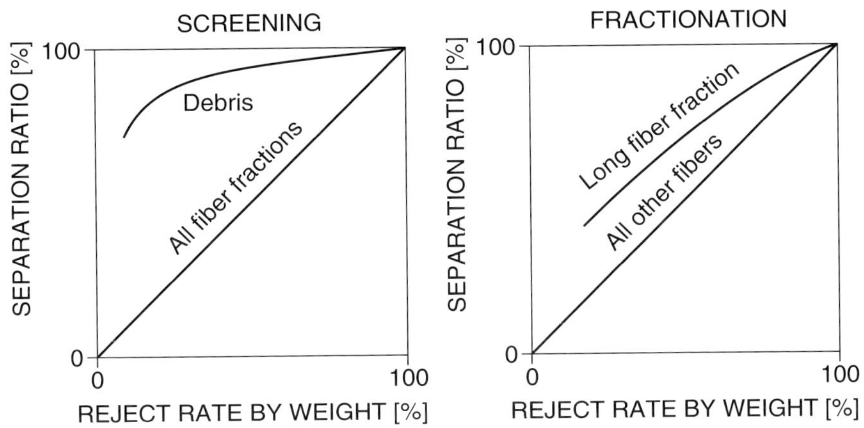

Figure 43. Separation diagram showing the main differences between screening and fractionation goals.

goal of fractionation is to change the composition of the suspension regarding fiber length or fiber flexibility. The requirements can vary between the following:

- Short fiber fraction ("accept flow") with lowest possible long fiber content
- Long fiber fraction ("reject flow") with highest possible yield and lowest possible short fiber content
- Long fiber fraction-free "accept" flow.

Considering the long fiber fraction (14 mesh fraction) as an example means that the curve or operation points for this fraction should lie as far as possible above the curve of equal distribution, i.e., the diagonal or 45º line.

4.3 Machinery and design versions

Different types of screening machines have use at different points in the system depending on raw material, debris particles, flake content, and stock consistency. They are used for coarse and fine screening and at all stages in screening systems[16,25].

All screening machinery has in common a (usually welded) housing with connections for the inlet, accept, reject, and sometimes light-weight reject with a rotor with clearing devices and a screen. The housing can be pressurized or at atmospheric pressure. The screen axes can be vertical or horizontal.

Screen surfaces can be flat (disk screens) or cylindrical (cylindrical screens). Hole perforations are about 0.8–1.5 mm diameter for cylindrical screens and 2.0–3.0 mm diameter for flat screens. Slotted screens are generally in the cylindrical form with slot widths from 0.1–0.4 mm. Both hole and slotted cylindrical screens can have profiles on the inlet side depending on application.

Disk screens usually have a smooth structure on the intake side. They are fitted with radial bars. The screens are also useful for breaking down residual flakes and sometimes for fractionation. Table 5 summarizes the various kinds of screens and their main operating conditions.

Table 5. Overview of screening machine types and their main operating conditions.

Operating parameter	Screen openings	Rotor circumf. speed	Consistency range		
Screen type			MC		LC
	[mm]	[m/s]	< 6%	< 4.5%	< 1.5%
Disk	Hole 2.0–3.0	20–30	•		
Cylindrical	Hole 0.8–1.5	10–30		•	
	Slot 0.1–0.4	10–30		•	•

4.3.1 Coarse screening

Disk and cylindrical screens are useful for coarse screening. Both types operate under pressure. Disk screens have use primarily for pulp slurries with a high trash and flake content. The reason is that these machines are more insensitive to a high content of materials hindering the screening process and have a good deflaking efficiency. For relatively clean suspensions and demands for high accept cleanliness, cylindrical screens find common use.

Disk screens

Recovered paper may contain components that are difficult to slush such as in department store waste or mixed recovered paper. Since the flake content of such components is high after slushing, coarse screening exploits the high disintegration potential of disk screens. This minimizes fiber losses and ensures operational stability of the downstream screening stages.

Figure 44 shows a disk screen with the cover removed. The rotor turns at a peripheral speed of about 20–30 m/s with several curved blades that keep the screen openings free by suction pulses and transport the trash and flakes radially outward. The spacing between rotor and screen plate is about 2–4 mm. Much of the deflaking effect is due to baffle bars near the rotor periphery. Figure 45 shows how to calculate the deflaking potential for this unit. The deflaking potential is the effectiveness in flake defibering.

Figure 44. Disk screen (opened) with vaned rotor, screen plate, baffle bars, and conical housing.

Disk screen machines also have some use in the second screening stage with a cylindrical screen in the first stage. The objective here is to minimize fiber losses again because of their higher deflaking potential and ensure reliable operation of the second, third, or both stages.

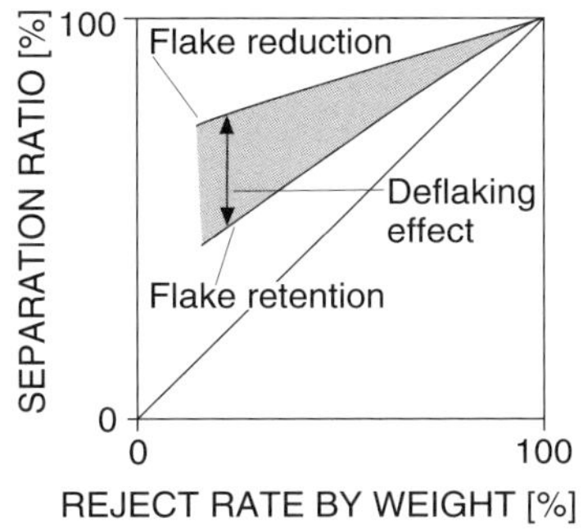

$$\text{Flake reduction} = \frac{\text{Flake concentr. inlet} - \text{Flake concentr. accept}}{\text{Flake concentr. inlet}}$$

$$\text{Flake retention} = \frac{\text{Flake concentr. reject}}{\text{Flake concentr. inlet}}$$

Figure 45. Assessment of flake reduction, flake retention, and deflaking effect in screening.

Cylindrical screens

Screen designs

For pulp slurries with low flake contents of about 5% or less, coarse screening occurs in cylindrical screens. These also have a certain deflaking effect although considerably less than that of disk screens. Cylindrical screens operate under pressure at stock consistencies up to about 4.5%. As coarse screening machines, they can have baskets with holes and slots wider than usual.

Holes in screen baskets are machined by milling and drilling. Slotted baskets can be machined by milling and sawing. Today, most are mounted together from individual bars (bar-type screen baskets). Coarse screening machines of this type are normally identical or similar to fine screens. The exceptions are machines where the screen rotates while the pulsing blades are stationary (Fig. 46). These machines only have use in coarse screening.

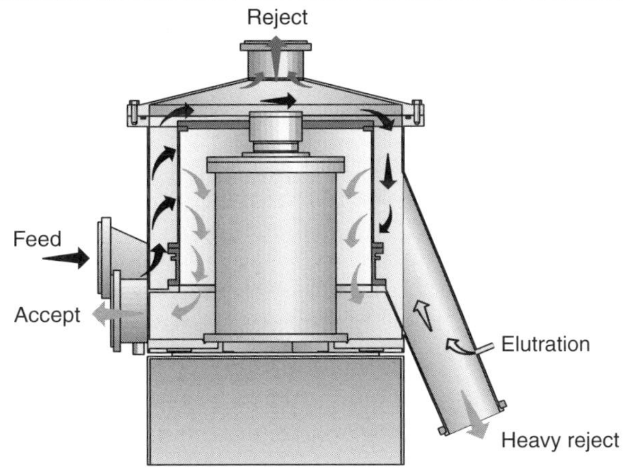

Figure 46. Coarse screen with rotating screen cylinder.

Many different types of screen designs equipped with a basket are available. The rotor can be positioned at the basket inlet or accept side. Flow through the basket can be centrifugal or centripetal. The most common arrangement is with the rotor on the inlet side usually with centrifugal flow through the basket. The spacing between rotor and basket is about 2–20 mm.

Rotors

A primary difference between coarse and fine screens is the rotor design. The rotor must keep the screen free of debris particles and flakes while fluidizing the suspension to mobilize the fibers and contaminants. The rotor design must also ensure reliable screen operation with high debris content by preventing plugging by spinning components or by material jamming between rotor and screen. Often it should also have a sufficient deflaking effect to ensure reliability of the downstream screening stages and systems. This requires additional energy demand.

Screen rotor shapes vary widely according to application. They may have blades immersed in the suspension or have the form of a closed rotor with integral contouring. Figure 47 shows three typical rotors for coarse MC screening. The insensitive lobed rotor for coarse screening with hole screens operates very reliably with only slight thickening effect and high specific throughput. The versatile step rotor uses both hole and slotted screens to ensure high reliability with good screening and only slight thickening effect. The gentle-acting bump rotor used with slotted screens gives high cleanliness efficiency but with correspondingly high fractionation and thickening effects.

Figure 47. Coarse screening rotors.

Wear

Due to high contaminant content in the pulp slurry, coarse screens are subject to more wear than fine screens. All their components therefore require special design such as abrasion-resistant surfaces. For this reason, the use of very fine slots in such screens can be critical. In addition, the stock preparation aggregates preceding coarse screening must ensure the best possible removal of abrasive materials[22,26].

Final stages of coarse screening systems

The final stages of coarse screening systems must remove all debris particles enriched in the previous stages. Only at this point is the screening system open to permit the discharge of the debris. Any contaminants not removed here transfer to the production level either directly (forward flow) or indirectly after being recirculated one or more times (reverse flow)[27].

Final stage coarse screening machines must handle high trash content with minimum fiber losses and highest cleanliness efficiency. For optimum effect, these requirements are contradictory. They require optimization according to each individual case[28].

The following illustrations show examples of final stage coarse screening machines used particularly for mixed (brown) recovered paper grades. Figure 48 shows a schematic of a nonpressurized machine with bladed rotor that continuously throws the trash-enriched suspension against the screen, removes the retained material, and returns it to the screen. The accepts are continuously ejected while the retained material is diluted with water, the fibers are washed to the accept chamber, and the debris is transported to the reject outlet and removed with a low moisture content. Figure 49 is a photograph of this machine with the cover opened.

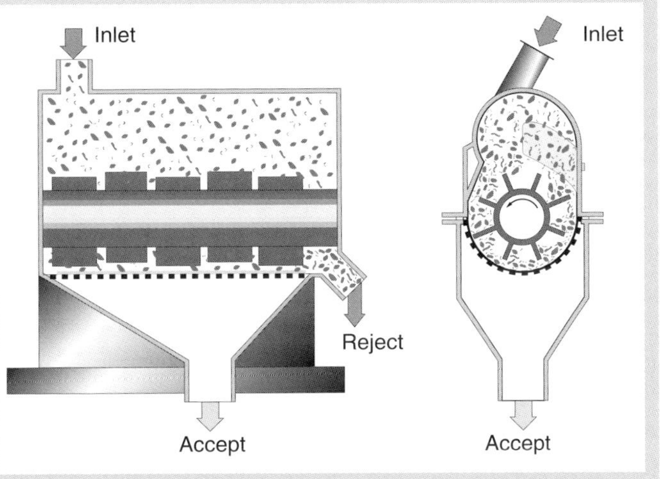

Figure 48. Schematic of a final stage screen with rotor and screen plate operating at atmospheric pressure.

Figure 49. Final stage screen with cover opened.

Figure 50 shows a schematic for a final stage screen pressurized in the lower half and operating at atmospheric pressure in the upper half. In the lower section, pre-screening uses a disk screen. Its reject is diluted with water and fed to the upper part of the machine. Most fibers are removed here in the nonpressurized basket screening zone. The final rejects have a low moisture content. Figure 51 shows a photograph of this unit.

Both machines have a high deflaking potential that helps to reduce fiber losses. The high reject consistency is also favorable for disposal. In addition, the cleanliness efficiency of both machines is usually sufficiently high to allow direct forward transfer of the accepts. This relieves the load on the coarse screening system.

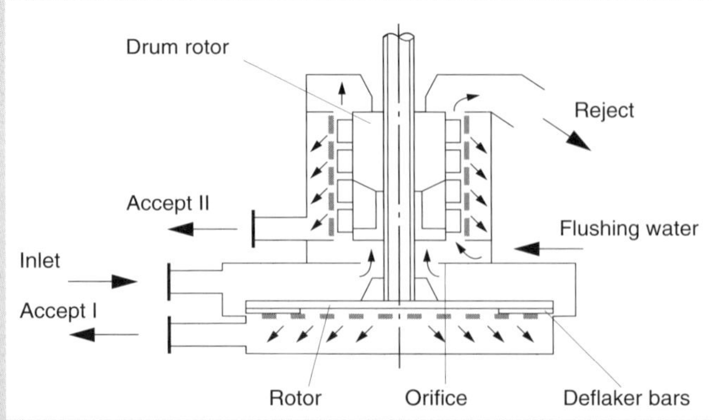

Figure 50. Schematic of a final stage screen with flat and cylindrical screen using ambient pressure outlet.

This discussion does not consider the classical vibrating screen that also operates at atmospheric pressure. Because of the large screen area required for high contaminant content in the final screening stage and the high production capacity of many recovered paper processing lines, vibrating screens are not cost-effective for this application. They are only suitable for coarse contaminant separation.

Figure 51. Final stage screen.

Pressurized screens are also useful in the final stage in continuous or batch operation with or without a wash cycle. They usually have the same screen hole sizes as in the previous stages. Their disadvantages are lower deflaking potential and corresponding limitation of admissible debris content.

4.3.2 Fine screening

Machines

Fine screening occurs in the LC range and partially in the MC range following coarse screening. For fine screening recycled fiber pulp for writing and printing papers with a high proportion of short fiber pulp, slotted baskets with slot widths of 0.1 mm or more are useful. For brown stock, slot widths of 0.15 mm or greater have use for domestic old corrugated containers (DOCC) or 0.25 mm for American old corrugated containers (AOCC) depending on fiber type and length. The precision of these screen baskets regarding slot width and profile angle must be very high. Operating results strongly depend on this exact geometry. Extremely abrasive particles therefore must be removed from the suspension as much as possible before fine screening. This occurs best with hydrocyclones operating in the LC range. This is why fine screening should occur after such an LC cleaner stage.

Besides these active measures, passive protection against abrasion is also possible. Screen basket surface treatments are available that substantially reduce abrasion and increase service life accordingly. As a result, operating behavior during the basket life only varies within admissible limits.

Among the results of abrasion are the change in profile angle that Fig. 52 shows. Although a smaller profile angle improves cleanliness efficiency, the specific throughput decreases significantly with greater fractionation and thickening effects.

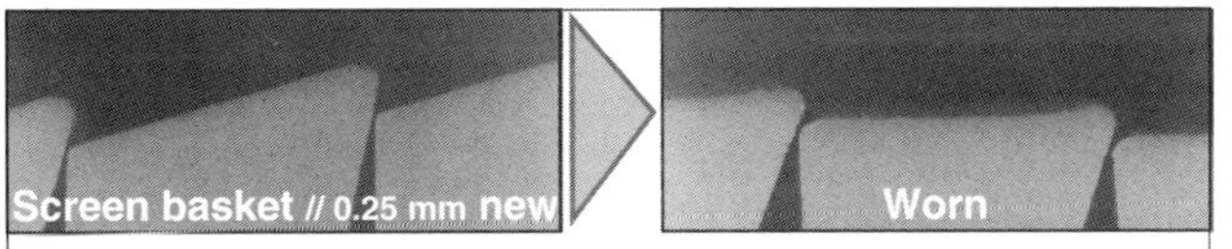

	Screen basket // 0.25 mm new	Worn
Thickening factor $C_E = \frac{C_r}{C_i}$	1.12	1.45
Reduction in stickies area	69%	80%
Reduction in number of stickies	49%	66%

Figure 52. Surface profile comparison of a bar-type screen basket when new and after long service in an abrasive environment and the effect on operating results[16].

Bar-type screen baskets are manufactured by welding, brazing, clamping, or some combination of these operations for affixing the individual bars to their mountings at the correct spacing as Fig. 53 shows. The bar spacing corresponds to the slot width that must be absolutely precise over the entire screen basket surface. With efficient manufacturing procedures, tolerances of ± 0.01 mm are now possible.

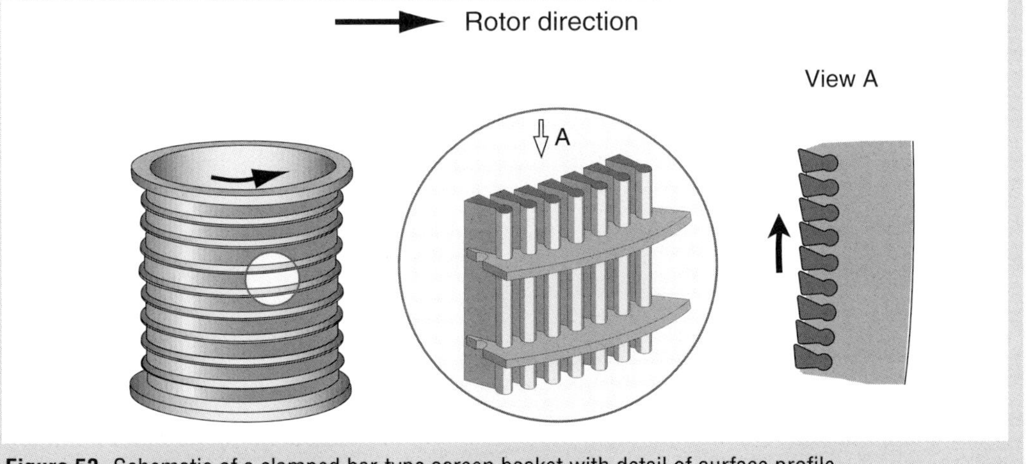

Figure 53. Schematic of a clamped bar-type screen basket with detail of surface profile.

The bar-type screen basket manufacturing procedure allows for easy production of bars with a complex profile on the inlet side. This ensures efficient screening with high hydraulic throughput simultaneously. The open screen area of bar-type baskets is also larger than with machined baskets to increase maximum throughput of a screen even more.

For fine screening in the MC range, rotors such as the bump rotor or foil rotor of Fig. 54 have use. In the LC range, foil rotors are generally employed.

Figure 54. Foil rotor used for fine screening.

Final stage screens in fine screening systems

For quality reasons, only slotted screens have use in the final stage of fine screening systems. This is due to the high demands on fine screening efficiency at this stage. An optimum compromise is also necessary here between cost-effective screening quality and efficient fiber recovery.

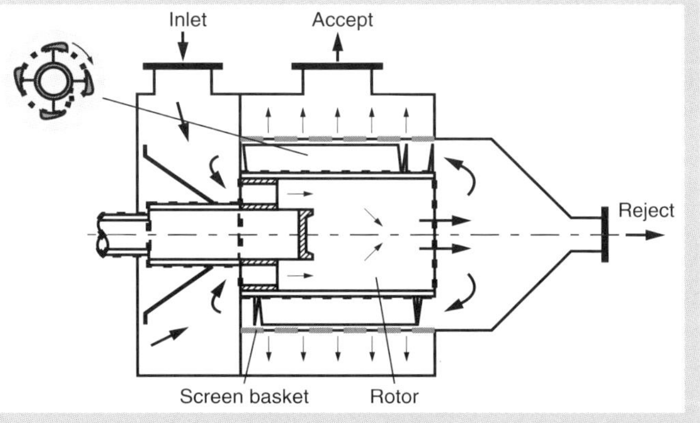

Figure 55. Final stage screen for continuous operation with slotted basket in an LC screening system.

Two main types of final stage fine screening machine are available. One type operates continuously and has a high cleanliness efficiency but relatively high fiber losses (Fig. 55). It is similar to a normal screening machine but has a special rotor design. Because of good internal axial mixing, it operates very reliably at small reject flow.

The other kind of final stage fine screen operates in batch mode with a washing function. Toward the end of each cycle, the suspension in the casing that has a high debris content is diluted with flushing water. This washes out the fibers while the debris is removed in batches. Naturally, the screening efficiency decreases with increasing reduction of

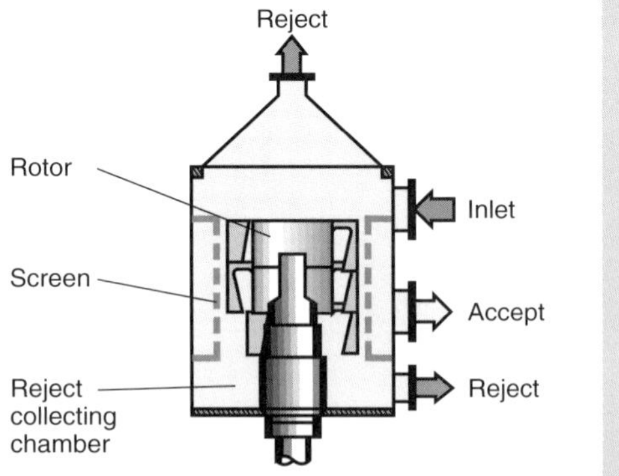

Figure 56. Final stage screen for batch and continuous operation with slotted basket in an LC screening system.

fiber losses during the wash cycle that removes the screen residues. Figure 56 shows a final stage machine of this type for batch operation. As a special feature, this machine can also operate continuously. Figure 57 shows the system for a batch fine screening machine, i. e., washing operation mode of this tail.

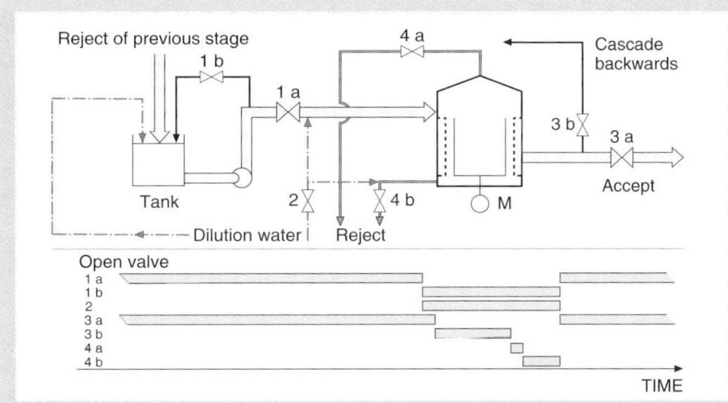

Figure 57. Control and operation diagram for a final stage screen on batch operational mode.

4.3.3 Fractionation

Fractionation can use disk screens or cylindrical screens[13,14]. The machines are the same as for screening with some differences regarding equipment and operating conditions. With disk screens and cylindrical screens, the holes are smaller than for screening. Slotted screen baskets have narrower slots, zero-angle profile, or both.

The mass flow split in screening for reject rate is about 5%–25%. The split in fractionation is higher. Depending on the requirements, it is about 30%–40%.

4.4 Screening results

Screening is always a compromise between machinery outlay, cleanliness efficiency, fiber losses, and operating reliability. It is a balance between quality demands and cost effectiveness. The primary relationships involved are explained here first for individual screens and then for screening systems as a whole. The main results discussed are separation of stickies because stickies are the most significant challenge in screening. Solving this problem therefore involves all screening parameters. Its study provides a good overview of screening as a whole.

4.4.1 Individual machines

Screening results depend on machine parameters and system operating conditions. The results are defined as cleanliness efficiency, reject thickening, and fractionation that is also reflected in freeness development. The most important machine parameters are as follows:

- Rotor suction and pressure pulse characteristics: severe or gentle
- Rotor periphery speed: high speed increases the suction and pressure pulse amplitude if other features remain the same
- Screen basket hole diameter or slot width and surface profile
- Housing geometry that influences the uniformity of basket through-flow.

The following system parameters are critical for screening results:

- Pulp characteristics: consistency and fiber type
- Flake content
- Debris content and type
- Throughput requirement that determines the through-flow velocity through the screen openings
- Reject rate.

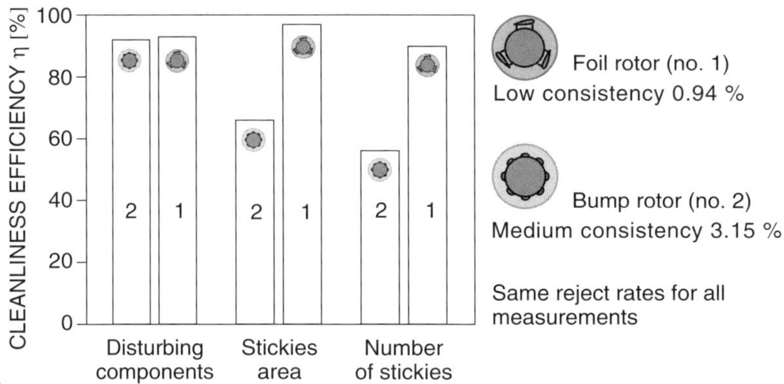

Figure 58. Effect of stock consistency and rotor type on slotted cylindrical screen cleanliness efficiency[16].

Screening effect

Figure 58 shows the influence of stock consistency and rotor type on cleanliness efficiency of stickies. The content and area of stickies here were determined by the pick-up method[29]. Clearly, the sensitive stickies are more difficult to remove due to the greater shear forces involved in MC screening. No difference is evident with the other contaminants.

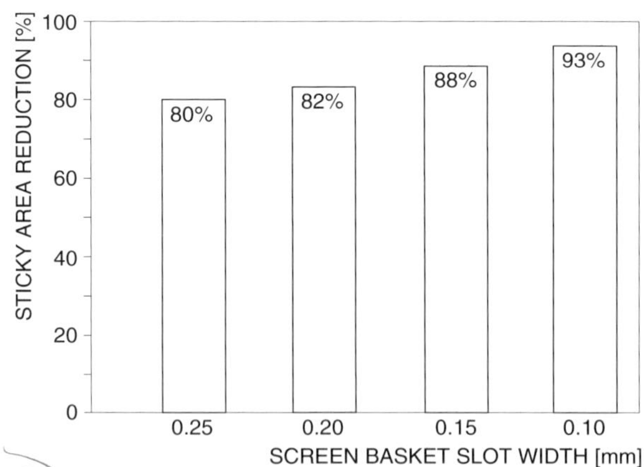

Figure 59. Effect of bar-type screen basket slot width on cleanliness efficiency for stickies removal.

Figure 59 shows the influence of bar-type basket slot width on removal of stickies. The profile angle is constant. This has great importance since slot width alone is not the only influence as noted earlier.

Figure 60 shows the influence of profile angle in a bar-type basket on screening results. The displacement of particle size spectrum between inlet and accepts is given for two different profile angles with a slot width of 0.1 mm.

Figure 61 shows an example of the influence of mean screen through-velocity on removal of stickies. The cleanliness efficiency remains at a high level over a wide velocity range. This means that this type of screen has a wide operating range with good efficiency[30].

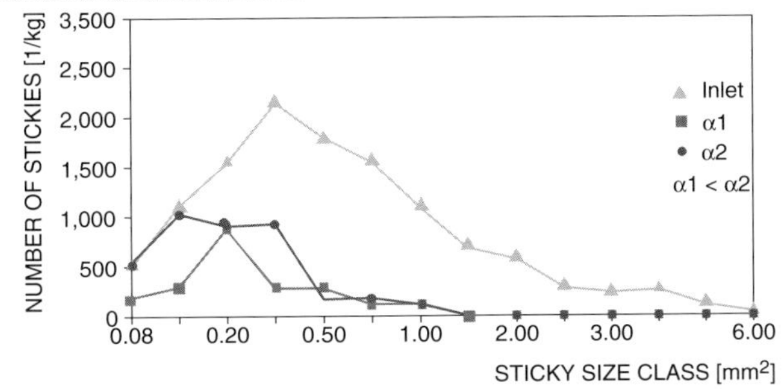

Figure 60. Size spectrum displacement of stickies for bar-type basket screens with equal slot width but different profile angles, α.

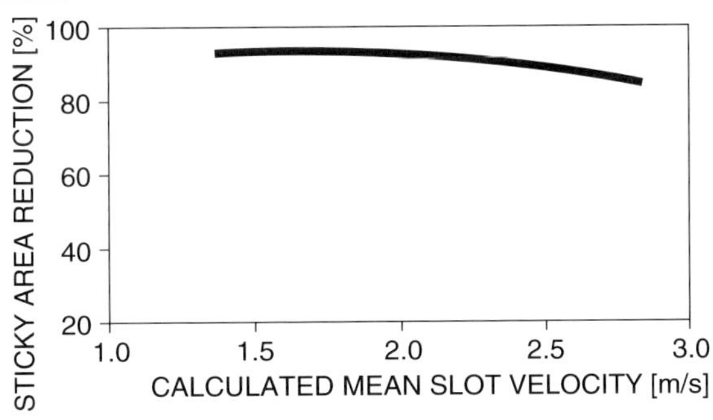

Figure 61. Effect of mean screen through-velocity on removal of stickies.

Thickening factor

The two following diagrams show how screening conditions such as slot width and mean through-velocity influence the thickening factor of the reject. Figure 62 plots the thickening factor against slot width for a certain type of screen basket. The disproportionately steep gradient of this curve is due to the fact that a slot width of 0.1 mm is almost the same order of magnitude as the fiber thickness (about 0.02–0.05 mm)[20].

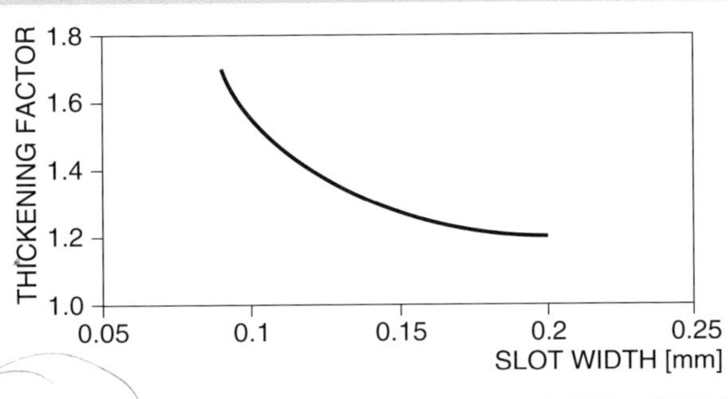

Figure 62. Effect of slot width on reject thickening factor[16].

Figure 63 shows the influence of mean through-velocity of a bar-type screen basket on the reject thickening factor. CSF (Canadian Standard Freeness) of the accept increases with increasing thickening factor similarly.

Cleanliness efficiency and thickening factor link closely. Figure 64 shows this for a screen with baskets of varying slot width of 0.09–0.15 mm. The cleanliness efficiency under otherwise largely similar conditions, e.g., specific throughput, improves with increasing thickening effect despite the corresponding rise in mean through-velocity with reduced slot widths.

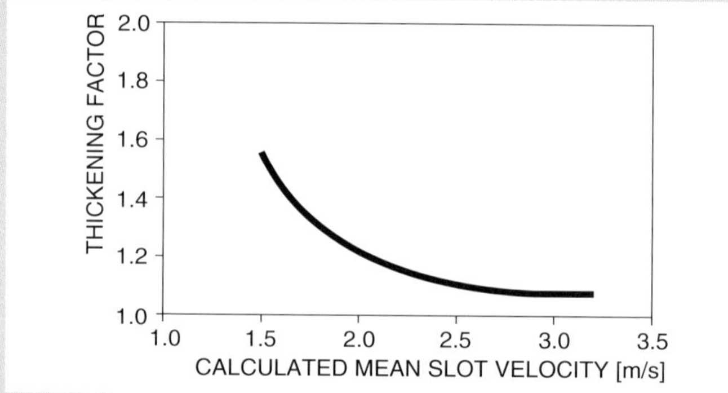

Figure 63. Effect of mean screen through-velocity on reject thickening factor.

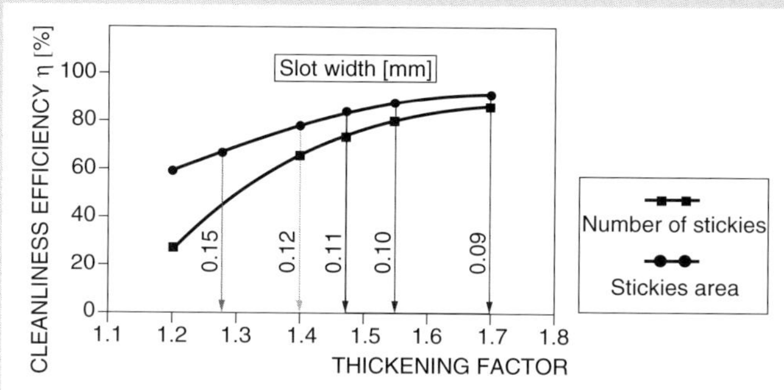

Figure 64. Cleanliness efficiency vs. thickening factor for baskets with different slot widths in the same screen[16].

4.4.2 Screening systems

The debris concentration in the pulp slurry at the screen inlet and the required pulp quality vary widely according to the application range. For this reason, screen designs and operating modes vary widely as does their integration in the screening system[25].

Each screening subsystem comprises several stages with a final stage screen. The interconnection of individual screen accepts and rejects in the various stages likewise varies according to requirements either with forward flow or full or partial cascade as Fig. 65 shows.

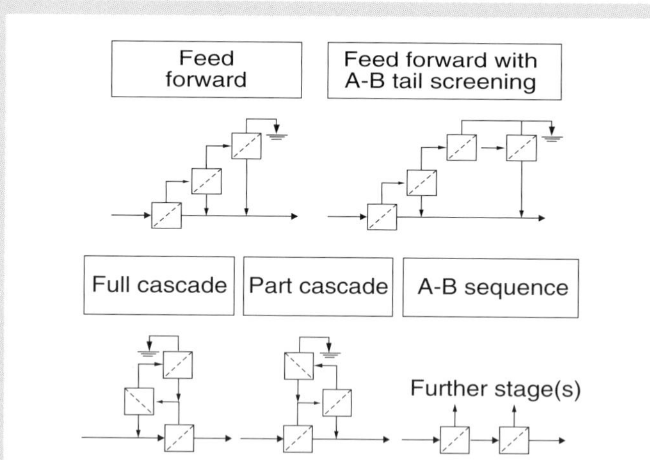

Figure 65. Screening system diagrams for recycled fiber pulp processing.

Cleanliness efficiencies in a cascade system are theoretically higher than with forward flow if the machines operate with the same parameters including specific throughput, stock consistency, and reject flow. The difference depends on particle size distribution in the inlet. A higher proportion of debris particles separable by probability means a greater advantage for the cascade systems[31–34].

In practice, effects such as comminution processes in screens and pumps counteract or even reverse this advantage if the proportion of debris particles that can be broken down is sufficiently high. In addition, a compromise often is necessary in selecting parameters between given machine sizes. This distorts the advantages and disadvantages of forward and backward flow of secondary or tertiary accept streams even more.

For optimal cleanliness results at the same fiber losses, the screens should be arranged in series (A-B arrangement). This can be very advantageous in the primary, intermediate, or final stage of a screening system. Arranging MC and LC slotted screening systems in series also gives excellent results. These arrangements are also useful in processing virgin pulps.

Screening deficiency

A screening deficiency can occur in systems with backward flow of accepts such as that from the final screening stage to the previous stage. Figure 65 shows this for the full and partial cascade. Figure 66 shows this when the final stage screen characteristics are inferior to those of the previous stage. The curve section m_i shows the debris particle size distribution in a system inlet flow with screens A and B. With ideal screening related to size in stage A, all small particles with size $d_p < d_{p,A}$ pass through with the accept (curve section $m_{a,A}$). All larger particles with size $d_p > d_{p,A}$ are separated. They pass with the reject from screen A into the inlet of screen B.

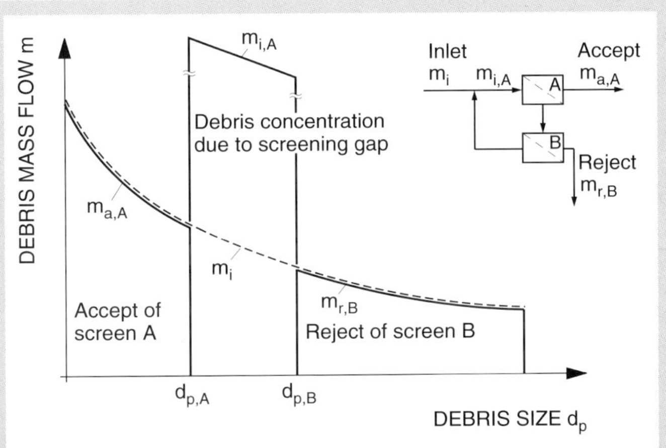

Figure 66. Explanation of screening deficiency.

If the characteristics of screen B are inferior to those of screen A $(d_{p,B} > d_{p,A})$, the larger debris particles, $d_p > d_{p,B}$, are separated as rejects (curve section $m_{r,B}$). All smaller debris particles $(d_p < d_{p,B})$ pass through with the accepts and are again returned to the previous stage A. This screen rejects all particles $d_p > d_{p,A}$ and passes them back to screen B. All particles of size $d_{p,A} < d_p < d_{p,B}$ remain trapped in the circuit and are enriched theoretically to an infinite extent (curve section $m_{i,A}$).

In practice, this often leads to a system collapse. Screens also have a certain comminution effect so that trapped particles are reduced during circulation to a size $d_p < d_{p,A}$ and are then removed from the system with the accepts of screen A. This means the following:

- Increase in debris content in the internal circuit depends on the effectiveness of the comminuting process
- Debris content in the accept increases
- Debris particle size in the accept corresponds to the characteristics of the finer screen A.

Screening effect

Figure 67 summarizes test results on numerous systems with different screening concepts. As the figure shows, a larger available screen area with more or larger machines gives increased cleanliness efficiency. These test results all lie within a continuous band for both MC and LC screening.

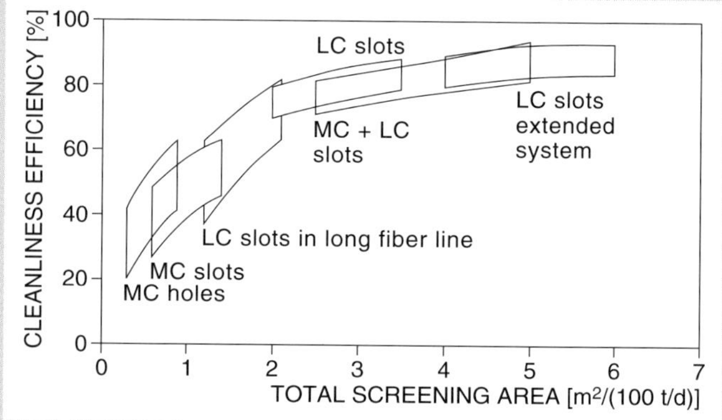

Figure 67. Cleanliness efficiency vs. available total screening area of different systems[16].

The flattening curve shows that continuing improvement of cleanliness efficiency can only occur at the expense of disproportionately high screening area. In addition, optimum efficiency is only possible with LC screening. Screening systems should therefore operate at the upper limit of this band. An engineering challenge is to extend this upward.

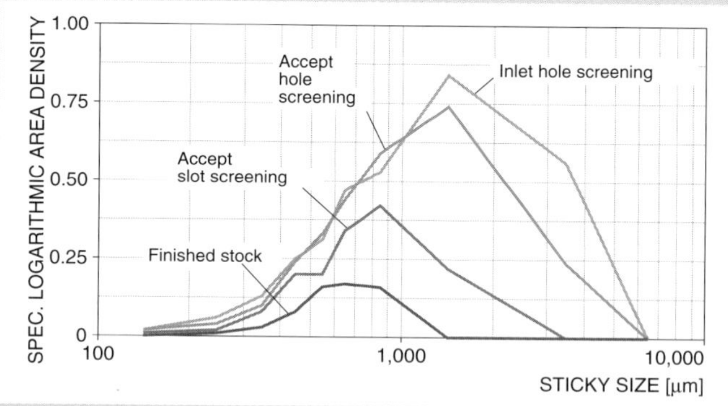

Figure 68. Change of size spectrum of stickies influenced by screening stages[29].

Figure 68 shows how the size spectrum of stickies changes after each screening stage. Clearly evident are the effects of coarse screening that passes only very few large debris particles and fine screening. The size data for stickies given here use the theoretical diameter in a melted condition according to the pick-up method for measuring macro stickies[29]. This means the stickies had different shapes in the suspension. Their dimensions relevant to screening were also different.

4.4.3 Thickening

Figure 69 shows the influence of slot width on thickening in each stage of a screening system. Narrowing the slot width from 0.15 to 0.1 mm increases the thickening factor in the second stage to that in the third stage with 0.15 mm slot width. Narrowing the third stage to 0.1 mm slot width increases the thickening effect to almost three times the first stage inlet stock consistency.

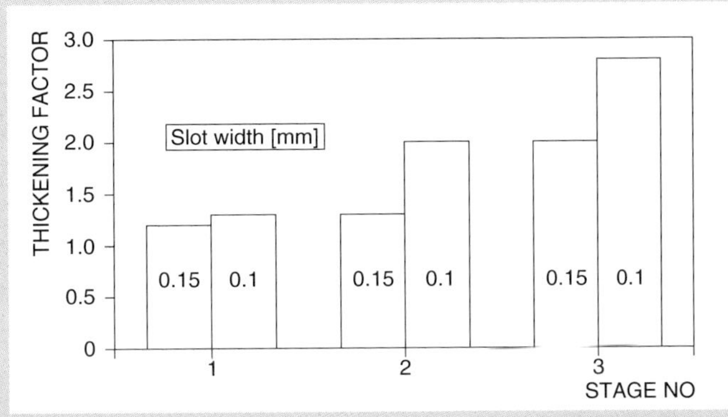

Figure 69. Effect of bar-type screen basket slot width on the reject thickening factor of different stages[16,25].

Similarly critical thickening or fractionation effects must also be considered when screening mixtures with a high proportion of stiff and thick fibers. Greater fractionation normally accompanies increased thickening. This is an additional operational disadvantage in the next stage. Designers of screens and screen systems strive to avoid excessive thickening.

4.4.4 Deflaking

With extremely high flake contents, the first stage can use a coarse disk screen followed by another disk screen in the second stage and a tail screen as Figs. 48 and 50 show. This ensures adequate deflaking capacity to break down the flakes into single fibers. With medium flake content, a cylindrical screen has frequent use in the first stage. Since these screens often have a low deflaking effect, the second stage often comprises a disk screen with higher deflaking potential for flake disintegration. Due to the necessarily high reject flow from the first stage, this must have a large size.

Figure 70 shows the importance of deflaking for reducing fiber losses and for operational reliability of the second and third screening stages. These are the results of theoretical flake content calculations in the different stages for various systems. The inlet flake content into the primary stage is assumed to be 5%. The total mass flow ratio of reject to accept is 0.3/0.7 with a constant cleanliness efficiency for flake screening of 80% for all three stages. The deflaking effect, i. e., the flake disintegration effect, in the first and third stages is 0%.

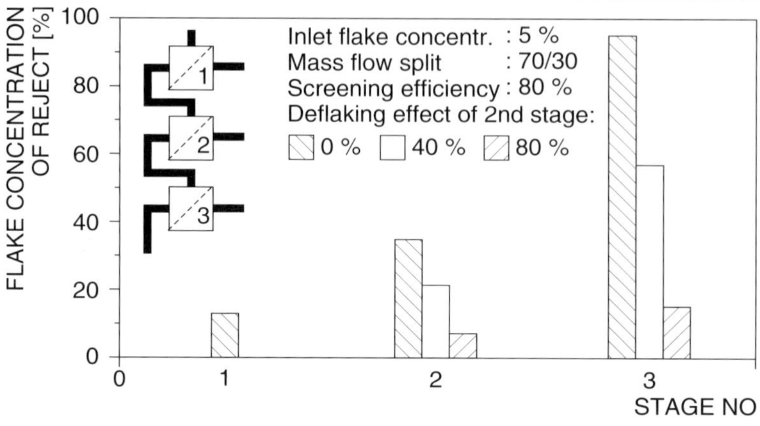

Figure 70. Theoretical example of reject flake concentration in the first, second, and third stages of a coarse screening system as a function of deflaking potential in the second stage with zero deflaking potential in the first and third stages.

If the second stage also had no deflaking effect, the flake content in the third stage would be almost 100%. That would mean the machine was completely full of flakes and no longer operable as a screening machine. Fiber losses would be extremely high. The flakes must therefore be disintegrated as much as possible. A deflaking effect of 40% in the second stage reduces flake content in the third stage to less than 60%. A deflaking effect of 80% reduces the flake content below 20%. This shows how an excessively high flake content in the first stage inlet of a screening system with inadequate deflaking in the following stages can cause system collapse.

Despite incomplete deflaking, disk screens have the special property that they screen simultaneously. If a deflaker were used with high debris content of the suspension at this point instead of a disk screen, the flake content would be lower but the debris would not be removed. More sensitive debris particles would be broken down additionally. High contaminant loads of the following stages would still remain.

With low debris concentrations, a deflaker is better because the deflaking effect is better and therefore the fiber loss is smaller. The system concept depends on whether coarse screening will occur at extremely low flake content with accordingly more gentle screening or with high flake content and necessarily more "aggressive" screening machines for residual deflaking.

5 Centrifugal cleaning

5.1 Objectives and system integration

Centrifugal cleaning removes particles from the suspension that affect paper quality or cause excessive wear in subsequent processing machinery. These contaminants may include heavy weight (HW) particles such as sand, metal pieces, and shives or light weight (LW) particles including plastic foam or other plastic material. For their efficient removal, such particles must differ in density from water and in size and shape from other suspension components.

Separation uses hydrocyclones such as Fig. 71 shows. The centrifugal field generated in these machines forces the heavy particles outward while the light ones move to the center. The flow streams containing primarily heavy or light particles are separated from the main flow.

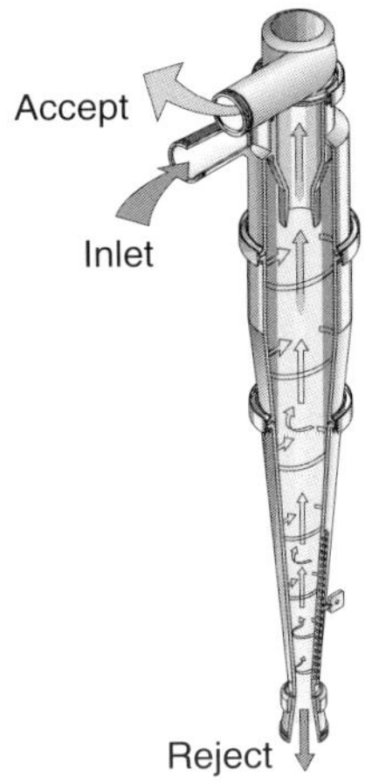

Figure 71. Cleaner for HW particle removal at low pulp consistencies.

The following text discusses the main requirements and special aspects of hydrocyclones for recycled fiber pulp processing. Compared with virgin pulp processing, the quantity of particles for separating is normally much larger. Their type and composition also vary considerably.

Centrifugal cleaning can remove particles with very different characteristics provided they meet the necessary differentiation requirements for separation. For processing of recovered paper, these characteristics include the following:

- Density or specific gravity: Higher than water (metal, glass particles, and sand) or lower such as plastic foam to allow centrifugal separation (Some particles such as certain stickies are unsuitable since their density is the same or almost the same as water.)

- Size: Ranging from slushing screen perforation diameters of 8–20 mm to 5 μm or less, e.g., fillers (Centrifugal cleaning can remove smaller particles than screening, i.e., as low as 10 μm.)

- Deformability, e.g. of stickies: Allows them pass through screen openings (Hydrocyclones can remove these particles if their density varies sufficiently from that of water and their size is sufficiently large.).

Centrifugal cleaning is an essential separation process that complements other separation methods such as screening. Hydrocyclones have use for various pulp consistency ranges. The contaminant and flake content or other contaminants hard to separate can also vary widely. For example, strips and tapes may twist together into long strings due to the rotational flow in hydrocyclones.

Figures 27 and 28 in the section on "Screening and fractionation" show how hydrocyclones can integrate into recovered paper processing systems. In the test liner system of Fig. 27, coarse HW cleaning is used after slushing. Fine HW cleaning occurs in the LC range. The deinking stock preparation system of Fig. 28 for improved newsprint production includes an additional HW cleaner in the LC range in the approach flow system.

Hydrocyclones can only operate efficiently with a design and operation according to specific requirements. This is why various types of these machines find use in recycled fiber pulp processing.

The inner wetted surfaces of hydrocyclones where the suspension flows at high velocity are subject to high wear rates due to abrasive components in the suspension. Adequate service life is still necessary. Knowing the degree of wear during operation is always helpful.

5.2 Principles

5.2.1 Hydrodynamic flow principles

The flow in a hydrocyclone is three-dimensional:

- Circumferential flow component generates the centrifugal force.
- Flow component in the axial direction is for transportation of the solid particles undergoing separation.
- Radial flow proceeds from the periphery to the center.

Hydrocyclones operate on the counterflow or unidirectional flow principle according to the flow directions of the reject and accept relative to the inlet. In counterflow cyclones, the inlet and accept connections are at the top. The reject outlet is at the bottom. In unidirectional flow cyclones, the accepts and rejects connections are opposite the inlet end. The following description of cleaners represents all hydrocyclones used in recovered paper processing.

As Fig. 72 shows, both HW and LW cleaners can be counterflow or unidirectional flow. Counterflow HW cleaners have the accepts connection on top and the rejects connection at the bottom of the cone. Counterflow LW cleaners are primarily arranged in the opposite way.

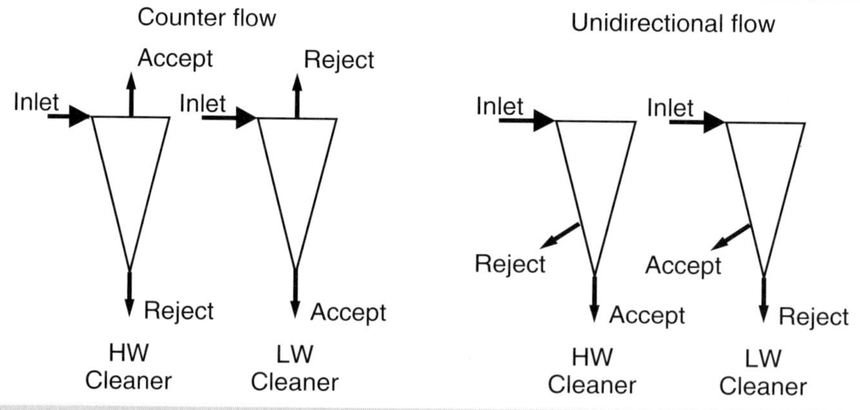

Figure 72. Principles of counterflow and unidirectional flow hydrocyclones for HW and LW cleaning.

Unidirectional flow HW and LW cleaners have both outlets at the bottom of the cyclone cone. The only difference between the units is that HW cleaner reject is removed from the suspension near the wall. In LW cleaners, this is where the accept leaves. This exploits the centrifugal force principle that HW collects on the outside and LW collects on the inside.

The HW particles follow a spiral path near the wall to the bottom of the hydrocyclone for removal through the reject opening with a small proportion of the incoming suspension (about 3%–15% by volume or 3%–30% by mass). In counterflow HW cleaners, the accept is taken out as the main flow at the top of the cyclone.

The LW particles tend to migrate to the center of the cleaner. In counterflow LW cleaners, light debris is removed with a portion of the total flow (about 3%–15% by volume or 1%–15% by mass) through the reject opening at the top of the cleaner. Most suspension leaves the cyclone through the accept opening at the bottom. The rejects are removed continuously or intermittently at ambient pressure (continuous only) or under pressure (both techniques). Reverse flushing of fibers occurs only with continuous reject removal.

5.2.2 Hydrocyclone theory

Free and core vortices

Particles are removed in the centrifugal field of hydrocyclones due to density difference between the particles and suspension[35–38]. Their shape and size also play important roles. The rotary flow necessary for creating a centrifugal field is generated by directing the suspension tangentially through an appropriately shaped inlet. The inlet geometry should optimally convert pumping energy into rotational flow[39]. The inlet flow is sometimes supported by a rotor at the top of the cyclone to ensure adequate tangential flow velocity at reduced throughput or high stock consistency.

The inlet velocity, $v_{t,i}$, in a cyclone with radius, r_i, results in centrifugal acceleration, a_c, toward the cyclone wall:

$$a_c = v_{t,i}^2 / r_i \tag{1}$$

Assuming flow without friction, the free vortex equation is then the following:

$$v_t \cdot r = constant \tag{2}$$

The tangential velocity, v_t, increases with decreasing radius, r. Correspondingly, the centrifugal forces also increase toward the center. Since flow without friction does not exist in practice, the flow is first turbulent, i.e., with intrinsic friction, and also is exposed to frictional resistance on the cyclone wall. The formula actually applying is therefore the following:

$$v_t \cdot r^n = constant \tag{3}$$

where $n = 0.5$ is commonly assumed in practice.

The free vortex or potential vortex attaches itself to a central core vortex defined by the following:

$$v_t = \omega \cdot r \tag{4}$$

where v_t is the tangential velocity
ω the constant angular velocity at radius, r.

The center of the core vortex usually contains air if any is in the suspension. Figure 73 shows the typical distribution of tangential velocity over the cyclone radius with ring and core vortices.

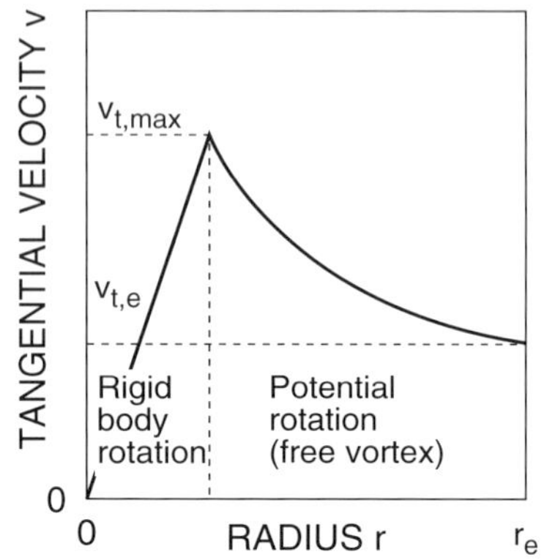

Figure 73. Tangential flow velocity vs. radius in a hydrocyclone with free vortex (potential rotation) and core vortex (rigid body rotation).

Acting forces

In vortex flow, the following forces act on a particle of volume, V_p, and density, ρ_p, moving with the suspension of density, ρ, at tangential velocity, v_t, on a circular path with radius, r:

- Centrifugal force F_c acting radially outward:

$$F_c = V_p \cdot \rho_p \cdot v_t^2 / r \tag{5}$$

- Buoyancy force F_b acting radially inward:

$$F_b = V_p \cdot \rho \cdot v_t^2 / r \tag{6}$$

- Flow resistance F_r to the radial motion of the particle at velocity, $v_{p,r}$:

$$F_r = c_w \cdot A_p \cdot \eta_s \cdot v_{p,r} \tag{7}$$

where c_w is the resistance factor
A_p the projected area of the particle
η_s the apparent dynamic viscosity of the suspension
$v_{p,r}$ the radial particle velocity.

Friction on the cyclone walls decelerates the suspension. To maintain adequate centrifugal acceleration, the cyclone radius is usually reduced conically toward the bottom. Most hydrocyclones therefore comprise a cylindrical top inlet section and a conical bottom section.

Maximum possible centrifugal acceleration

Ignoring friction, the maximum possible velocity, $v_{t,max}$, in a counterflow cyclone can be derived from the pressure differential, Δp, between inlet and accept:

$$v_{t,max} = (2\Delta p/\rho_s)^{1/2} \tag{8}$$

From this equation, the maximum centrifugal acceleration, a_c , in the cyclone can be derived by inserting the radius, r, at which the maximum velocity, $v_{t,max}$, occurs:

$$a_c = v_{t,max}^2/r \tag{9}$$

For example, at a pressure differential of $\Delta p = 2$ bar and radius of $r = 50$ mm, the centrifugal acceleration, a_c , is about 8 000 m/s^2. This is approximately 800 times the acceleration due to gravity. The actual centrifugal acceleration may be significantly less than the theoretical value. This is due to wall friction, internal fluid friction, and unfavorable flow conditions caused perhaps by cyclone overloading.

Particle separation

As particles approach the central vortex finder pipe for stabilizing cyclone flow, the centrifugal forces acting on them in the ring vortex increase. The diameter of the core vortex is about the same as that of the vortex finder pipe. Its length is adjusted according to need. Due to the increase in centrifugal forces toward the vortex finder pipe periphery, the separation effect also increases toward the center. HW particles with small diameter, unfavorable hydrodynamic shape, small density difference, or a combination of these factors are therefore separated effectively and propelled toward the walls. Particles whose separation probability is 50% due to their density, shape, and size are cut-size particles.

HW particles collect near the walls and are removed at the bottom of the cyclone with the suspension surrounding them. The suspension freed of these heavy particles lies further inward and is drawn off from the center, at the top of counterflow cyclones, and at the bottom of unidirectional flow cyclones.

LW particles are propelled inward in the centrifugal field of the free vortex subject to the above mentioned conditions whereby the separation forces increase radially toward the center. The LW particles collecting in the center are removed from the top of counterflow cyclones and from the bottom of unidirectional flow cyclones through the central pipe. The suspension largely freed of light particles rotates near the cyclone walls from where it is removed.

In unidirectional flow HW particle cleaners, the HW-enriched suspension near the wall must be separated from the cleaned suspension at the bottom of the cyclone in a confined space. The same applies to LW cleaners where the LW-enriched suspension is in the center of the cyclone.

In counterflow LW cleaners suspension is fed at the top or base of the cleaner and the accept is removed at the bottom or the apex. High volumetric reject flow assures a high efficiency of such cleaners. Since the accepts and reject ports are reversed in comparison with the conventional “forward” cleaner for HW removal, these LW cleaners are often called reverse cleaners.

5.2.3 Assessment of cleaning results

Hydrocyclone cleaning results are assessed by the cleanliness efficiency, η, that is also used for assessing screening machines:

$$\eta = (c_i - c_a)/c_i \tag{10}$$

where c_i is the particle concentration related to total solids at the inlet
c_a the particle concentration in the accepts.

The separation diagram previously explained in the section on "Screening and fractionation" also has use for assessing cleaner efficiency.

5.2.4 Separation efficiency and cleaner operation

As indicated by the above formulas, a high cleanliness efficiency is possible in different ways:

- High pressure differential between the inlet and accepts: This results in a high tangential velocity and high centrifugal acceleration. Since the pressure loss directly affects cyclone pumping energy demand, it must be limited for cost reasons. In a given cyclone, the pressure loss will rise with increasing throughput, the dwell time will be shorter, and the turbulence will be greater. A maximum cleanliness efficiency therefore exists at a certain throughput.
- Small cyclone diameter: This has practical limitations. For example, the inlet and outlet connections require dimensioning to prevent plugging for operational reliability. The minimum possible diameter therefore depends on the maximum particle size or trash length such as wires in each processing section. In addition, the low capacity of a small hydrocyclone demands many units in parallel. This sets cost limits on the minimum cyclone diameter.
- Low stock consistency: This limits freedom of particle motion the least. At higher stock consistencies, the fiber network becomes more dense and increasingly restricts particle motion. This is therefore a matter of fiber content, since ash content has little effect on freedom of particle motion.
- Low flake content: In the same way as high stock consistency, flakes restrict particle motion. Since more flakes are removed in an HW cleaner, fiber losses increase.
- High suspension temperature: This reduces the water viscosity and therefore reduces flow resistance to radial particle motion.
- Appropriate hydraulic loading of the hydrocyclone and appropriate split into accept and reject flows: The cyclone should operate within the hydraulic

design limits regarding throughput and reject flow. A higher reject flow improves separation but increases fiber losses.

- Appropriate reject removal: This should be continuous as much as possible. With intermittent operation, flushing water flows should be as low as possible so they do not flush back the HW already separated. This again means higher fiber losses.

The following comments apply to particle properties:

- Particles with large density difference compared with water separate more efficiently whether they are LW or HW.
- Particles whose density is very near 1 g/cm^3 may separate as heavy or light particles depending on suspension temperature or not at all. For example, this applies to some hot melts ($0.95 < \rho < 1.05$ g/cm^3) since the density of water decreases from 0.995 to 0.971 g/cm^3 when its temperature rises from 30°C to 80°C.
- A larger particle at comparable density gives a larger difference for centrifugal and buoyancy forces (inward acting for LW and outward acting for HW) and a better separation effect. For example, HW cleaners hardly separate filler particles 3 μm in diameter, but they separate coating particles 30 μm in diameter very well.
- Particles of comparable density with favorable hydrodynamic shape $(c_w \cdot A_p)$ separate more effectively, i.e., spherical particles separate better than flat ones.

Table 6 gives a qualitative overview of the primary ways to enhance cleanliness efficiency.

5.3 Machinery and cleaning effect

According to the stock consistency ranges of their use, cleaners are HC cleaners (or high density (HD) cleaners), MC cleaners, and LC cleaners[40–51]. Table 7 summarizes the primary design features and operating data of hydrocyclones used in these three consistency ranges.

Table 6. Parameters affecting hydrocyclone cleanliness efficiency.

...results in... / Increase of...	Cleanliness efficiency	Degree
Pressure difference	↑ ↓	●
Volumetric throughput	↑ ↓	◐
Cyclone diameter	↓	●
Consistency	↓	●
Flake content	↓	◐
Suspension temperature	↑	○
Reject rate	↑	◐
Flushing water flow	↓	◐

↑ Increase ↓ Decreace ● High ◐ Medium ○ Low

Table 7. Primary features of cleaners for various pulp consistency ranges.

Cleaner type / Characteristic		HC Cleaner with rotor	HC Cleaner without rotor	MC Cleaner	LC Cleaner HW	LC Cleaner LW
Preferred flow principle	–	Counter flow	Counter flow	Counter flow	Counter flow	Uniflow
Consistency range	%	2–4.5 (6)	(1–5) 2–4.5	1–2	0.5–1.5	0.5–1.5
Throughput per cleaner	L/min	100–10 000	80–10 000 (20 000)	600–3 000 (10 000)	100–1 000 (2 000)	100–500 (5 000)
Spec. energy demand /stage	kWh/t	0.5–3	0.5–3	1–4	2–10	2–50
Pressure drop inlet - accept	bar	0.1–1.0	0.4–2 (3)	1–2	0.7–2 (4)	0.8–2
Approx. g-factor	–	< 60	< 60	< 100	< 1 000	< 1 000
Length Diameter (max) Reject opening	mm mm mm	3 000 300–700 80–120	2 000–5 000 100–500	3 000–5 000 100–700 40–80	 75–300 10–40	 110–450 40–60
Reject rate by mass/stage	%	0.1–1.0	0.1–1.0	0.1–1.0	5–30	3–20 (0.2)
Reject discharge (not final stage)	–	Batch Continuous	Batch Continuous	Continuous Batch	Continuous	Continuous
Reject thickening (not final stage)	–	–	–	–	1.5–7	0.2–1.0
Final stage reject discharge	–	(if any) Batch	(if any) Batch	Batch	Continuous Batch	Continuous

( ) in special cases

As an example of the different separation effects in these three pulp consistency ranges, Fig. 74 shows typical size limit ranges for sand particles.

This discussion does not include machinery combinations allowing simultaneous coarse screening and centrifugal cleaning or centrifuges.

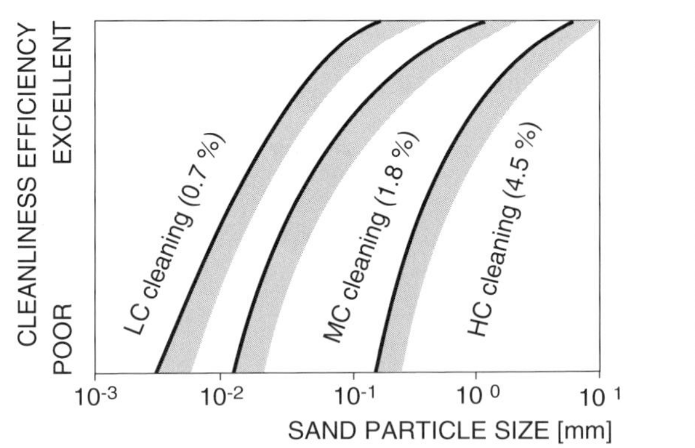

Figure 74. Sand separation performance of cleaners for various pulp consistency ranges.

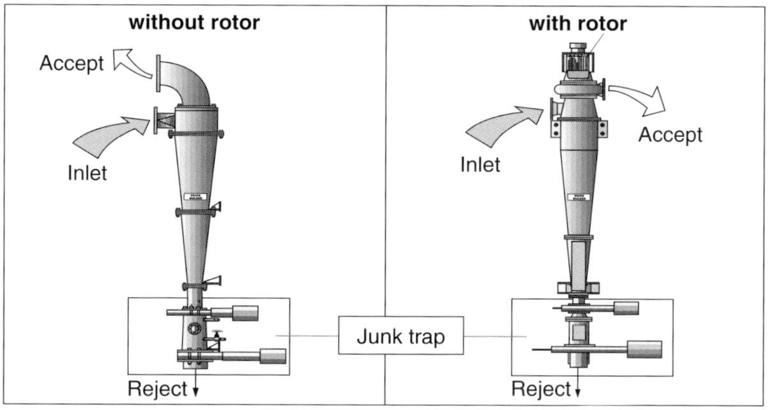

Figure 75. HC cleaners with and without rotor for intermittent operation.

5.3.1 HC cleaners

The HC cleaners of Fig. 75 have use for pre-cleaning. They remove coarse heavy particles from the suspension to protect excessive wear or damage in subsequent machinery such as refiners, screens, and pumps. An important requirement for HC cleaners in the pre-cleaning position is high operational reliability. Small reject orifices and reject chambers tend to plug due to the variable nature and size of the coarse contaminants to be removed.

In recycled pulp preparation lines, HC cleaners have use between slushing and coarse screening. Particles for separation must be larger than about 1 mm with density significantly higher than 1 g/cm^3. Examples are staples and paper clips, pieces of wire, and coarse glass and sand particles. Due to the particle size, all flow cross sections in HC cleaners must have adequate size to prevent plugging especially in the reject outlet.

HC cleaners generally operate at pulp consistencies of about 2%–5% or sometimes 6%. Their cleanliness efficiency decreases with rising apparent suspension viscosity that depends on stock consistency and type.

HC cleaners operate at centrifugal accelerations in the range of 40 g and higher. Some have an additional rotor in the inlet zone to promote acceleration of the suspension. This rotor must operate at adequate speed to ensure good separation at low throughputs or high stock consistencies.

Modern HC cleaners work on the counterflow principle and have three main connections for the inlet, accept, and reject. The pressure differential between inlet and accept is about 0.1–1.2 bar for machines with a rotor and 0.4–2 bar or sometimes 3 bar for machines without a rotor.

The heavy particles are often removed periodically through a heavy particle sluice. This is a junk trap comprising an upper and lower slide valve and an intermediate sluice chamber. To minimize fiber losses with the reject, HC cleaners are flushed through the bottom cone or heavy particle sluice that is a valved reject chamber.

In HC cleaning systems such as Fig. 76, the rejects from the first stage (usually comprising one or more HC cleaners in parallel) are continuously removed. This significantly improves the stage efficiency[52]. These rejects are diluted to a stock consistency of about 1.5% or less. They then go to a sedimentation tank from which the larger heavy particles are removed through a heavy particle sluice. The next cyclone stage and the succeeding pump are therefore protected against plugging and abrasion.

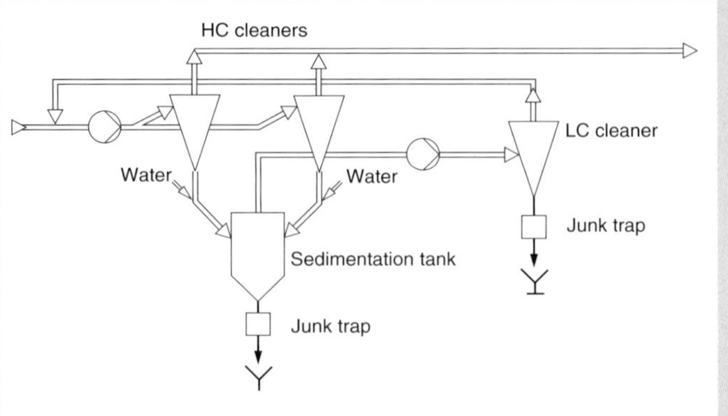

Figure 76. System showing HC cleaners in the first stage and LC cleaners in the second stage.

The second stage that treats the reject of the first has one or more hydrocyclones in parallel that remove the residual small heavy particles very efficiently. This stage also uses a heavy particle sluice and reverse flushing for periodic reject removal. Reject from the entire HC cleaning system is about 0.1%–1% of inlet mass flow depending on dirt content. Figure 77 compares the cleanliness efficiency of a single stage HC cleaner with that of the two-stage system described above with the same HC cleaner installed in the first stage.

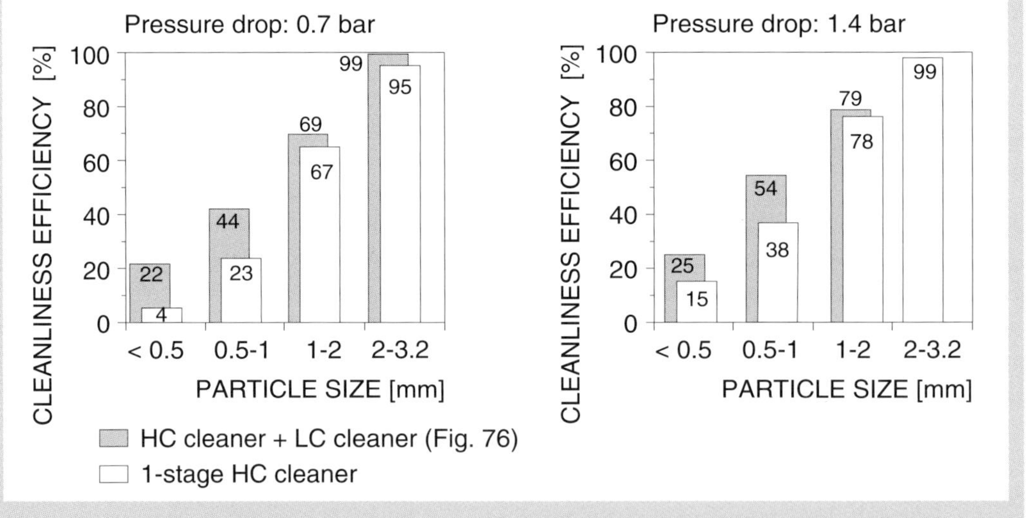

Figure 77. Cleanliness efficiency of a single stage HC cleaner compared with the system of Fig. 76.

5.3.2 MC cleaners

Centrifugal cleaners operating at pulp consistencies up to about 2% are the MC cleaners of Fig. 78. They comprise medium size hydrocyclones without rotor and often operate as single-stage cleaners with a junk trap. This type of cleaner also protects subsequent machinery from wear and ensures reliable operation by removing heavy particles such as sand, glass particles, and paper clips. Reject flow is about 0.1%–1% depending on the contamination of the recycled fiber pulp.

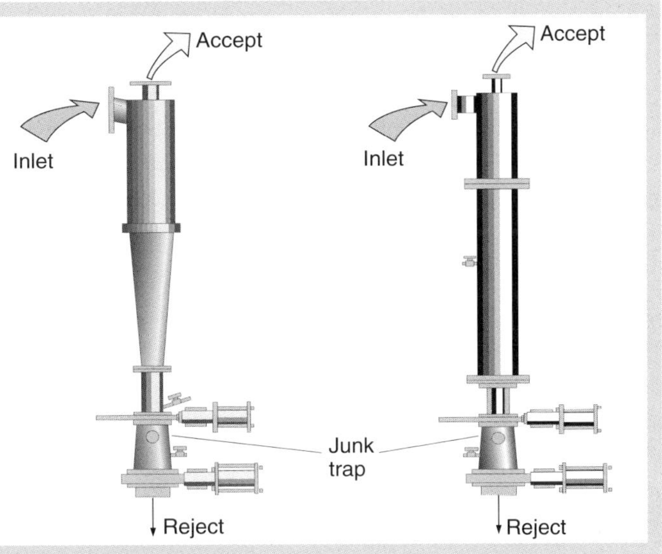

Figure 78. MC cleaners for batch operation.

5.3.3 LC cleaners

LC cleaners have use for heavy or light particle cleaning. They normally operate at stock consistencies of 0.5%–1.5%. They are usually smaller than the hydrocyclones described above and operate up to 1 000 g. For this reason, their separation efficiency is often significantly higher. Due to the size of the particles they can remove, they also have use for separating very small heavy particles and small light particles. The specific energy demand per ton of pulp is high because of the low consistency range. This energy is necessary for pumping the pulp slurry to the inlet of the cleaner.

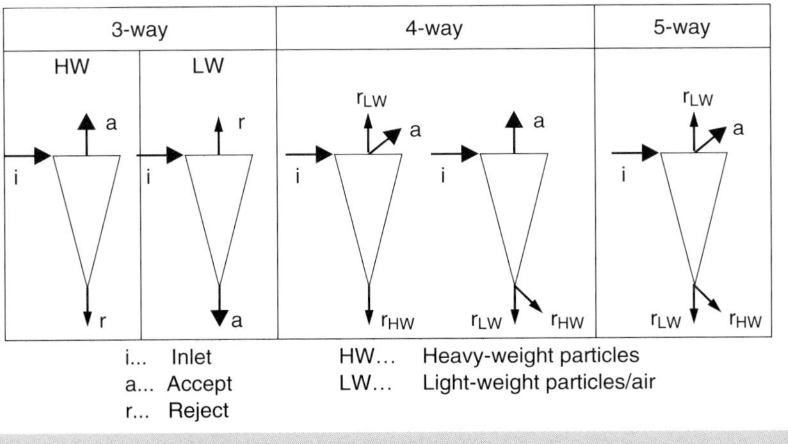

Figure 79. Number and arrangement of outlets on LC cleaners.

The wet surfaces of this type of cleaner are generally smooth but can sometimes have a spiral design or internal ribs at the end of the reject outlet. Cleaner body materials comprise plastic, steel, or ceramics.

Number of connections

Similar to other hydrocyclones, LC cleaners normally have three connections such as the three-way cleaners of Fig. 79. Combination cleaners for simultaneous removal of heavy particles, light particles, and air have one or two additional outlets. Four- and especially five-way types are more complex to operate. Their efficiency and reliability is often much lower than three-way machines. Figure 80 shows a four-way combination cleaner.

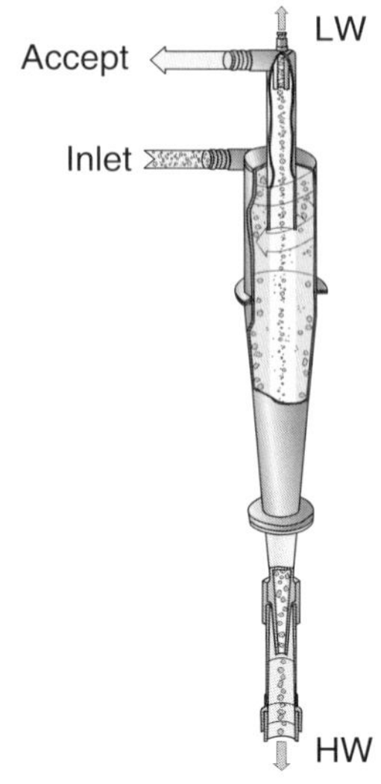

Figure 80. Combination cleaner.

Heavy particle cleaners

Heavy particle cleaners have recently become more important with the availability of screen baskets with slot widths down to 0.1–0.15 mm. These screens collect much more sand (particle size > slot width) and are considerably more sensitive in operation to abrasion. They therefore require safeguarding upstream. Heavy particle cleaners can do this very effectively. They adress smaller particles than the HC and MC cleaners.

Most heavy particle cleaners operate on the counterflow principle. Figure 81 shows heavy particle cleaners for final stage cleaning with continuous reject outlet

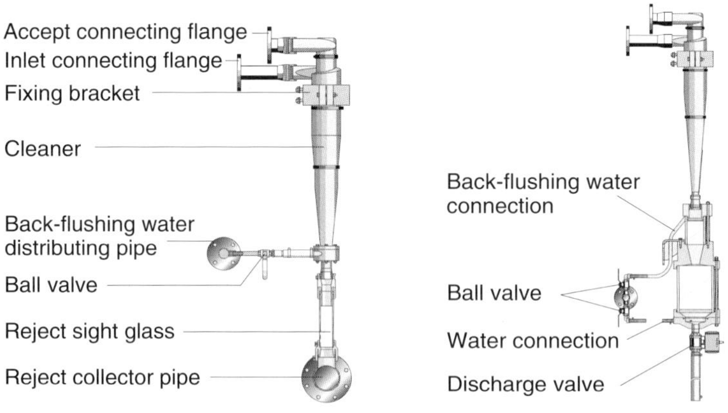

Figure 81. Heavy particle cleaners for continuous and batch operation of reject outlet.

under pressure and for batch operation. In suspensions with high heavy particle content, the continuously operated cleaner of Fig. 81 is also applied in previous stages for fiber loss reduction. Figure 82 shows an example of the cleanliness efficiency of a given LC heavy particle cleaner at varying operating conditions of consistency and pressure difference resulting from varying throughput.

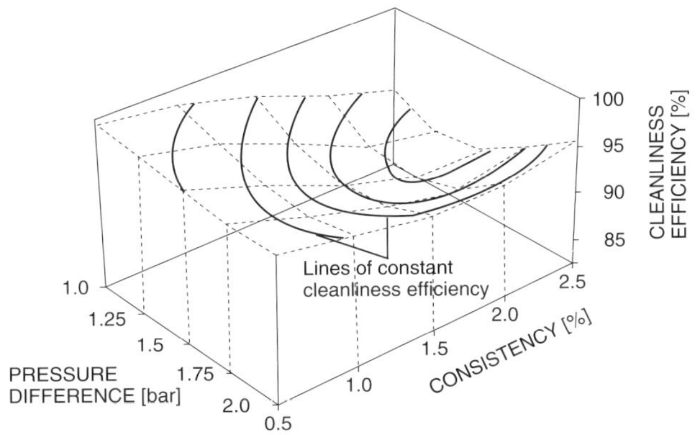

Figure 82. Cleanliness efficiency of a heavy particle cleaner vs. consistency and pressure difference.

Light particle cleaners

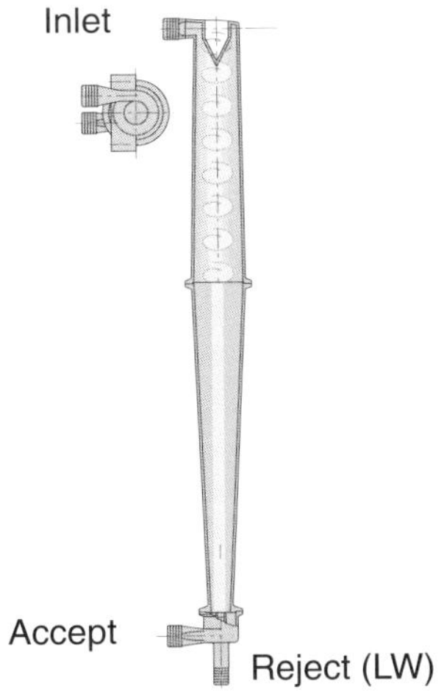

Figure 83. Section through a light particle cleaner with unidirectional flow.

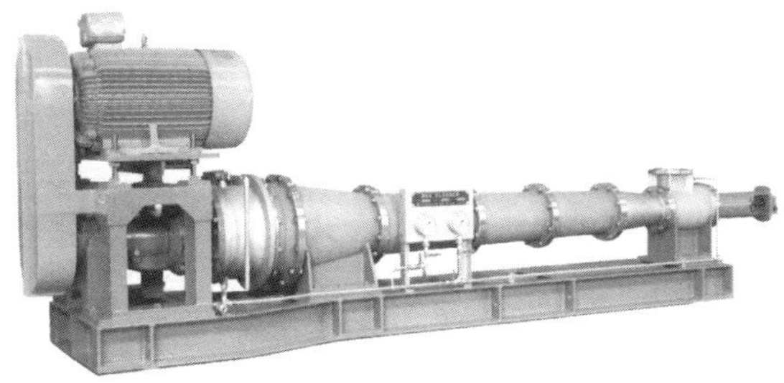

Figure 84. Horizontal light particle cleaner for large throughput.

Due to improved screening technology with narrow basket slots in pressure screens, light particle cleaners now play a less important role in recycled fiber pulp processing. Their main application today is for stocks with high wax content and for removing plastic foam. Light particle cleaners usually operate on the unidirectional flow principle.

Figure 83 shows a cross section through a light particle cleaner, and Fig. 84 gives a picture of a horizontal light particle cleaner for high throughput. An impeller accelerates the suspension entering the cleaner through the head. This causes a pressure gain of 0.5 bar instead of a pressure loss between inlet and outlet of the cleaner under normal operating conditions.

Cleaner batteries

For high throughput with minimal structural outlay, individual cleaners connect in parallel to form cleaner batteries of various types as Fig. 85 shows. The suspension is fed via a distributor pipe or chamber to the cleaner units whose accepts and rejects then flow to a collector pipe or chamber.

Open reject outlets previously had collection and disposal in a common tray. Today, rejects are almost always collected in a closed pipe under pressure with a common large-section regulating valve that controls the total reject flow of the cleaners of the individual stage. This allows use of larger cleaner reject orifices that reduce the risk of plugging compared with an open, non-pressurized system where a smaller reject outlet is necessary to limit the reject flow.

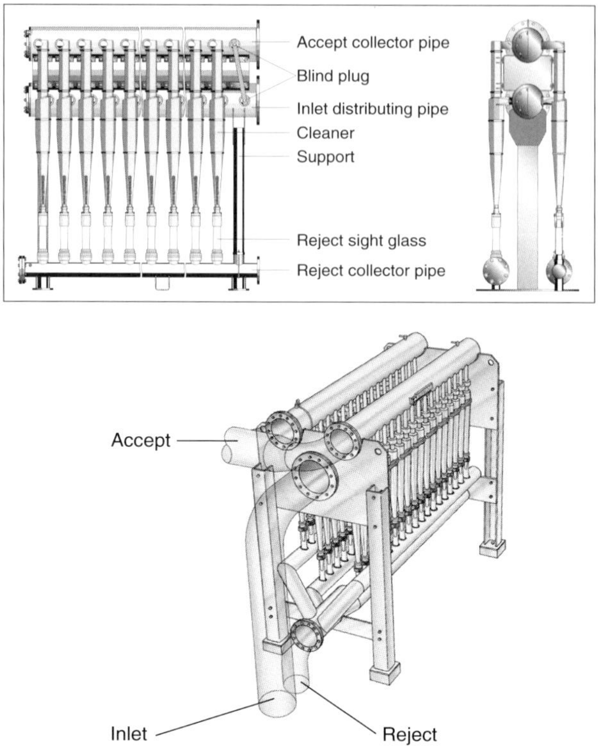

Figure 85. Cleaner battery arrangements showing two rows and four rows.

5.3.4 Cleaner systems

Rejects from the first-stage LC cleaners are processed in as many as four stages depending on system size and the debris content of the incoming suspension. The usual arrangement is cascade as in Fig. 86. Rejects from the first cleaner stage at production level flow to the second stage at recovery level whose accepts return

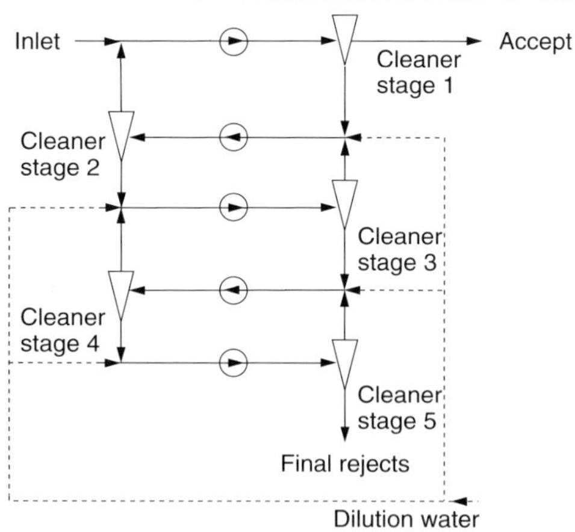

Figure 86. Cleaner cascade system.

to the first stage inlet. Rejects from the second stage go to the third cleaner stage whose accepts return to the second stage inlet. The rejects are fed forward to the next higher stage if any until reaching the final cleaner stage. Typical feed consistencies range from 0.6%–1.3%.

The stock consistency of heavy particle cleaner reject is higher than the inlet consistency. The thickening factor for continuous operation depends on the reject flow ratio, stock type, and type of cleaner. A higher separation efficiency usually corresponds to a higher thickening factor. For cleaners with low volumetric reject rate such as an open, nonpressurized reject outlet, the thickening factor is about 3–7. For pressurized cleaners with higher volumetric reject rates, the thickening factor is about 1.5–3. This is not solely due to fiber consistency increases since high thickening factors are also the result of an increased amount of fillers and heavy particles contained in the reject.

For efficient handling in the downstream cleaner stages, the rejects require dilution. To reduce flow rates and cleaner size, the stock consistency of the dilution water should be as low as possible.

Since flow decreases as it goes through the subsequent stages, the number of cleaners in each stage can be smaller. In heavy particle cleaner systems for example, each stage has only about 25%–45% of the previous stage capacity. The number of cleaners necessary also depends on the thickening factor, reject flow ratio, and dilution water stock consistency.

Cleanliness efficiencies of cleaners depend heavily on throughput. For large changes in output, some cleaners of the first stage should be switched in or out of the system accordingly. The active cleaners can then operate in the optimum range.

5.3.5 Final stage cleaners

Final stage cleaners must remove contaminants from the reject with the lowest possible fiber loss. Flushing water is therefore injected appropriately into the reject outlet to flush

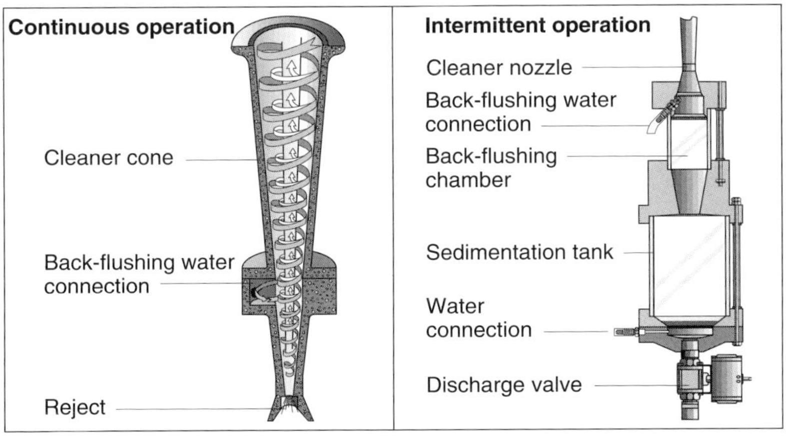

Figure 87. Flushing system for fiber recovery in final stage cleaners for continuous and batch operation.

fibers back into the cyclone. Figure 87 shows some typical flushing systems for continuous or intermittent reject removal on heavy particle cleaners. Since minimum fiber loss and maximum cleanliness efficiency are contradictory goals, they require optimization according to each individual case.

5.3.6 Centrifugal cleaners with rotating housing

The special feature of the centrifugal cleaner shown schematically in Fig. 88 is rotation of its housing[44,50]. This design has use for light particle cleaners. Due to flow conditions in the cyclone and the high centrifugal acceleration, light particles with almost the same density as water will separate. Figure 89 shows such a machine under construction.

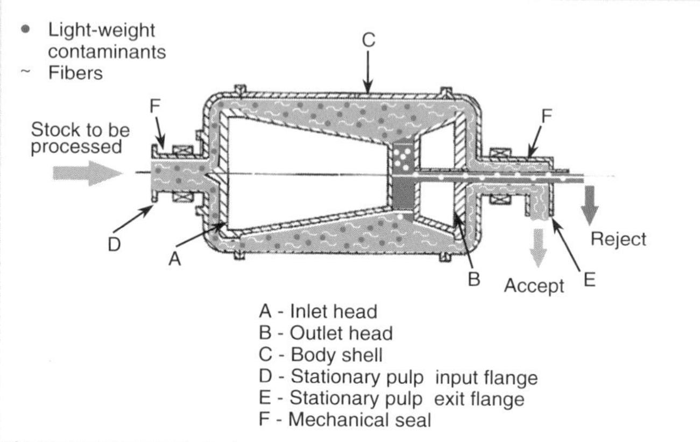

Figure 88. Schematic arrangement of centrifugal cleaner with rotating housing.

5.3.7 Abrasion

Hydrocyclone housings are subject to extremely abrasive heavy particles (metal, glass, sand, fillers, and coating pigments) contained as contaminants in the pulp slurry and to very high flow velocities such as in LC cleaners. Abrasion is especially serious where these particles impact the inner walls with high kinetic energy, travel across the wall surfaces at high velocity, or both. Often the particles remain in almost stationary orbit at a given height in the cyclone. The inlet zone and hydrocyclone cone therefore have a high risk of abrasion.

Figure 89. Centrifugal cleaner with rotating housing under construction.

Abrasion is countered in two different ways. First, the endangered surfaces are protected with ceramic coatings or made entirely of ceramic. Other coating materials such as rubber or hard chromium plating have some use. Second, the design of the hydraulic contours of the hydrocyclone ensure optimum particle flow paths.

To avoid unpleasant surprises or even production stoppages, wear indicators are sometimes installed so that worn components can be replaced during operational stops. Figure 90 shows a simple kind of wear indicator comprising grooves in the cone as design failure points. Slight leakage through these grooves provides an early warning of excessive wear. Leakage can be stopped by inserting a colored marking plug before replacing the worn component.

Figure 90. Cleaner wear indicator.

6 Flotation

6.1 Objectives and system integration

Flotation is a separation process for suspension cleaning based on probability using air bubbles that attach themselves to the particles and transport them to the surface of the suspension. Two kinds of flotation have use in the paper industry. Deinking flotation uses selective flotation for removing ink particles. Microflotation or dissolved air flotation (DAF) has use for cleaning process water or effluent by separating all solids from the liquid as much as possible. The following text only discusses selective, i.e., deinking flotation. Dissolved air flotation is dealt with in Chapter 6.

Selective flotation has use in recycled pulp processing for removing contaminants from the suspension. The separation criterion is the different surface wettability of the particles to be removed and the fibers to be retained. Particles whose surface is or has been rendered hydrophobic, i.e., water-repellent, can be floated within a certain size range. These particles include printing ink, stickies, fillers, coating pigments, and binders.

The main application of selective flotation in the paper industry is deinking. This improves pulp brightness by removing ink particles and enhances cleanliness by removing dirt specks. Dirt specks are large particles visible to the naked eye such as toner particles from copy machines or ink-jet printers. The minimum size limit for most measuring methods is about 50 μm. Ink particles with an equivalent diameter smaller than this are part of the brightness assessment.

Flotation also partly removes fillers. It has a certain de-ashing effect that is often desirable. This effect can prevent so-called "ash collapse" (excessive build up of ash content) in the system loops without removing excessively organic fines material. Flotation is also beginning to find use systematically for removing stickies with varying efficiency. Stickies cover a wide variety of substances (See Chapter 11) whose size and surface chemistry are not always suitable for flotation.

Printing inks, fillers, and stickies to some extent can also be removed by wash deinking. Under specific dewatering conditions, wash deinking can adequately remove fines and fillers below about 30 μm. Flotation deinking removes a wide particle size range of about 10–250 μm very efficiently with a high yield of fines and fillers. Some systems even remove larger particles up to 500 μm at reasonable efficiency. Wash deinking was preferable in the United States. Flotation was more favorable in Europe. Today flotation is common globally with use of the two processes on a complementary basis. Their specific benefits are particularly useful when high demands are placed on pulp processing due to raw materials, final product requirements, or both.

Indispensable for efficient flotation is the detachment of the particles from the fibers so that they can move freely in the suspension. They must also be in the proper size range for flotation. Mechanical and chemical preparation is therefore very important for efficient flotation.

The chemical and physical characteristics of the particles must also be known. For example, the various printing and nonimpact reproduction techniques used according to paper grade have a very important effect on deinking flotation results (See Chapter 8). They also influence processing requirements and their integration in the stock preparation line. Figure 91 shows the integration of flotation and washing in a wood-free office paper processing line for deinked market pulp with three individual flotation process steps in series.

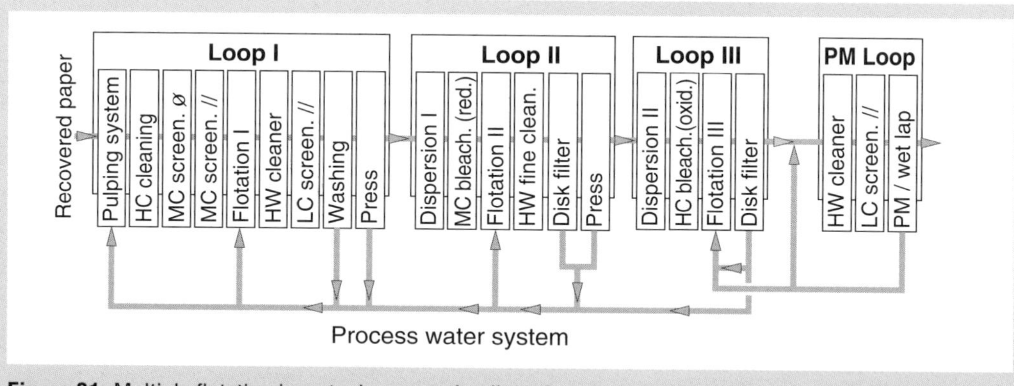

Figure 91. Multiple flotation in a stock preparation line of wood-free office paper for deinked market pulp where red. = reductive and oxid. = oxidative bleaching.

6.2 Principles

6.2.1 Prerequisites for good flotation results

As mentioned above, the particles must be able to move freely in the suspension without adhering to the fibers. They must also fall within a certain size range. The mechanisms involved in releasing ink particles from fibers are not yet completely known, but one can assume that they include fiber swelling during slushing, mechanical forces due to friction, and chemical bond loosening. Fiber swelling breaks ink apart by crack formation. When flakes disintegrate, the ink particles also break down. Mechanical forces remove the ink through friction between fibers. Firmly attaching particles such as oxidized inks require greater separation forces that mechanical dispersion can apply.

The size range of free ink particles in a suspension is very wide. Originally, the basic size of ink particles (carbon black and pigments) was 0.02–0.1 µm. Water-based flexographic printing ink agglomerates are 1–5 µm. Offset printing ink agglomerates can be up to about 100 µm. The oxidized ink agglomerates that adhere so strongly to the fibers often reach sizes of 500 µm and more. For efficient flotation of ink particles, their size range must be 10–250 µm. Any larger particles therefore require reduction in size. Dispersion shear forces are sometimes necessary. Smaller particles require agglomeration into larger ones. Flat particles are more difficult to float than cubic ones of the same volume. Reshaping is also beneficial in this instance[53].

For some recovered paper mixtures according to product demands, flotation is necessary multiple times as Fig. 91 shows.

Most commonly used printing inks are hydrophobic. This can be accentuated by applying calcium soap. It agglomerates small ink particles and also partly changes the nature of filler particles to make them float. A precondition for good agglomeration is sufficiently but not excessively hard water.

For further information, see the following chapters of this book: “Deinkability of printing inks” and “Deinking chemistry”.

6.2.2 Flotation process

The following mechanism removes particles from a recycled fiber slurry. Air injection forms bubbles to which the hydrophobic particles attach.The bubbles move to the surface of the suspension where they form a foam layer that is easily removable. The particles are either hydrophobic initially or made hydrophobic by adding suitable chemicals (surfactants).

Probability mechanism

Flotation is a probability process[53–55] in four stages each of which has its own probability. The particles must collide with an air bubble (collision probability), attach themselves firmly to it (attachment probability), rise to the foam layer as a stable particle-bubble aggregate (stability probability), and undergo removal with the foam (probability of particle transport with foam). The last two stages together are removal probability.

Number and size of bubbles

The probability of bubbles and particles colliding within a given volume clearly increases with the number of bubbles, i.e., with the volume of air injected for a given bubble diameter, and with the number of particles. A smaller bubble diameter, d, for a given air volume, V, gives a larger number of bubbles, z:

$$z = 6V/\pi d^3 \tag{11}$$

Small bubbles can catch and transport small particles, but large particles need large bubbles. Figure 92 gives a scale overview of size relations for contact and attachment between 1 mm diameter bubbles and 10–500 µm particles. For floating particles of various sizes, generation of a certain size range of bubbles is also necessary. For example for a given air to suspension ratio, small bubbles will offer a very high surface area to a large number of small particles. An adequate number of large bubbles must also be present to collect the bigger particles (usually at a smaller number).

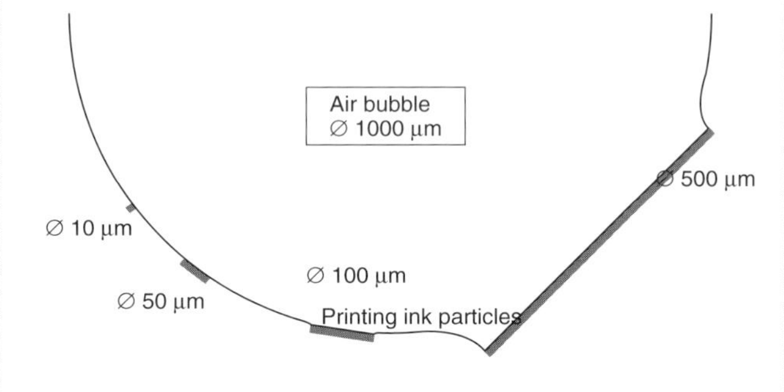

Figure 92. Size relations for contact and adhesion of various particle sizes to air bubbles.

Collision between bubbles and particles

An important criterion of collision probability is the relative motion between bubbles and particles. Only when the particles move relative to the bubbles they do approach sufficiently near for collision and attachment under the proper conditions.

Figure 93 shows the primary requirements for collision. The bubble and particle approach each other on their relative stream lines. For all particles within a critical radius, $r_{capture}$, collision will occur. The radius, $r_{capture}$, depends on bubble size, relative velocity between bubble and particle, suspension viscosity, and other factors. In this example, par-

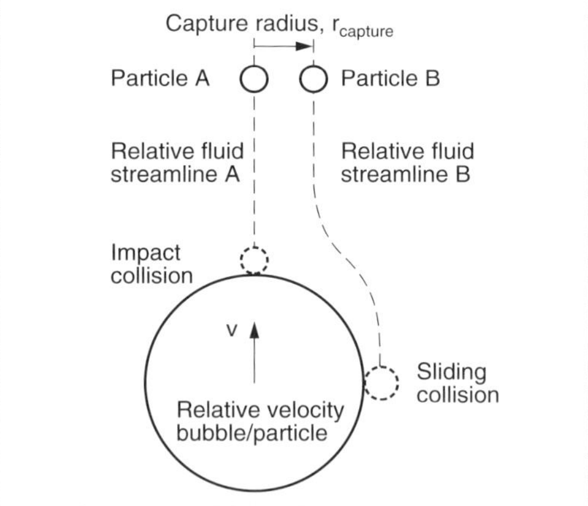

Figure 93. Stream lines and collision conditions between particles and bubbles[54–56].

ticle A moves on its relative stream line A until impact collision with the bubble occurs. The particle may slightly deform the bubble and rebound or attach itself. Particle B that approaches the bubble at critical capture radius collides tangentially with it (sliding collision) and may also rebound or attach.

Disk shaped particles of the same mass and material are more likely to collide than spherical particles because the disk diameter is far bigger than the sphere diameter. Disks are much less likely to collide with bubbles than spheres of the same diameter with far greater mass because they extend only in one plane. Collision therefore depends on their alignment to the bubble.

Attachment

For firm attachment to the bubble after collision, a three-phase limit must be fulfilled. The liquid film between particle and bubble must be displaced to form a new state of equilibrium between particle, air bubble, and liquid regarding surface energy. Figure 94 shows this mechanism schematically[56,57].

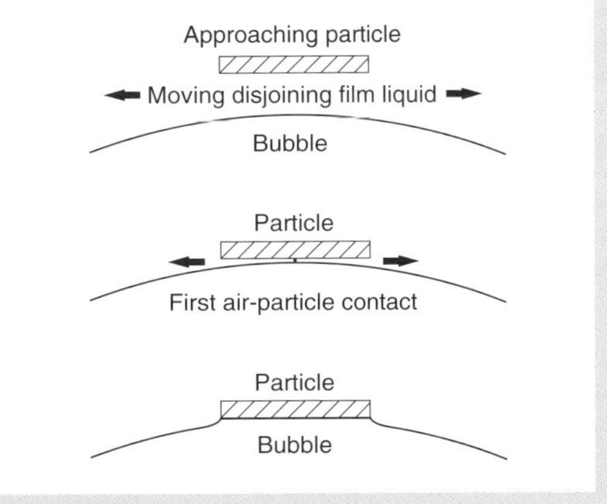

Figure 94. Mechanism of particle attachment to an air bubble[54–56].

Spherical particles attach themselves more easily than disk shaped particles. Although the latter ones offer a greater surface area for attachment, considerable liquid film volume must be displaced for attachment during the limited contact time. If the edge of a disk shaped particle contacts the bubble, the film volume displacement is much less as is the surface area.

During the brief interlude between collision and final formation of the three-phase limit, the adhesion forces are not fully developed. The particle can easily separate from the bubble again by flow forces.

Stable transport

Only when a state of equilibrium (lowest energy) has occurred can the particle adhere optimally to the bubble and resist the hydrodynamic flow forces trying to remove it. Due to flow resistance, the particle moves to and remains on the downstream side of the bubble.

The hydrodynamic flow forces acting on the bubble-particle aggregate cause tensile forces, centrifugal loading, and shear forces. Some smaller particles less than about 10 µm diameter never seem to come into true contact with their bubble (contactless flotation). Flow forces therefore remove them more easily [54].

Bubble formation and motion

Other important mechanisms are bubble generation, relative motion of bubbles and particles, and bubble ascent to the surface. Bubble generation occurs in various ways primarily connected with the processes of collision of bubbles and particles, transport of the bubble-particle aggregate to the surface, and removal in the foam layer. Table 8 shows principles of individual flotation phases.

Table 8. Principles of individual flotation phases.

Solution	No mixing energy (porous media)	With mixing energy	
Subfunction		**Static mixer**	**Dynamic mixer**
Air supply	Pressing	Sucking Pressing	Sucking Pressing
Collision of ink particles with air bubbles	Countercurrent flow of suspension and air	Static elements (Flow energy)	Rotating elements (Mechanical energy)
Discharge of air bubbles and ink particles	Long path to surface	Short/ medium path to surface	Medium path to surface

Bubble generation

In the simplest case, air is injected through a permeable body such as perforated sheeting, textile matting, or sintered ceramic. The bubble size and air mixing in the suspension are not influenced at the source by additional energy application such as flow energy. Bubble size here primarily depends on the surface tension of the suspension, size of air injection openings, and air injection volume.

Another way of generating bubbles involves high mixing energy application. This energy depends on pressure loss and type of mixer. It strongly influences the bubble size spectrum.

The precise determination of bubble sizes under operating conditions with corresponding stock consistency is difficult. Although much information is available on bubble size, shape, and behavior in water, practical conditions in stock suspension are very different especially if the consistency is more than about 1%. For this reason, bubble tests with water are not very conclusive[58,59]. The size spectrum of bubbles usually depends among other factors on the type of mixing element, mixing energy applied, and suspension properties such as pulp consistency and type and amount of surfactant[60].

Relative velocity

While air bubbles rise according to buoyancy and resistance forces, flotation particles follow flow paths dictated by hydraulic forces, inertia, and gravity. The relative velocity between air bubbles and suspension determines the collision probability of bubbles and particles. If bubbles are generated without applying mixing energy, collision must occur during ascent. Bubble ascent time should therefore be as long as possible to increase the probability of collision.

The ascent velocity of a 1 mm diameter air bubble in water is about 0.2 m/s. This increases to 0.3 m/s for 2 mm diameter as Fig. 95 shows[58]. Since flotation aeration is very intensive, entire bubble clusters are generated. Due to their flow resistance, these carry far more suspension with them than a single bubble. This pumping effect causes strong re-circulation since the suspension transported upward with the bubbles requires replacement by downward flow.

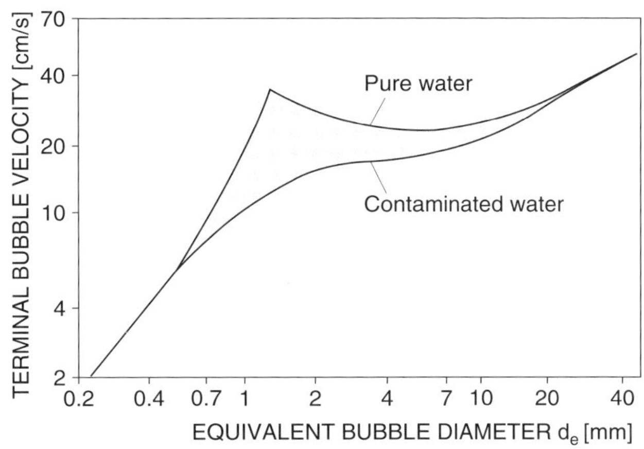

Figure 95. Bubble ascent velocity[58].

Individual bubbles contacting each other in a cluster often coagulate into larger bubbles. With high stock consistencies and long ascent paths, this often leads to bubble channel formation through which the air rapidly escapes.

Static or dynamic mixers create strong relative motion between bubbles and particles due to the air injection energy. This ensures good contact between the bubbles and particles. Ascent paths to the surface foam layer can be much shorter with this method.

Typical operating conditions

Flotation systems generally operate at stock consistencies of 0.8%–1.5% and temperatures of 40°C–70°C. The pH should be about 7–9 with water hardness of 5–30 °dH when using fatty acid surfactants. With other surfactants, this hardness requirement is not so important. The relative air load is mostly about 300% and more expressed as total air volume flow to total suspension volume flow. Certain flotation systems operate at air loads up to 1 000%.

Besides ink particles, the flotation foam also contains organic components such as fibers and fines as well as inorganic materials such as fillers and coating pigments. Losses of these fibers, fines, fillers, and pigments can be minimized by reducing reject foam volume or using a secondary stage flotation at recovery level where the reject (foam) of the primary stage is floated for fiber, fines, and some ash recovery. Research and development work is in progress for washing fibers and fines from the foam for higher yields in deinking flotation[61,62].

Reducing the amount of primary cell reject foam is only possible to a limited extent without influencing flotation results of brightness and cleanliness. A secondary flotation stage therefore has common use. It purely reduces losses while the first stage maintains quality requirements.

Secondary flotation usually operates at lower stock consistency than the primary stage. The solid composition in the pulp slurry is very different. It contains less fibers but considerably more fines and fillers. The secondary flotation operating parameters therefore require adjustment to these conditions.

Secondary flotation aims for at least the same brightness and residual dirt speck content as at the primary flotation inlet so that the accepts can be recirculated to that point without affecting quality. The secondary flotation rejects are optimized to minimize total losses and organic losses without reducing accepts quality. Secondary flotation is also selective between organic and inorganic components. This means considerably more inorganic than organic particles separate compared with the inlet composition.

6.3 Flotation cells and their operating principles

6.3.1 Characteristics of different flotation cells

Flotation deinking cells vary widely[63]. The following overview summarizes flotation cell technology in the past and the latest developments today. Table 9 shows their common features and differences. A primary difference between flotation cells is their aeration system. These can either comprise air-permeable bodies as mentioned above or static and dynamic mixers. Other variable features of flotation cells are the following:

- Number of aeration stages and units required for complete flotation
- Type of air supply (natural aspiration or compressed air)
- Foam removal methods (paddle or free flow over a weir, vacuum, or pressure)
- Use of secondary flotation stage as a standard
- Closed or open cells
- Cell design and arrangement of multiple cell units.

6.3.2 Flotation cells on the market

Table 9 lists most flotation cells for deinking flotation with their main characteristics. The special features of flotation cells on the market today are summarized below primarily using manufacturers' data.

Ahlstrom
The Ahlstrom flotation cell of Figs. 96 and 97 has a self-aspirating dynamic mixer. Foam from this round open cell is removed by free weir flow. Several cells can combine into a flotation line.

Beloit
The PDM 2 cell of Figs. 98 and 99 is an improved version of the PDM cell with air injection by static mixer. One to four of these long cells in series can form a complete flotation line. The PDM 2 cell operates under slight pressure that is used for foam removal.

Table 9. Flotation cells.

Characteristics		Aeration mixer			Number of aeration steps		Air supply		Foam discharge						Cell		
Supplier . Product name	Year of introduction	Porous medium	Static	Dynamic	Single	Multi	Ambient	Compressed	Scraper/weir	Free weir overflow	Vacuum	Pressure	Secondary stage	Cell inclosed	Shape of body	Main axis	Marketed today
Ahlstrom																	
Kamyr GSC	1991	•				•		•				•	•		cycl.	vert.	
Flotation Cell	1997			•		•			•						cyl.	vert.	•
Beloit																	
LinearCell	1983		•			•	•				•				tube	horiz.	
PDM Cell	1987		•			•		•				•		•	tube	horiz	
PDM 2 Cell	1995		•			•		•				•		•	tube	horiz.	•
Comer																	
Spidercel	1992		•	•		•	•		•					•	col.	vert.	
Cybercel	1996		•	•		•	•		•					•	col.	vert.	•
Denver																	
Flotation Cell	1950			•		•	•		•				•		rect	vert.	
Kvaerner																	
Flotation column	1994	•			•			•	•					•	col.	vert.	•
Shinhama																	
Hi-Flo Flotation Cell	1991		•			•	•			•		•	•		cycl.	vert.	•
Sulzer Escher Wyss																	
FZ Cell	1972			•		•		•	•				•		cyl.	vert.	
Uni Cell	1976	•				•		•		•			•		rect.	horiz.	
CF Cell	1984		•			•	•			•			•		cyl.	vert.	
CFC Cell	1992		•			•	•			•			•	•	cyl.	vert.	
Sunds																	
Swemac	1976	•				•		•	•				•	•	cyl.	vert.	
Thermo-Black Clawson Black Clawson																	
Ultracell	1981		•			•	•						•		cyl.	vert.	
IIM/BC Flotator	1991			•	•			•	•				•	•	rect.		•

Table 9. Flotation cells. (continued)

Characteristics / Supplier . Product name	Year of introduction	Aeration: Porous medium	Aeration mixer: Static	Aeration mixer: Dynamic	Number of aeration steps: Single	Number of aeration steps: Multi	Air supply: Ambient	Air supply: Compressed	Foam discharge: Scraper/weir	Foam discharge: Free weir overflow	Foam discharge: Vacuum	Foam discharge: Pressure	Secondary stage	Cell inclosed	Cell: Shape of body	Cell: Main axis	Marketed today
Lamort																	
Verticel	1978		•			•	•				•				col.	vert.	
Verticel DA	1986		•			•	•				•				col.	vert.	
Verticel DAD	1993		•			•	•				•						
Verticel MAC	1995		•			•	•					•		•	col.	vert.	•
Voith																	
Paddle Cell	1959			•		•	•		•				•		rect.	vert.	
Injector	1981		•			•	•		•				•	•	tube	horiz.	
Multi-Injector	1986		•			•	•			•			•	•	tube	horiz.	
E-Cell	1990		•			•	•			•			•	•	tube	horiz.	
Voith Sulzer																	
EcoCell	1995		•			•	•			•				•	tube	horiz.	•

cyl... cylindrical *rect... rectangular*
cycl... cyclone *horiz... horizontal*
col... column *vert... vertical*

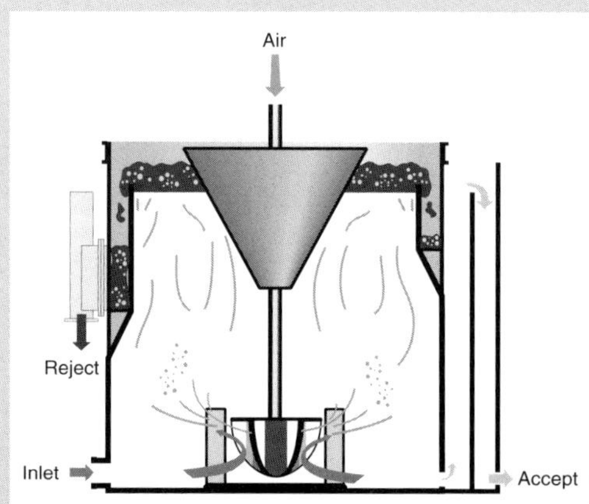

Figure 96. Schematic arrangement of the Ahlstrom flotation cell.

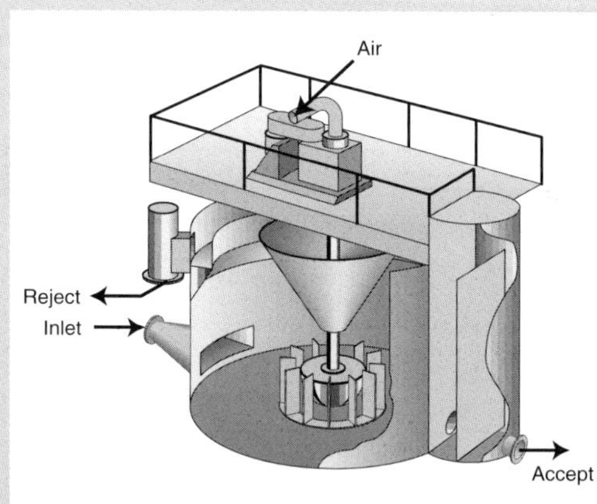

Figure 97. Cut-off view of the Ahlstrom flotation cell.

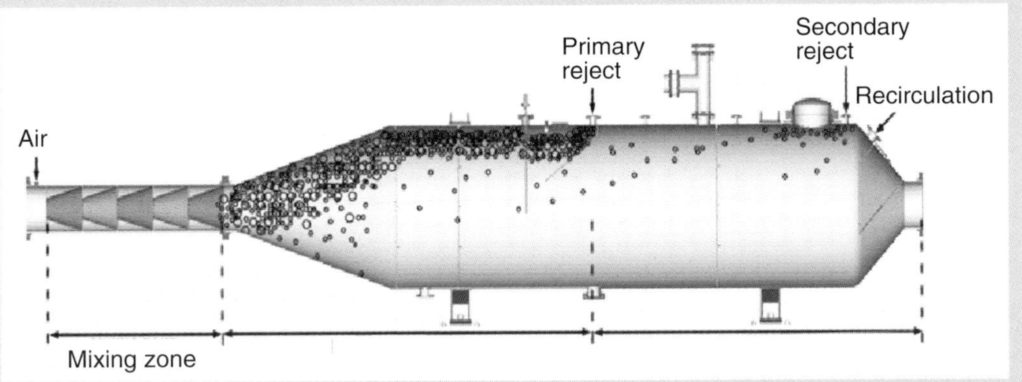

Figure 98. Schematic arrangement of the Beloit PDM 2 flotation cell.

Figure 99. Beloit PDM 2 flotation cell.

Black Clawson (Thermo Black Clawson)

The II M/BC Flotator cell in Figs. 100 and 101 is closed and rectangular. Dynamic mixers in the cell generate fine bubbles from air supplied through a manifold. The pulp slurry is recirculated and re-aerated several times. High specific air flows of up to 1 000% are used. Foam is removed by scrapers over a weir. Suspension dwell time in the cell ranges from 8-15 min. This restricts throughput with the given cell volume. A single cell is sufficient for adequate flotation results. Normally, several cells are arranged in parallel.

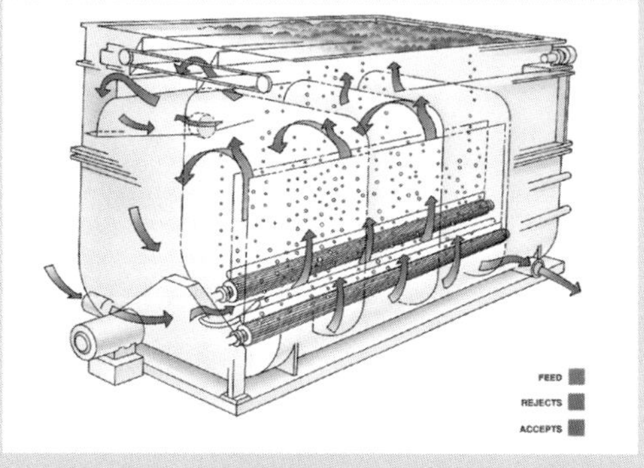

Figure 100. Schematic arrangement of the Black Clawson II M/BC Flotator cell.

Figure 101. Black Clawson II M/BC Flotator cell.

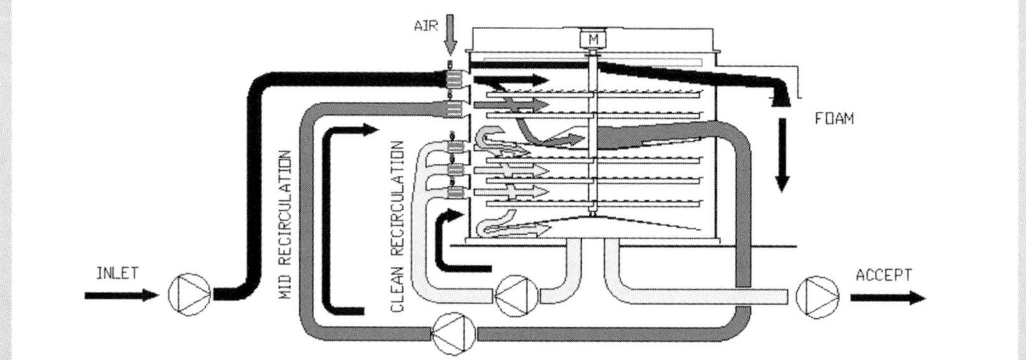

Figure 102. Schematic arrangement of the Comer CYBERCEL.

Comer

The CYBERCEL of Figs. 102 and 103 was introduced in 1996 as further development of the previous SPIDERCEL dated 1992. This tower-type cell has a static mixer for air injection and a mechanical agitator for improved bubble distribution. The suspension flows in the opposite direction to the ascending air bubbles. Some suspension is recirculated and aerated again before distribution to the middle and lower tower zones. The foam is removed by a paddle over a weir. Two of these cells in series are sufficient for adequate flotation with an optional secondary flotation cell.

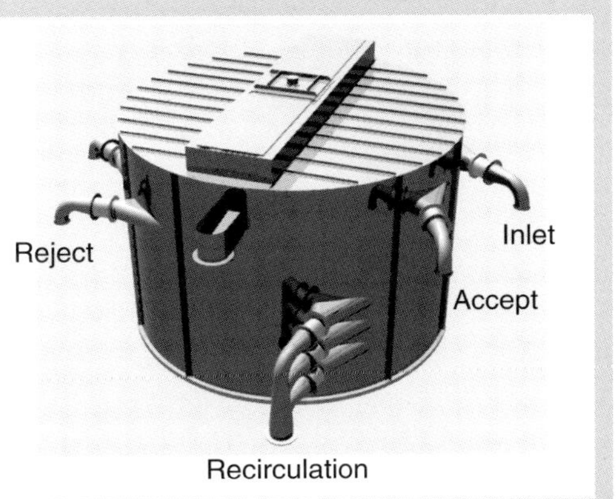

Figure 103. Comer CYBERCEL.

Kvaerner

These closed flotation columns of Figs. 104 and 105 operate on the counterflow principle with compressed air injection through porous sintered metal tubes at the bottom. As an option, the suspension inlet flow can also be aerated. Flow stabilizing fins are fitted to the inside of the column. Foam is removed over a weir. The flotation effect is given by a single pass through the column. No secondary flotation stage is used.

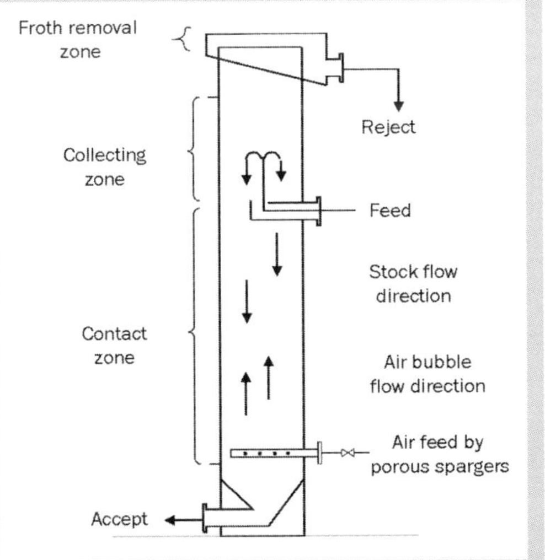

Figure 104. Schematic arrangement of the Kvaerner flotation column.

Figure 105. Kvaerner flotation column.

Lamort (Thermo Black Clawson)

The Lamort MAC cell in Figs. 106 and 107 also has a single flotation pass through the column stages. No secondary flotation stage is necessary. This circular enclosed and pressurized cell is subdivided into 3–5 superimposed elementary cells. Suspension flows from top to bottom where it is tapped off from each elementary cell and aerated before returning to the next lower cell. Aeration is by self-aspiration using static mixers. The air injected in each cell ascends through the upper cells to the foam layer on the surface that is removed under low pressure. Valves allow control of reject rate and composition.

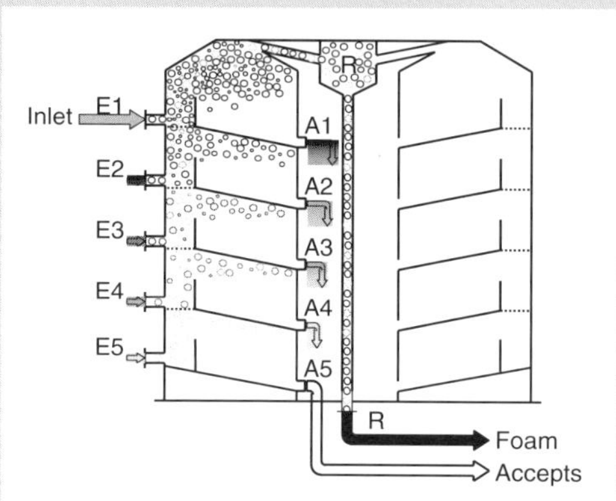

Figure 106. Schematic arrangement of the Lamort MAC flotation cell.

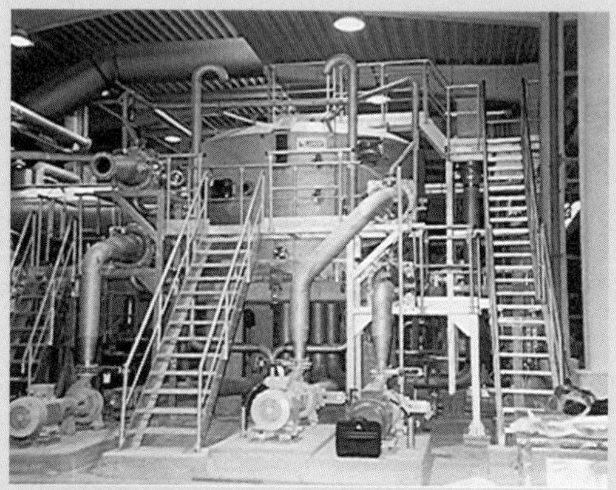

Figure 107. Lamort MAC flotation cell installation.

Shinhama

The Shinhama Hi-Flo cell in Figs. 108 and 109 has a flotation cyclone. Aerated suspension enters the cyclone by natural aspiration through the static mixer. The air bubbles with particles collect by centrifugal force in the center of the vortex flow from which they ascend to the top. Several of these units are connected in series usually with a secondary flotation unit.

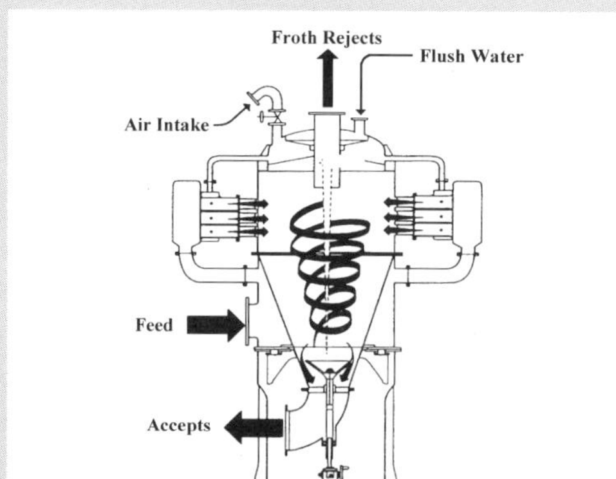

Figure 108. Schematic arrangement of the Shinhama Hi-Flo flotation cell.

Figure 109. Shinhama Hi-Flo flotation cell.

Voith Sulzer

The Voith Sulzer EcoCell of Figs. 110 and 111 combines the Voith E-cell body with the mixing and aeration principle of the Sulzer Escher Wyss CFC cell. The elliptical EcoCell tube contains individual cells in which the suspension is aerated each time it flows through them. Aeration occurs by a self-aspirating static mixer working on the same principle as the step diffuser. The foam is usually treated in a secondary stage.

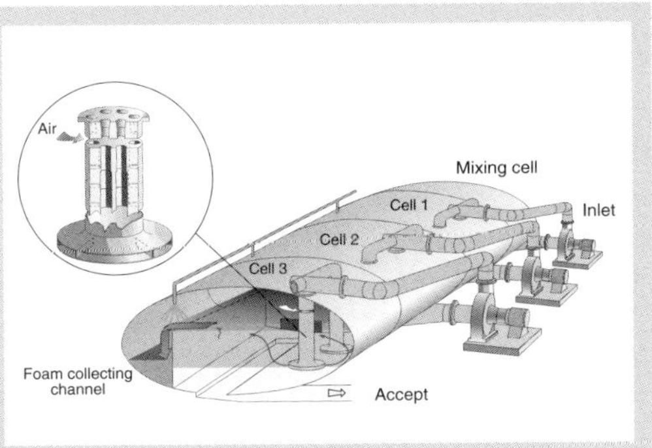

Figure 110. Schematic arrangement of the Voith Sulzer EcoCell.

Figure 111. Voith Sulzer EcoCell in double-decker arrangement.

6.3.3 Foam deaeration

Rejects from the primary cells contain large quantities of foam. To ensure stable operation of the secondary flotation cells, the rejects therefore require deaeration. Figure 112 shows a typical deaeration cyclone with mechanical defoamer. Deaeration makes the suspension easily pumpable and also separates ink concentrate for disposal.

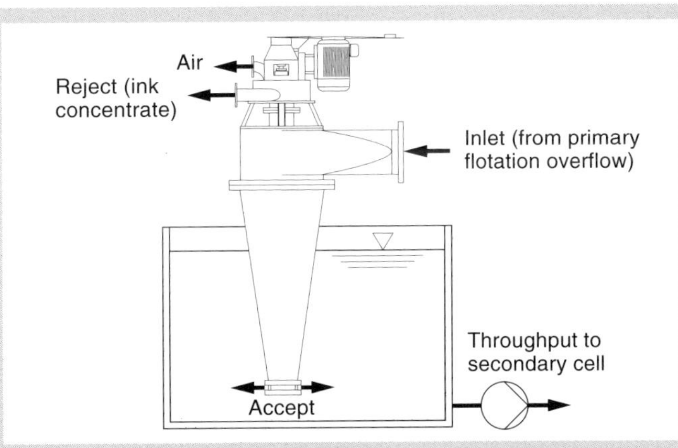

Figure 112. Deaeration cyclone with mechanical foam breaker.

6.4 Assessment of flotation results

Besides ink particles, stickies, and ash, flotation removes a certain amount of fibers and fines. It increases cleanliness by enhancing brightness and reducing dirt specks. In the future, flotation will have increasing use for removing stickies and some ash.

To remove stickies by flotation in the same size range as ink particles (10–250 µm), the stickies must have proper hydrophobic characteristics. Although semi-quantitative data is available only for macro stickies larger than about 100 µm, one can assume by analogy from test results on dirt specks (ink particles larger than 50 μm) and brightness measurements (ink particles from 50 μm down to 3–5 μm, or < 1 μm for water-based inks) that micro stickies smaller than 100 μm are similarly separable.

6.4.1 Main quality criteria of deinked pulp

The two most important quality criteria for deinking (brightness and dirt specks) depend heavily on the relative air volume injected. This is the ratio by volume of air throughput to suspension throughput. For flotation columns operating on the counter-flow principle with one single pass, this is the ratio of air input at the bottom of the column to suspension intake at the top. For cells connected in series, the air inputs in each cell are added and taken as a ratio to the total suspension flow. Flotation results are often expressed according to the number of cells applied.

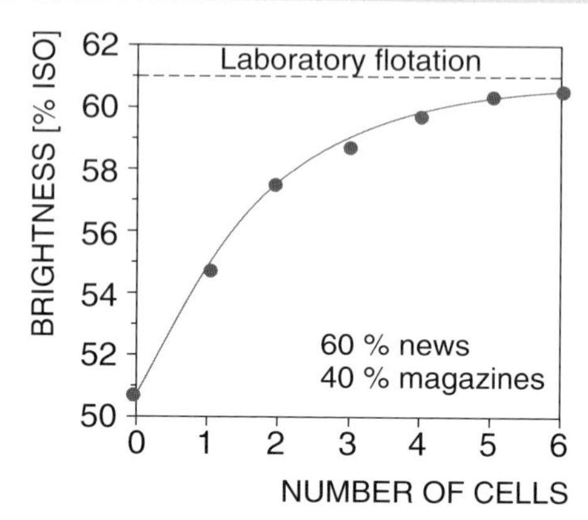

Figure 113. Brightness development vs. the number of flotation cells for a mixture of old newspapers and magazines.

Since optimum results depend considerably on fiber and solids losses during flotation, this is a particularly interesting comparison basis. It enables assessment of widely varying flotation result presentations. The following illustrations explain the principle technical relations involved. Figure 113 shows brightness as a function of cell numbers or of air injection volume if the amount of air sucked in is known. Clearly, the brightness curve flattens with increasing cell numbers or flotation time. This is because the number of ink particles not yet

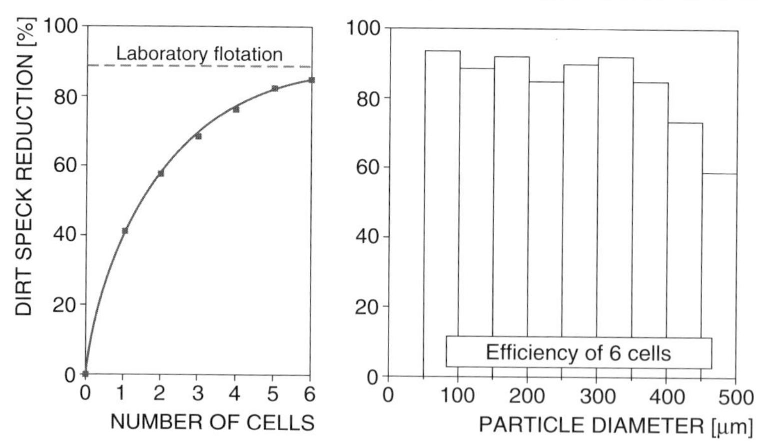

Figure 114. Dirt speck reduction vs. the number of flotation cells.

removed decreases correspondingly. According to the reduced bubble collision probability with constant air input, only few particles can be captured.

The same explanation applies to the dirt speck reduction curve in Fig. 114 that shows how the number of dirt specks removed declines with increasing flotation time and lower dirt speck count. This diagram also illustrates that the flotation efficiency decreases with increasing particle size.

Figure 115 illustrates how brightness depends on foam removal or solids loss. A higher solids removal usually means higher attainable brightness. Brightness can also decrease again as explained below. When assessing or setting the reject flow, realizing that the ink particles alone make up a considerable mass proportion is important. For example, the ink proportion of newspaper mass is about 1.5%, for magazines up to 5%, and for nonimpact printed papers as much as 6%.

Figure 115. Brightness development vs. total solids losses for a mixture of old newspapers and magazines.

6.4.2 Laboratory cells

Attainable flotation effect in terms of brightness increase and dirt speck reduction depends strongly on the recycled fiber pulps entering the flotation cell. To assess flotation cell results for a particular case, standard laboratory tests use the same conditions as the actual cell regarding solids loss so the results are reproducible and applicable in practice.

Several types of laboratory cell are available for such tests. Depending on operating condition variations in stock consistency and fiber losses, results vary accordingly.

Some methods strive to measure the maximum flotation limit that is not easy to standardize. Depending on the recycled fiber pulp such as from recovered paper with a high ratio of mechanical fibers and high fillers content, brightness may even decrease after a certain flotation time. This is because the brightness first increases due to the large number of ink particles removed. When only a few remain, very little additional improvement is possible. Simultaneously, bright fillers and pigments are still being removed so brightness decreases again. It may recover later if the darker mechanical fibers are removed rather than

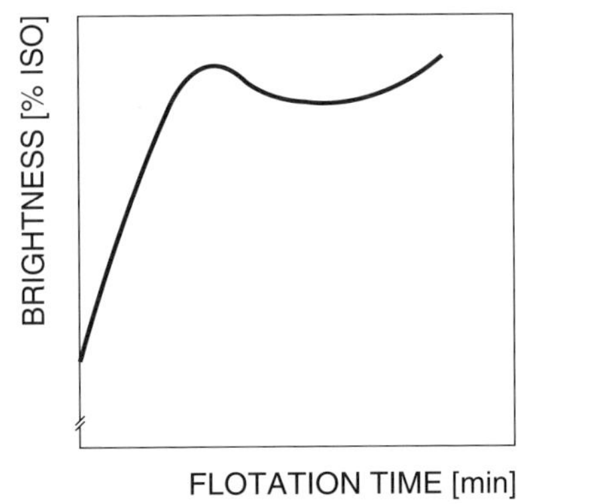

Figure 116. Brightness development vs. flotation time for a mixture of old newspapers and magazines.

chemical fibers. In the end, this depends on the brightness of the primary solids being removed. Fig. 116 illustrates such a case.

With other stock mixtures, brightness can improve continuously even with very long flotation times. This is because unbleached fibers float better than bleached ones. More of them are therefore removed with increasing flotation time. This then enhances brightness.

The same applies to dirt speck reduction. If dirt particles have not yet released from the fibers, they cannot float. If increasingly more fibers with hydrophobic ink particles attached to them float, dirt specks will decrease in the end.

To obtain practical information from laboratory tests, operating parameters such as stock consistency, solids loss, and flotation air volume or flotation time must be realistic and deliver reproducible results. Figure 117 shows a laboratory cell meeting these requirements. Aeration is automatic with a dynamic mixer. The air flow corresponds with conditions in practice. The standard stock consistency is 1%, but this can be higher or lower. The cell is operated batchwise. Foam is removed automatically by a scraper. Solids losses correspond with industrial practice both regarding quantity and composition. This kind of equipment therefore has use for assessing the plant condition and optimizing stock input, chemicals, and flotation time.

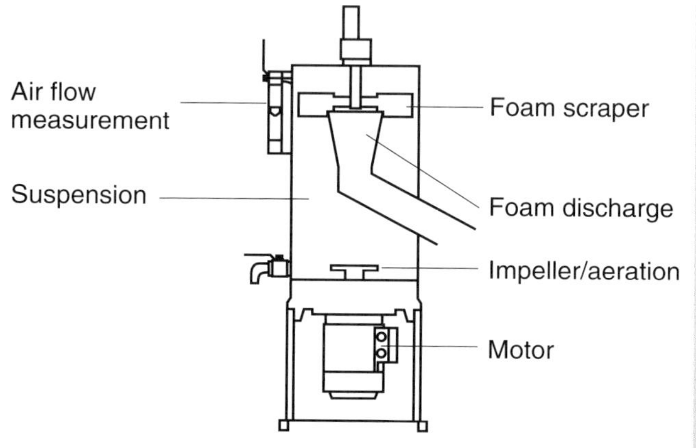

Figure 117. Laboratory flotation cell.

7 Dewatering

7.1 Objectives and system integration

The purpose of dewatering is to remove process water from the pulp slurry and increase the stock consistency simultaneously as Fig. 118 shows. This means separating its solid and liquid components to a certain degree that is given by the stock consistency ratio, c_t/c_i. Dewatering can have the following technical and economical objectives:

- Allow specific kinds of additional processing of the stock such as dispersing in the consistency range of 22%–30% or more
- Making subsequent processes at the production level (Fig. 3) more cost-effective such as bleaching where outlay for chemicals, energy, and space requirements varies widely according to stock consistency (5%, 15%, or 30%). Space also plays an important role in storage outlay.

- Allow making wet laps for economical stock transport or storage
- Achieve the highest possible stock consistencies at discharge level for energy-effective recycling or energy recovery of rejects and sludge
- Separate water loops regarding chemical and contaminant load and temperature at production level (This is done by pressing to high stock consistencies of 30% or more at the end of each loop.).

The section on "Washing" describes the dewatering processes and machinery where solids fractions require separation from one another rather than just the separation of solids from liquids. This includes not only washers but also fiber recovery filters.

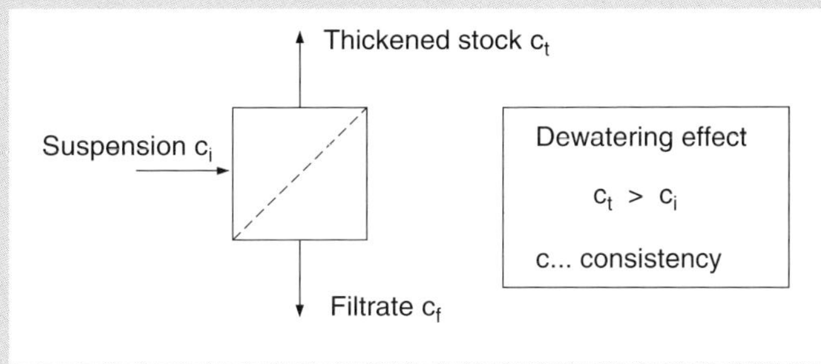

Figure 118. Dewatering terminology.

Figure 91 in the section on "Flotation" shows how dewatering processes integrate into recovered paper processing systems. In Table 10, the tasks, basic processes, and objectives of the two separation processes, dewatering and washing, relate to the typical machines for recovered paper processing.

Table 10. Processes involved in dewatering.

	Dewatering	
Function in process	**Solid/liquid separation**	**Solid/solid separation**
Target	Increasing consistency – enabling next process step – loop separation – web (sheet) forming	Fractionation (particle size) – fines/ash vs. fibers
Unit process	Filtration Pressing	Filtration (Pressing)
Desired result	No solids in filtrate	High particle size separation purity
Examples of machinery	Disk filter Belt filter Twin wire press	Washer Recovery filter

7.2 Principles

Dewatering is a filtration process with mechanical pressing if a further increase in stock consistency is necessary. During filtration, a filter cake or filter mat is accumulating on the filter base such as a wire or screen. The filter mat comprises some solids to be retained from the suspension such as fibers or rejects. Some solids go with the filtrate. This is primarily due to initial solids loss before a filter mat has accumulated on the filter base[64,65].

The classical save-all method can also be used. An auxiliary filter layer first accumulates with long fiber materials before filtering the suspension. A thicker filter mat build-up means the filter base and the initial solids losses have less effect on the overall filtering process and results. Filter bases used in recovered paper processing are wire mesh filters and perforated or slotted metal screens. Porous fine-filtering materials such as ceramics and membranes have use for water purification.

When the filtering process starts, only the filter base itself retains the solids. Since filter bases usually have openings that are larger than a significant share of the particles in the suspension, many particles first pass through with the filtrate. For this reason, the initial filtrate has a high solids content as Fig. 119 shows. Only the large particles are caught on the filter base surface where they start forming an auxiliary filtering layer. Due to the low flow resistance before this layer accumulates, the earlier throughflow rate is high.

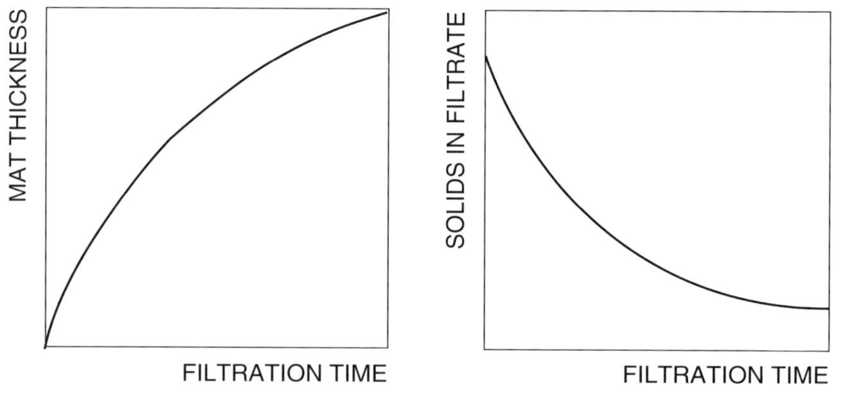

Figure 119. Filter mat thickness and filtrate solids content vs. filtration time.

As the layer accumulates, it increasingly catches smaller particles both on the surface by surface filtration and inside the filter mat by deep-bed filtration. As a result of this decreasing throughput and increasing flow resistance, dewatering performance decreases considerably. Even particles smaller than the available flow channels can be retained on their way through the filter mat by deep-bed filtration if they are retained mechanically or electrokinetically on fibers or other solids they encounter.

Filter dewatering performance primarily depends on the intake stock consistency, the suspension freeness and temperature, and the furnish composition.

The forces involved in dewatering include the following:

- Gravity: Free drainage as a function of suspension height above the filter
- Centrifugal force: Applied to the liquid column height instead of gravity, this increases the dewatering pressure accordingly

- Pressure differential: Created by generating a vacuum behind the filter perhaps with a suction box, pressurizing the suspension before the filter by wire tension over a curved surface, or a combination of both
- Mechanical pressing: A nip between two rolls similar to the press section in a paper machine or with a screw press.

Dewatering without further mechanical pressing is often called thickening. Thickening is also used for dewatering processes through a filter base where no stable filter mat can accumulate because it is continually removed. The suspension is therefore thickened on the filter intake side from where it is removed through the outlet. In some cases for example with certain types of belt filter presses such as sludge presses or screw presses, additional shear forces are induced in the filter mat. These break down its structure and therefore improve dewatering.

Dewatering capacity is throughput as a function of time and unit area. Increasing the suspension temperature improves dewatering capacity by reducing the water viscosity. Stock dewatering characteristics are normally expressed in the paper industry as Schopper-Riegler value (SR) or Canadian Standard Freeness (CSF). Another way is to measure the water retention value (WRV). Dewatering machine design also uses hand filter tests (handsheet tests) that indicate filtrate volume flow as a function of time at constant dewatering pressure.

Table 11 shows the main types of dewatering machines used in recovered paper processing with their separation objectives, stock consistency range at intake and outlet, separation efficiency, and type of dewatering forces involved.

Table 11. Operating conditions of different types of dewatering machines.

Machine	Separation		Consistency range [%]					Separation efficiency Solids/liquids	Dewatering by			
			Inlet		Outlet							
	Solid/liquid	Solid/solid	<2.0	>3.5	5–10	10–20	20–>30		Gravity	Vacuum	Wire pressure	Pressing
Drum filter	■		■		■	▒		◐	■			▒
Disk filter	■		■		■	▒		●	■	▒		
Belt filter	■		■		■	▒		◐	■	▒		
Belt pressure filter	■		■	■	■	▒		◐	■		■	
Twin wire press	■		■	■			■	◐	■		■	■
Screw press	■			■			■	◔	■			■
Belt washer		■	■		■		▒	◕		▒	■	▒

■ *Usual conditions*
▒ *Special application*

● *Good* ◐ *Average* ○ *Poor*

7.3 Machinery and effects

The following sections describe the various types of dewatering machines used in recovered paper processing. These machines are primarily distinguished here according to the type and shape of the filter base. They may comprise perforated or slotted metal screens or wire mesh screens formerly made of metal but now made with plastic. The latter are also differentiated either as disk filters, drum filters, or as continuous belt filters[66,67].

7.3.1 Belt and drum filters

Figure 120 shows the three principal forms of dewatering machines with continuous filter belts. The simplest of these is a belt filter onto which the suspension is fed and dewatered through the wire either by gravity alone or with vacuum assistance under the wire. Attainable dry content with vacuum assistance is about 10%. Without vacuum it is about 5%.

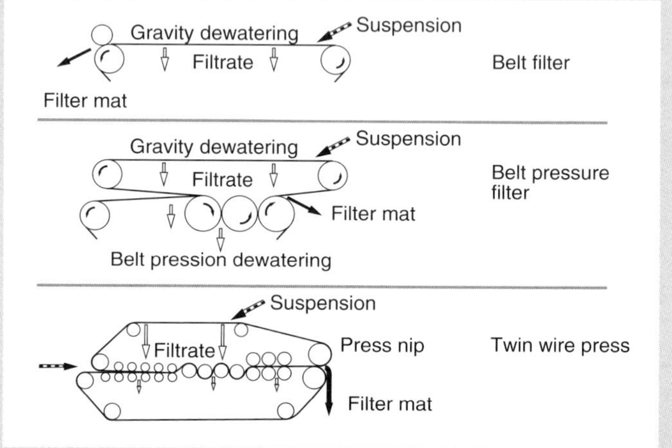

Figure 120. Principle of the belt filter, belt pressure filter, and associated dewatering machinery.

In belt pressure filters, the zone where dewatering by gravity occurs is followed by a zone where pressure is applied for additional dewatering effect. This is done by passing the filter mat between two belts over one or more rolls. Due to relative motion within the filter mat and the pressure caused by tension in the outer belt, the result is intensive dewatering especially with small roll diameters. The dewatering pressure, p, generated by this method has the following definition:

$$p = 2S/D \tag{12}$$

where S is the belt tension, N/mm
D the roll diameter or peripheral surface curvature diameter, mm.

Figure 121. Twin wire press.

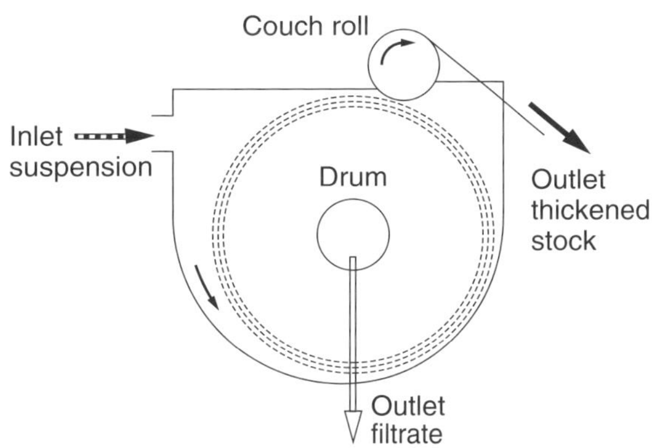

Figure 122. Schematic arrangement of a drum thickener.

A dry content of about 10% is attainable with belt pressure filters.

Belt presses carry the dewatered filter mat from the belt pressure filter zone to the press zone where it is mechanically pressed in one or more nips to a dry content of about 25%–50%. Figure 121 shows a twin wire press as one form of a belt press.

The drum thickener of Fig. 122 is a special kind of belt filter where the belt length is reduced to the circumference of a screening cylinder. Drum thickeners therefore comprise an open roll covered with the filter wire or filter cloth. The rotating drum is immersed in a vat full of suspension that is dewatered as it flows into the drum from the outside. The filtrate is removed from the inside at one end of the drum through a hollow shaft. The dewatering pressure is generated by the differential head between the suspension level in the vat and filtrate level in the drum. The filter mat is removed by a scraper bar or couch roll. The latter serves simultaneously as a press roll and therefore increases the dry content of the filter mat.

7.3.2 Disk filters

Disk filters shown in Figs. 123 and 124 mainly comprise a suspension trough in which numerous hollow disks (up to 34 at about 12 m length) with a diameter of 3.0–5.5 m rotate. They are mounted closely and equally spaced on a hollow shaft. Each disk is covered on both sides with plastic filter fabric. Due to their compact design, disk filters offer a much greater filter area than drum filters or thickeners of the same size. The filter disks are immersed about halfway in the suspension and rotate at about 0.5–2 rpm that corresponds to a peripheral speed of about 5–20 m/min. Due to this rotation, the available filter area is continuously renewed.

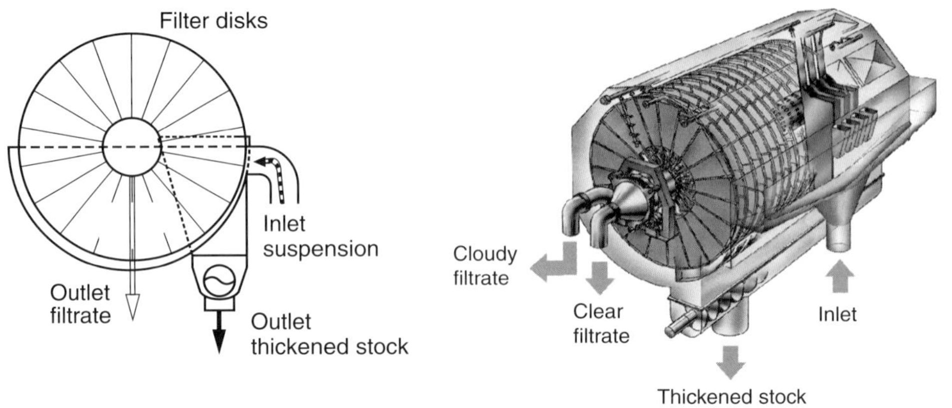

Figure 123. Principle and schematic arrangement of a disk filter.

Figure 124. View inside a disk filter.

Unless vacuum assistance is used, the dewatering pressure depends entirely on the hydrostatic head difference between the suspension level and the filtrate level. The following is the operating principle of the thickening filter. Due to the small head differential, the filter mat is continually removed from the filter screen by flow and gravity forces. As a result, the suspension is thickened in the trough and ejected over a weir due to the pumping effect of the filter disk surfaces. The filtrate has a high solids content due to the continuous filter mat removal and renewal and subsequently shows a greater specific screen throughput.

Dewatering pressure in vacuum-assisted disk filters initially also depends on the hydrostatic head difference. It then greatly increases by applying vacuum (about 0.25–0.5 bar or an absolute pressure of 0.75–0.5 bar) through the hollow shaft. As soon as the disks are immersed, a filter mat starts to form that remains firmly attached to the filter base throughout the dewatering process. The filter mat is removed from the disks in the upper quadrant by knock-off showers.

The stock consistency at the outlet is about 10%–12%. Since the initial filtrate has a high solids content such as 250–500 mg/L for deinking stock, it is generally drained off separately as cloudy or grey filtrate at about 50%–70% of total flow. As the filter mat thickens, the filtrate solids content becomes low at 50–100 mg/L for deinked pulp. This filtrate comprising about 50% to 30% of total flow then has use as shower water.

7.3.3 Screw press

The screw press in Fig. 125 is a primary type of dewatering machine using perforated cylindrical metal screens. It comprises a housing inside which is a perforated or slotted screen cylinder and screw rotor. The screw pitch angle can decrease toward the filter cake outlet, the screw rotor diameter can decrease, or the screw rotor shaft diameter can increase as Fig. 126 shows. As a result, the volume available for the predewatered stock as it is transported from intake to outlet decreases. This increases the dewatering pressure.

As it rotates, the screw rotor continuously cleans the screen cylinder surface simultaneously transporting and reshaping the filter cake. Thanks to this constant renewal of filter surface area, a high dry content and dewatering capacity is possible. Solids loss to the filtrate is high because the filter mat is continuously removed. Screw presses are more compact than belt presses but their specific energy demand and solids content in the filtrate are higher[68,69].

Screw presses also have use for reject dewatering. Figure 127 shows such a machine with the throttling device at the reject outlet open.

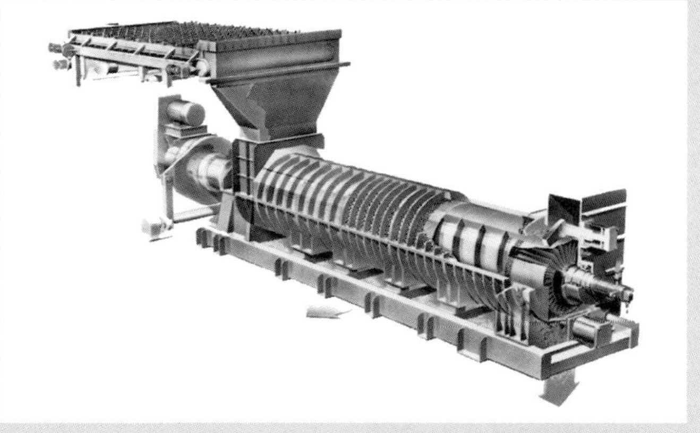

Figure 125. Cut-off section of a sludge screw press with pre-dewatering.

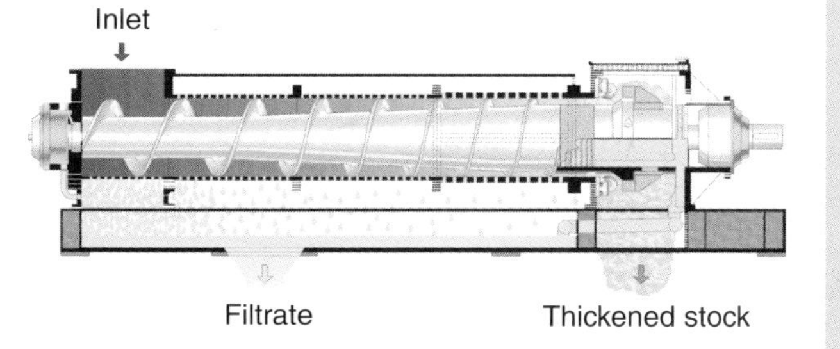

Figure 126. Cross section of a screw press for stock dewatering.

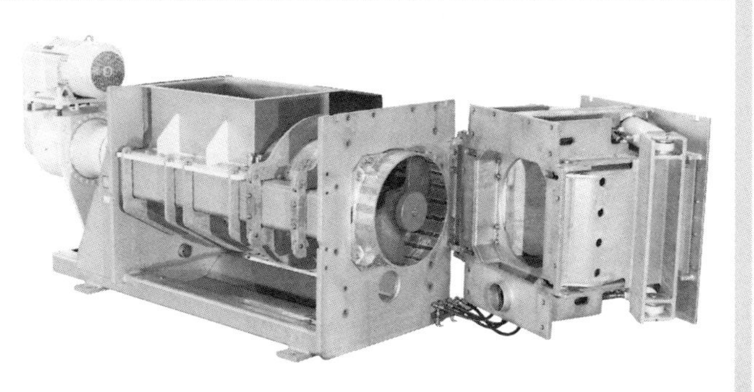

Figure 127. Screw press for reject dewatering with throttling device at the reject outlet open.

7.3.4 Other kinds of dewatering machinery

Another kind of dewatering machine using metal filter elements is the inclined or side-hill screen of Fig. 128. This comprises a curved metal screen with horizontal slots about 0.15–2.0 mm wide. Inclined screens are often vertical with the suspension distributed at the top by nozzles or an overflow weir. Dewatering occurs by flow over the horizontal slots without any filter mat formation. Inclined screens have no moving parts. Their specific dewatering capacity is low. They have primary use for thickening rejects and recovering fibers.

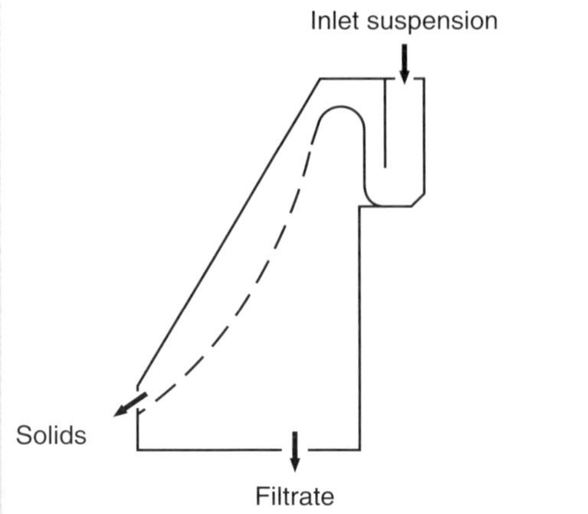

Figure 128. Schematic arrangement of an inclined dewatering screen.

A special version is the vibrating inclined screen. The entire machine is vibrated by a motor with an out-of-balance shaft. This improves screen cleaning and increases the dewatering capacity.

The extractor is another type of machine using a metal filter base. It comprises a suspension tank with a slotted screen cylinder and screw rotor inside. Unlike the screw press, the rotor is not tapered and therefore generates no pressure increase. The suspension is pumped upward on the Archimedian spiral principle. The filtrate passes through the screen perforations in the same way as the screw press. The filter cake is continuously removed. Stock consistencies of about 4%–8% are possible with this type of machine.

Another kind of extractor is used at discharge level (for explanation of discharge level refer to Fig. 3). Floating solids are firstly removed from the suspension via an overflow while the heavy particles are removed by sedimentation. The suspension enters the dewatering area at the bottom and is transported upward by the screw rotor. The water escapes through the screen cylinder perforations leaving the solids that are removed as mentioned.

8 Washing

8.1 Objectives and system integration

In recovered paper processing, washing is a filtration process for removing debris particles smaller than 30 µm from the suspension. Simultaneously, dissolved and colloidal contaminants are removed with the filtrate. The object in both cases is to remove substances negatively affecting the papermaking process or finished product quality. Materials for washing include fillers and coating particles, fines, micro stickies, and ink particles in wash deinking[66,67,70–72]. Dissolved and colloidal contaminants include inorganic and organic substances such as those contributing to COD and anionic trash.

The undesirability of contaminants depends on the raw material characteristics, processing system, and finished product requirements. When washing out contaminants effectively, not removing materials that are valuable for paper production is important. Fiber losses must certainly be avoided. In many cases, the quantity of fillers and fiber fines removed must also be restricted. Washing requirements therefore vary widely according to application. Table 12 shows the different washing tasks involved in processing recovered paper whose quality can vary enormously depending on its collection system. Figure 91 shows a typical washing system layout.

8.2 Principles

In contrast to dewatering that separates solids such as fibers from the fiber suspension, washing primarily separates different kinds of solids. The filtrate acts here as a carrier medium for the fine particles to be removed. Since separation of the dissolved and colloidal substances occurs in the same way during dewatering, this discussion does not cover it. For a better understanding of the washing process, knowing the particle size limits for efficient removal is helpful. Another critical parameter is filter pore size as Table 13 shows.

Table 12. Requirements on washers in recovered paper processing according to their application.

Final product	Desired removal of			
	Ash	Fines	Conventional printing inks	Aqueous printing inks
Newsprint	no	no	yes	yes
SC paper	partially	partially	yes	yes
LWC paper	yes	partially	yes	yes
Tissue	yes	partially	yes	yes

Table 13. Solid size ranges of recycled fiber suspensions and mesh widths of washing equipment.

Solids	Dimension [µm]
China clay	1–2
Kaolin	0.3–5
Coating grit	< 60
Ink particles	0.5–500
Fines fraction	< 200
Short fiber fraction (length)	120–400
Long fiber fraction (length)	> 400
Sieve/wire pore size	10–500

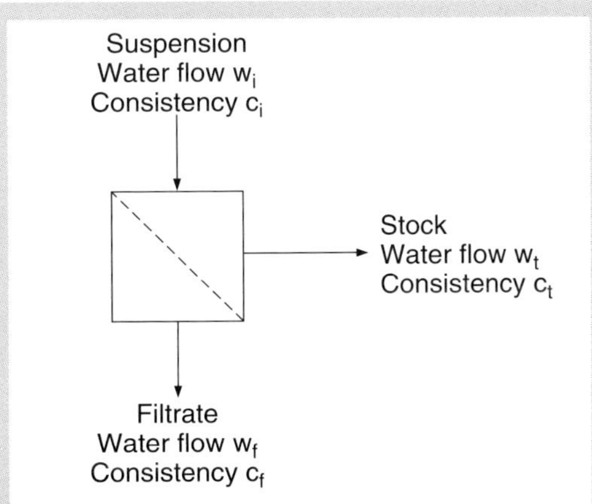

Figure 129. Washer balance of water flow.

Washing theoretically has an ideal separation effect when the quantity of small particles in the filtrate corresponds to the difference of stock water flow, w_t, for filter cake and filter mat and filtrate water flow, w_f, as Fig. 129 shows. The theoretical maximum washing effect therefore depends on the intake stock consistency, c_i, and output stock consistency, c_t. A lower intake stock consistency gives a higher liquid content and allows washing out more fine particles. A higher outlet stock consistency, c_t, of filter mat means more liquid separates from the suspension and more fine particles can theoretically be removed with the filtrate.

Although the following primarily covers filler (ash) separation, the relations involved also apply in principle to the removal of other substances such as ink particles or fines. Figure 130 shows relations in the ideal case where the washing effect corresponds to the ratio of stock water flow (filter mat) to filtrate water flow[71]. Definition of this theoretical maximum ash removal = water removed, W_f / water input, W_i.

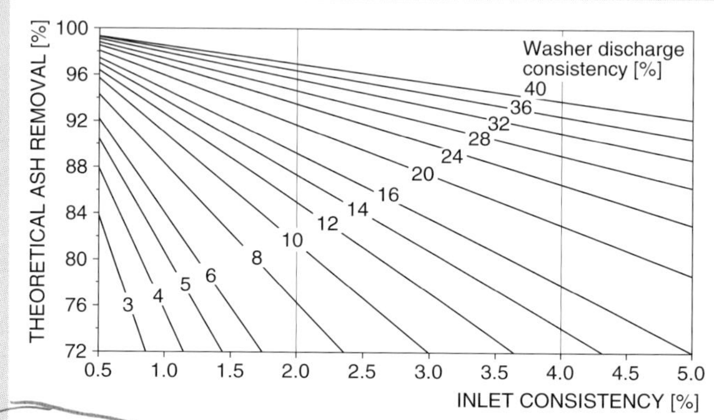

Figure 130. Theoretical maximum washing effect vs. inlet and outlet consistencies.

The washing effect never reaches this maximum because some separable particles are retained in the filter mat. Other solids retention mechanisms are explained in the section on "Dewatering." A separable particle in this context is one that is sufficiently small to pass through the capillaries between solid fibers in the filter mat. In the aforementioned ideal case, such particles flow along the stream lines without catching on the fibers and consequently leave the filter cake and flow with the filtrate. In reality, inertia, gravity, and electrokinetic forces divert the particles from the stream lines as Fig. 131 shows so that they become caught on the fibers. A longer path through the filter mat, i.e., a thicker mat, and finer capillaries give a higher probability of retention.

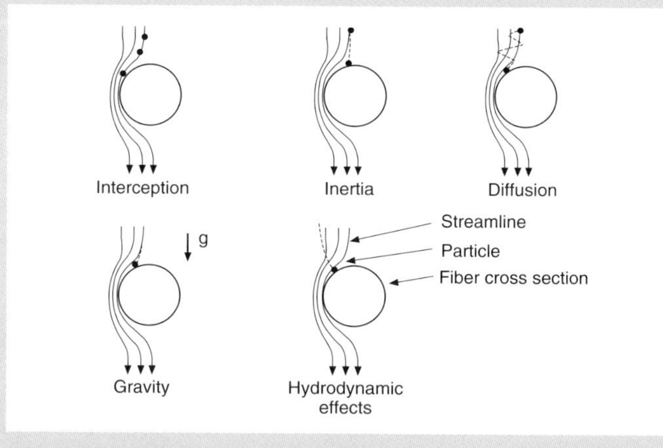

Figure 131. Stream lines and particle flow paths through a filter mat[67].

For separation by washing, the particles must be detached from the fibers. In addition, they must not exceed a critical size or the fiber mat will retain them. Particles with a diameter of more than 30 µm are hardly washed out as soon as a filter mat has accumulated[66]. This is a very important point in wash deinking and processing for coated paper grades. An efficient separation or comminution stage must therefore be inserted before the washing stage such as deflaking or dispersion.

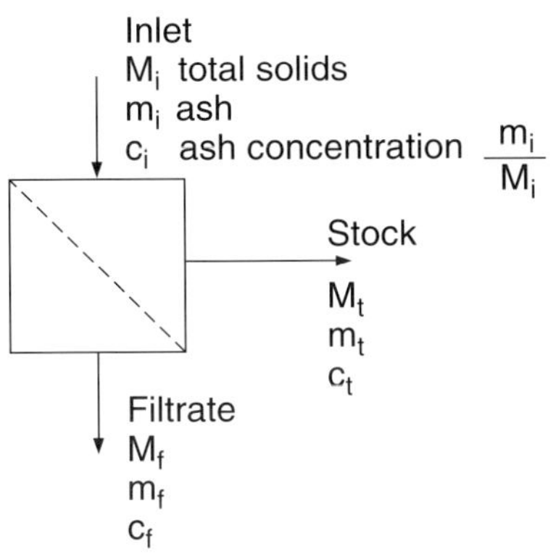

Rel. ash removal $T_a = \frac{m_i - m_t}{m_i}$
Rel. solids removal $T_s = \frac{M_i - M_t}{M_i}$
Ash removal efficiency $\eta_a = \frac{c_i - c_t}{c_i}$

Figure 132. Definitions of ash and solids removal in washers (M and m are expressed as kg/s and M includes m).

Figure 132 explains some important definitions regarding separation effects in the washing process. For ash removal alone: relative ash removal = ash removed/ash input.

For washing out fine organic components and ash: relative suspended solids removal = suspended solids removed/suspended solids input.

Suspended organic solids washed out unintentionally with the filtrate represent a reduction of yield. Besides the ash removal rate, another important criterion for washing efficiency is therefore the amount of other solids in the filtrate and especially the loss of fibers and fiber fines.

A separation diagram can represent the washing process and its results as Fig. 133 shows. The section on "Screening" describes this in greater detail. The solids mass flow ratio is defined as follows:

$$R = (M_f/M_i) \cdot 100\ [\%] \tag{13}$$

where R is the solids mass flow ratio
M_f the solids mass flow in the filtrate
M_i the solids mass flow in the inlet.

An important separation requirement is met if the long fiber and short fiber fractions are completely removed with the stock mass flow, M_t. The fines often should remain completely in the stock, too. They do this to a higher degree only with sizes larger than about 30 µm. Only the very smallest fines fraction, the ash particles and colloids are ideally distributed in equal concentration (referred to water split) in the feed suspension and filtrate. In practice, even these ultra-fines are retained to a certain extent in the stock. Finally: ash removal efficiency = (ash concentration of inlet suspension – ash concentration of washed stock)/ash concentration of inlet suspension.

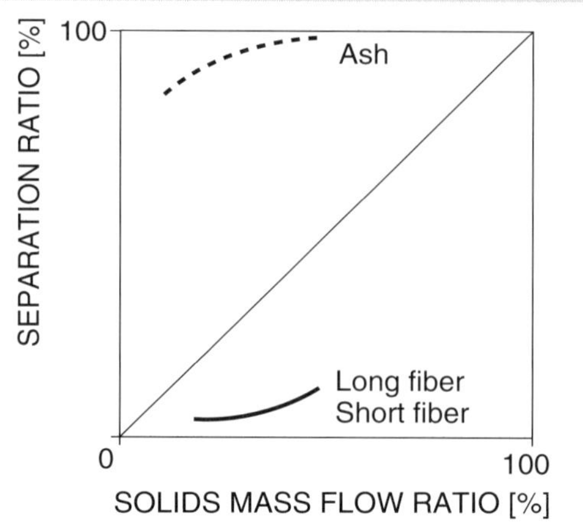

Figure 133. Separation ratio diagram for the filtrate in stock washing.

Washing (separation of defined solids fractions) is distinguished from dewatering (separation of water) by the following definition. In washing systems, the solids contained in the filtrate are subsequently separated, usually by dissolved air flotation (DAF) and removed from the system. In simple dewatering, the filtrate is recirculated. Or the solids in the filtrate are also separated in DAF units, but are reused in the process.

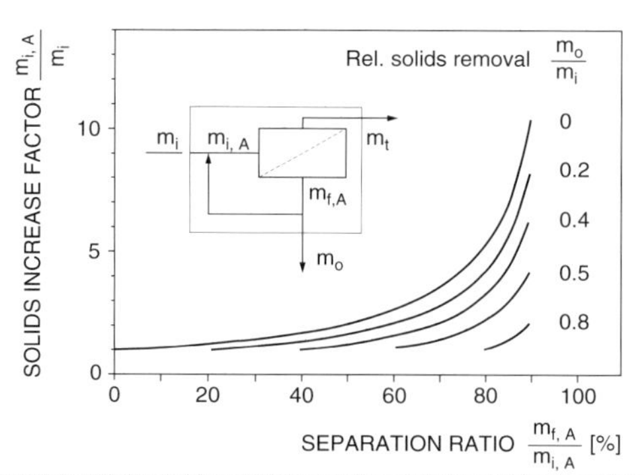

Figure 134. Solids increase factor at the washer inlet affected by washing efficiency and relative solids removal from the system.

With thickening, the filtrate should have the lowest possible solids content. In washing, a defined content of fine particles is necessary. Recirculating unpurified washing filtrate would increase solids content at the washer intake to a limit depending on the washing efficiency as Fig. 134 shows. In other words, a high washing efficiency would cause a very high solids content, $m_{i,A}$, at the point of washer inlet. As the diagram shows, this can be reduced by extracting some solids content from the system circuit. In this connection, the relative system overflow is m_o/m_i = ash mass flow of filtrate removed/ash mass flow of inlet suspension.

In many cases, defined ash quantities of different levels must be removed from the system. This is possible with the following:

- Exploiting the ash separation adjustment range of the different washers employed

- Washing only some suspension at production level
- Purifying only some filtrate.

These methods only have use to a very limited extent for wash deinking. This requires entire removal of ink particles from the filtrate for optimum brightness. To minimize fiber fines and ash removal, the filtrate can be cleaned by selective flotation. This is a method used for removing water-based inks[72].

8.3 Machinery

Washing machinery for separation of defined solids fractions has use for the primary suspension flow at the production level (See section on "General overview"). At the recovery level, washing has primary use for fiber recovery. The machines for washing comprise belt filters, belt pressure filters, disk filters, static filters such as inclined and curved screens, and pressure screens[66,74,75] as described in the section on "Dewatering." Figure 135 shows the washing efficiencies of these different machine types for total solids separation and ash removal for a recycled fiber stock with an ash content of 15%–17%.

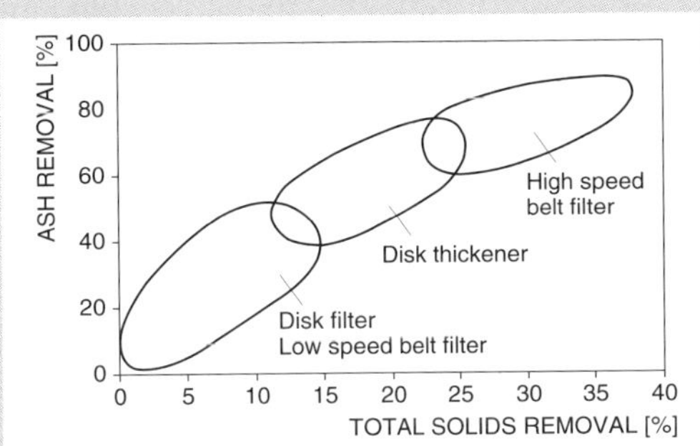

Figure 135. Classification of different washing principles according to total solids and ash removal.

8.3.1 Belt filter type machines

One type of belt filter is the gravity table. This is like a fourdrinier in which gravity alone dewaters the suspension. The washing effect depends on the amount of particles not retained similar to the wire section of a paper machine.

Among the fast-running belt pressure filters are the VarioSplit[70] and the Double Nip Thickener (DNT). The suspension is dewatered under pressure by the wire that runs with the stock over one or two rolls.

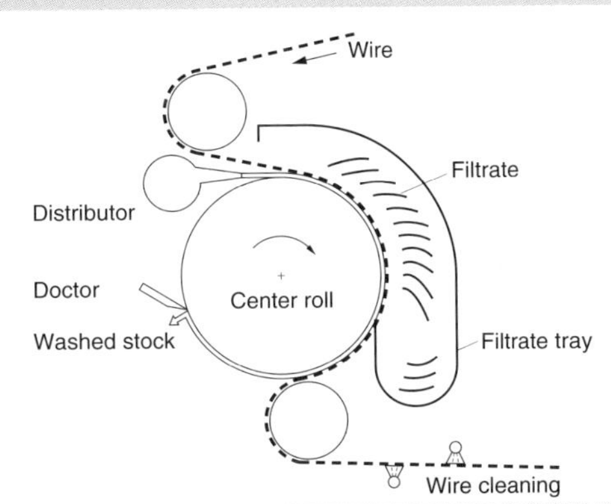

Figure 136. Dewatering principle of a belt filter type washer.

Figure 136 shows the operating principle of the VarioSplit. The suspension is injected by a nozzle into the gap between the wire and a closed dewatering roll with diameter of D. The dewatering pressure, p, is as follows:

$$p = 2S/D \quad (14)$$

The wire tension, S, generates the pressure as explained in the section on "Dewatering." The filtrate containing the particles washed out is then purified. The dewatered stock leaves the roll at a consistency of about 5%–10%. Figure 137 illustrates the schematic layout of this machine with two dewatering units that is shown in practice in Fig. 138.

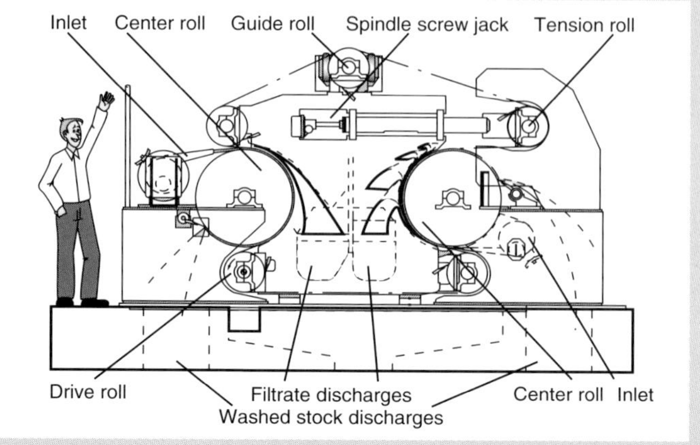

Figure 137. Schematic arrangement of a VarioSplit with two dewatering units.

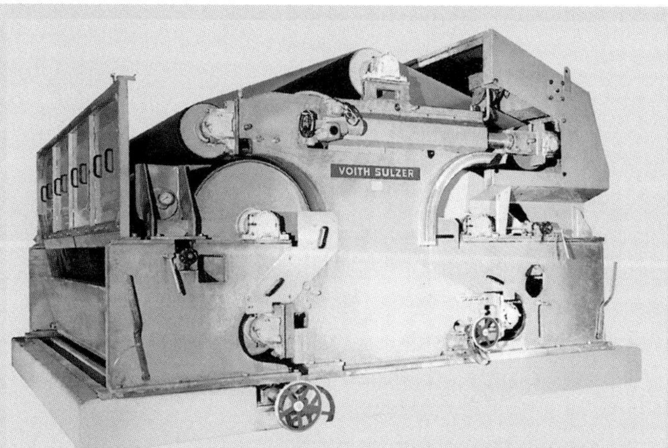

Figure 138. VarioSplit.

As Fig. 139 shows, the ash removal efficiency and solids losses for a given recovered paper mixture depend considerably on the basis weight of the web of the washed stock. Varying this parameter therefore allows adjustment of the washing effect over a wide range. The difference between the two curves is the organic component removal rate. This rate increases slightly more steeply with decreasing basis weight than the ash removal rate. This is due to the fact that retention of larger particles in the fines fraction decreases rapidly with decreasing filter mat thickness. Due to their small size, ash particle retention is low even at higher basis weights.

The VarioSplit operates at speeds of 350–1 000 m/min and inlet stock consistencies of about 0.7%–1.5%. With working widths from 1 x 2.5 m to 2 x 4.5 m, throughputs from about 115–500 t/d are possible.

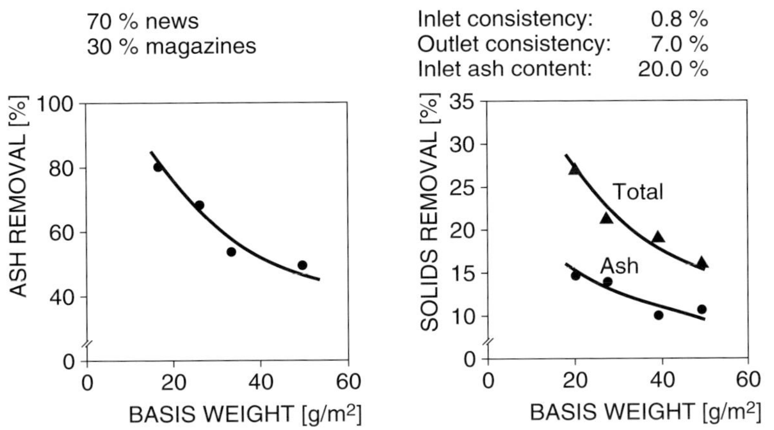

Figure 139. Ash and solids removal with the VarioSplit vs. basis weight of the washed web.

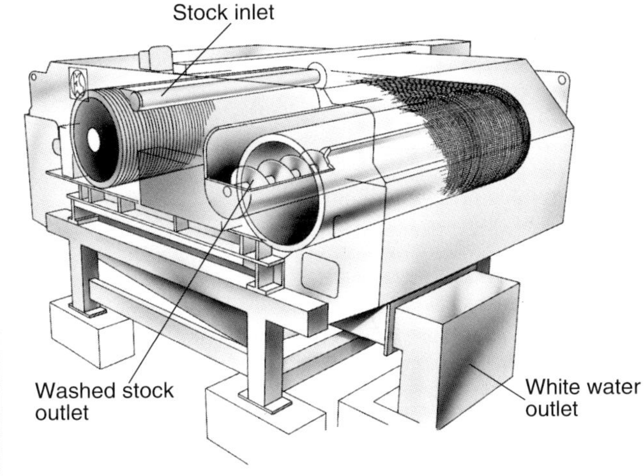

Figure 140. Schematic arrangement of a double nip thickener.

Figure 140 shows the schematic layout of a Double Nip Thickener in which the pulp slurry is also injected between the peripheral wire and the first dewatering roll. In contrast to the smooth VarioSplit roll, this unit has peripheral grooves. The pre-dewatered suspension then travels on the wire to the second dewatering roll that is smooth. Its final dry content here increases to about 5%–10% before removal. The relations shown in Fig. 139 between de-ashing efficiency, wire throughput, and web basis weight also apply in principle to the Double Nip Thickener. The production output of a unit is about 120-300 t/d depending on furnish and feed consistency at an operating speed of 800 m/min and working width up to 2 m.

The Fiber Roller also belongs to the belt filter family. In this machine, the suspension is injected between the stationary outside wire directly connected to the nozzle and a suction roll with chambers at various vacuum levels. The stationary wire is sealed after the nozzle outlet with a plastic cover so that dewatering can only occur inwardly. At stock consistencies of about 5% and above, shear forces between the stationary wire and rotating dewatering roll generate a fiber rolling effect. Vacuum is applied that causes additional dewatering. Stock consistencies of 20% and more are therefore possible. With an optional pressing roll, they can exceed 30%. The operating speed of the Fiber Roller is 100–200 m/min with a dewatered basis weight of 25–65 g/m^2.

8.3.2 Disk thickeners

Disk thickeners belong to the disk filter family but operate without vacuum[75]. Besides thickening at the production level, they also have use for fiber recovery in recovered paper processing lines. They are especially useful in cases where only a slight de-ashing or washing effect is necessary. In fiber recovery, the suspension has only a very small fiber content. The filter mat is corresponding thin. The filter wire fineness is therefore very important.

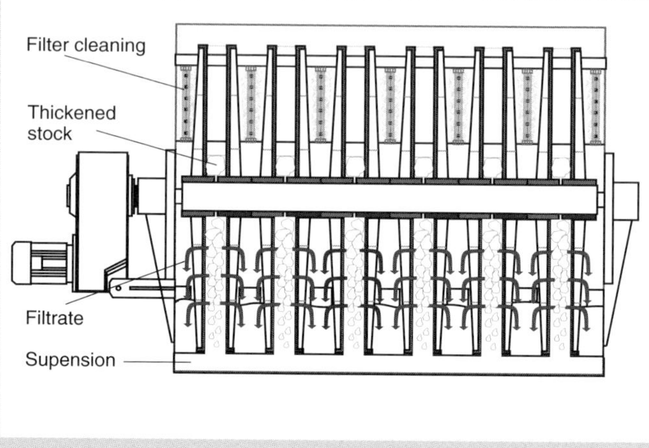

Figure 141. Schematic arrangement of a disk filter for suspension thickening, fiber recovery from a suspension, or for ash and fines removal.

Figure 141 shows the schematic layout of a disk filter used for suspension thickening, fiber recovery from a suspension, or for ash and fines removal. Speed variation can adjust the de-ashing effect. The dewatering efficiency depends on the hydraulic pressure differential between the inlet suspension level and the filtrate level. The suspension is dewatered through a fine-mesh screen. Material retained is removed by disk rotation on the opposite side of the intake.

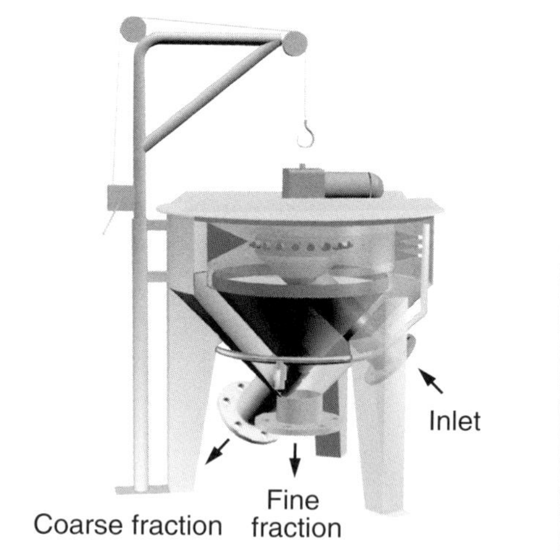

Figure 142. Schematic arrangement of a spray filter for fiber recovery.

8.3.3 Spray filters

The machine shown schematically in Fig. 142 operates on the spray filtration principle[73]. The suspension is sprayed at high velocity from a horizontal nozzle ring onto the rotating fine-mesh cylindrical screen. The high dynamic pressure forces a large quantity of suspension through the screen. Throughput and washing effect can vary as a function of screen mesh and jet characteristics. Since these parameters have a mutual connection, the washing effect adjustment range has corresponding limitations. One or more spray nozzles flush off the fibrous material retained on the screen surface.

8.3.4 Pressure screen type washers

The Dynamic Washer of Fig. 143 is essentially an extremely fine pressure screen[74] with holes or slots of about 0.2 mm. The bump rotor used in these machines has many hemispherical projections on its cylindrical surface. In contrast to conventional screens, the fibers are retained on the screen basket. Only ash, fines, and a very small proportion of short fibers can pass through the small perforations.

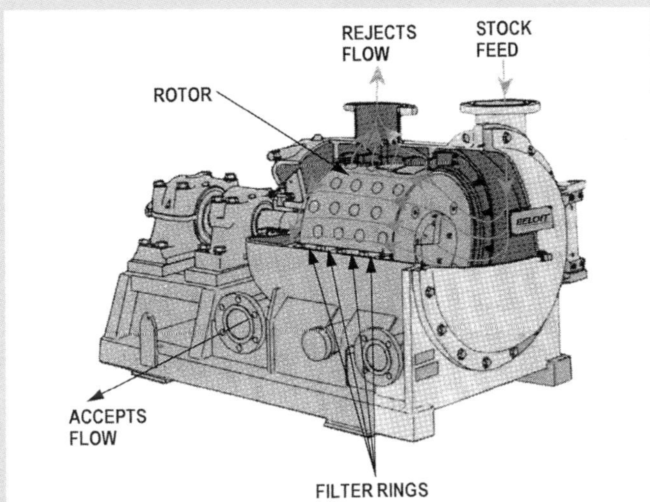

Figure 143. Schematic arrangement of a Dynamic Washer.

9 Dispersion and kneading

9.1 Objectives and system integration

Demands on dispersion machinery vary considerably depending not only on the recycled fiber furnish quality but also on the final product. The primary tasks of a dispersion system are as follows:

- Reduce dirt specks below the visibility limit and distribute them finely or make them floatable
- Break down stickies and distribute them finely or make them floatable
- Distribute wax very finely
- Break down coating and sizing particles
- Detach ink or toner particles from the recycled fibers
- Mix in bleaching agents
- Treat fibers mechanically for retaining or improving their strength characteristics
- Treat fibers thermally to increase bulk
- Ensure decontamination of microorganisms.

Table 14 shows how these various dispersion tasks relate to the paper grades produced from recycled fibers. The table covers white and brown grades.

Table 14. Dispersion requirements for paper grades produced from recycled fibers.

Grades / Task of dispersion	Newsprint	SC paper	LWC paper	Tissue	Test liner	Board (Filler)	Board (Top liner)
Dirt specks and stickies dispersion	■	■	■	■	■	■	■
Wax dispersion					■	■	
Coating grits dispersion		■	■	■			■
Ink/toner detachment	■	■	■	■			■
Bleaching agents/ stock mixing	■	■	■	■			■
Strength improvement		■	■	■	■		
Bulk increase				■		■	
Microbial decontamination				■	■	■	■

Dispersion does not remove contaminants (ink and toner particles and stickies) but breaks them down to a size where they usually no longer interfere or detaches them from the fibers if necessary to remove them in a subsequent process. This is the only way dispersion can be a method for reduction of stickies or dirt specks[76–78].

For solving problems with stickies or increasing the optical cleanliness of brown grades, dispergers can often be replaced by fine screens with slot widths as low as 0.15–0.25 mm. This saves energy and avoids excessive loop water heating.

Dispersion involves application of high shear forces to the fibers and the debris particles to be dispersed. To transfer the required amount of dispersing energy, the stock consistency must be about 22%–30%. Although this involves considerable investment and operating costs, it also provides some advantages. The dispersion should always be located at the point of water loop separation because this is where the pulp must be thickened to high consistency. Bleaching at high stock consistency and especially oxidative bleaching provides technical, operational, and economical advantages. The disperger can therefore have simultaneous use as a mixer for bleaching agents.

If demands on the finished stock are especially high, dispersion occurs at two points in the system. Sometimes, the same type of disperger has use. For other cases, a high-speed disk disperger in combination with a slow-running kneading disperger has use[79–82].

9.2 Principles

9.2.1 Particle shear strength

During dispersion, shear forces must be sufficiently strong to break down contaminants and distribute them finely. The forces must also remove particles adhering to the fibers.

Particle shear strength is therefore an important consideration. Figure 144 shows laboratory shear test results on an adhesive at 20°C, 50°C, and 80°C and on a toner at about 80°C as a function of test shear velocity[83].

Although shear velocities in the disperger are many times higher than test shear velocities, one can assume that shear strength test results also apply at higher speeds or at least indicate the relative shear strengths of different particle types. These test results refer to shear strength in the dry condition. Wetting considerably reduces this. For the materials tested here, wet shear strength is only about 30%–50% of the dry shear strength.

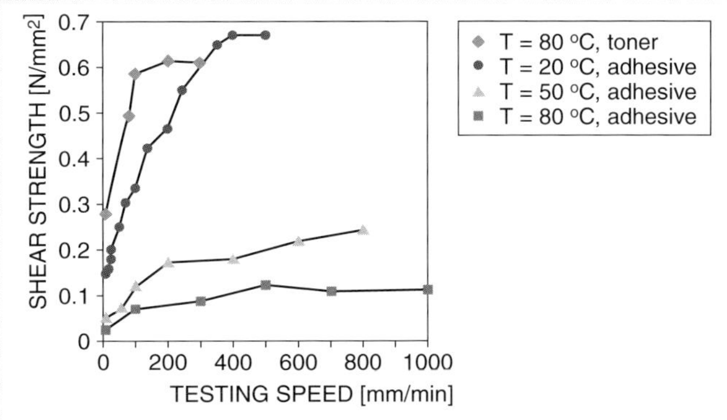

Figure 144. Shear strength of adhesive and toner (dry) vs. testing speed [83].

9.2.2 Dispersion mechanisms

Dispersion mechanisms in disk and kneading dispergers are different due to their different internal components and operating speeds. No generally valid model for dispersion mechanisms is available for either type of disperger. The following explanations may help clarify the dispersion processes occurring in these machines.

Disk dispergers

Fillings in a fast-running disk disperger have intermeshing teeth or refiner-type bars. The following analysis of this mechanism only concerns teeth as fillings. Here, the stock is forced radially through the small channels between the teeth and through the shear

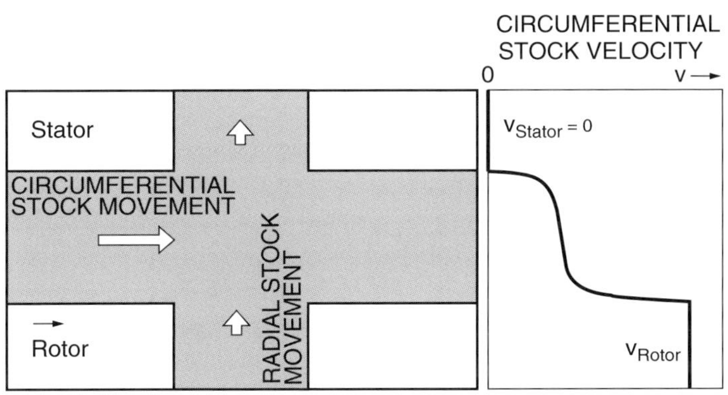

Figure 145. Model of the process in a high-speed disk disperger.

zones between rotor and stator. In Fig. 145, only two of the rings with teeth are considered. These are a rotor ring and an adjacent stator ring surrounding it. The stock is pumped through the fillings radially to the outside by the centrifugal force generated by the rotor. This fills the intermediate space between rotor and stator. Only when this intermediate space is completely full is the stock forced through the stator channels and only to the extent as it is replenished by the rotor action.

Stock leaves the rotor filling at peripheral velocity and impinges on a slower stock stream turning in the intermediate space. This is braked by the stator wall and stator channel entry. No complete analysis of the velocity profile between rotor and stator is now available.

The main dispersion effect occurs at the rotor outlet and stator inlet. This is due to the high shear forces there and the impact of the particles on the edges of the fillings. All particles are forced through the high shear force zone where intensive fiber-to-fiber and fiber-to-particle friction occurs. Particle impact on the edges of the fillings is more a matter of probability.

Kneading dispergers

The rotor fillings of kneading dispergers often run at a peripheral speed of about 5–15 m/s. This is slow compared with disk dispergers at 50–100 m/s. In some kneading machines, the speed is even lower. In addition, the rotor to stator element spacing of typically 10 mm is much wider than in disk dispergers where it is 1 mm or less. Efficient dispersion still demands shear forces as high as in disk dispergers, i.e., a high velocity gradient in the stock. How such high shear forces with intensive fiber-to-fiber friction occur in kneading dispergers is unclear. Adequate resistance against rotation or free movement of the stock for instance at the stator side seems important. This occurs to allow kneading and shear action.

Mechanisms in the kneading disperger seem to be very different from those in the disk disperger. This influences technical results. One can assume that stock treatment by the rotor elements in kneading dispergers is primarily by chance even if frequently repeated. This is possibly because the stock is not forced through an adequately high shear force zone.

9.3 Machinery design

9.3.1 Dispersion subsystem

In all dispergers, the amount of energy that can apply to the stock depends on the stock viscosity, temperature, type of fillings, and peripheral speed. The stock consistency necessary for efficient dispersion is ensured by dewatering in screw presses or twin wire presses. These machines are described in the section on "Dewatering."

To ensure the right dispersing temperature, the stock requires pre-heating. This normally uses direct steam heating that reduces the stock consistency by a few percent due to steam condensation. This must be considered before setting the dewatering stock consistency. Various types of heating screws have use operating at different rotational speed with steam injected to condense on the stock to increase stock temperature and with paddles or blades for stirring and transporting the stock. The crumbled stock should be continuously broken down to provide the largest possible condensation surface area.

At temperatures above 100°C, the heating and dispersion system operates under pressure. The stock then must be forced by plug screw feed into the pressure zone. More steam is necessary for heating above 100°C. Its thermal content that can transfer into mechanical energy is higher than at atmospheric pressure. To save costs, dispersion under pressure should therefore have use only in cases where the stock quality makes this indispensable. The effects of high temperature on the requirements on stock quality and especially all strength characteristics require consideration[84]. The effects on chemical and mechanical fibers are different.

Considering the high consistency of the stock and the high specific energy input, the stock temperature in a kneading or disk disperger can rise significantly.

Subsequent processes define which stock consistency and pressure conditions are necessary after the dispersing zone. This includes HC or MC bleaching or dilution to 5%. Figure 146 shows a typical dispersion system with dewatering screw (screw press), heating screw, and disperger.

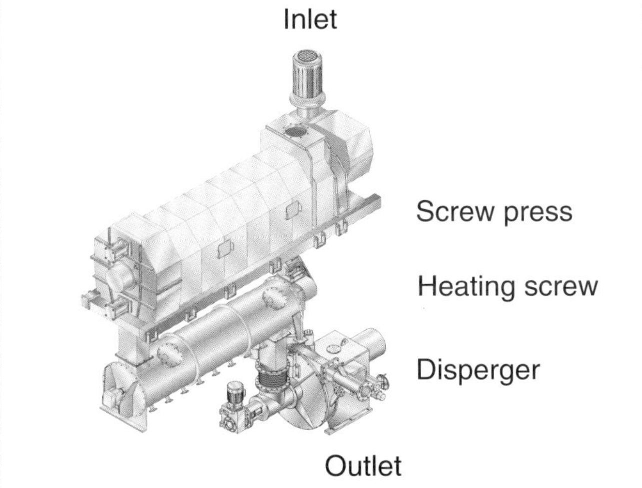

Figure 146. Dispersion system with dewatering screw, heating screw, and disperger.

Disk dispergers

Disk dispergers are compact machines that can have various kinds of fillings. Figure 147 shows a typical disk disperger design with central feed screw, stator and rotor disk fillings, and housing with outlet. This type of disperger normally uses cast fillings. Figure 148 shows a typical example.

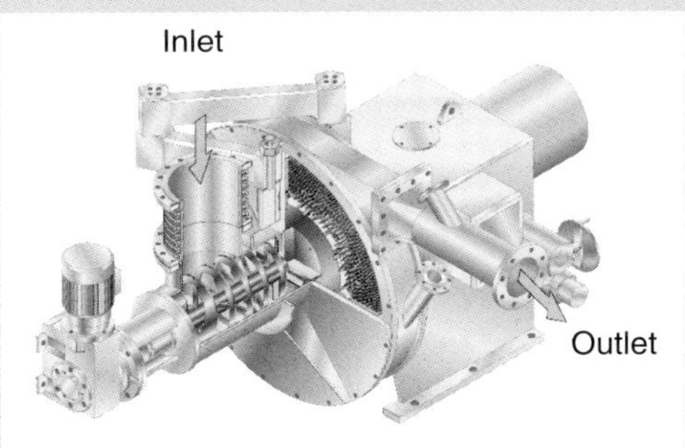

Figure 147. Disk disperger.

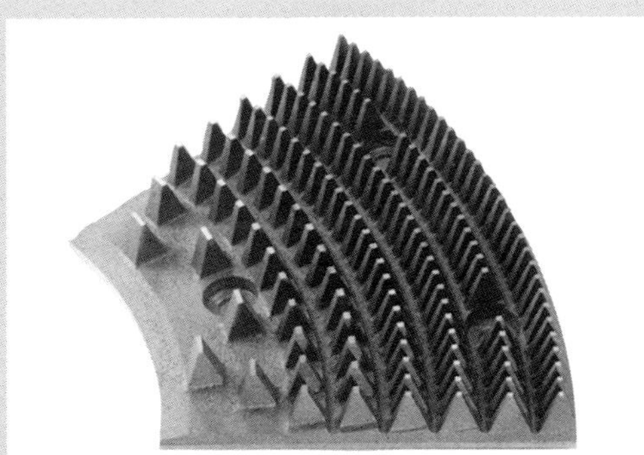

Figure 148. Disk disperger fillings.

The amount of dispersion energy applied is adjusted approximately by varying the inlet stock consistency. An important consideration is that the dispersion effect depends not only on the specific energy demand but also on the stock consistency to a degree at least for "difficult" particles. Fine adjustment and control of the specific energy demand occurs during operation by varying the axial spacing of the fillings. This spacing also requires adjustment to compensate for operational wear.

Kneading dispergers

Kneading dispergers or kneaders have a horizontal tubular housing with a stock feed chute at one end and an outlet at the other. Inside the housing are one or two rotating shafts fitted with kneading elements that also act as impellers. The kneading chamber has resistance elements.

Figure 149 shows an example of a single-shaft kneading disperger having a rotor with rhomboidal impellers inside a stator casing with cylindrical resistance elements. Regulation of specific energy demand on this type of machine occurs by throttling the outlet with a damper valve or by varying the discharge screw speed.

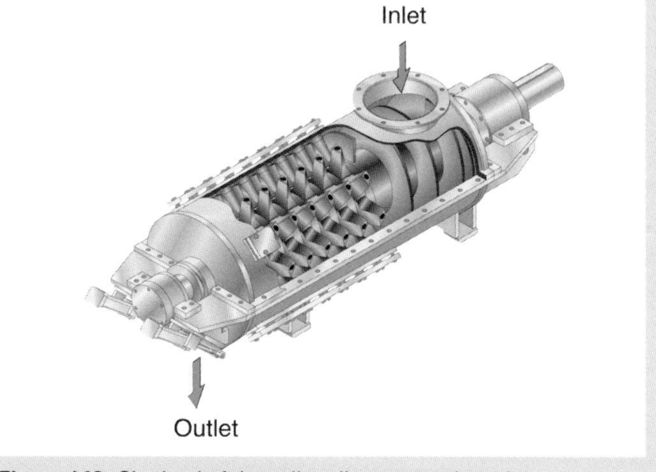

Figure 149. Single-shaft kneading disperger or kneader.

Figure 150. Twin-shaft kneading disperger or kneader.

In twin-shaft kneading dispergers, the impellers on the two shafts may provide any additional resistance required for intensive kneading. The twin-shaft kneading disperger in Fig. 150 shows the impellers on the counterrotating shafts. In the same way as single-shaft machines, power demand is regulated by throttling the discharge opening such as with a damper valve.

9.4 Dispersion results

The following results show the principle relations applying to dispersion. They also apply to disk or kneading dispergers. The next section discusses any slight differences.

Figure 151 shows the key influence of specific energy demand on reduction of stickies. Clearly, a minimum specific energy demand exists above which dispersion reduces almost all stickies below the visible limit[29].

At constant specific energy demand, reduction of stickies improves with increasing stock consistency. High temperature improves the result due to the fact that particles at softening temperature are easier to disperse. Peripheral velocity has no additional effect on dispersion of stickies after reaching about 50 m/s. High velocities cause greater wear and tear[85,86].

Figure 152 shows how specific energy demand influences dirt speck reduction[76,82,86]. High specific energy demand again brings improved results but only up to about 120 kWh/t. In some special applications in ink detachment, a specific energy up to 150 kWh/t is desirable.

Mechanical loading at excessively high stock consistency can cause fiber crinkling in kneading dispergers[87].

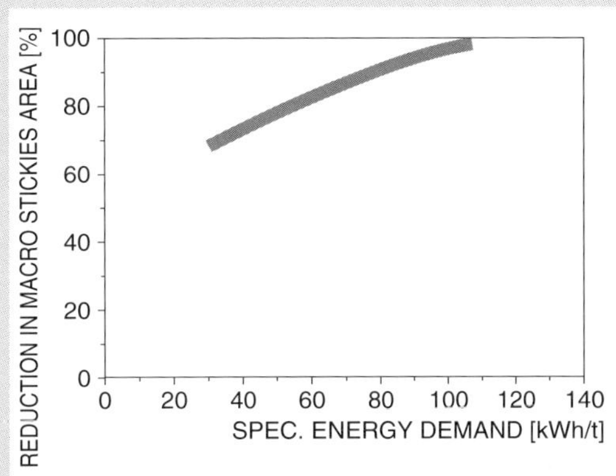

Figure 151. Reduction of macro stickies area in a disk disperger vs. specific energy demand.

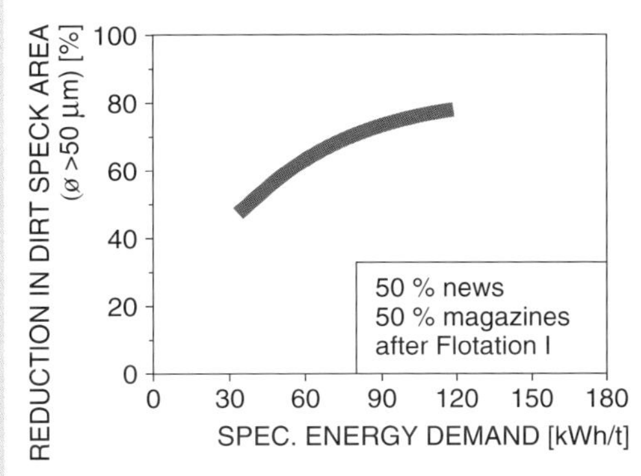

Figure 152. Reduction of dirt speck area in a disk disperger vs. specific energy demand.

9.5 Applications of disk and kneading dispergers

Table 15 summarizes the key characteristics of disk and kneading dispergers. The effects of the two disperger types are very similar. For some applications, one type may be preferable. Since practical applications usually involve several different goals regarding product quality and processing, the best type of machine depends on individual conditions. Each type of machine from different suppliers also comes in various versions with varying performance.

Table 15. Key data for disk and kneading dispergers.

<table>
<tr><th colspan="2">Type of dispersion machine
Criteria</th><th>Disk disperger</th><th colspan="2">Kneading disperger</th></tr>
<tr><td rowspan="2">Circumferential speed difference rotor/stator</td><td rowspan="2">m/s</td><td rowspan="2">50–100</td><td>7–13</td><td>0.5–1.5</td></tr>
<tr><td>1-shaft</td><td>2-shaft</td></tr>
<tr><td>Specific energy demand</td><td>kWh/t</td><td>50–80 (150)</td><td colspan="2">30–80 (120)</td></tr>
<tr><td>Stock consistency in</td><td>%</td><td>22–35</td><td colspan="2">22–35</td></tr>
<tr><td>Stock consistency out</td><td>%</td><td>4/15/35</td><td colspan="2">25–35</td></tr>
<tr><td>Temperature at entering disperger</td><td>°C</td><td>60–130</td><td colspan="2">40–95</td></tr>
</table>

( ) in special cases

Table 16. Typical operating data of disk and kneading dispergers.

Type of dispersion machine Criteria	Disk disperger	Kneading disperger
Positioning in systems with only 1 dispersion stage	Before post-flotation	Before post-flotation
Positioning in systems with 2 dispersion stages	Close to end of system prior to pulp storage	Before flotation and washing
Impact on freeness (CSF) - with heating - without heating	 Moderate decrease Strong decrease	 No impact No impact
Impact on strength - with heating - without heating	 Moderate increase Strong increase	 No impact No impact
Impact on dirt speck reduction (inks)	Good effect, especially with conventional inks	Good effect, especially with non-impact inks/toners
Macro stickies reduction	Good effect	Hardly any effect, leaves large stickies screenable

Disk dispergers generally reduce stickies more efficiently than kneading dispergers that tend to leave more stickies untouched and easier to screen. Disk dispergers also disperse conventional printing inks more efficiently. Kneading dispergers have certain advantages in treating toner particles from nonimpact inks.

The generally steeper decrease in freeness (increase in Schopper-Riegler value) achieved with disk dispergers gives improved strength characteristics. If this is not important, using a kneading disperger that leaves freeness practically unchanged may be better.

Kneading dispergers can operate unheated at normal process loop temperatures without noticeably decreasing freeness. Fast-running disk dispergers without heating rapidly decreases freeness and increases the Schopper-Riegler value. This operating mode has use for HC refining of recycled fibers.

Table 16 lists typical operating data of disk and kneading dispergers. The data primarily use a direct comparison between one supplier's disk and kneading dispergers but do also include test results from other suppliers.

10 Refining

10.1 Objectives and system integration

The purpose of refining in recycled fiber processing is to improve the main properties of recycled fibers[88–97]. Many of these fibers have high strength potential that refining can exploit. This will correct the negative effects of fiber treatment in previous stages of papermaking such as on swelling, flexibility, and fiber-to-fiber bonding potential.

Since the objectives of refining recycled fibers depend on the final product and the recovered paper grade used, refining applications vary widely. To restore fiber strength at least partially, the fiber surface is reactivated and the fiber fines lost during processing are regenerated.

Table 17. Application of recycled fiber refining for white and brown paper and board grades.

Paper/board grade	White			Brown	
	Standard newsprint	Improved newsprint	SC/LWC paper	Fluting/ Test liner	Board
Application	Partially required	Required	Required	Required	Required
Target	Strength improvement	Strength improvement Fiber coarse-ness reduction	Strength improvement Fiber coarse-ness reduction	Strength improvement	Strength improvement
Net refining energy demand [kWh/t]	< 50	50–100	~ 100	~ 100	~150 (200)

Strength enhancement is often a refining goal in packaging paper production. The AOCC raw material used in the United States can therefore be made sufficiently strong to eliminate the need for a sizing press. The DOCC grades used in Europe and elsewhere do require surface sizing. In either case, refining can still reduce the sizing outlay necessary to meet strength demands.

Refining is also useful for reducing unacceptably high R14 or R30 fractions in furnishes for supercalendered (SC) and lightweight coated (LWC) papers. The stock can be suitably prepared under appropriate refining conditions by eliminating the R14 component and reducing the R30 fraction to the desired level.

Table 17 gives an overview of the main refining applications in recycled fiber processing for white and brown grades. The white grades primarily comprise SC and LWC papers, improved newsprint, and some standard newsprint. Wood-free writing and printing papers and tissue are not normally refined or use only a brushing refiner. With brown grades, fluting and test liner definitely require refining.

Refining can increase certain strength characteristics. Since fiber properties change due to refining, other paper characteristics are also affected. With decreasing freeness or increasing Schopper-Riegler value for instance, bulk and drainage decrease. Optical characteristics such as brightness, opacity, and light scattering coefficient and tear strength are negatively influenced to a greater or lesser extent.

An optimum compromise is therefore necessary not only between cost-effectiveness and quality but also between the various quality parameters. This compromise can occur by varying the refiner fillings, energy input, refiner type, and the processing stages before refining. Recycled fiber pulp is mainly refined in the low consistency range of about 3%–6%. Sometimes, high consistencies up to 30% and above such as HC refining have use.

Recycled fiber pulp usually comprises a mixture of different fiber types according to origin (type of wood), pulping (chemical, mechanical, bleached, or unbleached), and previous history (during paper production, finishing, and use). This results in significant differences in refining resistance or potential between the various stock components.

In some cases, fractionation can reduce this problem so that the different fiber components produced can be treated separately. This kind of selective refining instead of full-stream refining further consolidates the coarse R14 fibers in deinked pulp for example. The separate components can be mixed again or allocated to the different layers in multilayer products.

Refining occurs at the end of the stock preparation line on the cleaned pulp to avoid damage of the sensitive refiner fillings. If refining were to occur before fine screening, the debris particles would become so small that they could no longer be screened efficiently.

Figure 27 shows how the refining stage is integrated in a processing line for test liner production using DOCC or AOCC. In this case, the long fiber fraction of the fine screening stage is HC refined. The other fraction is LC refined. Figure 153 shows where LC refining is used in deinking processing lines for SC paper.

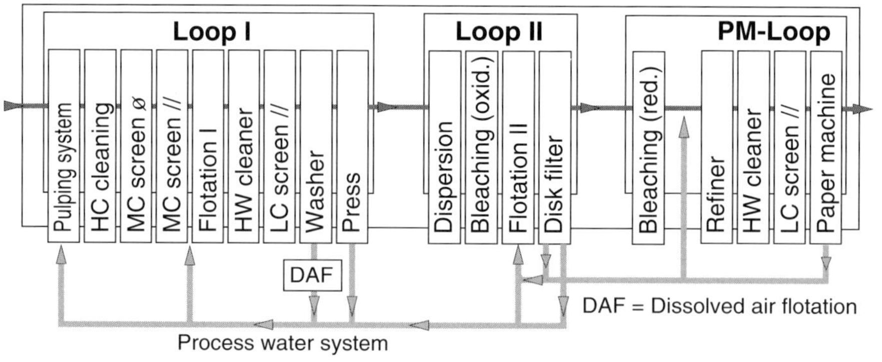

Figure 153. LC refining arrangement for deinked pulp in SC paper production with red. = reductive and oxid. = oxidative bleaching.

10.2 Principles

10.2.1 LC refining

The object of refining recycled fiber pulp is to improve interfiber bonding in the paper by modifying the structure of fibers. In LC refining, this occurs as the pulp passes between the bars of the fillings attached to the stators and rotors of disk and conical refiners.

Refining results depend to a high degree on the angle between the bars (cutting angle) and their edge geometry.

Small cutting angles and bar widths with sharp edges result in a fiber shortening effect. For a fibrillation effect with gentle refining, the bars must have large angles and widths with rounded edges. The final edge shape during operation depends on the wear resistance of the filling material, the specific edge load, and the mineral filler content in the suspension that can cause rapid wear.

A very important criterion affecting refining results is the specific edge load (SEL). This is calculated from the net refining energy divided by the cutting edge length (CEL). The net refining energy is the difference between power consumption and pumping power (no-load power) of the fillings at large spacing between them for a given flow measured with water or pulp slurry . The CEL is calculated from the cutting edge formed per second by the contra-rotating bar edges. A higher SEL means more fibers are cut and shortened. A lower SEL has a more pronounced fibrillation effect.

Many fibers in a recycled furnish have been previously refined and therefore have lower refining resistance. In addition, each recycling process weakens many fibers slightly. For this reason, recycled fiber refining should be very gentle. This low intensity refining requires a low SEL. This optimally exploits refining potential of the fibers that primarily should not be shortened any further. This is a very important requirement for high tear strength that rapidly diminishes with high SEL. Especially in recycled fiber processing for SC and LWC grades, any thermomechanical pulp (TMP) fiber bundles that were not fully defibered must be broken down.

At low refiner loading, no-load power also becomes a significant factor compared with total energy outlay. The amount of refining energy transferable at constant SEL therefore must be higher. This uses fillings with an extremely high CEL with very narrow bars and equally narrow grooves.

Due to material strength considerations, the bar width can only be decreased to a limited extent. This is also influenced by bar height since a lower height allows for more narrow bars. The bar height must be sufficient to allow adequate wear, filling service life, and throughput. This is particularly important because recycled raw materials are generally more abrasive than primary materials due to their high filler content. To some extent, lower SEL counteracts this. Another important aspect for consideration with groove width is risk of blockage that depends on the type of fibers and their condition. The highest risk of plugging occurs with long fibers.

10.2.2 HC refining

HC refining uses refiners with special high consistency fillings or disk dispergers. Disperger fillings are suitable in some cases for refining. The specific energy demand depends on the spacing of fillings. The freeness effect on the fibers strongly depends on the HC refiner inlet temperature. This can range between process temperature and normal dispersing temperature, i.e., 40°C–100°C. Freeness is only slightly reduced by high temperatures but drops steeply at low temperatures.

The HC refining process primarily comprises shear forces between fibers rather than cutting by the filling bars. SEL theory is therefore not applicable in this case.

Since the stock requires thickening to about 30% consistency or more, energy demand and hardware costs for HC refining are higher than for LC refining if the process does not need a dewatering stage to separate water loops.

10.3 Machinery and design versions

10.3.1 LC refining

For recycled fibers, the same types of LC refiners have use as for virgin pulp. These include double disk refiners with four internal components each as in Fig. 154 or more as in Fig. 155. The conical refiners of Fig. 156 typically have a single rotor and stator.

Figure 154. Double disk LC refiner with rotor swung free.

Figure 155. Multiple disk LC refiner.

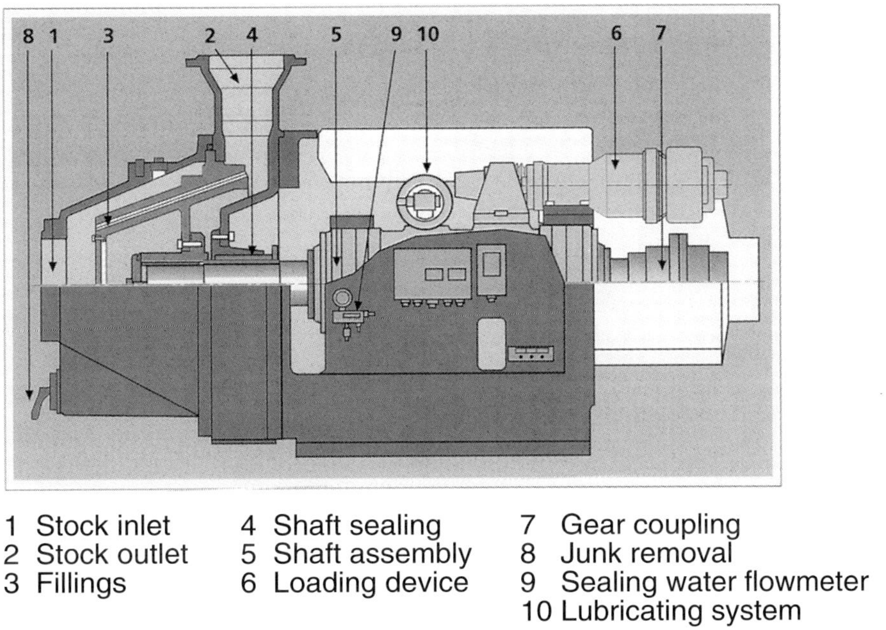

Figure 156. Conical LC refiner.

In all cases, the edge length must be as long as possible to keep the portion of no-load power at low specific edge load as small as possible compared with total refining energy demand. Figure 157 shows a typical example of this kind of filling. With tougher materials, bar widths of 2 mm are possible with bar heights of about 5 mm or more. The groove width is 2–3 mm depending on plugging tendency of the fiber stock.

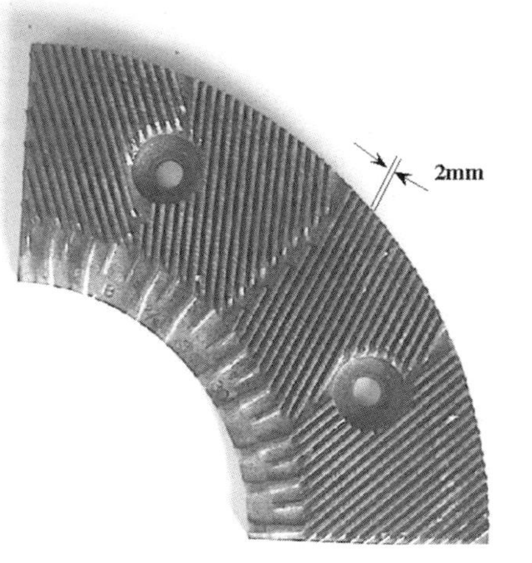

Figure 157. LC refiner fillings for low intensity refining.

Varying the refiner energy varies the specific refining energy demand. This means controlling the spacing between rotor and stator or the throughput.

10.3.2 HC refining

Figure 158 shows an HC refiner and Fig. 159 shows a disk disperger used in this application. Kneading dispergers are unsuitable for refining because their fillings are too widely spaced. Figure 160 shows the kind of fillings used in HC refiners comprising plates with bars with the rotor fillings operating at peripheral velocities of about 60–100 m/s. The rotor fillings of disk dispergers operate at peripheral speeds of about 50 m/s or more. The stock is pressed through the gaps between rings with intermeshing teeth. To illustrate this, Fig. 161 shows the disk disperger fillings in end projection and cross section.

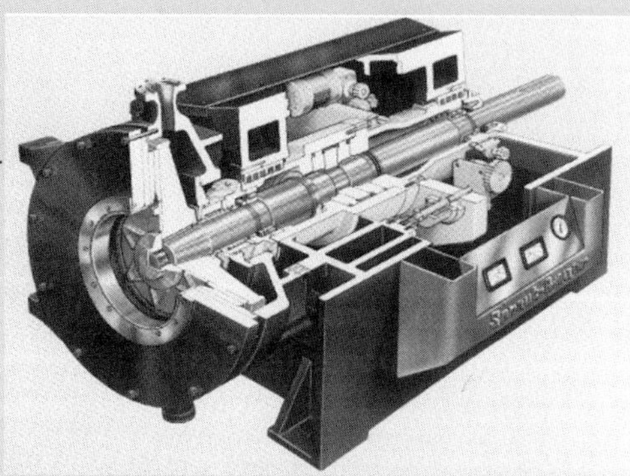

Figure 158. HC refiner.

Figure 159. Disk disperger used for HC refining with cover opened.

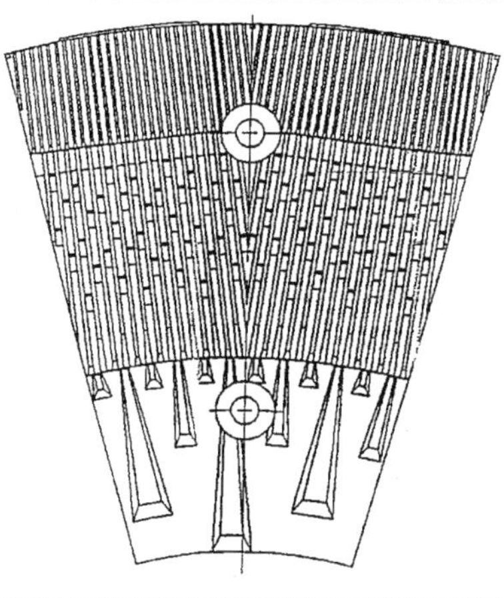

Figure 160. HC refiner fillings.

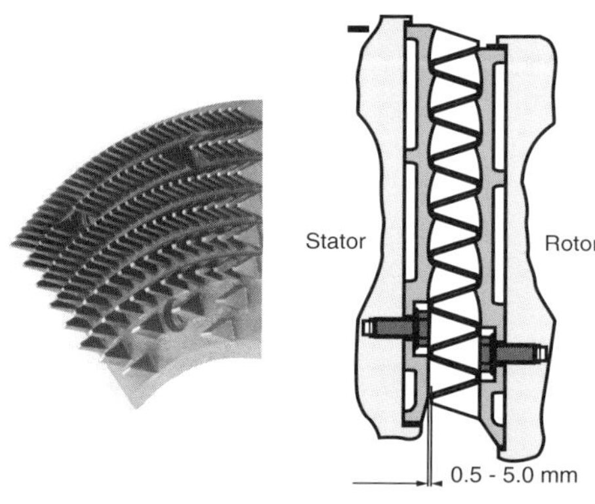

Figure 161. Disk disperger fillings used for HC refining.

10.4 Refining results

Since recycled fiber pulp is an extremely heterogeneous stock, the following examples are only a rough guide. They comprise LC and HC refining test results on recycled fiber pulp for packaging papers and improved newsprint or SC and LWC grades. The results of HC refining are as follows:

- High stretch making it suitable for sack paper
- Little decrease in freeness (increase in Schopper-Riegler value) with strength improvement
- Very little fiber shortening
- Very slight reduction in tear strength.

10.4.1 Refining of recycled fiber pulp for packaging papers

Table 18 shows refining test results for LC and HC ranges. The objective of the initial trials was to improve the strength characteristics of DOCC to manufacture test liner. The LC refining used three passes through a double disk refiner at a specific edge load of 1 500 Ws/km. Freeness decreased accordingly. The strength criteria of breaking length, burst, and short span compression test (SCT) improved by 20%–30%.

The HC refining tests used AOCC and DOCC stock that was fractionated. The long fiber fraction was HC refined in a disk disperger at about 40°C. This temperature is much lower than the normal dispersing temperature of at least 90°C. With a corresponding decrease in freeness (increase in Schopper-Riegler value), the breaking length, burst, and SCT values improved by 10%–20%.

Table 18. Results of LC refining tests on DOCC and HC refining tests on mixed AOCC and DOCC.

Mode of refining		LC refining		HC refining	
Criteria		**DOCC**		**AOCC (50%)/ DOCC (50%), long fiber fraction**	
Consistency	%	4.2		37	
Spec. edge load	Ws/km	1 500			
Spec. energy demand	kWh/t	3 x 30 (net)		127(total)	
		before	**after**	**before**	**after**
Freeness	CSF	360	180	620	520
SR value	SR	35	54	19	24
Breaking length	km	3.5	4.5	3.5	4.5
Burst	kPa	158	200	162	220
SCT	kN/m	1.55	1.87	1.53	1.69

10.4.2 Refining of recycled fiber pulp for writing and printing papers

Figure 162 gives qualitative information on the influence of LC and HC refining on fiber length, elongation, tear, and tensile strength for deinked pulp. Plotted vs. specific refining energy demand, these curves show the typical trends.

The refining test results shown in Table 19 were recorded on deinked pulp for SC paper production. The object here was to eliminate the R14 fraction and reduce the R30 component to a certain degree. LC refining used a comparatively low specific edge load of 1 500 Ws/km. HC refining was also done for comparison. Only LC refining practically eliminated the R14 fraction. The other relevant technical characteristics developed accordingly.

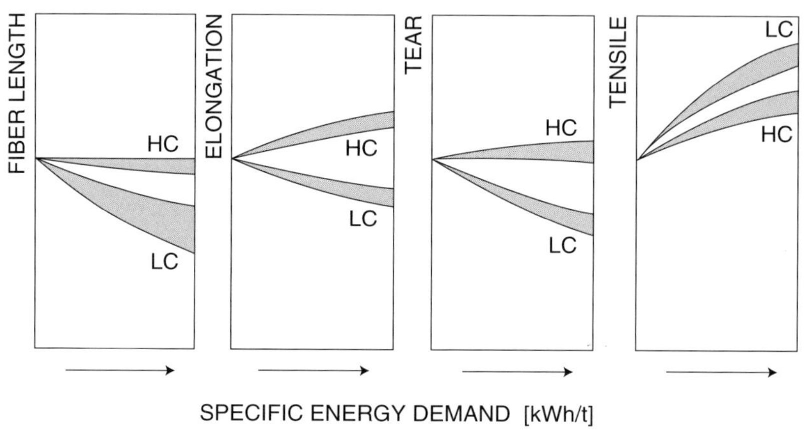

Figure 162. Effect of LC and HC refining on fiber length, elongation, tear, and tensile strength of deinked pulp.

Table 19. Comparative LC and HC refining test results on deinked pulp.

Criteria		LC refining	HC refining
Specific energy demand	kWh/t	148	155
Specific edge load	Ws/km	1 500	–
R14 fraction*	%	-100	-14
R30 fraction*	%	-40	-5
Schopper-Riegler value*	ΔSR	+20	+8
Freeness*	ΔCSF	-100	-50
Breaking length*	%	+15	+11
Burst*	%	-4	+3
Tear*	%	-43	-4
Young's modulus*	%	+40	+10
ISO brightness*	%	~0	~0

** Change in values on refining.*

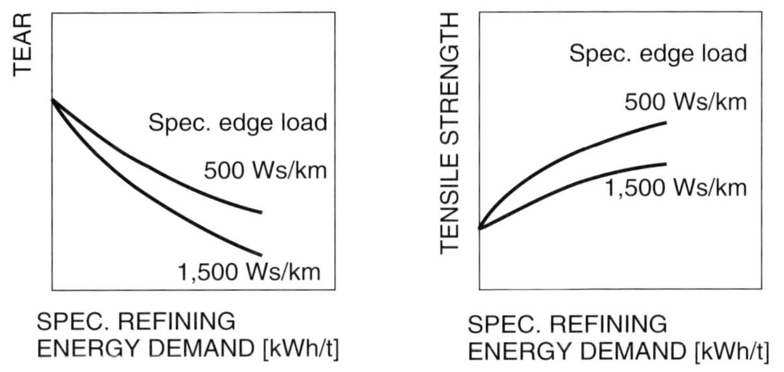

Figure 163. Effects of specific edge load in LC refining on characteristics of writing and printing papers.

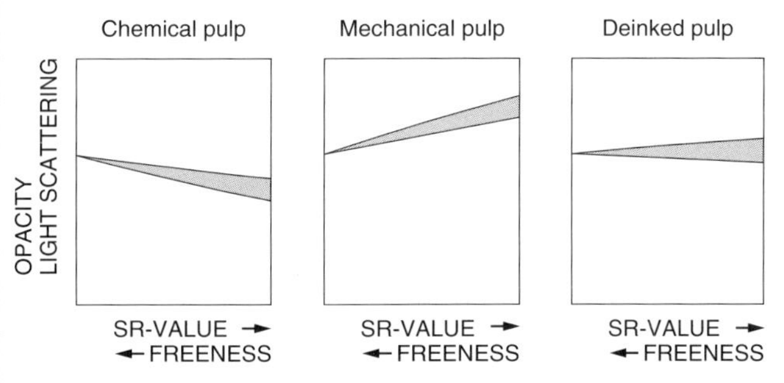

Figure 164. Effect of refining of deinked pulp on optical characteristics.

Figure 163 shows how certain characteristics change when the specific edge load decreases more. Today, these kinds of deinked pulp are refined at edge loads down to 500 Ws/km, i.e., very gently.

Figure 164 shows the effective trends of refining on opacity and light scattering coefficient. The typical development of these parameters is plotted vs. freeness and Schopper-Riegler value for chemical and mechanical pulp and for deinked pulp. The deinked pulp is a mixture of approximately equal parts of these fiber types.

11 Mixing and storing

11.1 Objectives

Mixing and storing are two different tasks. The main objective in storing low consistency pulps is usually to prevent solids from floating to the top, settling at the bottom, or flocculating excessively.

The purpose of mixing is to integrate several mass flows as completely as possible. The various components may be different kinds of fiber suspension. Other suspensions or liquids may also require mixing.

Another important point with recycled fiber processing is that the suspension may contain components that can wind themselves around rotor blades to interfere with the operation.

11.2 Principles

11.2.1 Mixing

To mix all the component flows in a chest efficiently, the entire volume requires complete stirring[98,99]. This large-volume agitation requires a correspondingly high energy input. This can be reduced by mixing in chests whose geometry allows good circulation throughout the entire chest volume. Due to economical reasons, the chest should not require excessive floor space. This means that the diameter to height ratio of the chest should be about 1:1–1:1.6 with a smaller active height. In other words, the chest should not be too low and not too high in relation to its base area.

For a given volume and geometry of the chest, the actual energy demand for complete agitation depends on stock type, consistency, and temperature of the suspension. Energy demand increases with increase in consistency and decrease of temperature. Unbleached softwood kraft pulp requires the highest energy input followed by bleached softwood kraft pulp, mechanical pulp, and hardwood pulp.

11.2.2 Storing and prevention of demixing

To prevent demixing, energy input can be limited to that required for hindering or correcting separation and reflocculation. Demixing takes a longer or shorter time according to the type of stock but can be prevented with relatively low energy demand. Application of mixing energy continuously or throughout the chest simultaneously is not necessary. Mixing only a small part of the suspension at regular intervals is sufficient. After some agitation time at different positions, the entire suspension will have been locally remixed.

11.3 Machine technology

11.3.1 Mixing

Different kinds of machinery are necessary for these two functions. Mixing usually occurs in chests whose height is equal to or greater than their diameter or length and width. The mixing propellers are similar to ship propellers as Fig. 165 shows.

The mixing effect can be adjusted somewhat by varying the propeller blade angle. Today, this is normally done by speed variation that ensures adequate mixing at the lowest possible energy demand. Suspension mixing propellers can be as large as 2 200 mm in diameter with power capacities to approximately 200 kW. Specific energy demand varies from 0.2–0.5 kW/m^3 depending on size and suspension conditions.

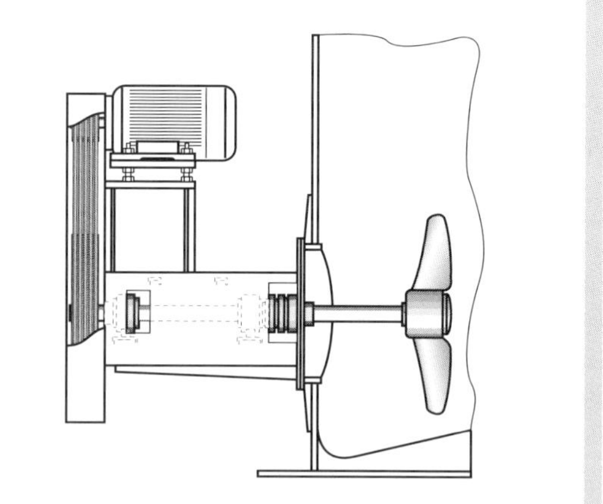

Figure 165. Schematic arrangement of a chest mixing propeller.

11.3.2 Storing and prevention of demixing

For suspension storage without mixing requirement, chests with a similar geometry as for mixing are used. Agitation preferably uses vertically suspended stirrer shafts on which propeller blades are attached in pairs at several levels. In large diameter chests, the stirrer shaft rotates not only about its own axis but also epicyclically around the chest centerline as Fig. 166 shows. The epicyclic motion is generated by the rotation of the shaft about its own axis at 60 rpm for a blade diameter up to 1 600 mm. An appropriate blade pitch angle ensures this.

Figure 166. Scheme of a chest stirrer shaft for preventing demixing.

With especially large chest diameters, two diametrically opposed stirrer shafts are installed. The epicyclic rotation speed depends on agitation requirements. For example, each zone of the chest may be agitated once every 20 s. Specific energy requirements for this kind of arrangement are about 0.02–0.1 kW/m^3. Stirrer shafts can either have fixed bearings at the top and bottom or be suspended from an upper bearing only with or without stabilization weight at the bottom.

11.3.3 Prevention of winding of components

Depending on their position in the process line such as for pulper dumping, storage chests are susceptible to some components winding themselves around the propeller blades and interfering with operation. Suitable shaping of the blades can counter this tendency. The blades have leading edges curved backward so the components in question are thrown out toward the blade end. Figure 167 shows the blade design of a stirrer shaft propeller for suspensions with a high content of such trash.

Figure 167. Stirrer shaft with backward curved blades for suspensions with a high content of spinnable trash.

11.3.4 Storing suspensions with high stock consistency

For storing high consistency stocks of about 10%–15%, demixing is not normally a problem. Dead zones may still exist where stock remains stationary sufficiently long to decompose. The main requirement in this case is to ensure continuous removal of the stock diluted close to the outlet. Storage towers with capacities of about 500 m^3 or more are conical at the bottom. Only this extraction zone has agitation propellers to form a pumpable suspension by mixing with dilution water.

The lowest possible storage volume depends on the stock consistency. For example, stock with a consistency of 15% requires only one-third of the storage volume required for a 5% stock consistency. High consistency stock has a semi-solid or crumbly nature. The bulk density of crumbly stock at 30% consistency is about the same as the storage density of a suspension with 10% consistency.

12 Acknowledgment

The author appreciates all the valuable contributions received during the preparation of this chapter. Special recognition goes to Herbert Britz, Axel Gommel, Anton Gmeiner, Rolf Hartmann, Harald Heß, Martin Kemper, Jochen Krebs, Almut Kriebel, Wolfgang Mannes, Volker Niggl, Peter Respondek, Samuel Schabel, Josef Schneid, Peter Schweiss, Harald Selder, Paul Sepke, Ronald Sigl, and Klaus Steinbild of Voith Sulzer Paper Technology Stock Preparation Division for their competent and encouraging support. Many thanks are also due to the following organizations for providing data, photographs, and schematic drawings of their equipment:

Ahlstrom: Figs. 20, 46, 85 b, 96, 97

Andritz: Figs. 121, 124, 125, 158, 160

Beloit: Figs. 98, 99, 143, 155

Comer: Figs. 102, 103

Kvaerner: Figs. 104, 105, 123 b, 126

Meri Voith Sulzer: Figs. 127, 141, 142

Shinhama: Figs. 84, 108, 109, 150

Sunds: Fig. 156

Thermo Black Clawson, Lamort: Figs. 88, 89, 100, 101, 106, 107, 140

Valmet

Voith Sulzer: Figs. 6, 13, 14, 15, 16, 17, 18, 24, 25, 26, 30, 37, 44, 47, 48, 49, 50, 51, 53, 54, 55, 56, 57, 71, 75, 76, 78, 80, 81, 83, 85 a, 87, 90, 91, 110, 111, 112, 117, 136, 137, 138, 146, 147, 148, 149, 153, 154, 157, 159, 161, 165, 166, 167

References

1. Moss, C., Tappi J. 80(4):69(1997).
2. Holik, H., "The next decade – a challenge for paper production technology" 1993 SPCI Conference Proceedings, SPCI, Stockholm, p.10.
3. Holik, H., Wochenbl. Papierfabr. 121(14):583(1993).
4. Personal communication.
5. Fischer, S.A., Tappi J. 80(11):141(1997).
6. Holik, H., "Towards a better understanding of the defibering process", TAPPI 1988 Engineering Conference Proceedings, TAPPI PRESS, Atlanta, p. 223.
7. Cleveland, F.C., in Secondary Fiber Recycling (R. J. Spangenberg, Ed.), TAPPI PRESS, Atlanta, 1993, p. 91.
8. Savolainen, A., Jussila, T., and Nikula, S., Tappi J. 74(11):147(1991).
9. Siewert, W., Das Papier 37(7):313(1984).
10. Krebs, J., Wochenbl. Papierfabr. 121(5):157(1993).
11. Himanen, J., "Fiberflow drum pulper technology for modern deinking applications", 1995 Paperex Proceedings, Indian Agro Paper Mills Association, New Delhi, p. 295.
12. Rosenfeld, K. and Mickley, G., Wochenbl. Papierfabr. 91(11/12):560(1963).
13. Musselmann, W., in Secondary Fiber Recycling (R. J. Spangenberg, Ed.), TAPPI PRESS, Atlanta, 1993, p. 207.
14. Meltzer, F. P., Progress in Paper Recycling 8(3):60(1998).
15. Schabel, S. and Respondek, P., Wochenbl. Papierfabr. 125(16):736(1997).
16. Rienecker, R., Schabel, S., and Schweiss, P., "Screening—a tool for stickies removal", Voith Sulzer Brochure St.SD.05.0003, Ravensburg, Germany, 1997, 14 pp.
17. Yu, C.J. and DeFoe, R.J., Tappi J. 77(8):219(1994).
18. Gooding, R.W. and Kerekes, R.J., J. Pulp Paper Sci. 15(2):J59(1989).
19. Heise, O., Tappi J. 77(2):78(1994).
20. Gooding, R., "The passage of fibers through slots in pulp screening", M.A.Sc. thesis, University of British Columbia, Vancouver, 1986.

21. Schweiss, P., "Application of the finite volume method and design example for a MultiScreen", Voith Sulzer Brochure St.SD.05.0001, Ravensburg, Germany, 1996, 10 pp.

22. Rienecker, R., Wochenbl. Papierfabr. 125(17):787 (1997).

23. TAPPI TIS 0605 – 04, "Screening symbols, terminology, and equations."

24. TAPPI TIS 0508 – 04, "Evaluating the performance of a fractionator using fiber length data."

25. Schweiss, P., Wochenbl. Papierfabr. 125(17):855(1997).

26. Vitori, C.M. and Philippe, I.J., "New technology for improved performance and longer wear life in contour surface slotted screen cylinders", TAPPI 1989 Pulping Conference Proceedings, TAPPI PRESS, Atlanta, p. 707.

27. Pothmann, D. and Bülow, C., "Reducing fiber losses in waste mill discharge", TAPPI 1988 Pulping Conference Proceedings, TAPPI PRESS, Atlanta, p. 331.

28. Siewert, W.H. and Flinn, P.J., "Final stage screening strategies", TAPPI 1989 Contaminant Problems and Strategies in Waste Paper Recycling Seminar, TAPPI PRESS, Atlanta.

29. Heise, O., Holik, H., Dehm, O., et al., "A new stickies test method – statistically sound and user friendly", TAPPI 1998 Recycling Symposium Proceedings, TAPPI PRESS, Atlanta, p. 213.

30. Vitori, C. M., Pulp Paper Can. 94(12):109(1993).

31. Steenberg, B., Svensk Papperstidn. 56(20):771(1953).

32. Almin, K.E. and Steenberg, B., Svensk Papperstidn. 57(2):37(1954).

33. Kubat, J. and Steenberg, B., Svensk Papperstidn. 58 (9):319(1955).

34. Doshi, M.R. and Prein, M.G., "Effective screening and cleaning of secondary fibers", TAPPI 1986 Pulping Conference Proceedings, TAPPI PRESS, Atlanta, p. 67.

35. Bloor, M.I.G. and Ingham, D.B., Trans. Instn. Chem. Engrs. 51:36(1973).

36. Trawinski, H., Aufbereitungstechnik 36(9):410(1995).

37. Svarovsky, L., "Hydrocyclones", Technomic Publishing Inc., London, 1984, 198 pp.

38. Bradley, D., "The Hydrocyclone", Pergamon Press, Oxford, 1965, 330 pp.

39. Sjögren, K. and Rüppel, G., Wochenbl. Papierfabr. 122(3):83(1994).

40. Julien Saint Amand, F., "Principles and technology of cleaning", 2nd Advanced Training Course on Deinking Technology, CTP, Grenoble, 1993, C. R. no. 3324.

41. Lindquist, S. and Moss, Ch.S., Progress in Paper Recycling 3(1):35(1993).

42. Merriman, K., in Secondary Fiber Recycling (R. J. Spangenberg, Ed.), TAPPI PRESS, Atlanta, 1993, p. 101.

43. Ferguson, J.W.J., Tappi J. 81(1):125(1998).

44. Berger, J., Wochenbl. Papierfabr. 113(21):810(1985).

45. Julien Saint Amand, F., Perrin, B., and Bernard, E., Wochenbl. Papierfabr. 120(8):295(1992).

46. Olson, C.R., Hall, J.D., and Philippe, I.J., Progress in Paper Recycling 3(1):24(1993).

47. Bormett, D.W., Lebow, P.K., Ross, N.J., et al., Tappi J. 78(8):179(1995).

48. Borchardt, J.K., Matalamaki, D.W., Lott, V.G., et al., Progress in Paper Recycling 4(4):16(1994).

49. Borchardt, J.K., Matalamaki, D.W., Lott, V.G., et al., Progress in Paper Recycling 4(1):44(1994).

50. Julien Saint Amand, F. and Bernard, E., Progress in Paper Recycling 3(3):33(1993).

51. Neeße, Th., Dallmann, W., and Espig, D., Aufbereitungstechnik 47(1):6(1986).

52. Janke, H., Wochenbl. Papierfabr. 121(7):241(1993).

53. Schmidt, D.C. and Berg, J.C., Progress in Paper Recycling 6(1):67(1996).

54. Schulze, H.J., Wochenbl. Papierfabr. 122(5):160(1994).

55. Thompson, E.V., Review of Flotation Research by the Cooperative Recycled Fiber Studies Program, Department of Chemical Engineering, University of Maine, Paper Recycling Challenge, Vol. II, Doshi, Appleton, USA, 1997, p. 31.

56. Nguyen, A.V., Schulze, H.J., and Ralston, J., Int. J. Mineral Process. 51:183(1997).

57. Nguyen, A.V., Stechemesser, H., Zobel, G., et al., "Elementary step of three-phase contact line expansion in bubble-particle attachment: an experimental approach to flotation theory", Proceedings of the XX IMPC, Max Planck Institute for Colloids and Interfaces, Freiberg, Germany, 1997, p. 31.

58. Clift, R., Grace, J.R., and Weber, M.E., "Bubbles, Drops and Particles", Academic Press, New York, 1978, 380 pp.

59. Hunold, M., Krauthauf, T., Müller, J., et al., J. Pulp Paper Sci. 23(12):555(1997).

60. Rao, R. and Stenius, P., J. Pulp Paper Sci. 24(5):156(1998).

61. Robertson, N., Patton, M., and Pelton, R., Tappi J. 81(6):138(1998).

63 Britz, H., Gommel, A., Holik, H., et al., German pat. DE 19,611, 864 (Nov. 12, 1997).

63. Eriksson, T.P. and McCool, M.A., "A Review of Flotation Deinking Cell Technology", Paper Recycling Challenge, Vol. II, Doshi, Appleton, USA, 1997, p. 69.

64. Rushton, A., Ward, A.S., and Holdich, R.G., "Solid-Liquid Filtration and Separation Technology", VCH Verlagsgesellschaft, Weinheim, Germany, 1996, 538 pp.

65. Britt, K. and Unbehend, J., Tappi 68(4):67(1980).

66. Horacek, R. G. and Forester, W., in Secondary Fiber Recycling (R. J. Spangenberg, Ed.), TAPPI PRESS, Atlanta, 1993, p. 163.

67. Bliss, T. and Ostoja-Starzewski, M., "Suspended Solids Washing Overview", Paper Recycling Challenge, Vol. II, Doshi, Appleton, USA, 1997, p. 85.

68. Enenes, T.H. and Helle, T.J., Pulp Paper Sci. 22(5):J93(1992).

69. Kappel, J., Fluch, H.-W., and Brunnmair, E., "Modern dewatering systems for sludges from the pulp and paper industry", 50th ATIP Conference Proceedings, ATIP, Paris, 1997, p. 8.

70. Siewert, W. and Horsch, G., "Operating results and experience with the Escher Wyss VarioSplit", TAPPI 1984 Pulping Conference Proceedings, TAPPI PRESS, Atlanta, p. 611.

71. Julien Saint Amand, F., "Ink removal by flotation and washing, hydrodynamic and technical aspects", CTP, Grenoble, 1997, C. R. no. 3660, p. 38.

72. Matzke, W., Gehr, V., and Heise, O., "High yield deinking process technology for handling water-based printing inks", COST Workshop, CTP, Grenoble, 1998, p. 5.

73. Moss, Ch., Paper Technology 38(8):59(1997).

74. Patel, R., "Dynamic washer and filter", 1994 PPI Recycling Conference Proceedings, Pulp Paper International, London, Session 7.

75. Holle, D., Wochenbl. Papierfabr. 113(19):711(1985).

76. Borchardt, J.K. and York, G.A., Progress in Paper Recycling 7(4):24(1997).

77. Carré, B., Galland, G., Vernac, Y., et al., "The effect of hydrogen peroxide bleaching on ink detachment during pulping and kneading", Wastepaper 95, Fourth International Wastepaper Technology Conference Proceedings, CTP, Grenoble, 1995, p. 89.

78. Kakogiamos, A., Johnson, D.A., and Thompson, E.V., Tappi J. 81(4):159(1998).

79. Alexander, A.C., Kurtz, R., and McBride, D.H., in Secondary Fiber Recycling (R. J. Spangenberg, Ed.), TAPPI PRESS, Atlanta, 1993, p. 197.

80. Cochaux, A., Carré, B., Vernac, Y., et al., Progress in Paper Recycling 7(4):89(1997).

81. Carré, B., Vernac, Y., and Galland, G., "Comparison between low-speed kneader and high-speed disperser: Effect of temperature, energy consumption and chemistry on optical properties", Doc CTP 1815, CTP, Grenoble, 1997.

82. Niggl, V. and Kriebel, A., Das Papier 51(10):520(1997).

83. Anon.,"Scherfestigkeit von Störstoffen, Teil 1 und 2", Voith Sulzer Internal Reports 3/57/95 and 3/87/95, Ravensburg, Germany, 1995.

84. Drehmer, B. and Back, E., Progress in Paper Recycling 5(4):49(1995).

85. Mannes, W., Wochenbl. Papierfabr. 125(18):938(1997).

86. Kriebel, A., Mannes, W., and Niggl, V., Voith Sulzer Technology Mag. (5):22(1998).

87. McKinney, R. and Roberts, M., "Curl in recycled fiber. The impact of dispersion and kneading", 1996 EUCEPA Symposium, EUCEPA, Paris.

88. Levlin, J.-E., "On the beating of recycled pulps", 1976 EUCEPA Symposium, EUCEPA, Paris.

89. Guest, D.A., Refining waste paper – Theory and practice, Current and future technologies of refining, PIRA, Leatherhead, UK, 1991.

90. Lumiainen, J., Tappi J. 75(8):92(1992).

91. Lumiainen, J., Paperi Puu 74(4):319(1992).

92. De Fox, R.J., Progress in Paper Recycling 2(1):31(1992).

93. Demler, C. and Silveri, L., Appita J. 49(2):87(1996).

94. Fletcher, R.S., Hong, A.J., and Sasahi, K.R., Pulp Paper Can. 99(2):T44(1998).

95. Rihs, J., "Refining of recycled fibers for brown and white grades", TAPPI 1992 Papermakers Conference Proceedings, TAPPI PRESS, Atlanta, p. 239.

96. Linck, E. and Selder, H., Das Papier 48(10A):V121(1994).

97. Selder, H., Das Papier 49(6):335(1995).

98. Schnell, H., Das Papier 38(8):393(1984).

99. Luiga, A., "What is mixing in paper pulp and what parameters are of importance when designing agitators", 1998 Asia Paper Conference Proceedings, Asian Paper, Singapore, Session 3, Paper no. 4.

CHAPTER 6

Design of recycled fiber processes for different paper and board grades

CHAPTER 6

Michael Schwarz

Design of recycled fiber processes for different paper and board grades

1 Introduction

The technology of processing recycled fibers is more complex than that of virgin fibers (chemical and mechanical fibers) because the composition of the raw material precludes any defined quality equivalence to primary pulps[1–3]. Besides contaminants, recovered paper always contains some fillers and coating pigments that can vary according to type and composition. Due to the technical outlay and stock losses involved, complete removal of these components is not cost-effective[4,5].

Recycled fiber (RCF) processing concepts must therefore consider the wide variations of the furnish quality. The fiber mix, fillers, and additives must balance to suit paper machine characteristics and product requirements. By appropriate balancing of the various components of RCF stock, optimum paper quality is often possible at lower production costs than when using virgin fiber furnish exclusively. Cost results can vary widely from one individual case to another.

This chapter covers some typical examples of RCF processing concepts considering these aspects. The approach flow system of the paper machine is included in full. Fiber recovery and reject processing are only considered regarding deviations from virgin fiber processing.

Depending on the paper or board grades produced and on the product and process conditions, RCF yield can fall as low as 60%. The residues as thickened rejects or sludge can be partly re-used or require disposal in several ways. This includes incineration for energy recovery, material recycling in other industries, and landfilling (See Chapter 12).

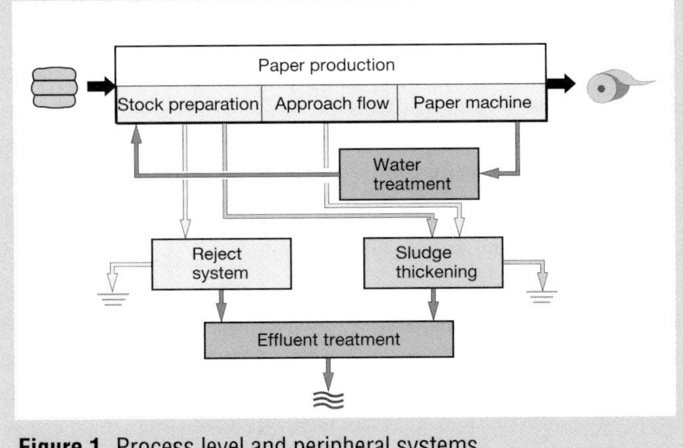

Figure 1. Process level and peripheral systems.

Integral parts of all RCF processing lines are the peripheral systems for reject and sludge treatment with loop water cleaning. These systems are discussed in detail for each concept of Fig. 1.

For simplification, the cases presented are white and brown grade systems. White grade technology strives for the best possible optical quality parameters such as brightness and cleanliness. The brown grade systems of Table 1 primarily concern strength parameters although certain cleanliness requirements are necessary.

Table 1. Key parameters for cleanliness.

	"White" system parameters	"Brown" system parameters
Production level	Optical properties - Brightness - Dirt specks Stickies Ash content	Stickies Flakes Sand Fines Ash content
Process water management	Suspended and colloidal solids Chemical oxygen demand Cationic demand Conductivity pH	Suspended and colloidal solids Chemical oxygen demand Cationic demand Conductivity Biological activity

2 White paper grades

The raw material for the processes discussed below comprises graphical post-consumer recovered paper grades including a small proportion of unprinted pre-consumer grades. This kind of deinking material uses primarily old newspapers and magazines and therefore contains mechanical fibers. In contrast to this, recovered office paper predominantly contains wood-free writing and printing papers. The unbleached chemical fiber content of the furnish must be as low as possible since delignifying bleaching processes are too demanding for cost-effective stock preparation (See Chapter 9). The description of a process layout for writing and printing papers examines the processing subsystems, the peripheral systems for water treatment, and reject and sludge thickening. The systems presented focus on differences compared with the standard process layout valid for virgin pulps.

2.1 Standard newsprint and improved paper grades

Figure 2 represents key system data and arrangements of process stages employed for the production of newsprint and improved paper grades. The raw material processed is deinking material from household collections comprising approximately equal parts of old newspapers and magazines. The main product is newsprint although this system can also produce pulp in adequate quality for improved paper grades. Other pulp components used for these paper grades generally include thermomechanical pulp (TMP). For basis weights below 45 g/m^2, the strength potential of the RCF and TMP mixture is usually not adequate. Chemical fibers must also be added. For lightweight coated

(LWC) and supercalendered (SC) grades, certain quantities of chemical pulp may require addition for the same reason. Using TMP as a furnish component does not influence the process stages of the recovered paper line. Water clarification must be adapted to the increased content of colloidal substances caused by TMP. Chemical pulp as the third component has no effect on RCF processing[6].

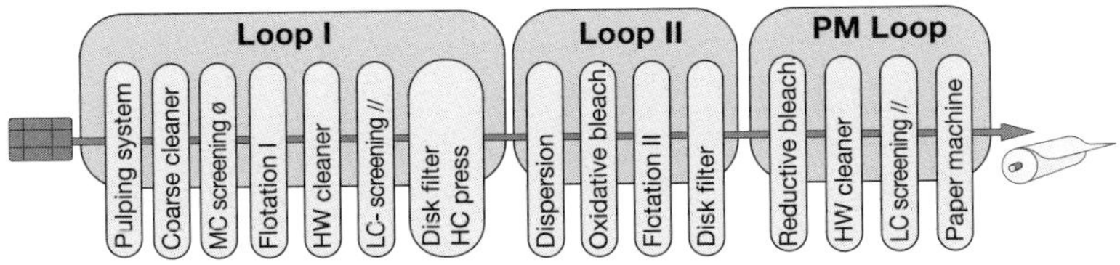

Product: Newsprint, and improved grades, approx. 40-90 g/m²
Furnish: News, magazines
Yield: Approx. 83 %
Effluent: Approx. 8 L/kg
Ash: Feed approx. 22 %, finished paper approx. 13 %
Brightness: Finished paper approx. 63-66% ISO

Figure 2. Concept for newsprint and improved paper grades.

2.1.1 Production level

Table 2 shows an overview of the tasks performed by each process stage[7]. Chapter 5 gives a detailed description of the machinery used in these stages.

Table 2. Function of main process stages.

Technological demands / Process stage	Optical cleanliness	Debris and sticky reduction	Ash content
Pulping PreClean System	Detachment and dispersion of printing ink particles	Saving screenability, pre-screening by PreClean	Ash enrichment by process water
Coarse cleaner		Heavy particles	
MC hole screening		Coarse flat debris	
Flotation I	Brightness increase, dirt speck reduction	Stickies reduction	Ash reduction
HW cleaner	Dirt speck separation	Fine sand and debris removal	
LC slot screening		Stickies removal by finest slot width	
Dispersion	Printing ink dispersion	Stickies dispersion	
Bleaching (oxidative)	Brightness of fibers		
Flotation II	Brightness/ dirt specks dispersed residual ink	Stickies reduction	Ash reduction
Bleaching (reductive)	Brightness/color		

Pulping is gentle with a low degree of contaminant break down. Operation is either as a batch in the pulper or continuously in a drum pulper. Major heavy particles are removed directly after the dump tower by the protector system. The high removal efficiency of the protector system uses a two-stage arrangement. A continuous reject flow from the high density cleaners of the first stage is fed to a sedimentation tank. It is adjusted to low consistency and then passed to the low consistency (LC) cleaners in the next stage. Both cleaners and the sedimentation tank have junk traps.

Medium consistency (MC) screening uses three stages with 1.2 mm screen holes. The first and second stages are fed forward. Only the third stage is arranged in cascade. The first stage operates at approximately 3.5% consistency, but the following stages are diluted. MC screening is followed by flotation I, the principal stage for removing printing inks and other fibrous and nonfibrous components present in the recovered paper processed. When the flotation stage is operated at approximately 1.2% stock consistency, small light-weight particles and stickies are also removed to a certain degree with the printing inks. Operation of the subsequent heavy-weight cleaners is optimum at the flotation discharge stock consistency. The cleaners provide efficient protection for the baskets of the following LC screening stage. This avoids damage or premature wear of the 0.15 mm slot screens used there. Experience from many plants has shown that expensive light-weight cleaners are no longer necessary when using the latest fine slot screening technology.

Thickening with disk filters and subsequently with wire press or screw press technology to approximately 30% completes the first loop. Dispersion then detaches residual ink and reduces stickies in size[8]. High consistency (HC) bleaching with peroxide occurs at this higher stock consistency under optimum conditions. After the HC bleaching tower, the stock is diluted before entering flotation II.

Before storage of the finished stock, it is dewatered to approximately 10% in a disk filter. For better loop separation, installation of an HC press is recommended here. By increasing stock consistency to 30% at the end of loop II, undesirable carryover of chemicals and contaminants can be effectively reduced. This improves the operating efficiency of the entire plant. Reductive bleaching is optional here.

In the approach flow system, the cleaner plant combined with deaeration and 0.35 mm slot screens does policing work only.

Figure 3 summarizes the stock consistency sequence throughout the process. In the first loop, the lowest stock consistency is 0.8% due to the requirements of the cleaners and the following LC screening. The second loop is set to the higher flotation stock consistency so that a smaller size disk filter can be used in this loop. A higher filler content of the RCF furnish requires more precise balancing of the ash content throughout the process. This is because the ash content affects not only the overall system yield but also the dimensioning of individual process machinery. If the filler content of the loop water increases, more water must be added for diluting to the desired stock consistency than with a lower filler content. The hydraulic and solids loadings on the process machinery are then higher.

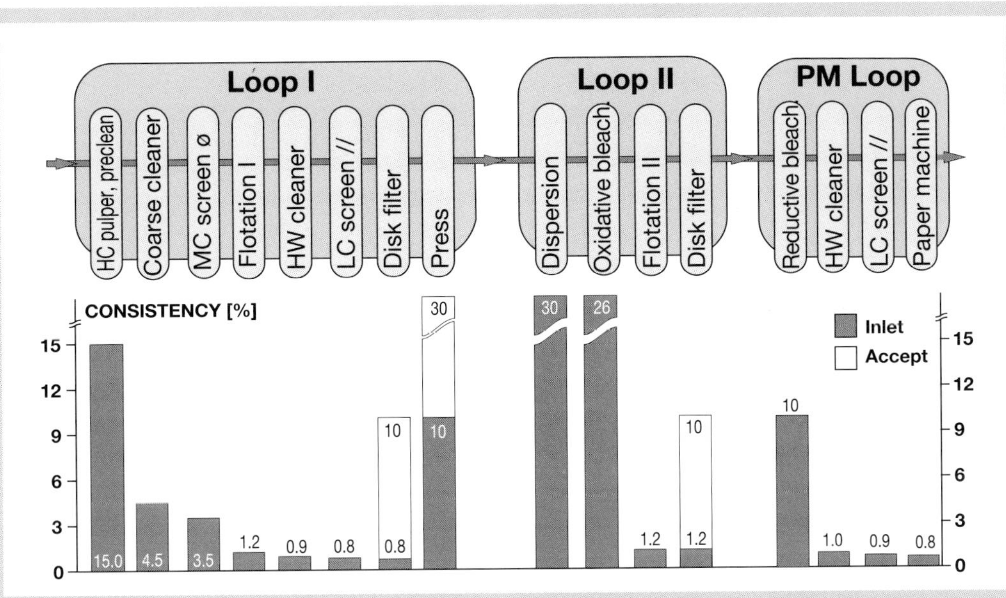

Figure 3. Newsprint and improved paper grades showing sequence of consistency level at different process steps.

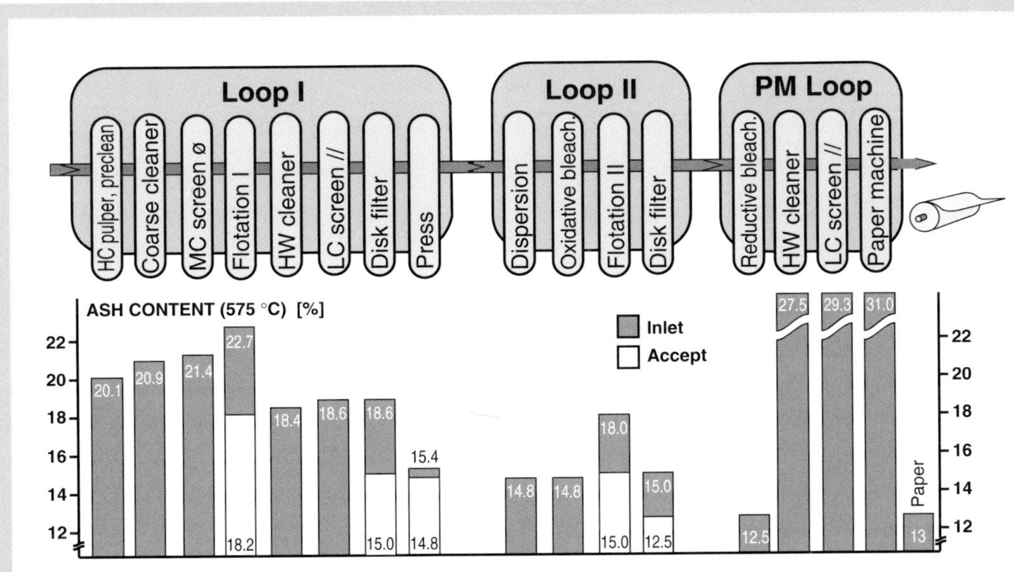

Figure 4. Newsprint and improved paper grades showing sequence of ash content at different process steps.

Figure 4 shows a typical example of the development of the ash content. By adding dilution water containing fillers, the ash content increases until the first flotation stage intake. Flotation then reduces the ash content substantially by selective removal of fillers and fines. A further reduction occurs during thickening and dewatering at the end of the first loop. Additional ash removal occurs in the second flotation stage. In the approach flow section, peak values are normally measured due to the high ash concentration of the white water of the short circulation even without adding fresh fillers. Stock consistency in the approach flow section should be assessed from two different viewpoints: fiber consistency and total consistency. Besides the filler content of the recycled fiber pulp at the end of the deinking line, retention on the forming section decisively influences stock consistency.

Since ash retention is always lower than fiber retention, fiber consistency is the decisive parameter for dimensioning screening machinery and cleaners. Because the filler content only marginally affects viscosity and screening characteristics, the total stock consistency and fiber consistency must always balance separately for machinery layout in the approach flow section.

An important parameter affecting optical cleanliness is the content of stickies. This also has a substantial effect on the runnability of the paper machine. Examining each process stage of Fig. 5 for its influence on content of stickies gives the following picture.

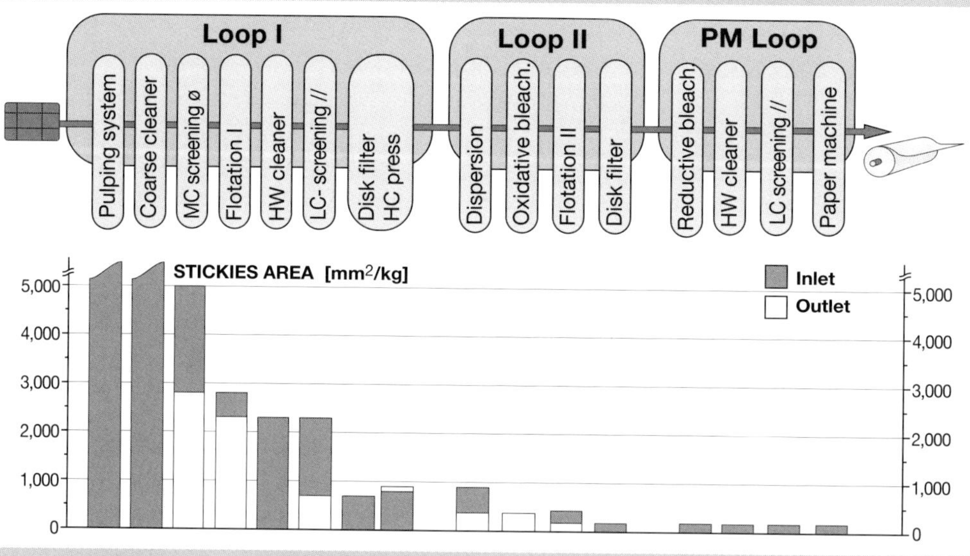

Figure 5. Newsprint and improved paper grades showing reduction of macro stickies.

Macro stickies are removed by screening, cleaning, and flotation. The reduction of the area of stickies of 2 000 mm^2/kg o.d. pulp to less than 1 000 mm^2/kg clearly indicates the importance of LC fine screening. Dispersion detaches the small stickies still adhering to the fibers and activates the surfaces of the stickies by shear forces. Afterwards in flotation II, the stickies can then be removed efficiently.

Before dispersion, the RCF pulp must be cleaned as efficiently as possible. Only the residual amount of stickies should be dispersed. The dispersion of pulp that still contains a high amount of macro stickies can have a significant drawback. After dispersion the stickies and their fragments tend to agglomerate in an agitated suspension. Agglomeration can easily happen in storage tanks or in process water loops where there is plenty of opportunity for the tiny sticky particles to collide and adhere to each other. Because of their small size, the particles can escape into the filtrate during washing and thickening operations. If the filtrate is not sufficiently clarified, larger sticky particles can be formed here only to be mixed into the pulp stream at the dilution stages. Note that after dispersion only a second flotation stage would be effective in removing dispersed stickies. The other mechanical separation methods are no longer suitable. At the end of the system, the stickies content is less than 100 mm^2/kg. Each figure refers to the area of macro stickies measured according to the pick-up method. Figure 6 shows the procedure.

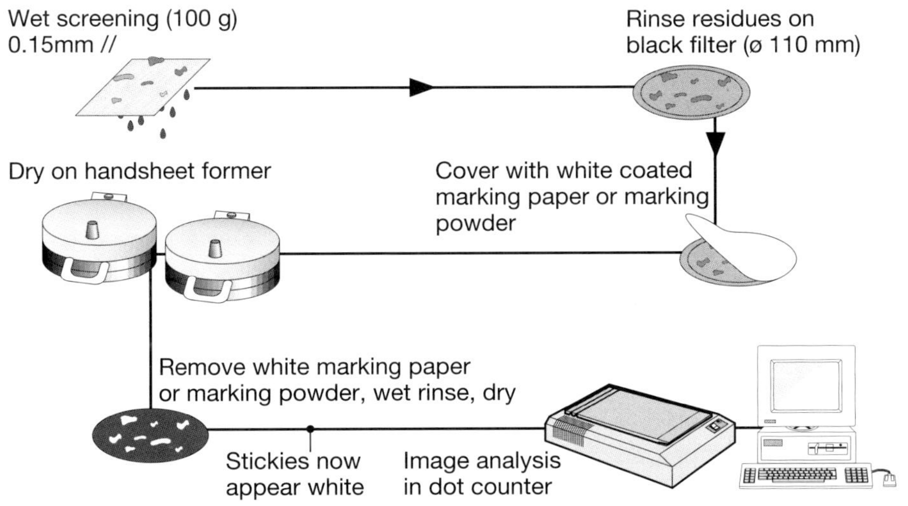

Figure 6. Pick-up method for the determination of macro stickies.

Another measure of optical cleanliness uses the number and area of visible dirt specks. Figure 7 shows how the dirt speck area is reduced in each process stage. The values shown comprise all visible dirt specks > 50 µm in diameter. The main process stages responsible for removing dirt specks are flotation I and II. Dispersing with a specific energy demand of about 75 kWh/t has also a significant influence on dirt speck reduction. Dirt specks not removed in the foregoing screening and flotation stages are separated from the fibers during dispersion and largely reduced in size below the visibility threshold. The finished stock therefore has a residual dirt speck area of about 100 mm^2/m^2.

Figure 8 illustrates brightness development. Most printing ink is removed in the first flotation stage. Residual ink particles still adhering to the fibers are then separated during dispersing for efficient removal in the second flotation stage (flotation II). Brightness loss due to ink fragmentation is compensated by the effect of oxidative bleaching during the same process stage.

Alkali and sodium silicate are added at the heating coil intake or shortly before the disperger. Peroxide addition is immediately before the disperger. This separate addi-

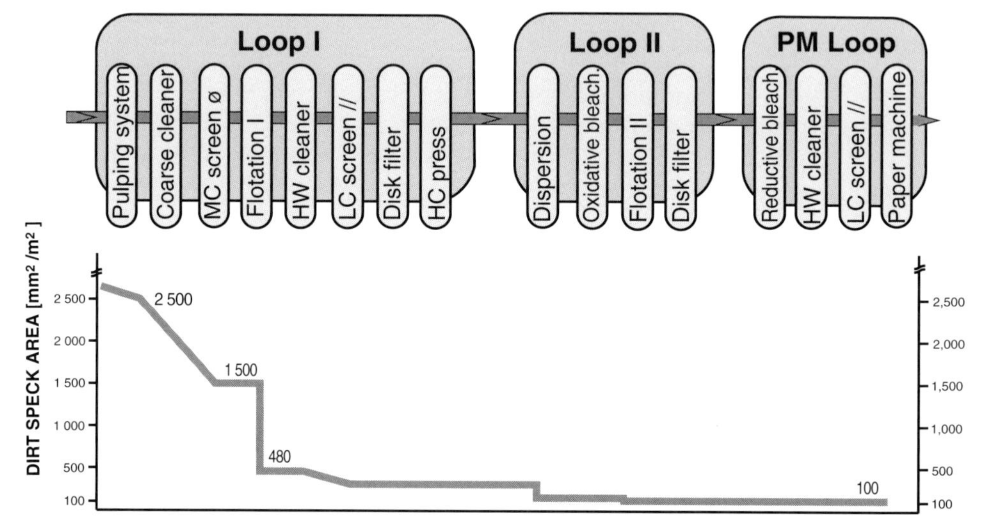

Figure 7. Newsprint and improved paper grades showing reduction of dirt speck area.

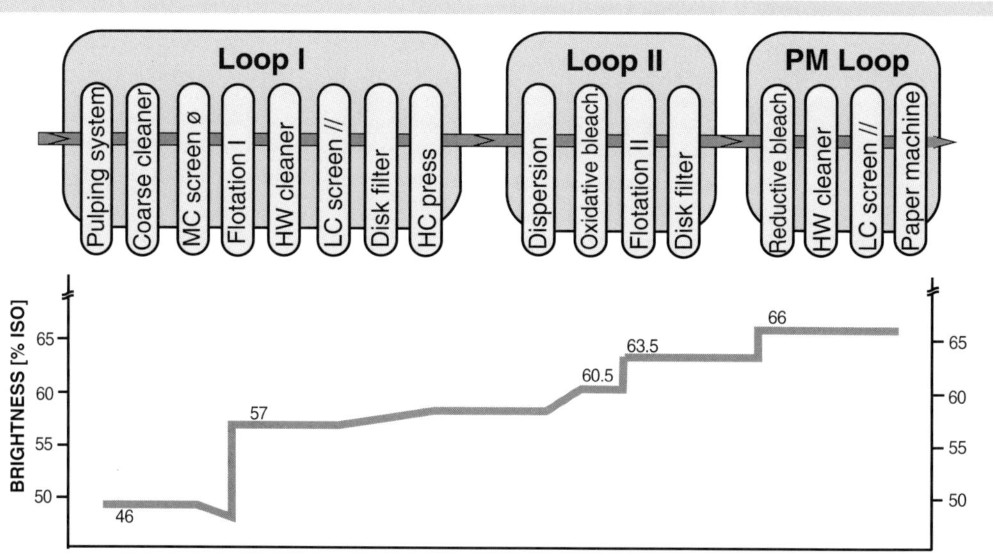

Figure 8. Newsprint and improved paper grades showing development of brightness through process stages.

tion of chemicals decisively combats undesirable silicate deposits. The actual bleaching process occurs in the HC bleaching tower where high stock consistency and high temperatures contribute to optimum exploitation of the bleaching agents with short dwell times. With an efficient process layout today, chelating agents such as ethylenediaminetetraacetic acid (EDTA) or diaminetriethylenepentaacetic acid (DTPA) are entirely unnecessary (See chemical additive guidelines in Table 3). An additional brightness increase occurs in flotation II.

Not until after thickening at the end of loop II does reductive bleaching occur in the MC range. If formamidine sulfinic acid (FAS) is used as a bleaching agent (See Chapter 9), no special precautions are necessary regarding chemical additives. For dithionite, acidification is necessary to destroy residual peroxide. The air and oxygen content also requires reduction by appropriate measures such as an MC pump with its deaeration effect .

Table 3. Consumption figures for different processes (dosages are for neat chemicals based on oven-dry furnish).

	Newsprint and improved grades	High-grade writing and printing papers	Tissue system	Market DIP
Raw material	Wood-containing recovered paper	Wood-containing recovered paper	Mainly wood-free recovered paper	Mainly wood-free recovered paper
Energy demand				
- Electrical power	300–380 kWh/t	400–500 kWh/t	400–500 kWh/t	650–750 kWh/t
- Heating energy	450–900 MJ/t (= 0.2–0.4 t steam/t)	650–1 100 MJ/t (= 0.3–0.5 t steam/t)	650–1 100 MJ/t (= 0.3–0.5 t steam/t)	900–1 300 MJ/t (= 0.4–0.6 t steam/t)
Additives				
- Slushing	0.5%–1.0% H_2O_2 0.5%–1.0% NaOH 1.0%–2.0% Na_2SiO_3	0.5%–1.0% H_2O_2 0.5%–1.2% NaOH 1.0%–2.0% Na_2SiO_3	0.0%–1.0% NaOH	0.0%–1.0% NaOH
- Flotation 1	0.3%–0.6% Soap	0.4%–0.8% Soap	0.3%–0.6% Soap	0.3%–0.6% Soap
- Flotation 2	0.2%–0.4% Soap	–	–	0.2%–0.4% Soap 0.0%–0.15% Surfactant 0.0%–0.3% NaOH
- Flotation 3	–	–	–	0.0%–0.15% Surfactant
- Bleaching	*Oxidative bleaching:* 1.0%–2.0% H_2O_2 0.5%–1.2% NaOH 1.0%–1.8% Na_2SiO_3 *Reductive bleaching:* 0.4%–1.0% Dithionite 0.2% NaOH	*Oxidative bleaching:* 1.0%–2.5% H_2O_2 0.5%–1.5% NaOH 1.0%–2.0% Na_2SiO_3 *Reductive bleaching:* 0.4%–1.0% Dithionite 0.2% NaOH	*Oxidative bleaching:* 1.0%–2.0% H_2O_2 0.5%–1.2% NaOH 1.0%–1.8% Na_2SiO_3 *Reductive bleaching:* 0.4%–1.0% Dithionite 0.2% NaOH	*Oxidative bleaching:* 1.0%–2.0% H_2O_2 0.5%–1.2% NaOH 1.0%–1.8% Na_2SiO_3 *Reductive bleaching:* 0.4%–1.0% Dithionite 0.2% NaOH
- Flocculants (internal water clarification and sludge thickening)	0.5–1.0 kg/t	3.0–5.0 kg/t	1.8–2.8 kg/t	3.5–5.5 kg/t
Losses				
- Solid waste	150–220 kg/t	300–350 kg/t	300–370 kg/t	350–400 kg/t
- Inorganic content	55%–65%	45%–55%	50%–60%	50%–60%

2.1.2 Total plant concept

Figure 9 documents the layout of the peripheral plant components showing reject system, sludge dewatering, and loop water treatment. Coarse rejects from the pulping section, heavy particle separation, and MC screening have a dry content of about 60% after drainage. The fine rejects are prethickened with the flotation and loop water cleaning sludge and then drained to about 60%. The filtrates are purified by dissolved air flotation (DAF). In this case, three separate microflotation stages are established of which DAF 1 is implemented in the conventional long circulation loop of the papermaking process.

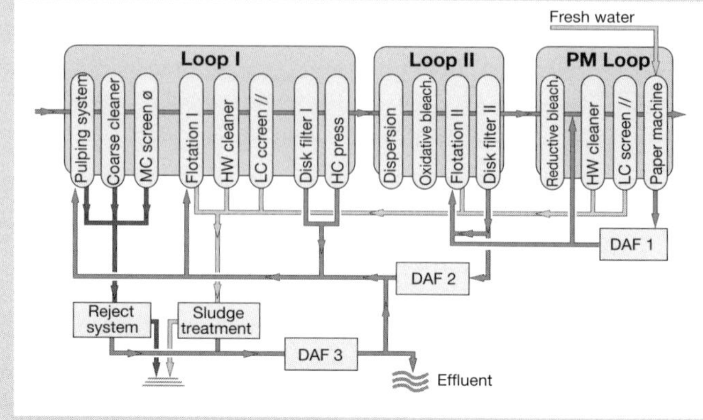

Figure 9. Newsprint and improved paper grades showing total plant concept.

2.1.3 Rejects and sludge treatment

Figure 10 classifies the wide variety of contaminants entering the system with the recovered paper. During pulping (drum pulper and pulper), the coarsest components are removed. They range from large light weight pieces of foamed plastic or foil to all kinds of larger, heavy particles with high density.

During coarse cleaning, small particles with high density such as sand, staples, and glass fragments are removed. To keep stock losses with the rejects as low as possible, the particles separated in high density cleaners are washed out with high efficiency. Depending on the system characteristics, the rejects in the following stages have a high content of organic materials including fibers.

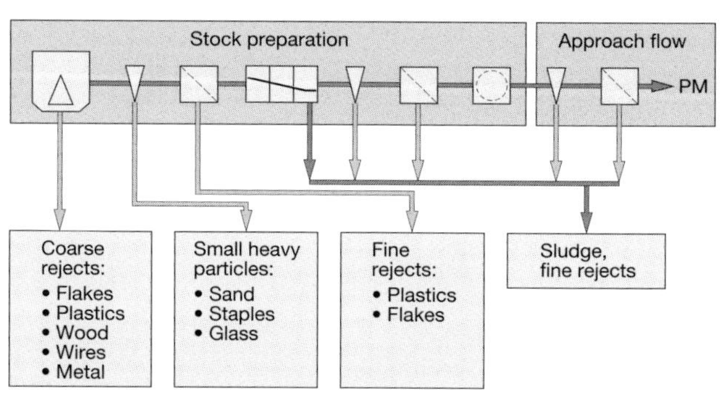

Figure 10. Classification of rejects and sludge.

During MC screening, coarse particles mainly comprising light weight plastics and flakes are removed. In cleaners, LC screens, and approach flow systems, fine rejects are separated. Since their composition is similar to that of the flotation sludge, they are commonly treated with the flotation sludge.

Figure 11 gives an overview of how reject machines and sludge dewatering equipment are integrated. Coarse rejects from the pulping area with rejects from the MC screening are dewatered in a screw press (Compactor). If an intermediate conveying system must be installed for coarse rejects, a generously dimensioned open screw conveyor (Screenex) with dewatering zone is preferable for this purpose. Rejects from the protector system (high density cleaner) are best dewatered in sedimentation equipment (Sedimator). Fine heavy particles separated in LC cleaners cannot be separated efficiently in such a machine since the sedimentation of these small particles leads to very low sedimentation velocities. These rejects are collected with the rejects of the LC screening stages. These fine but still fibrous rejects can be dewatered with the flotation sludge. By forming a fiber matrix, the small quantity of fine fibrous rejects helps to facilitate the treatment of flotation waste in sludge presses.

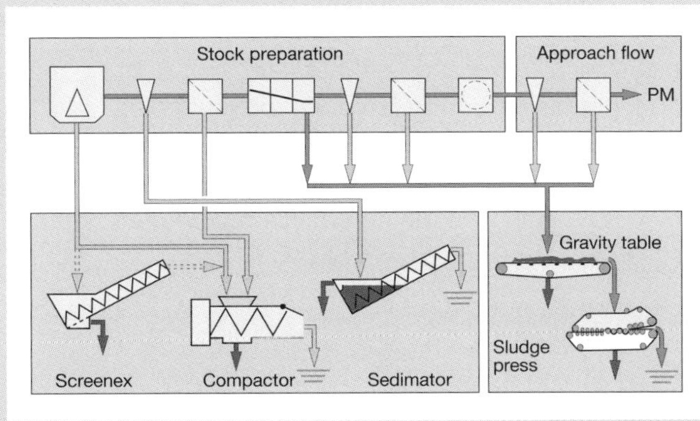

Figure 11. Machines for reject and sludge treatment.

The common practice now is to use two-stage treatment with mild pre-thickening from about 1%–2% to about 12%–16% dry content followed by equally mild dewatering to a final dry content up to 60%. Normally, flocculation agents (organic polymers or inorganic adsorbers) must be used since cake filtration that requires the accumulation of a natural filtering layer is not possible here. The sludge is pre-thickened in rotary thickeners of the drum or disk type or in gravity table machines. The particular advantage of gravity table thickening is that it avoids violent agitation and prevents flake break-up by depositing material in the screen gently. With very little additives, this ensures optimum filtrate qualities and high output consistencies as Fig. 12 shows.

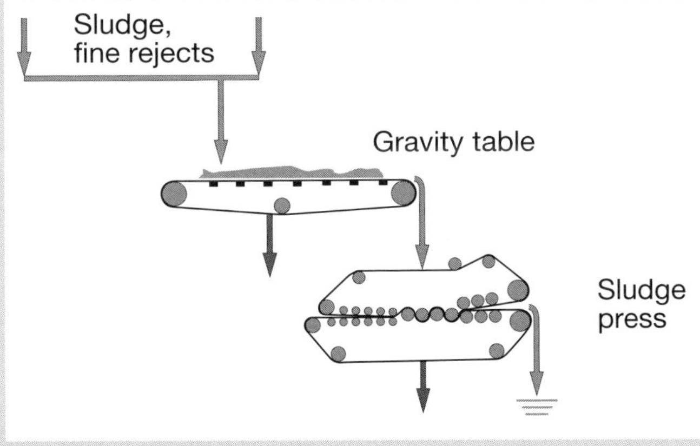

Figure 12. Sludge thickening.

The pre-thickened sludge is then dewatered in twin wire presses or screw presses. Both processes have many reference installations and many years of experience. In twin wire presses, dewatering can be continuously adjusted by varying the pressure to suit material characteristics. Additional advantages are low energy demand

and good accessibility for maintenance. The main advantage of screw presses for sludge thickening is their compact, totally enclosed design. Particularly important for large systems is that they require little space, are easy to install, and give a high dry content. Pre-dewatering also uses gravity tables.

Worth mentioning in this context is the importance of laying out the sludge dewatering subsystem comprehensively. Only by selecting the most suitable machinery in optimum combination with the respective peripheries of the main system is maximum efficiency and cost-effectiveness possible[11,12].

2.1.4 Water management

Figure 13 shows the principle of optimum water management for a typical process for the production of improved newsprint using pre-flotation and post-flotation. Three loops separated from each other by drainage stages contribute to sound water management. Fresh water is supplied to the paper machine only. Clear filtrate from the save-all is fed as make-up water to the stock preparation line on the counterflow principle. Besides very little overflow and filtrates from the reject handling machinery, the only effluents from the entire plant are the filtrates from the sludge thickening machines[13].

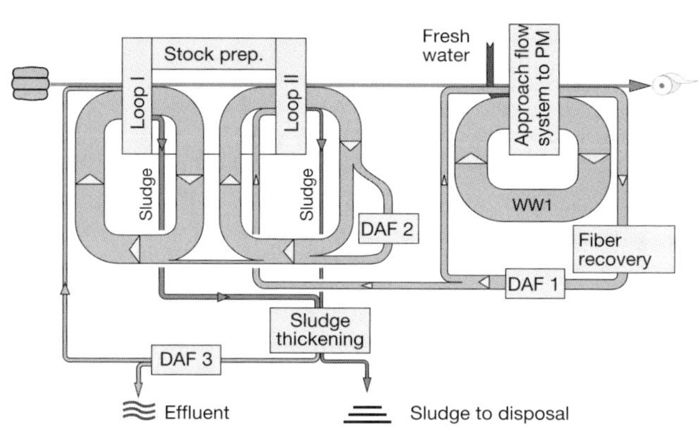

Figure 13. Water management.

Some filtrate from the second loop in stock preparation is clarified in an additional dissolved air flotation stage (DAF 2). The sludge thickening filtrate returned to the first loop always passes through the third dissolved air flotation unit.

The balance data for production-related parameters such as ash content, COD, or cationic demand show the great importance of optimum water management. Concentrations of flow-related substances such as COD can decrease significantly from the first loop in the stock preparation line, through the second loop, and to the approach flow system. This is essential for sheet formation and product quality. In this process example, the COD concentration in the first loop totals 4 300 ppm (4 300 mg/L), in the second loop 2 400 ppm, and finally in the approach flow 1 100 ppm. This means that the respective ratios of loop concentrations are approximately 100:50:30 as Fig. 14 shows.

Until now, no reliable method has been available to quantify directly detrimental substances that impair sheet formation and product quality. Dissolved and colloidal substances are distributed and mixed according to flows and their concentrations. COD may be used as the master parameter that indicates the amount of detrimental substances.

Adequate water management must not be limited to only one stable plant condition as Fig. 13 shows. Some sludge dewatering filtrate is recirculated to save fresh

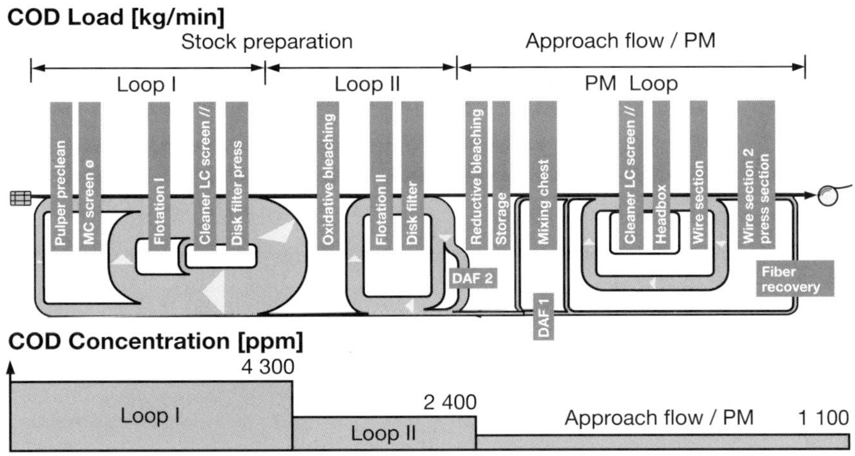

Figure 14. COD concentration of the process water loops.

water. The remainder is drained off at DAF 3 to the effluent treatment. If more filtrate is recirculated from sludge thickening, less make-up water is transferred from the second to the first loop. Less make-up water is required by the paper machine, and less fresh water is consequently required by the entire system.

The ratio of recirculated to discarded filtrate at DAF 3 therefore determines the specific fresh water consumption of the entire system from pulper to press section. This is implemented in the control system by regulating the water system buffer and storage tank levels in cascade from the first loop in the stock preparation line to the paper machine where fresh water is fed to the system. Depending on the concentration of anionic trash, additives consumption, and product quality demands, the degree of loop closure can be continuously adjusted to requirements during operation without the need for complex control algorithms.

As already mentioned, adding an HC press after the disk filter at the end of loop II is strongly recommended. This will significantly reduce undesirable carryover of chemicals and other detrimental substances. The left columns of Fig. 15 show the concentration of COD in different loops when considering only the usual thickening to 12% by a disk filter. If an HC press is

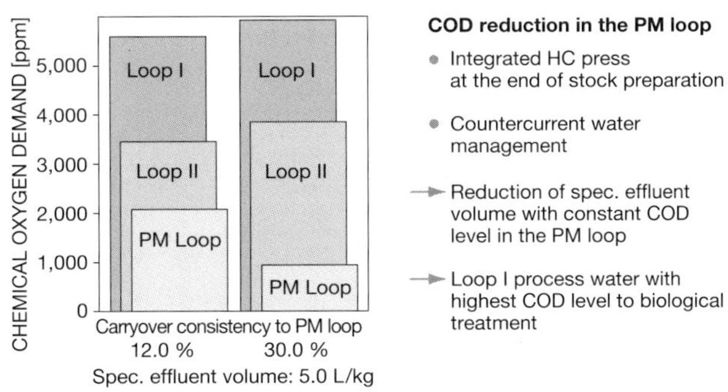

Figure 15. Newsprint system showing COD affected by water loop separation.

installed for consequential water separation, the transfer consistency to the paper machine loop increases to 30%. The columns in the right part of Fig. 15 represent the figures achieved by these modified conditions. The concentration in the paper machine loop where the paper quality is finally determined decreases to approximately half the level without the HC press.

Note that the concentration in loops I and II increases. The detrimental substances leave the process to a smaller extent through the finished paper as the values indicate. At a constant effluent volume from the stock preparation loops, the concentrations of the effluent and the prior loops must therefore increase accordingly. The concentration level of detrimental substances primarily depends on the specific effluent volume. For 5–10 L/kg, Fig. 16 shows the relations. The values refer to the process example discussed in detail with only one disk filter at the end of loop II.

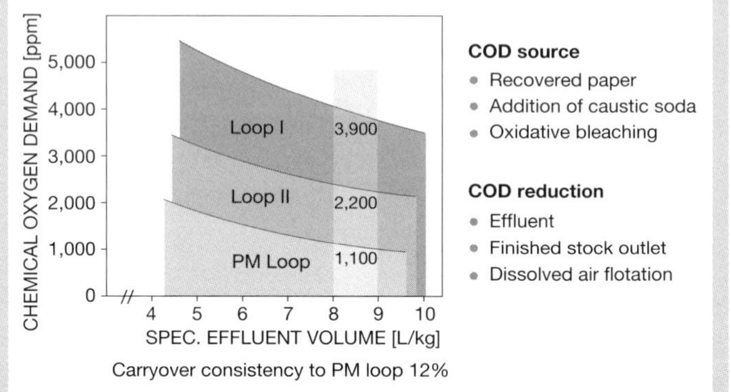

Figure 16. Newsprint system showing COD concentration vs. specific effluent volume.

2.1.5 Process water clarification

State-of-the-art is the dissolved air flotation (DAF). This is the only technology allowing effective treatment of large water volumes economically. Anionic trash and fines agglomerate into physically treatable flocs by using chemicals (flocculants). Figure 17 shows this in a simplified form as a cationic polymer "catching" the particle like a chemical "octopus." These agglomerates can then be disposed in the clarification tanks by flotation.

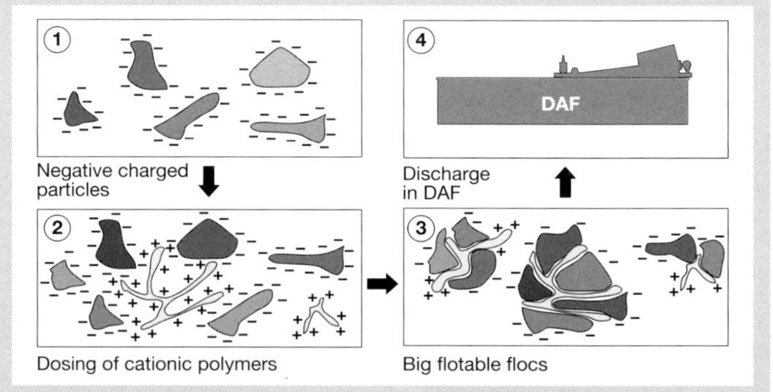

Figure 17. Flocculation by polymer additives.

With dual dosing, different chemicals can be combined to minimize the costs for agents and also catch colloidal substances. Combining salt additives such as alum or poly-alum chloride (PAC) with polymer and inorganic adsorbents such as bentonite is possible.

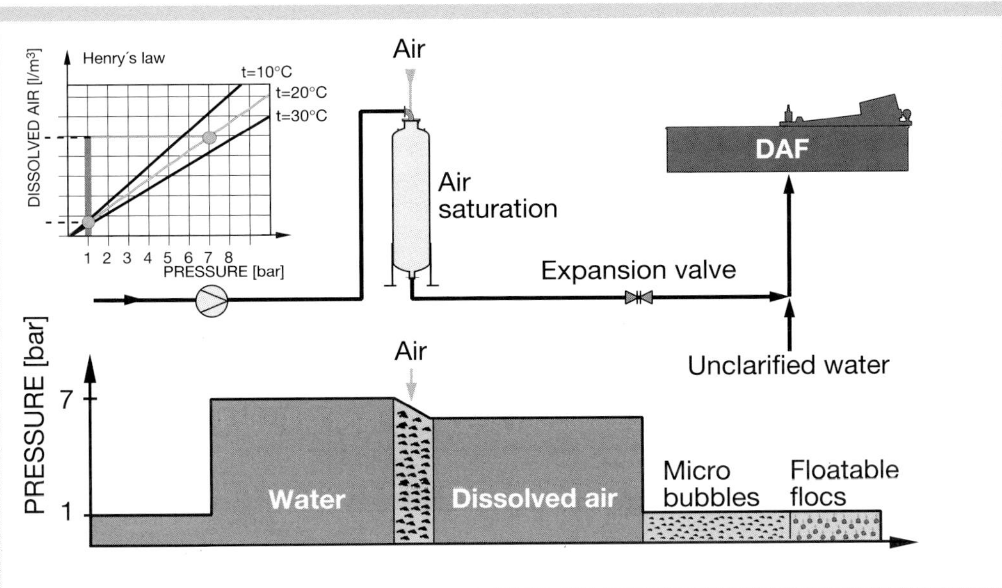

Figure 18. Principle of dissolved air flotation.

Before entering the clarification tank, some untreated water or a corresponding amount of clarified water is saturated with air after raising the pressure to 7 bar. No air bubbles may remain after aeration since these would collect smaller bubbles and reduce the flotation efficiency. After depressurizing the saturated air and water mixture to atmosphere pressure, very small air bubbles appear within a very limited size range as Fig. 18 shows.

After mixing the aerated water with the unclarified process water at the inlet of the clarification tank, flocs already formed attach to the air bubbles, rise to the surface, and form a stable layer of sludge that is removed by a paddle. Process water equally distributes around the periphery. The flow velocity reduces steadily in the outward direction. Clarified water is drawn off below the rim as Fig. 19 shows. Essential for efficient clarification is an optimum interplay of chemistry and physics. In other words, the adequate flocculants must be injected in the right quantity at the right place with optimum machine technology[13,14].

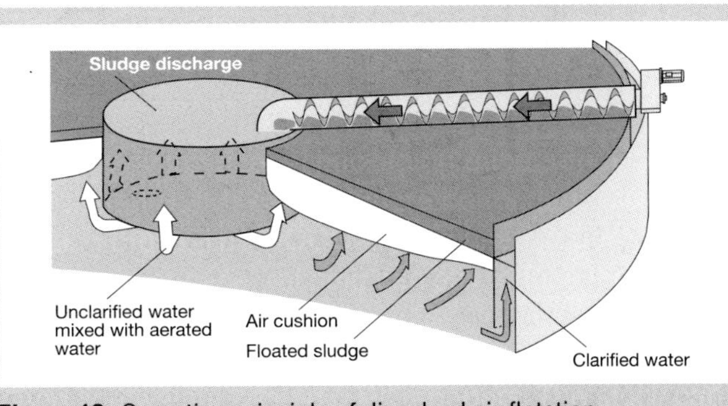

Figure 19. Operating principle of dissolved air flotation.

2.2 High-quality writing and printing paper grades

Figure 20 illustrates the process layout with peripheral systems. As in the standard system, two processing loops exist. In contrast to the standard system, an additional MC slotted screening stage is after MC hole screening. For optimum brightness and more efficient de-ashing, a washer replaces the disk filter at the end of the first loop. This allows controlled reduction of fines and fillers. Stock is then recovered from the filtrate of this washer in a separate DAF (DAF 3).

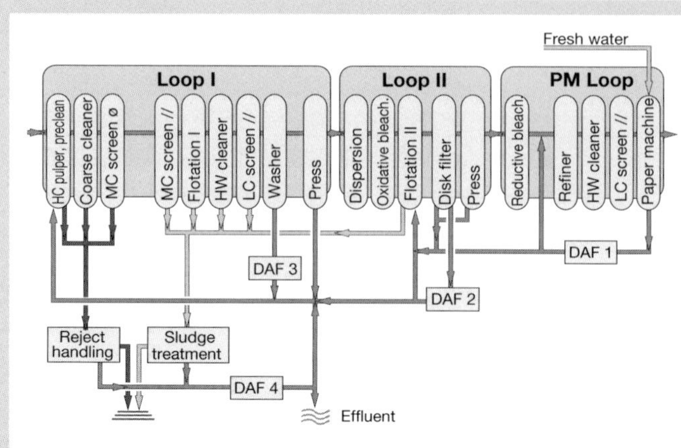

Figure 20. Concept for high-grade writing and printing paper grades.

In the case of high-grade writing and printing, bleaching always occurs in two stages. Peroxide bleaching not only increases brightness but also reduces the fiber mottling effect by brightening individual brown fibers. Reductive bleaching with dithionite or FAS then contributes to an additional increase in brightness and color stripping (See Chapter 9). Thanks to optimized bleaching, no need exists for color compensation in the machine chest that would otherwise be necessary.

Dispersing uses a high specific energy demand of more than 100 kWh/t. A refining stage is also necessary after reductive bleaching. This post-refining reduces the 14 and 30 mesh fraction and adapts coarse TMP fibers in the newsprint furnish to the more refined TMP stock for manufacturing SC and LWC papers. This greatly improves smoothness and especially printability[15–19].

Due to the chemical fiber content, deinked pulp (DIP) prepared in this way has significantly better strength characteristics than TMP or groundwood. For lightweight SC and LWC grades made exclusively of virgin pulp stock, minimum strength requirements must occur by adding chemical fibers to the TMP. When using DIP as an additional component, the chemical fibers contained therein largely make adding virgin chemical fibers unnecessary. Total raw material costs can therefore decrease by optimizing the furnish.

2.3 Market pulp

The most demanding process is the production of market DIP from RCF. A prerequisite is the exclusive use of wood-free grades as furnish using sorted office waste. The mechanical fiber content should be below 3% as much as possible and must never exceed 5%. To meet the highest quality demands, the stock preparation line includes an additional loop compared with Fig. 20 as Fig. 21 shows.

Loop I is similar to that in the process example for writing and printing paper grades. Loops II and III include two additional full-flow flotation stages. Two-stage dispersion occurs in the second and third loops that are separated from each other by a thickening stage to 30%. Only this repeated sequence of dispersing and flotation fulfills the high cleanliness requirement at the end of the system.

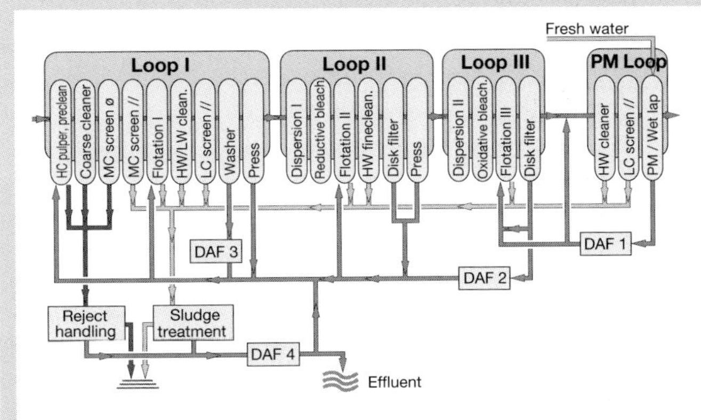

Figure 21. Concept for market DIP from wood-free mixed office waste (MOW).

Table 4 shows a comparison of input and final values achieved for the parameters brightness, dirt specks, and content of stickies. The change of technical parameters in individual process stages is detailed in Table 5.

Table 4. Efficiency of a MOW system.

Parameter	Untreated stock	Finished stock requirements
Brightness*	60% ISO	80%–85% ISO
Dirt speck area	2 800 mm^2/m^2	< 10 mm^2/m^2
Stickies	12 000 mm^2/kg	< 20 mm^2/kg
Ash content	21%	5%

*without UV luminance

Table 5. Efficiency data of unit processes for MOW processing.

Technolog. parameters		Brightness change	Dirt speck area reduction	Stickies area reduction	Ash reduction
Unit process		% ISO	%	%	%
Slot screening MC/LC		0	20	80	0
Flotation I		+3	65	30	30
Washing		+3	-20	0	85
Centrifugal	heavy weight	0	30	30	0
cleaning	light weight	0	0	30	0
Dispersion I		-1	65	70	0
Reductive bleaching		+8	10	0	0
Oxidative bleaching		+10	10	0	0

An additional advantage of the three-loop system for recycled fiber processing is that it allows a reductive-oxidative bleaching sequence (dithionite-peroxide). The first reductive bleaching stage in loop II is less susceptible to heavy metals than oxidative bleaching. It is also insensitive to catalase that is destroyed by heat treatment in the dispersion stage (See Chapter 9).

Peroxide bleaching at the end of the system therefore occurs under optimum conditions with clean stock and loop water from which any heavy metals have already been removed. Residual peroxide does not require decomposition. It helps with stock conservation by restricting bacteria growth.

When planning RCF lines of this kind for manufacturing a chemical pulp substitute, careful account is always necessary for cost-effectiveness. Besides the high machinery investment costs incurred, the low yield of about 65% also requires consideration. Careful design and the latest technology contribute to long-term profitability [20].

2.4 Tissue

The main goal in tissue production from RCF is to reduce the filler content of the furnish. Ash removal also involves fiber fines removal. At high system de-ashing rates, ash removal is less selective, i.e., the inorganic part of the sludge removed can be below 50%. Total stock yield may then fall to 60% or even less. Figure 22 summarizes test results on numerous tissue stock preparation lines. Using these statistics, residual ash adherence to the fibers is 1%–1.5%. This does not influence the tissue manufacturing process or the product quality. For cost-effective tissue stock preparation lines, the ash content of the finished stock should therefore only be decreased as far as absolutely essential.

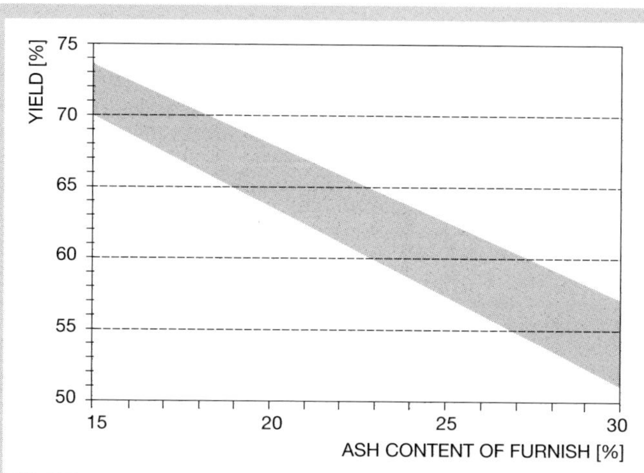

Figure 22. Stock yield vs. ash content in a tissue system.

Figure 23 shows the total system for a tissue stock preparation line. Since the furnish comprises wood-free recovered paper grades, the content of offset printed paper is low. Only a deinking flotation stage is necessary. After MC hole screening, slotted screening is also included in the MC section. Since no LC screening stage exists, only small-size machinery is necessary for thickening to 30%

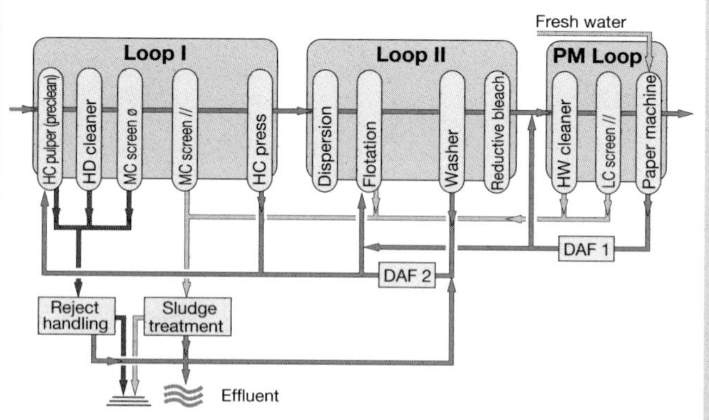

Figure 23. Tissue system (standard and wood-free furnish).

before dispersing. This reliably breaks down toner particles. After flotation under neutral conditions, a washer is also part of the line. This removes fillers while also functioning as a thickener. At 8%–10% stock consistency, reductive bleaching can easily be integrated afterwards. Stock is recovered from the washer filtrate of the DAF 2. The washer sludge is thickened with the fine rejects and flotation sludge[21,22].

The microflotation stage of DAF 1 does not perform cleaning but merely recovers fibers from white water II. For this reason, the floating material normally returns to the mixing chest. Since microflotation requires no auxiliary filtering stage, the flotation sludge exclusively comprises solids from the white water. Using the DAF 1 stage for controlling ash content is therefore also possible. For this purpose, the microflotation sludge is treated partly or totally with the other sludge.

In the example shown in Fig. 24, the stock consistency in loop I after MC hole screening decreases. LC slotted screening with extremely narrow slots follows the cleaners. The advantages of this arrangement are highly efficient stock cleaning at the beginning of the line and the possibility of including a washer in loop I because of low stock consistency. With the second washer in loop II, this results in an ash removal with this system higher than that of the standard process described above. For optimum brightness, two-stage bleaching (oxidative and reductive) is best.

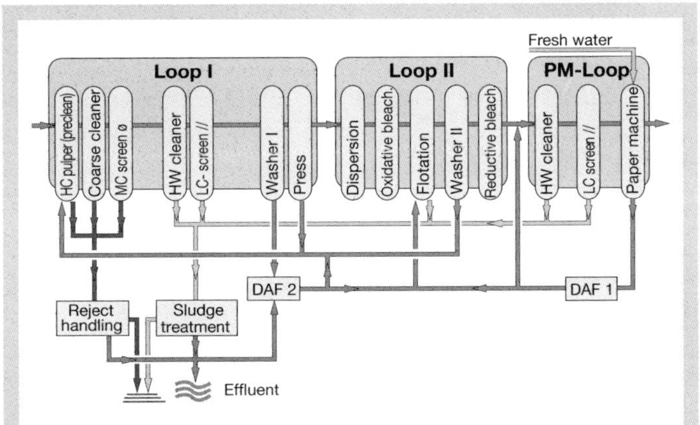

Figure 24. Tissue system (high-grade, wood-free furnish).

If the furnish also contains mechanical fibers, two flotation stages should be provided. Figure 25 shows an example of such an arrangement. In the same way as for writing and printing paper grades, the first flotation stage is already integrated into loop I. After the cleaning stage and LC fine slot screening, the first washer provides thickening and filler removal. The remainder of the line is identical to the high-grade tissue system for wood-free furnish described above.

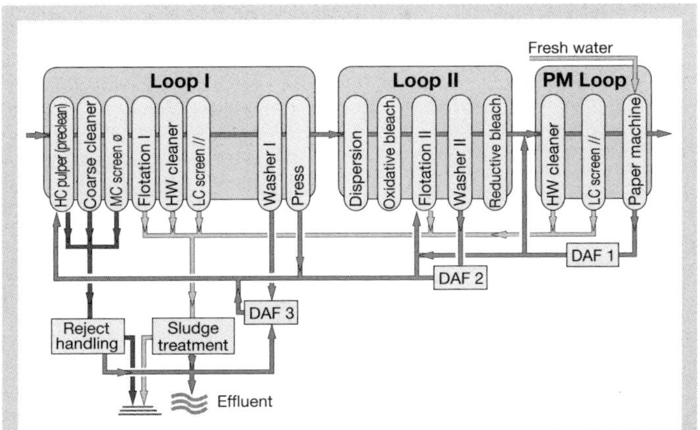

Figure 25. Tissue system (high-grade, wood-containing furnish).

3 Packaging paper and board grades

During conception and design of stock preparation lines for graphic paper grades, the major objective is to attain optimum optical characteristics (Compare Fig. 2). The development of adequate strength characteristics is a prime objective in the stock preparation process for packaging grades. On packaging paper grades, the requirement for optical characteristics has increasingly become a focal point in processing recovered paper. The requirement for the best possible runnability of the paper machine points is similar. Only securing sufficient basic cleanliness can avoid problems with deposits and associated downtimes for cleaning.

The cleanliness problems are intensified by the fact that the quality of the available raw material will decrease more. This applies primarily to Central Europe where the proportion of recycled fibers used in liner and fluting (corrugating medium) grades has been almost 100% for many years. This trend is now becoming apparent globally with various chronological delays. As a consequence, manufacturers of packaging papers and board must deal with an increasing proportion of detrimental substances, an increase in ash content, and a decline in fiber quality.

An additional problem encountered is the disposal of a growing volume of solid rejects, sludge, and highly loaded effluents resulting from the impaired quality of the recovered paper grades. This makes adapting the entire papermaking process to these new conditions necessary. Recycling process technology must be upgraded. The additional development of established technologies, introduction of necessary additional process stages, consequent water management, and advanced handling of rejects and sludge are indispensable in meeting the challenge of future packaging paper production.

Figure 26 describes in detail a process example for recovered paper processing for the production of liner grades. The processes for the production of corrugating medium are very similar to those for liner. The critical differences are the less stringent demands put on the optical cleanliness of the medium. This essentially results in a lower screening requirement. The slot widths in the fine screening stage can be adjusted wider than in the production of test liner.

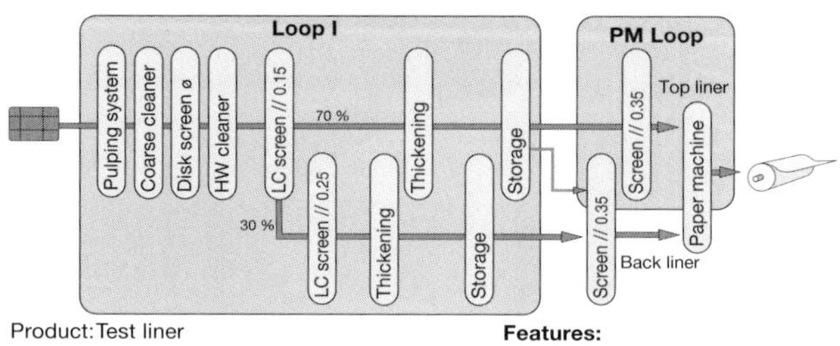

Product: Test liner
Furnish: Recovered household/ supermarket grades
Yield: Approx. 93 %
Effluent: Approx. 2.5 L/kg

Features:
- 2-loop system
- Storage possibilities
- Loop separation
- Screening carried out in loop I, in approach flow police function only

Figure 26. System layout for test liner production.

3.1 Test liner

Available raw material is low quality recovered paper from households and supermarkets as post-consumer material. The system illustrated as an overview in Fig. 26 and in more detail in Fig. 27 relates to Central European conditions where the strength potential of the untreated stock is low. The process design is simpler and contains fewer process stages than the stock preparation systems for the production of graphic paper grades previously discussed. Basically, technically sophisticated processes are also desirable in packaging paper and board production, but the requirements of the final product do not permit sufficient leeway for integration of additional fiber treatment and cleaning stages. Each individual machine must therefore be examined for its economic profitability even more closely than with graphic paper grades.

The continuous pulper operates at a stock consistency between 4.5%–5.5%. The vertical design permits safe ragger operation. The ragger removes strings, flat, and other spinning contaminants. Various types of contaminants accumulate in the pulper trough. An efficiently operating de-trashing system is therefore a prerequisite for trouble-free operation of the pulper system. Even on recovered paper grades that are easily slushable as in this example, a dump tower is essential following the pulper.

No defined slushing time that would be the same for all types of fibers is guaranteed by continuous operation of the pulper. The actual dwell time in the pulper sometimes deviates considerably from the mathematical average dwell time. The specific slushing operation is as short as possible for reasons of efficient operation. This results in the flake content immediately after the pulper remaining at 20%. Only after a storage in the dump tower for at least 30 min at a temperature of at least 50°C can the stock be deflaked and homogenized to such an extent that screening with small holes and narrow slots becomes possible.

The protector system contributes to an efficient separation of heavy-weight particles using the principle of centrifugal force. The special feature of this cleaning system is its two-stage cleaner arrangement. A continuous flow of rejects enriched with heavy particles is withdrawn from the first stage high density cleaners. After dilution to the low consistency range around 1%, this partial stream is cleaned very efficiently in the clean-

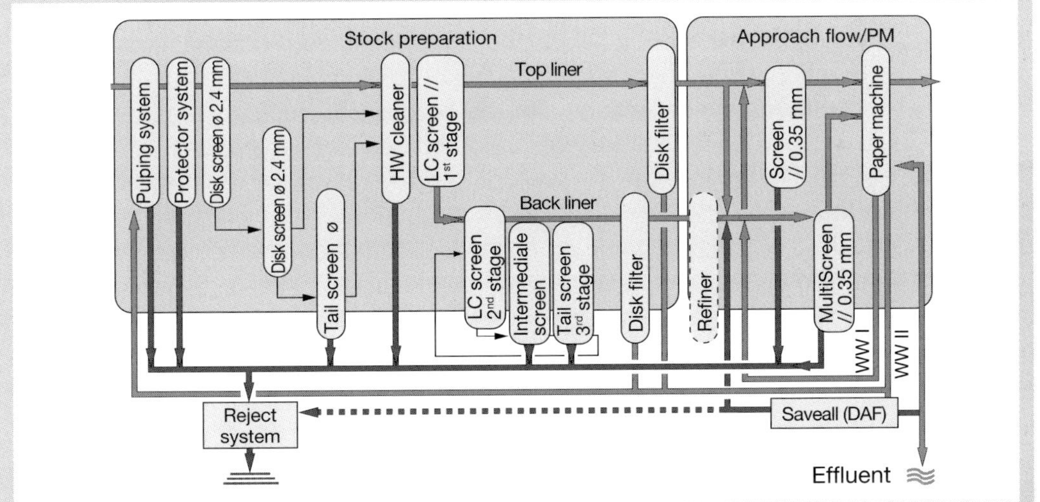

Figure 27. Test liner production showing total plant concept.

ers positioned in the second stage. The stock suspension at approximately 3.5% consistency then passes the hole screening system. Disk screens reduce the residual flake content. Disk screens are more robust than cylindrical screens and have a much higher deflaking potential (See Chapter 5). The holes of the disk screen are about 2.4 mm. If the deflaking potential in the final stage is sufficiently high, cylindrical screens can be used alternatively in the first stage and possibly in the second stage. To provide a recycled fiber quality equivalent to stock having passed disk screens with 2.4 mm holes, installation of 1.6 mm holes in cylindrical screens is recommended.

The first and second stage is always fed forward while the accept of the tail screen can be led forward or cascaded back in front of the second stage as Fig. 27 shows. The flake content after the pre-screening may not exceed approximately 4% to ensure good runnability of the following fine screening and cleaning stages.

The stock consistency is then reduced to approximately 1.2% and fed to the full-stream cleaner plant. Modern cleaner plants work at a very high efficiency. Avoiding an increased concentration of the separated fine heavy particles by discharging the complete rejects of the cleaner plant is necessary. This means that the cleaner rejects are thickened in a sedimentation or filtration unit. The overflow from sedimentation or the filtrate from filtration is then released to the effluent treatment plant.

The decisive process contributing to a clean stock is fine screening in the LC range. Modern screens used here have slots of 0.20 or 0.15 mm width in the first stage. The accept of the first stage is fed forward to the top liner of a two-ply test liner while the rejects are screened with a 0.25 mm slot width and used as pulp for the back liner. Note that in this line not only the contaminants but also the residual flakes accumulate. Handling the amount of nonfibrous materials requires the intermediate screening that Fig. 27 shows.

The basic rule is that the screening principles within a subsystem may not be changed. This means that all screening units within a screening system must have slotted or hole screens. In this instance, an intermediate screening unit with hole screen is intentionally used for the accepts in the second screening stage. This machine does not function as an additional screening stage. Its only purpose is to separate some rejects with heavy concentration at this location. For this reason, the accepts from this intermediate screening will also be fed to the final stage that has slots as do the first and second stages. Because of the lower reject content, this final stage can operate at a clearly lower fiber loss than would be possible in an arrangement without intermediate screening. With this method, reject is discharged at two locations from the fine screening subsystem: the intermediate screening stage and the final stage.

Instead of this intermediate screen, a deflaking machine before the tail screen can be considered. In this case, the reject particles are partly disintegrated. The removal efficiency of the fine screening is therefore lower. In addition, the lifetime of the internal deflaker components becomes less due to the high amount of dirt.

After screening, the accept streams are thickened and stored. The consistency in the storage tower is normally about 4.5%. A loop separation of the stock preparation loop from the paper machine is therefore achieved. In addition, this thickening has use for de-ashing and fiber fines removal. This is possible by filtrate treatment in a dissolved air flotation unit.

In the approach flow system, a brushing refiner can be placed before screening is performed with 0.35 mm slots. For higher paper machine speeds above 1 000 m/min, deaeration must be considered. In contrast to graphic paper systems, a combination of deaeration and cleaners is not feasible for two reasons. First, the headbox consistency is increased with modern paper machines to 1.5%. Because of the former technology used today and the consistencies above 1.2%, the cleaning efficiency is drastically lower. Second, the stock is already cleaned sufficiently after having passed the full stream cleaner plant in the stock preparation line.

For this reason, no further cleaner stage in the approach flow system is considered. A disk filter or a dissolved air flotation unit may be provided as a save-all. If a disk filter is used, the entire thickened stock must be returned to the mixing chest since the sweetener must not be lost. If DAF is provided in place of a disk filter, the possibility exists to discharge some floating material as the broken line in Fig. 27 shows. The wire section of the paper machine is consequently used as a washer. The white water II especially contains mainly fines and fiber particles that do not contribute to sheet formation but that could negatively affect retention and consistency. A targeted fine reject removal results in upgrading the low-grade furnish.

One must consider that efficient cleaning and screening is only feasible when a certain amount of material can be rejected. The yield of such a system is therefore only 93% or even less. This is primarily due to the coarse rejects of the furnish. The fine rejects of the fine screening are approximately 1.5% by weight. For cleanliness, the macro stickies content is perhaps the most important parameter. Figure 28 shows how macro stickies are reduced in such a system.

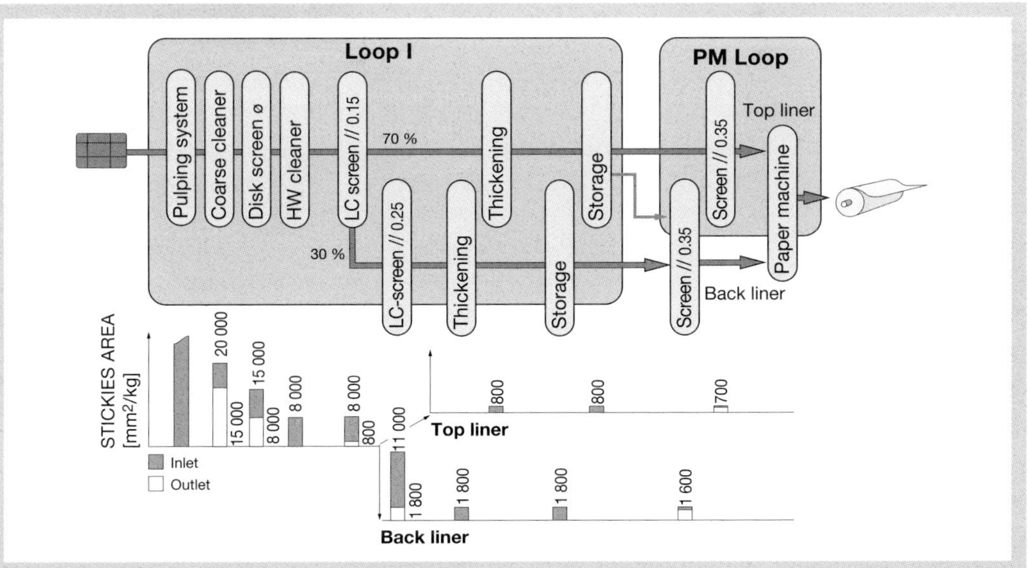

Figure 28. Test liner production showing reduction of macro stickies.

The measurable macro stickies are about 20 000 mm^2/kg after slushing with big variations depending on the recovered paper processed. The accept from the 0.15 mm slot screening for the top liner decreases to less than 1 000 mm^2/kg at the end of the stock preparation system. The accept for the back liner is still less than 2 000 mm^2/kg although the stock entering the 0.25 mm slot screening is loaded with macro stickies of more than 11 000 mm^2/kg.

In discussing dissolved substances and other fine particles that cannot be treated in screening units, the following picture is the result.

The state-of-the-art is to open the system to approximately 4 L/kg effluent corresponding to approximately 5.5 L/kg fresh water as Fig. 29 shows. Experiences from various plants show that under these condtions dissolved substances are washed out in a sufficient proportion (COD enriches only to approximately 7 500 ppm).

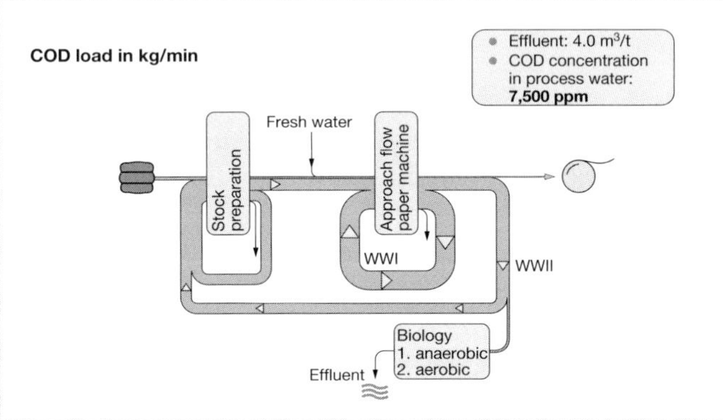

Figure 29. Packaging papers showing limited effluent volume.

If the effluent is further reduced, the detrimental substances accumulate dramatically. This results in impairment of strength properties, reduced runnability of the paper machine, corrosion, and odor problems. Figure 30 shows these conditions. With completely closed systems, the COD concentration reaches 30 000 ppm and more.

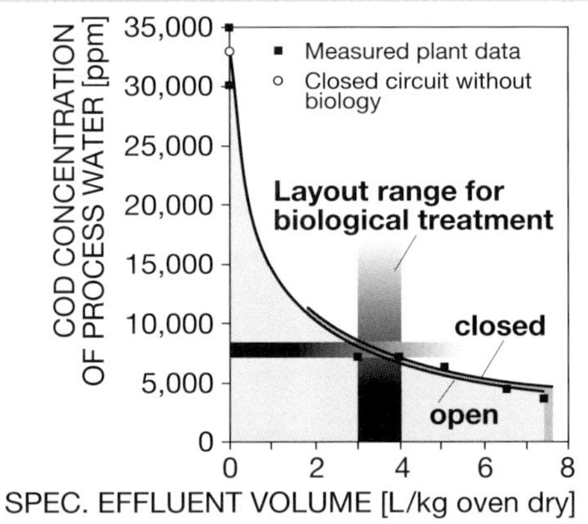

Figure 30. Packaging papers showing COD vs. specific effluent volume.

The COD concentration of the loop water and simultaneously of the effluent increases at an accelerating rate with reduction of the specific effluent volume. This increase is limited only by the fact that even in zero effluent systems a small water volume leaves the system that must be supplemented by fresh water. Besides the small volume of water that is discharged with the thickened rejects, this is essentially the water that is transported with the moist paper web from the press section into the dryer section and that subsequently evaporates in the dryer section. In packaging paper production, this is approximately 1.5 L/kg paper.

Compared with writing and printing paper grades, the required optical properties for packaging paper grades are less demanding. Integrating a biological treatment into the process without impairing product or process technology is possible with packaging grades as Fig. 31 shows (See Chapter 12). Using biology as a "system kidney" for dissolved organic substances results in reasonable loop closure down to zero effluent without running into trouble with micro stickies and dissolved substances[23,24].

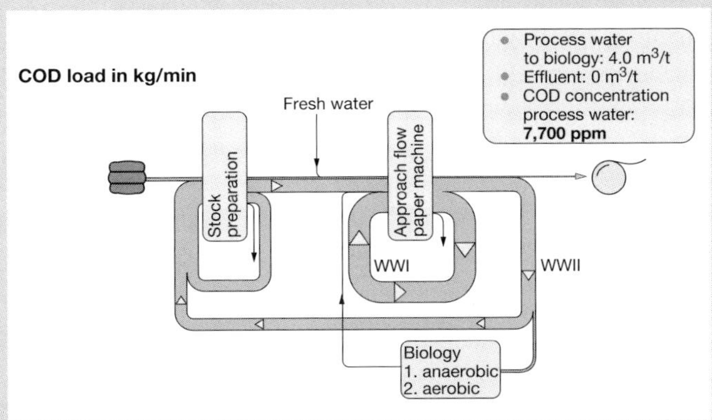

Figure 31. Packaging papers showing zero effluent with integrated biological process water treatment.

3.2 Fractionation systems

The process discussed so far does not provide for fractionation (See chapter 5). The reason for this is the fact that no significant fiber length displacement is possible by single-step fractionation on low grades such as mixed recovered paper and department store waste. Even if long fiber components can be separated through two-stage fractionation, little improvement of fiber quality will occur because the refining potential of the fibers can be too low if they have been recycled several times. With better raw material qualities, a targeted long fiber treatment could well be significant. The following text will discuss possible designs involving fractionation systems or combinations of screening and fractionation.

Fractionation is a specific application of screening with rather different objectives. The purpose of fractionation is to generate two pulp fractions with clearly differentiated qualities such as fiber length distribution or shives content. This separation is a prerequisite for a selective processing of each fraction.

As Fig. 32 shows for most virgin and recycled fiber grades, considerable overlap exists between the

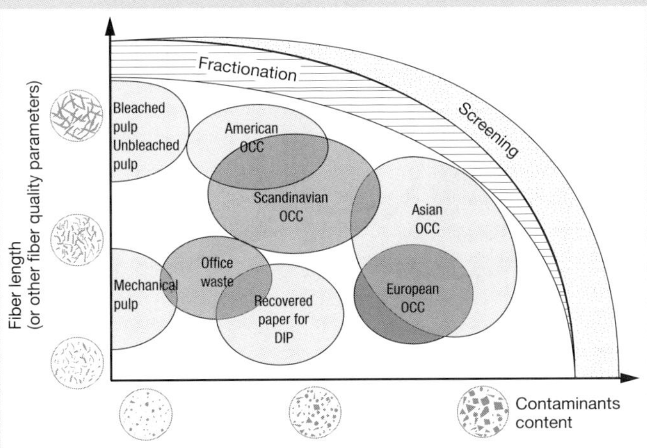

Figure 32. Fiber stock characteristics globally.

screening and fractionation processes. For example, bleached and unbleached chemical pulps with low contamination have advantageous fiber properties such as length, bonding potential, and flexibility. Mechanical fibers despite their extractive contents are typically low in contaminants but have poorer fiber properties such as low strength potential. Recovered paper as a raw material for graphic or packaging paper and board production covers a wide range of contamination and fiber qualities. While American and Scandinavian old corrugated container (OCC) still has outstanding fiber qualities with a medium contaminant content, the fiber properties of Central European OCC are deteriorating with a continuing increase in contaminants. The main problems here are stickies and waxes whose characteristics are so varied that separation is very difficult.

Consequently, many stock preparation strategies use combined fractionation and screening either in individual machines or subsystems. This provides efficient removal of contaminants and generates fiber fractions that can be optimally treated.

The fractionation of OCC grades is a well-established process for producing fluting, test liner, and folding boxboard (white lined chipboard). In such processes, the recovered paper is prepared in a very simple way. Pulping is followed by coarse screening. The coarse screening system uses plate screens with holes, and the tail screen is not pressurized. The tail screen primarily operates in a washing mode to minimize fiber losses. The fiber fractionation stage using hole screens follows this. The prerequisite for good screening and fractionation results is use of relatively clean recovered paper with an adequate long fiber content as a raw material. The long fiber fraction can then be dispersed and refined for homogenization of the stock and for strength improvement, respectively.

In Asia and Europe, the amount of contaminants of the OCC grades has increased over the last few years with a deterioration in fiber quality. The cleanliness of the short fiber fraction attainable with the process described above is no longer adequate to meet current product requirements with demands for higher runnability on the paper machine. As a result, the conventional fractionation process must be modified.

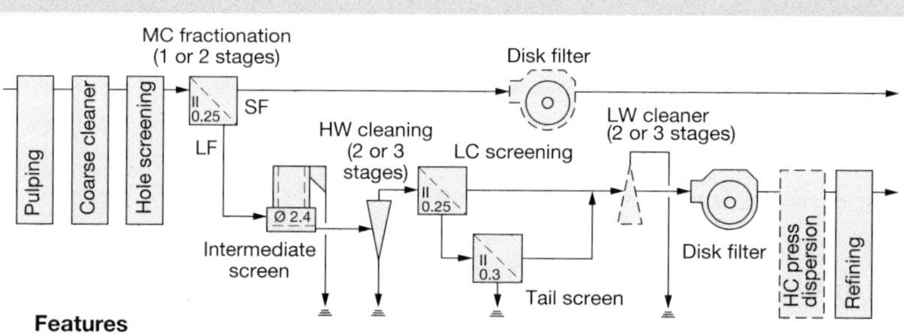

Features
- Single or double slot fractionation with narrowest possible slots
- Intermediate screening and deflaking of LF fraction with CombiSorter
- HW cleaning for fine sand removal and reduced wear of subsequent LC screens
- Continuous operated tail screen ensures high stickies removal
- Optional LW cleaning for wax removal (AOCC)
- Optional dispersion for high-quality products

Figure 33. Fractionation system for OCC furnish (AOCC=American OCC).

The papermaker's challenge is to integrate the fractionation process into the stock preparation system with the greatest possible contaminant removal.

Figure 33 shows a modern system concept. The stock that has already been coarsely screened is fractionated in one or two stages depending on production output and quality demands. The narrowest possible slot widths are selected driven by operational reliability. This guarantees the highest attainable cleanliness in the short fiber (SF) fraction and the desired long fiber (LF) enrichment in the other fraction. The SF fraction is then thickened and stored for use as top liner in test liner production.

Since the enriched LF fraction also contains a high proportion of flakes and contaminants, an intermediate screening stage is necessary. This guarantees a reliable operation of the subsequent heavy-weight cleaning and the stickies-selective LC screening while simultaneously deflaking the pulp. Another point worth noting is that the tail screen in the LC screening system should operate with a continuous reject discharge and with a portion of the reject stream recirculated to handle the accumulated amount of contaminants. This is the only way to ensure an acceptably low contaminant content in the tail screen accepts.

If the recovered paper furnish contains wax, light-weight cleaners are recommended. Since the wax separation efficiency in light-weight cleaners is not sufficient, the LF fraction should then be treated in a disperger after thickening. This contributes to the necessary homogenization for optimum product quality. Simultaneously, the dispersion stage disperses printing inks and residual contaminants such as wax and stickies. The treatment of the LF fraction is completed by refining to improve strength properties. The prepared pulp is suitable for the back liner in test liner production.

3.3 Boxboard

The demands upon folding boxboard (white lined chipboard) and liquid packaging board are defined by their purposes as packaging and advertising medium. Particular emphasis is put upon high bending stiffness, ply-bond strength, smoothness, brightness, and cleanliness. In the food sector, microbial and chemical cleanliness are additional requirements. Multilayer board structure provides the possibility to select the most suitable raw material for each layer.

For example, the top layer can use bleached kraft pulp. A recovered paper prepared under deinking technology may be integrated into the top under-layer. The filler includes chemithermomechanical pulp (CTMP) or mixed recovered paper. The back layer again uses bleached kraft pulp[25].

The process technology of the individual process lines depends on the raw materials used. When using old newspapers and magazines for example, stock is prepared with the double flotation procedure already described for use as top under-layer.

The challenge and peculiarity of carton production is in the combination of the different stock components in sheet formation with the objective of securing the desired characteristics such as bending stiffness and smoothness. The individual stocks are prepared separately and are supplied over several individual approach flow sections to the respective formers. This strict separation of the fiber main streams is especially influenced negatively by mixing white waters in the wet section of the board machine.

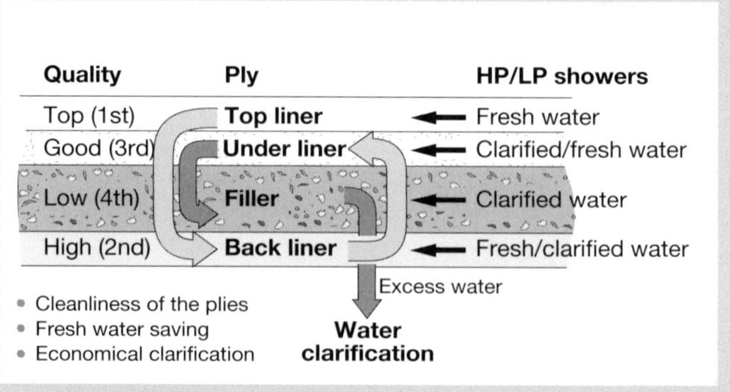

Figure 34. White water flows in multi-ply board machines.

Varying contamination levels and contaminant contents of the individual stocks require targeted control of water coming into the board machine required for the individual layers from "clean" to "less clean." Only with optimum water management can the normally demanded multitude of grades be controlled at the board machine without jeopardizing quality. Figure 34 shows schematically how fresh water can provide the maximum advantage in a cascade arrangement.

Basically, each layer has its own loop. The surplus water is always supplied to the next qualitatively lower layer. Providing individual fiber recovery in each layer would be too expensive. In lines that convert virgin pulp, only one fiber recovery is often provided in the filler line. In processes that process recycled fibers, at least two and in exceptional cases three separate recovery units are recommended. Besides the central fiber recycling in the filler line that is qualitatively the lowest standard, the layers that are supplied with bleached virgin fibers in grades having white top layers undergo additional separate filtration if suitable.

References

1. Göttsching, L. and Stürmer, L., Wochenbl. Papierfabr. 106(23/24):909(1978).
2. Howard, R.C. and Bichard, W., J. Pulp Paper Sci. 18(4):J151(1992).
3. Ferguson, L.D., Paper Technology 33(10):14(1992).
4. Göttsching, L., Wochenbl. Papierfabr. 103(19):687(1975).
5. Ortner, H. and Bergfeld, D., "Recycling of waste paper, present status – Future development", ASEAN Symposium on the Pulp & Paper Industry, ASEAN, 1991.
6. Weidhaas, A., Wochenbl. Papierfabr. 118(11/12):473(1990).
7. Siewert, W.H., Wochenbl. Papierfabr. 123(16):90(1995).
8. Niggl, V., Wochenbl. Papierfabr. 121(9):324(1993).
9. Borschke, D., Das Papier 52(2):51(1998).
10. Schwarz, M., Wochenbl. Papierfabr. 125(14/15):692(1997).
11. Stark, H., Papier aus Österreich (11):11(1992).
12. Schwarz, M., Wochenbl. Papierfabr. 123(18):792(1995).
13. Stark, H. and Schwarz, M., Papier aus Österreich (6):3(1994).
14. Hamm, U. and Göttsching, L., Wochenbl. Papierfabr. 123(10):444(1995).
15. Demel, J. and Möbius, C.H., Allg. Papierrundschau 111(24/25):746(1987).
16. van Raven, A. and von Troil, S., Wochenbl. Papierfabr. 119(2):891(1991).
17. Bergmann, W., Wochenbl. Papierfabr. 116(18):788(1988).
18. Novak, G. and Malesic, J., Das Papier 46(3):109(1992).
19. Bürger, G., Das Papier 47(5):233(1993).
20. Matzke, W. and Selder, H., Wochenbl. Papierfabr. 118(19):835(1990).
21. Siewert, W.H. and Horsch, G.R., "Operating results and experiences with the Escher Wyss VarioSplit", TAPPI 1984 Pulping Conference Proceedings, TAPPI PRESS, Atlanta, p. 611.
22. Kleuser, J.,"Deinking systems for tissue, writing and printing papers", 1990 APPITA Annual Conference, APPITA, Rotorua, New Zealand.
23. Börner, F., Dalpke, H.-L., Geller, A., et al., Wochenbl. Papierfabr. 110(9):287(1982).
24. Borschke, D., Schwarz, M., and Selder, H., Das Papier 50(7/8):444(1996).
25. Hanke, H., Braun, E., Siewert, W.H., et al., Das Papier 48(10A):74(1994).

CHAPTER 7

Deinking chemistry

Anders Lassus

Deinking chemistry

1 Introduction

The purpose of deinking is to remove printing inks and other substances that might effect the papermaking process or final properties of paper. Unsatisfactory separation gives a paper with low brightness and dirt specks. Larger ink particles > 50 μm are visible to the naked eye as black or colored spots in the paper. Smaller particles lower the brightness of paper and increase the need for bleaching.

In a deinking process, the ink should first detach from the fibers. Flotation or washing then separate ink and fibers from each other as Fig. 1 shows. A sequence of these two processes is also possible.

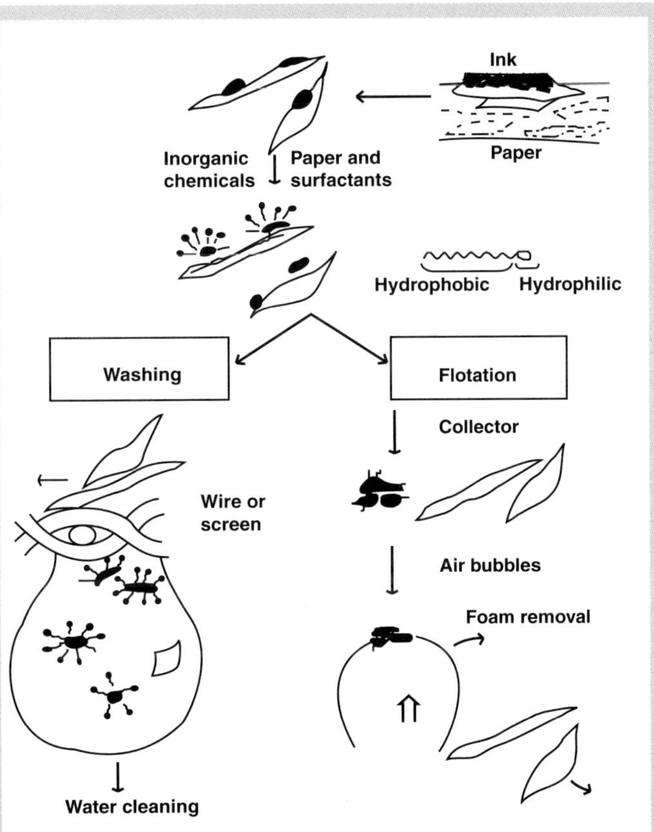

Figure 1. Deinking by flotation or washing[1].

Flotation is the most common method used. It provides a high yield of fibers, i.e., a low amount of rejects, and less efforts to purify the white water produced by the following process stages. The accept and the reject that comprises ink particles and stickies can be separated from each other according to the different surface properties of the particles in the recycled fiber pulp slurry.

In washing, the particles separate primarily according to their size. Small particles such as fillers and coating pigments are easily washed out. Washing the pulp has primary use when making low ash tissue paper.

Effective deinking of recycled fiber pulp based on household collected news and magazines can increase brightness by 15%–20% ISO. After pulping wood-containing recovered paper, the brightness is about 40–45% ISO. After deinking, the brightness is about 60% ISO. Bleaching can improve the brightness even more (See Chapter 9).

The efficiency of ink separation depends on the ink properties. Letterpress printing ink is easy to remove from fibers. Offset printing ink can cause problems that appear after the paper ages and the alkyd binder of the offset ink cures. Rotogravure printed paper is easier to deink than offset printed paper. Flotation deinking does not efficiently remove water-based flexographic printing ink, but washing can remove this water-based ink. Flexographic printing of graphic papers is not common in Europe. Chapter 8 describes the deinking of different printing inks and paper grades.

Deinking also depends on the paper quality used. Deinking of coated paper is generally more successful than deinking of uncoated paper. Recovered graphic paper contains the following:

- Fibers, fillers, coating pigments, starch, synthetic coating binder, and other paper chemicals
- Printing inks
- Adhesives and plastic films from converted paper
- Metal, sand, glass.

Removal of thermoplastic adhesives, plastics, and coating binders is as important as the removal of printing ink because these materials can cause problems in the papermaking process. These tacky substances are called stickies (See Chapter 11).

The requirement for chemicals in a deinking process varies considerably depending on raw materials, quality specifications of the deinked pulp, and the type of process. The flotation process is more chemical intensive than the washing process. Recovered papers of high quality such as (wood-free) office papers usually require less chemicals than household collected (wood-containing) recovered paper. The quality specifications for tissue paper can sometimes tolerate a few specks in the final product. The lowest chemical doses therefore occur when producing tissue from wood-free office paper with a washing process. The highest chemical input occurs when producing wood-containing graphic papers from household collected recovered paper using the flotation process.

2 Pulping

Recovered paper is disintegrated in water at a consistency of 5%–18% depending on the type of pulper used. Consistency in pulping is important for the chemical doses since these are calculated on the amount of air-dry paper. A higher stock consistency in a pulper or a drum pulper therefore means less water and a higher concentration of chemicals. This gives a higher chemical efficiency, the possibility to optimize the chemical doses, or both.

When necessary, surface active chemicals are added to the pulper to decrease adhesion of the printing ink to the fibers. Due to the influence of the chemicals and the mechanical shear forces, ink particles of different sizes are released to the aqueous phase.

In the pulping stage, the pH is high at approximately 9–10. Adding caustic soda gives this high value. Because of the added caustic soda, the fibers swell so the printing ink is removed more easily. A possible mechanism is that some binders in a printing ink can saponify and hydrolyze at high pH. This could improve the water solubility of the ink. Actual proof for this theory is not available[2]. When using fatty acid soaps as collector chemicals, high pH is very important.

Water glass, sodium silicate, is a water soluble polymer that can bind multivalent ions. It adsorbs on the surfaces of colloidal particles. The water glass improves dispersion of the printing ink and prevents it from redepositing on the fibers. Water glass is also a pH buffer that keeps the pH at a desirably high level.

The hydrogen peroxide added to the pulper works as a bleaching agent by reversing the yellowing of mechanical fibers caused by the caustic. Peroxide bleaching is most successful at high temperature and consistency. High temperature softens the thermoplastic stickies that occur in recovered paper as contaminants and makes their removal more difficult. The maximum stock consistency in the pulper depends on its design. The consistency is frequently not sufficiently high for effective bleaching. When drying and storing offset printed paper, chemical cross-linking occurs between the alkyd binders of the ink. The peroxide can break these bonds so the ink removes more easily[3].

Some metal ions such as iron, manganese, and copper can catalyze the decomposition of peroxide. Adding organic chelating agents such as ethylenediaminetetraacetic acid (EDTA), diaminetriethylenepentaacetic acid DTPA), or diethylenetriaminepentamethylenephosphonic acid (DTMPA) that react with the metallic ions can prevent the decomposition. The use of chelants in deinking is not common today (See Chapter 9).

Peroxide can also decompose due to microbiological action. Bacteria in the process water generate an enzyme called catalase that decomposes the peroxide. Different techniques are available to attack this problem. One method is to use biologically active com-

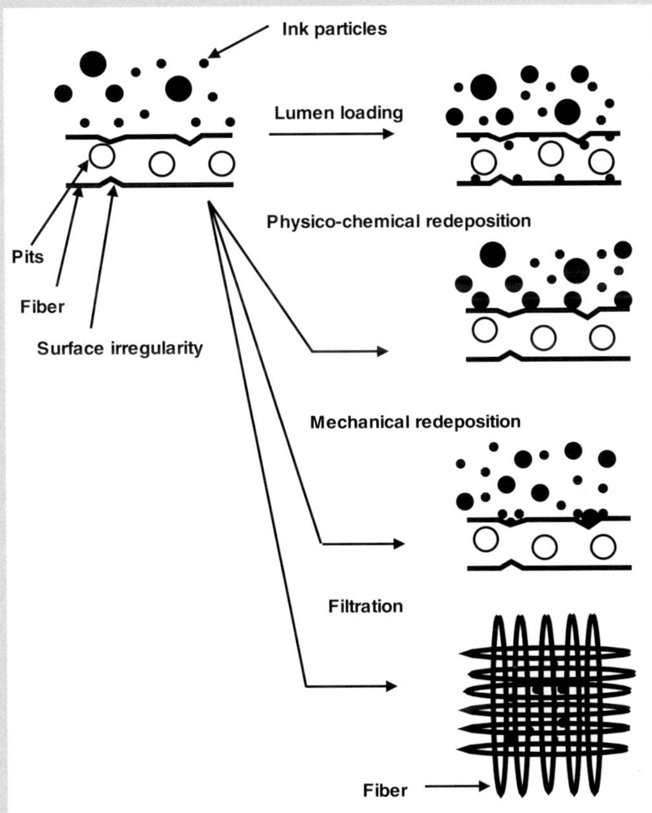

Figure 2. Different mechanisms of deposition of ink particles onto fiber[4].

pounds such as glutardialdehyde. Another method is to use elevated doses of hydrogen peroxide when necessary, i.e., so-called shock dosage.

3 Ink redeposition on fibers

When ink particles have detached from the fibers and dispersed into the aqueous phase, a risk exists that they will redeposit on the fiber surfaces if the colloidal stability of the ink particles is not sufficiently high. Three types of redeposition are lumen loading, chemical redeposition onto the fiber surface, and mechanical redeposition onto the fiber surface. Ink particles can also become entrapped in the fiber network by a filtration mechanism as Fig. 2 shows.

Lumen loading means that small ink particles pass through the pits of the fiber wall and enter the lumen. This phenomenon can occur independently from the surface properties of the particles. Hydrophilic and hydrophobic particles can redeposit by lumen loading.

Larger particles can redeposit on the fiber surface. This can occur by a physicochemical adsorption or by mechanical entrapment into fiber surface irregularities. To avoid redeposition, the ink particles and the fibers must be hydrophilic. The chemicals used in the pulper such as the surfactants, caustic, and water glass provide a hydrophilic character to the ink particles. If there is a partly hydrophobic character of the fibers it can contribute to ink redeposition. Such increased fiber hydrophobic character can be due to residual deinking chemicals deposited on the fibers or wood extractives originating from mechanical pulp.

4 Flotation

After pulping, the detached and dispersed printing ink has a certain particle size distribution. The composition of the ink and the hydrodynamic regime of the pulper determine the size distribution. If the ink disperses easily, the average particle size is considerably smaller than if the ink is difficult to disperse. Different separation processes function at different particle size ranges. The washing process is successful when the ink particles are very small. Flotation requires larger particles. Different literature sources provide information about the optimum particle size of the various processes[5–7].

Surface chemical properties of an ink also influence the degree of ink separation. Hydrophilic particles favor the washing process. Flotation requires hydrophobic particles. For successful flotation, the ink particles must flocculate to larger agglomerates. The surface properties of the agglomerates also must change, i.e., the hydrophobic character must increase. The flotation process will not separate hydrophilic particles.

Flotation is a separation method using the different surface properties of particles. The principle of the method is as follows. Air is introduced into a diluted fiber suspension of 0.8%–1.5% consistency. The water-repelling ink particles attach to the air bubbles and rise to the surface. The hydrophilic fibers remain in the water phase. The foam that contains the ink can be removed mechanically, by overflow, or by a vacuum extraction.

The foam also contains some fibers, fines, and ash (fillers and coating pigments). Flotation removes the fillers and pigments of the papers to some extent. This means the ash content of the reject is high. Practical experience has shown that the ash content of the raw material is very important for the amount of rejects and the foaming properties. The foam is richer in ash when the recovered paper contains magazines or other papers with a high content of fillers or coating pigments[8]. The binders of coated papers also increase the amount of foam. Flotation does not remove the sticky contaminants in the pulp slurry sufficiently. Flotation removes about 50% of the stickies[9,10]. For efficient removal of macro stickies, special screens and cleaners are available (See Chapter 11).

Size of bubbles covers a large spectrum and depends primarily on the type of air injector and the hydrodynamic parameters. One theory states that flotation is a first order process. The number of particles removed is proportional to the number of particles in the suspension. A flotation rate constant can use the time of flotation[11]:

$$dN/dt = -kN \rightarrow N/N_o = \exp(-kt) \tag{1}$$

where N is the number of particles per liter pulp slurry
t the flotation time
k the flotation rate constant that depends on type of contaminants, chemicals, temperature, type of pulp, consistency of the pulp, bubble size distribution, and flotation cell design.

In industrial flotation cells, using the assumption of a constant flotation rate based on the air/pulp slurry ratio requires care. This is because changing the air flow at constant bubble size distribution modifies the hydrodynamic behavior of the suspension especially at high air contents. Excessively high turbulence in the flotation cell might adversely influence the flotation result.

Julien Saint Armand and Perrin have experimentally determined the influence of the size of ink particles and flakes on the flotation rate constant[11]. The most effective flotation occurs when the particle size is about 20–40 μm. This is about the lower limit for particles that are visible to the naked eye.

The most effective flotation also occurs when the contact angle between water and the particle to be removed is 90°. The remaining particles (fiber and fiber fines) are fully wetted, i.e., their contact angle is 0°[12].

Chemicals that collect ink into greater particles while simultaneously giving these agglomerates a hydrophobic surface are collectors. The collector must also attach to the air bubbles. Before addition of the collector, the ink particles for removal from the fibers are too small and their surface properties are too hydrophilic for flotation. Collectors are added to the pulper or immediately before flotation. Adding in portions is also possible. One part can be added to the pulper, and the other part can be added to the flotation cells.

To fulfill these demands, the collector must be difficult to dissolve in water and be able to form hydrophobic surfaces. Different types of collector chemicals are available. They are fatty acids, semi-synthetic, and synthetic collectors.

5 Fatty acid collectors

5.1 Fats and fatty acids

Each natural oil or fat contains a mixture of fatty acids in proportions that depend on the source. Sometimes, one particular fatty acid predominates such as lauric acid in coconut oil. Nature seldom provides fatty acids in the free, unreacted state. In most oils and fats, they occur as triglycerides. These are esters in which one molecule of glycerol chemically combines with three molecules of fatty acid. Structurally, natural fatty acids have straight chains of carbon atoms with a terminal carboxyl group. These acids usually have an even number of carbon atoms. The most abundant fatty acids are those with 16 and 18 atoms.

At one end of the chain, the terminal carbon atom has three hydrogen atoms attached to it as Fig. 3 shows. At the other end, the terminal carbon atom forms part of the carboxylic (COOH) group that gives the weakly acidic properties of these chemical compounds. Most or all the intermediate carbon atoms each have two hydrogen atoms attached. In the case when all the intermediate carbon atoms have two hydrogen atoms attached, the fatty acid is saturated. If only one hydrogen atom is attached to each of two adjacent chain carbon atoms, the fatty acid is unsaturated. If two or more pairs of carbon atoms have only one hydrogen attached, the fatty acid is polyunsaturated.

Fatty acids result from breaking down the triglyceride molecules of natural oils and fats into their constituents by hydrolysis. The products are glycerin and fatty acids. After hydrolysis or splitting, the crude fatty acids can be purified by vacuum distillation. The number of double bonds can be decreased by allowing the fatty acid to react with hydrogen gas. This process is called hydrogenation.

The most common fatty acids obtained from natural fats and oils are the following:

- Linoleic acid, C:18:2 (18 carbon atoms, 2 double bonds)
- Linolenic acid, C:18:3 (18 carbon atoms, 3 double bonds)
- Oleic acid, C18:1 (18 carbon atoms, 1 double bond)
- Palmitic acid, C16:0 (16 carbon atoms, no double bonds)
- Stearic acid, C:18:0 (18 carbon atoms, no double bonds).

An isomer of 18:3 occurring in coniferous wood is pinolenic acid to distinguish it from linolenic acid. The difference between these two species is the position of the double bonds. In pinolenic acid, the double bonds are at carbon atoms 5, 9, and 12. In linolenic acid, they are at 9, 12, and 15.

The properties of a fatty acid or a fatty acid mixture depend on the average chain length and the number of double bonds. The following describe the composition:

- Iodine value
- Titer.

The iodine value describes the number of double bonds. The iodine value of pure linoleic acid is 181. The iodine value of oleic acid is 90, and the iodine value of stearic acid and palmitic acid is 0. The titer is the crystallization point of the acid, i.e., the temperature at which the fatty acid solidifies. Titer increases with an increased chain length and an increased degree of saturation. The titer of linoleic acid is -5°C, oleic acid is 14°C, palmitic acid is 63°C, and stearic acid is 70°C[14].

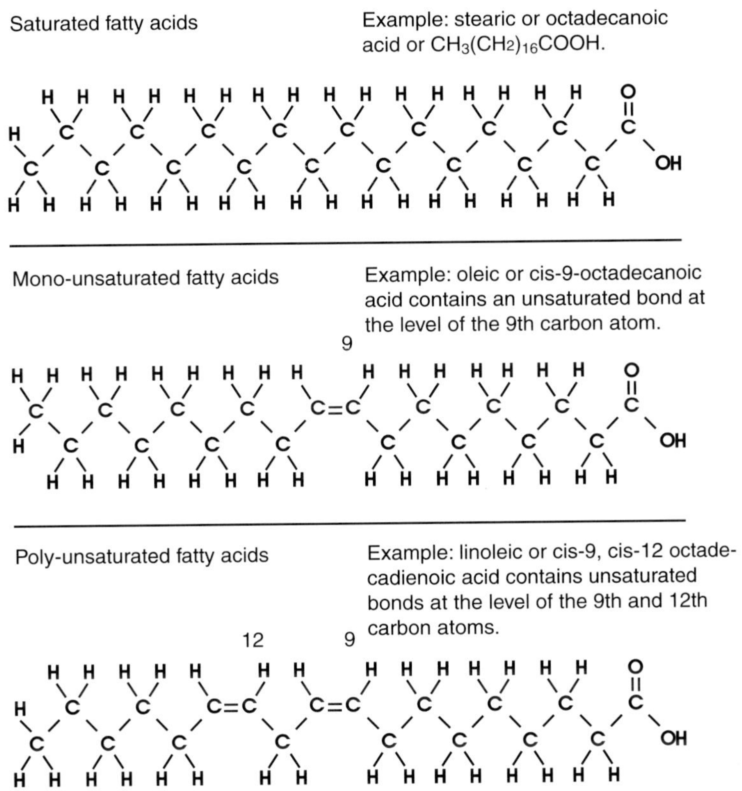

Figure 3. Common C18 fatty acids[13].

5.2 Soap

Water alone is not a very effective detergent. During prehistoric times, adding ash from burned plants improved the cleaning effect by making the water alkaline. Modern soap making developed from this method.

Soap is an anionic surfactant that consists of a long hydrocarbon chain to which a water soluble (hydrophilic) group attaches. The water soluble group tries to orient itself toward water. The hydrocarbon group is water insoluble (hydrophobic) and tries to orient away from water. Figure 4 shows the sodium soap of stearic acid.

$$\underbrace{CH_3(CH_2)_{16}}_{\text{Hydrophobic}}\underbrace{COO^-\ Na^+}_{\text{Hydrophilic}}$$

Figure 4. Sodium soap of stearic acid.

A compound such as this containing hydrophilic and hydrophobic portions will concentrate at the boundary surfaces of water. Such surfaces are the interface between air and water and the interface between a solid surface and water.

Soap is usually manufactured either by direct saponification of a triglyceride or by neutralization of a fatty acid. The reaction rate of fatty acids with alkali is much faster than the corresponding reaction rate of triglycerides.

Triglyceride saponification

This reaction involves the cleavage of three ester groups usually in a stepwise fashion[15]:

$$\begin{array}{lllll} RCOOCH_2 & & & & OH \quad OH \quad OH \\ \quad\ \ | & & & & \ \ | \qquad\ \ | \qquad\ | \\ RCOOCH & + 3NaOH & \rightarrow & 3\,RCOONa + & CH_2\text{-}CH\text{-}CH_2 \\ \quad\ \ | & & & & \\ RCOOCH_2 & & & & \\ \textit{Fat} & \textit{Caustic soda} & & \textit{Soap} & \textit{Glycerin} \end{array} \tag{2}$$

Fatty acid neutralization

Fatty acid neutralization involves a rapid abstraction of fatty acid protons by a hydroxide anion followed by sodium cation transfer to give the products of the neutralization reaction[15]:

$$\begin{array}{lllllll} RCOOH & + & NaOH & \rightarrow & RCOONa & + & H_2O \\ \textit{Fatty acid} & & \textit{Caustic soda} & & \textit{Soap} & & \textit{Water} \end{array} \tag{3}$$

Water solubility of soap

The sodium soaps of fatty acids are water soluble. The viscosity of the liquid strongly depends on the temperature. Soaps made of unsaturated fatty acids can remain in liquid form at lower temperature and at higher concentrations. Soaps made of hardened (hydrogenated) fatty acids need an increased temperature to be in liquid form. For example, a 5% sodium stearate solution requires at least 70°C to be a low viscosity liquid. At room temperature, the 5% sodium stearate is a solid gel. Sodium oleate at 20% concentration is a liquid at room temperature.

When sodium soap is used in hard water, its cleaning effect is remarkably less. The reason is simple. The sodium soap reacts with the calcium ions of the hard water and forms calcium soap that is practically insoluble:

$$\begin{array}{lllllll} 2\,RCOONa & + & Ca(OH)_2 & \rightarrow & (RCOO)_2Ca & + & 2\,NaOH \\ \textit{Na-soap} & & \textit{Lime} & & \textit{Ca-soap} & & \textit{Caustic soda} \end{array} \tag{4}$$

Chemical analysis such as titration can determine water hardness. The hardness is expressed as parts per million calculated as $CaCO_3$. In Europe, expressing water hardness in German hardness units, °dH, is common. One °dH is 10 mg CaO/L and corresponds to 17.8 ppm $CaCO_3$.

5.3 Handling fatty acid and soap

Fatty acid and soap addition to the deinking process can take the following forms:

- Fatty acid liquid
- Fatty acid powder
- Soap powder
- Soap dissolved in water
- Fatty acid saponified in a mill dosed as a soap solution.

Unsaponified fatty acid, liquid

The chemical is added as fatty acid and is saponified in the pulper by the sodium hydroxide present. The primary disadvantage with this alternative is the risk for incomplete saponification. To ensure an even distribution of the chemical in the pulper, using fatty acids that have a liquid form at the process temperature is important. This excludes the stearic fatty acid blends with a higher melting point. The dosing point is restricted to the pulper. Adding the chemical to the flotation cells is not possible.

Unsaponified fatty acid, powder (beads)

The fatty acid is in spray dried form usually packaged in small paper bags of 10–25 kg size. The bags are added unopened to the pulper where they disintegrate and the fatty acid saponifies. The main advantage with this system is the simplicity.

No special storage, handling, and dosing equipment is necessary. An advantage compared with the liquid, unsaponified fatty acid alternative is the possibility to use saturated fatty acids. Unsaturated fatty acids should not have use in this application because caking of the beads can occur. The dosing point is naturally restricted to the pulper. As with the liquid unsaponified fatty acid, a risk for uncompleted saponification that could lead to waxy spots of fatty acids in the final paper does exist. Dosage can change by using bags of different sizes. Bag handling systems are usually more labor-intensive than chemical handling in bulk.

Soap, powder

An alternative to fatty acid beads is soap powder in paper bags. As with the fatty acid beads, the bags are used unopened. In the pulper, the bags disintegrate and the soap dissolves. Compared with fatty acid beads, the dissolution rate is faster. The possibility also exists to change the fatty acid composition over a broader range since the risk for lump formation is smaller. The dosing point is naturally the pulper.

Soap pellets, dissolved in hot water

Soap pellets can be stored at room temperature in a soap silo. Before use, the soap pellets are dissolved in hot water. The hot soap solution is then dosed into the deinking process. A handling system for soap pellets in bulk requires the following:

- Storage silo for the soap pellets
- Dissolving tank
- Storage tank for the soap solution.

The soap dissolving tank is usually a small tank of 2–4 m^3 equipped with a mixer. First, the tank is filled with hot water. The soap pellets are then added. The dissolving time depends on the size and composition of the pellets. Small pellets of 4–5 mm diameter can dissolve in 10 min. Since the soap solution can solidify to form a highly viscous gel at lower temperatures, all storage and handling systems for the soap solution must be insulated and heated. The temperature of solidification for a saturated soap of 5% concentration is about 50°C. The soap solution can be added to the pulper or to the flotation unit. Split addition is also possible where some soap goes to the pulper and some goes to the flotation unit.

The density of soap pellets in bulk is 400–500 kg/m^3 depending on the shape and size of the pellets. Problems that can occur with soap pellets are caking of the hygroscopic pellets and dusting. Caking problems are overcome with mechanical solutions and formulation of the raw materials for the soap. Dusting is avoided when the soap is carefully pelletized.

Fatty acid, saponified in a mill

Saturated fatty acid blends of stearic type solidify at a temperature of 50°C–60°C and therefore require storage at elevated temperature. A typical storage temperature is 75°C. The density of the fatty acid blend is about 850 kg/m^3.

A saponification system requires the following:

- Storage tank for the fatty acid
- Storage tank for caustic soda
- Saponification reactor
- Storage tank for the soap solution.

Continuous saponification systems and batch systems are available. In a batch system, the mixing tank is first filled with hot water. The fatty acid and the caustic are then added. The fatty acid reacts with the caustic and forms soap. After the reaction, the soap solution is transferred to the soap storage tank. The saponification of a fatty acid emulsified in hot water is instantaneous and complete. The rate determining step is the insolubility of the fatty acid in water. Effective mixing is therefore necessary to ensure that the two reacting components come into close contact with each other, i.e., that the fatty acid droplet size is as small as possible.

Most important for a saponification unit is the temperature control. If insufficient heating or insulation exists, the fatty acid or the soap solution may solidify.

Soap vs. fatty acid

When comparing doses and prices of different fatty acids and soaps, one must remember that 1.0 kg of fatty acid compares to 1.08 kg of soap. Soap products can also contain 15%–18% moisture. One kilogram of product contains 820–850 g soap corresponding to 760–790 g fatty acid. The raw material for soap is fatty acid or neutral fat. When soap is produced from neutral fat, the conversion will take much longer since the reaction involves cleavage of three ester bonds[15]. During commercial manufacture of soaps, the saponification process is terminated before 100% conversion of the fat has occurred. This leaves trace amounts of glycerides in the soap that are a mixture of tri, di, and monoglycerides.

5.4 Dosing point and dosage

The soap can be added to the pulper or the flotation unit. Very strong opinions exist regarding the optimum dosing point. Both dosing points find frequent use. If all soap is added in the flotation step, a need arises to add some surface active compound in the pulper.

The main advantage with addition to the flotation operation is better control of the foaming properties in the cells. The level of the flotation cells are more easily controlled, and the consistency of the reject is high. A hypothetical risk with dosing is that the reaction between soap and calcium ion will be incomplete. Some soap then might not function as a collector. The required optimum dosage would be high.

The most important argument for addition in the pulper is that one can take advantage of the dispersive effect of the surface active soap. This dispersive effect of the soap probably depends very strongly on the water hardness. In a system with soft water, the dispersive effect is stronger. In systems with hard water, one can assume that most soap converts to inert calcium soap with low detergency. Recent studies at the Institute for Surface Chemistry in Stockholm, Sweden have shown that the reaction between soap and calcium is slow. Figure 5 describes surface tension of a soap solution as a function of time for four different calcium cloride concentrations. In these experiments, the sodium oleate concentration was 50 mg/L with initial pH adjustment to 9.

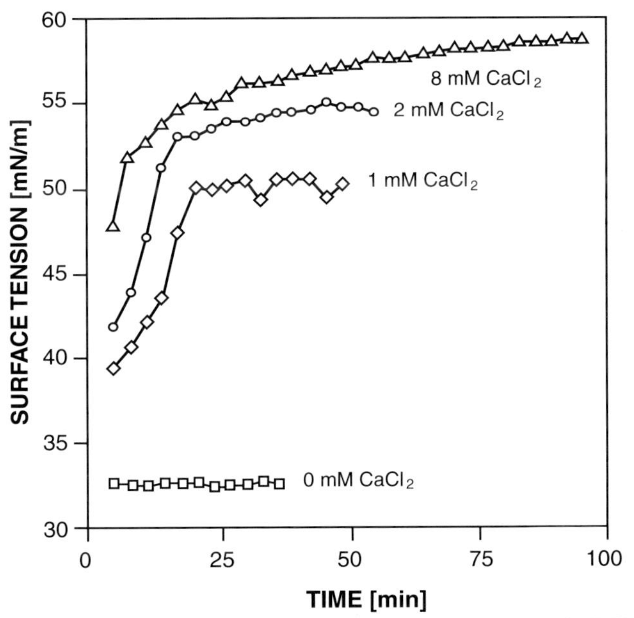

Figure 5. Surface tension vs. time for four different calcium chloride concentrations[16].

The temperature was 22°C. Note that at 50 mg/L sodium oleate is at a concentration below that of the critical micelle concentration[16].

The slow reaction rate supports addition to the pulper. Sufficient dwelling time is then available for the reaction to occur. Optimum performance occurs when the soap addition is divided with some to the pulper and some to the flotation unit. A quick rule is the following:

- Addition of soap to the pulper gives collecting properties
- Addition of soap to the flotation unit gives foam control.

In a system with both pre- and post-flotation, addition to the post-flotation stage usually provides no advantage. The residual chemicals remaining after the pre-flotation are sufficient for the post-flotation stage. A minor dosage to the post-flotation can give some foam controlling properties.

The soap dosage naturally varies for different raw materials. A typical dosage for a mixture of old newspapers and magazines is 6–7 kg soap per ton of recovered paper.

5.5 Mechanism of function

Soap requires a minimum water hardness to function well. Normally, the threshold limit is about 10 °dH. At locations with soft water, this is possible by adding calcium as calcium hydroxide or as calcium chloride.

Several different theories exist regarding the mechanisms by which calcium soap formation improves flotation[17]. One suggests that the sodium soap reacts with the calcium to form insoluble calcium soap as Fig. 6 shows. The calcium soap then precipitates on the ink surfaces to give the surfaces a more hydrophobic character. This also agglomerates the ink particles into larger floating particles[18].

If the water is sufficiently hard as in Central Europe, no extra addition of calcium is necessary. Notice also that recovered paper can contain calcium carbonate that will increase the hardness. If the paper contains considerable carbonate, lowering the dosage of calcium or not adding any is possible.

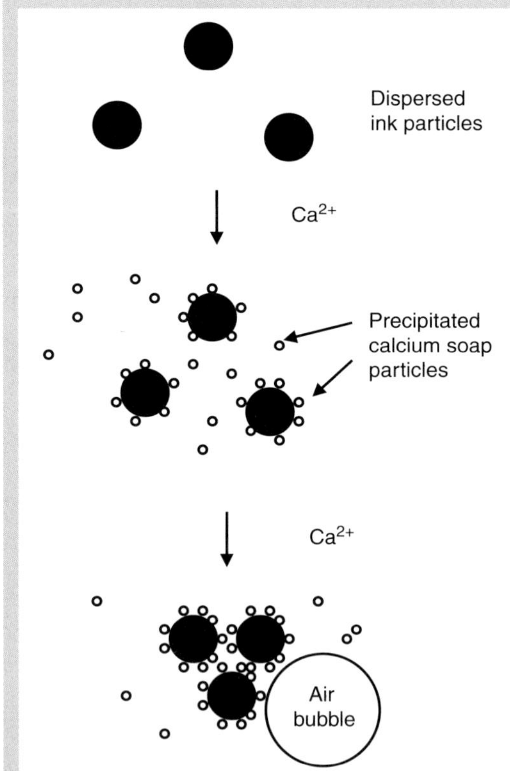

Figure 6. Hypothesis for the collecting mechanism of calcium soap[18].

5.6 Fatty acid composition

Composition of the fatty acid is important for proper function of the flotation stage. The collector usually contains a blend of different fatty acids. The most common of these are C18:0, stearic acid; C16:0, palmitic acid; and C18:1, oleic acid.

Change in the composition can directly influence the amount of foam, ink selectivity, amount of reject, and carryover of chemicals. A more saturated or hardened product gives less reject. The foam is dark, and the ink selectivity is good. A hardened product is especially preferable when a lower carryover is desirable. One should also consider that the precipitated calcium stearate has a harder consistency than calcium oleate. Calcium oleate deposits on the paper machine are therefore more tacky.

Optimum foaming properties in the flotation cells require some unsaturated fatty acids in the mixture. If the fatty acid is excessively hardened, the amount of reject is too small. This can have a negative effect on the brightness. A typical iodine value for a deinking fatty acid blend is therefore 10–25. Chain length can also influence the amount of foam. Shorter fatty acids give more foam than longer fatty acids. The proper amount of foam varies with the type of flotation cells. Some cells especially when the foam will be removed by free overflow require more foam than other cells. The composition of the recovered paper (age of paper and ratio between newspapers and magazines) also influences the optimum fatty acid composition.

5.7 Soap collectors

Good performance is the primary soap characteristic. The final brightness of the pulp is good. Normally, flotation will work well with no problems connected to reject removal. The soap also overcomes variations in raw material. Soap is usually the most “forgiving” collector. Some disadvantages associated with soap are the following:

- Handling: Fatty acids and soap solutions need heated, insulated equipment for storage and dosing.
- Higher dosage than for semi-synthetic and synthetic collectors: Carryover to the paper machine may disturb the papermaking process.
- Need for calcium in soft water: A risk exists for scaling and deposit problems since different calcium salts such as silicate, oxalate, carbonate, soap, and rosinate precipitate on the process equipment. An increase in water hardness can also affect the yield in the flotation. Normally, the stock losses depend on the hardness in the system, i.e., a higher hardness gives higher losses.

5.8 Biodegradability

The soap is removed from the process as a calcium soap. Most soap is removed with the reject, but some remains in the process water and goes to the effluent. The biodegradability of the calcium soap is very fast. The decomposition rate of the soap depends on the fatty acid composition. Obviously, the decisive factor is the water solubility. Fatty acids with shorter hydrocarbon chains form calcium soaps that dissolve more easily than fatty acids with longer hydrocarbon chains. The biodegradability of the shorter fatty acids is therefore faster than for fatty acids with longer hydrocarbon chains. Calcium oleate is more easily biodegraded than calcium stearate as Fig. 7 shows[19].

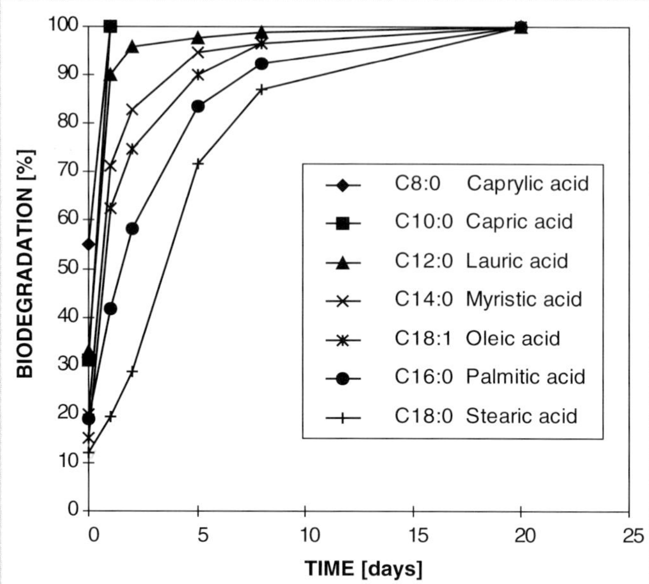

Figure 7. Biodegradation of calcium soaps[19].

6 Synthetic collectors

To avoid the disadvantages associated with use of soap, so-called synthetic collectors are available. The handling of these chemicals is easy because they are liquids at room temperature. The dosage calculated on the recovered paper is lower for synthetic collectors than for soap. Since the water hardness does not influence their function, no extra calcium addition is necessary. The risk of calcium salt deposits in the process is therefore lower. The greatest disadvantage of the synthetic collectors is their performance in the flotation stage. Often the final brightness of the pulp is lower compared with a process using a fatty acid, soap, or a semi-synthetic collector.

In the pulping stage, these chemicals have a dispersive effect on the ink. In the flotation stage, they can act as collector chemicals. The mechanisms for different synthetic collectors vary. One can choose surfactants with low water solubility that gives the ink particles an oily surface. Synthetic collectors containing ester bonds such as fatty acid ethoxylates can hydrolyze in an alkaline environment. Before hydrolysis, the chemical acts as a dispersant. After the hydrolysis, one part acts like a typical fatty acid and the other part is completely water soluble.

The raw material sources of synthetic surfactants can be of either natural (fat and vegetable oil) or petrochemical origin (mineral oil). Similar to soap, the synthetic surfactants consist of a hydrophobic and a hydrophilic part. Depending on the electric charge of the molecule, surfactants are anionic, nonionic, or cationic. In deinking formulations, nonionic surfactants are most common. Anionic surfactants do not have extensive use primarily due to two disadvantages:

- Water solubility: The good water solubility of anionic surfactants can cause an enrichment of the surfactant in the water.
- Foaming properties: Anionic surfactants usually give a high level of foam that can lead to high yield losses. The anionic fatty alcohol sulfate and fatty alcohol ether sulfate especially give high fiber losses[20].

The primary problem related to use of cationic surfactants is their selectivity. Cationic surfactants easily adsorb onto the anionic fibers and therefore give high fiber losses.

The most important nonionic surfactants in deinking are ethylene oxide (EO) and propylene oxide (PO) copolymers, alcoxylated fatty alcohols, and alcoxylated fatty acids. For the alcoxylation process, ethylene oxide or a mixture of ethylene oxide and propylene oxide is used.

The use of ethoxylated nonylphenols has decreased during the last few years because their biological degradation is unsatisfactory. Toxic phenols form during the biodegradation. The EO/PO block copolymers are also unsatisfactory in this aspect since their biodegradation is slow. Ethoxylated fatty alcohols and fatty acids biodegrade easily[21]:

$$\underset{\text{Ethoxylated fatty alcohol}}{R\text{-}O\text{-}(CH_2\text{-}CH_2\text{-}O)_n\text{-}H} \qquad \underset{\text{Ethoxylated fatty acid}}{R\text{-}\overset{\overset{O}{||}}{C}\text{-}O\text{-}(CH_2\text{-}CH_2\text{-}O)_n\text{-}H} \tag{5}$$

Changing the number of ethylene oxide units and the composition of the hydrophobic component influences the foaming, solubility, wetting, and dispersive properties. When using a propylene oxide and ethylene oxide mixture, the foaming properties are adjusted by the ratio between the propylene oxide and ethylene oxide. The ethylene oxide gives a surfactant with more foaming. The propylene oxide gives a less foaming product.

The hydrophilic lipophilic balance (HLB) describes how a surfactant interacts with water soluble and water repelling substances. When the HLB is high, the influence from the hydrophilic part of the molecule is greater than from the hydrophobic part and vice versa. The HLB scale has an empirical gradation on which oleic acid has the value 1.0 and sodium oleate has the value 18.0. HLB provides an estimation in advance for the water solubility properties of a surfactant as Table 1 shows.

Table 1. Effect of HLB on water dispersibility/solubility[22].

HLB	Water solubility	Suitable application
1-4	Insoluble	Defoamer
3-6	Poor	W/O emulsifier
6-8	Miscible	Wetting agent
8-10	Opaque dispersion	O/W emulsifier
> 13	Clear solution	O/W emulsifier, solvent

7 Semi-synthetic collectors (emulsions)

To combine the advantages of synthetic collectors and traditional fatty acid collectors, so-called semi-synthetic collectors are available. These are water miscible emulsions. The components are fatty acids, fatty acid derivatives, surfactants, and water.

Advantages possible when using a semi-synthetic collector compared with fatty acid are the following:

- Emulsion performs well in soft water
- Optimum dosage using active, dry content is about half the dosage of the soap, i.e., the carryover to the paper machine is lower.

The fiber losses in the flotation stage depend on the water hardness in flotation. A lower hardness normally gives lower stock losses as Fig. 8 shows.

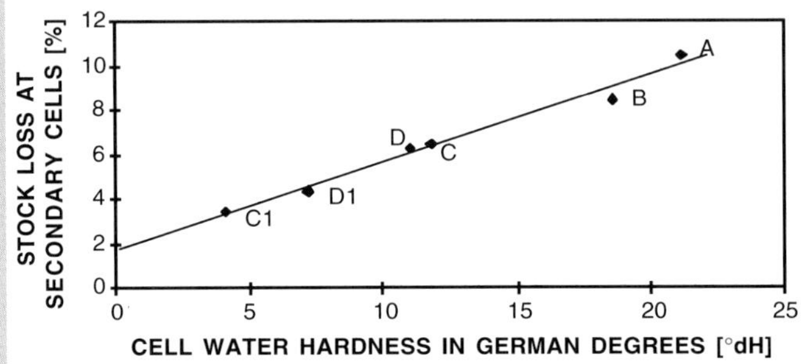

Figure 8. Stock losses across flotation deinking cells vs. water hardness for various industrial deinking plants[17].

When the emulsion is stable, the handling of the emulsion is as easy as the handling of synthetic collectors. Special arrangements such as the following can be useful:

- Storage tank should have a mixer. The mixing should be gentle. Air must not be mixed into the product since the product usually will foam easily.
- Storage and dosing system should not allow any air in the emulsion, e.g., filling pipes should be below the liquid surface.
- All dosing pumps are not suitable for dosing an emulsion since high shear forces can break an emulsion.

When the emulsion works well, achieving the same brightness as with a fatty acid collector is possible. The emulsions are more sensitive to variations in the process than soaps. The amount of pigments and the water hardness especially influence the foaming properties. If the raw material contains a high portion of fillers and coating pigments, a risk can exist that an emulsion will give an excessively abundant foam. This will lead to problems with reject handling and level control in the flotation cells.

The normal dosing point for a semi-synthetic collector is the pulper since the chemical requires a certain residence time to perform well in the flotation operation.

7.1 Combination of soap and emulsion

A combination of soap and an emulsion (dual system) can provide some advantages for a deinking plant:

- Compared with soap:
  - Lower carryover of fatty acid to the paper machine
  - Possibility to run at a lower water hardness in the deinking plant
  - Higher fiber yield in the flotation.
- Compared with emulsion:
  - Better control of foam levels in the flotation cells
  - Higher brightness of the deinked pulp
  - No problems with reject handling.

A system with high ash content emphasizes these advantages, i.e., when 100% emulsion can give excessive foaming. In a dual system of emulsion and soap, the emulsion is dosed to the pulper. The soap can be dosed either to the pulper or to the flotation unit. An alternative to a dual system consisting of emulsion and soap would naturally be a combination of a synthetic collector and a soap collector. In Europe and especially in the Nordic countries, the dual system of emulsion and soap is the most common combination. The soap/emulsion ratio is chosen depending on what properties are prioritized. If high brightness is desired, a high soap/emulsion ratio is used. When low carryover is the main criterium, a lower ratio is normally preferred.

8 Choice of collector

Different collectors have different advantages and disadvantages depending on priorities:

- Final brightness of the deinked pulp
- Carryover of deinking chemicals to the paper machine
- Easy handling of the chemical
- Requirement for extra calcium
- Chemical costs.

Most important to the choice of chemical are the process conditions such as water hardness, pH, and temperature. Naturally, the process design and the different type of process equipment also influence the performance of the chemical. The composition of the recovered paper is also extremely important. Paper from different geographic areas will contain different types of printing inks and additives. Soap or fatty acid in bulk is not economically possible at every mill. The mill should be a certain size to justify investment in a saponification unit or a system for handling soap pellets. High water hardness and high pH normally favor a soap based chemistry. Lower ash content of the raw material can enable an emulsion based chemistry.

9 Carryover of fatty acids and wood resins

Some deinking chemicals can carryover to the paper machine where they can have adverse effects. The most common problems caused by fatty acid soaps are deposits on papermaking equipment. Monitoring calcium soaps throughout the deinking and papermaking process is essential for deinking process control, and development.

In a paper mill using recycled fiber pulp, other sources of extractives than the deinking chemicals exist as Table 2 shows. In a mill using both deinked pulp and mechanical pulp, most extractives originate from the wood raw material. Wood-containing recovered paper also contains residual wood resins. Other potential sources for fatty acids and fatty acid derivatives in the recycled fiber pulp are residual paper chemicals and printing ink binders.

Table 2. Different sources of fatty and resin acids.

Source	Typical main composition
Deinking chemicals	
- Soap	C18:0, C16:0, C18:1
- Emulsion	C18:0, C16:0
Recovered paper	
- Wood-containing paper	Resin acids, C18:2, C18:3, C18:1, C16:0
- Paper chemicals	Resin acids
Mechanical pulp	Resin acids, C18:2, C18:3, C18:1, C16:0

Some typical differences in the composition of fatty acids and fatty acid derivatives exist due to the origin of these extractives. The residual deinking chemicals usually contain saturated fatty acids. The primary components of wood resins are unsaturated fatty acids and various resin acids. The volume in this series entitled *"Forest Product Chemistry"* describes the composition of wood resin.

The fatty acids in wood resin are in ester form whose primary components are triglycerides and free fatty acid. In fresh wood, most fatty acids are in triglyceride form[23]. Deinking chemicals consist primarily of fatty acids in their free or saponified form. In the process, the water hardness and the pH determine the form of the fatty acids. At higher pH, the fatty acids can be in calcium soap form. At lower pH, the fatty acids appear in their free form[16].

As Table 2 shows, deinking collectors usually contain considerable stearic acid, C18:0, but very little polyunsaturated acids such as linoleic acid. Wood resin contains very little stearic acid but a larger amount of linoleic acid, C18:2. Consequently, stearic acid and linoleic acid are good markers for deinking chemicals and wood resin, respectively.

When analyzing samples taken from the deinking plant and paper production, sample preparation has extreme importance. Wood resins and deinking chemicals are present in the fibers and in colloidal and dispersed form in the aqueous phase. If paper sheets or pads are analyzed, only some extractives are analyzed. An alternative to forming a sheet is to evaporate the water and analyze the solid residue. This procedure gives information about the total extractives, i.e., the amount in the solid phase and the aqueous phase of the sample. Freeze drying of the pulp samples is a gentle method for evaporation of the water.

Fatty acid calcium soaps are slightly but not sufficiently soluble in solvents such as chloroform and tetrahydrofuran. Efficient extraction of such soaps for further analysis requires conversion of the soaps to free acids that are highly soluble in many solvents. This is possible by treatment with acid or an acidic salt[24,25].

Figure 9 shows a procedure for the analysis of wood resin and residual chemicals. The alkaline treatment after the extraction is necessary to hydrolyze the esterified part of the wood resin. Esterified fatty acids can be determined as the difference between an alkaline-treated, hydrolyzed sample and an untreated sample. Analysis of silylated or methylated extracts on a standard gas chromatographic column gives detailed information on individual fatty acids, resin acids, and free sterols.

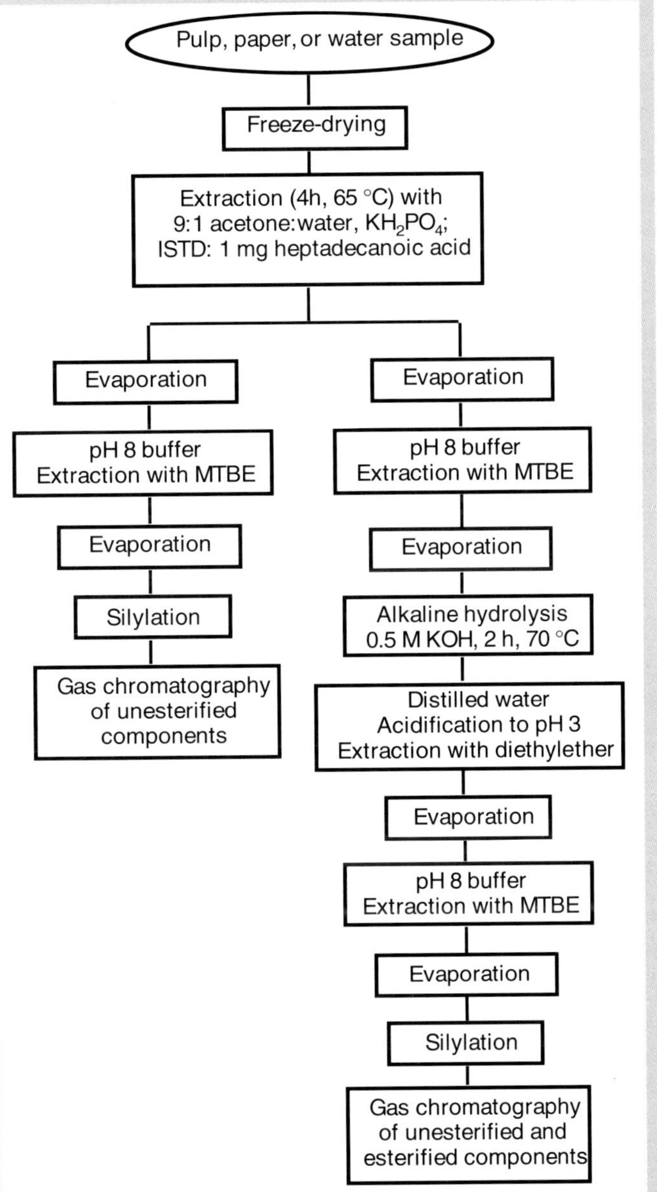

Figure 9. Procedure for determination of fatty and resin acid calcium soaps and wood resin[25]. (ISTD: internal standard, MTBE: methyl tertiary butyl ether.)

9.1 Case study

Samples of pulp and process waters were taken from a European deinking plant. The samples were stored in a refrigerator for a few days until they were freeze dried and analyzed[26]. The process is a normal two-loop design with pre- and post-flotation. The mill uses a dual system soap and emulsion.

Stearic acid

The main results were as follows:

- Almost all stearic acid enters with the deinking chemicals
- Stearic soaps are removed primarily in the flotation steps but to some extent over the first disk filter and first screw presses
- Dewatering stages in loop 2 do not remove measurable amounts of stearic soaps
- 80%–85% of the deinking chemicals added are also removed from the deinking process. This corresponds to similar studies in other European deinking mills[27].

Linoleic acid

The total amount of linoleic acid (both free and esterified, i.e., liberated by alkaline hydrolysis) was determined on the same samples as above. Most linoleic acid is present in triglycerides and steryl esters that are totally insoluble in water even at pH 10–11. The results showed the following:

- Dispersed wood resin enters with the recovered paper and the paper machine white water
- Wood resin is removed to about 50% over both flotation stages
- About 80% of the dispersed resin entering with the recovered paper is removed in the deinking process.

10 Influence of pH

The deinking process usually runs at high pH. The pH in the pulping stage remains at 9–10 and drops to 8–9 in the flotation stage. If the deinking plant is a two-loop process, the second process water loop normally runs at even lower pH. A pH of 7.5 is typical for the post-flotation in the second loop.

If the paper machine is running at an acidic pH, a need exists for acidification of the deinked pulp. This is normally accomplished with sulfuric acid. If the deinked pulp contains calcium carbonate originating from fillers and coating pigments, the acid dissolves the carbonate. This leads to an increase of the water hardness and the risk of problems related to deposits. These deposits often consist of different calcium salts such as calcium carbonate, calcium silicate, calcium oxalate, and calcium soap. Deposits can occur when a change occurs in the chemical or physical conditions such as temperature or pH. Another disadvantage related to dissolving the minerals is a reduction in yield. The recent tendency has been to run the deinking plant with a more even pH profile. The pH in the pulper is carefully optimized, i.e., overdosing of caustic and water glass is avoided. Simultaneously, the pH in the papermaking has in many mills successfully been increased to neutral or pseudo-neutral.

Some advantages could be possible if the entire deinking process runs in a neutral environment. The most important are the following:

- Less dissolving of water soluble components in the recycled fiber pulp gives an increase in yield and a reduction in COD and BOD of the effluent
- Less yellowing of mechanical fibers
- Savings in chemical costs.

In practice, running without alkaline chemicals is difficult. The most success with neutral deinking has been in applications with raw material of high quality composition such as (wood-free) office paper. When using normal household collected (wood-containing) paper, the lack of caustic and peroxide most often gives a significant drop in brightness of the deinked pulp. The situation can improve to some extent by keeping the dosing of water glass at a normal level. This means running at a slighly elevated pH[28].

Other disadvantages related to neutral deinking are the following:

- More anionic trash in the deinked pulp that can influence the papermaking negatively
- Risk for plugging of screens related to insufficient disintegration of wet strength papers.

11 Enzymes in the treatment of recycled fibers

The use of enzymes in deinking has had considerable interest recently[29]. One can distinguish between two different approaches: Enzymes that react with the fibers and enzymes that react with the printing ink. The first group consists of so-called cellulases and hemicellulases. The exact function of these enzymes is not completely understood. Different mechanisms have been proposed[30]:

- Peeling effect on the cellulose microfibrils
- Hydrolysis of accessible cellulose chains
- Removal of cellulosic microfibrils or fines
- Cleavage of hemicellulose releasing lignin from carbohydrate-lignin complexes.

One potential disadvantage with these enzymes is that their fundamental function uses degradation of fiber components. Careful dosage control and optimization of reaction time is necessary to minimize yield losses.

Another approach to enzymatic deinking is to attack the printing ink. Lipases and esterases can break down the carrier or binder in vegetable oil-based printing inks.

Normally, the enzymes are added in the beginning of the process replacing part or all the traditional deinking chemicals. Washing or flotation then removes the released particles. The replacement of some deinking chemicals can provide certain advantages. If the ink detachment can occur in a neutral environment, the advantages mentioned in the previous section will result.

Mixed information exists about the performance of enzymatic deinking. A negative factor can be the tendency of enzymes to reduce the ink particles to much smaller particles than other methods do. The ink particles can be too small for effective flotation. If the disintegration is continued excessively, they can even be so small that they penetrate into fibers. This reduces the washing efficiency.

The total picture of the impact of enzymes seems to be that they are efficient when low residual ink area is desirable. The brightness of the deinked pulp seems to be on a lower level or in some cases on a par with conventional deinking methods.

Enzymes in deinking have undergone extensive study in laboratory and pilot plant scale. A break-through in commercial applications has not emerged. The decisive factors in the future will be cost and performance. What will be the cost of enzymatic treatment related to the quality improvement such a treatment can achieve? Enzymes in deinking could be an interesting new approach.

12 Deinking by washing

Washing is achieved by rapid dewatering of the pulp through a screen or wire. The washing process typically removes fines and fillers from the pulp very efficiently. Note the difference between washing and dewatering (See Chapter 5). Special equipment is available for washing. Normal dewatering equipment such as disk filters or wire presses do not give much washing effect since a thick fiber layer forms. Ink particles easily become entrapped in this layer.

To obtain successful washing by deinking, the ink must be well dispersed and the particles sufficiently small. Chemistry in the pulping is similar to flotation deinking exept that washing chemicals or dispersants have use instead of collectors. Water glass, hydrogen peroxide, and caustic soda have frequent use in washing of wood-containing recycled fiber pulps.

The dispersant usually consists of a mixture of synthetic surfactants. When choosing the dispersing agent, the following are important:

- Dispersant wets the paper effectively
- Ink particles detach from the fiber
- Ink particles remain dispersed in the water phase, i.e., a so-called redeposition phenomena is prevented
- No foaming problems occur in the process such as in the washers
- Dispersant is easily biodegradable.

The white water from the washing stage containing ink, fillers, and stickies is usually cleaned in separate equipment normally by dissolved air flotation. The dispersant must therefore not impair cleaning of the water. This means the following must occur:

- Dispersant must not be too water soluble. This could cause an enrichment of the dispersant in the process water
- Dispersant is not too dispersive. The HLB should not be too high
- Dosage of dispersant is optimized. Excessive dosage is avoided.

When choosing the optimum HLB for the dispersant, one must compromise between high dispersing efficiency and function of the process water treatment. Closure of the system determines the proper HLB.

13 Treatment of process water

In the production of pulp and paper, a trend exists to lower use of fresh water. Processes are becoming more closed. This is naturally also important in the production of deinked pulp. When closing the system, having efficient purification of the process water is important (See Chapter 6).

Another important driving force for cleaning process water is that it can lead to a substantial quality improvement of the pulp. This is especially true when using recovered paper printed by water-based inks. In washing by deinking, cleaning the process water is necessary since this is the only practical means of ink removal from the process.

Other advantages of internal water purification are as follows:

- Energy savings because the need for heating the process water is decreased
- Savings in the effluent treatment[31,32]
- Savings in chemical costs in the pulping (Dosages can be lower when the process water is cleaner.).

The process water is cleaned by adding chemicals that flocculate the colloidal particles into larger agglomerates or flocs. These can then be removed by microflotation. Another common name for this process is dissolved air flotation (DAF). The function of DAF is the following (See also Chapter 6).

Cleaned water that is saturated with air is added to the microflotation. The pressure is about 5–7 bar. The ratio between the already cleaned, air saturated water and the water to be cleaned depends on the amount of contaminants in the water. If the water is very contaminated, it might require as much as 20% air saturated water based on the total amount of water. When the water saturated with air goes into the microflotation cell, many small air bubbles form due to the sudden pressure drop. The flocs attach to these bubbles and rise to the surface from which the sludge can be removed.

Most detrimental substances in the process water have a negative surface charge, i.e., their zeta potential is negative. Repulsive forces exist between the particles. Neutralizing the negative charge by adding cationic chemicals can decrease these forces. After neutralization, joining the particles is possible by using long chain polymers that can be nonionic, slightly cationic, or anionic.

Using an active mineral with a very large specific surface such as bentonite is also common. The particles to be flocculated adsorb on the mineral surface. The mineral particles then join by using a polymer. Several substances in the process water are water soluble when the pH is high. At a neutral pH, these substances precipitate for removal by flocculation and microflotation. Water cleaning is therefore less efficient at higher pH.

References

1. Elsby, L., personal communication.

2. Ben, Y. and Dorris, G.M., "Chemical changes in linseed oil under repulping and flotation conditions", 1997 CPPA 4th Research Forum on Recycling Proceedings, CPPA, Montreal, p. 269.

3. Blechschmidt, J. and Ackermann, C., Wochenbl. Papierfabr. 119(17):659(1991).

4. Johansson, B., "Colloidal chemistry of ink particle agglomeration and flotation", Lic. thesis, Department of Pulp and Paper Chemistry and Technology, Royal Institute of Technology (KTH), Stockholm, 1997.

5. Know-How Wire, Jaakko Pöyry Magazine (May):32(1988).

6. Paraskevas, S., "Ink removal–Various methods and their effectiveness", TAPPI 1989 Seminar on Contaminant Problems and Strategies in Wastepaper Recycling, TAPPI PRESS, Atlanta, p. 41.

7. Marchildon, L., Lapointe, M., and Chabot, B., Pulp Paper Can. 90(4):90(1989).

8. Stenius, P. and Heimonen, J., "The effect of coating components and fillers in the deinking of paper", 1993 Oxford Fundamental Research Symposium Proceedings, PIRA, Leatherhead, UK, Vol. 3, p. 1707.

9. Biza, P. and Lippert, G.V., Wochenbl. Papierfabr. 120(11/12):439(1992).

10. Asp, F., Engstrand, P., Holmbom, B., et al. "Evaluation of contaminant flows in a deinking process and estimations of the value of the raw material", 1994 PPI Paper Recycling Conference Proceedings, Pulp Paper International, London.

11. Julien Saint Amand, F. and Perrin, B., "The effect on particulate size on ink and speck removal efficiency of the deinking steps", 1991 CPPA 1st Research Forum on Recycling Proceedings, CPPA, Montreal, p. 39.

12. Stenius, P., Svensk Papperstidn. 84(4):14(1981).

13. Anon., Product catalog "Radia oleochemicals", Fina Chemicals, Brussels, 1993, p. 13.

14. Sonntag, N.O.V., Bailey's Industrial Oil and Fat Products, John Wiley & Sons, New York, 1979, p. 16.

15. Spitz, L., "Soap Technology for the 1990s", American Oil Chemists Society, 1990, p. 61.

16. Johansson, B., Wickman, M., and Ström, G., Nordic Pulp Paper Res. J. 11(2):74(1996).

17. Turvey, R.W., "Technology of Paper Recycling, Chemical Use in Recycling," Chapman & Hall, London, 1995, 139 pp.

18. Larsson, A., Stenius, P., and Ödberg, L., Svensk Papperstidn. 87(18):R158(1984).

19. Brüschweiler, H., Felber, H., and Schwager, F., Tenside Surfactant Detergents 28(5):348(1991).

20. Hornfeck, K., Wochenbl. Papierfabr. 113(11):646(1985).

21. Janicke, W., Tenside Surfactant Detergents 25(6):345(1988).

22. Griffin, Q.C., in Encyclopedia of Chemical Technology (R. E. Kirk and D. F. Othmer, Eds.), Wiley & Sons, New York, 3rd edn., Vol. 8, 1979, p. 900.

23. Assarsson, A., Svensk Papperstidn. 99(9):291(1966).

24. Dorris, G.M., Douek, M., and Allen, L.H., J. Pulp Paper Sci. 10(1):TR1(1983).

25. Sundberg, K., Hemming, J., Lassus, A., et al. "Determination of fatty and resin acid calcium soaps", 1997 9th Int. Symposium on Wood and Pulping Chemistry, CPPA, Montreal, 108 pp.

26. Holmbäck, Å., "Analys av fettsyratvålar och vedharts i en returpappersanläggning", M.Sc. thesis, Åbo Akademi University, Lab. of Forest Products Chemistry, Turku, Finland, 1995.

27. Lassus, A., personal communication.

28. Hanecker, E., "Sicherstellung und Erhöhung der Wirksamkeit des Deinkingprozesses durch Anpassung des Chemikalieneinsatzes und die Altpapierzusammensetzung", PTS-FB 14/96, Papiertechnische Stiftung, Munich, 1996, 107 pp.

29. Welt, T. and Dinus, R.J., Progress in Paper Recycling 4(2):36(1995).

30. Mörkbak, A. and Zimmermann W., Progress in Paper Recycling 7(2):14(1998).

31. Eisenschmid, K. and Steinbeis, B., Wochenbl. Papierfabr. 117(14):645(1989).

32. Eisenschmid, K. and Steinbeis, B., Wochenbl. Papierfabr. 118(16):709(1990).

CHAPTER 8

Deinkability of printing inks

CHAPTER 8

Katharina Renner

Deinkability of printing inks

1 Definition of deinkability

Deinking is a term describing a process for detaching and removing printing inks from recycled fibers to improve optical characteristics of pulp and paper using recovered printed material. The detachment of the printing ink from the fibers of the disintegrated recycled pulp uses chemical and mechanical conditions. Removal of the detached ink particles then uses flotation or washing. Deinking is the most important process stage in processing of recovered paper as a raw material for the production of deinked pulp (DIP) primarily used for manufacturing graphic papers, hygienic papers, or top liners of white lined chipboard.

The deinkability of recovered paper describes the degree of printing ink removal by a deinking process that is measured first by the improvement of optical parameters. These properties include luminance and brightness of the undeinked and deinked pulp. Besides the printing ink, the printing process, the print substrate, and the aging of the printed product, the conditions and parameters of the deinking processes in terms of the chemical, mechanical, and hydraulic regimes influence the deinkability of printed material.

Deinking primarily uses recovered graphic paper products such as old newspapers and old magazines as wood-containing products. It also uses wood-free paper from offices. In Germany for example, the deinking material obtained from household paper collections comprises equal parts of newspapers and magazines. This accounts for about 80% of the deinking base material. The remaining 20% consists of other wood-containing papers and wood-free papers and a certain proportion of unwanted brown packaging materials and colored papers that impair the cleanliness and brightness of the deinked pulp.

Newsprint and magazine papers use different furnishes and additives and are printed by different processes. Among the classical printing processes used for printed products are letterpress, offset, and rotogravure. The flexographic process also has use for newspaper printing occasionally in North America. Except in Italy, flexography for newspaper printing has only very small use in Europe.

Included among the nonimpact printing processes that are used especially in the office sector and in small volume applications are electrophotography, electrography, electrosensitive processes, magnetography, thermography, and ink-jet printing. Parallel to the growing versatility in office communication, the heterogeneity of recovered paper concerning types of paper, printing processes, and printing inks is also increasing.

Depending on the printing ink composition, the deinkability of different printed products can vary considerably.

Evaluation of deinkability primarily uses the optical parameters of brightness and color or shade. Traditionally, the simplest and fastest way to evaluate deinkability is to characterize the difference between the brightness of the deinked pulp and the brightness of the pulp before deinking. This is brightness gain.

On a laboratory scale, the deinkability factor, DEM_f, developed by Papiertechnische Stiftung (PTS) in Munich, Germany, and calculated from the ISO brightnesses (R_{457}) of the unprinted deinked pulp (US) (i.e., made from unprinted paper subjected to the deinking process), the printed undeinked pulp (BS), and the deinked pulp (DS) can characterize deinkability of a printed material[1]:

$$DEM_f = \frac{brightness(DS) - brightness(BS)}{brightness(US) - brightness(BS)} 100[\%] \quad (1)$$

DEM_f can be 0%–100%. A value of 100% describes complete removal of the printing ink.

Assessing the deinkability of color print samples using DEM_f is possible only to a limited degree since the ISO brightness insufficiently describes the optical properties of colored deinked samples. The color difference of a sample to a reference sample as a vector in the L*a*b* color system (CIELAB) was used to develop another deinkability factor (See section "Optical parameters and color of deinked pulp"). The Department of Paper Science and Technology (IfP) in Darmstadt, Germany, developed the deinkability factor, DEM_{Lab}[2]:

$$DEM_{Lab} = \left(1 - \frac{\sqrt{(L^*_{US} - L^*_{DS})^2 + (a^*_{US} - a^*_{DS})^2 + (b^*_{US} - b^*_{DS})^2}}{\sqrt{(L^*_{US} - L^*_{BS})^2 + (a^*_{US} - a^*_{BS})^2 + (b^*_{US} - b^*_{BS})^2}}\right) 100[\%] \quad (2)$$

This parameter describes the color difference between unprinted deinked pulp (US) and deinked pulp (DS) in relation to the color difference between unprinted deinked pulp (US) and printed undeinked pulp (BS). As with DEM_f, 100% corresponds to complete deinkability.

Another way to evaluate deinkability is to determine the effective residual ink concentration (ERIC) value proposed by Jordan and Popson[3,4]. Measurements of reflection use a sample surface (handsheet, filter pad) in the (invisible) IR range with a wavelength of 950 nm and at 457 nm in the visible wavelength range. At 950 nm wavelength, only black printing inks should absorb light. Other materials such as the yellow lignin do not absorb light in the IR range. Results of ERIC measurements are expected to provide the amounts of residual ink. The ERIC value only provides information on the influence of the residual ink on the change of light absorption at a wavelength of 950 nm. It gives no information on the actual quantity of residual ink.

Besides evaluation of deinking results by optical parameters, the image analysis method is finding increasing use to quantify the optical heterogeneity of deinked pulp. The method measures the number and area of ink speck particles with results in histograms for the size distribution of particles. Using these parameters provides conclusions regarding the cleanliness of paper made of deinked pulp. The particle size distribution can also assess effectiveness of the individual stages of processing recovered paper.

Suitable for use as image analysis devices are systems that use a camera or scanner to inspect the sample surface (handsheets, filter pads, or membrane filter samples) to be measured and subsequently analyze it. Higher resolution of an image analyzing system gives more detailed information about the particle size distribution. More than 90% of all dirt particles in recycled fiber processing are below a diameter of 50 µm. This is invisible to the naked eye[5–8].

2 Composition of inks

2.1 Inks for conventional printing processes

The basic components of the printing inks used on the conventional lithographic (offset), relief (letterpress and flexographic), and rotogravure printing processes are vehicles, colorants (pigments or dyes), additives such as drying agents, and other components. The vehicle functions as a carrier (e.g. solvents) and as a binder (e.g. resins). The printing process concerned and the requirements of the printer determine the selection of the components and their proportions. Table 1 shows the composition of printing inks.

Table 1. Composition of printing ink[9].

Ink components	Proportion
Vehicles - Carrier - Binder	 15% – 80% 5% – 50%
Colorants (pigment or dyes)	5% – 30%
Additives	1% – 10%

Conventional printing processes use inks with different compositions for technical reasons. Depending on the printing process concerned, their consistency can vary from low viscosity to thick, elastic, and viscous. Table 2 shows typical recipes for black printing ink systems.

Table 2. Composition of different black printing ink systems[10–12].

System	Offset sheet-fed	Offset coldset	Offset heatset	Rotogravure	Flexo for newsprint
Pigment	Carbon black				
	15%–25%	15%–25%	15%–25%	5%–10%	12%–20%
Carrier	Mineral oil 15%–25%	Mineral oil about 60%	Mineral oil 25%–35%	Toluene 50%–70%	Water 65%–80%
Boiling range	250°C–300°C	250°C–400°C	200°C–300°C	70°C	100°C
Binder	Various resins 25%–50%	Colophony derivatives 10%–20%	Colophony derivatives 25%–35%	Colophony derivatives 20%–30%	Polyacrylate 5%–15%
Additives	5%–10%	1%–5%	5%–10%	2%–10%	2%–10%

Carbon black has use for all types of black inks. Differences exist in the recipes especially for the vehicles and additives. The ink manufacturer can make appropriate adjustments within these recipes to satisfy special requirements due to technical conditions and print substrates.

2.1.1 Carriers and binders

Binders are partially responsible for forming the ink film. They envelop the pigments, carry them across the inking and printing unit, and bind them firmly to the print substrate after the drying process.

Carriers dissolve the solid binders and keep the ink in the press fluid. A dissolved binder in a carrier is a varnish. This controls the consistency of the ink[9]. After application of the ink to the substrate, the carriers should evaporate as much as possible without residue or absorb into the substrate. This allows the binder to form a film. Some binders occur naturally, and others are synthetic[9].

Carriers and binders differ greatly in their physical and chemical characteristics depending on which printing process applies the resulting inks and which characteristics are desirable. Table 3 shows binders and carriers for different printing processes. The subdivisions of binders and carriers illustrate the variety of inks possible. The inks will obviously have different deinking characteristics.

Table 3. Binders and carriers for different printing processes[9].

Printing process	Binders	Carriers
Offset (web-fed rotary for news)	Hydrocarbon resin	Mineral oil
Offset Letterpress	Drying oils Alkyd resin Phenol-modified colophony resin Hydrocarbon resins	Mineral oil
Rotogravure	Colophony derivatives Hydrocarbon resins	Toluene
Packaging flexo and Packaging rotogravure	Cellulose nitrate Polyvinyl acetate Polyamides	Ethanol Isopropanol Ethyl acetate Benzine
Water-based flexo and water-based packaging rotogravure	Acrylic resins Maleic resins	Water (Alcohols)

Important base materials among the natural ink binders are colophony, asphalt, and bitumen. Drying oils have a vegetable origin (linseed oil and soybean oil). They come from oil seeds with purification by refining. Alkyd resins are modified drying oils. Another natural base material for ink binders is cellulose.

Most synthetic binders are polymerization products. Important synthetic binders are polyvinyl chloride, polyvinyl acrylamide copolymers, acrylic resins, and hydrocarbon resins. For the water-based ink systems that are becoming increasingly important, acrylic dispersions or saponified acrylic or maleic resins have use.

The most important carriers in inks are alcohols (isopropanol and ethanol), esters (ethyl acetate), glycols and glycol derivatives (methoxypropanol), ketones (acetone), hydrocarbons (mineral oils, toluene, xylene), and water.

The ink drying process has a decisive influence on binder and carrier selection and on the ink composition and deinkability of the printed product. Following are the different types of drying:

- Physical drying
  - Absorption into an absorbent print substrate
  - Evaporation of the solvent
- Chemical drying
  - Oxidative drying
  - Drying by high energy radiation (UV radiation and electron beam curing).

Absorption and evaporation belong to the category of physical drying because no material changes occur except the absorption of carriers by the print substrate or evaporation of solvents in the air. Oxidative and high energy radiation drying belong to the chemical drying category. The binder molecules here become enlarged by oxygen uptake or polymerization. Offset printing uses a combination of different drying methods as Table 4 shows.

Table 4. Ink drying processes[13,14].

Vehicles and reaction	Drying process	Print process	Substrate
Drying oils, alkyd resins + oxygen	<u>Chemical:</u> Oxidative polymerization	Letterpress Sheet-fed offset	Paper, board
Mineral oils penetrate substrate	<u>Physical:</u> Absorption	Rotary printing Letterpress Sheet-fed offset (coldset)	Newsprint Uncoated paper
Mineral oils evaporate in special dryers	<u>Physical:</u> Evaporation and part absorption	Web offset Heatset	Coated paper Uncoated paper
Resins harden by radiation	<u>Physical/ Chemical:</u> Oxidation comb. with absorption (IR drying) Polymerization (UV drying)	Sheet-fed offset Web offset Flexo (packaging, labels)	Paper, board, foils
Solvent or water evaporates under heat or in air stream	<u>Physical:</u> Evaporation	Toluene-based rotogravure Water-based rotogravure and flexo	Paper, board, corrugated board

Contact of the ink film with the print substrate differs according to the drying process. With ink absorption, the ink penetrates or absorbs into the paper pores as Fig. 1 shows according to the density and caliper of the paper or the pigment coating layer. Chemical reactions such as oxidation do not occur.

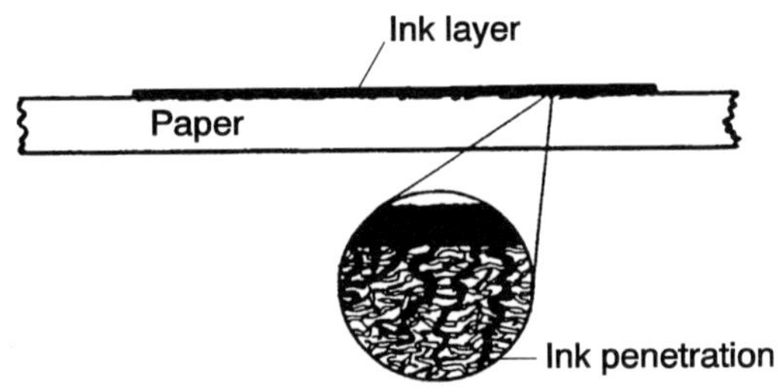

Figure 1. Ink penetration into paper[15].

Inks for web offset printing without thermal drying (coldset) including newspaper printing are usually formulated for absorptive drying. They therefore do not contain large proportions of oxidative drying oils or mineral oils with especially low boiling points. The mineral oils contained therein absorb quickly on the paper and leave the resins and the pigments bound on the paper surface. Due to the incomplete ink drying, the readers' fingers turn black upon contact with a freshly printed newspaper. Printed materials of this type in a fresh, unaged state are usually easily deinkable. Using surfactants, the inks will emulsify in water for removal during the flotation process.

For drying by evaporation, the solvents transform into a gaseous state. Cold or hot air blown onto the printed surface can intensify evaporation as Fig. 2 shows. The decisive factors for sufficient drying by evaporation are evaporation rate of the solvent and the ability of the binders to surrender the solvents quickly and completely. Rotogravure inks use highly volatile solvents such as toluene that evaporate rapidly on the substrate. During this evaporation, the dissolved binders solidify and become firmly attached to the paper surface. They dry primarily without chemical changes and are most easily deinkable in the alkaline environment of the deinking process.

Figure 2. Ink drying by evaporation[15].

For drying of web offset heatset inks, the mineral oils evaporate as Fig.3 shows. The ink resins melt and adapt to the substrate surface. After leaving the dryer, chill rolls cool the hot ink film. The print of web offset heatset inks in a fresh, unaged state is also easily deinkable. Oxidative drying vegetable oils or alkyd resins are present only in small amounts in heatset inks. Because these substances do not evaporate, the ink film remains moist and sticky.

Water-based flexographic inks dry by absorption and evaporation of their water content. Due to its physical characteristics, water evaporates considerably faster than mineral oil. Some water therefore evaporates during the transport of ink from the inking roller all the way to the time of transfer to the print substrate.

1st Stage: **Penetration** ➡ Oil penetrates into the paper, leaving a more solid layer of resin and pigment at surface
Separation of resin and oil

2nd Stage: **Influence of heat** ➡ Evaporated oil above the ink layer restricts further drying
Evaporation of a part of the oil

3rd Stage: **Hot air application** ➡ Residual oil is retained by resin and thick ink layer
Evaporated oil is blown away
Reduced oil content of ink

4th Stage: **Cooling** ➡ Residual oil is encapsulated by ink layer
Consolidation of ink layer

Figure 3. Drying stages of web offset heatset inks[16].

Conventional flexographic print is poorly deinkable. In the most commonly used alkaline flotation deinking process, the binder of the water-based flexographic inks (usually acrylate resins) dissolves in a finely distributed manner. This releases very small ink particles that remain equally with the fiber stock and the filtrate. Due to their small particle size, wash deinking will more effectively remove flexographic inks. Ink manufacturers are trying to improve the flotation deinkability of flexographic inks for newspaper printing. They are having some success at the test stage with so-called third-generation flexographic inks[17].

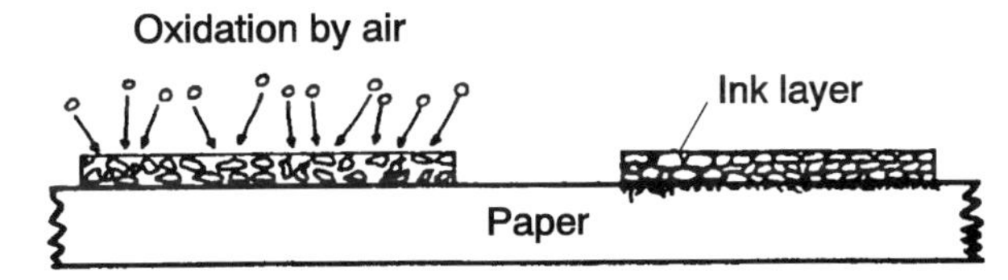

Figure 4. Ink drying by oxidation[15].

Besides absorbing mineral oils, high quality offset inks also contain oxidative drying components such as linseed oil or fatty alkyd resins. These components absorb oxygen from the air that causes them to oxidize and cross-link to large, insoluble molecules with a high mechanical strength (Fig. 4). The pigments attach very firmly to the substrate using the oxidized parts of the binder. This makes the print resistant to abrasion and chemically resistant. For this reason, oxidation drying inks are more difficult to deink. The development of deinkable inks therefore gives preference to the use of nonoxidizing binder components.

Drying by high energy radiation takes fractions of seconds. The chemical polymerization forms large molecules that are insoluble in most solvents and offer outstanding mechanical characteristics. With deinking methods used today, this also means that printing inks "cured" by high energy drying cannot easily detach from fibers. Note also that a difference exists in the deinkability of uncoated and coated papers. With coated papers, the size of the ink particles after pulping can be partly too large for removal by flotation deinking.

Besides the absorption and drying behavior of the binders and solvents, the physical and chemical characteristics are also determining factors for deinkability. Table 5 provides an overview of some characteristics of the vehicles that influence deinkability of the printing inks[18]. A problem is that the recovered paper is pulped in water during the deinking process with the addition of deinking chemicals. The chemicals are sodium hydroxide, hydrogen peroxide, sodium silicate, and anionic or nonionic surfactants (See Chapter 7). The binders can react with the chemicals used and influence their deinkability.

The ink manufacturer can use this data as a guideline when formulating ink recipes. The ink manufacturer must achieve the desired deinkability without sacrificing the important printing and application characteristics.

Table 5. Binder and solvent characteristics that influence deinkability[18].

Binder and solvent characteristic	Deinkability	
	good	poor
Polar		•
Covalent	•	
Good cross-linking		•
Poor cross-linking	•	
Alkaline soluble		•
Alkaline insoluble	•	
Poor pigment wetting		•
Good pigment wetting	•	

2.1.2 Colorants (pigments and dyes)

A distinction exists in the colorant component of printing inks between pigments and dyes. Pigments in the application medium are practically insoluble, inorganic or organic, and chromatic or achromatic colorants. Dyes in the application medium are soluble organic colorants that have rare use in conventional printing inks.

Pigments and dyes in the ink communicate the color impression. They influence the runnability of the inks and determine the fastness characteristics of the printed products such as light fastness, alkali fastness, and resistance to water, acids, and solvents. A distinction exists between black, color, and toner pigments. Toner pigments or dyes have use where a certain ink shade is necessary. The following shows different chemical categories of pigments and dyes[9]:

- Insoluble inorganic pigments
  - Colored mineral pigments such as natural chalk
  - Mineral pigments produced artificially from metallic salts such as magnesium and calcium carbonate or aluminium hydroxide
  - Metal powder including pulverized alloys of copper and zinc to produce gold and silver inks
  - Carbon blacks produced by a controlled, incomplete burning of petroleum, coal tar products, or natural gas or by thermal splitting of acetylene (They are used to produce black inks).

- Insoluble organic pigments
  - Natural organic pigments are not used for printing inks
  - Artificial organic pigments from coal tar and petroleum distillation (50 000 color compounds are available of which 300–500 are commercially available).
- Soluble organic dyes or aniline dyes
  - Natural organic dyes have no significance
  - Artificial organic dyes such as azo dyes.

As the list indicates, the primary distinction is between inorganic and organic pigments that are insoluble in water, solvents, or both and the organic dyes that are soluble in these solvents. The pigments have primary use in offset and letterpress inks, and the dyes are additionally used in individual cases for rotogravure and flexographic inks[12].

Besides chemical constitution, the physical parameters such as particle size, particle shape, and particle surface characteristics influence the quality of pigments. Carbon blacks differ in particle size, color depth, and color shade. This provides the ink manufacturer with a wide choice of suitable types. Table 6 shows examples of the mean particle size and specific surface area for carbon black pigments and several color pigments.

Table 6. Particle sizes of ink pigments[19,20].

Pigment	Mean particle diameter [μm]	Specific surface area [m^2/g]
Channel carbons:		
Type high color	0.013	980
Type long flow	0.025	520
Flame soot:		
Type long flow	0.022	155
Type medium flow	0.027	110
Pigment yellow 13	0.50	73
Pigment red 57	0.60	49
Pigment blue 15	0.80	57
Pigment green 7	0.60	51

Note that the particle size measured from the pigment powder does not correspond to the particle size in the application medium. For deinkability, the particle size in the application medium is the decisive factor. This is because the inks detached from the fibers at the time of deinking are then present in the suspension as solids with different particle sizes and surface structure. The layer thickness of the ink application is very thin. Depending on the printing process concerned, it can be 0.5-10 μm. The common range is 0.8-4 μm.

As Table 7 shows with the pulping of recovered paper, a flexographic ink disperses during processing into much smaller agglomerates than letterpress, offset, or rotogravure inks. The spectrum of particle sizes depends greatly on the pH of the recovered paper processing operation.

Table 7. Particle size of different inks after recovered paper disintegration[21].

	Ink particle diameter	
Inks for	Uncoated paper	Coated paper
Letterpress	2 – 30 μm	10 – 100 μm
Offset	2 – 30 μm	50 – 100 μm
Water-based flexo	0.3 – 1 μm	0.7 – 2 μm
Rotogravure	2 – 30 μm	5 – 30 μm
Laser print, photocopy	40 – 400 μm	40 – 400 μm

The optimum ink particle size for effective ink removal using the flotation process is 10–250 μm. Small particles adversely influence deinkability in the flotation process. Wash deinking can effectively remove ink particles having a particle diameter less than 10 μm. Flotation deinking is more effective in removing particles larger than 10 μm (See Fig. 16 in Chapter 10).

When considering deinkability under conventional alkaline deinking conditions, pigment characteristics are important. Hydrophobic pigments can be more easily deinked than hydrophilic pigments as Table 8 shows. Alkali-resistant pigments remain unchanged during deinking. This includes the chemically inert carbon black. Pigments that are not alkali-resistant dissolve partially during deinking. This alters the color shade of the deinked pulp and of the white water.

Table 8. Relationship between deinkability and colorant characteristics[17].

Colorant characteristics	Deinkability	
	good	poor
Hydrophilic		•
Hydrophobic	•	
Alkali-resistant	•	
Not alkali-resistant		•

2.1.3 Additives

Additives are useful to adapt certain ink characteristics such as drying, gloss, and abrasion resistance. Examples of such additives are drying agents, oils, waxes, plasticizers, lubricants, and dispersing agents. Drying agents are a necessary ingredient in all oxidative drying inks where they act as catalysts for the polymerization of the self-drying oil components. They accelerate the process.

Additives can be incorporated at the time of ink production or at the press to adapt the ink to the technical environment and the print substrate. The additives are always strongly application specific. Their type and amount are therefore proprietary. Because of this secrecy, someone doing deinking cannot know whether such additives have adverse effects. No information is presently available to what extent these additives influence deinkability.

2.2 Inks for nonimpact printing processes

Nonimpact printing is a common but imprecise international term used to describe a large group of printing processes. The term originated to make a clear distinction from printers having a mechanical impact such as type shaft, spherical head, belt type, and matrix needle printers. The electronic printing process describes all electronically driven printing processes including the mechanical processes not discussed here. Table 9 lists the printing processes belonging to the nonimpact printing group.

Table 9. Nonimpact printing processes[22].

Printing process	Area of use
Electrophotography	Basic technique of the majority of office copiers and intelligent printers especially laser printers
Electrography	Plotters, ionographic printers
Electrosensitive processes	Limited to small formats (small reels)
Magnetography	Used in high-speed printers and copiers
Ink-jet printing	Used in text and graphic printers, video printers, for printing packaging and commercial products
Thermographic printing	Used in color printers and digital color copiers

The nonimpact printing processes offer possibilities that were not previously possible in classical printing techniques for economical reasons. Their most important area of application is office communication. With the growing range of possibilities in the office communication sector has come an increasing flexibility regarding types of paper, printing processes, and printing inks.

All nonimpact printing processes visualize the image information that is available in electronic form and transfer this to a substrate using colorants. These liquid or dry colorants vary considerably in their physical and chemical characteristics. This is why individual printing processes require a detailed explanation. A common opinion is that paper printed by the nonimpact process is difficult to deink because the particles produced during disintegration of recovered paper are too large for removal by conventional deinking processes such as flotation or washing. Due to adaptation of deinking process steps and chemistry, this is no longer a valid consideration.

2.2.1 Toner printing

Electrophotography and electrography are electrostatic printing processes. With magnetography, these three processes are all toner printing processes. Electrophotography is the leading printing technique among all nonimpact processes. It achieved its initial breakthrough with office copiers and is continuing its advance with laser and light emitting diode (LED) printers.

The electrophotographic process of Fig. 5 comprises the steps of charging, exposure, development, transmission, fixing, and cleaning. The heart of the electrophotographic process is the photoconductor that is negatively charged through ionization of the surrounding air (1). After this step, the charged photoconductor is illuminated by the reflected light from the original (2). At the places where the light is projected, the charge disappears. This provides an invisible latent image from the original. On the remaining charged spots on the photoconductor, toner will transfer (3). Now the image is visible. The positively charged toner particles transfer in the electric field to the plain paper (4). After this step, the toner is melted onto the paper (5). This fixing of the toner onto the paper occurs primarily with heat and pressure by transporting the paper between two heated rollers. Another fixing possibility is noncontact fixing with IR or UV radiation as Fig. 6 shows. In the final step, the remaining toner on the photoconductor is cleaned (6). The photoconductor is illuminated again for re-use[23].

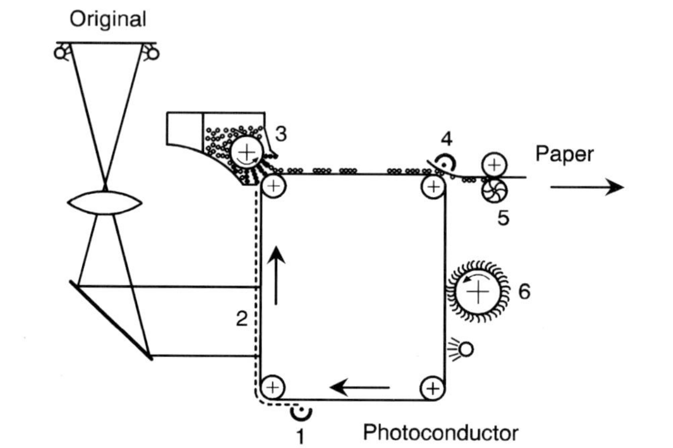

Figure 5. Electrophotography[23].

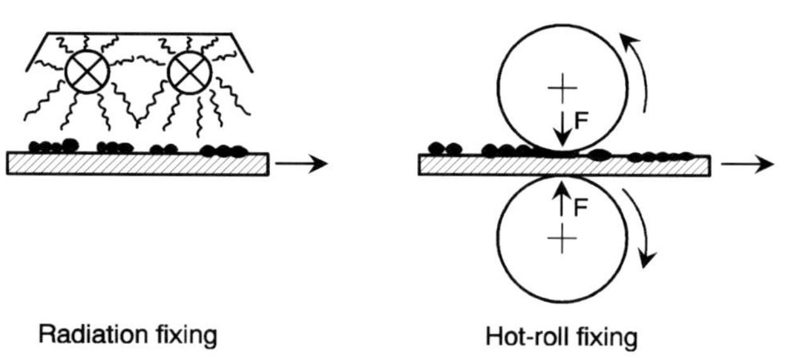

Figure 6. Toner fixing methods[23].

The ink film produced by fixing can be efficiently detached from a fiber surface during pulping, but it does not fragment again into fine toner particles. Ink agglomerates that result can have particle sizes up to about 500 µm. Due to the large particle diameter of the ink agglomerates, their removal by conventional flotation is problematical and not at all possible by washing.

Electrophotography makes the following demands on the toner[24]:

- Good electric or magnetic characteristics
- Sufficiently color development for high optical density
- High fixing capacity for a resistant, rub-proof print
- Environmentally-friendly, i.e., little volatile and no toxic components.

The actual composition of the toners are proprietary to the particular manufacturer and have been formulated for optimum performance in specific equipment. Most toners contain the general components listed in Table 10[25].

Table 10. Toner components[25].

Component material		Function
Colorant	Pigment or dye	Color
Base binder resin	Blend of monomers: styrene acrylates, polyesters, epoxy-type resins	Fusing of toner particles
Modifier resin	Natural material: abietic acid, resin, wax	Enhance strength of toner/paper interface
Charge control agents	Quaternary ammonium salts, sulfonates, zinc complexes	Provide toner particle with electrostatic charge necessary for development

Depending on the particular process, a distinction exists between single-component dry toner, two-component dry toner, and liquid toner as Fig. 7 shows[26]. Single-component toners have use in devices for medium and small print volumes. Two-component toners have primary use in large devices for office applications. Due to their solvent emission and the need for special types of paper, liquid toners are disadvantageous for most applications in the office area and therefore have no significance. For special applications, dry toners cannot simply replace liquid toners because extremely fine resolutions are possible only

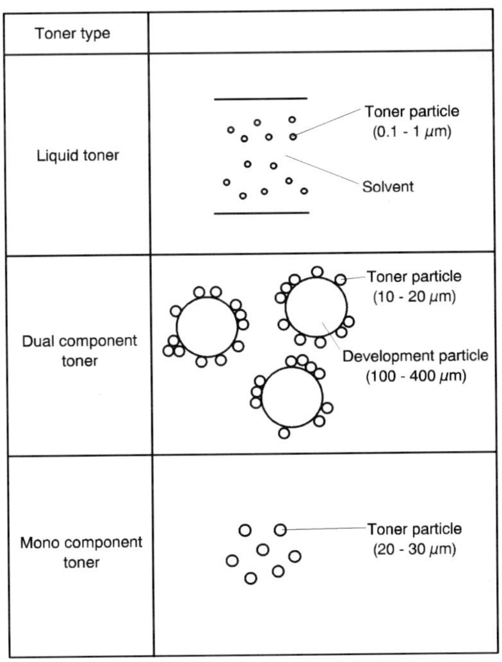

Figure 7. Toner types[26].

with liquid toners due to the very small particle sizes of 0.1–1 μm. Compositions of the different types of toner are described in the literature[22,24,26].

Single-component toner

The particles of the single-component dry toners combine coloring and the magnetic, electrical, and fixing characteristics in one component. They consist of iron oxide, thermoplastic resin, pigment (primarily carbon black), and additives. The toner particles have a diameter of 20–30 μm. This limits resolution.

Two-component toners

Two-component dry toners are powder compounds of carrier particles with a diameter of 100–400 μm that can be magnetized and coloring fine toner particles of 10–20 μm. The carrier particles consist of pure iron with a resin layer or a surface oxide layer. They may also be iron oxide with admixtures of copper, zinc, magnesium, barium, cobalt, nickel, or manganese. These components determine the magnetic and electric characteristics of the two-component toners.

Liquid toners

In two-component toners, the carrier is a dielectric liquid in which the chargeable toner particles of <1 μm are present in a finely dispersed and suspended manner. The most commonly used carriers are isoparaffin hydrocarbons with a flash point of 40°C–65°C. The transfer of the toner particles occurs electrostatically onto an image-charged photoconductor with subsequent transfer to paper or directly onto a special paper, e.g., coated with zinc oxide, that serves as a photoconductor.

Electrography and magnetography differ from electrophotography only in the recording of the latent image. Papers printed by this process have similar deinkability. The toners for magnetographic printing processes are very similar to the single-component toners for electrophotography. The magnetic toner particles of 10–20 μm contain ferromagnetic pigment particles as a finely grained powder and other components such as resin that are important for characteristics such as stability and fixing.

2.2.2 Ink-jet printing

With ink-jet printing, ink is sprayed from electronically controlled fine jets as tiny particles onto a paper. The inks are usually water-based products with alcohol additives. They contain primarily azo dyes and additives such as wetting agents and buffers to provide ink stability. For most inks, water is the solvent. Higher alcohols or glycols act as moisteners. Easily volatile solvents such as alcohols or esters ensure sufficiently fast drying. The ink must also contain a binder (polymers or resins) that ensure ink fixing after evaporation of the volatile portion. When printing a paper, the ink should run precisely and not flow excessively, a problem which may occur with water-soluble inks primarily on certain papers. Table 11 shows the typical composition of a water-based ink.

The deinkability of ink-jet prints differs from that of prints produced by electrostatic processes. Due to the aqueous composition, the ink disintegrates into very small particles during pulping. The tiny ink particles are very difficult to remove by flotation deinking. They remain in the pulp suspension and can reattach to the fibers. Wash deinking improves this result only marginally. Recovered paper blends that contain a certain amount of ink-jet printed products usually undergo a shade displacement during stock processing. Bleaching with reductive bleaching agents can effectively destroy the chromophoric groups of azo dyes used as ink dyes to counteract the shade displacement[27].

Table 11. Typical composition of a water-based ink[24].

Ink component	Proportion
Demineralized water	80%–83%
Polyethyleneglycol	8%–10%
Diethyleneglycol	3%–7%
Dye (mainly azo dyes)	2%–4%
Wetting agents	0%–2.5%
Additives (e.g. biocides)	Traces

2.2.3 Thermographic printing

In thermography, printing uses a heat-induced conversion process, i.e., the physical characteristics of the ink are thermally altered. A distinction exists between direct thermal printing and transfer thermal printing. A specially coated paper is necessary for direct thermal printing. Under the influence of heat, the leuco dyes embedded in the coating transform into a visible dye. The leuco dyes have a particle size of about 1 μm. They are water soluble and do not attach to the fibers. The addition of alkali effectively destroys these dyes[28]. Wash deinking removes such fine particles more effectively than the flotation deinking process. Older facsimile machines use thermal printing papers.

Thermal transfer printing permits printing on normal paper. For this application, an additional process step is necessary in which an image coating transfers by heat from a carrier (ribbon or film) to the print substrate. The inking layers activate by external heating elements or internal heating. For internal heating, the carrier layer contains fusible colorants. Typical components of an ink for thermal transfer printing are as follows[24]:

- Polymers such as polyamide resin or styrene and alkyl alcohol copolymers
- Oils to lower the polymer fusion point
- Oil gelling agents with a low melting point
- Softeners, thickeners, dispersion agents, and solvents
- Pigments and possibly dyes
- Components for controlling electrical conductivity.

2.2.4 Electrosensitive processes

In electrosensitive processes, a current passes into electrosensitive materials and triggers a physical or chemical process for optical information recording as Fig. 8 shows. Such a process can be the evaporation of a top layer in a laminate (metallic paper printers, electroerosion processes) or the color change of a coating (electrochromic or electrolytic processes).

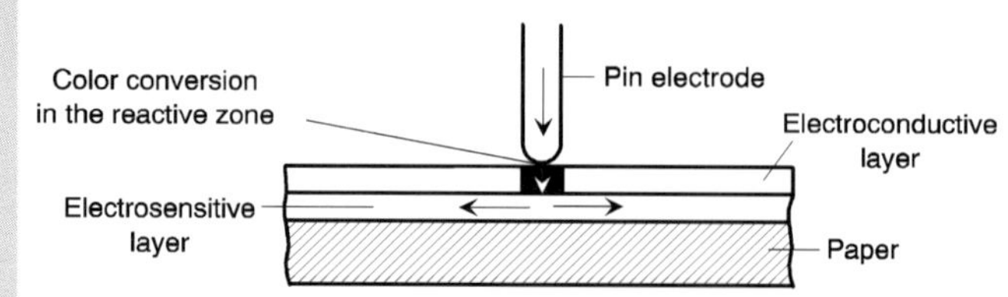

Figure 8. Electrosensitive printing process[22].

The advantage of this process is the very simple design of the printing head (pin electrode rail) and the single-step operation. Disadvantages are the low speed, need for coated papers, and release of reaction products from the coating. Electrosensitive recording processes have primary use in tape and test protocol printers that work with coated paper.

3 Deinkability of printed products

The deinkability of conventionally printed products depends on the type of ink applied on the paper substrate, the paper itself (coated or uncoated), and the age of the printed product at the time of deinking.

Newsprint has an ISO brightness of 60% compared with improved newsprint that is about 62%, uncoated rotogravure paper (supercalendered paper) that is about 67%, coated wood-containing paper (lightweight coated paper) that is 70%–72%, and wood-free copy paper that is more than 80%. As a consequence of printing, the brightness of the processed printed products is always significantly lower. The brightness after deinking improves with higher brightness of the unprinted substrate. The aging of the ink film and the type of ink can sometimes invalidate this correlation. A closer examination of the printed products and inks is therefore appropriate.

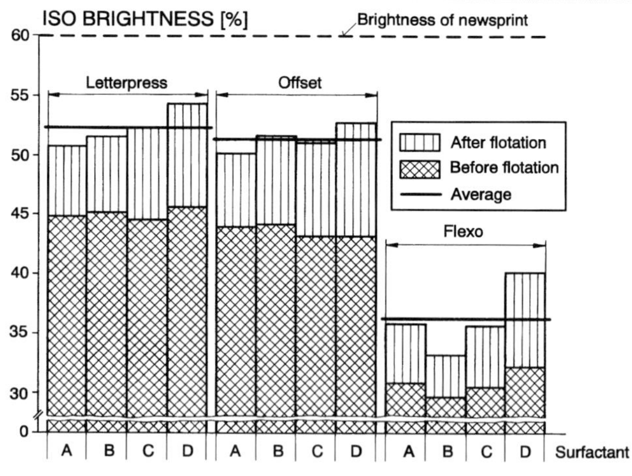

Figure 9. Brightness of newspapers before and after flotation influenced by printing process and type of surfactant[29].

3.1 Newspapers

Figure 9 shows results of laboratory tests on newspapers printed by the letterpress, offset, and flexographic processes[29]. For newsprint printed by the offset and letterpress processes, the average defiency of brightness of undeinked pulp is about 15% ISO compared with unprinted paper at 60% ISO.

For newsprint printed by the water-based flexographic process, one can assume that the ink transfer is approximately identical to that in the conventional newspaper offset and letterpress processes. Despite this, the brightness of the flexo printed newspaper is about 30% ISO lower than that of unprinted paper. This is a result of the greater dispersion of the ink particles and the consequential larger specific particle area. Under alkaline deinking conditions, the binder of flexographic inks is dispersed. This causes release of very small, hydrophilic pigment particles that do not float to the same extent as offset and letterpress ink particles. This gives a consistent darkening of the pulp slurry and the paper produced from it.

A comparison of the brightness before and after flotation shows the degree to which removal of ink can influence the gain of brightness. The average brightness that is possible after a one-stage flotation of letterpress-printed newspapers is slightly higher compared with (aged) offset-printed newspapers as Fig. 9 shows. Commercial letterpress-printed products often provide better deinking results than newsprint printed by the offset process. The letterpress process today has no significance for newspaper printing. Considerably poorer deinking results result with newspapers printed with water-based flexographic inks (33%–40% ISO). In this example, establishing the influence of the flotation agent (surfactant) used on the ink removal is possible.

3.1.1 Vegetable oil vs. mineral oil inks

For health and environmental reasons, newsprint ink vehicles were partly changed to vegetable oil based on renewable raw materials. In the middle 1980s, the World Health Organization (WHO) established that highly aromatic mineral oils with their polycyclic benz(a)pyrene compounds could cause cancerous skin tumors. The volatile organic compounds (VOC) in the inks also polluted the atmosphere and attacked the ozone layer. The United States enacted the Clean Air Act in 1990 to achieve a reduction of VOC emissions[30].

Some ink manufacturers hoped to eliminate any health risk by developing water-based inks (newsprint flexographic inks). Simultaneously, the industry launched newsprint inks using highly hydrogenated or solvent-refined naphthenic mineral oils. These were more expensive and required new printing conditions. Another desire was new base materials for ink production that would satisfy the health and environmental regulations. Recent years have therefore seen a partial change in the base materials for vehicles used in inks for high-speed web offset printing. Today, newsprint inks containing pre-treated mineral oils are used as are others using renewable materials. Renewable base materials are tall oil (a by-product of kraft pulp production), linseed oil, rape seed oil, ricinus oil, and soybean oil.

It is questionable whether vegetable oils in offset inks compared with mineral oil inks give good deinkability. In the fresh, unaged condition, most offset printed papers can be efficiently deinked. As stated before in the section dealing with ink composition, drying of the unsaturated vegetable oils occurs by oxidation, i.e., by absorbing oxygen from the air with molecule enlargement, and by polymerization. After a certain storage time the deinkability suffers. This can be attributed to the existence of oxidizable binders and cross-linking reactions in the ink.

The modified drying oils or alkyd resins also dry by oxidation, but their unsaturated fatty acids cross-link faster than the vegetable oils. They therefore contribute to development of an elastic, more resistant ink film. The chemical oxidation here compared with the absorption of chemically inert mineral oils that occurs within minutes is a very slow process taking hours or even days.

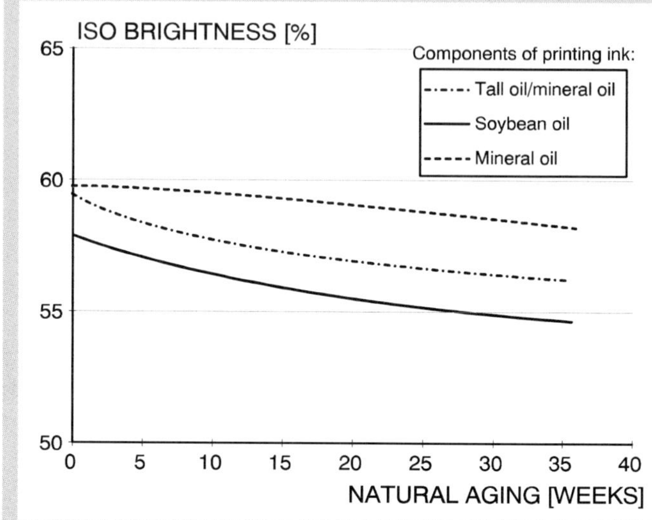

Figure 10. Brightness of DIP influenced by different types of binders[30].

Examination of offset samples printed on one side of the same newsprint with mineral oil, a tall oil and mineral oil blend, and soybean oil in the ink composition show that the offset ink containing 100% mineral oil provided slightly better brightness results in deinking in an unaged condition[30,31]. Figure 10 shows the brightness after deinking for all three inks resulting from natural aging. The offset ink using mineral oil showed only minor changes in brightness after natural aging and flotation. Compared with the inks using mineral oil, the ink containing soybean oil as binder showed poorer results.

3.1.2 Water-based flexographic inks for newsprint

Water-based flexographic printing has primary use in package printing. Due to economic advantages, it finds use in newspaper printing in Italy and to a much lesser extent in the United Kingdom and the United States. Saponifiable synthetic dispersions are the binders. Because these dispersions are soluble in an alkaline environment, the conditions of the flotation deinking process are insufficient to deink water-based flexographic printing satisfactorily. In contrast to offset and rotogravure inks, the inks do not have a hydrophobic nature. The particles of the flexographic ink are also very small (See Table 7). This results in impaired flotation efficiency.

In the example in Fig. 9, a maximum brightness gain in flexographic newspapers of not more than 8% ISO was possible by flotation. This is a poor result. Using washing instead of flotation, the brightness gain was about 20% ISO[29]. Results were similar for deinking tests for various European newspapers[32].

Figure 11 shows the results of deinking tests with newspapers printed by the offset and flexographic processes. The flotation of offset newspapers gave acceptable brightness of 52%–57% ISO. In contrast, flotation of Italian and English flexographic newspapers produced a very low brightness of 22%–29% ISO. Thickening after flotation had little effect in newspaper offset.

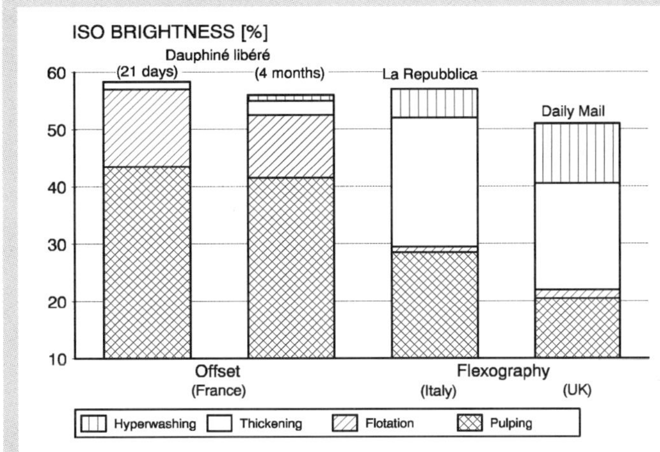

Figure 11. Deinking trials with newspapers printed by offset and flexographic printing processes[32].

In flexographic printing, thickening did increase the brightness by about 20% ISO. This brightness gain is due to removal of the small flexographic ink particles by dewatering. A resultant discoloration of the process water does occur. In an industrial deinking process, this would cause problems in the process water loop because this water is intended for re-use. Another problem in the deinking of flexographic newspapers is the phenomenon of reattachment of the finely dispersed flexographic ink particles to the fibers.

Hyperwashing limits the extent to which detached ink particles are still present in a pulp slurry. Figure 11 shows that especially for the flexographic printed newspapers a washing stage leads to a further brightness increase by removal of detached ink particles. This will also influence the condition of the process water to be recirculated.

In recent years, several studies have tried to discover ways to improve the deinkability of water-based flexographic inks[17,32–34]. Laboratory tests have established a strong relationship between pH of the pulp suspension and deinkability of flexographic inks[32–34]. Flotation of conventional water-based flexographic inks achieves higher levels of brightness in a slightly acid pH range of 5.5–7 than with flotation in an alkaline environment[32].

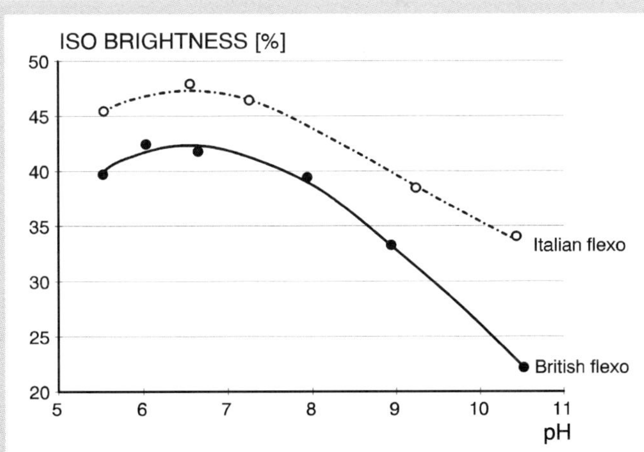

Figure 12. Effect of pH at flotation on the brightness of deinked pulp using flexographic printed newspapers[32].

Figure 12 shows the flotation results of flexographic newspaper at different flotation pH. Due to this relationship between the brightness achieved and the pH, using a two-stage flotation for deinking flexographic printed material is recommended. The first stage should remove the flexographic ink particles in an acidic environment at pH 5.5. This decomposes the calcium carbonate present in most pulp slurries. The second stage should occur in an alkaline medium to remove the conventional ink particles. Pilot tests gave a brightness of 55% ISO using a two-stage flotation even for 100% flexographic newspapers. Patents have been applied for this process. Figure 13 shows a flow sheet for a possible two-stage flotation process.

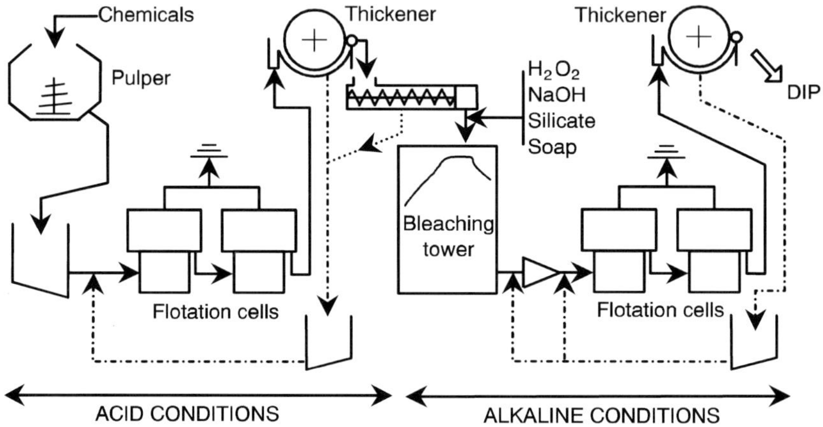

Figure 13. Two-stage flotation process adapted to mixtures of newspapers containing flexographic printed newspapers[32].

The pH dependency of the flotation result should improve deinkability of flexographic newspapers by using adequate surfactants in the flotation process[33–35]. Such surfactants can have an anionic, nonionic, or cationic nature. With the hydrophobic particles of conventional inks such as offset inks, attachment of the surfactant to the ink particles does not depend on its ionic nature. For water-based flexographic inks, one must

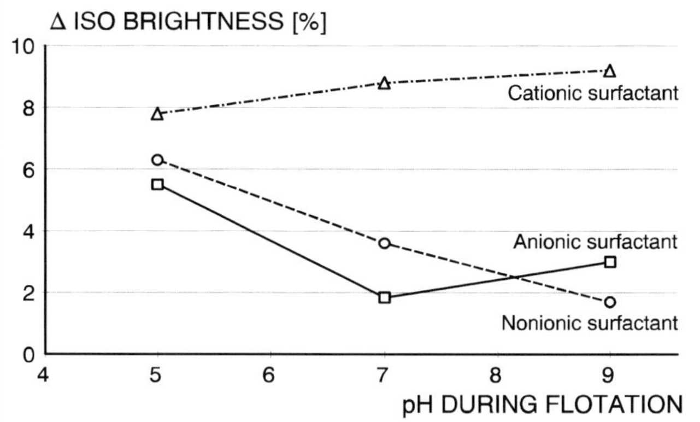

Figure 14. Brightness gain of flexographic printed newspapers influenced by pH during flotation[34].

assume that hydrophilic, negatively charged particles are present in the alkaline environment. Their surfactant load has extreme importance for attachment of the surfactant to the particle surface. Due to the hydrophilic character of these ink particles, no great tendency exists on the part of the particles toward attachment to the air bubbles. For this reason, making the ink particles water repellent by surfactants is necessary. After the ink particles have received this treatment, less problems will be experienced with approaching and attaching to the air bubbles. The attachment of the surfactant to the hydrophilic, negatively charged particles of water-based inks also depends on the applied flotation pH as Figs. 14 and 15 show[32–35].

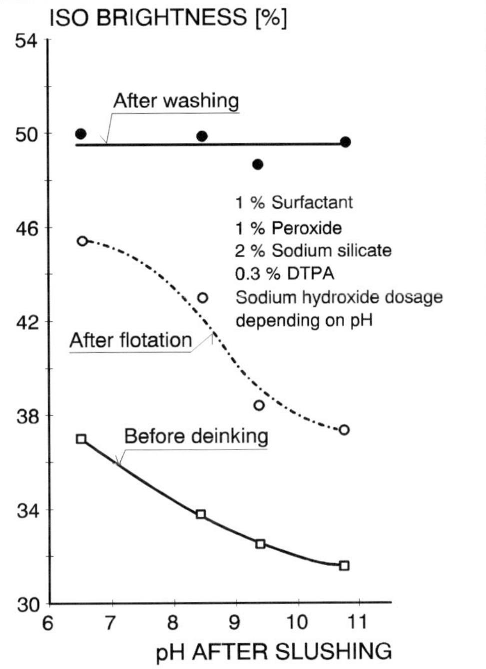

Figure 15. Brightness of deinked flexographic printed newspapers influenced by pH and the deinking process[29].

Figure 14 shows that a cationic surfactant retains its attachment capacity throughout the entire flotation pH range of 5–9. This gives a brightness increase of 8%–9% ISO[34]. With anionic and nonionic surfactants, a certain brightness gain is possible at flotation as the pH decreases. This is due to a greater tendency toward attachment of these surfactants in the acidic environment. The brightness gain never reaches the level of the cationic surfactant.

Figure 15 documents the results of tests using a constant chemical formulation but with different alkalinity adjusted by sodium hydroxide. With decreasing pH, the brightness increases before and after flotation. The results of the washing remain nearly unchanged but at a higher level[33].

Ink manufacturers are trying to optimize the composition of water-based flexographic inks to improve

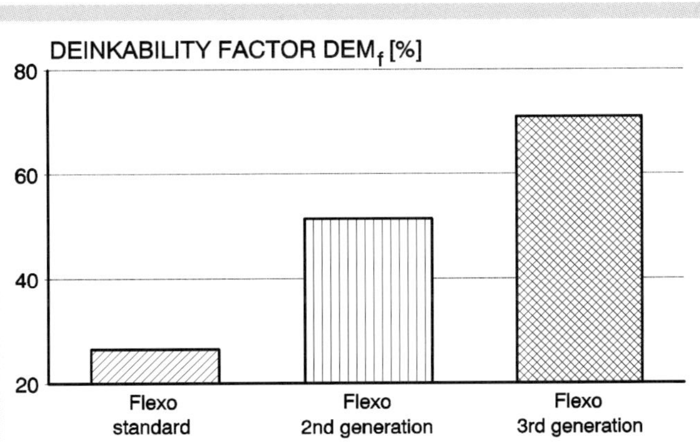

Figure 16. Deinkability of different flexographic newsinks[17].

their deinkability. This involves using binder systems in the ink composition that do not completely disintegrate with conventional alkaline deinking formulations. The ink formulas for these deinkable, water-based inks are available in patent specifications[36,37]. In one recipe for a deinkable water-based ink, the water-diluted cationic resin is an acrylic resin. As an example of this printing ink development, Fig. 16 shows improvement in the deinkability of new, flexographic inks. Conventional water-based flexographic ink had a DEM_f of only 25%. A DEM_f of about 70% was possible using the improved ink formula of the so-called third generation.

3.2 Magazines

Magazines, catalogs, advertising supplements, and direct-distribution advertising matter are printed on uncoated and coated papers that are primarily wood-containing supercalendered (SC) and lightweight coated (LWC) paper grades and improved newsprint. LWC papers have primary use for catalog production. SC and LWC papers are used for magazines. The wood-containing and wood-free coated and uncoated graphic paper grades find use in advertising supplements. Magazine papers are printed by rotogravure and offset processes. Offset printing is primarily web-fed and sheet-fed heatset.

3.2.1 Coated and uncoated magazine paper

Fewer problems usually result with the deinking of coated papers than with uncoated papers. The ink detaches with the pigment coating during pulping. In contrast, deinkability of uncoated paper is influenced by the absorption of the ink into the fiber network of the paper. Depending on the ink composition and its absorption or drying behavior, the ink penetrates the paper to a lesser or greater degree.

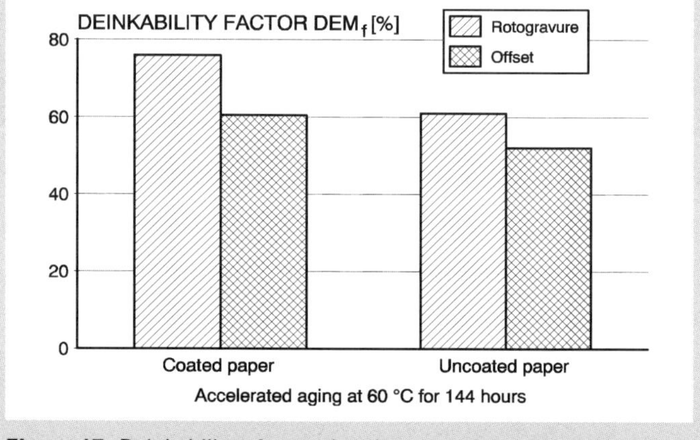

Figure 17. Deinkability of coated and uncoated magazine papers printed by offset and rotogravure processes[38].

Several examples of deinking results obtained from uncoated and coated print products have interest. Figure 17 shows the deinkability factor by flotation of uncoated and coated magazine papers considering the printing process used. Before deinking, the print samples were subjected to accelerated aging to simulate the natural aging of printed paper until its processing. The highest deinkability factor with an average DEM_f of 76% occurred with the rotogravure printed coated magazine paper[38]. The rotogravure printed uncoated magazine paper had a DEM_f of 60%. For both uncoated and coated magazine papers, the deinkability factors of the rotogravure printed papers are greater than those of papers printed by the offset process.

Figure 18 shows the brightness of printed and unprinted samples for various paper grades and printing processes. Compared with offset and letterpress printed newsprint, these results indicate that the difference in brightness between unprinted and printed magazine papers is greater with rotogravure printed papers. This is a result of the greater ink application of the rotogravure process[29]. The average ink application to the paper in this printing process is about twice as much as in the offset process. Table 12 shows this by presenting the mass of dried ink applied to the paper and the thickness of ink film for different printing processes.

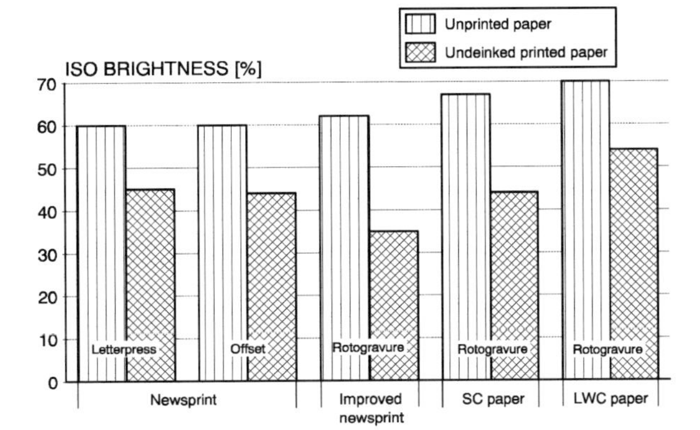

Figure 18. Brightness of printed and unprinted samples influenced by paper grade and printing process[29].

Table 12. Ink application to paper for various printing processes[39,40].

Printing process	Ink film thickness [μm]	Ink application [kg dry ink/t paper]
Letterpress	4 – 6	15 – 25
Offset	1 – 3	10 – 20
Rotogravure	3 – 12	20 – 50

The furnish of the paper also influences deinkability of uncoated and coated papers. In a series of tests, wood-free and wood-containing uncoated and coated papers were printed by the heatset offset process. A good ink detachment occurred for the deinking of the wood-free uncoated and coated print samples[41]. Brightness increased in both cases by more than 17% ISO as Fig. 19 shows. The bright-

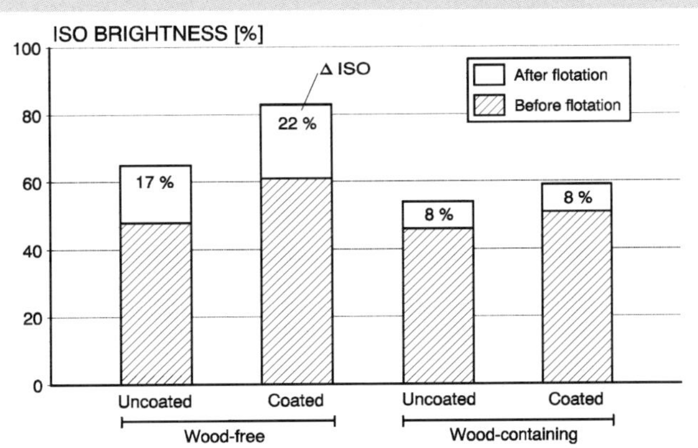

Figure 19. Brightness of heatset offset printed magazine papers before and after flotation[41].

ness gain for the wood-containing uncoated and coated papers was only 8% ISO. Poorer ink detachment occurred here for the wood-containing papers than for the wood-free papers. In the case of wood-containing fibers, an interaction occurs between the lignin and the printing ink resulting in an intensified attachment of the ink on the fibers. Analyses of ink particles showed that the deinking of wood-containing coated print samples gave a considerably greater number of ink particles with diameters below 30 μm before and after flotation. This gave considerably poorer deinking by flotation.

The deinkability of SC and LWC papers was investigated in the following test series[42]. The rotogravure printed samples represented catalog printing, i.e., the solid printed samples with a high ink application. The deinking results of the print samples were very unsatisfactory as Fig. 20 shows. The average brightness of 48% ISO achieved for the LWC papers after flotation with an initial brightness of 70%–72% ISO for the unprinted LWC paper is a poor deinking result. A better flotation result occurred for catalog printing of the same print image on SC papers. This is surprising since one usually assumes better ink detachment for coated papers. In this case, the brightness increase due to flotation proved to be greater for the SC papers than the LWC papers. The flotation result depends in this case significantly on the particle size spectrum of the printing ink influenced by the mechanical forces at slushing with the coating layer. This resulted in a greater brightness of the SC papers after flotation although the brightness of the unprinted SC papers at 67% ISO is below the brightness of the LWC papers.

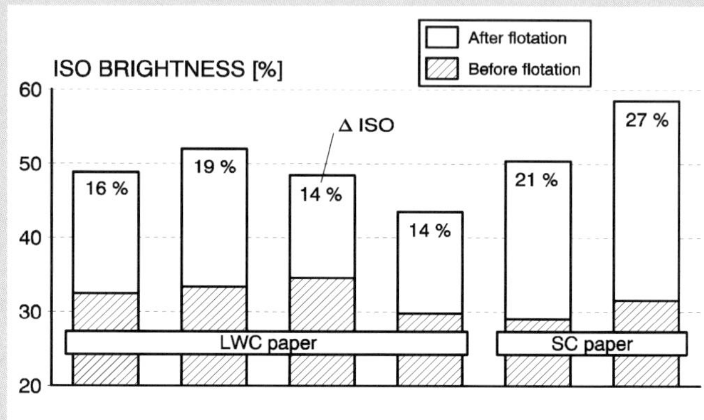

Figure 20. Brightness of rotogravure printed magazine papers (catalog) before and after flotation[42].

3.2.2 Toluene- vs. water-based rotogravure inks

The conventional rotogravure ink used in a high speed printing process contains toluene as a solvent. A printed product using conventional rotogravure ink gives good deinkability depending on the print substrate. This is because the brightness achieved after flotation of DIP is sufficient for production of newsprint. As a result of legislation such as regulations governing residual toluene in printed products, water-based rotogravure inks could become important in the future. In preparation for this change from the toluene- to water-based ink systems, many organization have performed tests on a laboratory scale[2,43,44].

Under conventional alkaline deinking conditions, the water-based inks are less deinkable compared with a toluene rotogravure print. The same reasons are valid for water-based rotogravure inks as for water-based flexo inks: the very small particle size and the hydrophilic character. The deinking results obtained with printing of an identical subject produced under commercial operating conditions using water-based and toluene rotogravure inks showed brightness gains of only 4.2% ISO and 23.9% ISO, respectively, as Fig. 21 shows. This major difference in deinking means considerable development work remains for the ink producers and deinking technology. The production and use of water-based rotogravure inks is still discouraged by their high price and costs of the necessary technical conversion of the printing presses.

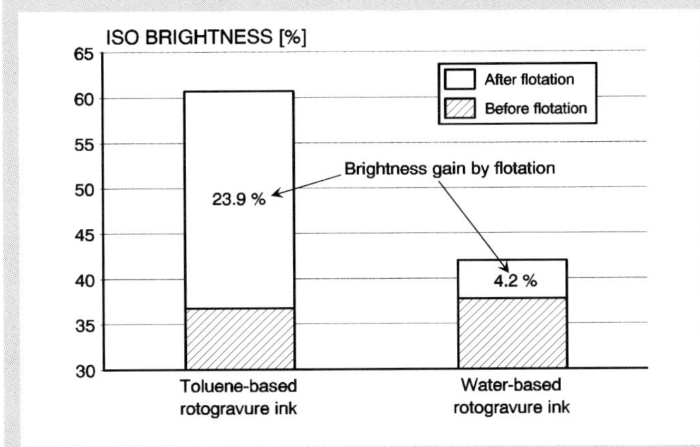

Figure 21. Brightness of toluene- and water-based printed advertisement material before and after flotation[45].

3.3 Nonimpact inks

Laser printed office papers usually provide toner particle size distributions after pulping with larger dimensions than papers printed by conventional processes. The dry or fluid inking agents transferred to the paper by these printing methods vary considerably in their physical and chemical characteristics. Large ink particles give an insufficient removal of ink during flotation. An ink particle size of 10–150 μm is optimum for good ink removal by the flotation process. This is the particle size range that is mainly present in the pulping of news and magazines printed by conventional processes.

For nonimpact printed papers from laser printers and copiers, experience has shown that a particle diameter of 70 μm to more than 400 μm is present after pulping. Figure 22 shows examples of the particle size distribution of different prints produced by laser printers and copiers[46]. The toner particle size after pulping for one commercial unit can have an average diameter of about 350 μm.

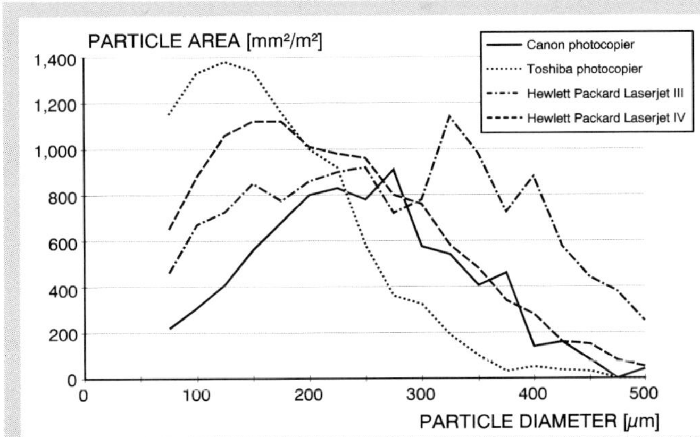

Figure 22. Particle area distribution of disintegrated copies of different copy and printer units[46].

To process these paper grades effectively into bright, clean pulps, a combination of different process steps is necessary. The following processes can be considered as possible treatment steps for the removal of large dirt particles:

- Toner fragmentation by kneading
- Toner fragmentation by dispersing
- Toner particle removal with cleaners after chemical pre-treatment in the pulper (agglomeration to larger particles)
- Flotation after ink particle fragmentation and detachment by dispersing or kneading.

Kneading and dispersing both efficiently realize ink particle detachment and fragmentation (See Chapter 5). Kneading is becoming increasingly important for recycling of office papers. It has a greater benefit for fragmentation of laser ink particles than dispersing. Dispersing in a high-speed disperger is more efficient for homogenization of particles of offset or rotogravure printing inks. With dispersing and kneading, the size of the residual ink particles decreases accompanied by a loss of brightness. Post-flotation, peroxide bleaching, or both can compensate for this brightness loss. Table 13 shows a comparison of kneading and dispersing technologies.

Table 13. Comparison of kneading and dispersion technologies [47,48].

Kneading	Dispersing
One kneading step: positioned before post-flotation Two kneading steps: generally positioned in the middle of the system before post-flotation and washing	One dispersion step: positioned before post-flotation Two dispersion steps: Generally positioned before post-flotation and close to the end of the system
Results in only a negligible change in freeness	Causes a moderate change in freeness in the disperger
Supports a better size reduction of nonimpact inks	Supports a better size reduction of conventional inks

In the kneader at high consistencies of 25%-30%, compressive and shearing effects apply. Also important for the kneading process are the applied temperature and kneading time. A distinction is necessary in the kneading process between cold and hot operation. The cold operation gives a poorer dirt size reduction but lower operating costs[48,49].

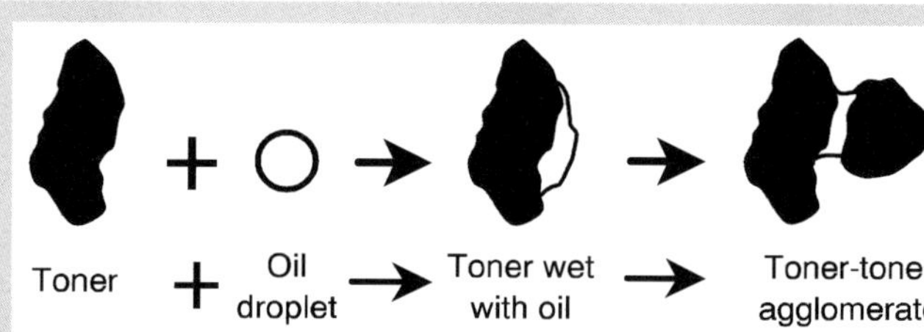

Figure 23. Liquid bridge agglomeration process[51].

Another possibility for removing large toner particles is use of forward cleaners. Forward cleaning is regarded as an effective method for removing ink particles from nonimpact printing processes[50]. This method can also combine with chemicals for pre-treatment of the toner particles. Special chemicals cause the toner particles to form larger aggregates to facilitate their removal with a forward cleaner. For example, toner particles are supposed to agglomerate to a diameter of more than 1 mm by addition of an oil in water dispersion into the pulper as Fig. 23 shows. Adding surfactants or other chemicals was not necessary[51].

The ink-jet printing process is the most important nonimpact printing process besides laser printing. As mentioned in the section entitled "Inks for nonimpact processes," the ink is water soluble and contains dyes as colorants. When disintegrating ink-jet printed office paper, the ink dissolves. Flotation deinking cannot remove it. In wash deinking, the filtrate from the pulp slurry removes the ink, but sufficient cleaning is not possible by dissolved air flotation (DAF) of the process water. Bleaching contributes to the decomposition of chromophores of the azo dyes of the ink. Reductive bleaching with dithionite or formamidine sulfinic acid (FAS) is more effective than oxidative bleaching with peroxide[27].

3.4 Effect of aging

As the age of a print product increases, its deinkability may suffer. Besides the time factor, environmental factors such as temperature, humidity, components of the air, and UV radiation have roles in the aging of paper and ink.

To determine quickly the influence of aging on deinkability by flotation or washing of printed products, printed samples are subjected in laboratory tests to accelerated aging at 60°C and 0% RH in a circulating air drying oven for 144 h according to the German standards DIN 54606-1 and -2. An approximate value is that accelerated aging of 1 h under these conditions corresponds approximately to a natural aging of one day for oil-based offset printed newspaper. Although the conditions applied for accelerated aging are more extreme than in natural aging, they provide an estimation of deinkability tendencies resulting from aging processes in a reasonable time.

The section on "Carriers and binders" made reference to the different mechanisms of ink drying and their relation to deinkability. Especially with offset inks does increasing age cause poorer deinkability. The offset process has greatest importance in newspaper printing. Surveys concerning the composition of recovered paper in deinked material from German homes revealed that newspapers account for about 40%[52]. The deinkability of mixed recovered graphic paper therefore depends on the age of offset printed newspapers, a major constituents.

Offset inks contain proportions of binder components that dry by oxidation. This includes alkyd resins or vegetable oils that use oxygen from the air for oxidizing and cross-linking. The oxidizing parts of the binder cause the ink pigments to attach very firmly to the fibers. Consequently, the print becomes chemically and mechanically more resistant to deinking. The flotation of offset inks in an unaged state using binders that dry by oxidation can result in similar brightness as offset inks that use penetrating binders. Figure 24 shows the difference in brightness by flotation of offset ink with and without components that dry by oxidation compared with natural aging before deinking. After flotation, the brightness corresponding to an offset ink without oxidizing components is greater. The loss of brightness due to the aging period is considerably less than for offset ink that contains components that dry by oxidation.

Figure 25 shows the brightness after flotation of different newsprint samples printed in an offset press. The newsprint used differed in ash content between 0.5%–12%. The ink was a type that dried by oxidation. For freshly printed paper, the resulting brightness difference was only 4% ISO. While all brightness values decreased with age, the low ash paper values decreased the most.

The deinkability of newspapers printed by the letterpress process shows a lower effect of age[53,54].

Besides being influenced by the printing process used (heatset offset or rotogravure), the effect of aging on deinkability of magazines is a function of use of uncoated or coated paper. Offset and rotogravure inks printed on coated paper detach easily even after aging resulting in an adequate deinkability. The aging of uncoated magazine paper causes a significant reduction in the deinkability of offset inks and has less influence on the deinkability of rotogravure inks.

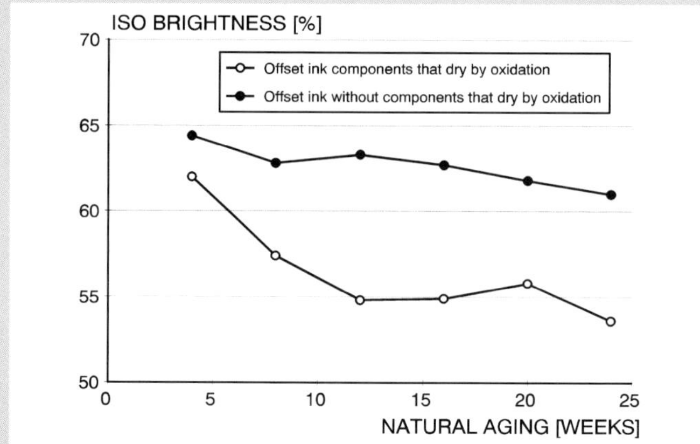

Figure 24. Brightness of wood-containing deinking material after flotation deinking influenced by natural aging[53].

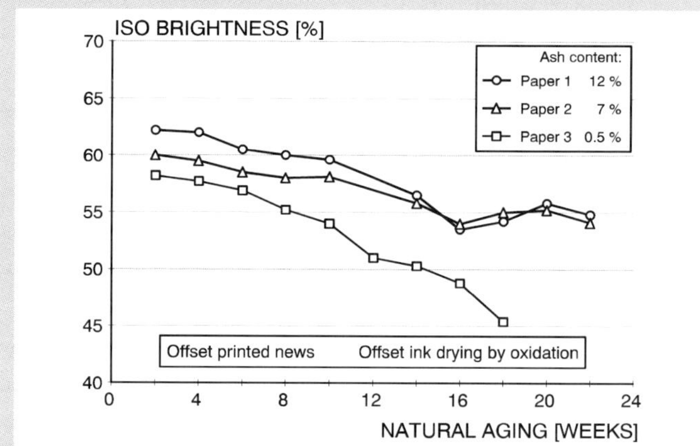

Figure 25. Brightness after flotation deinking influenced by natural aging of newsprints with different ash contents[53].

4 Evaluation of deinking results

4.1 Yield of deinking processes

To determine the deinking process yield, the pulp mass remaining in the accept is calculated from the mass of solids (fillers, pigments, fines, fibers, ink particles, and stickies) removed by deinking and the mass of recovered paper used (all values are on an oven dry basis):

$$Yield = \left(1 - \frac{removed\ mass\ by\ deinking}{input\ mass\ to\ deinking}\right)100[\%] \quad (3)$$

For flotation, the removed solids for consideration are in the foam of the flotation process. With a washing process, the rejected solids of the wash water (filtrate) must be used to calculate the yield. The yield for one flotation step is about 85%–90% depending on the recovered paper mixture used. The yield of one washing stage is about 75%–85% for a wood-containing, European recovered paper mixture.

The total yield of a multiple deinking process depends primarily on the number of flotation and washing steps. In addition, the applied chemistry with its pH and the air volume used can influence the flotation yield. Dilution and thickening determine the yield of a washing stage. From the calculated mass of the reject and its ash content, the selectivity of the process may be demonstrated in relation to fiber and minerals removal.

Figure 26 uses an example of the laboratory flotation of recycled fibers from a household collection (old newspapers and magazines) to illustrate the change of yield depending on the duration of flotation.

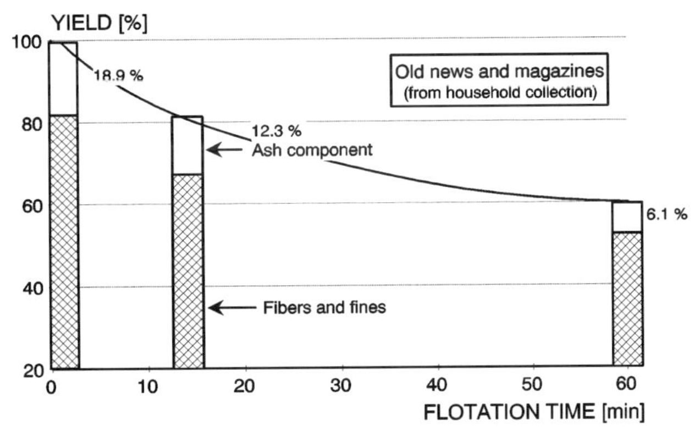

Figure 26. Influence of flotation time on deinked pulp yield[55].

4.2 Optical parameters of deinked pulp

The parameter most frequently used to evaluate the optical quality of deinked pulp is the spectral reflectance factor R_{457} used as a measure of ISO brightness. The measuring instruments are spectrophotometers. The ratio of light reflected from a paper sample at a given solid angle to the light that under identical measuring conditions is reflected from a totally white material at the same solid angle is calculated. Two reflected light intensities are therefore compared. The instrument is calibrated to a known white reference and the measured value of the sample read as a reflectance factor. Two main methods are available for determining the brightness of paper or pulp:

- TAPPI standard method[56] used especially in North America
- ISO standard methods[57,58] used primarily in Europe.

Both methods use the principle of measuring the reflectance factor of the paper or pulp sample at an effective or more precisely the centroid wavelength of the light of 457 nm given as spectral reflectance factor R_{457}. The difference between the methods

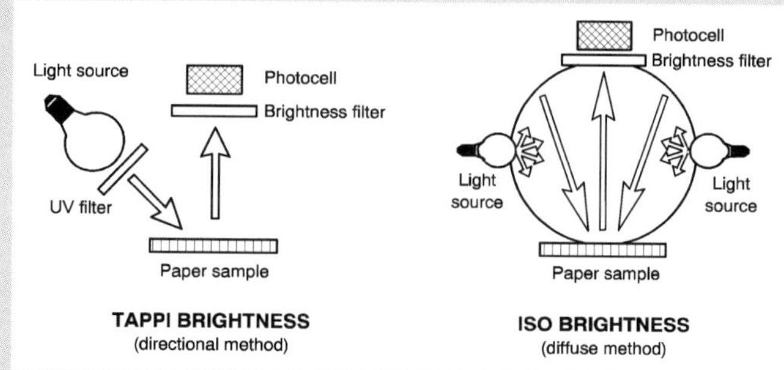

Figure 27. Differences in brightness measurement methods[59].

is due to the geometry of the incoming light beam that according to TAPPI is 45°. For the ISO measurement, diffuse illumination is necessary as Fig. 27 shows. Measuring in both cases uses 0°.

When using deinked pulp to manufacture graphic papers of particular shade specifications such as the color recommendations for SC and LWC paper, the colorimetric characteristics of the individual base material components and therefore also of the deinked pulp must be known. The spectral reflectance factor R_{457} alone is insufficient to describe the optical impression of colored samples. For example, blue pigments that have their strongest reflection in the wavelength range of 440–480 nm always reveal very high spectral reflectance values. The sample to the observer is clearly not very "bright" but appears blue.

With the aid of colorimetry, simulating the color sensitivity of the human eye by physical measurements is possible. The pure colors can be allocated to particular light wavelengths within the visible range of 380–780 nm. One way to assess colored samples is therefore the evaluation of the reflection curves in the wavelength range of 400–700 nm that spectrophotometers can make.

Another possibility to evaluate colored samples from the recovered paper processing operation is to determine their color. A color is defined unequivocally by the tristimulus values X, Y, and Z. The tristimulus value Y and the reflectance factor R_{457} are the most frequently used parameters to evaluate the optical quality of deinked pulp.

For a clear representation of the color, the measured tristimulus values X, Y, and Z are converted into the chromaticity coordinates x, y, and z. Because $x + y + z = 1$, defining a color by stating two chromaticity coordinates, x and y, is possible. Since the chromaticity coordinates in an x-y diagram (CIE chromaticity diagram for the standard illuminant C and 2° CIE standard colorimetric observer) are represented, only minor changes in the numerical values often appear for deinked pulp due to the physical properties of x and y. This makes an interpretation of the results difficult.

The CIELAB color space offers an alternative to the original CIE color space. Using the CIELAB color space, one can determine the color difference of colorimetric values. The advantage of this system is that it uses equidistant color differences corresponding to the color sensitiveness of the human eye. It uses the three coefficients, L*, a*, and b* to identify colors that can be calculated from the tristimulus values X, Y, and Z as Fig. 28 shows. The L* value stands for lightness, the a* value identifies the red or green axis, and the b* value represents the yellow or blue axis. The color difference, $\Delta E*$, of two different samples is the difference between two color vectors and can be calculated by the following formula:

$$\Delta E^*_{ab} = \sqrt{(\Delta a^*)^2 + (\Delta b^*)^2 + (\Delta L^*)^2} \quad (4)$$

Calculation of the color difference produces a relationship between the optical proportion of the deinked pulp, the unprinted but chemically treated pulp, and the printed but untreated pulps with regard to the CIELAB color space as described in the section "Definition of deinkability." This uses calculation of the deinkability factor, DEM_{Lab}.

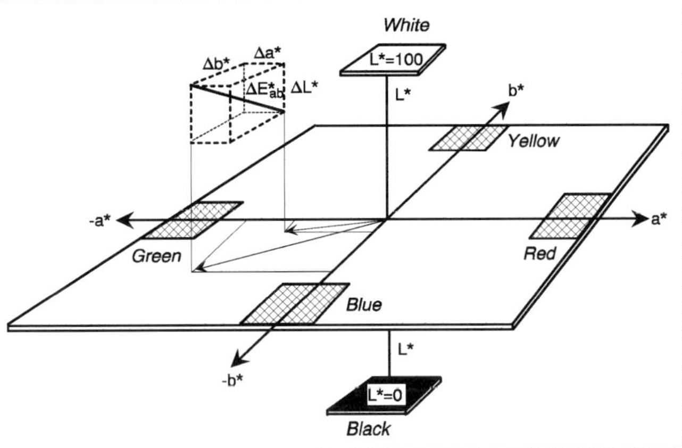

Figure 28. CIELAB color system[60].

When measuring ISO brightness, controlling the UV content of the light source is important since the measured reflectance values of deinked pulp can be influenced by the content of optical brighteners. Figure 29 gives an example for the brightness measurement of deinked pulp with and without UV radiation[61]. Higher reflection values occurred after the removal of the UV cut-off filter of the spectrophotometer especially in the wavelength range of 400–500 nm. The difference between the two spectral reflectance curves has a maximum at about 430 nm. This relates to the fluorescent optical brightener content of the deinked pulp. In the example shown for

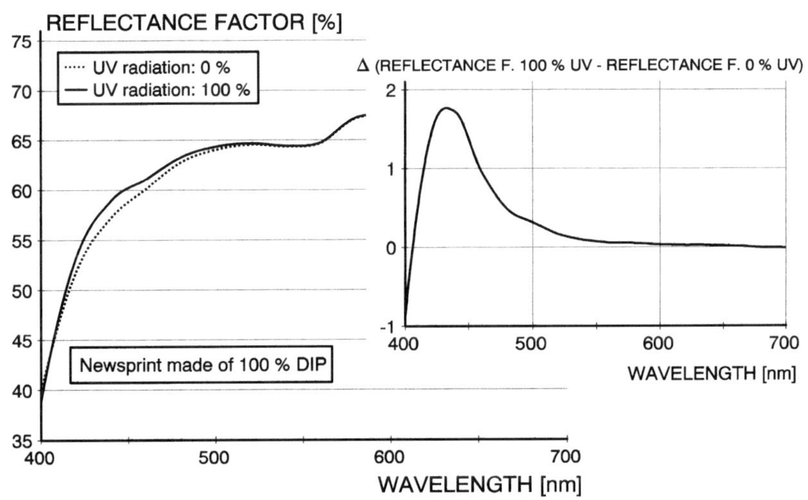

Figure 29. Brightness measurement of deinked pulp with and without UV radiation[61].

commercial deinked pulp, the maximum difference was 1.3 percentage points. Experience has shown that a high proportion of UV radiation in light can cause the spectral reflectance factor R_{457} to increase by several percentage points. For ISO brightness, the rule is that an increasing proportion of UV gives a greater absolute value.

Figure 30 shows that different UV proportions also change the color. In the example shown, the maximum UV radiation lowers the b* value by about one point, and the a* value increases almost to the same extent.

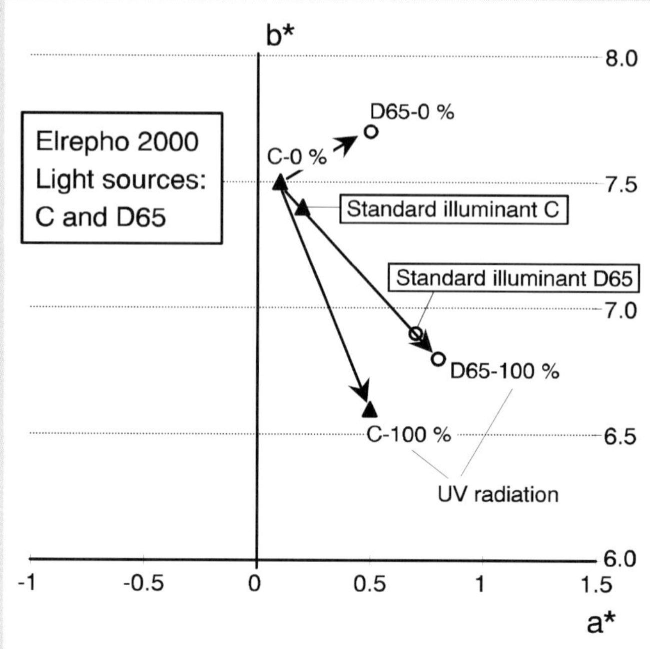

Figure 30. Changes of color influenced by UV radiation[61].

4.3 Particle analysis by image analysis

Evaluation of the quality of deinked pulp and the effectiveness of ink removal can use optical parameters such as brightness, luminance, or color. Image analysis systems are useful for recording optical inhomogeneities. These are dirt specks including ink particles in the deinked pulp or in the paper produced that form an optical contrast to the brighter sheet background. They are primarily residual ink particles that cause an increasingly grey pulp as the particle size becomes smaller.

Image analysis with different specifications is useful to determine the number, diameter, and area of the residual dirt particles in the macroscopic and microscopic range. Numerous personal computer image analysis systems working as scanners with a resolution not less than 40 µm particle diameter provide high speed measurement of particle numbers and areas in the visible (macroscopic) area. These devices are "low cost" image analyzers with primary use commercially for the quality control of deinked pulp.

More powerful image analysis systems that can also capture particles invisible to the naked eye to a size as small as 2–3 µm have primary use in research.

With the exception of a TAPPI method for determining the equivalent black area (EBA), no standardized evaluation methods currently exist for the measurement of dirt specks with these image analysis systems. Several methods have been developed for particle analysis by visual comparison[7,62–66].

The first step in an image analysis evaluation is preparation of the sample. A distinction occurs here not only between laboratory samples and commercial samples but also for the image level to be assessed. The surface measurement of a handsheet, filter pad, or paper with finite thickness only allows a qualitative dirt particle analysis. This is because it only records the dirt particles on the surface of the sample. Quantitative dirt particle analyses can use measurement on very thin fiber mats such as membrane filter samples where the z-direction of the sample to be measured is almost zero. Only single, separate particles occur at the measuring level.

Although quantitative analyses are very lengthy due to the sample preparation, they precisely record the number and area of the particles through a process. Image analysis investigations of the size distribution of dirt particles in deinked pulp have shown that considerably more than 90% of all particles are smaller than 20 µm after pulping (Fig. 31). They are therefore below the visibility limit[66,67]. The frequency or distribution in relation to the particle area shifts toward larger particles. Particles of less than 10 µm diameter and between 10–20 µm account for about one-third each of the recorded total area. For quantitative dirt particle recording, this shows that measurements in the microscopic range are essential.

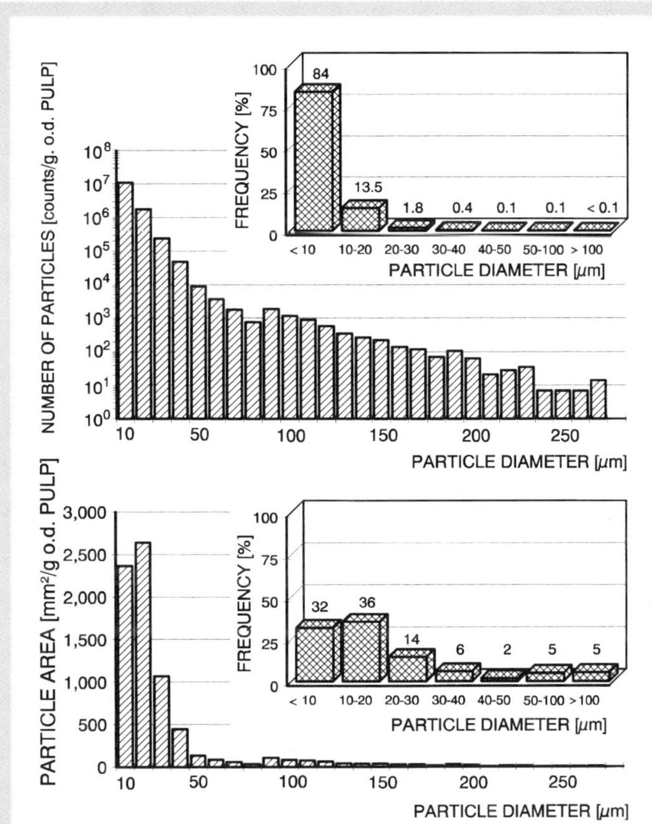

Figure 31. Size distribution of dirt particles after pulping[66].

In commercial practice for off-line process control, qualitative analyses in the macro area with particle size > 50 µm have gained widespread use. They use filter pads, handsheets, or the produced paper. Comparative dirt particle measurements performed with a powerful interactive image analysis system (IBAS) and a "low-cost" analyzer (e.g., Dot Counter) on filter pads, handsheets, and membrane filter samples have clearly shown that both analyzing devices categorize the deinkability of recovered papers similarly. This includes the process stages of disintegration and flotation with respect to visible particles[67]. As soon as the following process stages of the deinking process such as dispersing reduce the size of the dirt particles remaining in the pulp suspension into the microscopic area (< 50 μm), no additional useful information is available from a "low-cost" image analysis device as Fig. 32 shows. Major differences exist between a dot counter and IBAS measurements for the number of measured dirt particles. Independent of the method of sample preparation, the interactive image analysis system will record higher values for the dirt particle number per square meter sample area than the dot counter.

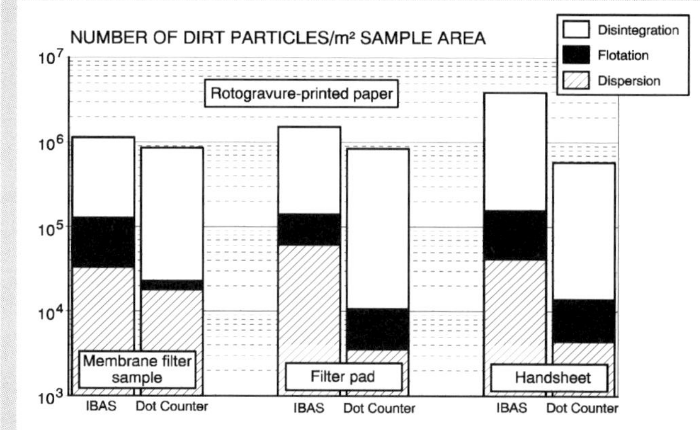

Figure 32. Dirt particle analysis in the macroscopic area[67].

Besides sample preparation, the steps of image acquisition, image processing, and image recognition are important for quantitative dirt particle recording as Fig. 33 shows. The recognition and recording of dirt particles by image analysis uses the difference in contrast between the dirt particles and their background, a fiber monolayer. The image captured by the camera is digitally converted into pixels whose size depends on the field of view determined by the measuring range used. Identification of the dirt particles uses grey value differences. A value for this with a maximum of 255 is assigned to each pixel in accordance with its reflectance. Image processing functions enhance the contrast between the measured objects and the image background. The image segmentation converts the digitized grey value image of the camera to a binary image. In this way, all pixels with a grey value below a certain threshold value are identified as dirt particles and receive the value of 1 in

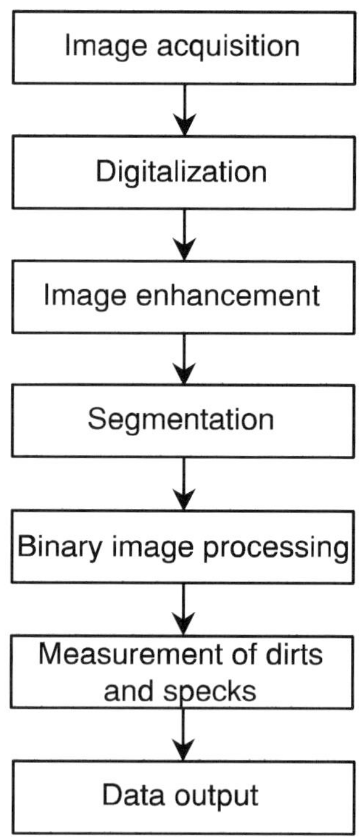

Figure 33. Dirt particle analysis procedure[62].

the binary image. The image analysis of the sample concludes with measurement of the dirt particles and the data output.

The total size and total number of particles and their division into individual size categories are relevant. The measuring parameters usually relate to 1 m^2 paper sample or 1 g of o.d. pulp. Due to its optical influence, the dirt particle area has special importance.

Figure 34 gives an example of the development of brightness and dirt particle area at different measuring points of a recovered paper processing plant[67]. As the dirt particles decrease, the brightness improves. The greatest brightness increase and therefore the strongest reduction in the dirt particle area occurs during the first flotation stage. After dispersing, a slight loss of brightness occurs. As a result of particle fragmentation the particle number increases, many particles are below the resolution limit of the image analysis system. They can therefore not be recorded completely resulting in a slight reduction of the total particle area.

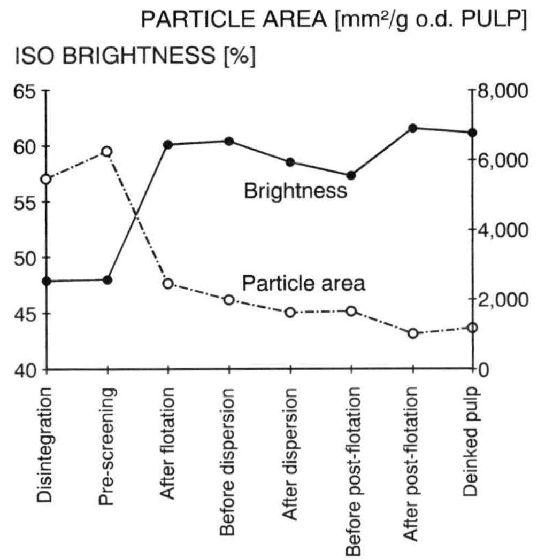

Figure 34. Development of brightness and particle area in the deinking process[8].

The diameter of the circle of equivalent area is calculated from the individually measured particle areas. As a general rule, a larger particle has smaller influence on brightness but contributes to the optical heterogeneity of the pulp or paper in terms of dirt specks.

References

1. *PTS-Merkblatt PTS-RH 010/87, "Prüfung von Altpapier – Kennzeichnung der Deinkbarkeit von bedrucktem Altpapier im Flotations-Deinking-Verfahren", Papiertechnische Stiftung, Munich, 1987, 8 pp.*
2. *Renner, K., Putz, H.-J., and Göttsching, L., "Untersuchungen zur Deinkbarkeit von wasserbasierendem Tiefdruck", INGEDE Report 1791 IfP, Darmstadt, Germany, 1992, 23 pp.*
3. *Carré, B., Galland, G., and Julien Saint Amand, F., Progress in Paper Recycling 4(1):80(1994).*
4. *Jordan, B. and Popson, S., J. Pulp Paper Sci. 20(6):J161(1994).*
5. *Jordan, B.D., Nguyen, N.G., and Trepanier, R.J., Tappi J. 76(10):110(1993).*
6. *McCool, A. and Taylor, C.J., Tappi J. 66(8):69(1983).*
7. *Trepanier, R.J., Pulp Paper Can. 95(12):T547(1994).*
8. *Renner, K., Putz, H.-J., and Göttsching, L., Das Papier 49(10A):V48(1995).*
9. *Anon., "Ink-Report", Gebr. Schmidt Druckfarben, Frankfurt a.M., Germany, 1995, 59 pp.*
10. *Ackermann, C., Putz, H.-J., and Göttsching, L., Pulp Paper Can. 95(8):T307(1994).*
11. *Kübler, R., Wochenbl. Papierfabr. 116(10): 386(1988).*
12. *Anon., "Werkstoff Druckfarbe", Hartmann International, Niedernhausen Ts., Germany, 1977, 13 pp.*
13. *Cosma, R., Verpackungs-Rundschau 36(2):99(1985).*
14. *Kübler, R., Allg. Papierrundschau 33(14):383(1984).*
15. *Williams, L.R., "Paper & Ink Relationships", Mennonite Press, Newton, USA, 1988, 150 pp.*
16. *Frank, E., "Untersuchungen zur Deinkbarkeit von Heatsetdrucken", 5th PTS Deinking Symposium Proceedings, Papiertechnische Stiftung, Munich, 1992, p. 54.*
17. *Buchweitz, J., Wochenbl. Papierfabr. 121(18):747(1993).*
18. *Schröter, K. -D., Wochenbl. Papierfabr. 119(1):8(1991).*
19. *Matzke, W., Allg. Papierrundschau 42(29):752(1993).*

20. Hermanies, E., "Polygrafische Technik", VEB Fachbuchverlag, Leipzig, Germany, 1978, 628 pp.

21. Borchardt, J.K., Progress in Paper Recycling 5(1):81(1995)

22. Schneider, J., "Technologien und Anwendungspotentiale der unkonventionellen Druckverfahren", 1988 FOGRA Symposium Proceedings, Elektronisches Publizieren, FOGRA, Munich, 17 pp.

23. Jeurissen, H., "Dry toner imaging process of copiers/printers", 1996 COST Workshop on the Fundamentals of Deinking, TNO, Delft, The Netherlands, paper no. 6.

24. Rosenberg, A., "Toner, Tinten und Beschichtungen", 1988 FOGRA Symposium Proceedings, Elektronisches Publizieren, FOGRA, Munich, 10 pp.

25. Olson, C.R., Hall, J.D., and Philippe, I.J., Progress in Paper Recycling 2(2):24(1993).

26. Leach, R.H., Armstrong, C., Brown, J.F., et al., "The Printing Ink Manual", 5th edn., Society of British Printing Ink Manufacturers Ltd., Padstow, UK, 1989, 839 pp.

27. Török, I., Putz, H.-J., and Göttsching, L., Wochenbl. Papierfabr. 119(11/12):395(1991).

28. Matzke, W. and Selder, H., Wochenbl. Papierfabr. 118(19):835(1990).

29. Putz, H.-J., Török, I., and Göttsching, L., Das Papier 43(10A):V48(1989).

30. Fuchs, B. and Thoyer, B., "Why are soybean based newsinks so successful in the U.S.A.?" ifra-Special Report 1.12, IFRA, Darmstadt, Germany, 1995.

31. Tuovinen, J., "Welche Einflüsse haben die Druckfarbenkomponenten auf das Deinkingergebnis?" 5th PTS Deinking Symposium Proceedings, Papiertechnische Stiftung, Munich, 1992, p. 110.

32. Galland, G. and Vernac, Y., Pulp Paper Can. 94(6):T181(1993).

33. Putz, H.-J. and Török, I., World Pulp Paper Technol. (3):85(1992).

34. Putz, H.-J., Schaffrath, J., and Göttsching, L., Pulp Paper Can. 94(7):T193(1993).

35. Järrehult, B., Horacek, R.G., and Lindquist, M.-L., "Deinking of wastepaper containing flexographic inks", TAPPI 1989 Pulping Conference Proceedings, TAPPI PRESS, Atlanta, 1989, p. 391.

36. Henkel-Hartmann, German pat. P43 22 999.9 (Jan. 1, 1995).

37. Michael Huber Munich GmbH, German pat. P41 15 731.1 (Nov. 11, 1992).

38. Förster, W., Großmann, H., and Baumgarten, H.-L., Wochenbl. Papierfabr. 120(8):281(1992).

39. Auhorn, W. and Melzer, J., Wochenbl. Papierfabr. 110(11/12):392(1982).

40. Frank, E., personal communication.

41. Hanecker, E., Klein, R., and Großmann, H., Wochenbl. Papierfabr. 125(8):418(1997).

42. Ackermann, C. and Putz, H.-J., "Verbesserung der Deinkbarkeit von Zeitschriften und Magazinen durch praxisbegleitende bundesweite Untersuchungen in Zusammenarbeit mit Zeitschriftenverlegern", INGEDE Report 4194 IfP, Darmstadt, Germany, 1997, 74 pp.

43. Galland, G. and Vernac, Y., ATIP Revue 47(5):154(1993).

44. Hooper, C., Daneault, C., and Dorris, G.M., Pulp Paper Can. 97(8):T269(1996).

45. Personal communication.

46. Vidotti, R.M., Johnson, D.A., and Thompson, E.V., Progress in Paper Recycling 3(3):39(1994).

47. Niggl, V., "Dispergierung – der Prozeßschritt zur Verbesserung der optischen Eigenschaften", Customer Information Seminar: Stock Preparation, Voith Sulzer Paper Technology, Ravensburg, Germany, 1997, 7 pp.

48. Cochaux, A., Carré, B., Vernac, Y., et al., Progress in Paper Recycling 6(4):89(1997).

49. Galland, G., Julien Saint Amand, F., and Vernac, Y., ATIP Revue 48(5):132(1994).

50. Sutman, F.J., Philippe, I.J., and LeBlanc, P., Tappi J. 77(5):87(1994).

51. Snyder, B.A. and Berg, J.C., Tappi J. 77(5):79(1994).

52. Renner, K., Putz, H.-J., and Göttsching, L., "Erfassungssysteme für Altpapier – Bestandsaufnahme und Bewertung neuer Erfassungsstrategien zur Qualitätsverbesserung von Altpapier", INGEDE Report 4094 IfP, Darmstadt, Germany, 1997, 73 pp.

53. Großmann, H., Phan-Tri, D., and Hanecker, E., Wochenbl. Papierfabr. 120(5):172(1992).

54. Nikki, M., Lucander, M., and Kokkonen, R., Wochenbl. Papierfabr. 110(14):507(1982).

55. Weithaler, R., "Vergleichende Bewertung von Laborflotationszellen für das Deinken von Altpapier", M.Sc. thesis, Darmstadt University of Technology, Darmstadt, Germany, 1994, 64 pp.

56. Tappi T452 om-92, "Brightness of pulp, paper, and paperboard (directional reflectance at 457 nm)".

57. ISO 2469, "Paper, board and pulps – measurement of diffuse blue reflectance factor (ISO brightness)".

58. ISO 2470, "Paper, board and pulps – measurement of diffuse reflectance factor (ISO brightness)".

59. *Riviello, A.E., Scamehorn, J.F., Christian, S.D., et al., "The use of image analysis to measure brightness of paper", TAPPI 1995 Pulping Conference Proceedings, TAPPI PRESS, Atlanta, p. 165.*

60. *Anon., "Das CIELab-Farbsystem", Special Issue, Gebr. Schmidt Druckfarben, Frankfurt a.M., Germany, 1994.*

61. *Renner, K., Putz, H.-J., and Göttsching, L., "Einfluß optischer Eigenschaften von Deinkingstoff auf optische Eigenschaften von altpapierhaltigen graphischen Papieren", INGEDE Report 3593A IfP, Darmstadt, Germany, 1996, 92 pp.*

62. *Riempp, G., Török, I., Ackermann, C., et al., Das Papier 47(3):128(1993).*

63. *Riempp, G., Török, I., Ackermann, C., et al., Das Papier 47(4):186(1993).*

64. *Klein, R., Wawrzyn, H., Grefermann, A., et al., Wochenbl. Papierfabr. 123(13):596(1995).*

65. *Fischer, S., Wochenbl. Papierfabr. 112(1):15(1984).*

66. *Göttsching, L., Putz, H.-J., and Renner, K., "Beeinflussung von Druckfarbenpartikelgröße und -verteilung beim Deinken von Altpapier", AiF-Report No. 8126, Department of Paper Science and Technology (IfP), Darmstadt, Germany, 1992, 100 pp.*

67. *Renner, K., Putz, H.-J., and Göttsching, L., Das Papier 49(7):423(1995)*

CHAPTER 9

Bleaching of deinked pulp

Christiane Ackermann

Bleaching of deinked pulp

1 Importance of DIP bleaching

The bleaching of deinked pulp (DIP) is an important and often vital process stage in the processing of recovered paper. It contributes decisively to improving the optical characteristics of deinked pulp. Bleaching is not always routine. It only occurs in cases where the quality of the recycled fiber pulp that is to be processed has certain demands. Recycled fibers intended for the production of packaging papers and board with the exception of white top liner for white lined chipboard are therefore usually not bleached. For economic reasons, bleaching only occurs when the intended use of the paper manufactured with recycled fibers requires improved optical characteristics such as brightness and luminance. Bleaching in processing of recycled fibers is usually therefore also integrated into a process concept in which ink is removed in one or more stages by flotation, washing, or both from the pulp slurry. The deinked pulp is usually used in the production of printing papers such as supercalendered paper and office paper and of hygienic papers.

Many bleaching chemicals and processes have use globally for DIP. Besides the chlorine-free bleaching chemicals, hydrogen peroxide, and sodium dithionite that have use without restriction for all types of fibers, chlorine-containing chemicals such as chlorine dioxide and sodium hypochlorite also have use especially in the United States. Due to their lignin-degrading effects, these types of bleaching chemicals are used only for wood-free recycled fiber pulp. The oxygen bleaching stage that comes from chemical pulp bleaching and is usually supported with hydrogen peroxide has become more common in commercial application. This is especially true for processing mixed office paper. To date, the use of ozone has not played any significant role in the bleaching of deinked pulp. Ozone is nevertheless the only chlorine-free bleaching chemical that can almost completely destroy optical brighteners in deinked pulps.

The bleaching of deinked pulp closely relates to other stages of stock processing. The combined application of flotation or wash deinking and bleaching produces synergistic effects that contribute to optimum brightness. High optical demands on the deinked pulp can only be satisfied if single-stage or multistage bleaching processes coordinate with the entire stock preparation technology. The choice of bleaching chemicals and processes depends on the quality of the recovered paper. Besides fiber composition, this also includes different types and proportions of contaminants and detrimental substances still present in the deinked stock.

1.1 Wood-containing DIP

Like mechanical pulp, wood-containing deinked pulp is usually bleached in a nondelignifying manner. The degradation of lignin such as occurs in chemical pulp bleaching is not desirable but does occur to a limited degree as a secondary reaction. With the aid of oxidative, reductive, or both types of chlorine-free bleaching agents, the chromophores of the mechanical fibers present in wood-containing DIP are modified to achieve an improvement of the optical characteristics of the deinked pulp.

An essential objective of bleaching wood-containing deinked pulp is to compensate for any yellowing of the fibers that occurs during their natural aging influenced by light, heat, and alkaline pulping conditions. Mechanical fibers are much more subject to these effects than chemical fibers. The result is that a greater bleaching gain is possible in cases with larger proportions of mechanical fibers in deinked pulp. With the conventional bleaching processes, no delignification will occur.

Color stripping used for example with colored papers is another objective of DIP bleaching. For economic efficiency, the destruction of the substantive, basic or acid dyes can occur only with reductive bleaching chemicals such as sodium dithionite.

Peroxide is the oxidizing bleaching agent frequently used for bleaching wood-containing DIP. For economic reasons, typical use is 1%–2% peroxide (based on a concentration of 100%) on o.d. pulp. Additional improvements in optical demands of the deinked pulp are possible with two-stage bleaching using a combination of peroxide and sodium dithionite or formamidine sulfinic acid (FAS).

When colored, recovered paper requires deinking. A reductive post-bleaching is necessary to reduce color shades. For this reason, the processing of recovered paper for hygienic papers frequently uses only a reductive chemical. For production of newsprint, adding peroxide during pulping is usually sufficient to satisfy the optical demands. This operation is not bleaching but a compensation for or the avoidance of alkali yellowing.

Some applications for wood-containing deinked pulp include uses in high quality graphic papers such as supercalendered (SC) and lightweight coated (LWC) papers. Multistage bleaching sequences must then occur when processing the deinked pulp to compete with paper grades manufactured exclusively from virgin mechanical and chemical fibers. Bleaching then frequently uses two stages with the combination of peroxide and dithionite or peroxide and FAS.

1.2 Wood-free DIP

In Europe, the availability of wood-free recovered paper for deinking is limited compared with that of wood-containing recovered paper grades. The application of the wood-free recovered paper primarily has use in the tissue sector. Wood-free recovered paper grades have greater use in North America. Consequently, mixed office waste (MOW) is an important category of recovered graphic papers in North America.

Previously, these recovered papers were usually treated with chlorine-containing bleaching chemicals such as elemental chlorine, chlorine dioxide, or sodium hypochlorite. Hypochlorite especially had wide use. When the proportion of mechanical fibers is less than 5%, considerable brightness occurs and dyes are effectively destroyed.

The chlorine-containing bleaching chemicals are being replaced by chemicals that also have use for bleaching chemical pulp. The use of oxygen for bleaching and homogenization has been studied extensively mostly in combination with peroxide. Initial results seemed promising. Larger scale pilot and industrial experiences have been mixed or negative, i.e., no or very little benefit from the addition of oxygen was found.

Primarily because of the considerable costs involved, little use of ozone for bleaching wood-free recovered pulp has occurred to date. In this application, this very reactive bleaching chemical can give brightness improvement up to 20% ISO. The depolymerization of cellulose that accompanies the treatment is detrimental to the strength properties. In the destruction or transformation of dyes, optical brighteners, or unbleached fibers from kraft pulp, ozone is the most effective of all oxygen bleaching chemicals.

Due to their lignin-degrading effect, the use of oxygen and ozone is confined almost exclusively to wood-free recovered pulp. Only the peroxide-supported oxygen stage with the suitable choice of reaction parameters can also be used for wood-containing deinked pulps. The brightness increase that can occur in such applications is modest. The increase in the COD load of the white water is considerable.

Wood-free deinked pulp is finding use as a substitute for hardwood chemical pulp to produce high quality printing papers. Besides multistage flotation and dispersion, the stock preparation processes for these chemical pulp substitutes in such cases include multistage bleaching sequences that combine conventional and unconventional bleaching chemicals.

1.3 Special aspects of DIP bleaching

Unlike the virgin fibers of chemical and mechanical pulps, residual printing ink and other impurities contaminate deinked pulp. Deinked pulp also primarily uses blends of mechanical and chemical fibers that were already bleached during their original production. Compared with virgin fibers, recycled fibers have altered morphological conditions, are more highly fibrillated, and have a higher fines content. The morphology of the fibers has a major influence on the reaction of the bleaching agent with the fiber-water system. Additional secondary effects will occur from the different

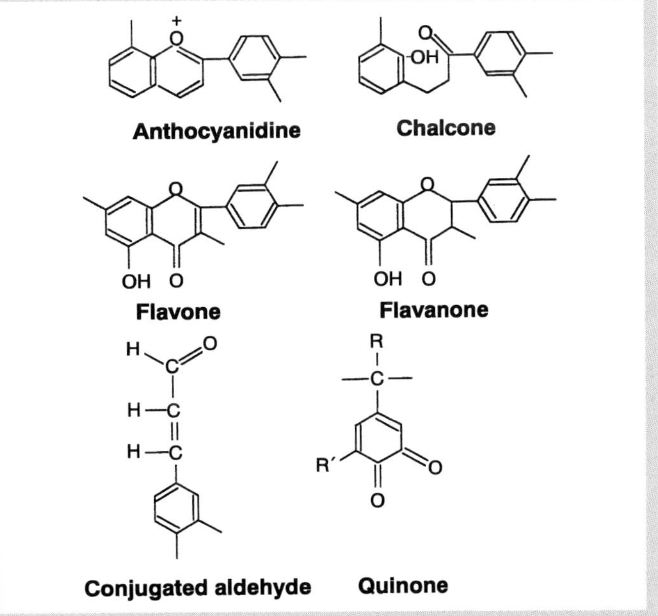

Figure 1. Chemical structure of some colored extractives and groups in wood and lignin[1].

impurities contained in the recycled fiber pulp. Examples of such contaminants are the chemicals used in previous papermaking and in the deinking process that may disturb the bleaching reactions. For these reasons, virgin fiber bleaching mechanisms cannot be applied without reservation to the bleaching of deinked pulp.

Fibers from wood or annual plants consist primarily of three polymers. Although the carbohydrates of cellulose and hemicelluloses do not cause any discoloration, the lignin structures with their chromophores give a bright yellow or dark brown shade to the fibers. This discoloration is the result of light absorption in the visible wavelength range of 380–780 nm[1–3].

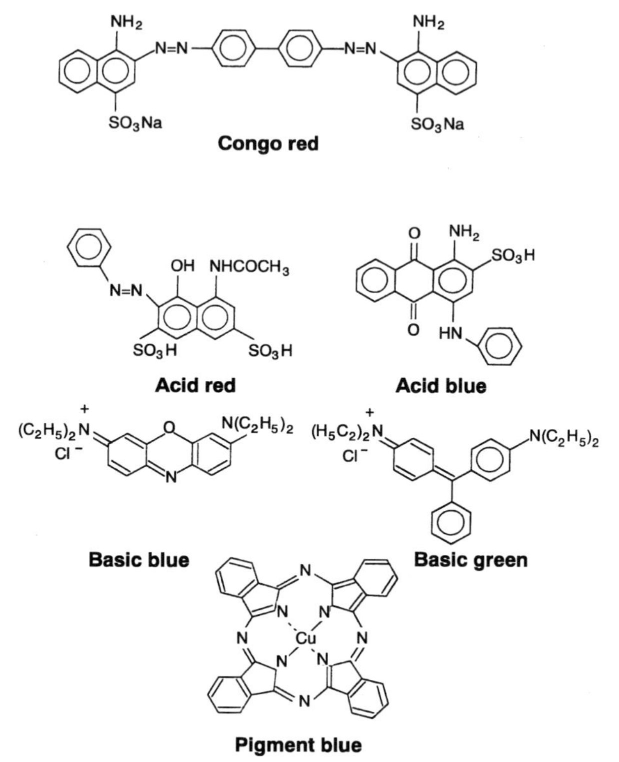

Figure 2. Examples of the chemical structure of dyes[1].

Figure 1 shows typical chemical structures of several wood chromophores. These have a system of conjugated double bonds that give the components their color. As a general rule, more conjugated carbon to carbon double bonds (C = C) give a wider wavelength range for light absorption, i.e., the chromophores will be darker.

Dyes and printing inks can also contribute to shading of the recycled fiber pulp. Figure 2 shows examples of chemical structures of dyes. Most dyes contain aromatic, carbonyl, or azo groups and carbon to carbon double bonds.

Improving the optical characteristics of deinked pulp requires elimination of the causes of discolorations. In this context, one must note that the bleaching potentials of individual pulp components differ considerably[1].

1.3.1 Chemical fibers

The primary characteristic of chemical fibers is their lignin content as far as their bleachability is concerned. This can range between 0% (fully bleached chemical pulp) and approximately 5% (unbleached chemical pulp). The residual lignin that remains in the fibers after digestion has a brownish color due to its condensed condition. This results primarily from extended conjugated structures, many aromatic groups, and several double bonds of carbon atoms. Only a small number of carbonyl groups are present. The bleachability of chemical fibers is therefore primarily a function of the level of condensation of the aromatic systems of lignin. Only electrophilic agents such as oxygen, chlorine dioxide, or chlorine can oxidize these. Nucleophilic agents such as hydrogen peroxide or hypochlorite can only attack side chains. This interrupts the conjugation and therefore the coloring there. Consequently, lignin-preserving bleaching chemicals cannot eliminate the brownish discoloration of the residual lignin in the chemical pulp.

Even with the extensive use of peroxide, the brightness of an unbleached kraft pulp would still be considerably lower than 50% ISO. The only possibility of color stripping such fibers is use of lignin-degrading bleaching chemicals such as oxygen or chlorine dioxide. In contrast, semi-bleached chemical pulp with a residual lignin content of about 0.5% can be bleached with nondegrading chemicals. The resulting brightness increase from an initial level of about 70% ISO has little significance.

1.3.2 Mechanical fibers

In contrast to unbleached kraft pulp, mechanical fibers are considerably brighter. The primary reasons for this are the natural, noncondensed condition of the lignin resulting from mechanical pulping and the presence of several chromophoric structures that are responsible for the yellow coloring of mechanical pulp. Because carbonyls are the primary factors here, mechanical pulp as opposed to chemical pulp can be bleached effectively with largely nondegrading, lignin-preserving chemicals such as peroxide or dithionite.

Under optimum bleaching conditions, brightness levels of as high as 84% ISO are possible with yield losses of less than 5%. Higher brightness levels are not possible even by using greater quantities of bleaching agents. This is because not all chromophores can be bleached and new chromophores that have a coloring effect form during bleaching. If mechanical fibers are treated with lignin-degrading bleaching chemicals, acceptable brightness gains can only occur at the expense of considerable yield losses.

1.3.3 Impurities

Besides fibers, fillers, or coating pigments, deinked pulp also contains variable proportions of residual printing inks, dyes, adhesives, and other impurities. These not only determine the optical properties of the pulp but also influence the success of a bleaching process. Although most impurities cannot be bleached and can even disturb the bleaching process, dyes can usually be effectively color stripped. Reductive bleaching chemicals such as dithionite or FAS and all lignin-degrading chemicals destroy these colored compounds. In contrast, color pigments such as phthalocyanine cannot be bleached reductively or by oxidation with hydrogen peroxide.

1.4 Conclusion

Using the current knowledge regarding bleachability of DIP gives the following general conclusions:

- If the discoloration of DIP is due to conjugated carbonyl structures, all lignin-preserving chemicals are suitable for bleaching.
- If the discoloration of DIP is due to conjugated azo groups, conjugated carbon to carbon double bonds, or condensed aromatic structures, only delignifying bleaching chemicals produce acceptable bleaching results.

Because the composition of DIP varies in most applications, bleaching frequently represents a compromise. If deinked pulp contains more than 15%–20% mechanical fibers, the use of delignifying bleaching chemicals is not possible at least for economic reasons. In the case of bleaching with peroxide, dithionite, or FAS, mainly carbonyl structures are stripped. This limits the bleaching effect. If DIP consists primarily of chemical fibers, using lignin-degrading chemicals such as oxygen or ozone that also react with other chromophoric structures is possible.

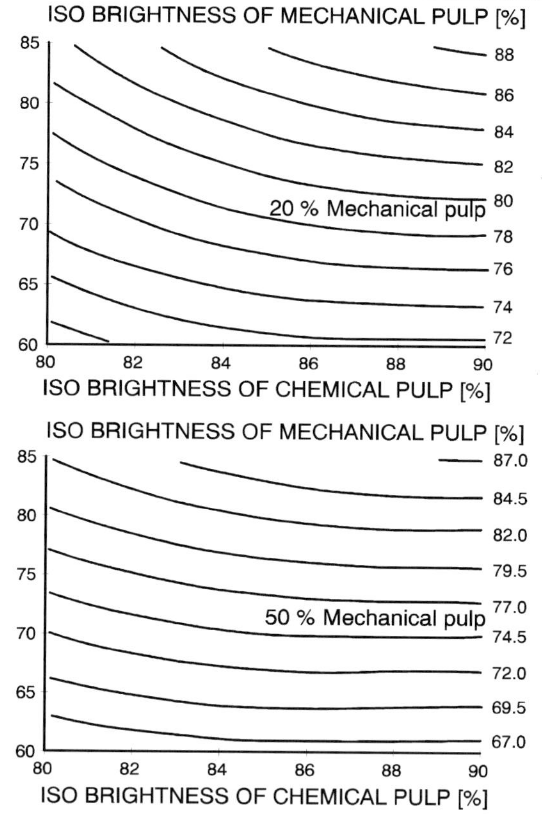

Figure 3. Brightness measured on mixtures of chemical and mechanical pulps of various brightness levels[1].

Figure 3 shows another limitation on bleaching of DIP compared with bleaching of pure virgin fibers. A greater proportion of mechanical fibers in the deinked pulp gives less optical influence of the chemical fibers on the brightness of the entire pulp mixture. With a 50% share of mechanical fibers for example, the brightness of the stock mix will not vary significantly with an increase in brightness of chemical pulp in the range of 80% to 90%. This indicates that the optical result of the bleaching of deinked pulp depends highly on the result of bleaching of the mechanical fibers. Accordingly, bleached chemical fibers cannot significantly improve the brightness of wood-containing deinked pulp. The explanation for this relates to a physical phenomenon in paper. The theory published in 1931 by Kubelka and Munk[4] describes the relationship between reflectance factor, R_∞, light absorption coefficient, k, and light scattering coefficient, s, as follows:

$$R_\infty = 1 + (k/s) - [(k/s)^2 - 2(k/s)]^{1/2} \quad (1)$$

With homogeneous stock mixtures, the reflectance factor of the individual components does not behave in an additive manner. Their light absorption, kb, and light scattering powers, sb, are as follows:

$$sb = \Sigma_i s_i b_i \quad (2)$$

$$kb = \Sigma_i k_i b_i \quad (3)$$

where b_i is the basis weight of the stock component i.

Consequently, any improvement in brightness of wood-containing deinked pulp will occur primarily from bleaching chemicals that react with the chromophores of the mechanical fibers. That explains again the preference of the lignin-preserving bleaching chemicals for bleaching wood-containing deinked pulp.

2 Lignin-preserving bleaching

With lignin-preserving, i.e., nondelignifying bleaching, the color stripping of the fiber stock uses a modification of the colored organic molecules into colorless molecules. Chemical reactions with the bleaching chemicals alters the wood chromophores. This involves only limited degradation of the organic substances into soluble compounds. All conjugated carbonyls are attacked. Oxidation can more effectively color strip them than reduction. Oxidation irreversibly destroys them. The resulting carboxylic acids dissolve partially as salts and enter the process water system as COD load. After the oxidative destruction of the chromophores and their extraction with the process water the brightness reversion of the pulp will be decreased.

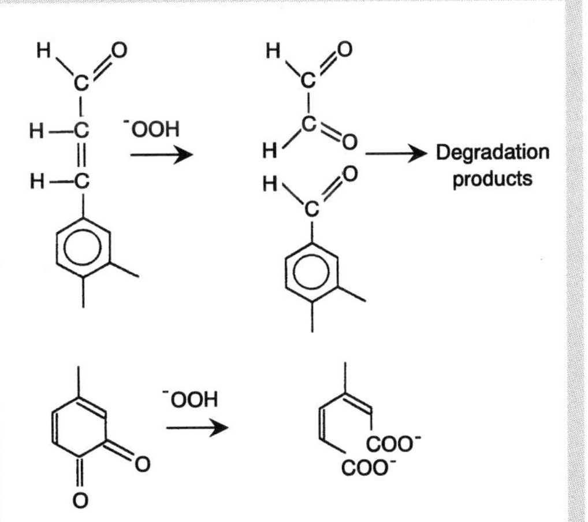

Figure 4. Reactions occurring during peroxide bleaching[1].

Reductive bleaching does not remove the quinoid structures, it only reduces them to aromatic compounds which might be reoxidized by oxygen. A reduction under neutral conditions has only a small impact on the solubilization of organic material. It does leave reversible reduction products in the pulp. For example, atmospheric oxygen or light can reoxidize corresponding ortho-phenols. The result is a reoccurrence of the brownish coloring. Figure 4 shows several reactions that occur when bleaching chromophores with peroxide.

2.1 Peroxide bleaching (P)

Although discovered in 1818 by Thenard, hydrogen peroxide (H_2O_2) only became a chemical product with a wide range of industrial applications following the development of new production processes (electrochemical processes) in the 1910s.

At the beginning of the 1940s, the demand for higher levels of brightness in wood-containing papers resulted in the first use of peroxide for mechanical pulp bleaching. Peroxide gained importance with the increased use of recovered paper for production of newsprint. Initially, peroxide had use to compensate for the yellowing that occurs during the flotation deinking of wood-containing recycled pulp in an alkaline environment. In the 1980s due to the dramatic increase in the proportion of recycled fibers in paper, separate peroxide bleaching after deinking became the state-of-the-art.

Initially, sodium peroxide, a yellowish powder that decomposes into sodium hydroxide and hydrogen peroxide, provided an inexpensive source of hydrogen peroxide. Its production by oxidizing electrochemically produced metallic sodium with oxygen was too expensive compared with production of hydrogen peroxide by the anthraquinone process. This had undergone constant development since the 1950s with the result that production of sodium peroxide ceased at the end of the 1970s. In addition, the alkalinity that occurred at the time of dilution was too high for optimum bleaching of deinked pulp. Sulfuric acid addition was necessary to prevent the yellowing of wood-containing fibers. Today, only hydrogen peroxide has use for pulp bleaching.

2.1.1 Chemistry

The chemistry of peroxide bleaching is well understood. The bleaching effect uses the dissociation of hydrogen peroxide in water to form a hydronium (H_3O^+) and a perhydroxyl ion (HO_2^-):

$$H_2O_2 + H_2O \leftrightarrow HO_2^- + H_3O^+ \tag{4}$$

The perhydroxyl anion acts as a nucleophilic bleaching agent. To achieve a high bleaching effect, increasing its concentration is necessary. This is possible by increasing the peroxide concentration and by adding alkali (sodium hydroxide). This is activation of the hydrogen peroxide:

$$H_2O_2 + OH^- \leftrightarrow H_2O + HO_2^- \tag{5}$$

Figure 5 shows the dependence of the bleaching result versus the amount of sodium hydroxide used[5]. The curves reveal that the nonlinear relationship between peroxide and sodium hydroxide dosage requires an exact definition of the ratio of peroxide to sodium hydroxide to obtain maximum brightness increases. At low alkaline dosages, i.e., before the maximum bleaching result occurs, the peroxide bleaching liquor does not have sufficient activation. In the highly alkaline range, i.e., after exceeding the optimum, a loss of brightness occurs with additional sodium hydroxide due to the increased yellowing reactions of excessive hydroxide ions on the lignin structures of mechanical fibers.

With H_2O_2/NaOH proportions at which optimum bleaching results occur, the ratio between the rates of bleaching and yellowing reactions is most favorable. An indication of an optimum peroxide activation is the residual peroxide content at the end of the bleaching process. The yellowing reactions dominate over the bleaching reactions if the residual peroxide content is less than 10% of the original peroxide dosage. For this reason, peroxide bleaching is an excess bleaching reaction.

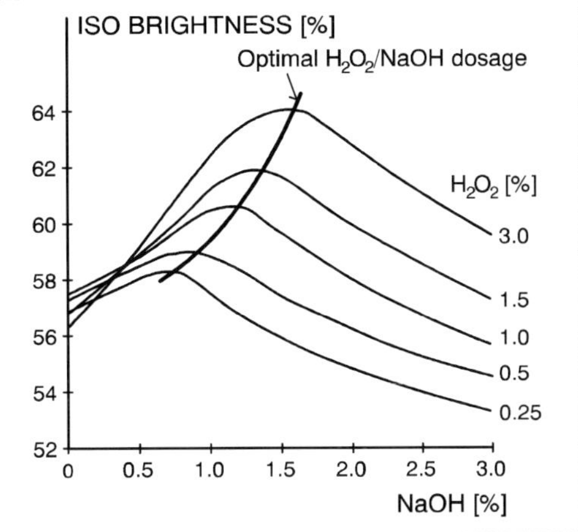

Figure 5. Effect of H_2O_2/NaOH dosage on brightness[5].

With increasing peroxide dosage, the curves in Fig. 5 move toward higher brightness levels. In a parallel development, the required amount of NaOH to obtain the maximum bleaching effect also increases. Doubling the quantity of peroxide used does not automatically require doubling the quantity of NaOH. For example, an increase in the amount of peroxide from 1.5% to 3.0% only requires a 0.2% increase in the amount of NaOH.

Bleaching should occur under optimum H_2O_2/NaOH conditions not only for economic reasons. Because of the development of the COD load of the process water with increasing alkalinity it is also necessary to keep the alkali amount to a minimum[6,7]. The solubility of oxidized lignin or hemicelluloses increases with an increasing pH value. This is especially true when processing wood-containing pulp.

The bleaching result improves with addition of peroxide. The brightness does not increase indefinitely because peroxide reacts only with certain chromophores. Figure 6 shows the development of brightness with increasing quantities of peroxide. The greatest increase occurs with peroxide additions of less than 1.5%. Larger charges lower the gradients of the brightness development considerably. For economic considerations, peroxide applications of more than 2% are not recommended.

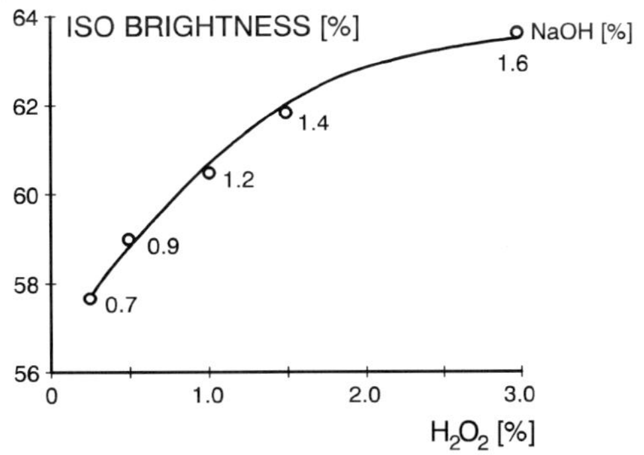

Figure 6. Effect of peroxide dosage on brightness by optimized NaOH addition[5].

The presence of catalytic ions (heavy metal ions) or enzymes (catalases) in recycled pulp slurries can cause decomposition of the peroxide. This causes loss of the activation energy necessary to break the oxygen to oxygen bond. Hydroxide radicals form that decompose via a chain reaction into water and oxygen:

$$H_2O_2 \rightarrow 2HO\bullet \rightarrow \rightarrow \rightarrow H_2O + \frac{1}{2}O_2 \quad (6)$$

or

$$H_2O_2 \rightarrow H\bullet + \bullet OOH \rightarrow \rightarrow \rightarrow H_2O + \frac{1}{2}O_2 \quad (7)$$

These undesirable reactions usually occur very slowly, but catalysts, increased temperature, and a highly alkaline medium can accelerate them. Besides the economically inefficient peroxide consumption, oxidative damage to the fibers are due to the radicals formed at the time of decomposition. To stabilize the peroxide, deactivating the catalytic substances in the recycled pulp slurry is necessary. Complexing, adsorption, and denaturation are the common reaction mechanisms as discussed in the following.

2.1.2 Peroxide stabilization

Water glass

Water glass, a form of sodium silicate solution, is an important and effective peroxide stabilizer. Produced from soda (Na_2CO_3) and silica (SiO_2), the soluble silicate can have different chemical compositions:

$$Na_2CO_3 + SiO_2 \rightarrow (SiO_2)_x + Na_2O + CO_2 \quad (8)$$

Their identification is defined by the ratio, x, of their components. Usually, the chemical composition of the silicates is expressed as a weight ratio and a molar ratio. Because the molecular weights of SiO_2 and Na_2O are almost identical at 60 and 62, respectively, sodium silicate has only a slight difference between the molar ratio and the weight ratio reflected by a factor of 1.031. With a given ratio that for most manufacturers uses the basis of weight, the density of a solution is defined by its concentration or solids content and expressed in degrees on the Baumé scale. A greater ratio of SiO_2 to Na_2O therefore gives a greater value of x and a lower density of silicate solution. As a standard, water glass solutions with densities of 38° Bé with a SiO_2/Na_2O ratio of about 3:1 are used.

The chemistry of the stabilizing effect uses a precipitative reaction of the silicate ions of the water glass with the calcium, magnesium, or both ions of the pulp suspension among other items. The resulting insoluble silicates serve as adsorption agents for detrimental substances such as organic impurities (anionic trash). Silicates also react with heavy metal ions and form complexes causing these ions to become inactive.

Although the reaction mechanisms do not currently have a complete explanation, the following positive effects are due to water glass:

- Peroxide stabilization
- Buffering effect to ensure a constant peroxide activation[8,9]
- Wetting and washing effect with improvement of the printing ink detachment
- Improved flotation yield by suppression of a co-flotation of fillers and fines[10].

The stated effects are not confined to the actual bleaching but have consequences for the entire deinking process. Figure 7 shows the influence of water glass on the flotation result where water glass is added during disintegration of the recovered paper in the pulper[5]. The improvement in brightness that occurs with increased addition of water glass is mainly due to its dispersing characteristics favoring ink particle detachment, improved collection and flocculation of dispersed ink particles, and other factors. The selectivity of ink removal during the flotation process is also enhanced because considerably fewer fillers and fines are removed as Fig. 8 shows. If the flotation process is intended especially to remove fillers, reduced use of water glass during disintegration is indicated. This is accompanied by a considerable fines loss that also reduces the DIP yield and therefore increases the amount of deinking sludge for disposal.

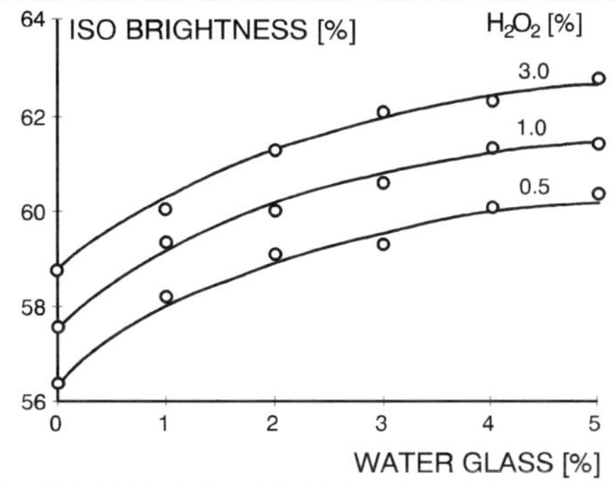

Figure 7. Effect of water glass on brightness[5].

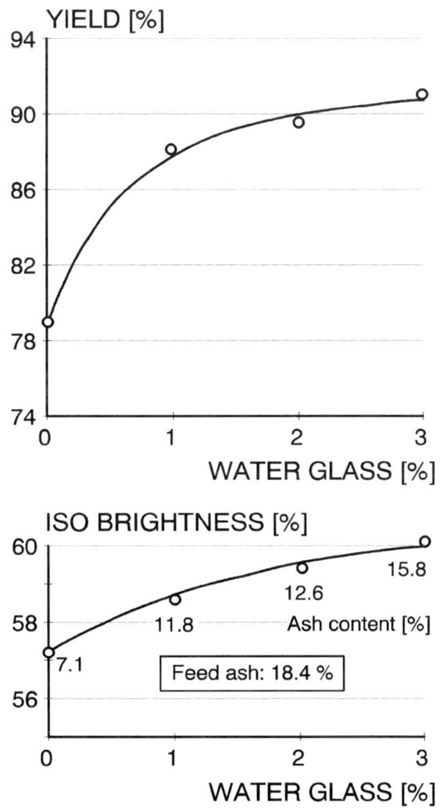

Figure 8. Effect of water glass on flotation yield and brightness[10].

Undesired reactions of water glass are the possible precipitation of alkaline earths, the formation of deposits in pipes and on machine parts used in the stock preparation process or paper manufacturing, and the impairment of the effectiveness of retention and flocculation agents. To eliminate these negative effects, the quantity of water glass used for processing of recycled fiber pulp has been reduced considerably. In the past, up to 5% was frequently added. Presently, a maximum of 2%–3% on o.d. pulp is used.

Chelating agents

Previously, silicate-free stabilizers as chelating agents had common use as a partial substitute for water glass. Such substances can only replace water glass for stabilizing the peroxide. They form coordination bonds with heavy metal ions. This masking prevents the decomposition of the peroxide.

The complexes formed consist of a central atom, M, and one or more ligands, L. The central atom and ligands in the complex are always in a molar relationship with one another. The complexes are therefore defined compounds from which a complex forming constant, k, can be calculated. Complexes with a $k > 2$ are usually stable. Although the central atom is usually a metal atom or ion belonging to the group of transition metals, ligands can have very different natures. Accordingly, distinguishing between the following types of chelating agents is possible[11]:

- Organic and inorganic
- Ionic and nonionic
- Single group and multigroup.

The characteristics of the ligands primarily determine the stability of the complexes. Because long-term stable complexes are necessary for bleaching deinked pulp, multigroup ligands or so-called chelaters are especially suitable for this application. Such chelating agents have several coordination points in one molecule. In this way, they firmly hold the central atom in the coordination complex as Fig. 9 shows.

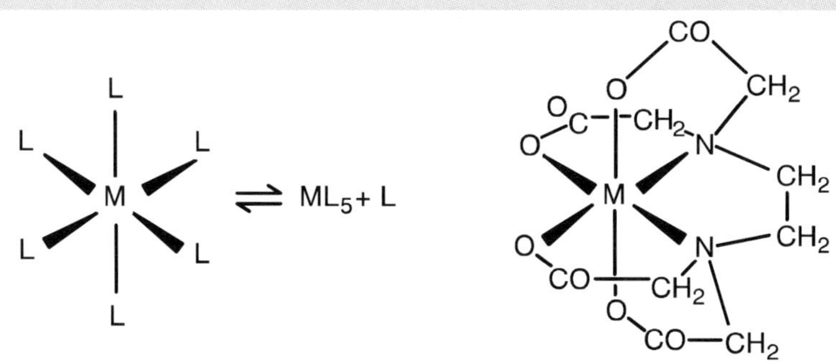

Figure 9. Complex structure of a single group and a multigroup chelating agent[11].

Chelating agents not only belong to different material categories but also are natural or synthetic products. Typical agents used for the bleaching of deinked pulp are the synthetic EDTA (ethylenediaminetetraacetic acid) and DTPA (diethylenetriaminepentaacetic acid) chelates that belong to the aminopolycarbon acids. Phosphonic acid-based chelates have no use for bleaching deinked pulp. Their use is limited to the bleaching of mechanical pulp where they act partly as a substitute for EDTA or DTPA.

Especially in cases where recycled fiber pulps and process water have a higher heavy metal ion content, the use of such chelating agents stabilizes the bleaching system. This leads to stable brightness increases during bleaching and an acceptable residual peroxide content[12]. Some controversy exists in publications about the need to use these substances for bleaching deinked pulp. A certain water hardness is necessary to ensure the success of the flotation process by supporting the formation of calcium soap flocs. With an increasing presence of polyvalent cations such as Ca^{2+} and Mg^{2+}, the reaction balances in the complexing change in favor of chelating of calcium ions. No specific effectiveness for complexing heavy metal ions exists[5]. These arguments increasingly lead to the conclusion that managing the bleaching of deinked pulps without such chelating agents is possible.

Ecological aspects are also part of the discussion. EDTA and DTPA are biologically almost nondegradable. They are only partly eliminated by adsorption on the activated sludge flocs of biological purification plants. If the sodium salt of EDTA and DTPA enters receiving waters, metal ions that are already bound in sediments can be remobilized. This is especially serious when drawing drinking water from surface water. For this reason and as a precautionary measure, many European countries are attempting to limit the use of chelating agents.

When using chelating agents as intended, they form stable complexes with heavy metals or alkaline earth metals immediately at the time of application. Chelating with the alkaline earth metal occurs especially if sufficient heavy metals do not correspond with the metered amount of agents. Consequently in this application, EDTA and DTPA do not enter the effluent as a sodium salt. In addition, evidence exists that almost half the quantity of chelating agents used does not enter the effluent but remains bonded on the fibers. The half detected in wastewater occurs in stable complexes. Because the concentration of calcium or magnesium ions in running water usually exceeds that of other metals by a factor of one thousand, the different concentrations mean that a stable development of calcium or magnesium complexes can occur even with better bonding of heavy metals compared with alkaline earth metals.

Work is presently occurring to find substitutes using natural, biologically or otherwise degradable compounds that produce similarly good results. Protein modules, low to medium molecular weight oligomers, and polymers are the chemical basis for this new generation of chelating agents[13]. Their market launch has been hindered by their considerably higher costs and poorer effectiveness.

Catalase

Excessive, uneconomic peroxide consumption can also be the result of the enzyme catalase that enters the process water loop of the stock preparation process with the recovered paper. Catalases are produced by all aerobic microorganisms and animal and vegetable cells. In the cell, they act as protective enzymes against the peroxide and some other substances produced in the respiratory chain. Peroxide has a strong toxic effect in the cell. The cell therefore tries to destroy the peroxide by the catalase as quickly as possible. Accordingly, catalases are also specific peroxidases that suppress an overproduction of peroxide and therefore prevent the destruction of the cells[14]. This natural reaction process can calculate the microbial activity in pulp slurries or filtrates. After an excess dosage of peroxide, the rate of decomposition is measured. This indirectly provides information on the presence of microorganisms[15].

In the deinking process, microorganisms find good growth conditions. Because the process water loops are largely closed, the microorganisms that enter the loops with the recovered paper have sufficient time to multiply rapidly. Practice has shown that a natural balance occurs after several days that then leads to a typical germ count for the deinking system.

Catalases are extremely effective catalysts for the decomposition of peroxide in oxygen and water. The conversion rate is 10^6 to 10^7 molecules of peroxide per molecule of enzyme per minute[14,16]. The rate of decomposition increases rapidly with an increasing catalase concentration as Fig. 10 shows. An increase in the concentration of a commercial catalase from 10 to 44 mg/L will accelerate the rate of decomposition after a reaction time of 10 min from 5% to 50%. The tests used a pulp of 15% consistency with addition of 5% peroxide[17].

Figures 11 and 12 clearly show that catalases remain stable over a wide range of temperature and pH. Only at temperatures above 70°C and pH values below 5 will the tetramerous haemin molecules become denatured. This irreversibly prevents their catalytic effect. Despite their heat sensitivity, enzymes become reactive again after a temperature treatment. This is mainly because the individual microorganisms can survive temperatures over 100°C and subsequently produce enzymes again. Spores can also resist such temperatures and provide the substrate for new microorganisms.

Autoclaving above 120°C will guarantee the destruction of microorganisms and therefore all enzymes. Such conditions do not occur in a paper mill processing recovered paper. The slushing of recovered paper containing mechanical fibers in a pulper or drum pulper usually occurs at temperatures of 40°C–50°C and a pH of 9 to about 10.5[18]. Under these conditions, catalases can completely decompose the peroxide used to prevent alkaline yellowing.

Because a large pH displacement is not technically possible in the strongly

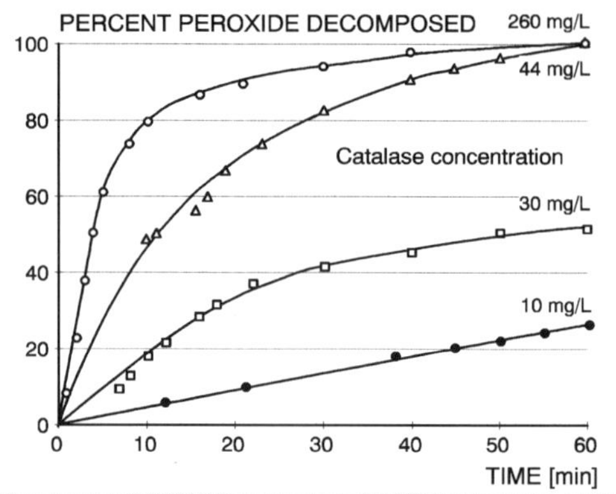

Figure 10. Decomposition rate of hydrogen peroxide for different catalase concentrations[17].

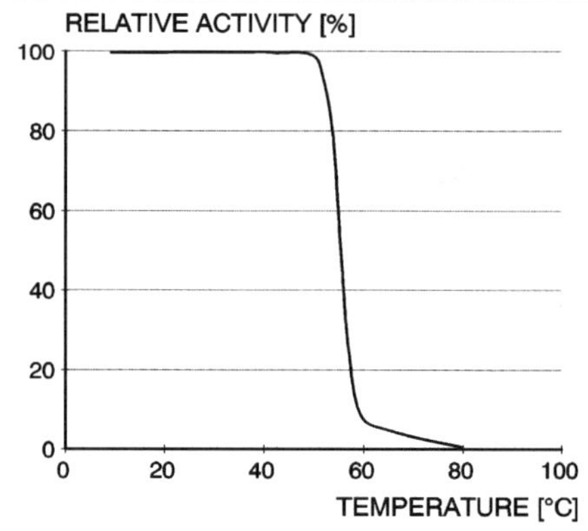

Figure 11. Effect of temperature on catalase activity.

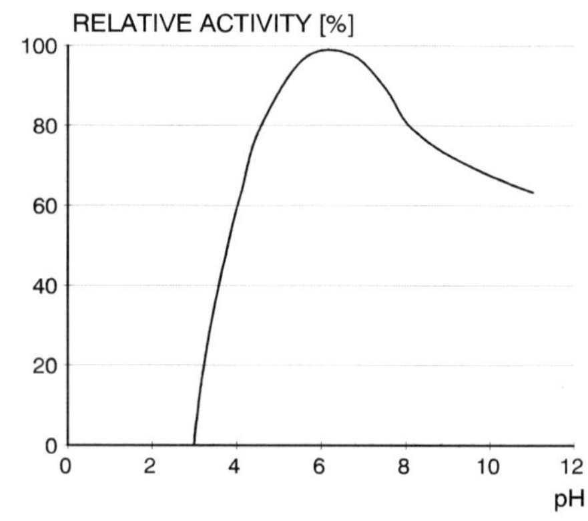

Figure 12. Effect of pH on catalase activity.

alkaline (to avoid yellowing of the mechanical fibers) or the strongly acidic area (to avoid decomposition of calcium carbonate), in commercial applications the catalases are thermally denatured or biocides are used. Since bacteria proteins are destroyed at temperatures above 70°C, thermal treatment during dispersion or kneading is an effective method to combat catalases. A poor separation of the process water loops can adversely influence the use of hot dispersion as a disinfecting stage. The duration of circulation of the considerable quantities of white water in the loop also permits renewed bacterial development. Occasionally, increasing the temperature during pulping is also attempted. Although this can effectively limit the growth of bacteria, it does have the disadvantage that tacky impurities (stickies) become more fragmented and therefore more difficult to remove in subsequent process stages such as screening.

If denaturation occurs with biocides, glutar-dialdehyde is useful as an inhibitor[19]. Another possibility to destroy the enzyme structure is a temporary excess of peroxide in the form of a shock dosage that also activates a "purifying" effect. This has primary use for treating process waters.

2.1.3 Process parameters

Process parameters such as stock consistency, temperature, reaction time, and the chemicals used considerably influence the bleaching result. Figure 13 shows the importance of the bleaching process at high consistencies[5]. The shift of consistency into a high range results in considerably greater brightness increases. This is due to two primary causes. First, a higher consistency gives a higher relative concentration of bleaching chemicals that come into direct contact with the fibers. Second, the displacement of the aqueous phase also eliminates dissolved matter that can adversely influence the bleaching reaction.

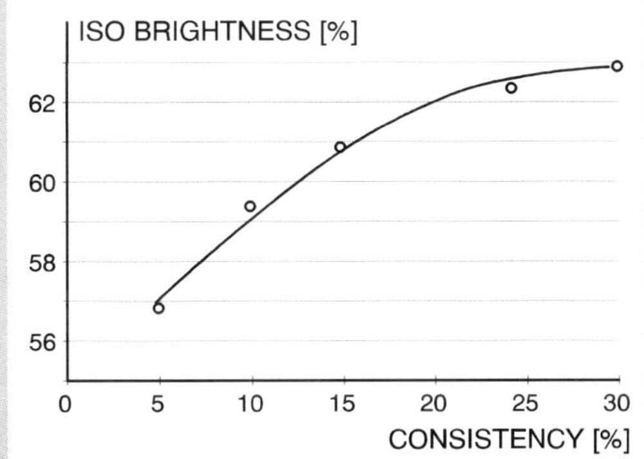

Figure 13. Effect of pulp consistency on brightness[5].

Above 30% consistency, mixing problems occur. The bleaching chemicals will no longer mix homogeneously into the pulp without use of highly effective mixing methods. The consequence is a leveling of brightness increases. For this reason, the consistency in separate bleaching stages is primarily limited to 25%. Because peroxide also has use in other stages of recovered paper processing such as during slushing, the application cannot always occur under optimum conditions. The consistency at bleaching usually is 10%–25%.

While no practical or acceptable alternatives to high consistencies exist for economic reasons, temperature, reaction time, and alkalinity can optimize the reaction. A close relationship exists in chemical reactions between the reaction time necessary to obtain the reaction product and the temperature at which the reaction partners interact.

In many technical applications, a high temperature is therefore useful to accelerate the reaction. This applies also for bleaching with peroxide. The temperature cannot increase indefinitely because undesired secondary reactions such as yellowing reactions or peroxide decomposition occur more rapidly. If bleaching occurs at excessively low temperatures, the reaction time requires extension to obtain similar bleaching results. For economic reasons, the temperature should therefore be 40°C–70°C. Typical reaction times are 1–3 h when bleaching occurs in a separate process stage.

Higher temperatures usually require a lower NaOH dosage to ensure optimum peroxide activation. Helmling *et al.*[5] stated that the alkalinity must be reduced by about 0.1% with a temperature increase of 10°C to guarantee optimum matching of "temperature activation" and "alkali activation." In contrast, the influence of the stock consistency on the H_2O_2/NaOH relationship is insignificant.

2.1.4 Point of application

On an industrial scale, the treatment of recycled fiber with peroxide uses different stages of pulp processing:

- During disintegration of the recovered paper in the pulper or drum pulper
- When storing the disintegrated, nondeinked pulp in a storage tower
- With a dispersion stage, after a deinking stage
- As post-bleaching, at the end of the stock processing line.

Pulper bleaching

The most important and oldest peroxide application is at the pulping stage. In industrial deinking operations, these invariably occur in an alkaline environment. To prevent yellowing of the mechanical fibers of the recycled pulp, peroxide is used with the alkali. The intended purpose of the addition of peroxide at this stage in the stock processing is therefore not actual bleaching but simply the compensation for or prevention of an alkaline yellowing. The main arguments in favor of the application of peroxide at this stage of the deinking process are the already existing alkaline conditions, the presence of water glass, and the temperature level of 40°C–50°C.

These conditions bring several considerable disadvantages. The consistency in the low consistency pulpers that were initially used was too low and prevented an optimum use of the bleaching chemistry. The knowledge of the effect of pulp consistency stimulated the development of pulping units. Today, only high consistency pulpers or drum pulpers are incorporated into the design concepts for processing recovered graphic papers. Although the increase in consistency to 15%–18% created considerably more favorable conditions for the use of peroxide, other conditions prevent an optimum use of the bleaching agent. The average pulping time of 15 min is therefore insufficient to obtain maximum brightness gains. The presence of a wide range of different impurities including printing ink also catalyses the undesired peroxide decomposition. The consequence is an increased addition of stabilizing chemicals such as water glass and possibly chelating agents.

The process efficiency is also adversely affected because parts of the stock are treated as contraries. These are removed in the subsequent cleaning and screening stages as rejects. They have therefore undergone an unnecessary chemical treatment.

Even if the use of peroxide at this stage of the process is not an ideal solution, no deinking systems currently exist that manage without it especially since no additional bleaching stages are then necessary in some cases. Seldom is more than 1% of the bleaching agent on o.d. pulp used. If additional peroxide treatment of the recycled fiber pulp during processing occurs, the initially added amount is reduced to less than half.

At the beginning of the 1980s, one machine manufacturer[20] proposed a concept to slush the recovered paper at a lower consistency and at reduced temperatures without the addition of peroxide. The intent was to achieve a more efficient use of the peroxide at the beginning of the stock processing. The chemical treatment occurred only after the cleaning and screening stages. For this purpose, the cleaned undeinked pulp was thickened to about 23%, heated, and subsequently stored for several hours at 45°C–50°C in a tower after the addition of a bleaching recipe. In this way, undesired decomposition reactions of the peroxide were supposedly avoided while simultaneously creating favorable conditions for peroxide treatment before a deinking stage. The concept did not prove successful. Instead, the practice of adding peroxide before or during a dispersion stage became more common[21–24].

Dispersion bleaching

Dispergers or kneaders have use in deinking systems to detach residual ink after a deinking stage or for fragmentation of visible contaminants in recycled fiber pulp (See Chapter 5). Mechanical dispersion of the primarily dark particles usually occurs with a loss of brightness. Recycled pulp is dispersed with high consistencies of 25%–30% and temperatures of 60°C–95°C. These process conditions and the high shear environment are ideal for admixing of peroxide. The so-called dispersion bleaching acts against the "graying" of the pulp resulting from the dispersion of the ink particles by improving the brightness of the fiber furnish if only by a few points.

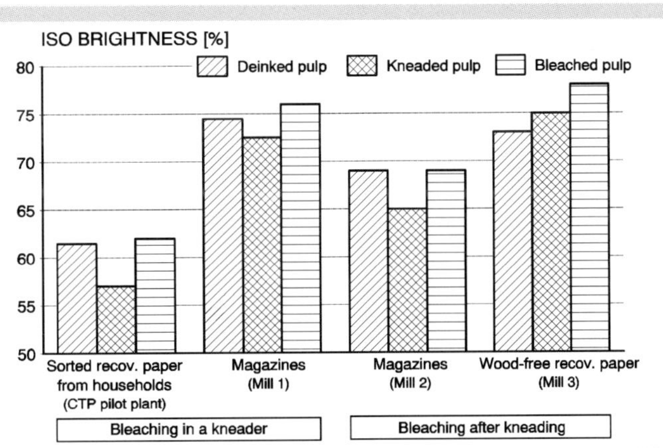

Figure 14. Peroxide bleaching in a kneader and after kneading[17].

The results of pilot plant and industrial tests shown in Fig. 14 prove that peroxide bleaching can compensate for the loss of brightness caused by dispersion[17]. Under optimum bleaching conditions, the brightness development has secondary importance whether bleaching occurs during dispersion or in a post-bleaching process. The addition of bleaching chemicals during dispersion can give positive side effects that then improve the entire stock processing operation. Adding chemicals during dispersion may improve the detachment of residual ink particles from the fibers so that a subsequent deinking stage can remove a larger quantity of such particles[25].

Use of peroxide at the dispersion stage has two important disadvantages. The residence time is too short to achieve maximum bleaching results. For this reason, many machine manufacturers install a high-consistency bleaching tower after the kneader or disperger that then allows the necessary storage time of the pulp[26,27]. The second reason is that excessively high temperatures lead to a thermal decomposition of the peroxide. This means that yellowing of the pulp can occur after only short reaction time. An additional mixer that mixes the alkaline bleaching solution homogeneously into the deinked pulp under optimum temperature conditions will help to stabilize the bleaching process.

For this reason, dispersion bleaching has use primarily in systems where the process occurs between two deinking stages. Peroxide is added directly before or during the actual dispersion treatment at temperatures of approximately 80°C. In this case, the dispersion units (feeding screw, disperger, or kneader) only ensure the homogeneous mixing of the bleaching chemicals. After dispersion (a process that lasts only a few minutes), the deinked pulp is diluted to a consistency of about 12% and pumped into a bleaching pipe. The bleaching reaction that produces acceptable results then occurs at temperatures of approximately 60°C and reaction times of 30–60 min. If dispersion occurs at temperatures of 60°C–70°C, the pulp can be fed directly into a downflow high-consistency tower. The residence time is 30–120 min. The dosage of bleaching agent is generally 1%–2%.

Post-bleaching

The bleaching of deinked pulp at the end of stock processing is post-bleaching. The process differs only slightly from that of virgin pulp bleaching that is a common practice in the treatment of mechanical pulp. The deinked pulp is thickened and heated with steam to a temperature of approximately 60°C. After adding bleaching chemicals, the pulp is stored in a bleaching tower at a consistency of about 15% for 1–3 h.

In peroxide bleaching there may be a trend toward even higher consistencies up to 40% (ultra-high consistency)[27]. Besides reduced steam consumption, the expected benefit is primarily savings in the introduced chemicals such as sodium hydroxide. This then results in reduced COD loads of the white water for recirculation. Reduced alkaline demand will also produce a lower load of anionic trash. Shortening of the reaction time by 50% when bleaching using a consistency of 30% instead of 10% is an important advantage of high-consistency bleaching[28].

2.1.5 Experiences from technical applications

A summary of the many, varied experiences from technical applications of peroxide bleaching for processing of deinked pulp is the following:

- Separate bleaching with peroxide after a deinking process (flotation or washing) is usually recommended[29,30]. The achievable brightness gains are superior compared with those obtained by peroxide bleaching before a deinking stage as Figs. 15 and 16 show. A possible alkaline yellowing during pulping due to the absence of peroxide is compensated by bleaching after deinking. The addition of peroxide at the pulping stage is necessary only in the case of high proportions of mechanical fibers in the pulp when even small amounts of peroxide contribute to brightness stabilization.
- To achieve greater brightness gains, increasing the dosage of peroxide above 2% on o.d. pulp is not recommended. Alternatively, splitting the addition of bleaching agent into some before and some after flotation gives additional improvement of the bleaching results[31]. Most deinking plants take advantage of the synergistic effects of flotation and bleaching by adding peroxide in two stages, i.e., during pulping and during dispersion or post-bleaching.
- A precondition for maximum brightness gain at the post-bleaching stage is the optimum removal of ink particles at the deinking stages[32,33]. A higher content of ink particles will give a poorer bleaching effect.

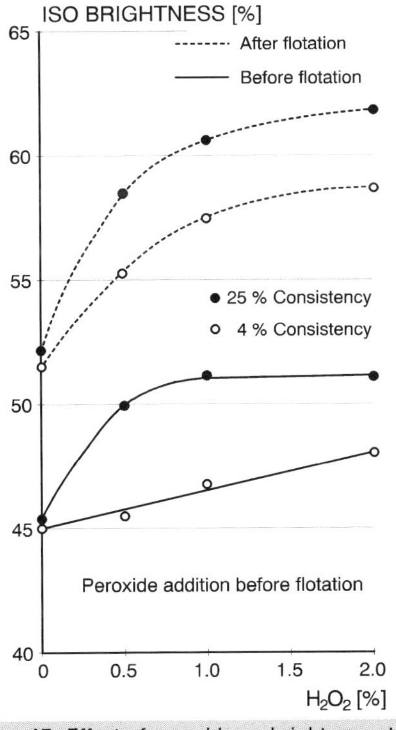

Figure 15. Effect of peroxide on brightness at different pulp consistencies[38].

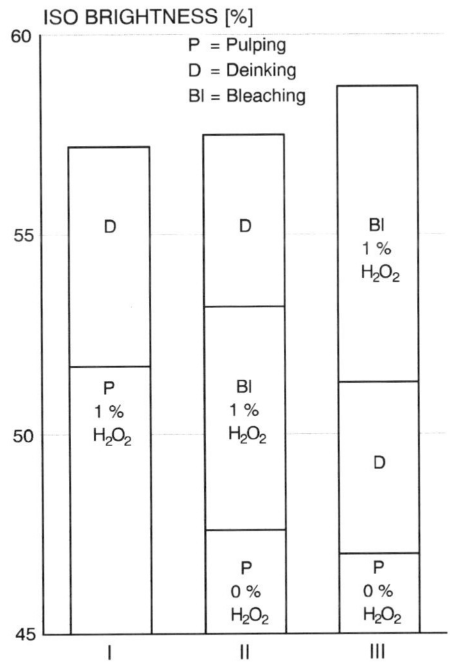

Figure 16. Brightness gain in a flotation deinking process using different ways for bleaching (75% old newspapers, 25% old magazines)[30].

Peroxide is a highly effective bleaching chemical for improving the brightness of deinked pulp. The bleaching result is sufficient for many applications for using deinked pulp to satisfy the optical requirements of the paper manufactured. A precondition for successful bleaching is stabilization of the peroxide to possible competitive reactions that cause its decomposition. The effects on chromophores of the lignin that use oxidative reactions are limited. If higher optical quality demands are made on the deinked pulp, additional bleaching with reductive bleaching chemicals will be necessary.

2.2 Dithionite bleaching (Y)

Common reductive bleaching chemicals used in treating recycled fiber pulp are sodium dithionite (sodium hydrosulfite) and formamidine sulfinic acid (FAS). FAS is finding increasing use primarily with wood-free recycled pulp. A reductive bleaching is essential especially not only for bleaching but also color stripping of colored recovered paper and carbonless paper. Many dyes used to color paper can be effectively destroyed by both reductive bleaching chemicals. Balancing of the optical parameters can occur even with highly varying recovered paper qualities.

In 1873, the German chemist, Schützenberger, succeeded in preparing the sodium salt of dithionic acid in an impure form. He also introduced the commonly used but incorrect term “hydrosulfite” for this chemical compound. Several years later in 1884, Bernthsen identified the correct formula as $Na_2S_2O_4$ and named it sodium dithionite.

Following research by Bazlen and Wolf, dithionite was produced for the first time in 1906 by BASF in Germany in powder form. Its maximum concentration of active substance is about 90%. Initially, the product was obtained by the zinc dust process. Zinc is converted into zinc dithionite in an aqueous solution with sulfur dioxide and then by sodium hydroxide to sodium dithionite and zinc oxide. A later process developed by Janson uses the sodium amalgam from the mercury cell electrolysis of sodium chloride solution and converts this directly with sulfur dioxide to sodium dithionite. The advantage of this process is that it produces a material free of heavy metals that is therefore stable. Due to the mercury content of the wastewater, chlorine-alkali electrolysis in Europe is converting from the amalgam process to the diaphragm process. For this reason, BASF developed the formiate process in which sodium formiate produced directly from carbon monoxide and sodium hydroxide is converted with bisulfite to sodium dithionite.

In the 1930s, dithionite was used for the first time for bleaching mechanical pulp. The sodium dithionite initially used was added only in small amounts directly as a powder. With bleaching times of only a few minutes and temperatures of approximately 20°C, the bleaching effect was limited. Satisfactory bleaching results were only possible by exclusion of atmospheric oxygen.

With the introduction of continuous bleaching processes such as tower bleaching, the use of solutions of dithionite proprietary blends became necessary to ensure their homogeneous mixing with stock. Another option is the use of cooled, alkalized solutions of sodium dithionite. At temperatures below 10°C, these solutions are so stable that they can be stored for long periods. This eliminates the complicated process of dissolving the powder.

Another source of sodium dithionite is the in situ production from Borol by the Ventron process. The borohydride bleaching is essentially dithionite bleaching. For this, sulfur dioxide, sodium borohydride, and sodium hydroxide are mixed in a reactor. Besides sodium dithionite that is present in the bleaching solution with about 85% active substance, sodium bisulfite and sodium perborate are produced. The advantage of in situ production is that the preparation of the bleaching solution can be controlled so no loss of active substance occurs due to the transport and storage of the naturally unstable dithionite solution. This form of dithionite preparation is especially interesting when long transport distances exist.

2.2.1 Chemistry

The reductive effect of the sodium dithionite during bleaching is due to its chemical transformation in aqueous solution into sodium bisulfite ($NaHSO_3$) and into the sodium salt of the unstable sulfoxylic acid ($NaHSO_2$). In the presence of reducible substances, the reaction occurs as follows:

$$Na_2S_2O_4 + H_2O \rightarrow NaHSO_2 + NaHSO_3 \qquad (9)$$

Although the resulting sodium bisulfite as a weak reducing agent barely influences the bleaching reaction, the sodium salt of the sulfoxylic acid is the actual bleaching agent.

In the presence of oxygen, the sodium salt oxidizes immediately into acid bisulfate ($NaHSO_4$):

$$NaHSO_2 + O_2 \rightarrow NaHSO_4 \qquad (10)$$

This useless consumption of bleaching chemicals is accompanied also by a shift of the pH into the acidic area. This then promotes the self-decomposition of the dithionite by forming thiosulfate and sulfur deposits. Dithionite solutions can also decompose as a result of the exclusion of air especially at higher temperatures. The sodium salt of the sulfoxylic acid then transforms into sodium thiosulfate ($Na_2S_2O_3$) according to the following:

$$2NaHSO_2 \rightarrow Na_2S_2O_3 + H_2O \qquad (11)$$

To stabilize the sodium dithionite solution, adding alkali as NaOH or Na_2CO_3 that is at least 5%–7% on a dry basis is recommended. Adding chelating agents can prevent the precipitation of calcium or magnesium salts in the solution. The use of chelating agents serves exclusively to sequester these polyvalent cations. The presence of heavy metal ions does not cause dithionite decomposition. The maximum concentration of the dithionite solution can be 120–200 g/L depending on the temperature.

Despite numerous investigations conducted primarily by Polcin and Rapson[2,3] in the 1960–1970s on mechanical pulp, the bleaching reactions of dithionite on the fibers are still not completely understood. During bleaching, only some wood chromophores react with dithionite. Primarily the simple structures (quinonoids such as ortho-quinone and semi-quinone and coniferaldehyde structures) are destroyed. Flavones can transform into less-colored chalcone chromophores. The treatment is indifferent toward conjugated ketones and double bonds of the coniferalcohol type and probably also toward extensively complicated quinone structures.

According to research by Steenberg and Norberg[34], only some reduced substances dissolve. This ensures a low COD load of the white water or wastewater due to the bleaching process. In addition, the reduction products are unstable and revert easily to their original structures. This is the cause of the tendency toward yellowing of the dithionite-bleached mechanical pulp. By using chelating agents, limiting the tendency toward yellowing of the bleached fibers is possible[35,36].

2.2.2 Process parameters

Compared with peroxide, the reducing dithionite has a higher kinetic reaction rate that is evident primarily in significantly shorter reaction times. Tests have shown that the bleaching reaction takes a few minutes. The bleaching effect is positively influenced primarily by rising temperatures that favor diffusion of the dithionite ions into the fiber cell wall[37].

A separate bleaching stage is essential due to the sensitivity of dithionite to oxygen. For this reason alone, a combination with the pulping stage of the stock processing operation such as occurs with peroxide bleaching is not possible. Low air contents occur with low pulp consistencies. For this reason, the optimum consistency range of dithionite bleaching was confined to 3%–5% for a long time. After admixing the bleaching chemicals, the pulp is directed into an upflow bleach tower so that the fibers reacting with the dithionite do not come into contact with atmospheric air. Only at the top of the tower can the ambient air react with still unconsumed dithionite and transform it into bisulfate and bisulfite.

Because of the short reaction time, the homogeneous mixing and distribution of the bleaching chemicals in the pulp is very important. In the low-consistency range, this task uses low-consistency mixers that have been the state-of-the-art for decades. Not until the development of an air-evacuating mixing pump was work also possible in the medium-consistency range with consistencies of 10%–15%. Fluidizing MC pumps that evacuate the air from the pulp slurry and simultaneously mix the bleaching chemicals are now standard components in the dithionite bleaching process. Depending on the required retention time, the bleaching performance continues in a similar upflow bleach tower or pipe.

By shifting the stock consistency into the more economical MC range, dithionite bleaching became established practice in recycled fiber pulp processing. Dithionite bleaching is now an important system module in the deinking process when a color correction by stripping dyes is necessary in addition to fiber brightening.

Increasing the temperature also accelerates the reaction in dithionite bleaching. As Fig. 17 shows, temperatures above 60°C produce superior bleaching results. The stability decreases with longer reaction time because the bleaching agent is completely consumed after a short reaction time[38]. Thermal reactions then dominate and lead to yellowing of the pulp. For furnishes with a higher mechanical fiber content, the pulp can be subject to brightness reversion when heated to very high temperatures. In addition, dithionite transforms into thiosulfate very rapidly at temperatures above 80°C accompanied by sulfur precipitation. For these reasons, the reaction times in dithionite bleaching of wood-containing DIP are usually 15–60 min. To limit energy consumption, a reaction temperature of about 60°C is common.

Higher temperatures offer additional brightness advantages especially with dithionite dosages of more than 1% on o.d. pulp as Fig. 18 shows. Below 1% dithionite addition, the brightness differences of recovered pulp treated at 40°C–80°C are minor. After a reaction time of 1 h at approximately 95°C, the bleaching results are lower compared with values at 80°C. This is due to heat-yellowing and fast dithionite reaction. To prevent brightness losses, reducing the reaction time at such high temperatures to 5–10 min is recommended[38]. Then the bleaching is normally performed in a pipe.

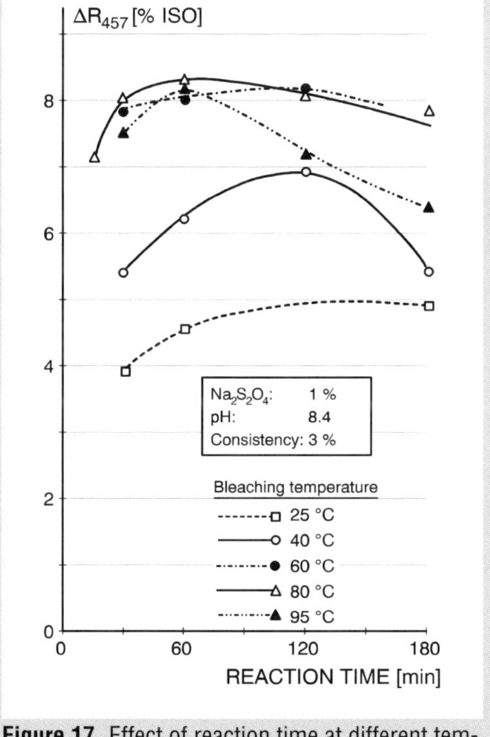

Figure 17. Effect of reaction time at different temperatures on brightness gain[38].

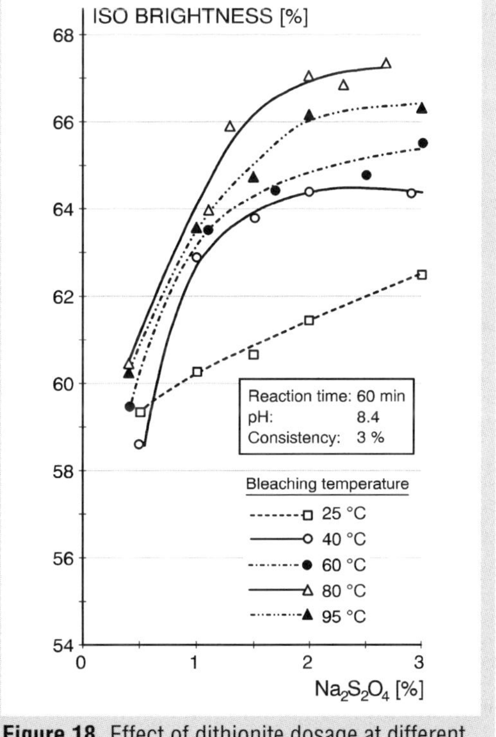

Figure 18. Effect of dithionite dosage at different temperatures on brightness[38].

As with peroxide bleaching, pH has a dominant effect on the reaction process in dithionite bleaching. With a dithionite amount of 1% depending on the reaction temperature, optimum results occur with initial pH of 6–7 as Fig. 19 shows. The pH for optimum results increases with increasing temperatures. Above 60°C, maximum brightness gains occur at approximately pH 7.

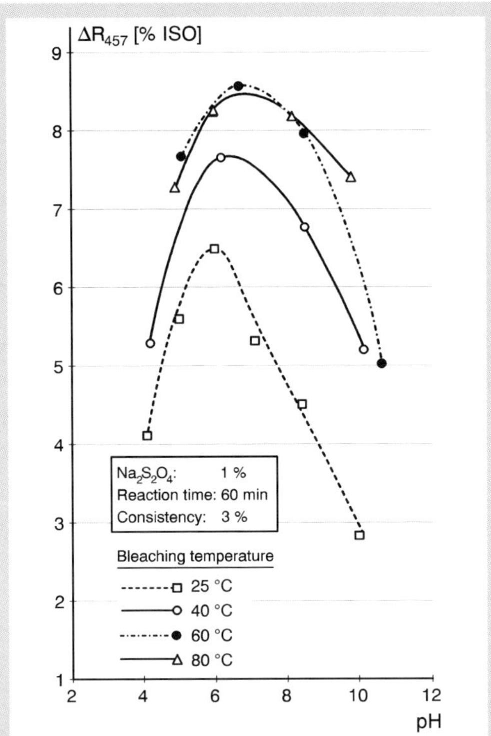

Figure 19. Effect of pH at different temperatures on brightness gain[38].

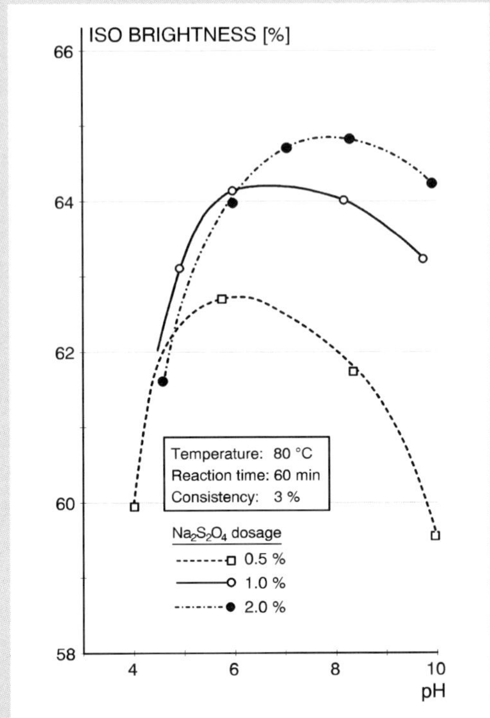

Figure 20. Effect of pH and dithionite dosage on brightness[38].

Larger dithionite additions and higher reaction temperatures require higher initial pH settings as Fig. 20 shows. Because of the interaction between high temperatures and added amounts, the pH falls rapidly and therefore prevents an optimum reaction. The optimum pH range in the dithionite bleaching of recycled fiber pulp as opposed to mechanical pulp bleaching is not in the weakly acid range but in the neutral to weakly alkaline area[39–41]. The refixation of detached ink particles is probably also responsible for the fact that the bleaching of deinked pulp produces poorer results compared with those obtained with mechanical pulp in the acidic range.

Under practical operating conditions, the amount of dithionite used in bleaching deinked pulp is usually 0.5%–1% on o.d. pulp. The reactions that occur primarily in the range of medium-consistency generally produce brightness increases of 4%–7% ISO in wood-containing deinked pulp. A better bleaching result is possible with wood-free deinked pulp if colored paper is part of the pulp.

Superior results frequently occur under laboratory conditions compared with industrial dithionite bleaching. The reasons for this may be the following[42]:

- Poor mixing efficiency of the bleach liquor with the pulp especially at high consistency in the mill environment

- Presence of air in the pulp slurry
- Presence of metal ions such as iron and aluminium
- Unstable dithionite liquor
- Presence of residual chemicals due to carryover especially of peroxide (See section "Two-stage bleaching" of this Chapter).

2.3 FAS bleaching (FAS)

Formamidine sulfinic acid (FAS) is a low-odor, crystalline reducing agent used in the bleaching of wood-containing pulp only since the 1980s. A 1983 patent[43] proposes bleaching conditions for mechanical pulp and recycled fiber pulp that can apply in a single-stage and a two-stage process with other bleaching chemicals.

Since then, FAS has had increasing use for bleaching recycled fiber pulp. Especially in cases where colored paper or carbonless paper are part of the recovered paper mix, FAS constitutes an effective bleaching agent for their decolorization. In the past, such types of recovered paper were usually bleached with hypochlorite especially in North America. This can very effectively destroy a wide range of dyes. Because of associated ecological drawbacks such as high chlorine and organic wastewater loads, FAS in increasingly replacing hypochlorite with comparable results.

As early as 1910, Barnett investigated the reaction of peroxide with thiocarbamide. Since then, the synthesis of FAS has been the subject of many publications. FAS contains thiocarbamide sulfur dioxide according to the X-ray structure analysis. The material forms at low temperatures under acidic to neutral pH conditions in a reaction from two peroxide molecules and one thiocarbamide molecule.

2.3.1 Chemistry

FAS is only slightly soluble in water. Although only about 27 g/L enter the solution as sulfinate, the solubility under alkaline conditions increases up to 100 g/L. The sodium salt of the formamidine sulfinic acid has a higher solubility, but its aqueous solution decomposes very rapidly. For this reason, alkaline bleaching solutions are prepared only shortly before their addition in a continuous process. Consumption must then occur as soon as possible.

In the reaction of the sulfinate with reducible compounds (chromophores), urea and sodium hydrogen sulfite are the primary products formed:

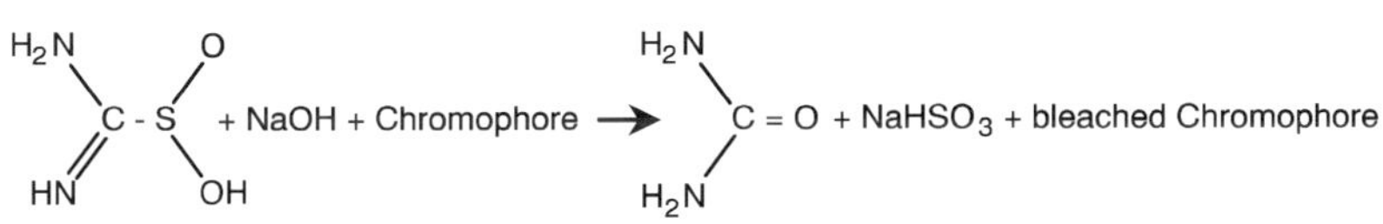

(12)

During the continuing course of the reaction, sodium hydrogen sulfite transforms to sodium hydrogen sulfate:

$$NaHSO_3 + \frac{1}{2}O_2 \rightarrow NaHSO_4 \tag{13}$$

The optimum ratio between sodium hydroxide and FAS is approximately 0.5 on a dry weight basis. Under optimum bleaching conditions, the pH drops during bleaching from about 9 to 7–8. With ratios in excess of 0.5, a risk of yellowing exists especially in cases where the deinked pulp contains high proportions of mechanical fibers. When treating a deinked pulp with a high calcium carbonate content, the amount of alkali can be reduced by two-thirds[44].

Like all reducing bleaching chemicals, FAS is oxidized by atmospheric oxygen although it is considerably less susceptible than dithionite. This fact permits a wider range of application possibilities for bleaching not only as a separate bleaching stage but also in combination with other stages of the stock processing operation.

Compared to dithionite, FAS also has a lower sulfur content. Combined with the mostly lower quantities used, this has an especially positive effect on the sulfate load of the white water loop. Consequently, the corrosion susceptibility of the equipment and instruments is lower as is the production of unpleasant odors due to hydrogen sulfide.

Under alkaline conditions, the sulfinate ion has a stronger reduction potential (>1 100 mV) than dithionite (about 800 mV). Chromophores can therefore also be destroyed that are not acted upon by dithionite. Azo dyes are destroyed irreversibly. Quinone structures are also reduced. FAS can therefore also have use to brighten unbleached kraft pulp by about 10% ISO.

2.3.2 Process parameters

The most important process parameter of FAS bleaching is the reaction temperature that is the most effective means of controlling the bleaching result. As Fig. 21 shows, increasing the temperature from 40°C to 90°C provides almost double the bleaching result of a wood-free deinked pulp after a reaction time of 30 min. At lower temperatures, the same figure also shows that increasing the reaction time is necessary to obtain an acceptable result. Increasing the reaction time alone is insufficient to compensate fully for the lower brightness level that occurs with lower temperatures.

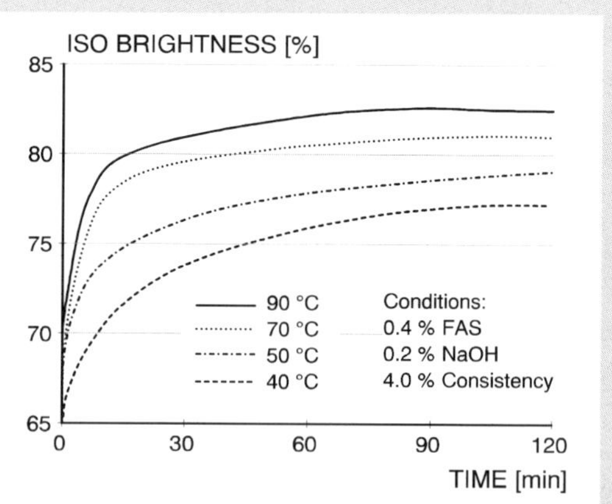

Figure 21. Effect of temperature and retention time on brightness in post-bleaching of wood-free deinked pulp[44].

If the temperature is set at an optimum value, only small quantities of FAS are necessary to perform effective bleaching as Fig. 22 shows. Only 0.2% FAS is sufficient to achieve a brightness after bleaching at a temperature of 80°C that is identical to that with a threefold larger addition of FAS at a reaction temperature of 50°C.

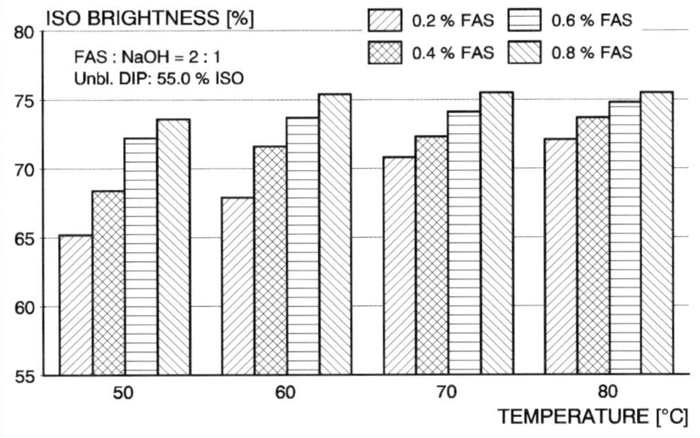

Figure 22. Effect of temperature on brightness in post-bleaching of deinked pulp with different amounts of FAS[44].

The influence of pulp consistency on the bleaching result is considerably less than with peroxide bleaching. Daneault and Leduc[45] developed a mathematical model for the effect of different parameters in FAS bleaching using numerous tests. Although the bleaching result improves as the consistency is reduced, they showed that the effect is considerably less than that occurring with temperature increase. The brightness of mechanical pulp increases by only 2% ISO with bleaching at 5% instead of 20% consistency. This dependence on consistency is only a result of the presence of atmospheric oxygen. The bleaching efficiency will therefore be greater with more effective exclusion of air.

Using tests with mechanical pulp, the following relationship held in the ranges investigated for the process parameters of temperature, consistency, and dosage of FAS [45]:

$$R_{457} = 55.05 + 0.08T - 0.12C + 5.7F - F^2 \tag{14}$$

where R_{457} is the ISO brightness,%
T the temperature,°C (from 40°C–80°C)
C the consistency,% (from 5%–20%)
F the FAS dosage,% (from 0.25%–2% dry basis).

One can assume that this relationship can also apply to the bleaching of wood-containing deinked pulp. The model illustrates that increasing temperature and FAS quantities or reduced consistency positively influence the bleaching result. Reaction time has no major importance.

Bleaching conditions in the FAS application can vary widely. A very simple method is to add FAS and sodium hydroxide directly to the pulper. The so-called pulper bleaching thereby effectively color strips colored and carbonless papers without requiring additional bleaching systems. In cases with greatly varying proportions of colored paper in the recovered paper mix, addition of the bleaching agent during disintegration has a balancing effect and therefore prevents undesired color deviations[44]. Besides the presence of air, a disadvantage is the low temperature that seldom exceeds 50°C.

Using a separate post-bleaching, one can largely avoid the presence of air and achieve an optimum temperature setting. The consistency conditions vary widely. Post-bleaching can therefore use low- and medium-consistencies in upflow or downflow bleach towers or pipes. The bleaching effect then is primarily a function of the temperature that is controlled by feeding steam to a mixer or pump.

Optimum temperature conditions for FAS bleaching can be readily achieved in a dispersion stage. The addition of the FAS solution best combines the effects of intensive mixing at a temperature that is favorable for FAS bleaching. Less air and water are present, and the mixing guarantees a quick and close contact between the fibers and the bleaching agents. The chemicals then react in a chest or a bleach tower with short retention times of 30 min[46]. A higher temperature in the disperger gives a shorter post-reaction time. With dispergers that can operate at temperatures above 100°C, a few minutes are sufficient for post-reaction. Furthermore, aerobic losses of FAS are minimized with increasing temperature. For best economics, the combination of FAS bleaching with a dispersion stage is attractive because no systems besides the disperger are necessary.

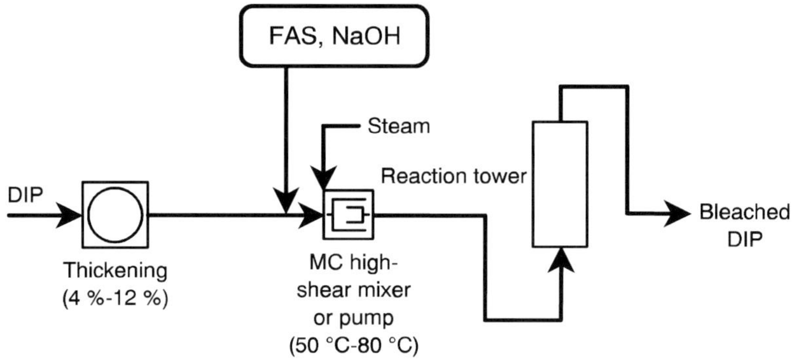

Figure 23. Typical application of FAS in post-bleaching in LC and MC ranges[44].

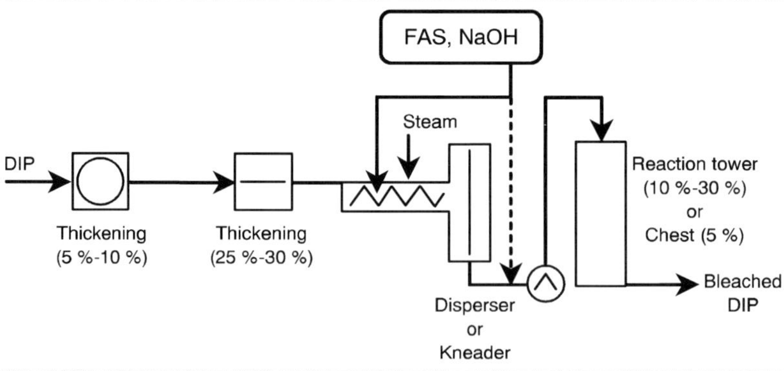

Figure 24. Typical application of FAS in dispersion bleaching[44].

Figure 23 shows a typical application of FAS for post-bleaching deinked pulp in the LC to MC range. After thickening, the pulp is mixed with the chemical solution, and steam is added in a high shear mixer. The pulp then enters an upflow tower. If the bleaching is integrated into an existing dispersion stage, the process goes according to Fig. 24. After thickening, the bleaching solution is mixed with the deinked pulp in a screw while adding steam. After treatment in a disperger or kneader, the pulp after dilution is fed to a tower, chest, or pipe where it reacts at lower temperatures[44].

Sometimes, a difference exists in bleaching results between lab and mill scale as in the case of dithionite bleaching. The most important effects result from the different laboratory equipment and conditions used. Often the laboratory set-up does not allow a high mixing efficiency at high consistencies and high temperatures. The presence of air cannot be prevented.

2.4 Combined bleaching stages

The use of deinked pulp depends largely on its optical characteristics. When using deinked pulp to produce high-quality graphic papers such as supercalendered or lightweight coated paper, the results of one-stage bleaching are usually not sufficient to meet the optical requirements. To increase brightness, two or sometimes three bleaching stages are necessary with oxidative and reductive chemicals. This destroys different coloring structures and gives a better bleaching result than the addition of larger amounts of chemicals during a single-stage bleaching process. The improvement in the optical characteristics results not only in better brightness but also in reduced yellow tint especially when treating deinked pulp with mechanical fibers[47].

Different combinations of bleaching stages are possible with the chemicals and technical processes used. Besides the application of different bleaching chemicals, bleaching sequences involving the use of only one chemical are also common.

2.4.1 Two-stage bleaching

Typical two-stage bleaching combinations for processing deinked pulp are as follows:

- Peroxide-peroxide
- Peroxide-dithionite
- Peroxide-FAS
- FAS-peroxide.

Oxidative-oxidative bleaching

Peroxide is frequently applied during disintegration in the pulper and subsequently in a dispersion stage or a separate post-bleaching process. Numerous tests have shown that a split addition between disintegration and dispersion or between pulping and post-bleaching should receive preference over the single addition of peroxide at the disintegration stage. Note that an increased dosage of peroxide during pulping is accompanied by a considerably reduced brightness increase in the second peroxide stage.

Extensive work[31] tracked the brightness increase during post-bleaching in relation to the addition of peroxide during pulping. Figure 25 shows the bleaching results of a recovered paper mix of 50% old newspapers and 50% old magazines treated with peroxide during pulping and in a post-bleaching process. A more intensive first peroxide treatment gives a poorer bleaching effect in the second treatment stage. The brightness increase is therefore reduced dramatically in post-bleaching when adding 2% peroxide during pulping. The bleaching effect of peroxide remains limited even in cases where it is added at different stages.

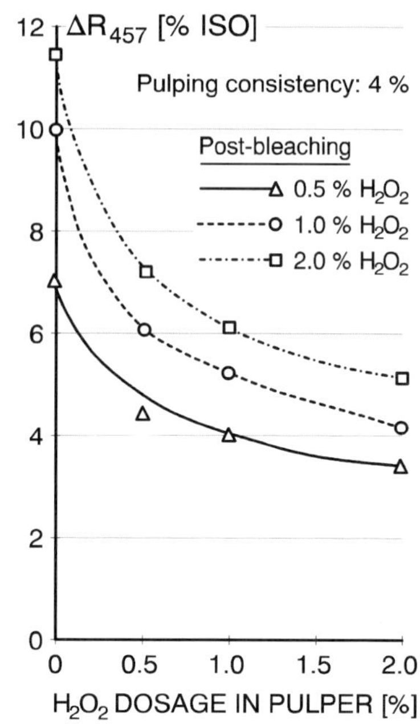

Figure 25. Brightness gain by peroxide post-bleaching vs. peroxide dosage during pulping[31].

The use of peroxide during pulping prevents yellowing caused by alkali when wood-containing deinked pulp is processed under alkaline conditions. The bleaching conditions during pulping are more unfavorable not only with regard to stock consistency and temperature but also with regard to the load of detrimental substances. In many applications, the addition of peroxide in the pulper is therefore reduced to a minimum. Later in the process, additional bleaching treatment uses a considerably larger dosage of peroxide under more favorable process conditions.

Oxidative-reductive bleaching

With the combined use of different chemicals, bleaching usually uses the sequence of oxidative followed by reductive. After the peroxide addition in the pulper, dispersion bleaching with dithionite or FAS is performed or post-bleaching with one or both reducing bleaching chemicals occurs in a separate process stage in a tower or pipe.

When using dithionite in the reductive bleaching stage, the previously mentioned sequence has various advantages compared with the opposite sequence:

- Prevention of alkali yellowing in the alkaline pulping of wood-containing recovered paper due to the addition of peroxide. Dithionite is ineffective in this pH range.
- A separate bleaching stage is essential due to the susceptibility of dithionite to atmospheric oxygen.
- If the dithionite stage is managed as post-bleaching, the final deinked pulp has a favorable pH for the paper machine.

With an oxidative-reductive bleaching sequence, care is necessary to ensure that after the peroxide treatment using an excess of bleaching chemical no residual peroxide enters the dithionite stage. Otherwise, dithionite would be consumed for the reduction of the residual peroxide. Previously, sodium bisulfite was frequently used to destroy and prevent a carryover of residual peroxide. This approach is being replaced increasingly by closed process water loops that ensure the unconsumed chemicals are reintroduced into the process for re-use[48]. Because flotation deinking usually occurs between the oxidative and reductive bleaching stages, the strong dilution with subsequent thickening prevents a carryover of process chemicals from the previous process stages. Simultaneously, this causes a more favorable pH range for dithionite bleaching even without additional acidification. With a low COD concentration of white water, an absolutely safe operational condition is thus guaranteed for a two-stage bleaching sequence.

In peroxide post-bleaching, the brightness gain is reduced with increasing addition of hydrogen peroxide at the pulping stage. Reduction of the bleaching effect in dithionite post-bleaching with increasing peroxide addition during pulping is only 30% as Fig. 26 shows. To achieve higher brightness gains, multistage bleaching with different bleaching chemicals is recommended.

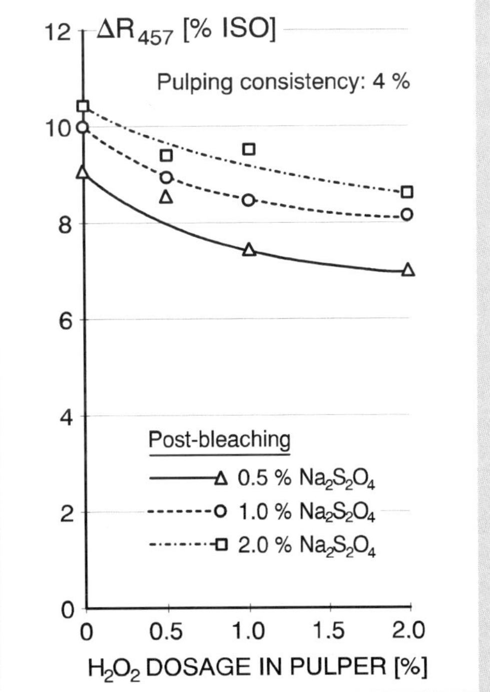

Figure 26. Brightness gain by dithionite post-bleaching vs. peroxide dosage during pulping[31].

An increasingly common practice is performing the second bleaching stage with FAS instead of dithionite. The FAS treatment then occurs in the disperger or in a separate processing step at the end of the deinking process. The main objectives with the use of FAS are balancing any optical variations that continue to exist in the deinked pulp and simultaneously increasing brightness. This bleaching sequence offers a constant quality of the final pulp especially in cases of strongly varying proportions of colored papers. The shade correction effect is greater when lower proportions of mechanical fibers are in the recycled pulp to be bleached. In cases with higher proportions of mechanical fibers, special care is necessary to ensure that the residual alkali is largely consumed after the total consumption of the FAS to prevent alkali yellowing. When papermaking under neutral conditions follows this, no additional acidification is necessary.

Reductive-oxidative bleaching

In contrast to dithionite, FAS can be used in the pulper. This is why the reductive-oxidative bleaching sequence seems more attractive than in the case of dithionite bleaching at least from an economic point of view. During pulping, no optimum temperature conditions can be set for FAS. Efficiency compared with that in an optimized post-bleaching is therefore adversely influenced. The FAS-peroxide bleaching sequence has primary use for the processing of wood-free recovered paper for production of hygienic papers.

For wood-containing pulp, the first bleaching is done after pulping. No clear preference exists for the oxidative-reductive sequence. Tests conducted at the French research organization CTP in Grenoble[48] show that both bleaching sequences produce nearly identical brightness results. Table 1 shows the brightness gains measured after two-stage post-bleaching with peroxide and dithionite. In the difference of brightness gains, both bleaching sequences show an average of 0.6% ISO. No significant difference was therefore established.

Table 1. Comparison of two-stage bleaching with different bleaching sequences[49].

Peroxide [%]	Dithionite [%]	ΔR_{457} [% ISO] Peroxide-Dithionite	ΔR_{457} [% ISO] Dithionite-Peroxide
1	1	5.9	7.3
1	2	7.0	7.6
2	1	8.3	7.9
2	2	7.3	8.7
5	1	8.8	9.6
5	2	9.5	10.0

If recovered office paper (MOW) is processed, the dithionite-peroxide bleaching sequence not only produces an exceptionally improved brightness but also an excellent cleanliness of the pulp[48]. Prerequisites for this include a three-loop system with two thickening and two dispersion stages and thickening after the peroxide stage to prevent peroxide carryover. Simultaneously, the pH of the stock is matched to the paper machine requirements. Both bleaching stages occur with a dispersion. Dithionite mixing is performed in a MC pump. Before admixing the chemicals, air is sucked off by a vacuum pump integrated in the MC pump.

One advantage of this bleaching sequence is that the deinked pulp enters the peroxide stage with a high level of cleanliness. This ensures maximum bleaching efficiency. Any catalases that may be present are destroyed in the first hot dispersion stage. Residual peroxide at the end of the process also ensures control of bacterial growth at pulp storage and the paper machine. Besides the considerable capital costs, higher COD loads of the white water loops as a result of an additional thickening stage after peroxide bleaching at the end of the process represent a disadvantage of this system. In addition, deposit and retention problems can occur due to excessive use of water glass.

The efficiency achievable with different bleaching sequences depends less on the particular bleaching sequence than on the stock processing technology, system integration, and white water circulation and purification. For every installation, one must decide in which sequence and with which chemicals bleaching will occur.

2.4.2 Multistage bleaching

In cases of high optical demands on wood-containing DIP, three-stage bleaching is most effective. Three-stage bleaching sequences usually use pulper bleaching, disperger bleaching, and a separate post-bleaching. Typical bleaching sequences are as follows:

- Oxidative-oxidative-reductive
- Oxidative-reductive-oxidative.

The addition of peroxide during pulping is restricted to small amounts of less than 0.5% used solely to prevent alkali yellowing. The actual bleaching occurs only after a deinking stage. A higher bleaching efficiency occurs due to the more favorable conditions. In this case, bleaching again uses peroxide, dithionite, or FAS with the disperger stage frequently used to mix the chemicals. After storage and a second deinking stage, the pulp is bleached alternately with a different chemical. If peroxide is added during dispersion, the post-bleaching uses dithionite or FAS.

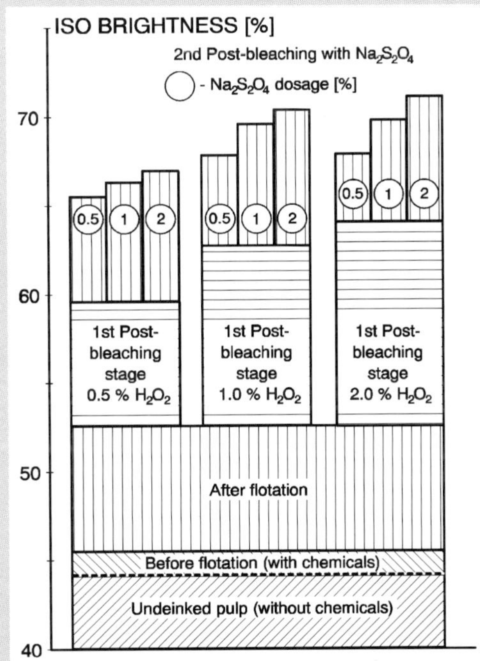

Figure 27. Effect of multistage bleaching on brightness[31].

Despite a triple application of bleaching agents, the effect on the optical characteristics of wood-containing deinked pulp remains limited. Even with greater amounts of bleaching chemicals added at the second and third bleaching stage, Fig. 27 shows that it is almost impossible to bleach wood-containing deinked pulp to a brightness more than 70%–72% ISO. For a superior optical level, other chromophores must be destroyed. This requires use of lignin-degrading bleaching chemicals.

3 Lignin-degrading bleaching

All lignin-degrading bleaching chemicals are oxidants used primarily in chemical pulp bleaching. Among the most important types used in the past have been elemental chlorine (Cl_2) with its compounds chlorine dioxide (ClO_2) and sodium hypochlorite (NaOCl). Since the 1980s when the discussion on potentially harmful ecological effects of chlorine and hypochlorite began, a global trend has existed toward replacement of these traditional bleaching chemicals with oxygen (O_2) and ozone (O_3).

Use of elemental chlorine has declined as a bleaching chemical due to the formation of potentially toxic chloro-organic compounds in the wastewater of the bleaching plant and in the fiber stock. Substitution by oxygen has already largely occurred in sulfite and kraft pulp bleaching. Oxygen is used as a delignification process.

The lignin-degrading bleaching processes exclusively use oxidative reactions in which the chromophore groups are dissolved away from the fiber structure and then removed. Consequently, the treatment usually has a yield loss. The effectiveness of the chemicals increases with their oxidation potential. For example, the oxidation potential of ozone in an acidic environment is 2.07 V. The oxidation potential of elemental chlorine is 1.36 V. Peroxide in an alkaline medium has an oxidation power of only 0.88 V.

Besides phenolic groups, attack especially occurs on the carbon to carbon double bonds. The colored organic substances are split into more hydrophilic, smaller molecules that are soluble in water or alkali. With the exception of hypochlorite, lignin-degrading bleaching chemicals do not modify carbonyl groups. Figure 28 shows typical reactions that occur during delignification. Besides the removal of lignin and extractives, most dyes are also color stripped.

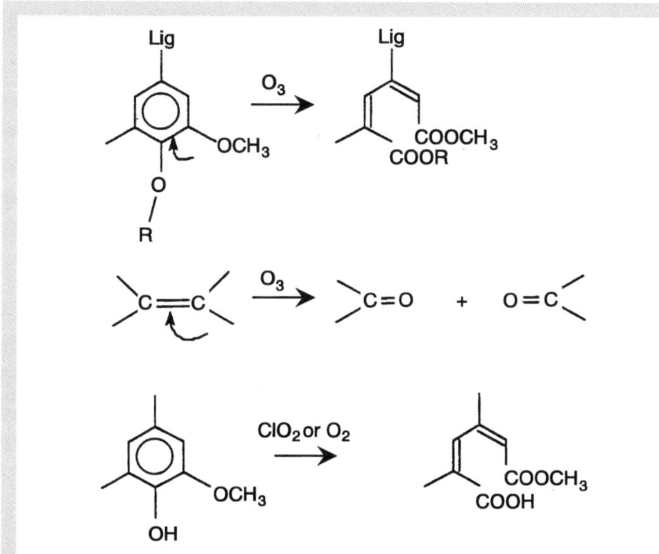

Figure 28. Main lignin reactions occurring during ozone, oxygen, and chlorine dioxide bleaching[1].

Consequently, the treatment of deinked pulp with lignin-removing chemicals is useful only if the largest proportion of the pulp consists of unbleached or semibleached chemical fibers. Bleaching will then reduce the residual lignin content and cause a brightness increase of the deinked pulp. If wood-containing deinked pulp is bleached using the aforementioned chemicals, an optical improvement will not occur. Depending on the bleaching conditions, a noticeable loss in brightness is possible. Finally, losses in yield must also be expected. These also burden the wastewater load of the deinking process[12,50].

3.1 Bleaching with chlorine-containing chemicals

In North America, wood-free deinked pulp is usually bleached with chlorine-containing chemicals. In the past, hypochlorite bleaching was common. Recycled fiber with a mechanical fiber content up to 5% is not only very effectively bleached but also color stripped. In cases with larger proportions of mechanical fibers, yellowing reactions on the mechanical pulp components accompany the treatment. Bleaching is then not possible with the usual single-stage process. It requires the prior removal of the residual lignin.

3.1.1 Hypochlorite bleaching (H)

An important precondition for achieving high brightness gains is a minimal lignin content. If too much lignin is present in the fiber stock, hypochlorite bleaching especially with low dosages results in discoloring. The pulp turns pink. This then requires using large quantities of chemicals to brighten the pulp. In conventional chemical pulp bleaching for this reason, a common practice was to carry out chlorination with subsequent alkali extraction of the chlorinated lignin compounds before the hypochlorite stage.

Depending on the requirements of the deinked pulp, the bleaching conditions can vary within certain limits. The quantity of active chlorine is 1%–6% on o.d. deinked pulp. In most cases, 2%–3% is used. The temperature is 50°C–55°C for a bleaching time of 2–3 h. Towers or chests are used as reaction vessels in which the deinked pulp is stored with consistencies of 3%–15%[51].

In technical scale hypochlorite is formed as sodium hypochlorite (NaOCl) by addition of gaseous chlorine (Cl_2) in sodium hydroxide (NaOH) according to the following reaction:

$$2NaOH + Cl_2 \rightarrow NaOCl + NaCl + H_2O \qquad (15)$$

The dissociation of hypochlorite solution is a function of pH. With chlorine dissolved in the strongly acidic range, almost no dissociation occurs. With an increasing pH, hypochlorous acid forms. In the alkaline range it dissociates completely as a salt of this acid. The pH falls during bleaching under alkaline conditions toward the neutral point as acid products such as carbonic acid and oxalic acid form that neutralize the base of the hypochlorite. Because cellulose is extensively degraded in the weakly acidic medium at pH 6–7, the pH at bleaching must be such that the recycled fiber pulp remains in the alkaline range at the end of the reaction. For this reason, bleaching usually starts at a pH of 9–10.

Treatment with hypochlorite usually uses a single-stage process. Depending on the composition of the recovered paper, brightness levels of 70%–80% ISO are possible. Sometimes, hypochlorite bleaching also uses two stages. A washing stage is then inserted between two reaction vessels to remove the oxidized lignin. This gives better use of the chemicals. Enhanced brightness levels at the end of the bleaching process reflect this[51].

3.1.2 Chlorine-containing bleaching sequences

If the proportion of mechanical fibers in the deinked pulp exceeds 10%, chlorination occurs before the hypochlorite stage to degrade the residual lignin content. This ensures the effectiveness of the subsequent hypochlorite stage even when using deinked pulp with a low mechanical fiber content.

The treatment with elemental chlorine uses low consistencies of about 3% and temperatures of 30°C–40°C. Following this, the deinked pulp is washed to remove the chlorinated, degraded lignin products. Only then is the pulp bleached with hypochlorite in the MC range of 12%–15%. Occasionally, small amounts of sodium hydroxide are added during washing to eliminate the oxidized lignin from the deinked pulp more effectively.

Previously, a complete extraction stage with alkali such as in chemical pulp bleaching was applied especially in cases where recycled fiber pulp with an even greater lignin content was bleached as Fig. 29 shows. Optimum results are possible with this bleaching sequence by adding greater amounts of the bleaching agent at the hypochlorite stage. This improves the brightness and its stability against color reversion.

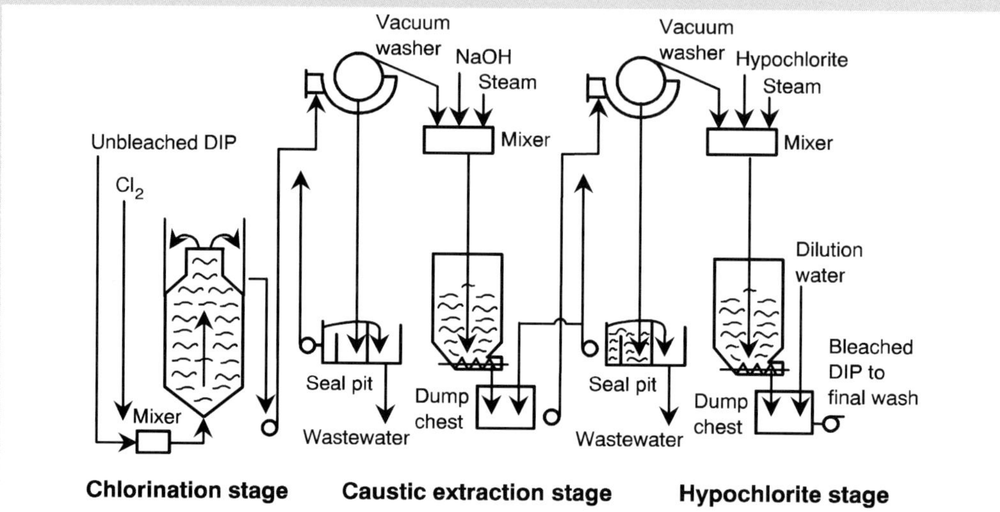

Figure 29. Typical chlorination-extraction-hypochlorite (CEH) bleaching sequence for deinked pulp with high mechanical fiber content[51].

Because of its negative environmental aspects, the described bleaching sequence has almost negligible importance for recycled pulp processing in the future. Instead, bleaching combinations with alternative chemicals such as FAS or with sodium dithionite will have increasing use for color stripping wood-free recycled pulp.

3.1.3 Chlorine dioxide bleaching (D)

Widely used for bleaching chemical pulp, chlorine dioxide has only minor importance for bleaching of wood-free deinked pulp. One reason for this is the expensive production process for the bleaching chemical that is profitable only with large deinking capacities.

The literature describes chlorine dioxide as an effective bleaching chemical for color stripping of colored recovered paper[52]. Especially in cases involving nonhomogeneous and large proportions of colored papers, the bleaching treatment efficiently eliminates most dyes. Brightening a strongly colored deinked pulp from mixed office waste having an initial brightness of 51% ISO is possible by single-stage bleaching with 1.1% ClO_2 on o.d. deinked pulp by 33% ISO while eliminating 90% of the dyes.

As Fig. 30 shows, the success of bleaching depends on the pH. An optimum bleaching result occurs in a low acidic medium at pH 6. While brightness levels of more than 80% ISO are possible at pH 4.5–7, an alkaline operation can achieve a brightness of only 70% ISO. The efficiency of bleaching also depends on the quantities of dyes in the deinked pulp. In most cases, the addition of 1% ClO_2 is sufficient for color stripping as Fig. 31 shows. With high dye amounts, a greater amount is necessary to obtain acceptable bleaching success. The AOX load of the wastewater is small.

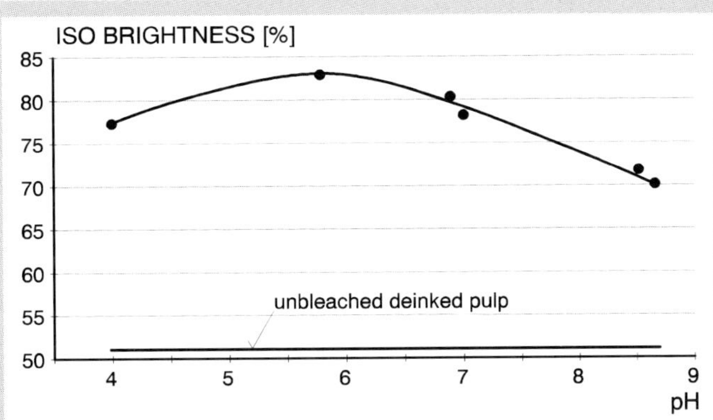

Figure 30. Chlorine dioxide decolorization effect of pH on brightness[52].

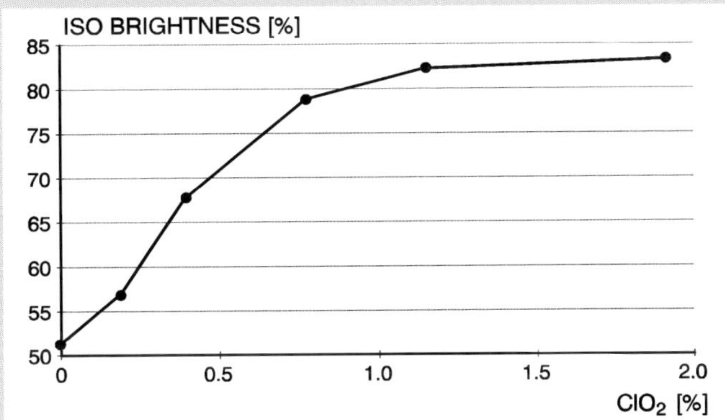

Figure 31. Effect of chlorine dioxide on brightness[52].

3.2 Bleaching with oxygen chemicals

For some years, a trend has existed not only in chemical pulp bleaching but also in recovered paper processing toward replacing chlorine-containing bleaching chemicals by oxygen-based chemicals. The driving forces behind this substitution are as follows[44]:

- Changed consumer demands: A growing demand exists for products manufactured without chlorine. The chloro-organic compounds (OX) (See Chapter 13) that occur in bleaching using chlorine-containing chemicals are becoming the target of increasing public criticism due to their perceived toxic characteristics.
- Demanding legislation: Legislation prescribes increasingly stringent emission levels such as AOX. Obtaining authorization to operate processing plants producing high amounts of AOX is becoming increasingly difficult.
- Changed composition of recovered paper: The partial replacement of chemical pulp by chemithermomechanical pulp (CTMP) increases the proportion of

high-yield fibers in the recovered paper grades that previously contained almost no mechanical fibers. This trend of increased mechanical fiber content is gaining impetus due to the growing use of recovered paper.

Besides the conventional, lignin-preserving chemicals, peroxide, dithionite, and FAS, additional substitutes are oxygen and ozone.

3.2.1 Oxygen bleaching (O)

In chemical pulp bleaching, an oxygen stage serves exclusively for removing residual lignin called delignification. The purpose of oxygen application in recovered pulp processing is to achieve an optical improvement of the pulp with an actual bleaching.

While oxygen delignification in chemical pulp bleaching has been applied on a industrial scale since the middle 1960s, results of the oxygen bleaching of recycled fiber pulp were reported for the first time in the second half of the 1980s[53,54]. The first tests with old corrugated containers (OCC) tried to improve pulp strength following the delignification that accompanies the treatment while simultaneously reducing the amount of wax introduced with the recovered paper. At a later stage, optical effects were also examined.

Today, oxygen treatment is used occasionally in processing wood-free deinked pulp as a chemical pulp substitute or as market DIP for the production of writing and printing papers with high standards of optical quality. In most cases, mixed office paper largely consisting of wood-free, bleached fibers is used. Depending on the quality of sorting, this can be contaminated to some degree with unbleached fibers, colored papers, and other impurities. The oxygen bleaching of widely varying pulp compositions then levels the varying optical properties of the deinked pulp[48].

Establishment of the oxygen stage for processing of deinked pulp for high quality graphical paper grades depends to a significant degree on economic criteria. The operation of the appropriate plants primarily in North America shows that their profitability links with the costs of cleaning, screening, and bleaching and the treatment of the resulting residues. Industrially, the efficiency is in doubt. The market situation of virgin hardwood pulp also has a major influence since both types of fiber stock compete as the raw material for the production of wood-free paper grades. The success of the process therefore connects to some degree to the political pressure for using recovered paper.

Chemistry

The complex reactions of the oxygen bleaching process are very fast radical chain reactions called auto-oxidation. With the participation of the oxygen and its radical reaction products, these reactions are responsible especially for degrading lignin and for forming hydrogen peroxide. The brightness enhancement that accompanies bleaching is due to the subsequent, relatively slow reactions of the hydrogen peroxide with the chromophores. Figure 32 summarizes the multitude of electrophilic (radical) and nucleophilic (ionic) reactions that can occur. They result in formation of hydroperoxides and many heterolytical and homolytical fragmentations. The hydroperoxides finally decompose. Although homolytical reactions for other oxygen chemical bleaching stages such as peroxide bleaching are unusable because

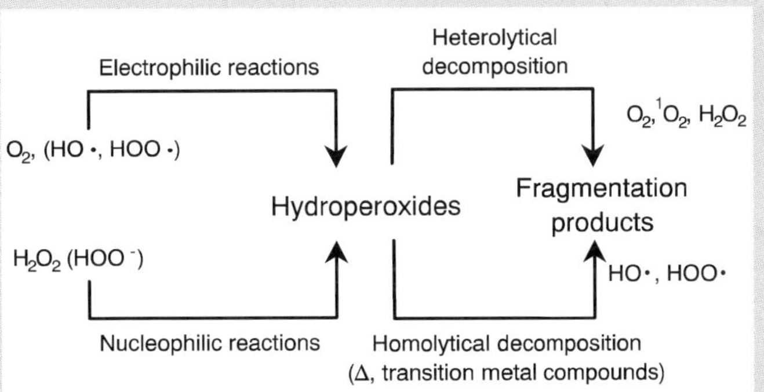

Figure 32. Scheme of reactions in oxygen bleaching[55].

they result in the decomposition of the bleaching agent, processes involving radicals are absolutely essential at the oxygen stage[55].

Due to its extremely low reduction and oxidation potential, the electrophilic character of oxygen is very weak. To start a reaction, raising the temperature or activating the system using a base such as sodium hydroxide is necessary. The activation uses an ionization of functional groups especially phenolic structures and structures with acidic hydrogen such as quinone methide in the lignin. To achieve practicable reaction times, temperatures in excess of 80°C are also necessary.

In contrast with chemical pulp bleaching, reactions on the contaminants introduced with the recycled fibers in the oxygen treatment of deinked pulp besides those on the chromophores of the lignin occur. These give an improved pulp cleanliness. In this context, the following effects are important[54,56–59]:

- Delignification of wood-derived dirt particles
- Color stripping of water soluble dyes
- Detackification of stickies
- Wet strength resin removal
- Improvement of laser and xerographic ink removal
- Fragmentation of adhesives such as hot melts and waxes
- Solubilization of binders that hold together flakes of coating, wet strength paper, and inks.

Industrially, the success of an oxygen stage for deinking plants is under critical review.

Process parameters

At atmospheric pressure, the low-cost oxygen dissolves only poorly in the bleaching solution. The process therefore uses pressure of 3–6 bar to increase the solubility. Originally conceived for the high consistency range, the oxygen stage is increasingly using pulp consistencies of 10%–15%. The precondition for application in this consistency range is an effective mixing of the oxygen into the recycled fiber pulp for which high-shear mixers are especially suitable. Usual reaction times are 30–60 min at about 100°C.

A commercial process introduced in 1994 provides for the oxygen bleaching of wood-free pulp as a separate process stage and with conventional bleaching stages. Figure 33 shows the process of the system. The cleaned and flotated medium consistency (10%) deinked pulp is heated by steam to 80°C–90°C and then introduced into a standpipe. With the exception of the oxygen, all chemicals are added before the pressure zone. Then the pulp slurry is pressurized in a MC pump with simultaneous degassing. After the addition of oxygen in a mixer, the deinked pulp is introduced into a pressurized upflow tower in which bleaching occurs. Depending on the optical characteristics required, the DIP is subjected to further deinking or bleaching stages after leaving the tower.

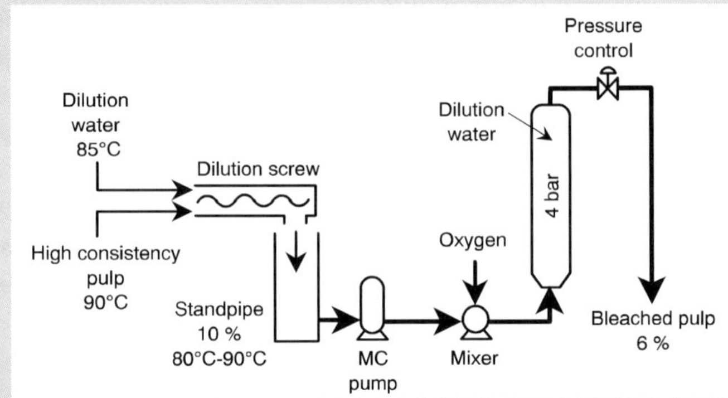

Figure 33. Process for the oxygen bleaching system[57].

To enhance the optical effect, peroxide is also frequently added to the bleaching solution. Figure 34 shows the interactions between the quantities of sodium hydroxide and peroxide and the brightness[56]. For maximum brightness gains with low peroxide applications, only small amounts of sodium hydroxide may be used. With greater peroxide additions, the sensitivity of the brightness gain to the amount added will decrease. Despite this, large alkali applications should be avoided since they have a negative effect on the COD load of the process water. In contrast to the promising results achieved in laboratory scale, this process has not been successfully established in mill scale.

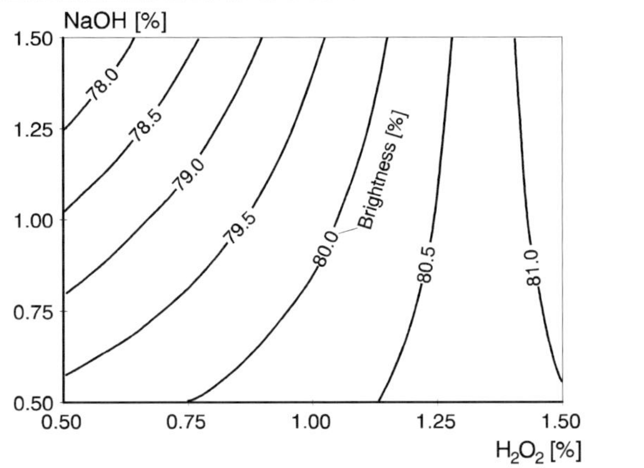

Figure 34. Two-dimensional brightness contour plot for a constant oxygen bleaching operation at 90°C, 10% consistency, 60 psig (4.2 bar) pressure, and 10 min retention time (ISO brightness of unbleached pulp: 72.8%)[56].

Water glass exerts the strongest influence on the bleaching process. Additions of up to 4% considerably improve the brightness but may also cause negative carryover effects to the paper machine especially in integrated mills. Literature reveals different results for the influence of magnesium sulfate. Although no advantages for optimization of the commercial process described above occurred with addition of magnesium sulfate[56], tests on the influence of additives at the oxygen stage indicate a stabilizing effect due to magnesium sulfate[60]. If the treatment uses addition of peroxide, slightly better bleaching results occur when also adding DTPA.

As the oxygen content increases, the optical effect improves. Greater brightness gains and an improved pulp cleanliness occur. Higher oxygen concentrations produce a stronger optical homogeneity especially for pulp mixtures having greatly fluctuating proportions of contaminants. Large oxygen quantities encourage the development of large gas bubbles particularly in medium consistencies. Some pulp is dragged along by their uncontrolled movement so the retention time is difficult to control. To keep the gas bubbles small, the pressure is often increased. The most effective method is to ensure an optimum distribution of the oxygen in the pulp with suitable mixing units. Oxygen levels of 5 kg/t pulp will then be sufficient to provide the desired bleaching effects.

Besides the effects of the technical parameters just described, the success of oxygen bleaching depends considerably on the mechanical fiber content of the recovered paper. Proportions of 10% limit the increase in brightness. If no peroxide is added at the oxygen stage, a decrease in brightness must be expected depending on the amount of sodium hydroxide. Tests carried out by Ackermann *et al.*[12] show that oxygen treatment can cause a dramatic loss in brightness in the bleaching of wood-containing deinked pulp due to yellowing of the mechanical fiber component. Only the addition of peroxide can prevent this brightness loss. Compared with a conventional peroxide bleaching without oxygen, better optical results are not possible.

Süss[61] was unable to determine any additional effects due to oxygen bleaching with the use of mixed office paper. Projects conducted by Patt *et al.*[59] with wood-containing deinked pulp originating from old newspapers and magazines also produced no significantly greater brightness after a peroxide-supported oxygen stage compared with conventional peroxide bleaching. Advantages accruing from the addition of oxygen did occur for the removal of visible dirt specks. When evaluating this bleaching effect, one must note that dispersing effects are also due to the intensive mechanical influences in the high-shear mixer that cause a reduction of visible dirt specks.

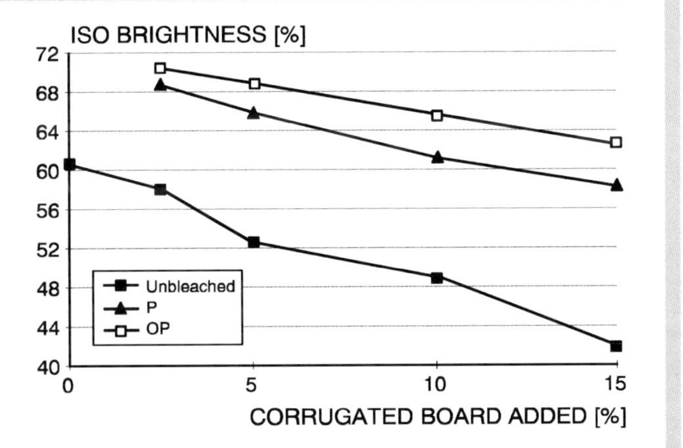

Figure 35. Brightness of unbleached and bleached deinked pulp influenced by increasing proportions of old corrugated board (OCC)[59].

Oxygen treatment represents an effective process step within recovered paper processing especially in cases where unbleached fibers negatively influence the optical cleanliness of wood-free deinked pulp as Fig. 35 shows. With increasing contents of unbleached fibers in recycled pulp, the advantage of the peroxide-supported oxygen stage compared with peroxide bleaching grows in importance[59].

Summary

Table 2 summarizes the characteristics of the oxygen treatment. The table shows that moderate bleaching conditions positively influence the result of the oxygen stage. Low alkali additions with the simultaneous application of peroxide produce the best bleaching results at temperatures of about 80°C and also reduce the COD load that is 5%–20% greater than with a conventional peroxide stage. The brightness gains are usually 4%–8% and in exceptional cases even 12% ISO.

Table 2. Effects of oxygen bleaching on deinked pulps[48].

Bleaching condition	High alkali application High temperature	Low alkali application Low temperature Peroxide addition
Wood-containing deinked pulp	High delignification High COD load Loss of brightness	Low delignification Relatively low COD load Increase of brightness
Wood-free deinked pulp	Delignification High COD load Low brightness increase	Low delignification Relatively low COD load High brightness increase

3.2.2 Ozone bleaching (Z)

Ozone bleaching is a recent development for bleaching recycled fiber pulp. In contrast to its use for chemical pulp bleaching where it serves exclusively to degrade residual lignin further, its application in the bleaching of deinked pulp destroys dyes and optical brighteners[59,62,63]. The strength gains that accompany the treatment have secondary importance. Due to its toxic nature and the high capital and operating costs involved, the use of ozone has not found considerable acceptance to date.

Chemistry

Ozone is a very strong oxidizing agent. Due to its mesomeric character, oxygen can react with different ionic mechanisms. The desired heterolytical decomposition of ozone in which different polyoxide stages occur results in the splitting of C = C bonds and the simultaneous formation of carboxyl and carbonyl groups. As part of this process, oxygen and hydrogen peroxide form in situ. In addition, ozone reacts with aromatic and especially phenolic structures and decomposes these to simple aliphatic acids. These reactions are not selective and last only a few seconds.

A disadvantage is the limited stability of ozone in aqueous systems that leads to homolytical decomposition. The presence of even the smallest amounts of base (pH > 3) or traces of heavy metal or transition metal compounds catalyze the decomposition of ozone. Besides forming hydrogen peroxide and oxygen, the radical reactions also result in an unwanted degrading of carbohydrate structures. Figure 36 shows the complex reaction processes[55].

Besides these reactions, treatment with ozone also negates the effectiveness of various substances contained in the recycled fiber pulp. As a result of its high reactivity towards conjugated bonds, ozone transforms dyes into colorless compounds[64]. Ozone is also the only oxygen-based bleaching chemical able to destroy optical brighteners that enter the recycled fiber pulp primarily via wood-free writing and printing papers[65,66].

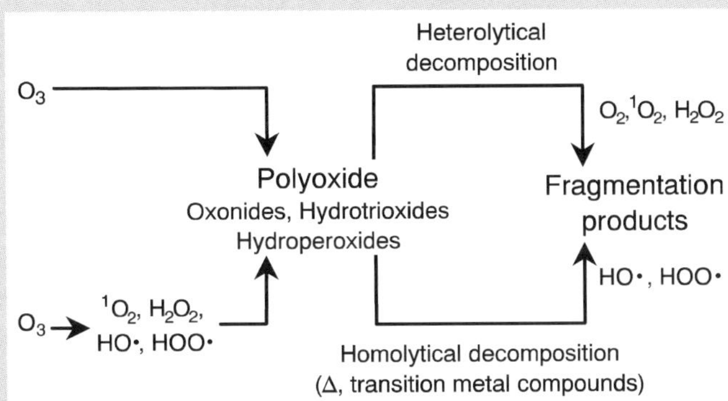

Figure 36. Reactions in ozone bleaching [55].

Process parameters

An ozone treatment uses a separate process stage. Due to the poor solubility of ozone in water, two different techniques have found widespread use.

To ensure that sufficient ozone can be brought to the fibers, bleaching usually uses high consistency conditions. The deinked pulp must therefore be thickened to consistencies of 35%–40% in a two-stage process. The separation of the water loops occurs in the first stage as Fig. 37 shows. A screw press is then used to obtain the consistency required for bleaching. Subsequent fluffing of the thickened pulp ensures the uniform treatment of all pulp components with the added ozone despite the atmospheric process conditions. This also prevents local excess ozone concentrations that lead to undesirable degrading of cellulose. After storage in a tower, thickening occurs again to separate the mostly acidic loop from the remainder of the process.

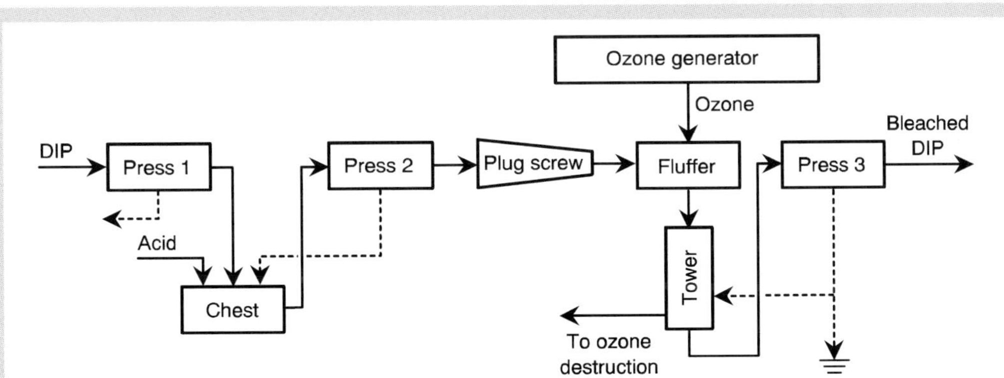

Figure 37. Ozone stage[68].

Alternatively, the treatment can also use the MC range. As in oxygen bleaching, efficient mixing units ensure that the bleaching chemicals quickly contact the fibers. For this purpose, the pulp is fluidized in a high-shear mixer and treated with ozone under approximately 10 bar. Due to the high level of reactivity of the ozone, the duration of the treatment depends on the ozone concentration in the air flow. The technically required reaction time is only a few minutes.

Due to the instability of ozone, the bleaching chemicals can only be produced on-site from oxygen by corona discharge. The produced ozone concentration in the gas can be 6%–12%.

Because of the wide range of possible reactions, the conditions for ozone bleaching have narrow limits. Chemical pulp bleaching occurs at pH <3 without heating at 20°C–60°C. The dosage of ozone is usually 1%. Due to the chosen conditions for ozone bleaching and the resulting unspecific reactions, a pre-treatment with chelating agents, an acidic washing, or both are not essential. The levels usually applied in chemical pulp bleaching can be used only to a limited degree for bleaching deinked pulp. Acidic pH conditions in particular can cause considerable disadvantages compared with alkaline deinking conditions[67]:

- Need to use acid to move the pH into the acidic range
- Increased risk of gypsum deposits
- Calcium carbonate losses
- Forming of secondary stickies due to the pH shock.

When ozone bleaching uses neutral or alkaline pH conditions, achieving a positive bleaching result is still possible. Tests by Kogan and Muguet[66] revealed that the radical reactions of the ozone caused by the increased concentration of OH ions can also lead to color stripping of dyes although less effectively than at a low pH.

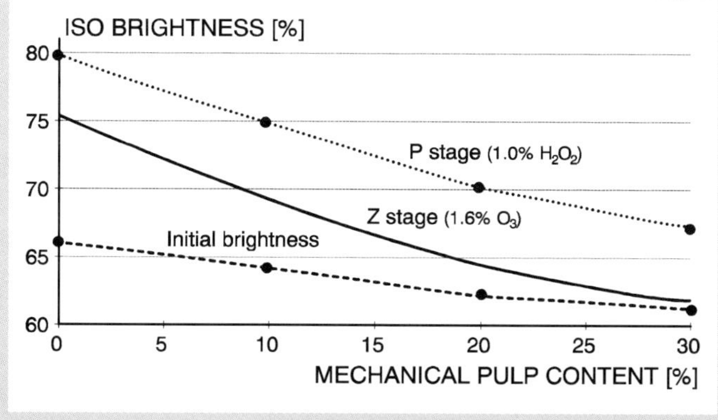

Figure 38. Brightness evolution vs. content of mechanical fibers[66].

Of major importance for the success of ozone bleaching is the composition of the deinked pulp. Proportions of mechanical fibers exceeding 20% prevent any optical gain as Fig. 38 shows. If deinked pulps recovered from old newspapers and magazines is treated with ozone, the higher ozone content can even cause a small loss of brightness as Fig. 39 shows.

A larger proportion of mechanical fibers is beneficial for the development of strength characteristics. As a result of the formation of new carboxyl groups, an increase in fiber bonding accompanies the ozone treatment. This results in a significant increase in strengths such as tensile index and burst index. This effect does not occur with wood-free deinked pulps[66,68,69].

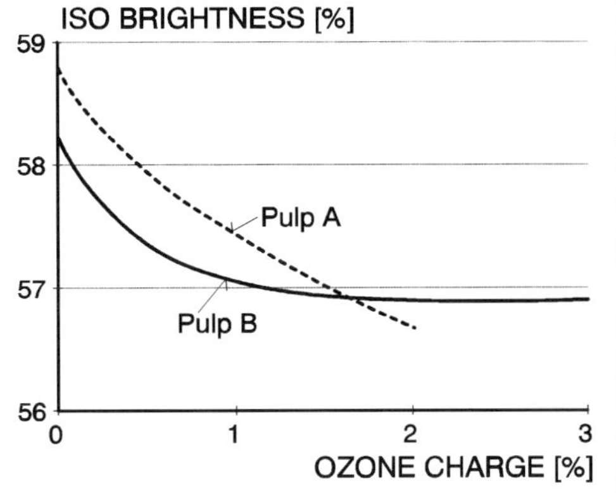

Figure 39. Effect of ozone on brightness of wood-containing grades based on 50% old newspapers and 50% magazines[68].

Besides the efficient color stripping of dyes, ozone treatment is the only oxygen chemical bleaching process that realizes an almost complete destruction of optical brighteners. This is especially important for using deinked pulp in papers destined to contact foodstuffs. For such applications, no chemical substances can migrate to the foodstuffs (bleeding resistance) regardless of whether they represent a potential health risk. Considering the considerable costs involved, this effect may not offer sufficient reason to introduce an ozone stage within recovered paper processing. By using chemical quenchers, achieving a comparable effect is possible in which the fluorescent materials are masked and therefore made ineffective[65].

3.2.3 Alternative bleaching processes

Searches for new, more efficient bleaching processes have provided laboratory results on alternative bleaching chemicals. These primarily include the peroxy acids including peracetic acid and Caro's acid.

Peracetic acid (CH_3COOOH) has a high oxidation potential and is therefore very acidic. Produced from concentrated acetic acid (CH_3COOH) and peroxide (H_2O_2), it exists in a state of equilibrium with acetic acid, peroxide, and water.

$$CH_3COOH + H_2O_2 \leftrightarrow CH_3COOOH + H_2O \tag{16}$$

Acid catalysts accelerate the reaction. Equilibrium peracetic acid is commonly available in concentrations of 5%, 10%, and 15%. Vacuum distillation can produce distilled peracetic acid from equilibrium peracetic acid. This requires storage at 4°C to prevent reversion to a new equilibrium. The price of peracetic acid in the equilibrium form is considerably higher due to its peroxide and acetic acid content. The complicated vacuum distillation requires high investment costs. The costs involved reduce the chances of industrial commercialization of this bleaching technology.

Caro's acid (H_2SO_5) is produced in situ from concentrated sulfuric acid (H_2SO_4) and peroxide. The reaction equilibrium will favor formation of Caro's acid by using highly concentrated starting chemicals. With very good cooling, yields of about 80% are possible.

$$H_2SO_4 + H_2O_2 \leftrightarrow H_2SO_5 + H_2O \tag{17}$$

The reactivity of peracetic acid in an acidic environment is due to the formation of hydroxonium ions (HO^+) that develop from the heterolytical splitting of the oxygen to oxygen bond. The reactions in neutral and alkaline environments are due to the peroxyacetate anion (CH_3COOO^-). This means that bleaching occurs by nucleophilic reactions as with alkaline peroxide bleaching. Like all peroxy compounds, peracetic acid also undergoes homolytical decomposition into acetic acid, peroxide, water, and oxygen. The pH largely determines the reaction processes. The presence of strong bases, heavy metals, or transition metals and increased temperature favor rapid decomposition.

At 60°C–80°C, peracetic acid and Caro's acid react as electrophiles[70]. They hydroxylate aromatic agents to facilitate the nucleophilic oxidative degrading in the following bleaching stages. Like ozone or chlorine and chlorine dioxide, they can destroy optical brighteners[71]. In addition, peracids oxidize alcohols and aldehydes to carbonic acids and can hydroxylate double bonds. This destroys the conjugation in chromophores.

Disadvantages are the acidic reaction conditions that cause gypsum to form and reduce the pulp yield when carbonate is present. The considerable costs for the bleaching agent and the unsuitable pH conditions for deinked pulp are the most important reasons negating industrial application.

References

1. Lachenal, D., Progress in Paper Recycling 4(1):37 (1994).

2. Polcin, J. and Rapson, W.H., Pulp Paper Can. 72(3):T103 (1971).

3. Polcin, J., Zellstoff Papier 22(8):226 (1973).

4. Kubelka, P. and Munk, F., Zeitschrift für technische Physik 32(11a):593 (1931).

5. Helmling, O., Süss, U., and Berndt, W., Wochenbl. Papierfabr. 113(17):657 (1985).

6. Berndt, W., Wochenbl. Papierfabr. 110(15):539 (1982).

7. Putz, H.-J. and Göttsching, L., Wochenbl. Papierfabr. 110(11/12):383 (1982).

8. Schlegel, M., "Neu- und Weiterentwicklung technologischer Verfahren zur Aufbereitung und Veredlung von Altpapierstoffen für grafische Papiere", Ph.D. thesis, Technische Universität Dresden, Dresden, Germany, 1979.

9. Bechstein, G., Das Österreichische Papier 12(4):16 (1975).

10. Süss, H. U., Nimmerfroh, N., Jakob, H., et al., Wochenbl. Papierfabr. 120(8):303 (1992).

11. Beurich, H.G., Wochenbl. Papierfabr. 113(18):689 (1985).

12. Ackermann, C., Putz, H.-J., and Göttsching, L., Das Papier 50(6):320 (1996).

13. Bast-Kemmerer, I. and Salzburger, W., Wochenbl. Papierfabr. 123(23/24):1096 (1995).

14. Aebi, H.E., "Enzymes 1: Oxidoreductases, Transferase", in Methods of Enzymatic Analysis, Vol. III (H.U. Bergmeyer, J. Bergmeyer, and M. Graßl, Eds.), Wiley VCH, Weinheim, Germany,1987, p. 273.

15. Prasad, D.Y., Tappi J. 72(1):135 (1989).

16. Bogner, B., Produktblatt 1-4, Boehringer GmbH, Mannheim, Germany, 1990, 4 pp.

17. Galland, G. and Vernac, Y., "Bleaching of deinked pulp", 2nd Advanced Training Course on Deinking Technology, CTP, Grenoble, 1995, Chap. 20.

18. Ferguson, L.D., Tappi J. 75(7):75 (1992).

19. Salzburger, W., Hartinger, S., Maltzeff, P., et al., Wochenbl. Papierfabr. 124(13):592 (1996).

20. Ortner, H.E., Tappi J. 63(10):83 (1980).

21. De Ceuster, J. and Carman, G., Paper Techn. Industry 18(4):126 (1977).

22. Boriss, P., Bradl, S., and Müller, R., Wochenbl. Papierfabr. 113(17):650 (1985).

23. Gilkey, M.N. and Mark, E.L., "Dispersion of sticky contaminant at medium consistencies", TAPPI 1985 Pulping Conference Proceedings, TAPPI PRESS, Atlanta, p. 567.

24. Siewert, W.H., Wochenbl. Papierfabr. 114(14):536 (1986).

25. Blechschmidt, J. and Ackermann, C., Wochenbl. Papierfabr. 119(17):659 (1991).

26. Sabbatini, M., "Optimization of tissue-paper manufacture from 100% secondary fibers",1983 EUCEPA Symposium Proceedings, EUCEPA, Paris, p. 129.

27. Matzke, W. and Kappel, J., "Present and future bleaching of secondary fibers", TAPPI 1994 Recycling Symposium Proceedings, TAPPI PRESS, Atlanta, p. 325.

28. Sharpe, P.E., "TCF bleaching of mixed office waste composition", TAPPI 1995 Recycling Symposium Proceedings, TAPPI PRESS, Atlanta, p. 157.

29. Galland, G., Bernard, E., and Sauret, G., "Contribution à l' amélioration du désencrage", 1983 EUCEPA Symposium Proceedings, EUCEPA, Paris, p. 167.

30. Bovin, A., "Improved flotation deinking by development of the air mixing chamber", TAPPI 1984 Pulping Conference Proceedings, TAPPI PRESS, Atlanta, p. 37.

31. Putz, H.-J. and Göttsching, L., Wochenbl. Papierfabr. 113(11):382 (1985).

32. Ackermann, C., Putz, H.-J., and Göttsching, L., Wochenbl. Papierfabr. 120(11/12):433 (1992).

33. Helmling, O., Süss, U., and Eul, W., "Upgrading of wastepaper with hydrogen peroxide", TAPPI 1986 Pulping Conference Proceedings, TAPPI PRESS, Atlanta, p. 407.

34. Steenberg, E. and Norberg, G., Deutsche Papierwirtschaft (3):111 (1977).

35. Schuster, G., Das Papier 32(7):299 (1978).

36. Auhorn, W. and Melzer, J., Wochenbl. Papierfabr. 110(8):255 (1982).

37. Melzer, J., "Kinetics of bleaching mechanical pulps by sodium dithionite", 1985 International Pulp Bleaching Conference Proceedings, CPPA, Montreal, p. 69.

38. Putz, H.-J., "Upcycling von Altpapier für den Einsatz in höherwertigen graphischen Papieren durch chemisch-mechanische Aufbereitung", Ph.D. thesis, Darmstadt University of Technology, Darmstadt, Germany, 1987, 202 pp.

39. Putz, H.-J. and Göttsching, L., "Die Bleiche von deinktem Altpapier", 1984 EUCEPA Conference Proceedings, EUCEPA, Paris, p. 279.

40. Nikki, M., Wochenbl. Papierfabr. 113(21):819 (1985).

41. Melzer, J., Tippling, P., and Jokio, P., Wochenbl. Papierfabr. 113(18):684 (1985).

42. Fluet, A., Dumont, I., and Béliveau, D., Pulp Paper Can. 95(8):37(1994).

43. Süss, H.-U. and Krüger, H., German Pat. DE 3,309,956 C1 (March 19, 1983).

44. Kronis, J.D., "Adding some color to your wastepaper furnish", TAPPI 1992 Pulping Conference Proceedings, TAPPI PRESS, Atlanta, Vol. 1, p. 223.

45. Daneault, C. and Leduc, C., Cellulose Chemistry Technology 28(2):205 (1994).

46. Faul, A.M. and Eisenschmid, K., Wochenbl. Papierfabr. 119(1):1 (1991).

47. Hache, M. and Joachimides, T., "The influence of bleaching on color in deinked pulps", TAPPI 1991 Pulping Conference Proceedings, TAPPI PRESS, Atlanta, p. 801.

48. Gehr, V. and Borschke, D., Wochenbl. Papierfabr. 124(21):929 (1996).

49. Galland, G., Bernard, E., and Vernac, Y., "Improvement of waste paper pulp bleaching", Commission of the European Communities, European Seminar Proceedings, CTP, Grenoble, 1991, p. 205.

50. Süss, H.U., Nimmerfroh, N., and Hopf, B., Wochenbl. Papierfabr. 122(20):802 (1994).

51. Heimburger, S.A. and Meng, T.Y., Pulp & Paper 66(2):139 (1992).

52. Quinnett, P.E. and Ward, L.R., "Chlorine dioxide for decolorizing deinked mixed office waste", TAPPI 1995 Recycling Symposium Proceedings, TAPPI PRESS, Atlanta, p. 149.

53. de Ruvo, A., Farnstrand, P.A., Hagen, N., et al., Tappi J. 69(6):100 (1986).

54. Markham, L.D. and Courchene, C.E., Tappi J. 71(12):168 (1988).

55. Gratzl, J.S., Das Papier 46(10A):V1 (1992).

56. Thomas, C.D., Hristofas, K., Yee, T.F., et al., Progress in Paper Recycling 5(1):37 (1995).

57. Thomas, C.D., Hristofas, K., Yee, T.F., et al., Progress in Paper Recycling 5(1):45 (1995).

58. Magnotta, V.L. and Elton, E.F., U.S. Pat. No. 4,416,727 (Nov. 22, 1983).

59. Patt, R., Gehr, V., Matzke, W., et al., Tappi J. 79(12):143 (1996).

60. Pauli, C., Patt, R., Gehr, V., et al., Wochenbl. Papierfabr. 121(20):852 (1993).

61. Süss, H.U., "Bleaching of waste paper pulp – changes and limitations", 1995 PIRA International Wastepaper Technology Conference Proceedings, Leatherhead, UK, paper no. 9.

62. Kogan, J. and Muguet, M., Progress in Paper Recycling 1(1):37 (1992).

63. Muguet, M. and Kogan, J., Tappi J. 76(11):141 (1993).

64. Karp, B.E. and Trozenski, R.M., "Non-chlorine bleaching alternatives: a comparison between ozone and sodium hypochlorite bleaching of colored paper", 1996 International Pulp Bleaching Conference Proceedings, TAPPI PRESS, Atlanta, p. 425.

65. Roy, B.P., Progress in Paper Recycling 4(1):74 (1994).

66. *Kogan, J. and Muguet, M., "Ozone bleaching of deinked pulp", TAPPI 1994 Recycling Symposium Proceedings, TAPPI PRESS, Atlanta, p. 237.*

67. *Muguet, M. and Sundar, M., "Ozone bleaching of secondary fibers", 1996 International Non-chlorine Bleaching Conference Proceedings, TAPPI PRESS, Atlanta, paper no. 8-1.*

68. *Kappel, J. and Matzke, W., "Chlorine-free bleaching chemicals for recycled fibers", TAPPI 1994 Recycling Symposium Proceedings, TAPPI PRESS, Atlanta, p. 231.*

69. *van Lierop, B. and Liebergott, N., J. Pulp Paper Sci. 20(7):J206 (1994).*

70. *Gierer, J., Holzforschung 36(2):55 (1982).*

71. *Dubreuil, M., Progress in Paper Recycling 4(4):98 (1995).*

CHAPTER 10

Papermaking potential of recycled fiber

CHAPTER 10

CHAPTER 10

Christiane Ackermann, Lothar Göttsching, Heikki Pakarinen

Papermaking potential of recycled fiber

1 Differences between virgin and recycled fibers

1.1 Introduction

The goal of a recycled paper or board manufacturer is to make a product that meets customers' specifications and requirements. At the present utilization rate, using recycled fibers (RCF) in commodity grades such as newsprint and packaging paper and board has not caused noticeable deterioration in product quality and performance. The expected increase in recovery rates of used paper products will require a considerable consumption increase of RCF in higher quality grades such as office paper and magazine paper. To promote expanded use of recovered paper, understanding the fundamental nature of RCF and the differences from virgin fibers is necessary.

To identify changes in fiber properties, many recycling studies have occurred at laboratory and pilot scales. Realistically repeating all the stages of the recycling chain is difficult especially when including printing and deinking. Some insight into changes in fiber structure, cell wall properties, and bonding ability is possible from investigations using various recycling procedures, testing methods, and furnishes.

Because mechanical pulps differ from chemical pulps chemically and physically, recycling effects on those furnishes also differ. When chemical fibers undergo repeated drying and rewetting, they are "hornified" and can significantly lose their originally high bonding potential. In contrast to the chemical pulps, originally weaker mechanical pulps do not deteriorate but somewhat even improve bonding potential during a corresponding treatment. Several studies[1–6] have shown good recyclability of mechanical fibers.

1.2 Effects of recycling on chemistry and morphology of fibers

1.2.1 Hornification of fibers during drying and recycling

A challenging aspect of pulp research has been to understand the cause of loss in the bonding strength of dried chemical fibers because mechanical fibers do not exhibit the same problem. Various hypotheses exist on causes of hornification. More detail is available in the volume *"Forest Products Chemistry"* of this textbook series. Jayme[7] introduced the colorful expression "hornification" in 1944 and quantified the phenomenon as the percentage reduction of water retention value (WRV) in pulp determined by centrifugation as Eq.1 shows:

$$Hornification = \frac{WRV_0 - WRV_1}{WRV_0} 100[\%] \tag{1}$$

where WRV_0 is the value of virgin pulp
WRV_1 the value of recycled pulp after drying and reslushing.

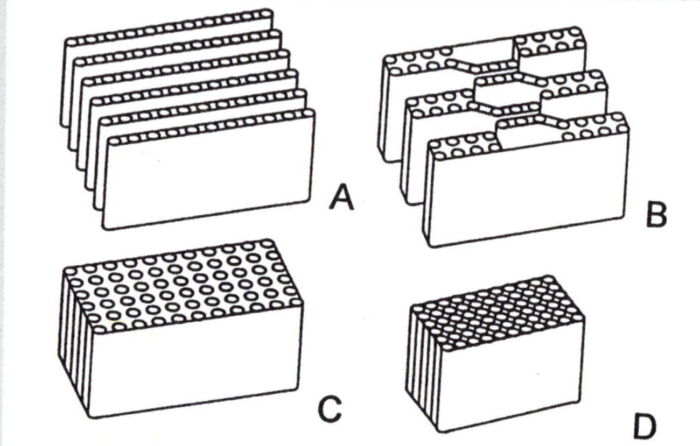

Figure 1. Changes in fiber wall structure[12].

According to the prevailing concept, hornification occurs in the cell wall matrix of chemical fibers[8]. During drying, delaminated parts of the fiber wall, i.e., cellulose microfibrils, become attached as Fig. 1 shows. Hydrogen bonds between those lamellae also form. Reorientation and better alignment of microfibrils also occur. All this causes an intensely bonded structure. In a subsequent reslushing in water, the fiber cell wall microstructure remains more resistant to delaminating forces because some hydrogen bonds do not reopen. The entire fiber is stiffer and more brittle[9,10]. According to some recent studies[2,11], hornification does not increase the crystallinity of cellulose or the degree of order in the hemicelluloses of the fiber wall. Earlier theories of hornification had used these ideas.

Weise and Paulapuro[12] did very revealing work about the events during fiber drying. They studied fiber cross sections of kraft pulp fibers in various solids by confocal laser scanning microscopy (CLSM) and simultaneously measured hornification with WRV tests. Irreversible hornification of fibers began above 30%–35% solids content and continued to 70%–80% solids content depending on the degree of beating. Hornification does not directly follow shrinkage since the greatest shrinkage of fibers occurs above 80% solids content. Figures 1 and 2 show the changes in fibers during different drying stages schematically.

In Figs. 1 and 2, stage A represents the wet kraft fiber before drying. In stage B, the drainage has started to cause morphological changes in the fiber wall matrix at about 30% solids content. The fiber wall lamellae start to approach each other because of capillary forces. During this stage, the lumen can collapse. With additional drying, spaces between lamellae continue shrinking to phase C where most free voids in the lamellar structure of the cell wall have already closed. Fiber shrinkage has exclusively occurred at right angles to the lamellar layers, and the fiber walls are thinner. Figure 2 shows that the width of the fiber has not yet changed. Toward the end of drying in stage D, the water removal occurs in the fine structure of the fiber wall, i.e., in its amorphous sections. Kraft fibers shrink strongly and uniformly during this final phase of drying, i.e., at solids contents above 75%–80%.

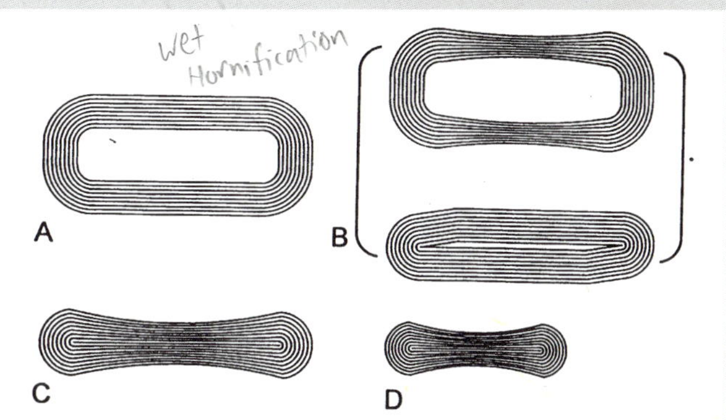

Figure 2. Shrinkage of a fiber cross section.

The shrinkage of stage D is reversible. Rewetting kraft fiber restores its swelling. Hornification has occurred in stages B and C. The original dimensions of the fresh fiber in a swollen state are not completely attained again by rewetting. The spaces between microfibrils in the lamellar structure of a cell wall do not fully open by reslushing or beating. They remain partly inaccessible to water.

Hornification already begins at solids levels typical on the wet press section of a paper machine. Hornification and its extent increase by beating chemical pulp. Intensity of drying also increases the hornification of fibers[13]. Weise [6] distinguishes "wet" hornification described above and "dry" hornification, i.e., the additional hornification that occurs by oven drying of kraft pulp sheet at 105°C to 100% solids content. In practical papermaking, properties of hornified low-yield pulps can often be restored by beating or refining. A later section of this chapter will discuss the role of pH and other chemical conditions in hornification and reswelling.

Scallan and Tigerstrom[8] observed that the elastic modulus of wet fibers of bleached kraft pulp can double during recycling. This loss of flexibility is evident by a drop of beating degree (Schopper-Riegler value) or an increase of freeness (Canadian Standard Freeness) in low-yield pulps. In recycling studies with high yield pulps[1,3,4], freeness changes have been very inconsistent. The changes are sometimes due to losses in fines content during experiments. The contradictory results can also depend on the specific energy input in the refining of the original thermomechanical pulp (TMP)[14]. The changes in pulp freeness are therefore not unique indications of hornification.

The different hornification propensity of different pulp types is most obvious in water retention measurements. For example, Korpela's[3] recycling experiments including reslushing with deinking chemicals and calendering of sheets show the typical behavior of various pulp types. Figures 3 and 4 show that WRV remained almost constant in mechanical pulps at an initially lower value but decreased considerably from much higher levels in softwood and hardwood chemical pulps particularly during the first cycle.

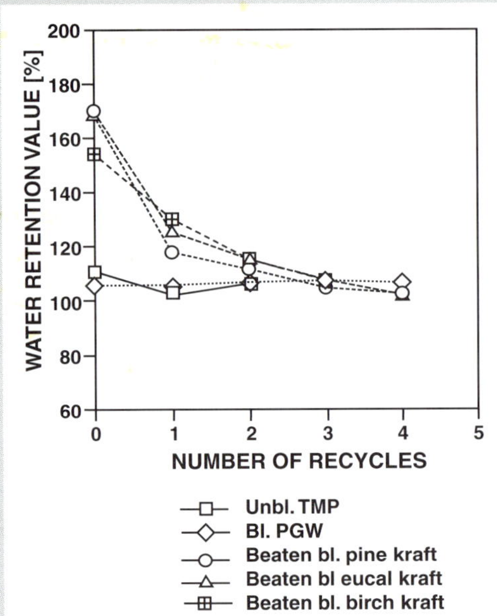

Figure 3. Effect of recycling on WRV of various mechanical and kraft pulps[3].

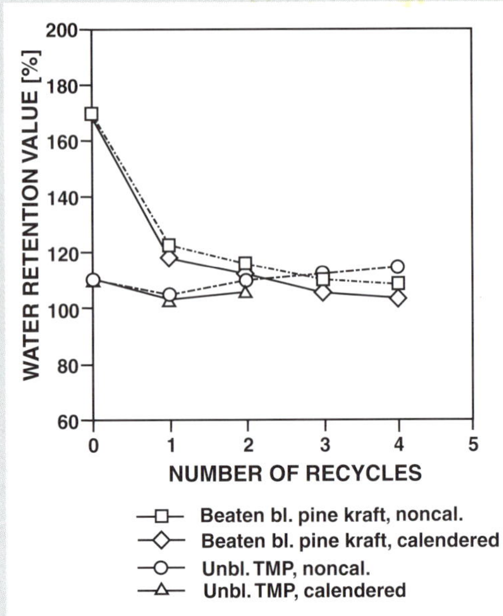

Figure 4. Effect of recycling on WRV of TMP and kraft pulp, as calendered and noncalendered sheets[3].

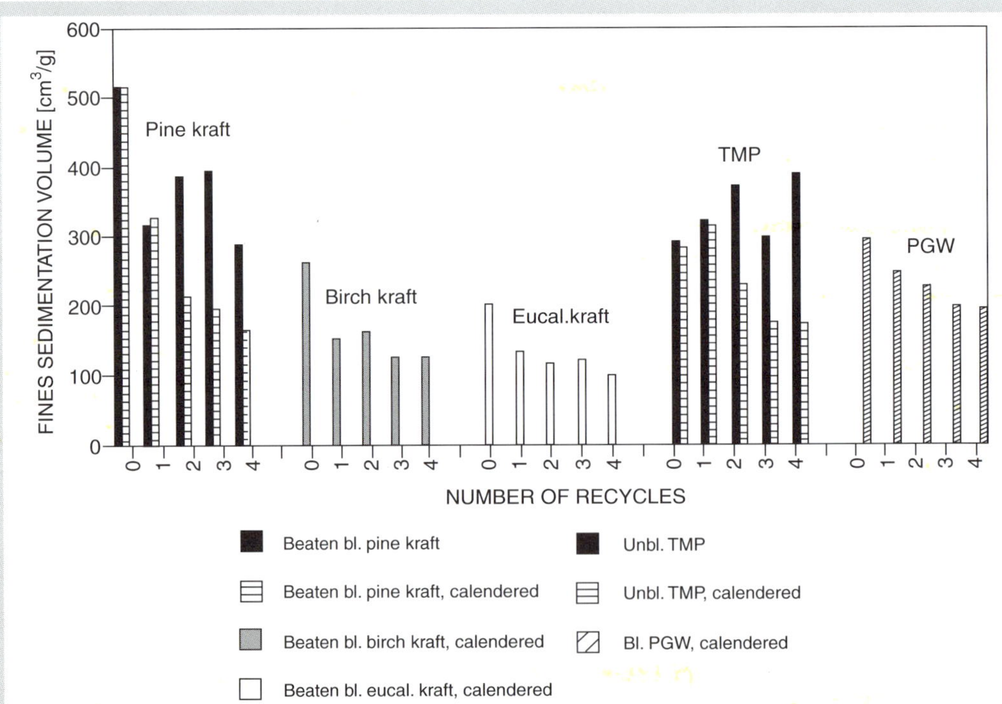

Figure 5. Effect of recycling on fines sedimentation volume[3].

The hornification of pulps also occurs in the fines fraction. Besides the water retention values of the whole pulps in the study quoted above[3], the effect of recycling on the sedimentation volume of pulp fines was also investigated as an indication of probable hornification or other morphological changes. When using Brecht's traditional classification for wood based fines[15], the sedimentation volume of highly fibrillated "slime stuff" types of fines is theoretically greater than that of "flour stuff" that has weak bonding ability. The sedimentation volume of pulp fines generally decreases during repeated recycling experiments. Figure 5 shows that the sedimentation volume of all chemical pulp fines decreased proportionally more than that of mechanical fines. This indicates hornification in chemical pulp fines. A morphological explanation could be that outer fibrils in fines particles are irreversibly twisted or worn out during recycling.

The results in Fig. 5 with TMP and kraft pulps also show that a hard nip laboratory calender produces fines or parts of fibers that sediment more rapidly than original fines. In practice, appropriate refining of recycled fibers can produce new "secondary" fines by peeling fiber walls. These have equal bonding ability to virgin fines as a subsequent section of this chapter will discuss.

A loosening or compacting of fiber structures can be examined to some degree with Simon's stain method[16]. Indication of probable chemical changes caused by hornification is difficult. The viscosity of chemical pulp dissolved in cupriethylendiamine (CED) solution is considered to be a measure of degradation of cellulosic material and an indication of its general strength potential[17]. Hornification decreases CED viscosity[18]. This could also be a measure of hornification. To conform this possibility, Alanko[1] examined it for TMP and kraft pulps recycled with deinking chemicals. The CED viscosity of beaten bleached kraft pulp decreased significantly[1]. The viscosity of TMP increased slightly during recycling in the same conditions. CED does not completely dissolve the lignin of mechanical pulps. That is why the viscosity behavior of pulps was different and perhaps not fully comparable. The directions of changes show that events such as occurred in the cellulosic structures of kraft pulp did not occur in TMP.

The main characteristic of the "milder" hornification in mechanical fibers described in the studies above is that they can significantly retain the ability to take up water after drying. Lower yield chemical fibers cannot. According to Scallan[8], lignin is a cross-linking agent in the lamellar microfibril structure within the cell. Hemicellulose preferentially coats the lamellae and acts as a "coupling" agent between lignin and cellulose microfibrils. This lignin-hemicellulose gel prevents formation of hydrogen bonds between microfibrils during drying and keeps the spaces between lamellae accessible for water. The pulping process of chemical fibers reduces the amount of the lignin-hemicellulose gel that makes the irreversible hydrogen bonding between microfibrils possible. In mechanical pulping, this gel primarily remains intact.

How much the lignin-hemicellulose material can be removed before the hornification phenomena appear is an interesting question. In the studies of Law[19], chemimechanical pulps with different chemical treatments and yields were recycled and strength properties were determined. A decrease of tensile strength was considered an indication of hornification. Depending on the wood species somewhere in the yield range of 80%–90%, the strength abruptly began to decline with recycling. A subsequent section of this chapter describes this study in more detail.

1.2.2 Changes in fiber morphology and flexibility

Changes in bonding potential are sometimes assumed to be due to changes in ratio of pulp constituents. To examine this relationship, Bouchard[11] undertook recycling experiments that included chemical and mechanical fibers with and without deinking chemicals. No direct relationship appeared between the removal of hemicelluloses or lignin and the loss or gain of strength properties during recycling. Infrared (IR) spectra of fibers also showed no significant chemical changes. Sodium hydroxide treatment during reslushing causes an increase in COD of process waters in deinking plants. Analyses of the repulping filtrates indicate that only negligible quantities of carbohydrates and lignin dissolve during recycling[20]. This corresponds to about 0.2%–1.0% yield loss for TMP and 0.1%–0.3% loss for kraft pulps. Such small losses are unlikely to have any effects on swelling or strength properties. Changes in bonding potential are therefore probably due to physical factors such as changes in fiber flexibility.

Effects of recycling on fiber length, coarseness, and distribution of fiber fractions have usually been negligible[1,3,21,22] when calendering and beating are excluded in experiments. Calendering can cause reduction in fiber length of all pulps depending on its type (soft or hard nip). As stated earlier[8], elastic modulus of wet chemical fibers doubled during drying and rewetting. This phenomenon has not been found or verified in a dry state. The specific strength and elastic modulus of dry fiber is usually indicated by the zero-span tensile test. Typical recycling experiments result in some loss of specific dry strength with chemical fibers by that test but show no effect or slight increase in that of high yield fibers[1,3].

Internal and external fibrillation physically influence the bonding ability of fibers. External fibrillation of beaten chemical pulp was quantitatively studied by image analysis[21] in a recycling study with and without use of deinking chemicals. The ratio of rough outline to smooth edge length of a fiber was the fibrillation index. This fibrillation factor decreased 20%–25% during four cycles. Similar loss of external fibrillation on TMP fibers was also found by visual inspection with confocal laser scanning microscopy[1]. Outer fibrils on the initial TMP fibers were long and curly. On the recycled fibers, those fibrils were partly removed. This kind of wear and tear of external fibrillation should also negatively influence bonding potential of recycled mechanical fibers. Other mechanisms compensate for these negative effects.

Internal fibrillation improves the flexibility and conformability of fibers. Flexibility or stiffness measurements can therefore be indirect methods for quantifying internal fibrillation of an individual fiber. In Alanko's recycling experiments[1], fiber flexibility measurements used the hydrodynamic method of Kerekes[23]. Measurements were made for long fiber fraction (18 mesh fraction) of TMP and kraft pulps that were recycled with deinking chemicals. Fifty fibers of each pulp were tested. Both TMP and kraft pulp showed an increase in flexibility. That of TMP almost doubled during four cycles. Microscopy also showed qualitatively that TMP fibers flattened in recycling. With the hydrodynamic method, calculated results assumed that fibers bent by water flow are round. The method therefore cannot distinguish between changes in geometry and changes in flexibility of fiber wall material. Although in this study kraft fibers also showed increased flexibility, the results do not necessarily mean that they start to bond better and improve sheet strength. In microscopic inspections, an increase of external damage was found.

Compressions, twists, and kinks occured in those recycled kraft fibers for which the hydrodynamic method of measuring the mechanical bending of a fiber is very sensitive.

The flattening of mechanical pulp fibers during recycling has been quantitatively found in some investigations[4,24]. Confocal laser scanning microscope (CLSM) is a powerful technique to quantify changes in fiber transverse dimensions. One study[24] examined three softwood pulps: a TMP, a chemithermomechanical pulp (CTMP), and a beaten bleached kraft pulp (HS). Fibers were recycled once and five times that only involved reslushing without chemicals, sheet making, and drying (shrinking restrained). Fiber cross-sectional images were generated by CLSM and the changes quantified by image analysis. Figure 6 shows the fiber transverse dimensions such as wall cross-sectional area, cell wall thickness, lumen area, and aspect ratio (D_{min}/D_{max}).

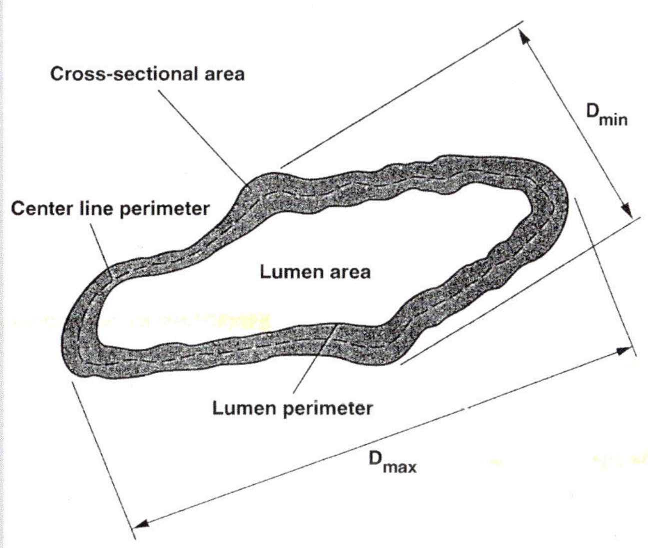

Figure 6. Cross-sectional dimensions of a fiber in analysis by CLSM[24].

All the parameters decreased between the first and the fifth cycle. Table 1 presents relative changes for each pulp. The values used are based on measurements made on about 500 dry fibers. For all pulps in this investigation, the decrease in wall cross-sectional area was about 8%.

Table 1. Relative reductions in cross dimensions between one and five times recycled fibers[24].

Parameters	TMP [%]	CTMP [%]	Kraft pulp [%]
Cross-sectional area	8.3	7.9	7.7
Wall thickness	7.9	5.0	3.0
Lumen area	14.0	31.0	0
Aspect ratio	9.6	9.6	3.9

The main observation in this CSLM study is that mechanical fibers became flatter during the recycling. Chemical fibers were almost all collapsed and flat already after the first cycle. No lumen area reduction therefore exists for kraft fibers in Table 1. Wall thickness reduction occurred in all fiber types although fiber coarseness measurements (mg/m) did not show any significant changes.

Two mechanisms can cause reduction in lumen area (flattening) of mechanical fibers upon recycling: decrease in rigidity of cell wall material by internal fibrillation or geometric decrease in fiber wall thickness. Table 1 has already shown the reduction in cross-sectional wall area of fibers. This same study[24] also plotted the lumen area of TMP fibers against wall thickness as Fig. 7 shows. Since the decrease in lumen area occurred in all classes of fiber thicknesses (in thin earlywood and thick latewood fibers), this suggests that the recycling has decreased the rigidity of the fiber wall and supports the first of the two mechanisms mentioned above.

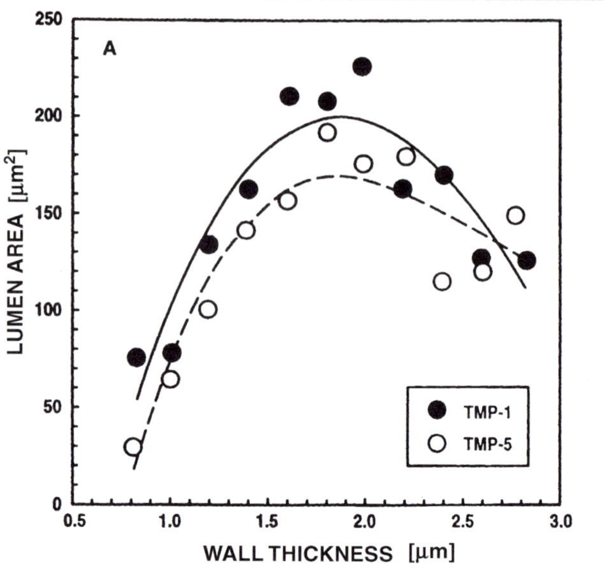

Figure 7. Lumen area vs. wall thickness for TMP fibers recycled one and five times[24].

This internal fibrillation and consequent flattening of mechanical fibers was the result of broke-like recycling treatment without refining or calendering. The forces that are probably adequate to soften fiber cell material during such a light treatment can be the compressive ones produced by imposed shrinkage in fiber bonds during drying. The shrinkage forces induced by surface tension of water can be enormous at 100–1 000 bar when the drying reaches nanometer size pores in the amorphous microstructures of the cell wall[25]. Fibers shrink about ten times more in their transverse direction than in the axial direction. If two fibers contact at right angle, the radially shrinking fiber will compress the other one into axial microcompressions in the bonded area. These microcompressions can manifest themselves in internal fibrillation of the cell wall depending on fiber type. In low-yield pulps for example, these compressive forces generated during the end stage of drying cannot obviously open hydrogen bonds formed within fibers during the earlier hornification stage. Microcompressions and their formation in fiber bonds are difficult to detect with conventional microscopy technology or other quantification methods.

In practice, appropriate beating of recycled fibers affects their morphology and can improve their bonding potential by increasing fines content and internal and external fibrillation. Possibilities to overcome the effects of hornification of chemical fibers by refining is the subject in the section of this chapter on papermaking. Recycled mechanical fibers also seem to have bonding potential that beating can enhance. In particular, the cell wall thickness of the long fiber fraction decreases with high-consistency refining. This implies that some fibrillar material from the outer cell wall layer peels off[22]. This effect occurred more in thicker latewood fibers than in thinner earlywood fibers. In some recycling investigations, high temperature in calendering conditions also increased permanent flattening of mechanical fibers[26]. Freezing of single fibers also brings very similar effects to drying[16].

1.2.3 Effects of chemical conditions on fiber flexibility

The increase in mechanical fiber flexibility presented above did not use any chemicals. Sodium hydroxide has frequent use in repulping of recovered paper to enhance defiberizing and ink detachment. Several studies have shown that alkaline NaOH treatment contributes to fiber swelling by ion-exchange reactions[1,8,20,27,28]. Acidic groups within the fiber wall, e.g., carboxylics in polysacharides, dissociate. Fiber swelling increases in response to the osmotic pressure generated by counter cations present in the solution (Donnan effect). The level of swelling increases in cation order: $Al^{3+} < H^+ < Mg^{2+} < Ca^{2+} < Li^+ \leq Na^+$.

In conventional papermaking, the most common cations are hydrogen, calcium, or aluminum. To some extent, alkali treatment of deinking replaces those ions by sodium that contributes to fiber swelling in the same proportion. Gurnagul[20] examined the response of different fiber types to alkaline treatment. In his experiments, TMP and low-yield, unbeaten kraft pulps (bleached and unbleached) were pretreated with hydrochloric acid (HCl) and sodium chloride (NaCl) before drying. In other words, the pulps were ion-exchanged to the hydrogen and sodium form. The level of fiber swelling was determined by the fiber saturation point method. The acidic group contents of pulps were tested by conductometric titration with simple inorganic electrolytes that have access to the bulk of the fiber wall. This measurement was made only for hydrogen form pulps. Table 2 shows the results.

Table 2. Influence of alkaline recycling treatment on swelling level and acidic group content in different pulp types pretreated to hydrogen and sodium form[20].

Pulp type	Sodium hydroxide content [%]	Fiber saturation [g water/g pulp]		Acidic groups [mmol/kg pulp]
		Na form	H form	H form
TMP				
Never dried	–	0.89	0.78	98
Dried and rewetted	0.061	0.96	0.82	163
Dried and rewetted	0.176	1.00	0.84	178
UNBL. KRAFT				
Never dried	–	1.42	1.33	85
Dried and rewetted	0.061	1.19	1.07	92
Dried and rewetted	0.176	1.22	1.09	91
BL. KRAFT				
Never dried	–	1.33	1.32	31
Dried and rewetted	0.061	1.05	1.02	30
Dried and rewetted	0.176	1.03	0.97	30

As expected, alkaline treatment produced an increase of swelling in TMP fibers. The swelling remained smaller in hydrogen form TMP than in sodium form TMP. Fewer sodium ions were present within the fiber wall in the alkaline treatment so that osmotic swelling effect was smaller.

In both kraft pulps, Table 2 clearly shows the hornification effect. The difference in the level of swelling between sodium and hydrogen form pulps is significant in the case of unbleached kraft but negligible with bleached kraft pulp because of the low content of charged acidic groups in this pulp.

In theory, these differences in charge content between various chemical pulp grades should be recognized for their hornification propensity. For example, hardwood kraft fibers have a considerably higher carboxyl acid content than softwood kraft fibers and should be more capable to avoid hornification when dried in sodium form[29]. In recycling studies[3,28,30], the results have not been consistent in this respect probably because of differences in the design of experiments. With low-yield recycled fibers, this effect of osmotic pressure caused by sodium ion probably cannot open a hornified fiber wall matrix without mechanical refining action.

Deinking and peroxide bleaching trials[31,32] in which caustic soda was replaced by ammonium or divalent cation (Ca^{2+} and Mg^{2+}) hydroxides have verified the general importance of the metal ion order in cell wall swelling. Despite high pH during reslushing of old newspapers and magazines, strength and density of pulp sheets remained lower and freeness was higher than with normal sodium hydroxide treatment.

1.2.4 Changes in surface chemistry of fibers

Besides influencing fiber swelling, the charge of fibers is important for adsorption of wet end chemicals, drainage on wire and press sections, and fiber bond formation during drying. When the amounts of dissociated carboxyl groups increase, the surfaces of fibers become more hydrophilic. When determining acidic group content in fiber surfaces of pulps, the molar mass of the titration electrolyte used is also important. Higher molecular weight polymers have less access to the wall of fibers particularly in the case of hornified recycled pulp. In the recycling experiments of Alanko[1] for instance, carboxyl group contents of TMP and beaten, bleached kraft pulp were measured by polyelectrolyte titration with a cationic polymer, Polybren, which has a small molar mass (~ 8 000). Figures 8 and 9 show changes after four cycles with different deinking chemistry.

When TMP was recycled four times in pure water, virtually no change to the fresh pulp carboxyl group content (initial point) occurred. The use of alkali caused an increase, but the highest carboxyl content on TMP fibers occurred when hydrogen peroxide was part of the treatments as Fig 8 shows. Stable metasaccharic acid end groups that contribute to the acid group content have probably replaced reducing end groups of polysaccharide chains. Alkaline saponification of ester and lactone derivates in TMP can also increase the number of carboxyl groups[1,27].

Alkali and peroxide did not increase the charge of bleached kraft in this study as Fig. 9 shows. The general charge level of bleached kraft is lower than that in mechanical pulp due to the removal of hemicelluloses during chemical pulping and bleaching as Table 2 also shows.

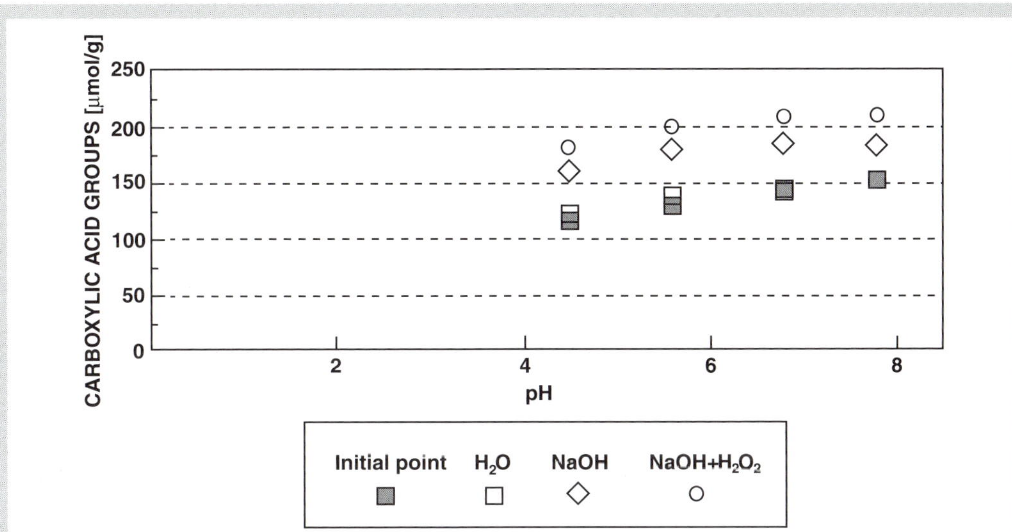

Figure 8. Carboxyl acid group content in long fiber fraction(30 mesh fraction) of fresh TMP and TMP recycled with different chemistry[1].

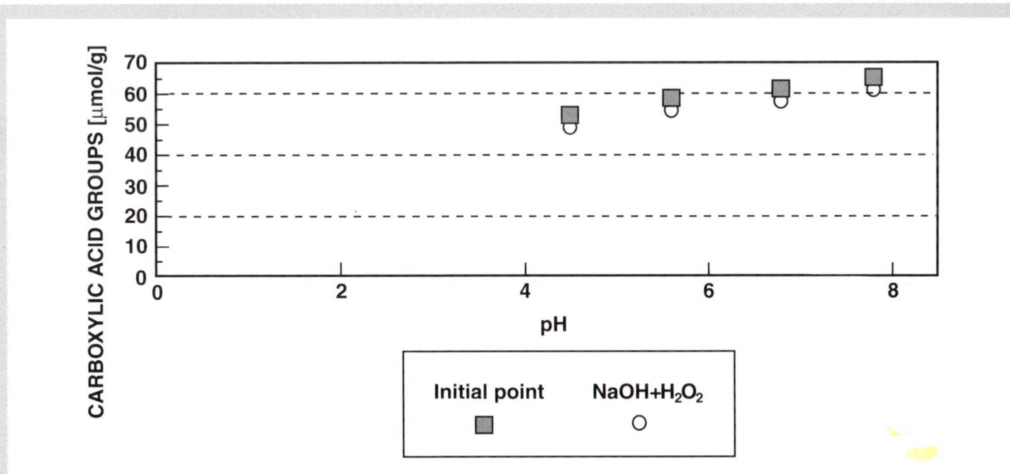

Figure 9. Carboxyl group content in fresh and recycled bleached kraft pulp with deinking chemicals[1].

Surfaces of TMP sheets in Alanko's study[1] were also analyzed by electron spectroscopy chemical analysis (ESCA). This technique indicates the ratio of carbon and oxygen and composition of bonds between carbon and oxygen in the outermost 5–10 nanometers of fibers. Using a curve-fitting program, Gaussian curves were fitted for differently bound carbon (classified into four categories) from which the total oxygen to carbon atomic ratio can also be determined. Chemical changes of TMP surfaces cycled with a blend of NaOH and H_2O_2 were followed by ESCA analysis as Fig. 10 shows.

Carbon spectra show that the number of aliphatic carbon to carbon bonds represented by the outermost right peaks in the pictures decreased markedly after the first cycle. This indicates removal of wood extractives from the TMP fiber surface due to the first alkali treatment. The general ratio of oxygen to carbon in the spectra increased during subsequent recyclings. Theoretical formulas give an oxygen to carbon ratio of 0.83 for cellulose and 0.33 for lignin. In this study[1], the oxygen to carbon ratio of 0.66 occurred with the fifth cycle. This increase can be explained by the removal not only of extractives but also of some lignin from the surface of fibers. The increase of carbon to oxygen bonds in Fig. 10 also indicates an increase of the cellulose content and a decrease of the lignin content.

The oxidizing effect of peroxide on lignin or carbohydrates can probably contribute to the increase of the oxygen to carbon ratio. Similar ESCA results of changes in surface chemistry resulted in another investigation[33] in which a reduction of extractive content was also confirmed in alkaline deinking.

The chemistry of the fiber surface and especially the content of extractives have much practical importance when considering bonding strength, wetting of dry sheets, or

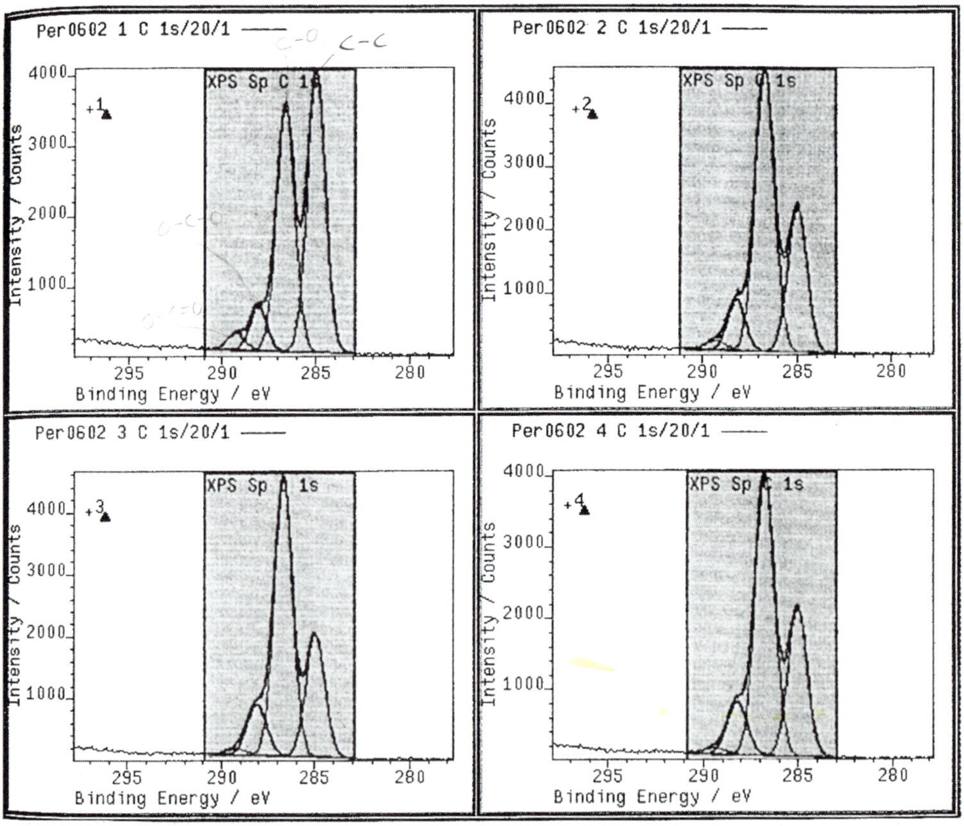

Figure 10. ESCA carbon spectra of TMP for the first, second, third, and fourth cycles. The peaks from right to left in the separate figures indicate bonds of C - C, C - O, O - C - O, and O - C = O.

friction of final paper. For example, an assumption is that films of lipophilic extractives on surface layers interfere with the formation of interfiber bonds[34,35]. Surface energy studies made by measuring the contact angle of water on individual fibers have been very contradictory due to the heterogeneity of recycling procedures and recycled fibers[5, 33, 36]. When measuring the wood extractive content in commercial paper grades, an obvious difference has consistently occurred between newsprint made of deinked pulp and that made of virgin mechanical fibers[37]. Most likely, the dissolving effect of regular alkaline deinking conditions has been the main factor to diminish the extractives content in RCF-containing paper.

With low-yield pulps, chemical pulping or bleaching processes already have very effectively reduced extractives and lignin from fiber surfaces. The changes in surface energy of chemical pulps during recycling are therefore negligible or at least uncertain[5,36].

1.3 Effects of recycling on pulp sheet properties

In the previous survey of the morphological and chemical changes that fibers undergo during recycling treatments, apparent hints existed on effects on pulp sheet properties. One particularly concerned the development of bonding ability that is the most crucial pulp property in this respect. Very contradictory mechanisms affect this property depending on the recycling procedures and the yields of pulps used. Although recycled pulps are often blends of different fibers and pigments, the influence of recycling on pure pulp grades is discussed in the following to clarify basic differences between pulp types.

1.3.1 Effects on strength properties in recycling

Probably the most often quoted studies in the literature when discussing the influence of repeated drying and reslushing on sheet properties are the experiments of Howard and Richard[4]. In that work, broke-like recycling without calendering was applied to various pulps. Pulp fines were recovered between cycles, and drying of sheets occurred at room temperature. Figure 11 shows typical behavior of mechanical pulps and indicates improved fiber bonding ability.

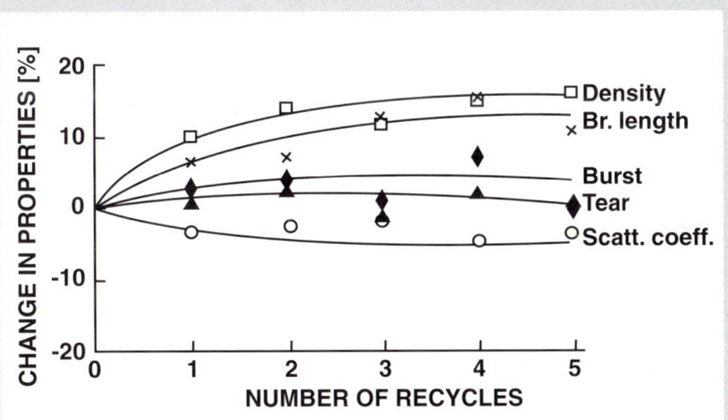

Figure 11. Changes in stone groundwood (SGW) sheet properties during recycling[4].

These results again imply that a mild, broke-like recycling is already producing improved bonding ability in mechanical fibers because of their increased flexibility caused by enhanced internal fibrillation. As described in a previous section of this chapter, the large shrinkage forces during drying cause microcompressions in bonds of crossing fibers. In the case of mechanical pulps, this produces internal fibrillation. Properties of chemithermomechanical pulp (CTMP) showed a similar trend to stone groundwood (SGW) as Fig. 11 shows, but it was even more pronounced. Tensile strength increased and light scattering fell 10% during the first cycle. Microscopic inspections of sheet cross-sections revealed that the number of collapsed or flatter fibers grew more in CTMP than in SGW. In this case, note that chemical pre-treatment of CTMP was associated with reduction in cell wall rigidity. Hornification had not yet appeared. Unfortunately, the yield of CTMP pulping was not determined. Behavior of regular TMP was somewhere between SGW and CTMP[4].

The actual chemical pulp of Howard's study behaved as presented in Fig. 12 in which the growth of tear strength was remarkable. Fibers were probably well saved from damage in this recycling procedure (no calendering and drying at room temperature). Strong hornification was apparent as indicated by a drop of burst and tensile strength and greater light scattering as Fig. 12 shows. The decrease of fiber bonding has been so apparent a mechanism in this recycling trial that sheet density also clearly decreased despite probable fiber flattening due to lumen collapse.

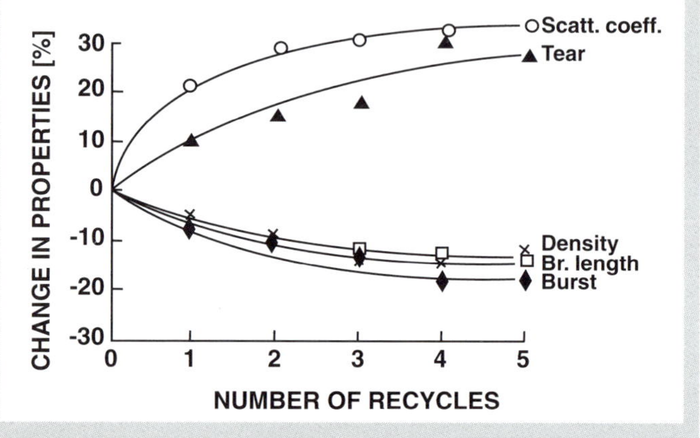

Figure 12. Changes in beaten bleached kraft sheet properties during recycling[4].

Tear strength is generally a good indicator of runnability on paper machines. It probably indicates the force required to pull out or break fibers at the tip of an advancing web rupture. Development of tear strength in softwood kraft during recycling is therefore of particular interest because its reinforcement ability is often necessary in printing paper grades con-

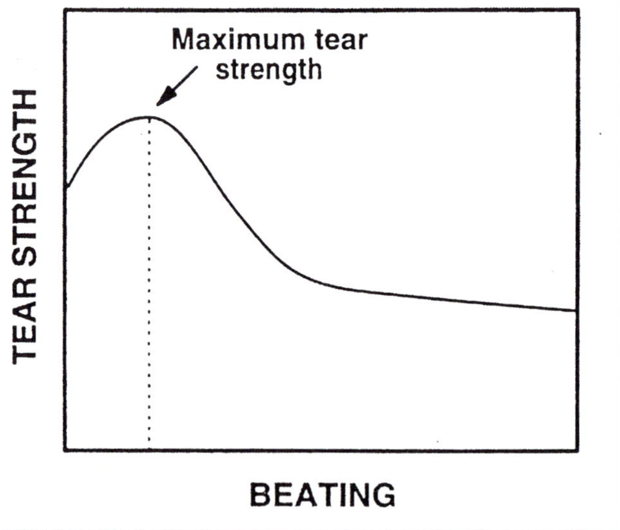

Figure 13. Typical plot between tear strength and beating (bonding) for softwood kraft pulp[38].

taining deinked pulp. In recycling studies[1,3,4,11,21], considerable disagreement has occurred on the direction of this development. Chaterjee *et al.*[38] found that the original beating degree of kraft pulp has an influence on that development. Figure 13 presents a typical curve of tear strength vs. beating degree or bonding ability for softwood kraft.

If the original virgin kraft pulp had a high freeness that corresponds to a beating level lying to the left of the maximum in the curve, the tear will decrease in recycling as bonding ability decreases due to hornification. If the tear value of beaten original pulp lies to the right of the peak, the decrease in bonding (hornification) would increase its tear value until it reaches the maximum as Fig. 13 shows. This hypothesis agrees well with the tear development in beaten kraft pulp of Fig. 12.

In the same way, improvement of fiber bonding in mechanical pulp during recycling has not consistently happened in all experiments[3,14,33,39]. That exceptional behavior has been explained by fiber cutting in calendering, loss of fines, effects of sheet or ink aging phenomena, and the method of drying of sheets. For example, free shrinkage of the sheet during drying produces an increase in tear and stretch but a considerable loss in tensile strength compared with drying under restraint. In one study[14], the specific energy consumption (SEC) in TMP manufacturing significantly influenced the development of properties in recycling. Tensile strength and density in high SEC pulp sheets grew. In low SEC sheets, it primarily declined. In the design of recycling experiments, refining energy input or original freeness level of pulps therefore requires consideration.

Different yields of CTMP corresponding to the extent of chemical pre-treatment influence development of the sheet properties during recycling. Law[19] recycled TMP, CTMP, and chemimechanical pulp (CMP) made from spruce and aspen without deinking chemicals but with the retention of fines. In comparison with CTMP, the CMP process had stronger sodium sulfite liquor, higher temperature, and longer treatment time and gave yields of 85% for spruce and 80% for aspen CMP. Spruce and aspen CTMP had yields of 92% and 90%, respectively.

The influences of repeated drying and reslushing depended on the yield of pulps and wood species. TMP and CTMP spruce pulps showed some increase in sheet density and tensile strength. Recycling of CMP exhibited a decrease of 15% in density and 30% in tensile strength. When the raw material was aspen, CTMP already showed a declining tendency in those properties. These results indicate that the CMP suffered noticeable reduction of bonding potential in the recycling process, but mildly treated CTMP suffered hardly at all. Chemical pre-treatment to a yield of 90%–80% removes ligno-hemicellulosic gel from cell wall lamellae to such an extent that it leaves the structures or some structure exposed to an irreversible hornification during drying. The exact yield boundary depends on wood species.

Differences in recycling behavior between pulps made of different wood species probably exists. In a study of TMP and CTMP made of spruce, pine, poplar, and birch, the pulps were recycled with a broke-like procedure without fines recovery [39]. Fines were lost, and the tensile and burst strengths decreased in all pulps. Spruce had the least decrease. Pine had slightly more, and hardwood pulps had the most decrease. The conclusion is that the bonding ability of fiber fractions in TMP and CTMP made of hardwood suffered more in recycling than those of softwood. Recycling studies made by Liebe [30] with various unbleached and bleached kraft pulps showed very similar development of sheet properties independent of bleaching method (ECF and TCF) with hardwood pulps being more prone to hornification than softwood.

1.3.2 Effect on other properties

Sheet properties that depend on bonding ability such as modulus of elasticity and Scott bond (delamination resistance) have very logically responded in recycling experiments. They have increased with increasing sheet bonding and vice versa. The stiffness of sheets has very consistently declined with all pulp types in recycling trials. According to paper physics, the resistance to bending action increases with growing modulus of elasticity or thickness of the sheet. The reason for less rigid sheets in the case of recycled low-yield pulps is probably the loss in fiber bonding (E-modulus). For mechanical pulps, the reason is the loss of sheet thickness due to flattening of fibers.

A dried (hornified) chemical pulp provides better dimensional stability for the sheet than undried pulp, i.e., drying decreases the hygroexpansion ability of fibers. Stürmer *et al.*[13] have studied the effect of drying history of recycled fibers on their hygroscopic behavior as Fig. 14 shows. A recycled kraft pulp sheet dried rapidly (pulp 3 in Fig. 14) absorbs significantly less water from humid air than a recycled sheet slowly dried at room temperature (pulp 1 in Fig. 14). As stated earlier, more intense drying brings still more hornification that further reduces access of water into the cell wall matrix by irreversibly closing the micropores. This prevents fiber swelling. In contrast, the initially lower hygroexpansive ability of mechanical fibers hardly changes during recycling[25].

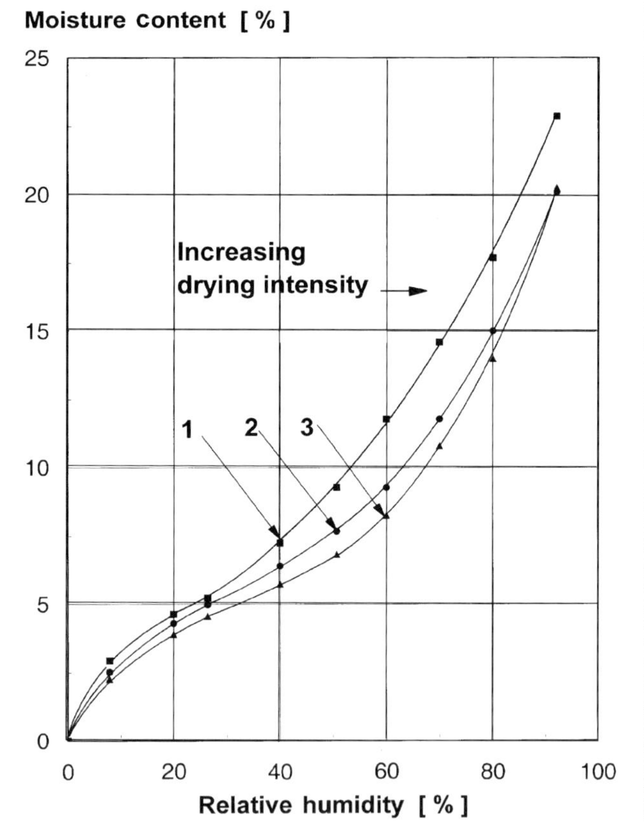

Figure 14. Absorption isotherms for recycled kraft sheets of different drying history. Drying intensity has been increased in the order 1, 2, and 3[13].

Optical properties except light scattering depend on bonding ability to a smaller extent. Ink removal and pulp bleaching govern brightness, light absorption coefficient, color shade, or number of ink specks as Chapters 8 and 9 of this book discuss. Accumulation of ink residues and probable age distributions of recycled fiber generations will be topics of sections 2 and 3 of this chapter.

Those properties influenced by the presence of hydrophobic extractives in the sheet can change if mechanical fibers have been subjected to alkaline deinking. Those wood extractives are primarily dissolved and washed out in a deinking plant. Porosity measured as air permeability has generally increased with chemical pulps on recycling. With mechanical pulps, changes have occurred in both directions. Factors such as recycling procedures or original freeness of pulps probably play significant roles in the development of porosity. Development of surface roughness has been more complex and brought contradictory results depending on the fate of fines in recycling arrangements. The easy densification of sheets made from recycled mechanical fibers in calendering compared with virgin pulp is a common practical fact that also provides good smoothness value measured with air leakage methods. Measurements with optical or electron microscopy can reveal more open surface structures for recycled sheets than for virgin material in the case of TMP[33].

1.4 Validity of experimental simulations of actual paper recycling

In practice, recycled fiber pulps are blends of different fibers and other paper furnish materials. All occurrences in the use and recycling chain are impossible to simulate, but laboratory and pilot scale recycling experiments are the only way to try to answer the question, “How many times can we subject a fiber to the deterioration it endures during recycling?”

1.4.1 Design of recycling experiments

A review of the literature shows that almost all researchers have used different procedures to recycle their fibers and pulps. A suitable question concerns what extent the differences observed between results of studies are due to the material being recycled or the method used to accomplish the recycling.

Recycling trials can start with recovered paper or virgin pulps depending on the target. In comparison with fresh pulps, the chemical fibers in a paper sheet are already hornified and the fiber lumens more or less collapsed. The beating degrees of all pulps and the original energy input in the mechanical pulping process require consideration since the development of bonding ability can depend on them. In the case of high-consistency refining, some fiber curl has developed in the pulp. This “latency” has an influence on sheet properties and can require removal before recycling trials. Deep freezing commonly used for storing of fresh pulps can cause some hornification for low-yield fibers[30]. In flash drying sometimes used in nonintegrated recycled pulp mills, fibers have shrunk without any restraint during drying.

The determination of the pulp yield in pilot trials is difficult if the simulation of complete multiple cycles is performed. If printing and deinking stages are included, fiber and especially fines losses can be considerable. The proportion of fines has a great influence on sheet properties. The hot dispersing or kneading stages at high-consistency and temperature are frequently not included in laboratory or pilot recycling simulations. In practice, those processes cause some fiber curling and probably also fiber kinks and microcompressions. Fiber damage can also occur in screw presses and high-consistency pumps.

Refining of pulp and some strength increasing additives compensate for reduced bonding ability of recycled chemical fibers in paper manufacturing. Refining can be part of laboratory recycling experiments, but the results depend on the type of equipment and the refining procedure. A laboratory beater often refines pulps gently and generates well-bonding fibrillar fines. In practice, refining effects are very variable.

In sheet pressing, intensity and the final solids content of sheets on laboratory wet presses do not correspond to those on production scale wet presses. The drying is also more intense on paper machines than on handsheets drying in the laboratory. No investigations on the effects of these differences on recycling simulation results are available. Differences between laboratory or pilot calendering and production equipment also require consideration in recycling experiments.

1.4.2 Chemical conditions in recycling experiments

Many nonfibrous substances such as pigments, latexes, sizes, anionic or cationic additives, and inorganic salts and their accumulation with increasing recycling rates can affect the sheet properties in different ways. In deinking experiments, balancing of chemicals, COD in water circulation and carryover of deinking chemicals to test sheets should also be monitored and kept at a realistic level. For example, additives such as cationic starch are more adsorbed on fines and fillers fraction than on fibers.

As previously stated, swelling of fibers decreases if sodium ions in the cell wall structure received in the conventional alkaline reslushing are replaced with hydrogen or other cations. pH and cations therefore require consideration. With inclusion of papermaker's alum in sheet making for example, some loss in tensile strength can be expected. In certain geographical regions, the use of hard water for stock dilution already establishes calcium as the prevailing cation in the system.

Aging of paper should be included in the design of recycling simulations especially when printing and deinking are involved. If the aging occurs naturally, it will lead to time consuming and complex recycling procedures. An artificial, accelerated aging is therefore often used for recycled sheets in which they are kept at elevated temperature for several days. In that case, pH and other chemical conditions of sheet making can play a critical role particularly with mechanical pulps. If pH has been on the acid side with no buffering of $CaCO_3$ in the furnish of the sheet, cleavage of acetyl and methoxyl groups of hemicelluloses under moist heat aging can produce organic acids that attack the load-bearing cellulose chains in fiber walls. This causes a decrease of strength properties[40]. The degradation of cellulose due to this acid hydrolysis also occurs during the natural aging of mechanical fiber but more slowly.

When accelerated aging is applied to printed sheets, intensified interactions between ink vehicles and fibers depend on the type of ink used. Heat treatment accelerates oxidation of possible unsaturated carbon bonds of binders and oils in inks. This leads to stronger adherence of ink to the fiber surface. Ink is then difficult to detach from fibers in paper reslushing and can result in poor deinking and excessive flotation losses[41,42]. This is why the selection or design of ink also requires consideration in a complete recycling simulation.

The design of recycling and deinking experiments is a demanding task. One must know what particular phenomena are to be studied and must be cautious in generalizing specific results. The available pilot or laboratory facilities and procedures may always have some limitations, but an understanding of the fundamental phenomena may help with interpretation of the results.

2 RCF quality related to remaining contaminants and inks

Depending on the recovered paper grade, RCF pulp is often a heterogeneous material. Besides fibers from different types of wood and from different pulping and bleaching processes and the fillers and pigments, recovered paper also contains components that enter the recycling loop in the course of paper converting and printing. While no exact definition of these components according to their categories currently exists, they generally include any materials that cannot be used in subsequent paper production. They therefore include papers and boards of grades unsuitable for the particular recycling process, substances from paper conversion such as adhesives and printing inks, and nonpaper components such as plastics, metals, etc.

Chemical and microbial contaminants are also present. Especially in the case of packaging papers intended to come into contact with foodstuff and of hygienic papers, the concentration of certain chemical contaminants must be less than the threshold values defined by legislation and regulations. Furthermore, certain contaminants must not migrate to the wrapped foodstuff or to the skin.

All the previously mentioned components adversely influence the quality of recovered paper, and their effects can vary greatly. Some only affect the quality of the paper they are used to produce such as brightness or optical cleanliness. Others are highly detrimental to the paper production process, e.g., stickies. Their effects often extend into paper converting or printing. Depending on the application of the recycled fiber, specifying that certain contaminants and types of recovered paper are virtually absent or at least do not exceed certain proportions is therefore necessary.

Removal of contaminants begins at the stage of manual sorting of recovered paper. This process primarily removes larger nonpaper components and unusable papers and board (See Chapter 3). The objective of the following multiple-stage processing at the paper mill is mostly to eliminate detrimental components remaining in the recovered paper. For this reason, the degree of contamination of the recovered paper used and the specifications of the finished paper mainly determine the technical configuration of a stock preparation plant (See Chapter 6). A more highly contaminated recovered paper requires greater investment and operating costs for its processing than less contaminated recovered paper grades or less quality-demanding paper and board grades to be produced.

2.1 Classification of contaminants

Contaminants in recovered paper can be classified according to different criteria. The following describes three different classification possibilities:

- Type
- Origin
- Effect.

Type

Contaminants in recovered paper are usually classified from a standard list of recovered paper grades that also provides standard definitions of such grades[43]. Some unwanted materials in recovered paper are identified such as nonpaper components and unwanted paper and board components.

Any material in the recovered paper that may cause damage to the machines during processing, disturbances during production, or a loss of quality and therefore value of the final product is in the category of nonpaper components. Included among the nonpaper components are metal, twine, glass, textile, wood, synthetic paper, sand, building material, and plastics.

Unwanted papers and boards are all paper and board grades that are unsuitable or hazardous to use as raw material for paper and board manufacture or that make the entire batch of recovered paper unusable by their presence. Typical paper and board grades that constitute a hazard to production are carbon paper, parchment, waxed carton, and wet strength paper.

Some paper and board grades can contribute to a loss of quality of the recycled fiber. The presence of brown packaging papers and board in the recovered graphic paper negatively influences the optical properties of the deinked pulp. For this reason, they should be present only in very small quantities or not at all in deinking material.

The type and quantity of unwanted materials are specified for different grades of recovered paper. For quality, recovered graphic paper is the most important material in newsprint production. Specifications for recovered graphic paper define a maximum of 1.5% nonpaper components and unwanted papers and boards. Extensive random analyses on German wood-containing recovered graphic paper have shown that the actual proportion was much higher[44]. Although the contamination with nonpaper components was less than 1%, the proportion of unwanted papers and board and especially packaging papers and board was intolerably high at an average of 6%. These findings reflect a global trend toward increasingly contaminated recovered papers. This is happening not only with recovered graphic paper from household collection but also with other recovered paper grades[45].

This way of classifying contaminants in recovered paper primarily uses a desired sorting quality and does not consider detrimental substances introduced during paper converting or those that accidentally enter the recycling process with the paper loaded with chemical additives during preceding papermaking (See Chapter 13).

Origin

This classification according to the origin of contaminants covers not only nonpaper components that enter the recovered paper externally in the course of its collection but also materials introduced internally in physical or chemical combinations with the paper. These include printing inks and adhesives and chemical additives used in the preceding paper production process[46].

External contaminants	Internal contaminants		
	Soluble in water	Insoluble in water	
		Nonstickies (not thermoplastic)	Stickies (thermoplastic)
Sand	Starch	Fillers	Adhesives
Glass	Alum	Wet strength paper	Plastics
Wires	Soluble glues	Inks	Wax
Golf balls	Sizing agents		Latex
Sneakers	Redispersible adhesives		
Wooden board			
Watches			
Engine blocks			
Kitchen sinks			
Styrofoam			
Cans			

1. Redispersible adhesive types are excluded from the category of insoluble adhesives
2. Redipersible adhesives may be sticky even when they are water soluble
3. Wax may or may not be sticky, depending on temperature
4. Certain components of inks may be water soluble, while insoluble components may be sticky or nonsticky

Figure 15. Contaminants in recovered paper[46].

According to Fig. 15, external contaminants that should not be present in recovered paper include sand, metal, glass, or objects whose intended original use should exclude their presence. Some contaminants can have large dimensions and are usually encountered in bales. Their separation generally occurs during a manual pre-sorting of the opened bales in the mill but at latest during the pulping process by suitable trash removal devices. Small external contaminants are effectively eliminated during processing at the cleaning and screening stages. Any effect on the quality of the recycled pulp due to such contaminants can be largely minimized.

Some internal contaminants were added during preceding paper production to produce certain paper properties or influence the papermaking process. These include starch, wet strength agents, pigments or fillers, and coating components. Included among the internal contaminants are materials that are required for paper conversion and printing. The most important of these are printing inks and adhesives.

In this classification, the internal contaminants are water-soluble and -insoluble substances. Soluble contaminants often called "disturbing substances" accumulate especially in the process water system. Their handling or removal depends primarily on the specific volume of effluent taken from the process. The detrimental consequences are evident as deposits on the machines and equipment, increased growth of slime and microorganisms, more corrosion, and reduction in the efficiency of chemical additives used in papermaking, e.g., retention agents. The effects of soluble contaminants are usually countered by periodic addition of materials such as biocides. Removal of dissolved contaminants with microflotation systems is not very effective. Primary processes for removal of dissolved organic substances are biological purification stages, membrane technology, and chemical additions for precipitation. These not only contribute to better pulp quality but also enhance the stability of the production process.

For their tacky properties, the insoluble contaminants are sticky and nonsticky substances. If the recycled pulp will be used for the production of graphic papers, printing inks are among the most important nonsticky internal contaminants. When recycled fiber pulp is used for the production of hygienic papers, fillers and pigments also have negative effects on the properties of the final products. Their removal is an important process step in recovered paper processing.

Although printing inks and fillers or coating pigments mainly influence the quality of the recycled pulp, tacky contaminants or stickies cause problems in the production and conversion operations. Stickies result from adhesives of paper and board conversion such as hot melts, pressure sensitive adhesives or latexes, and coating binders. Their control became an important area of recovered paper research (See Chapter 11 on stickies).

Effects

This classification additionally considers the intensity of the effects of contaminants on the quality of recycled fiber pulp and on its processing and the papermaking process. A survey in 1992 by CTP in Grenoble among French recovered paper processors tried to evaluate contaminants by their effects and frequency of occurrence. Table 3 shows the results[47].

Table 3. Contaminants listed in order of their detrimental effects according to the paper grades[47].

White paper grades	Brown paper grades
1. Self-adhesive labels	1. Bitumen
2. Adhesives	2. Wax
3. Wet strength papers	3. Self-adhesive labels
4. Synthetic adhesives	4. Hot melts
5. Hot melts	5. Wet strength papers
6. Wax	6. Adhesives
7. Laser printing ink	7. Styrofoam
8. Dyed papers	8. Metal objects
9. Aqueous inks	9. Glass fragments
10. UV inks	10. Ropes
11. Varnishes	11. Sand and gravel
12. Natural glues	12. PE coated papers

According to this survey, the internal contaminants introduced especially through the paper product have very negative effects on the recovered paper processing operation. For both paper categories, tacky contaminants had especially severe effects. For the white grades, poorly deinkable printing inks were also critical. For brown grades, bitumen, wax, and self-adhesive labels caused the major problems.

2.2 Removal techniques for contaminants

The need for manual sorting of recovered paper before pulping depends on the quality of the recovered paper grades. Contaminants are removed during processing primarily by the following techniques:

- Cleaning
- Screening
- Flotation
- Washing.

The removal of contaminants uses differences in their characteristic properties compared with wetted fibers and the pulp suspension, respectively. The most important distinguishing characteristics are size, density, shape, stiffness, and surface properties. Chapter 5 of this book extensively covers the removal techniques mentioned above.

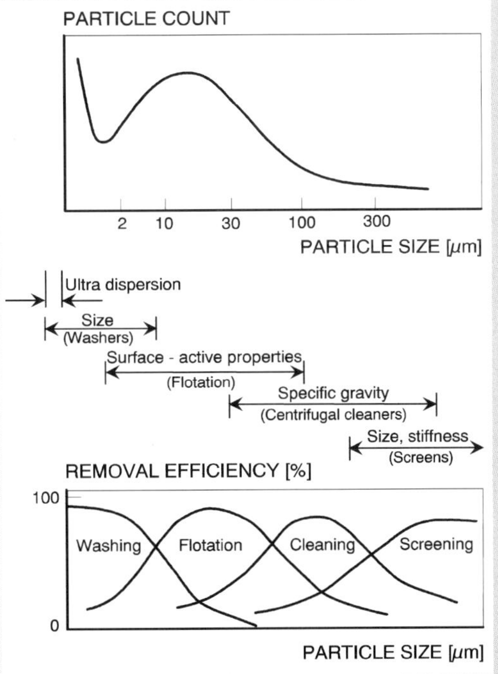

Figure 16. Particle size distribution and unit operation removal efficiency[50].

Size

Washing can remove contaminants smaller than the fiber components or smaller than the threshold of visibility. These include fillers and pigments and colloidal and dissolved detrimental substances. Larger contaminants can be effectively separated at the screening stages. This includes wet strength flakes, macro stickies, or even larger external contaminants such as plastics.

McCool[50] describes the removal efficiency of contaminants as a function of their particle size. According to Fig. 16 small contaminants less than 30 μm in diameter can be removed most efficiently by washing. No additional distinguishing properties for the fibers is necessary. Particles of 10–250 μm can be removed by the flotation process if their surface is sufficiently hydrophobic. The centrifugal cleaning process has no exclusive dependence on size. Even if particles of 100 μm can be efficiently separated by means of a cleaner, a different density than that of the fibers is a decisive precondition for the successful realization of the process. When applying this model, one must consider that the size limits of the different processes vary according to the separating criteria. This is an important factor especially for screening where increasingly smaller screening perforations are finding use.

Density

For separating contaminants by this distinguishing characteristic, their density must have a value other than 10^3 kg/m^3. By applying high centrifugal forces, heavy (density $>10^3$ kg/m^3) and light (density $< 10^3$ kg/m^3) contaminants can be effectively removed. These include most small external contaminants such as metal particles, glass fragments, or sand that have a higher density than fibers. It also includes latexes, plastics, and tacky particles with a density less than that of the wet fibers.

Shape and stiffness

The shape of contaminants has a major influence on the level of efficiency of the cleaning and screening stages. Spherical particles can be eliminated more efficiently by cleaning and slot screening than flat contaminants that are better removed by hole screening. In addition, the particles should not be too soft and deformable to prevent their passing through the openings of the screens despite having an apparently suitable size for separation.

Surface properties

This case uses the different wettability of fibers and contaminants in the aqueous medium as a separating criterion. The most important process here is selective flotation that removes hydrophobic particles from the hydrophilic fibers of recycled pulp slurries. This selective separating process is especially suitable for removing printing inks (deinking). Attempts are increasingly being made to separate fillers and stickies more selectively by this method[48,49].

2.3 Effects on RCF quality

This section discusses the effects of contaminants that are still present in the recycled fiber pulp even after adequate processing. They have a negative influence on papermaking, paper conversion, and paper quality. Besides residual printing inks, such contaminants primarily include stickies and dissolved and colloidal substances.

2.3.1 Effects of residual ink particles

Residual ink particles have a significant influence on the optical characteristics of the recycled pulp and the paper made of recycled fiber furnishes. Although this is usually of only secondary importance for the production of paper grades with low optical requirements, e.g., brown packaging paper or board, a certain content of residual ink particles in graphic paper grades can lower their quality.

The presence of residual ink particles leads to a reduction of brightness. Figure 17 shows that this reduction is nonlinear. The greatest drop in brightness occurs with very small amounts of small-size ink residues. The losses are greater with higher initial brightness of the paper. While a printing ink proportion of 0.1% in a newsprint causes a decrease of 16%, the brightness of a white paper falls by a total of 35%. A doubling of the amount of residual ink therefore results in only minor optical losses[51].

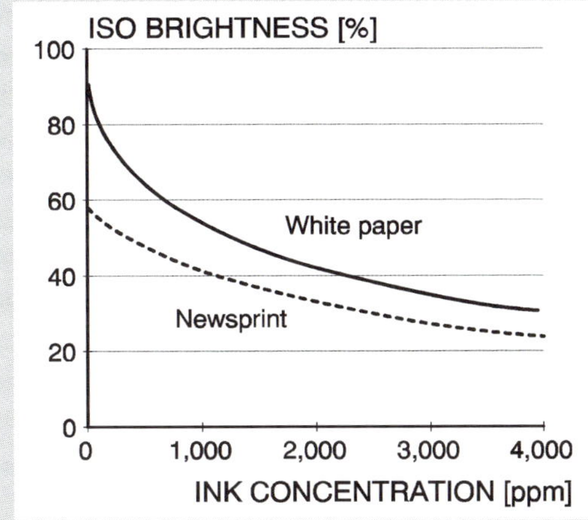

Figure 17. Brightness of newsprint and white wood-free paper vs. concentration of residual ink[51].

The reduction of brightness is also a function of the size of the residual ink particles. Smaller particles with their large specific surface area clearly influence greater brightness losses caused by higher light absorption than larger particles as Fig. 18 shows. In addition to this optical phenomenon, the so-called light-spreading effect also known as the Yule-Nielsen effect[52] causes further light absorption with a drop in particle size. According to this theory, light penetrates below the printing ink particles due to the scattering within the paper structure and is additionally absorbed there.

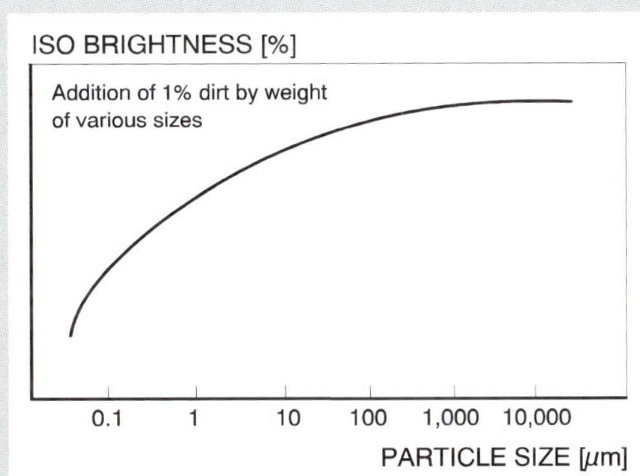

Figure 18. Brightness drops with a decrease in particle size[50].

Larger ink particles affect the optical homogeneity of the recycled fiber pulp and make it appear contaminated[53]. The particles that are visible as specks are due to ink residues as particles or agglomerates detached from the fibers but not removed and as printing ink still attached to the fibers. Difficult to deink printing inks such as nonimpact inks especially contribute to a greater speck concentration and negatively influence the optical quality of the finished product.

Besides optical effects, residual ink particles can also have a detrimental effect on runnability[54]. Experience has shown that conventional offset and rotogravure printing inks lead to deposits in the wet end of the paper machine. The effects of toner particles from nonimpact printing processes are more evident in the dry end as coated felts or fabrics and rolls. The consequences can be measured not only by the decreased efficiency of process stages but also by faults in the paper that can lead to web breaks.

2.3.2 Effects of colored fibers

Besides residual ink particles, colored fibers also cause an optical impairment of the deinked pulp and the paper produced from it. Colored fibers produced from unbleached brown or dyed recovered papers not only have a negative effect on brightness but also result in an increased optical inhomogeneity.

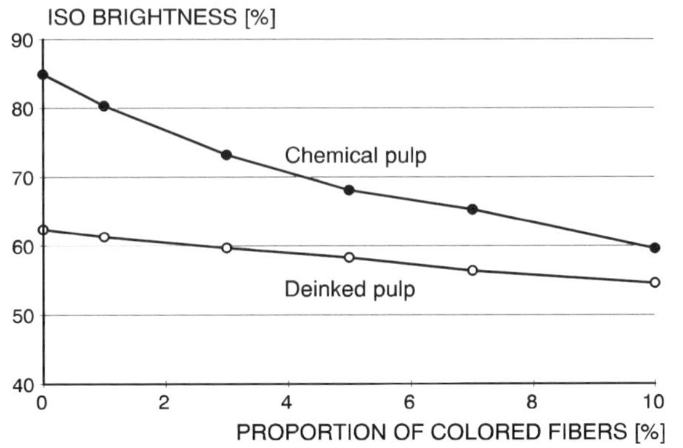

Figure 19. Effect of colored fibers on brightness of deinked pulp and chemical pulp[55].

Figure 19 shows the effects of small proportions of colored fibers on the brightness of deinked pulp and chemical pulp. Similar to the influence of printing ink particles, the reduction of brightness is considerably greater in brighter base materials. A proportion of 5% of corrugated board manufactured from unbleached fibers reduces the brightness of a bleached chemical pulp by about 17%. The brightness of the wood-containing, deinked pulp is reduced by only 4%. The presence of brown recycled fibers has especially negative effects in wood-free recovered papers for which the initial brightness is considerably superior to that of old newspapers and magazines. Achieving an effective decolorization of colored fibers is possible by oxidative bleaching (See Chapter 9 on "Bleaching of deinked pulp").

Colored fibers from dyed papers only result in losses in optical quality if the deinked pulp was not treated with suitable bleaching chemicals during stock processing. Undertaking reductive bleaching with dithionite or FAS generally gives an effective decolorization of such fibers.

2.3.3 Effects of stickies

Potentially tacky substances are now the most feared contaminants in recycled fiber pulps (See Chapter 11). Because of their tacky character, stickies deposit on wires, felts, rolls, and other moving parts especially on the paper machine. The detrimental influences can be manifest in different ways as Fig. 20 shows. Faults occur in the paper primarily as thin

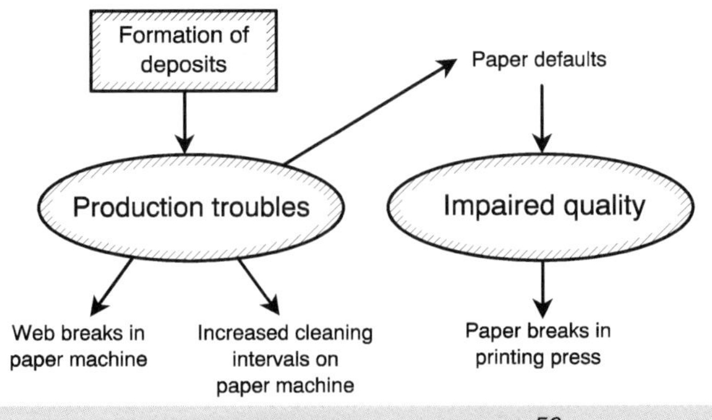

Figure 20. Effects of stickies in recycled fiber pulp[56].

areas or holes that can cause web breaks in a paper machine and a printing press. The result is longer nonproductive time and cleaning time that adversely affect the productivity of the processes.

Several reasons exist why tacky contaminants have more serious consequences: a greater use of pressure sensitive adhesive coated papers at home and in the office, the increasingly widespread use of mailing campaigns, and the common use of supplements and glued-in inserts in magazines. These ensure an increasing content of adhesives in recovered paper. In addition, increasingly more paper grades are being surface sized, coated, or finished in similar ways.

The stickies that are usually present in recycled fiber pulps do not necessarily need to cause problems. Macroscopic contaminants above a certain size (approximately > 150 µm) can be effectively removed from a recycled pulp slurry by using suitable screening techniques. Finely dispersed micro stickies (size approximately < 100 µm) pass through the screens with the accept and cause the aforementioned problems. These stickies are most feared because they form in part during pulp processing by secondary reactions with other components of the pulp or white water and are therefore difficult to control. One way to avoid such problems is to wash the recovered pulp followed by a purification of the washing filtrate. Another possible solution is to add talc to deactivate the tacky surface.

2.3.4 Effects of other detrimental substances

With the wide variety of detrimental substances, their effects on the quality of recycled fiber pulp can also vary greatly. Sometimes, these contaminants result in negative effects only after their concentration in the aqueous phase. The extent of this depends on the degree of process water loop closure and the adsorption tendency to fibers and minerals.

As usually understood, dissolved contaminants in terms of "disturbing substances" are the total of anionic oligomers, polymers, and nonionic hydrocolloids. They can not only harm the quality of the recycled pulp but also lead to serious technical difficulties. The following effects are due to such detrimental substances[57]:

- Reduced effectiveness of chemical additives such as retention agents, dyes, wet strength resins, and bleaching chemicals
- Poorer retention
- Deposits
- Reduced paper strength
- Specks and spots in the paper.

Such problems occur not only when using recycled fiber pulp but also when using virgin pulps with their contaminants. A potentially detrimental substance is introduced with mechanical pulp as pitch. Additives for filler processing such as polyphosphates and polyacrylates can also cause problems during paper production. Sodium silicate is a typical detrimental substance that is present in recycled pulp as a result of the deinking process[58]. Due to the chemical carryover, residual silicate enters the white water system and the paper machine with the pulp. There it primarily weakens the effect of retention agents such as polyethyleneimine or polyacrylamide. Residues of this peroxide stabilizer also negatively affect the efficiency of white water clarification by reducing the effectiveness of flocculants.

Most chemical contaminants contained in recycled fiber pulp that represent potential ecological hazards originate during the production of virgin fiber stock or enter the recycling process as papermaking or paper converting additives. Table 4 shows several chemical substances in recovered pulps and their origin as far as has been unequivocally established.

Table 4. Possible chemical components in recycled fiber pulps[54].

Component	Origin
Dioxin/furan	Chlorine bleaching (not common)
Polychlorinated biphenylene (PCB)	Carbonless papers (former production)
Pentachlorophenol (PCP)	Slimicides, biocides
Optical brighteners	Papers, coatings
Phthalate	Softeners in printing inks, additives
Formaldehyde	Wet strength agent
Heavy metals	Fillers, coating pigments, (printing inks)
AOX compounds	Chlorine bleaching, wet strength agents

The fact that these components are being considered increasingly important is due to the high level of awareness among consumers regarding potential environmental hazards and to the improved analytical techniques that permit their detection by economically efficient methods.

The most problematic substances that concern consumers are dioxins and furans that form as a result of the chlorine bleaching of chemical pulp and the use of dioxin-containing additives during the papermaking process. The global trend toward the substitution of chlorine in bleaching by oxygen-based bleaching chemicals has virtually eliminated the production of dioxins and furans during chemical pulp production. The result is that their content in recycled fiber pulp has also decreased significantly. From the paper industry's point of view, no potential hazard for consumers exists today (See chapter 13).

The same applies for polychlorinated biphenylenes. In contrast, pentachlorophenol continues to enter the recycling process as a result of nonuniform European regulations governing the production and use of PCP-containing biocides (See Chapter 13).

According to the current state of knowledge, optical brighteners that enter recovered paper with white writing and printing papers represent neither an ecological nor a toxicological hazard. Independent of any potential hazard, their resistance to bleeding must be ensured.

The concentrations of formaldehyde and softeners in recycled pulps are very low to the extent that they can be detected at all. Chapter 13 covers the question of heavy metal content and AOX compounds of recovered paper.

Theoretically, aromatic amines can form by the splitting of azo pigments from yellow printing inks during the reductive bleaching of recycled pulps via dithionite, formamidine sulfinic acids, or both. Such amines can be carcinogenic. Recovered paper processing mills have therefore investigated this problem with extensive tests. Under the applied bleaching conditions, the azo pigments are chemically inert. Aromatic amines were undetectable in the white water systems or the bleached recycled pulp[59].

Microbial contamination of recycled pulps caused by bacteria and fungi can be the cause of production disturbances and quality problems. Microbial loads are usually due to organic contamination that enters the stock preparation system via the recovered paper or during storage of the recycled pulp. In addition, tightly closed process water loop systems with a high concentration of biologically degradable, dissolved organic substances favor the growth of microorganisms. These lead to the covering of machine components and pipe systems with slime. Detaching slime clots disturb the sheet formation process and can even lead to losses of production due to web breaks. Because microbial respiration consumes the oxygen contained in the process water, anaerobic conditions with the production of organic acids occur that then cause odor emissions and corrosion problems. A microbial contamination of foodstuff by recycled fiber containing packaging paper can be excluded because microbes do not migrate from the paper on dry or moist food[125].

Previously, the measures for containing microbial activity were confined primarily to the addition of biocides. Today, thermal stock treatment is increasingly favored. Temperatures of 80°C can cause considerable decontamination. To reduce the use of biocides, thermal degermination such as DURING dispersion should be the preferred method[54,60].

3 Modeling of age distribution of RCF

Before their re-use for paper production, RCF of any recovered paper grade or recovered paper delivery to a paper mill can have traveled through the recycling loop one or more times including the following stages:

- Paper manufacturing
- Paper converting and printing
- Consumption of paper products
- Collection of used paper products
- Sorting of recovered paper
- Processing of recovered paper.

With each recycling loop, the recycled fibers or the paper manufactured wholly or partially from RCF experience changes in characteristics compared with virgin fibers. These changes can relate to morphological, chemical, mechanical, optical, or other physical characteristics:

- The drying process in the previous paper production causes fiber hornification and therefore a reduction in fiber-to-fiber bonding when using recycled chemical fibers to manufacture new paper. This can then have a negative effect on the strength characteristics, e.g., tensile strength or ply bond strength, of the paper produced.
- Under the influence of mechanical forces during recovered paper processing, e.g., refining, a reduction of fiber dimensions (fiber length and fiber diameter) occurs with fines formation. Consequently, certain characteristics of the fiber suspensions, e.g., freeness, and the paper produced can be influenced. This includes among others apparent density, porosity, pore size distribution, and tear.
- When deinking printed paper products, ink particles and other dirt specks cannot be fully removed. The amount of residual contaminants therefore increases with every recycling loop. Accordingly, cleanliness and optical characteristics such as brightness and color shade of the paper are negatively influenced.

Except the type of fibers concerned (chemical and mechanical fibers) and the upgrading measures in recovered paper processing, the change in the characteristics of RCF and paper made of RCF depends primarily on the number of recycling loops undergone by the fibers. For this reason, knowing how often on average the fibers were recycled for example in a national recycling system is important. Another important factor is the age distribution in relation to different recycling generations that the recycled fibers have for certain recovered paper grades. Such an evaluation could also be undertaken for different recovered paper grades, e.g., deinking grades or for different paper grades, e.g., newsprint or corrugating medium produced wholly or partially from RCF.

Even with fiber microscopy or chemical analyses, verification of how often RCF has undergone a recycling loop or even whether a fiber has been recycled is not possible. Quantification of the number of recycling loops of RCF can therefore be done only by statistical methods. In addition, this can vary greatly from one individual case to another because the recovered raw material does not necessarily reflect national averages. Local conditions are often dominant.

The following describes several statistical models to determine how often recycled fibers are being used anew for paper production or how many generations are included in a furnish for RCF-containing paper and their proportions:

- To facilitate understanding of the general principle concerned, a simple model is first presented as a one-parameter model. The simplified assumptions involved mean that it does not reflect the industrial realities.

- The two-parameter model that builds upon the first refers to a practice-oriented recovered paper mix, e.g., recovered paper for deinking, consisting of two components, e.g., old newspapers and magazines with their different origins at the time of the preceding paper production.
- Finally, the multiparameter model presented approximates the reality of industrial recycling.

In view of the importance of newsprint and the continually increasing share of recovered fibers contained in it in many countries, the two parameter and multiparameter models relate to this paper grade with special consideration given to the situation in Germany where newsprint is almost totally produced from recycled fibers.

3.1 One-parameter model

To become acquainted with the statistical procedure, a greatly simplified model in the first step is taken to calculate the number of recycling loops or generations of recycled fibers and their proportions in a furnish for the production of RCF-containing paper as Fig. 21 shows. This model uses the following simplifications:

- All paper grades in a country are produced with an identical recovered paper utilization rate, e.g., 50%
- No recovered paper exports or imports occur
- No paper exports and imports occur
- No yield losses occur during recovered paper processing.

An important simplification is that no paper import occurs since the import of paper produced from virgin fibers introduces young recycled fibers of the first generation into the recycling system of the importing country. The single variable in this model is the recovered paper utilization rate indicated with *a*. The designation $(1-a)$ corresponds to the fraction of virgin fibers introduced into the furnish during paper production.

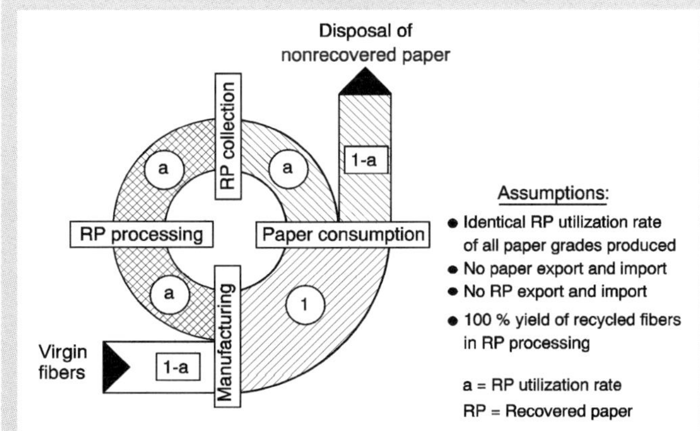

Figure 21. Simplified recycling loop of recycled fibers for one-parameter model[61].

Figure 22 uses the so-called age tree covering three generations G to illustrate the calculation of the age structure or number of generations of recycled fibers. Paper A is produced repeatedly from a mixture of recycled fibers and virgin fibers in the same proportions. Due to the origin of the used recovered paper (identical with used paper A), paper A consists of recycled fibers that have undergone the recycling system for a different numbers of times such as virgin fibers of generation G3 that have already been used three times for paper production.

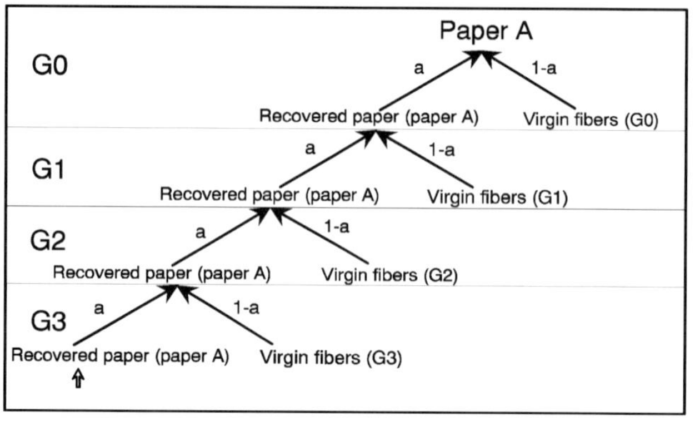

Figure 22. Structure of multiple recycling of paper[61].

In the following, the proportions, g_i, of the different fiber generations are given to calculate the age structure of the furnish for paper A:

$$g_0 = 1 - a \tag{2}$$

where g_o is proportion of zero generation = virgin fibers

$$g_1 = a(1 - a) \tag{3}$$

where g_1 is proportion of first generation of recycled fibers

$$g_2 = a^2(1 - a) \tag{4}$$

where g_2 is proportion of second generation of recycled fibers

$$g_n = a^n(1 - a) \tag{5}$$

where g_n is proportion of nth generation of recycled fibers.

By adding the elements of Eq. 2–5, the sum, S_n, of the proportions, g_i, of all generations of recycled fibers will be obtained including the zero generation (g_0) as virgin fibers.

$$S_n = g_0 + g_1 + g_2 + \ldots + g_n \tag{6}$$

$$S_n = (1 - a) + a(1 - a) + a^2(1 - a) + \ldots + a^n(1 - a) = 1(100\%) \quad 0 \le a < 1 \tag{7}$$

Figure 23 using Eq. 6 shows a histogram of the number and proportions of fiber generations present in the pulp for three different utilization rates of recovered paper. Here, generation G0 represents the proportion, g_0, of virgin fibers and all generations from G1 of the proportions of recycled fibers (g_1 to g_n) that have undergone the recycling system different numbers of times.

The lowest utilization rate of $a = 0.25$ (25%) reflects the situation in the United States in the 1980s. The utilization rate of $a = 0.5$ (50%) simulates the situation in Germany and Japan at the beginning of the 1990s. The rate for $a = 0.75$ (75%) applies to some other countries today. For mathematical reasons, the number of generations is infinite. The series in the diagrams is interrupted at a total pulp mass of about 99% so that all older, not represented generations have a total share of not more than 1%. For practical and also technical reasons, the reproduction of all older generations with a total stock weight less than 1% has not been included.

In place of the histograms, Fig. 24 shows the total number of fiber generations, N_{total}, in relation to the utilization rate. At $a = 1.0$

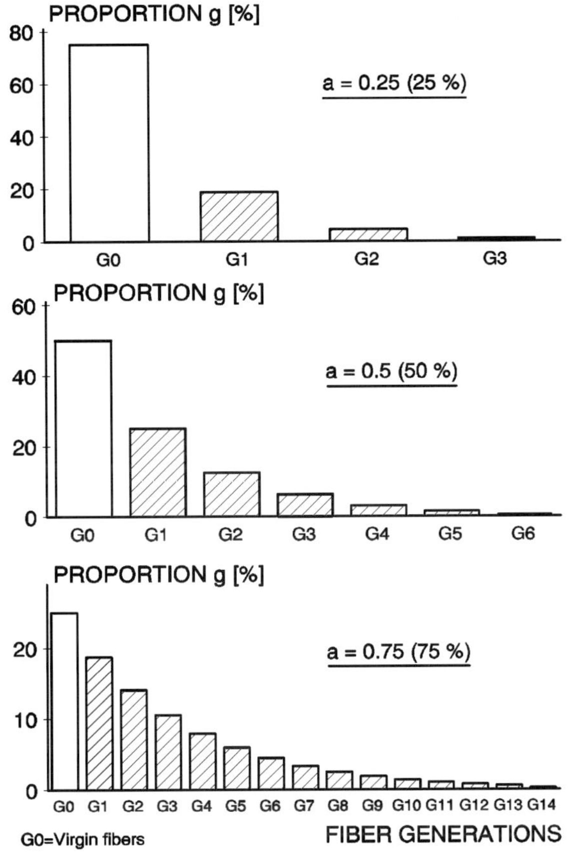

Figure 23. Number and proportions of different fiber generations vs. recovered paper utilization rate[61].

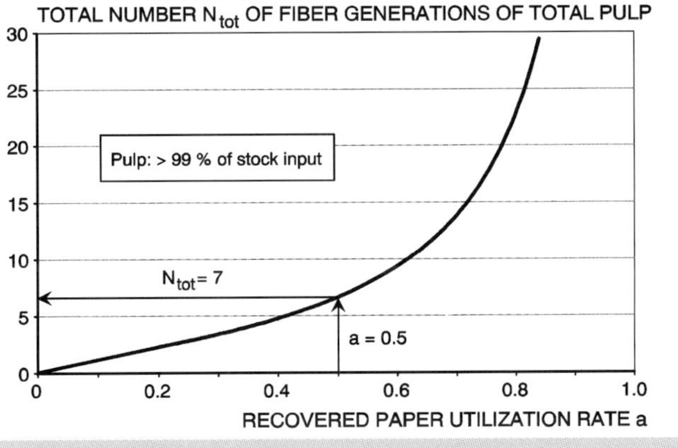

Figure 24. Total number of fiber generations of total pulp made from virgin and recycled fibers[61].

(100%), this function heads toward infinity. At $a = 0.5$ (50%), six generations of recycled fibers are contained in the pulp although the other 50% of the pulp comprises virgin fibers. Because of the greatly increasing number of generations with an increasing a, this representation of the total number of fiber generations can lead to a dramatic misunderstanding and become technically irrelevant due to the fact that fiber generations that account for only a very low percentage of the total pulp are included in this total number.

Instead of the total number of generations describing the RCF age structure of a pulp, an average age, $\bar{N}$, of the fiber generations as defined by Eq. 8 and Fig. 25 is more relevant from the technical point of view.

$$\bar{N} = \frac{a}{1-a} \quad 0 \le a < 1 \tag{8}$$

With a recovered paper utilization rate of $a = 0.5$ (50%) for example, the average age of all fiber generations of the pulp or the RCF-containing paper used in its production at $\bar{N} = 1$ is still very low due to the 50% share of virgin fibers in addition to the 50% share of recycled fibers with six generations. In this case, the proportions of virgin fibers and the first and second generation of RCF dominate in the pulp used for paper manufacture as Fig. 26 shows. Together, they constitute 90% of the pulp at a utilization rate of $a = 0.5$ (50%). With a utilization rate of $a = 0.67$ (67%), the average age increases to 2 fiber generations with 70% combined proportions of virgin, first, and second generation fibers.

The described one-parameter model was discussed from the point of view of a national recycling system. In the final analysis,

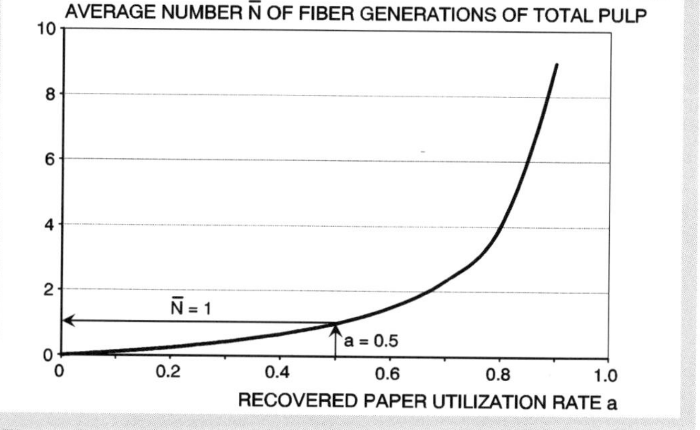

Figure 25. Average number of fiber generations of total pulp made from virgin and recycled fibers[61].

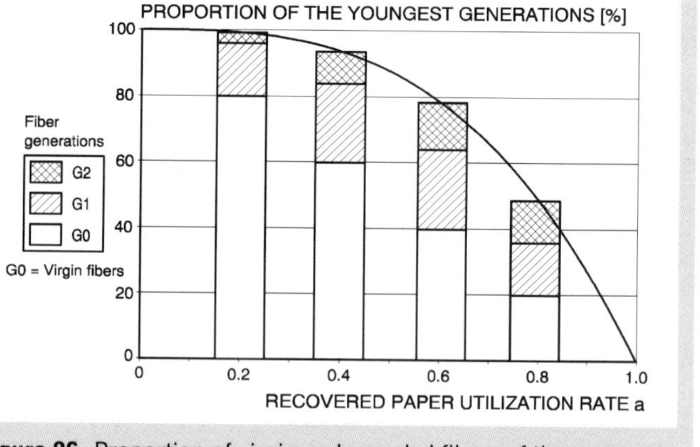

Figure 26. Proportion of virgin and recycled fibers of the youngest generations[61].

it is valid for a global recycling system to the extent that all paper grades produced worldwide have an identical recovered paper utilization rate.

3.2 Two-parameter model

To approximate the situation in industry in this case, one assumes that a certain paper grade in one country or in the world is produced from a recycled fiber mix based on two different recovered paper components and from a proportion of virgin fibers. Each of these two recovered paper components has its own origin or pulp composition especially in relation to the proportion and age structure of RCF in the furnish. This is the situation that occurs in most countries. When manufacturing newsprint from old newspapers and magazines, the following features apply:

- The old newspapers uses newsprint produced from virgin fibers and/or recycled fibers.
- The magazine papers [supercalendered (SC) and lightweight coated (LWC) papers] from which the magazines were produced consist exclusively or primarily of virgin fibers. For purposes of simplification, SC and LWC papers are not discussed separately in the following. Magazine papers are considered collectively instead.

Figure 27 shows the different origin of the two processed recovered paper components in relation to the fiber furnish used. Paper B represents magazine papers, and paper A represents newsprint. The simplifications of the one-parameter model apply especially regarding the assumption that no yield losses occur during recovered paper processing.

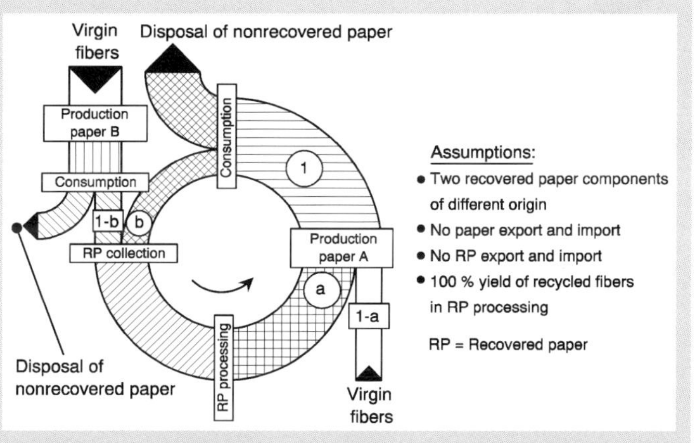

Figure 27. Two different recycling loops of recycled fibers for the two-parameter model[61].

Besides the utilization rate of recovered paper, a, taken thus far, the parameter, b, is also necessary. The latter parameter describes the proportion of the RCF-containing recovered paper in the stock for the paper that is to be produced. The expression $(1 - b)$ corresponds to the proportion of the recovered paper whose original paper consists exclusively of virgin fibers. If $b = 1$, the two-parameter model as a borderline case becomes the one-parameter model with the sole variable, a.

With the two parameters a and b, calculating the age structure of the recycled fibers in a pulp such as for newsprint is still relatively easy. The paper A that is to be produced uses the recovered paper utilization rate, a, that comprises an "old" RCF component, b, and a "new" RCF component, $(1 - b)$. For paper A with its proportions g_A, the following equations calculate the number and proportions of the different fiber generations:

$$g_{0,A} = (1 - a) \tag{9}$$

where $g_{0,A}$ is the proportion of the zero generation = virgin fibers

$$g_{1,A} = a(1 - a \cdot b) \tag{10}$$

where $g_{1,A}$ is the proportion of the first generation of recycled fibers

$$g_{2,A} = a^2 b(1 - a \cdot b) \tag{11}$$

where $g_{2,A}$ is the proportion of the second generation

$$g_{n,A} = a^n b^{n-1}(1 - a \cdot b) \tag{12}$$

where $g_{n,A}$ is the proportion of the nth generation.

The following equations apply for the total proportions of all fiber generations up to generation *n* for paper grade A:

$$S_{n,A} = g_{0,A} + g_{1,A} + g_{2,A} + \ldots + g_{n,A} \tag{13}$$

$$S_{n,A} = (1 - a) + a(1 - a \cdot b) + a^2 b(1 - a \cdot b) + \ldots + a^n b^{n-1}(1 - a \cdot b) = 1(100\%) \tag{14}$$

$$0 \le a < 1$$

For purposes of greater clarity and technical relevance, one can also calculate the average age of the fiber generations in the total pulp or in the paper grade A produced from it that is manufactured from recovered paper components with their different origin and from virgin fibers.

$$\overline{N}_A = \frac{a}{1 - a \cdot b} \qquad 0 \le a, b < 1 \tag{15}$$

If $b = 1$, Eq. 15 is identical to Eq. 8. Figure 28 shows the average fiber age of the produced paper grade from two different recovered paper components in relation to the proportion of recovered paper, a, and the additional variable, b. The curve with $b = 0.25$ represents a newly produced paper that consists of 75% recovered paper containing only first-generation recycled fibers, e.g., 25% old newspapers and 50% magazines both using paper produced from virgin fibers.The remaining 25% of the newly produced paper used a recovered paper component containing recycled fibers from several generations since the original paper, e.g., newsprint, was already produced with RCF. The curve $b = 1$ represents a newly produced paper whose recovered paper consists entirely of recycled fibers belonging to numerous generations (See one-parameter model). If $b = 0$, the recycled fibers consist exclusively of first-generation fibers so that with $a = 1$ the average number of fiber generations is $\bar{N} = 1$.

According to Fig. 28, a national recycling system that has only recovered paper grades at its disposal containing recycled fibers of many generations $(b = 1)$ would undoubtedly experience quality-related problems when the recovered paper utilization rate, a, exceeds 75% for example. This would be due to the disproportionately increasing average age of the recycled fibers. A national recycling system therefore has a major advantage if the used recovered paper also contains paper grades produced from virgin fibers. To the extent that the national recycling system is characterized by a high utilization rate, a, of recovered paper, a lower value of the variable, b, could be realized by imports of papers produced from virgin fibers. With this in mind, comparing the recovered paper utilization rate, a , on an international scale has little relevance as long as no distinction is made in the import ratio and the imported paper grades regarding their composition of virgin and recycled fibers.

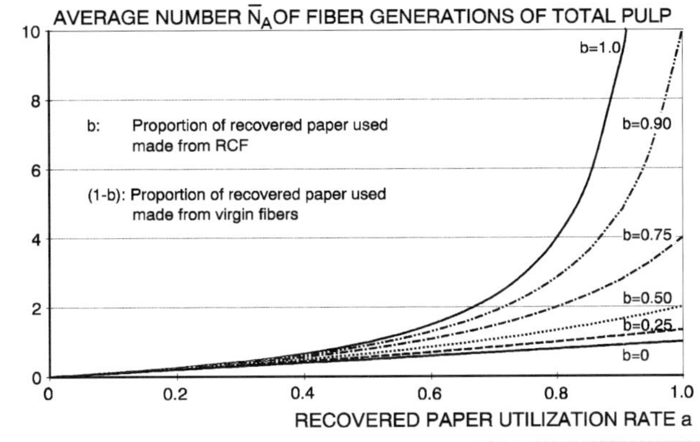

Figure 28. Average number of fiber generations of total pulp for paper A made from virgin and recycled fibers based on two different recovered paper components[61].

3.3 Multiparameter model

The next case describes the more practical example of newsprint production from recovered paper. Figure 29 shows the qualitative and nonquantitative characteristics of the paper and recovered paper components or pulp flows. Newsprint produced nationally uses virgin fibers and recycled fibers. The same applies for the nationally produced magazine papers and the imported newsprints and magazine papers. The composition of the recycled fibers from a recovered paper mix used to produce deinked pulp is more complicated. This domestic recovered paper mix consists of the following components:

- Old newspapers using virgin and recycled fibers, e.g., 45% by weight
- Old magazines consisting of SC and LWC papers using virgin and recycled fibers, e.g., 45% by weight
- Paper products manufactured from wood-free papers using virgin fibers, e.g., 10% by weight not shown in Fig. 29 to ensure greater clarity.

These four paper grades can therefore have been produced at home and abroad from mixtures of virgin and recycled fibers, only from virgin fibers, or only from RCF. This results in a complicated stock system organized by paper grades, fiber mixes, and the age of the fibers in the different paper grades.

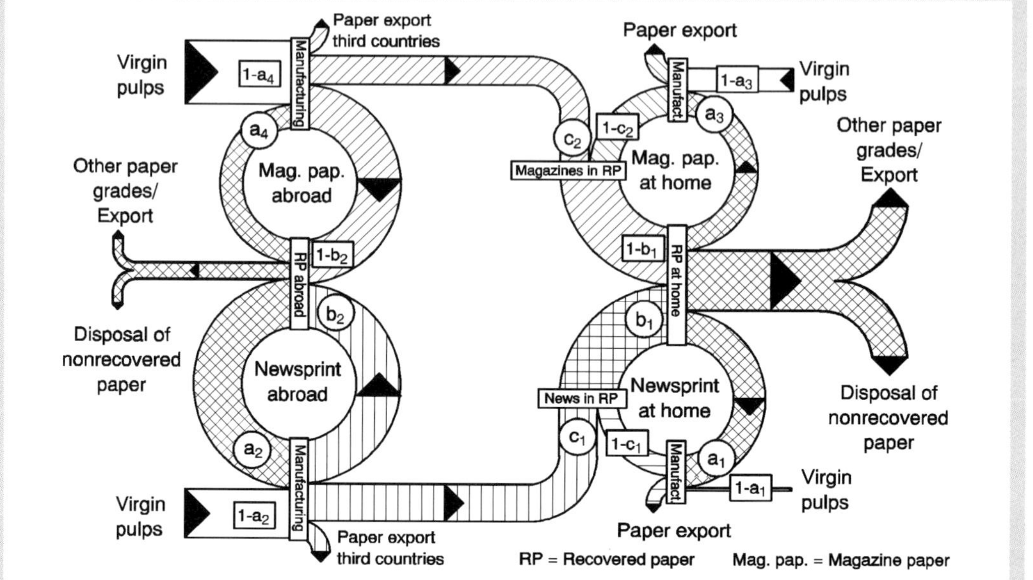

Figure 29. Recycling loops of newsprint and magazine papers produced at home and abroad[61].

Table 5. Utilization rates of different paper components of German deinking material for newsprint production.

Paper components of deinking material	Variable according to Fig. 29	Recovered paper utilization rates in paper production [%]	
		case 1	case 2
Domestic newsprint	a_1	90	115
Domestic SC paper	a_3	0	20
Domestic LWC paper	a_3	0	20
Domestic fine paper	-	0	0
Imported newsprint	a_2	20	40
Imported SC paper	a_4	0	10
Imported LWC paper	a_4	0	10
Imported fine paper	-	0	0

One must therefore consider that the newsprint used in domestic newspaper production originates at home and abroad, and the recovered paper utilization rate of the domestic and international newsprint varies as Table 5 shows. This same principle also applies for magazine papers manufactured from SC and LWC papers. For this reason,

18 variables are possible in the multiparameter model that can no longer be condensed in simple equations as was the case of the two-parameter model. Using special computer programs, one can calculate the average age of the fiber mix of virgin fibers and recycled fibers for the national newsprint. Figures 30 and 31 illustrate this for two examples. The calculation of the number, the average age, and proportions of the RCF generations contained in newsprint used the situation for 1994 (case 1) and for the future (case 2), both in Germany.

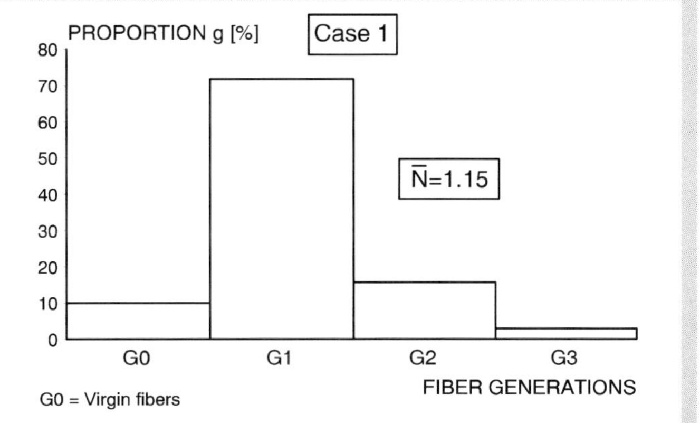

Figure 30. Number and proportions of fiber generations present in pulp for domestic newsprint made from virgin and recycled fibers in 1994[61].

The other parameters not described here relate to the proportions of the recovered paper components (old news and magazines made from SC, LWC, and fine papers) of the deinking material used at home and abroad and the ratios between domestic consumption of the four home-produced and the corresponding imported paper grades.

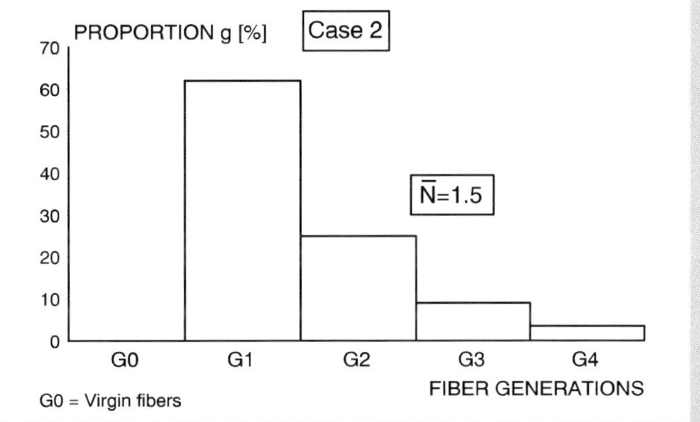

Figure 31. Number and proportions of fiber generations present in pulp for domestic newsprint made from recycled fibers in the future[61].

Figures 30 and 31 provide the following conclusions:

Situation in 1994 (case 1)

- Despite the high recovered paper utilization rate of 90% in domestic newsprint, no more than four fiber generations (three RCF generations and a 10% proportion of virgin fibers) exist as Fig. 30 shows.
- The average age of the mass-weighted proportions of virgin and recycled fibers does not exceed $\overline{N} = 1.15$ generations. The recycled fiber-containing newsprint is therefore younger than one would intuitively assume for such a

high recovered paper utilization rate. This low average age of the fiber furnish of newly produced newsprint is affected by domestic and imported magazine and fine papers made from virgin fibers that become recycled fibers of the first generation. Furthermore, the imported newsprint consists predominantly of RCF of the first generation, too.

Situation for the future (case 2)

- An increase in the recovered paper utilization rate in domestic (90% to 115%) and imported (20% to 40%) newsprint and in domestic SC and LWC paper (0% to 20%) results in four generations of recycled fibers in the pulp of newly produced newsprint as Fig. 31 shows.
- The average age of the pulp rises to $\overline{N} = 1.5$ generations. These conditions do not lead to any expectations of losses of quality in the newly produced newsprint. Note that countermeasures by intensified stock upgrading in its processing are taken by two-stage flotation deinking and dispersing for example.

4 Recycled fiber in papermaking

4.1 General aspects

Recycling experiments show that fundamental changes can occur in fiber and pulp characteristics as discussed earlier in this chapter. The conclusion of laboratory and pilot scale studies is that recycled fiber is not better or worse than virgin fiber, but it is different. The following sections discuss practical effects of fiber recycling related to the behavior of stocks during papermaking and the quality of products.

The most apparent quality differences with virgin fiber are ink specks and reduced brightness of unbleached deinked pulps since the ink removal processes do not perform perfectly. Strong hot dispersing can make all ink specks very small so they become invisible to the naked eye. Such smaller ink particles reduce pulp brightness and are more easily caught by fibrillated fibers. Aging of recovered paper also affects brightness and shade of the final recycled fiber. Bleaching can compensate for a decrease of brightness, but high brightness levels are difficult to achieve. Chapter 9 of this book discusses bleaching of deinked pulp.

Probably the most objectionable characteristics of recycled pulps for papermaking are their residual contaminants such as inks, adhesives, glues, plastics, etc., that can impact the runnability of paper machines by deposition on rolls and paper machine clothing. The content of stickies in pulps depends on the collection and processing systems of recovered paper. Various mechanical and chemical means for controlling detrimental effects of stickies are available. Chapter 11 of this book describes these in detail.

Lack of uniformity is an inherent problem of recovered paper when trying to predict the behavior or quality potential of recycled pulp. Although definitions for various recovered paper grades exist, collection systems take in different types of fibers and other paper furnish materials such as mineral pigments in variable amounts. In recovered paper from households and print rooms for deinked pulp (DIP) furnish for example, the proportions of mechanical and chemical fibers are different depending on geographical region.

4.2 Downgrading and upgrading of recovered paper

4.2.1 Downgrading

In theory, the processing of recovered paper should produce a RCF pulp that is suitable as furnish for the same paper grade as the recovered paper itself. For various reasons, one cannot completely apply this principle. Very often the use of RCF pulp must be confined to less demanding paper. This implies downgrading. The reasons for this include the following:

- The quality of fibers or the removal of contaminants in the recycling process are insufficient to permit the re-use of the RCF pulp in the original product.
- If recovered paper is a mixture of different types of printed products and fibers, the lower quality fibers in the mixture often restrict the end use of RCF. In such a case, the higher quality grades in the recovered paper mixture are downgraded.
- A higher grade of recovered paper can be used in a lower paper grade to compensate for shortcomings caused by other furnish components. For instance, producers of DIP-based newsprint often use controlled addition of old magazines to improve deinking performance and brightness.
- Coating of paper normally improves its quality. Coated paper grades are normally downgraded when recycled. From a recycling point of view, coating brings fine pigments and binders into RCF stock. These are less suitable as fillers and may negatively influence drainage. Coated papers can therefore force DIP producers to emphasize ash removal in a deinking operation.
- Certain paper grades will never come back into the production of the original type of papers. Many printed products have use as furnish for packaging papers and board, and certain tissue grades cannot be recycled. This does not necessarily imply downgrading of the RCF.

4.2.2 Upgrading

In principle, the upgrading of RCF is possible as illustrated by the following cases:

- The quality of RCF pulp can be improved beyond its state in the original product by bleaching or other chemical and mechanical treatments. Fractionation of DIP into long and short fiber classes that are subsequently treated separately and bleached for use in high quality printing paper grades is sometimes given as an example. This application is very exceptional in practice since papermaking potential of purely short or long fibers has several shortcomings. Fractionation of RCF has most often applied to old corrugated containers (OCC) or mixed (brown) recovered paper for multi-ply board applications.
- Small amounts of lower quality recovered paper sometimes come with better grades that still have the capacity to “hide” or compensate shortcomings in the

final RCF pulp. For instance, RCF pulp can be classified "wood-free" after bleaching although it contains 5%–10% mechanical fibers.

- RCF with inferior visual appearance can easily be "hidden" within the sheet by stratification such as in the multi-ply forming of various boards. This possibility has been discussed by producers of printing paper for some time, but no successful commercial application of this principle exists. The stratification might also bring benefits by hiding micro stickies.
- The use of relatively low quality RCF in the base paper furnish of coated grades can also be regarded as a similar "hiding" principle that will probably increase in the future. This will depend on the development of film transfer coating technology, i.e., roll coating, that has less demanding base paper quality requirements than today's dominant method of blade coating.
- A special case of upgrading is to use relatively low grade RCF in a better paper grade and to accept the resultant quality loss. For example, office paper that has a high and recognizable content of RCF up to 100% can be in demand due to "environmental" reasons.

4.3 Effects of deinking process yield on pulp characteristics

The target of all RCF deinking processes is to remove contaminants and printing inks that inevitably cause some losses of fibers, fines, and fillers. A direct relationship generally exists between flotation loss and brightness of final DIP. By intensifying flotation performance for example, the level of both variables increases in most cases.

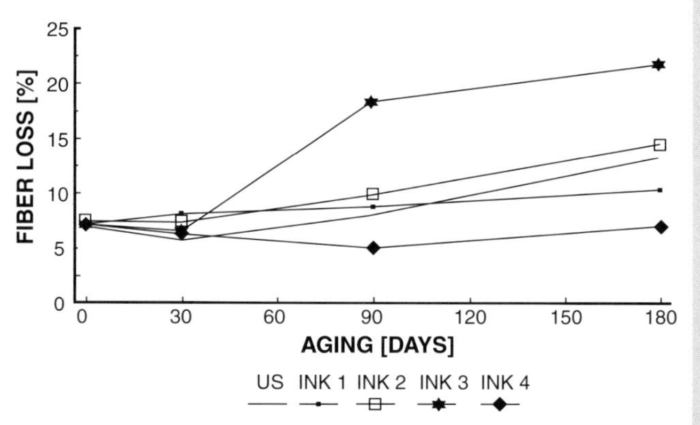

Figure 32. Fiber losses in flotation of different offset newspapers vs. aging[41].

Poor detachment of printing inks can affect not only pulp brightness but also yield of flotation. The aging of certain types of offset inks can significantly influence the deinkability of printed products[41]. Unsaturated bonds of carbon chains in those ink binders open due to oxidation and can interact with reactive groups of fibers[42]. This results in many ink specks attached on fibers and greater losses in flotation probably because fibers and fines have become more hydrophobic. In a pilot flotation of different TMP-based newspapers[41], losses increased from a level of 7% to 15%–20% with an ink prone to aging reactions as Fig. 32 shows.

Unfortunately, this trial did not determine whether these excessive losses were more concentrated in the long fiber fraction than in the fines. Aging propensity of inks is often part of recyclability evaluations of printed products as an important criterion.

Some investigations[3,36] have shown that the presence of residual printing ink has a very small influence on tensile strength or pulp drainage if the ink content is below 1% of the pulp. Only in DIP made of office paper could flat and flaky particles of toner inks strongly interfere with fiber-to-fiber contacts.

4.3.1 Losses of fines and mineral pigments in deinking process

Analyses of the compositions of the reject and the accept of a deinking process can reveal changes in the proportions of pulp components. The changes generally depend on raw material, design, and operation of the process. The results of a study made in several production plants can be summarized as follows[63]. In the accept of flotation deinking, the pigment content of the original furnish is significantly reduced. The proportion of fiber fractions (+ 200 mesh) is increased. The proportion of wood-based fines (-200 mesh) is approximately unchanged corresponding to their more than proportional removal rate.

Some recovered paper grades contain fillers and coating pigments in great quantities that have use as a replacement for fresh filler to a certain extent to improve optical properties of printing paper. A part of pigments are sometimes removed intentionally in a deinking process since an excessive ash content in a paper furnish reduces bulk and makes a sheet too limp for many end uses. High pigment content is also detrimental for softness of tissue, may cause a risk of blanket deposits in offset printing, and generally decreases strength properties in all paper grades.

A very interesting study on the present state of practical operations is that of the International Deinking Research Association (INGEDE) in Munich. This group characterized 20 deinking processes in European member mills producing pulp for various end uses (mostly for newsprint)[64]. The ash content varied widely depending greatly on the proportion of magazines in the recovered paper. The average ash content after reslushing of recovered paper was 21% (the maximum was as high as 30%) from which level it decreased to 13.5% during the deinking process. Reduction of calcium carbonate content was two times higher than that of kaolin probably due to differences in surface characteristics. An inverse relationship existed between strength properties and ash content in the final DIP, but the correlation was not high because other pulp components also varied considerably. For example, the chemical fiber content was 22%–60%.

The design of a process or appropriate flotation chemistry can considerably enhance the ability to remove ash. For example, suspended solids of white water in a process loop can be partly or totally removed by microflotation. A primary wash deinking stage can remove ash from the pulp flow almost completely. These operations also remove fiber fines as effectively as they remove fillers and coating pigments.

The loss of fiber fines can have a remarkable influence on sheet properties. When using (wood-containing) deinked pulp in printing paper for example, the lack of fines might become a disadvantage for several reasons. The fines are an essential factor in development of opacity and smoothness.

The desired fines yield generally depends on the end use of the pulp. In DIP processes for tissue grades, minerals and fines are almost completely eliminated by washing due to quality reasons. In packaging material recycling, the poorly bonding, hornified fines can accumulate and impair drainage so much that they may be removed like filler.

4.3.2 Fiber yield in deinking flotation

The high retention of the long fiber fraction is a desirable characteristic of all RCF processes. Flotation operating conditions or the hydrodynamic design of cells[65] can control this in deinking. In a cell, long fibers distribute themselves unevenly and concentrate into flocs depending on pulp consistency and turbulence. Shorter fibers and fines that cannot form networks concentrate in interfloc regions where they are hydraulically carried into foam with rising water and air bubbles that pass around the flocs. This is why the minerals and fines are more concentrated in the flotation reject. At low pulp consistency or under high turbulence, the flocs are broken. More long fibers become trapped by bubble flow into the foam and are therefore lost from the pulp yield. A highly flocculated pulp is less effective for ink removal since the air bubbles cannot contact the ink particles existing within flocs. A flotation cell should therefore first comprise a "highly turbulent" zone to maximize bubble to ink particle collisions followed by a more "quiet" flotation zone where the ink-laden bubbles migrate between the fiber flocs to become rejected.

A high calcium ion content is believed to make long fibers hydrophobic or to promote precipitation of ink on them to increase fiber losses[66]. The adhesion between large bubbles and long fibers has not been definitely shown. The hydraulic transport mechanism described above[65] therefore seems likely to be more dominant than possible surface chemistry related mechanisms. The surface chemistry mechanisms do explain the differences in yields between pigments and fibrous material during flotation. Screening and cleaning stages also require attention when improving fiber yield in recycling operations.

A traditional belief is that worn and contaminated fibers are selectively removed in deinking flotation. A comprehensive microscopic analysis[63] of long fibers in accepts and rejects was therefore made in full-scale flotation tests. The results showed no differences in fiber length or surface characteristics in fibrillation of fibers between accept and reject. The ratio of mechanical and chemical fibers was approximately the same for reject and accept. The differences in lignin content of fibers therefore do not play any significant role in flotation. Ink clearly contaminated the fibers in the reject. Determining the extent to which those fibers had been selectively floated or had simply adsorbed ink from flotation foam was difficult.

4.3.3 Effect of flotation on strength properties

Seeing the development of strength properties in industrial deinking flotation is interesting as changes in ratios of pulp components and fractions occur. In one study[67], pulps of three European plants using recovered paper collected from households were analyzed after reslushing and after the main flotation before the hot dispersion stage as Table 6 shows. The ratio of old newspapers and magazines differed considerably in their raw materials. On average, the mineral pigment reduction was about 40%, the increase in long fiber fraction was about 7% of total fiber furnish, and the proportion of wood-based fines increased slightly.

Table 6. Development of strength properties in industrial flotation processes[67].

Property	After disintegration	After flotation
Breaking length [km]	3.73	4.33
Burst index [kPam^2/g]	2.06	2.56
Tear index [mNdm^2/g]	911	1.006

A remarkable rise in bonding ability occurred due to loss of filler and coating pigments and growth of tear strength caused by the increased proportion of long fibers. The more than proportional loss of fiber-based fines did not negatively influence bonding potential. The higher amount of ink contaminants in an unflotated pulp may somewhat diminish strength properties compared with flotated pulp.

4.4 Improving papermaking potential of RCF by mechanical treatment

Pressurized or atmospheric hot dispersion or kneading stages are today an established part of RCF processing. These include dewatering up to consistencies as high as 25%–35% solids and an intense mechanical treatment at a chosen temperature. The operations effectively disperse stickies or ink specks, eliminate bacteria or their by-products, and provide intensive mixing for bleaching chemicals at high pulp consistency. Chapters 5 and 6 describe those functions in detail. An additional function of those process units can also be to modify pulp quality. For some paper and board grades, a refining stage at high or low-consistency is part of the RCF process specifically to improve the papermaking potential of hornified chemical fibers.

4.4.1 Effects of refining on mechanical fibers

Particularly when using recovered paper collected from households as furnish for high-quality printing grades, the stiff mechanical fibers originating from old news can create quality problems. The mechanical pulp component for those higher grades requires processing to give a good balance of strength, printability, and optical properties. In other words, the material should be an ideal blend of long, slender fibers with many finer particles contributing to bonding, surface smoothness, and light scattering. Good bonding ability in a pulp can also permit high filler loading of printing paper if necessary. For these grades, hot dispersion in several stages or combining with post-refining is therefore often recommended. This kind of operation starts to resemble the reject refining of TMP plants. An understanding of general refining mechanisms and the response of wood fibers to beating is therefore necessary when trying to improve quality of DIP by mechanical treatments.

Figure 33 shows a typical softwood tracheid representing 90%–95% of the fibers. The remainder primarily consists of very small (0.1 mm long) parenchyma (ray) cells. The S2 layer is the thickest part of the fiber wall. It determines whether the fiber belongs to thick-walled summerwood (latewood) or thinner-walled springwood (earlywood) fibers. Because of the much thicker cell wall, summerwood fibers have a considerably smaller lumen than springwood fibers. Springwood fibers collapse during drying more than summerwood fibers.

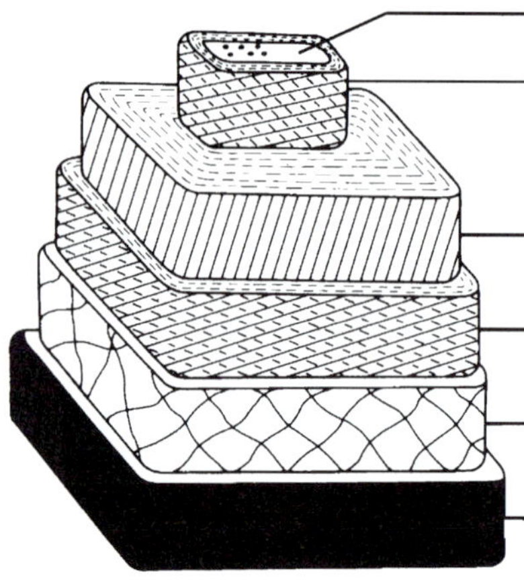

Figure 33. Schematic image of softwood fiber[68].

Stiff mechanical long fibers alone are not sufficiently flexible to form good bonds. If a laboratory sheet forms solely with the long fiber fraction, the fiber network will loosen by the so-called spring back effect after wet pressing and during drying. To counteract this, the flexibility of those fibers should be increased through a refining process. The more slender, refined fibers can then overlap other fibers and increase bonding areas to produce a denser and smoother sheet. Increased flexibility of coarse TMP fibers by refining is primarily due to loosened cell wall structure, i.e., internal fibrillation. Changes in the cross-sectional dimensions remain minor[69]. Selective refining of the separated coarse fiber fraction more effectively improves fiber flexibility than refining of the entire TMP[69]. In particular, the thick-walled summerwood fibers require strong refining intensity in that action.

Refining of mechanical fibers is a stochastic process in which their length shortens due to breaking and rupturing. The fiber wall thickness also decreases. Increased energy consumption simultaneously enhances both factors. During refining, the proportions of the middle-size fraction (48–200 mesh) and fines (–200 mesh) increase due to fragments of fibers and cell walls or wider, ribbon-like particles peeled from the S2 layer[22,68]. Generated in the beginning of the pulping process, chunky fiber fragments are often called primary fines. When refining proceeds to the S2 layer, so-called secondary fines with lamellar and ribbon-like shape occur. The secondary fines generally contribute to bonding ability. The primary fines give sheet smoothness and optical properties[68].

4.5 Effects of refining on deinked pulp

Improving the papermaking potential of DIP is a complex task because the material consists of chemical and mechanical fibers from old newspapers and magazines. Low-consistency refining is common particularly for post-refining to reduce the content of coarse fibers or shives. Due to fiber shortening, a decrease in tear strength is often inevitable in that type of refining. To maintain tear, moderate specific edge load (SEL) and energy consumption are recommended[70]. Low SEL does not develop tensile strength as fast as higher SEL refining. A decrease in coarseness of the long fiber fraction and in the light scattering of the sheet also indicates the generation of secondary fines in low-consistency beating.

Development of bonding ability in low-consistency refining also depends on the composition of a DIP furnish. In a beating trial, steady improvement of tensile strength for a prolonged time was characteristic of DIP containing a high proportion of chemical fibers[71]. With a furnish comprising more mechanical fibers, the growth of tensile strength leveled soon, but the drop in tear strength was considerable with further beating. That kind of furnish is easily over-refined at low-consistency.

The smoothness of paper measured by air leakage methods improves with low-consistency refining[72], but the improvement in final printing smoothness using the rotogravure process has not been consistently better[73]. The cut pieces of coarse fibers have probably not received adequate refining in low-consistency, or the secondary fines derived from the chemical pulp fraction have unfavorable effects on rotogravure printability (See section 4.8.2 of this chapter).

Any benefit to refining separately only long fibers of DIP instead of refining the entire pulp is unclear. Fractionation into long and short fiber fractions is possible with a properly profiled slotted screen. In a pilot scale trial[73] with DIP, the + 28 mesh fraction was separated almost completely from the remainder of the pulp for high-consistency refining. After each refining stage, the long fiber fraction was again combined with the short fiber fraction. The sheet properties of that blend were compared to those of the unfractionated pulp refined to the same freeness[73]. Surprisingly, the separate treatment of the long fiber fraction took more energy and the improvements in sheet smoothness and tensile strength were not significantly better those of the refined entire DIP.

Remember that the long fiber fraction of DIP is an uncontrolled blend of mechanical and chemical fibers that require different treatment to improve strength properties or printability. Chemical fibers also react in a particular way to refining at high-consistency and temperature as the following discussion shows.Today's fractionation technology cannot selectively separate those fiber components.

4.6 Effect of refining on hornified chemical fibers

The basic difference between recycled chemical and recycled mechanical pulps is the hornification effect in the cell wall structure of chemical fibers as described earlier. Hornification manifests itself in a reduced accessibility of water into the fiber wall when exposed to water. This reduced swelling ability is most evident in the water retention value (WRV) test. The impaired swelling can be seen in microscopic examinations as a reduction in cross-sectional fiber wall area and thickness[74]. The lumen and outer diameters of rewetted fibers decrease on drying. The entire fiber can collapse and flatten. Thick-walled summerwood fibers especially contract in this treatment. Figure 34 is a schematic drawing of morphological changes in a wet chemical fiber[75].

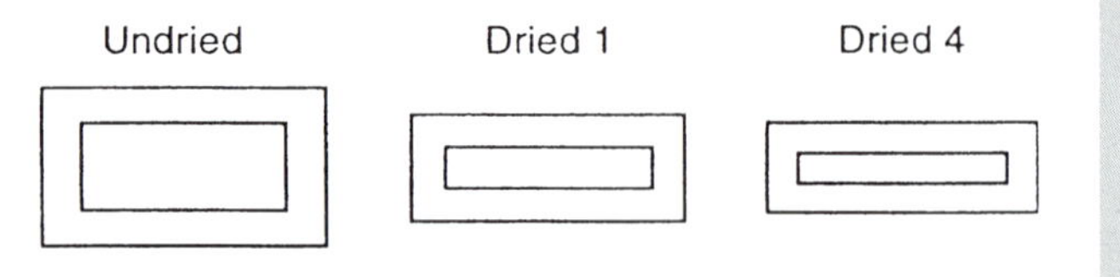

Figure 34. Schematic drawing of undried fiber and various times dried and wetted unbeaten fibers[75].

Hornification studies have primarily been made by recycling unbeaten chemical pulps. In papermaking, the recycled wood-free pulps have always undergone refining to some extent to improve the strength properties. Whether the original hornification is removed by refining or the restoration of swelling ability is a result of new changes in the fiber wall is unknown. Microscopic examinations showed that during continuing PFI mill refining the outer perimeter of a recycled kraft fiber first increases and then decreases. The fiber wall thickness steadily grows due to swelling toward the lumen[76]. That decrease of outer diameter is probably due to removal of material from surface parts of the fiber. A circular cross-sectional shape about the same as in virgin fiber will be partially regained by refining[76]. Figure 35 shows this decollapsing phenomenon of wet fibers schematically.

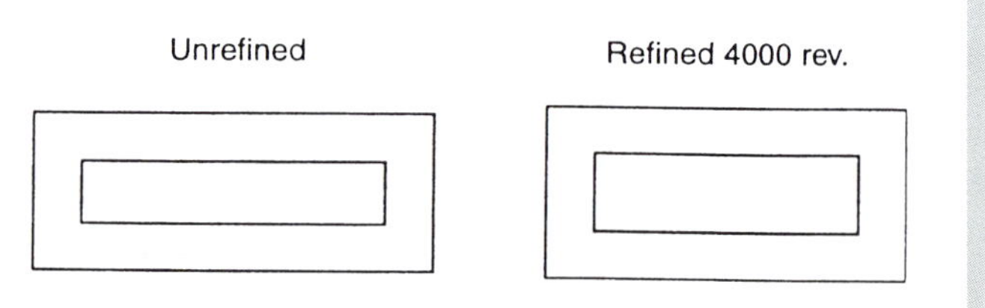

Figure 35. Schematic drawing of mean chemical fiber dimensions of dried, unrefined and dried, refined fibers[75].

Besides morphological changes in the fiber wall structure, the mechanical action of refining also peels secondary fines from the outer layers of recycled chemical fibers. When refining begins to affect the ordered structure of the S2 layer in a fiber as Fig. 33 shows, secondary fines will be created primarily containing lamellar, ribbon-like material. Secondary fines of beaten chemical pulps are found to a great extent to reside in spaces between the outer fibrils on fibers[77]. They therefore contribute to interfiber bond formation considerably. The primary fines of recycled chemical pulps such as parenchyma cells and the fines coming from the previous papermaking cycle are present already before refining. The WRV of primary fines falls with drying similar to that of long chemical fibers[3], but it cannot be restored by beating.

In one experiment[77], the secondary fines generated by refining from sheets that were subjected to harsh pressing and drying treatments proved surprisingly to be as effective in enhancing bonding and elastic properties of the pulp sheets as control fines refined from virgin fibers. Hornification of recycled chemical fibers does not influence the strength potential of the secondary fines gained by beating.

The sheet strength improvement with refining can considerably impair the drainage of recycled chemical pulp. In one study[78], the burst strength of virgin kraft pulp at 585 mL CSF was a reference. With each recycling, the reslushed stock was refined to the extent necessary to achieve the reference burst strength. At a constant burst strength, results showed that the sheet density, tensile strength, and other bonding related properties slightly increased in this type of recycling. Fiber length and tear strength considerably decreased. The biggest penalty was the decrease of freeness from 585 mL for the virgin pulp to 455 mL on the first cycle, 250 mL on the third cycle, and 90 mL on the fifth cycle to maintain the burst strength at the reference level.

Kneading or hot disperger stages of RCF processes cannot restore the loss of swelling ability in fibers by hornification. High-speed dispersing generally improves tensile and burst strengths depending on temperature[79]. Low-speed kneading causes a slight decrease in those properties. The air permeability of sheets also increases. Besides the beating effect, fibers are subjected to high-consistency and temperature, physical conditions that cause latency as discussed later. Low-consistency beating is normally included in wood-free RCF processes to remove curl or latency of fibers caused by high-consistency refining, hot dispersing, or flash drying of market pulp.

4.6.1 Effect of chemical conditions on refining results

Dry strength additives such as starch are useful to improve interfiber bonding without sacrificing as much dewatering ability of wood-free stock as with refining. The penetration of these additives into micropores in the cell wall structure is limited due their molecular size. They are therefore not a full substitute for beating in restoration of the swelling ability in hornified fibers. Chemical treatments with oxygen, peroxide, and alkali can sometimes be included in a recycling process. Alkali helps in the swelling of fibers by opening intrafiber bonds. In one study[80], pulp made of old corrugated containers (OCC) was soaked with 2% caustic soda for 4 h before refining in a PFI mill. Figures 36 and 37 show the results.

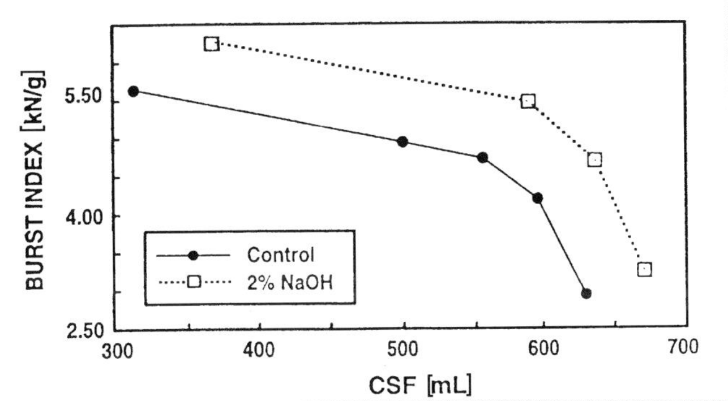

Figure 36. Effect of refining on burst strength.

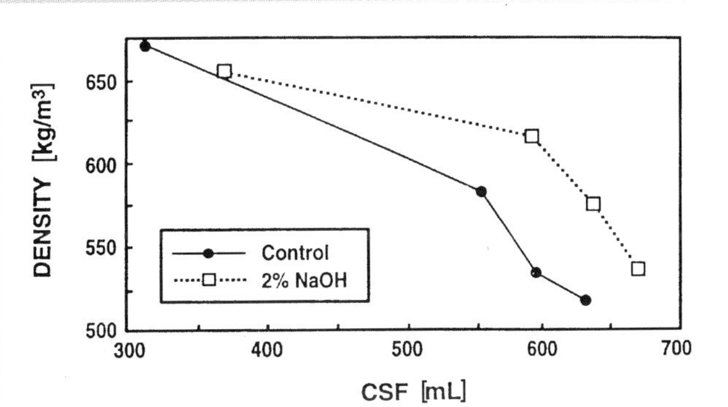

Figure 37. Effect of refining on sheet density[80].

The burst index curve of Fig. 36 is about 25%

higher for the NaOH treated handsheets than for the control made at pH 5.5. This swelling effect due to NaOH is also evident in sheet density particularly at higher freeness levels of pulp as Fig. 37 shows before the new generation of secondary fines created by further refining starts to level it.

As stated in section 1.2.3. of this chapter, alkaline conditions create osmotic pressure within a fiber wall microstructure. Acidic carboxylic groups dissociate in the presence of sodium cations. This chemically contributes to fiber swelling. In OCC furnish for example, more of these ionizable groups exist compared with bleached kraft fiber. Internal fibrillation due to refining also makes these charged sites in the cell walls more accessible for cations in the system. In contrast with sodium ions, multivalent cations or acid papermaking will hinder the contribution of charged groups to fiber swelling. In the study[80] quoted above, the handsheet forming at a reduced pH already induced a considerable loss in sheet strength from that achieved by soaking the stock in caustic soda.

4.6.2 Effect of elevated temperature on RCF during dispersing or refining

Hot dispersing and kneading are the stages of the recycling process that can cause changes throughout the fiber structure particularly in the properties of chemical fibers. Typically after thickening to 25%–35% consistency, pulp is steam heated for 2–3 min to an input temperature of 80°C–120°C. Higher temperatures are often preferable to kill bacteria and improve dispersion of contraries. The net energy input is generally moderate at 30–80 kWh/t. About 85% of the refining energy converts into heat so that pulp temperature might be 5°C–15°C higher than the input temperature for a moment.

Heat treatment changes the ability of a fiber to take up water[81]. The ability of cellulose and hemicelluloses to take up water and swell is an exothermic reaction. The fibers are fully saturated with water during treatment in a hot disperger or refiner. The maximum possible moisture content measured as WRV or fiber saturation point and the corresponding swelling of fibers decrease with increasing temperature. In other words, fibers deswell with increased temperature. Part of this deswelling temporarily remains as a "hysteresis" after cooling. The recovery of the swelling ability is slow. The temperature dependence of moisture sorption and desorption into fibers is described in more detail in the book on *"Forest Products Chemistry"* in this series. Figure 38 shows this deswelling at 90°C and the hysteresis during cooling for

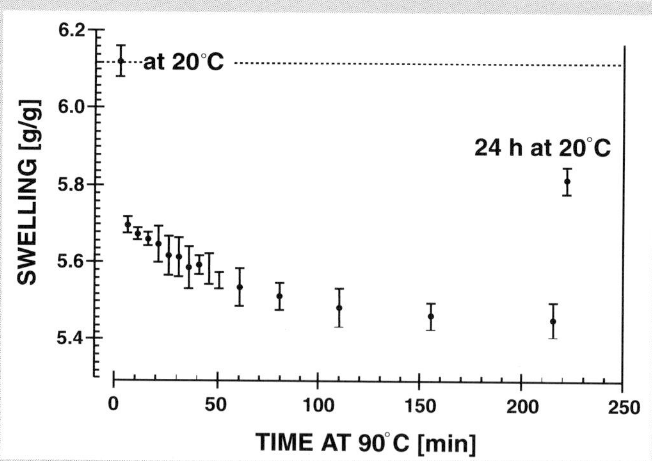

Figure 38. Water uptake or swelling in g/g for original pulp (dotted line), deswelling under mild stirring at 90°C, and swelling after subsequent recooling back to 20°C[81].

a regenerated cellulose gel. At higher temperatures above 100°C, the deswelling effect would be more evident.

This fiber deswelling has consequences for pulp freeness and strength properties since softening and flexibility of fibers depend on the swelling rate. The negative effects of temperatures exceeding 100°C on strength properties related to fiber bonding are available in the literature[79,81,82]. For example, a hot dispersion of OCC material between 60°C and 120°C was studied in pilot scale experiments[81]. Table 7 shows typical results.

Table 7. Effect of temperature in hot dispersing of OCC pulp[81].

Parameter	Unit	In pulper	After hot dispersing		
Input temperature	[°C]	50	66	98	120
Disperger energy demand	[kWh/t]	–	117	112	100
Schopper-Riegler value	[SR]	19	23	22	18.5
Sheet density	[kg/m^3]	590	630	620	600
Scott bond	[J/m^2]	160	240	220	190
Tensile index	[KNm/kg]	36	42	40	31
Burst index	[MN/kg]	2.4	2.8	2.7	2.2
Tear index	[Nm^2/kg]	13.4	12.7	13.4	13.2

Below 95°C in this trial, the refining effect of hot dispersing dominates and results in growth of bonding potential. When temperature increased to 120°C, a negative result in bonding ability occurred due to the delayed deswelling. The development of tear resistance had an opposite effect. This deswelling "hysteresis" in recooled pulps after a high temperature treatment resembles hornification caused by drying. One cannot be certain whether it is as irreversible as hornification because the water take-up ability of fibers recovers from the deswelled state with time[81]. Beating of hot dispersed pulp at an intermediate temperature after dilution is a practical means to speed the recovery.

4.6.3 Effect of high-consistency treatment on curl of RCF

Curl after refining of mechanical pulp at high consistencies and temperatures is common knowledge. This latency is easily removed from mechanical pulps by slight agitation at a low-consistency and an intermediate temperature. Curl is relatively stable with low-yield pulps. More mechanical work is necessary to reduce it[82]. Treatment of chemical pulps with low-speed kneaders brings more curl and microcompressions to fibers than with high-speed dispergers probably due to stronger longitudinal compression of fibers. In hot dispersion units, the elevated operating temperature increases chemical fiber curl considerably more than power input[82].

Page *et al.*[83] have presented an interesting theory describing curl creation and removal that explains the different behavior of various fiber types. When mechanical fibers curl at elevated consistency or temperature, the hemicellulose-lignin matrix in the fiber wall presumably flows under the curling stresses of treatment. The cellulosic lamellae (microfibrils) in the fiber wall are crystalline and resist the curling. If the fibers are diluted in cold water, the curl will be set since the hemicellulose-lignin matrix in mechanical fiber becomes rigid and the internal stresses balance. When increasing the temperature, the matrix softens and can flow again while the stresses of the cellulosic fibrils force the mechanical fiber back to its original, straight form.

In low-yield chemical pulps that contain very little lignin or hemicelluloses, created curl is quite insensitive to latency removal whether hot or cold. In hot alkaline kraft pulping, the cellulosic paracrystalline regions of fibers become less ordered, i.e., "amorphous." These amorphous regions can easily flow under the stress of high-consistency treatment, and the chemical fibers become curled. No internal stresses exist that restore the fiber to its straight form. Curled chemical fibers can straighten somewhat by low-consistency refining. Even mixing or screw press transfer in medium consistency (10%–15%) slightly increases curling in chemical pulps.

Curl and microcompressions in fibers and their control in RCF processes have remarkable effects on paper quality. Latency removal after hot dispersion increases tensile strength and sheet density. When intentionally increasing curling in fibers, wet web stretch and dry tear strength increase, but the sheet density and the strength properties related to fiber bonding decrease. In conclusion, control of fiber curl sometimes provides opportunities for quality development of the end product. In tissue paper grades for example, high bulk and softness are desirable properties to which curled fibers contribute. In some packaging grades, increased bulk, tear resistance, and higher stretch caused by curled fibers can also be quality improvements.

4.7 Effects of recycled pulp on wet end chemistry of paper machine

4.7.1 Carryover of deinking chemicals

Carryover means the transfer of organic and inorganic chemicals added in recovered paper treatment processes to the paper machine and its white water system. They can trigger the formation of deposits, so-called secondary stickies, on the paper machine, but they also contribute to sheet properties related to surface chemistry. This is especially true with surfactants and soaps of the deinking process. Besides the formation of stickies, an excessive carryover of deinking soaps can interfere with fiber bonding and can relate to low surface friction that can increase layer-to-layer slippage when winding a reel. If low kinetic friction (sliding) of newsprint coincides with a high density of a wound reel for example, it might lead to crepe wrinkles[84].

Calcium soaps of the flotation process also adhere to fibers. Retention of this surfactant on fibers is a combination of mechanical entrapment and adsorption. It is the result of the balance between van der Waals, electrostatic, and hydrodynamic forces. These retention phenomena were studied by Paprican[85] with pure fatty acid soaps and fibers. Fatty acids (FA) in sodium soap form were found to be present within fiber walls. When calcium ions were added to sodium soap-containing stock, no fatty acids entrapped within the cell wall structure were observed. Calcium soap forms large particles that can only reside outside the fiber for dimensional reasons as Fig. 39 shows.

Figure 39 presents dimensions of both fatty acid soaps in relation to fiber wall structure elements in a simplistic way. The spaces between microfibril lamellae are macropores. Those spaces between fibrils are micropores through which only small ions can move. The diameter of macropores in the cell wall may reach 30 nm to which sodium oleate (about 7 nm) in critical micelle concentration [critical micelle concentration (CMC) is about 0.48 mmol/L] has access. The corresponding calcium oleate particle with a hydrodynamic diameter of 230 nm cannot enter.

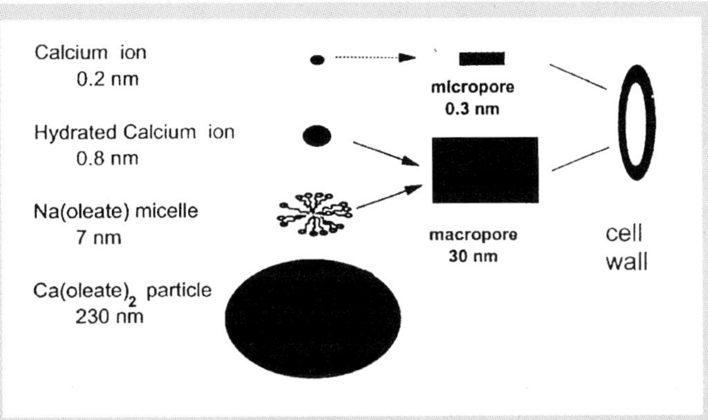

Figure 39. Simplistic view of cell wall porosity and the relative sizes of the chemical species of oleic acid involved in carryover of soaps[85].

Retention of calcium oleate on fibers was minor when the calcium ion addition was equimolar (about 7 mg/L). When calcium dosage increased tenfold (a level that is common in practice), as much as 60% of the soap was adsorbed within a kraft fiber bed in filtration tests[85,86]. The retention behavior of calcium soap can be explained by the so-called DLVO theory that describes the colloidal stability of electrostatically stabilized dispersions (See the book *"Forest Products Chemistry"* of this series). Calcium soap particles are negatively charged like fibers, and they normally repel each other. The excess calcium ions in the suspension suppress electrostatic repulsion to such an extent that other forces such as van der Waals or mechanical entrapment can significantly increase deposition onto

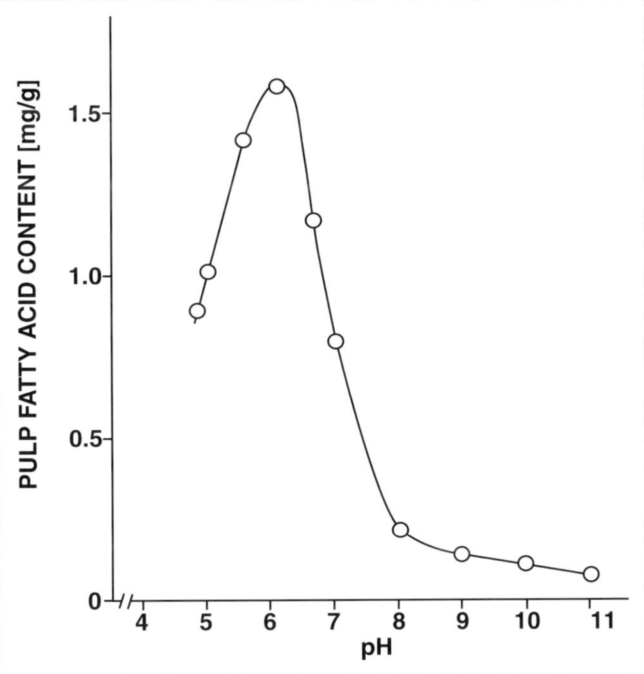

Figure 40. Amount of calcium soap and fatty acid deposited on TMP long fiber fraction vs. pH[86].

fibers. In addition, pH is also an important variable for the colloidal retention of calcium soap and fatty acids[86] as Fig. 40 shows. At pH values above 8, the retention is low, but it increases with decreasing pH to reach a maximum at pH 6. That behavior can also be explained by interactions between fibers and calcium soaps. At high pH, the acidic groups of fibers are more activated than at pH 6 and repel the soap particles.

The adhesion of calcium soap particles to fiber surfaces is not very strong. A dynamic adsorption-desorption balance exists with the surrounding liquid phase. The soap particles can be released into introduced "clean" water from fibers and conversely from "soapy" water onto clean fibers[85]. Analyses of some deinking plants have shown that fatty acids as calcium soaps normally follow the fiber streams in thickening processes[87]. The white water of a paper machine can therefore become contaminated by carryover originating from DIP although fresh water is added to the system at the paper machine.

The data in literature[87–89] about soap carryover with DIP is 0.6–13 kg/t. The average is about 2 kg/t. The design and operation of the deinking process, pH, and deinking chemistry influence the carryover. Sometimes, a switch from calcium soap chemistry to fatty acid emulsions or nonionic surfactants in neutral deinking has been explained by an intention to avoid carryover to the paper machine. Part of the great variation in carryover reported in the literature probably comes from different methods used for analyses. Chapter 7 of this book discusses fatty acid analytical methods.

4.7.2 Return of used wet end chemicals back to papermaking

Modern papermaking uses a large variety of wet end chemicals that can remain in the finished products and finally return with the recovered paper to the papermaking system. In particular, cationic polymers are attached to furnish solids although their exact location or mode of attachment are not exactly known. An interesting question is whether the chemicals added initially stay on fibers in the deinking process and influence the functioning of fresh chemicals added later in subsequent papermaking stages.

Probably the most common case is the use of cationic wet end chemicals as fixatives to bind anionic trash to the sheet. In a study[90], the release of a model substance (lignosulfonate) originally fixed by cationic additives was investigated during paper disintegration. While detachment of anionic substances was very small at neutral pH, the release was considerable above pH 9. This is a normal pH in alkaline deinking and bleaching. Wash-out of fixed anionic trash and old additives seems therefore to be possible in the deinking process if water management applies the countercurrent principle. The model substances were attached on fibers again when pH decreased[90]. This suggests that the interactions between fibers and still positively charged old cationic polymers can be restored when the repulsive effects caused by high pH decrease.

Significant amounts of cationic polymers such as starch or alkyl ketene dimer (AKD) size can often also remain adsorbed on fibers during recycling[91]. Some cationic long chain dry strength agents can lose their binding potential in a subsequent papermaking cycle even when they were well retained in the sheet[92]. Probably the mode of attachment of those long polymers to fibers, i.e., loops and tails, had changed significantly.

Because of the increasing use of cationic additives in papermaking and the good preservation of their functional effectiveness during recycling, progressive accumulation can also be expected in recycled fiber. If a system gradually becomes overcationized, these additives can also act as dispersing agents in suspensions due to cationic electrostatic repulsion. For example, the agglomerating effect of deinking collectors on ink particles can be disturbed by overcationizing as demonstrated by deinking uncoated supercalendered magazines that were strongly treated by cationic fixatives or alum during previous papermaking stages[93].

4.8 Reduction of water consumption in RCF processes

In today's integrated RCF lines, water is normally used countercurrently (See Chapter 6). Fresh water is added at the paper machine, and the effluent is discharged from the RCF plant. In the optimum conditions, as much as 90% of the material dissolved in water in the beginning of the process does not reach the paper machine[94]. This level depends on the general water consumption, the dewatering rate of thickeners, and the adsorption of dissolved materials into the pulp pad.

Due to environmental concerns and the need for efficient use of resources, internal water loops in papermaking are becoming increasingly more closed. When decreasing fresh water intake, the content of dissolved and colloidal substances in white waters grows dramatically. Accumulation of dissolved and colloidal substances can cause disturbances in control of retention, drainage, and even web breaks or quality defects if those substances start to agglomerate. A second major trend has been the increased use of recovered paper as furnish. It contains many substances that bring variable loads of organic and inorganic materials in dissolved and colloidal form into the system including those derived from glues and adhesives.

Total closure of the entire process water system can be implemented easier at test liner or corrugating medium producing mills in which the runnability of the paper machine is not as sensitive as in printing paper manufacture. Today with brown grades, effluent discharges of 0–2.5 m^3/t can be reached. In newsprint manufacturing, 8 m^3/t is a good achievement. In deinking process, discharges for removing detached inks and contaminants are always necessary.

Which dissolved and colloidal substances in RCF systems accumulate in water loops, which exit with the paper sheet, and what the differences are for selective to virgin fiber papermaking as water consumption is reduced are interesting considerations. One simple method to learn about the behavior of those substances is the so-called closed-cycle test in which a pulp sample is diluted and agitated. After filtration, the filtrate is re-used for dilution of the subsequent fresh sample of the particular pulp. This is repeated for selected times, and accumulation of various dissolved and colloidal substances is followed by chemical analysis.

A comprehensive closed-cycle test[95] compared the behavior of those substances in virgin and recycled pulps. The newsprint TMP was taken after the second main refiner, and the DIP deinked with the alkaline calcium-soap method was sampled from the thickener at the end of the process. The first dilution was made with distilled water and pH was adjusted to neutral. In this closed-cycle test, the water was filtered through the pulp pad two times in each cycle to prevent the permeation of ink and filler particles to the filtrate. The results[95] discussed below cover the accumulation propensity of some essential dissolved and colloidal substances and their composition.

4.8.1 Characteristics of anionic trash derived from recycled pulps

Anionic dissolved and colloidal substances are often called interfering or disturbing trash without any explanation of what they interfere with or what they disturb. Their behavior is connected to the negative anionic charge commonly appearing on useful and detrimental substances. The negative charge is normally due to dissociation of carboxyl and sulfonic acid groups of particles.

Figure 41 shows the development of cationic demand in DIP and TMP filtrates measured by the closed-cycle test[95]. The pulps bring negatively charged colloidal particles into the system. Unbleached TMP brings considerably more than the DIP. The enrichment behavior is different. Steady, continuous growth of charge in the TMP filtrate with the number of recycles indicates the very small size of the particles or molecules. The charge of DIP filtrate levels after five cycles. This behavior suggests that some colloidal charged particles of DIP are larger and start to be increasingly more retained in the pads of fresh pulp.

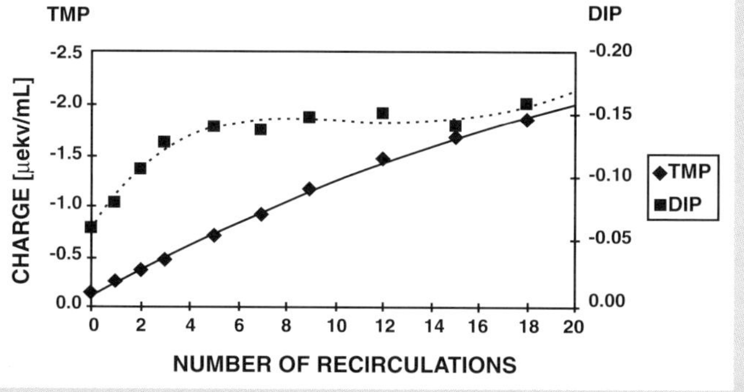

Figure 41. Charge of filtrates vs. number of pulp recirculations[95].

White water and pulp analyses in two European newsprint mills using both types of pulps[125] confirm their characteristic differences for anionic load. TMP was the main source of anionic trash that had very small particle size. Most charge was still present in water after filtration through a 0.45 µm membrane.

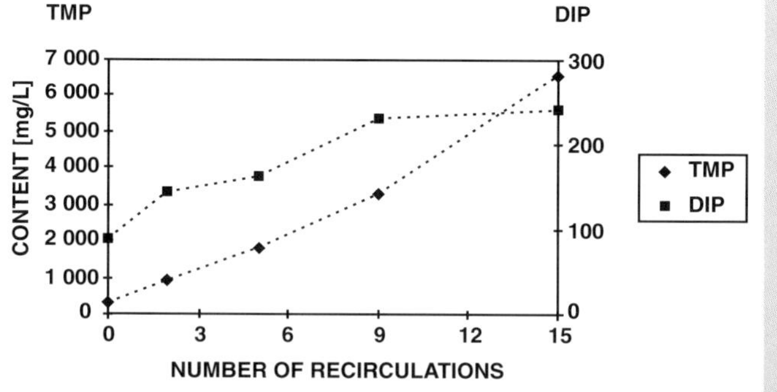

Figure 42. Total content of carbohydrates vs. number of pulp recirculations[95].

In DIP, some anionic particles are derived from

constituents of recovered paper such as coating binders, dispersants, starches, and ink binder resinates that do not exist in virgin TMP. No verified data about the exact particle sizes of those materials exists.

Most anionic load in mechanical pulps comes from carbohydrates (polysaccharides) that can be described as dissolved hemicelluloses. In the same closed-cycle experiment[95], enrichment of carbohydrates was analyzed during recycling as Fig. 42 shows. In the beginning, the carbohydrate content of TMP filtrate was four times higher than that of DIP filtrate but soon became 20 times higher and did not level off.

The reason for the poor retention of dissolved TMP carbohydrates in a pulp pad is the small size of polysaccharide particles below 0.1 μm. The chemical composition of dissolved TMP hemicelluloses is primarily mannose, galactose, and some glucose. Most DIP carbohydrate is glucose[125] that primarily comes from starch. In the mills that use old packaging material as furnish, surface sizing or converting starch is the major source of anionic load.

If alkaline peroxide bleaching is included in the DIP process, the cationic demand will increase unless the pulp is well washed. One reaction in this oxidative bleaching is the breakage of ester bonds of methylated galactouronic acids in hemicelluloses that then become soluble[96]. The reductive bleaching with dithionite does not directly cause the formation of dissolved substances.

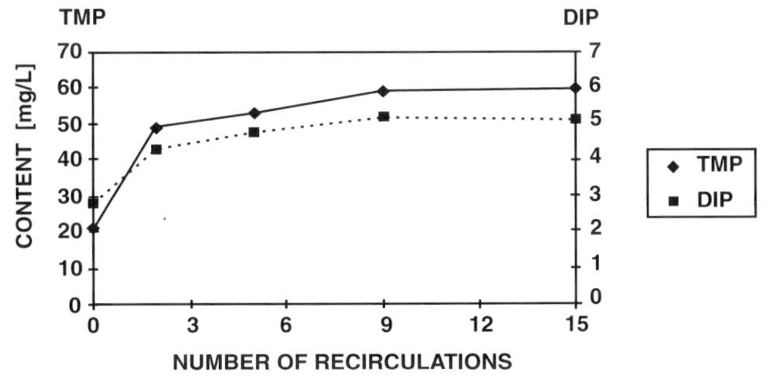

Figure 43. Content of dissolved colloidal extractives in filtrates vs. number of pulp recirculations[95].

Lipophilic extractives of wood are the most detrimental fraction of anionic trash since they can cause deposits on the paper machine alone or with other hydrophobic particles. Figure 43 shows enrichment of dissolved and colloidal wood resins in DIP and TMP filtrates during the quoted closed-cycle test[95].

The leveling of wood resin content occurred after three recirculations in both filtrates. Very little dissolved and colloidal lipophilic extractives were released to filtrates from DIP. The countercurrent water management washes out extractives, and deinking flotation also removes pitch droplets. To some extent, the analysis of composition of wood resins in TMP and DIP filtrates might also explain the difference in the extractive amounts[95]. In gas chromatographic analysis, the extractives were divided into the subgroups of triglycerides, steryl esters, sterols, resin acids, and free fatty acids indicated in Fig. 44.

In Fig. 44, the constitution of lipophilic extractives in TMP filtrates is very similar to that of extractives in fresh spruce wood. In DIP, triglycerides have decomposed to free fatty and resin acids probably during the aging of recovered paper. In their dissociated form at pH above 7, these acids follow the water fraction as micelles. Triclycerides preferably form larger extractive droplets that follow the pulp stream on thickeners[87]. Triglycerides especially are believed to be a major "culprit" to pitch problems on a paper machine[125]. When reducing water consumption in an integrated deinking line, an enrichment of free resin acids in particular can occur.

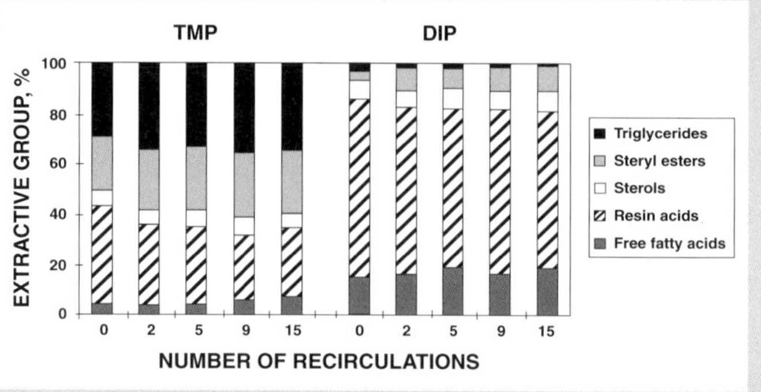

Figure 44. Distribution of wood extractive groups in DIP and TMP filtrates[95].

4.8.2 Chemical oxygen demand (COD) derived from recycled pulps

Chemical oxygen demand or total organic carbon (TOC) are often considered to be measures of released organic material in white waters. Their concentration closely depends on specific water consumption. In the previously referred closed-cycle test[95], the enrichment of COD in DIP and TMP filtrates was analyzed as Fig. 45 shows.

This experiment shows that DIP is "cleaner" in this respect, and COD of filtrate starts to level after fewer cycles than TMP filtrate. This indicates some differences in the molecular size of their COD materials. If an alkaline peroxide bleaching stage is included in the deinking process, the release of COD material will grow as a linear function of pH.

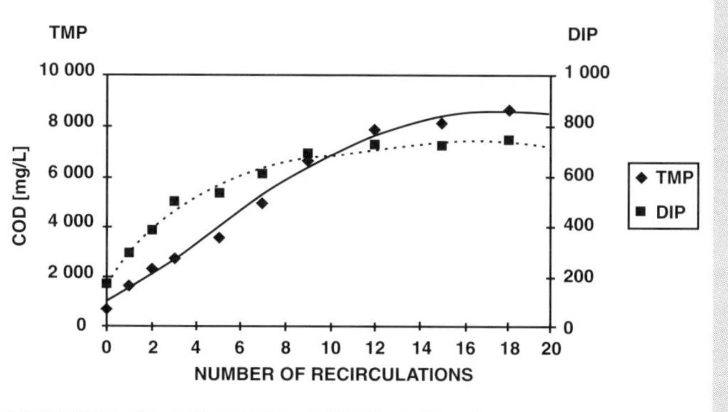

Figure 45. Accumulation of COD in TMP and DIP filtrates vs. number of pulp recirculations[95].

COD is a measure of dissolved and colloidal substances that can be chemically oxidized, but the value itself does not give very accurate information about organic components present. In the chemical analysis of DIP filtrates, COD constituents can fall into four categories: lignin derived molecules, small hemicelluloses, acetic acid, and very small particles or micelles of lipophilic extractives[97]. Lignin normally comes from mechanical pulp or in the case of OCC from brown packaging material. The carbohydrate particles are the largest. They can be removed by ultrafiltration and partly by microflotation. The size of lignin and extractive molecules is less than 0.01 µm. They are difficult to remove by conventional purification equipment. Some extractives in the COD load of DIP and especially free resin acids may originate from ink binders or deinking soaps.

When reducing fresh water consumption in RCF mills, the COD concentration will rapidly increase in white water. Values above 10 000 mg/L are not exceptional in closed OCC lines. An increasing load of organic dissolved and colloidal substances may also accumulate in the paper recovery chain. When disintegrating recovered paper in water, a COD level close to that in the previous papermaking is reached in the filtrate[97].

4.8.3 Effects of dissolved inorganics in recycled pulps

Besides the load of organic dissolved and colloidal substances, recovered paper contains various inorganic ions that accumulate heavily in closed water systems since their retention is generally poor. Most electrolyte load consists of the inorganic salts that existed in previous papermaking, but some can be added in recovered paper processing and bleaching. For example, alkaline deinking chemistry normally includes the use of sodium hydroxide and sodium silicate and sometimes a calcium ion addition to increase water hardness.

In papermaking, the performance of retention aids, dispersants, and other paper additives can be impaired when the concentration of dissolved salts exceeds a critical level. Negatively charged electrolytes such as sodium silicate directly function as anionic trash by consuming cationic additives. With a high dissolved salt content, the effect of high molecular weight retention agents based on the bridging mechanism is also weakened due to the balling-up of long polymer chains.

The most apparent results of high ionic strengths are inorganic deposits or scalings in undesirable places. Calcium carbonate and silicate can precipitate in the alkaline pH region of a deinking plant. Calcium oxalate forms deposits within a wider pH range. Oxalic acid is generated as a byproduct in oxidative bleaching of pulp. Gypsum can also occur in scalings if many sulfate ions exist in the system. Ions that can cause corrosion can also come with recovered paper. This includes chloride. They can also accumulate when fresh water consumption decreases.

An increase of calcium ion content will undoubtedly occur in the future since the amount of $CaCO_3$ in most recovered paper grades is growing. The accumulation tendency of calcium ions in DIP and TMP filtrates was analyzed by the closed-cycle test quoted earlier[95]. Figure 46 shows steady growth in both filtrates. The starting level of TMP was only 3 mg Ca/L since fresh water in the particular mill is very soft. For DIP, the corresponding value was 100 mg Ca/L because of the use of calcium soap as a collector.

Calcium carbonate is sparingly soluble above pH 8, but its decomposition will dramatically increase if pH declines. All acidity coming into a system dissolves $CaCO_3$ to release free calcium ions and CO_2 gas as foam. Typical sources of acidity in papermaking are the use of alum and dithionite bleaching residues. The pH of occasional, supplementary virgin pulps is also normally acidic. In recycled fiber systems, starch is often converted into organic acids by microbiological activities that also cause decomposition of $CaCO_3$. This is a problem in packaging paper and board manufacturing where considerable $CaCO_3$ today comes from mixed recovered paper. Surface sizing and converting operations bring starch to the furnish.

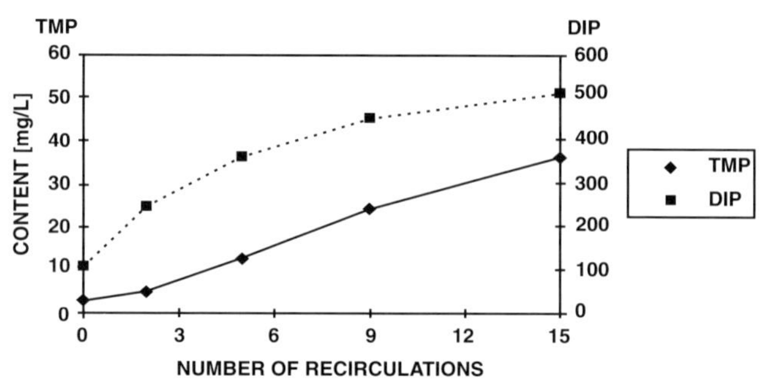

Figure 46. Dissolved calcium content in TMP and DIP filtrates vs. number of pulp recirculations[95].

High calcium ion concentration may create complex problems in papermaking. According to the earlier quoted DLVO theory, an increasing concentration of multivalent cations compresses the electric double layer around anionic charged colloidal particles. As the diffuse layer shrinks, other forces such as van der Waals forces overpower electrostatic repulsion and finally induce coagulation. In a recycled fiber system, this means that aggregation of anionically stabilized microstickies, soap, or pitch droplets is then more probable.

A study[98] examined the effect of calcium ion concentration on agglomeration tendency of colloids of groundwood (GW) and DIP process waters by a photometric dispersion analysis (PDA) instrument.

As Fig. 47 shows, the size of colloidal particles of both pulps starts to grow. The agglomeration of DIP colloids occurs at lower calcium content than that of the groundwood. When the pH was raised from 5.0 to 8.0 in the presence of calcium ions, the relative aggregation of DIP particles increased. This indicates interactions or reactions between calcium and organic colloids in suspension. Otherwise, the elevated pH would only enhance repulsion in the anionic charged system. The relative PDA values obtained in this study of Fig. 47 were converted to absolute particle dimensions by the Coulter N4MD instru-

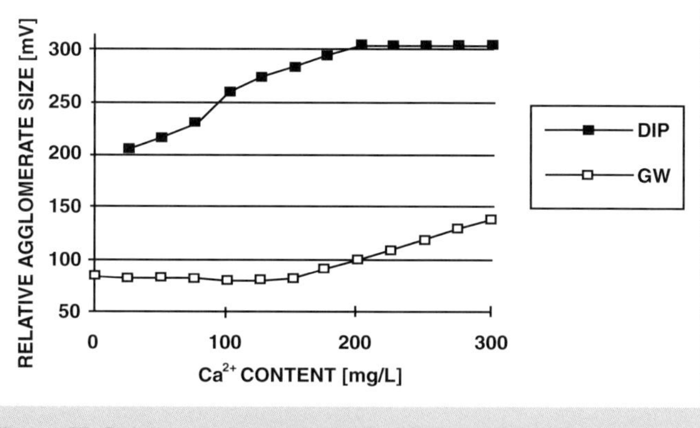

Figure 47. Relative agglomerate size by PDA of colloidal fraction from DIP and unbleached groundwood at pH 5.0 vs. calcium ion content[98].

ment. The PDA reading for GW at the start corresponded to a particle size average of 0.3 μm. This is typical for wood resin. For DIP, the value was about 3 μm. The latter reading suggests that agglomeration of colloids has probably started around mineral filler particles in the DIP water of this experiment.

Numerous water soluble adhesives used in paper converting are electrostatically stabilized suspensions. The charges remain unchanged in the particles that are obtained by disintegration of dried adhesive films. Those particles are also known as microstickies. The interfering effect of calcium ions on colloidal stability of a redispersed pressure sensitive adhesive film was studied by the PDA instrument[98]. Figure 48 shows that the size of microstickies grows during circulation (agitation) in the device but faster in the presence of calcium.

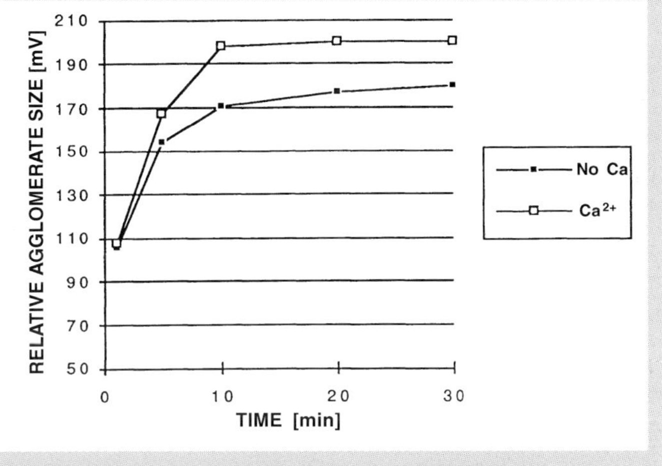

Figure 48. Relative agglomerate size of particles from dispersed pressure sensitive adhesive film as a function of circulation time without and with calcium (180 mg/L) at pH 5.0[98].

4.8.4 Removal of dissolved and colloidal material from recycled fiber circuits

As stated before, an increasing content of organic and inorganic dissolved and colloidal (DC) substances can negatively influence paper machine runnability and paper quality when fresh water use decreases. When trying to retain an increasing DC content in the paper web, the substances fixed or adsorbed in the web will finally return with recovered paper into systems. Internal purification of water circuits from DC substances therefore seems more desirable than fixing them in the sheet. The adaptation of these techniques is discussed below.

Today, microflotation [dissolved air flotation (DAF)] is a very convenient way to treat large water volumes. It removes fines, fillers, and some colloids very effectively by proper flocculation chemistry[97,99,100]. Originally, internal microflotations were adapted to eliminate those ink and stickies that were too small (< 5 μm) for removal by regular deinking flotation.

Small-size DC material (< 0.2 μm) is not removed by microflotation[97]. It can accumulate into significant concentrations when closing process water loops. The COD load in particular consists of particles of that small category. The best COD reductions by microflotation have therefore only been in a range of 20%–30%. The reduction of inorganic salts is poor with microflotation except in the case of aluminum since it can be bound on suspended solids or can form $Al(OH)_3$ flocs in alkaline conditions.

For selected water fractions, membrane technology is a way to reduce further the content of very small-sized DC substances. In a ultrafiltration trial[97] with a membrane cut-off number of 100 000 g/mol (corresponding roughly to 0.01 μm pore size), reducing COD of microflotated DIP white water by an additional 30% was possible. This reduction occurred by removal of anionic carbohydrates. Reduction of salts with ultrafiltration was moderate primarily in the region of 10% with very small ions such as Na and Cl almost zero. The use of ultrafiltration techniques with deinking white waters is reported for the removal of flexographic ink particles[101]. Recovered paper process waters are difficult due to fouling, i.e., permanent plugging of membranes.

Biological water purification processes offer very promising methods for internal removal of organic DC material from white waters in brown recycled paper systems. Already existing aerobic activated sludge processes can reduce COD in effluent waters of deinking mills by 80%–90%. Organic material is primarily converted into carbon dioxide. In anaerobic biological treatment, the conversion of COD into methane gas is not as complete. Investment and operation costs are less expensive, and the methane gas can be used for energy production. In the anaerobic water treatment process, the sulfate ions are reduced to sulfide ions. They can be further converted to elemental sulfur by a group of sulfide oxidizing bacteria to remove sulfur compounds from the system[102].

Interesting data from an anaerobic-aerobic treatment process have been published[102,103] about the experiences of a brown recovered paper based board mill that has run completely closed water systems (See Chapter 12). The COD concentration in the process increased to 35 000 mg/L primarily from the starch in converted OCC. The starch hydrolyzed into glucose that then decomposed into short chained volatile fatty acids due to the activity of bacteria. Those butyric, lactic, etc., acids caused odor problems in the product. Their acidity also decomposed the calcium carbonate coming from the recovered paper. Introduction of internal biological treatment[102] for white water considerably eased the situation as the values in Table 8 show.

Table 8. Effect of internal biological purification on white water of a closed OCC mill[102].

Parameter	Unit	Before white water purification	After white water purification
COD	mg/L	35 000	8 000
pH	–	6.25	7.25
Calcium	mg/L	2 650	420
Sulfate	mg/L	1 350	610
Maleic acid	mg/L	6 600	1 350
Butyric acid	mg/L	350	40
Acetic acid	mg/L	6 300	1 300

The aerobic stage of the purification process generates CO_2 gas. It reacts with dissolved calcium at appropriate pH resulting in precipitated $CaCO_3$ that was removed with effluent sludge. The anaerobic stage and sulfur specific bacteria reduced sulfate ion content as reported earlier. This biological treatment only produced a minor reduction in conductivity that indicates a persistent high salt content[102]. Evaporation or an essential demineralization process would be necessary to remove electrolytes. Table 9 characterizes some desalting processes.

Table 9. Comparison of various demineralization methods.

Parameter	Unit	Ion exchange (IE)	Reverse osmosis (RO)	Electro-dialysis reversal (EDR)
Organic contamination		Sensitive	Very sensitive	Less sensitive
Salt in feed	mg/L	< 700	400–50 000	500–4 000
Salt in outlet	mg/L	< 100	10–5 000	50–500
Process pressure		Low	High	Low

Electro-dialysis reversal (EDR) is an interesting technology since it concentrates salts from water into very small volumes by direct electric current. Reversal dialysis also inhibits membrane fouling since organic matters that do not pass through the membrane are flushed back into the process water.

4.9 Characteristics of deinked pulp fibers for runnability

4.9.1 Effect of DIP on dewatering in manufacture of printing grades

Wood-containing deinked pulp has common use as a partial or complete substitute for the mechanical pulp component in printing paper grades. Sometimes the belief has been expressed that the drainage of deinked pulp in newsprint sheet forming is inferior to corresponding virgin mechanical pulp due to its higher chemical pulp and filler content. In a trial on a pilot scale gap former at a speed of 1 500 m/min, the maximum dewatering capacity of the former was investigated as the ratio of DIP and TMP in a newsprint furnish varied[104]. In this particular DIP, the ratio of mechanical and chemical pulp was about 65/35 and the ash content was about 7%. The test was made by gradually opening the headbox slice until crushing in the structure of the formed sheet was observed. The result was that this maximum drainage capacity limit decreased when the proportion of deinked pulp increased as Fig. 49 shows.

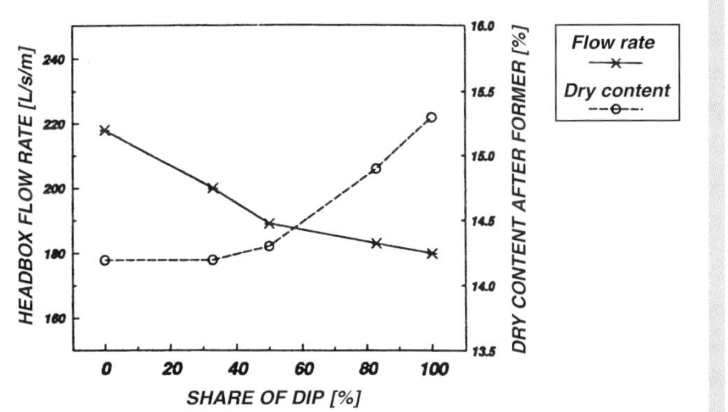

Figure 49. Maximum drainage capacity and final dry content on a gap former vs. DIP and TMP ratio[104].

Apparently, the drainage at the beginning of the gap former on the forming roll was poorer with DIP, but the dry content after the entire former section was about 1% higher. This suggests that the dewatering after the "dry line" was more effective. DIP-containing stock was probably compressing well and squeezing water out during the vacuum pulses of the blade shoes and suction boxes. After the wet press section, no remarkable differences occurred in solids content between the pulps, and the total runnability was good. Without the very intense refining that is not usual for a stock of mechanical printing paper, the total drainage probably does not present a runnability problem when the mechanical pulp is replaced by DIP.

4.9.2 Performance of recycled chemical pulp as reinforcement fibers

As previously stated, DIP made of household recovered paper in some geographical regions contains a large proportion of softwood kraft fibers (up to 50%). Economical advantages would result if that chemical pulp component could substitute for primary reinforcement fiber in printing paper grades. When DIP replaces the mechanical pulp component, the softwood kraft fiber content easily exceeds the limit of 30%–40% in a mechanical printing sheet (even in newsprint). This is theoretically adequate for the forming of a complete reinforcement fiber network[105]. Additionally, commercial wood-free recycled fiber is available in a deinked form from office papers.

Reinforcement fibers (RIF) added to paper will improve strength properties and runnability on a paper machine and in converting or printing operations. High fiber length, strength, and good bonding capacity are essential for high-quality RIF. Besides conventional strength properties, RIF should increase the fracture toughness of paper, i.e., its ability to resist crack propagation in the web. Fracture toughness measurement can use several techniques, but the property normally increases with increasing kraft pulp dosage in mechanical printing paper. Various opinions are available in the literature[105,106] on the best beating degree.

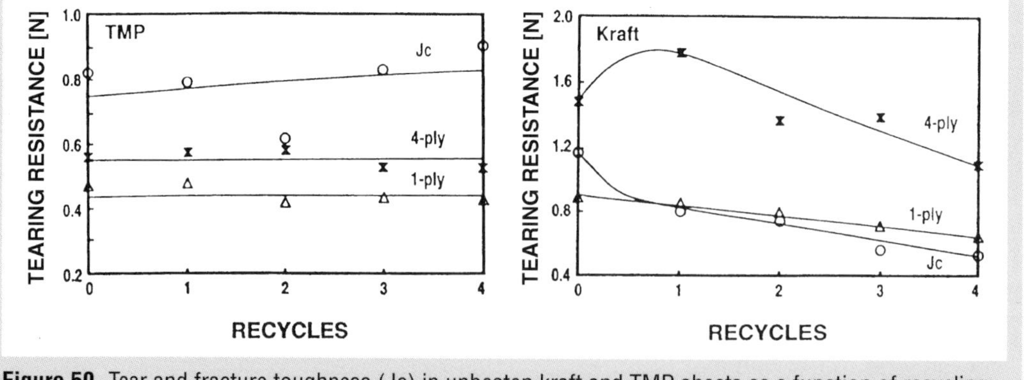

Figure 50. Tear and fracture toughness (Jc) in unbeaten kraft and TMP sheets as a function of recycling (number of cycles)[38]. Observe different scales in Y-axes.

A study[38] examined the fracture toughness measured as a so-called J-integral (Jc) and conventional out-of-plane tear strength as a function of recycling (number of cycles). The recycling of TMP and unbeaten, unbleached kraft used a broke-like treatment, i.e., only sheetmaking, restrained drying, and gentle reslushing.

The unbeaten kraft sheet lost up to 55% of its fracture toughness (Jc) and 27% of its tear strength after four cycles as Fig. 50 shows. When inspecting fracture paths for the virgin and the four times recycled kraft sheets, considerably more fiber pull-outs and less fiber breakage existed with the recycled fracture surface than with the virgin one. Apparently, the unbeaten kraft fibers became hornified during recycling and were easily pulled from the fiber matrix due to reduced bonding. Those properties of TMP did not change as a function of recycling as expected. In this study, the Jc value of TMP was even higher than that of four times recycled kraft as Fig. 50 shows. The entire level of Jc would have been considerably higher if the original kraft pulp had been beaten. The development of tear strength in recycling depends on the original beating degree of the kraft pulp (See section 1.3.1 of this Chapter).

The fracture toughness of a kraft fiber network seems to be very sensitive to hornification. The potential of wood-free fibers as reinforcement pulp therefore suffers significantly by recycling. The refining of recycled fibers derived from white office paper is a topic for later discussion.

4.10 Effects of growing RCF use on quality of printing paper grades

Technical and economic opportunities exist to re-use the increasing recovered paper amounts available from intensified recovery in printing paper grades. Gradually, larger quality differences compared with virgin fiber products will appear. Accumulation of contaminants and ink residues and their effects on quality have been discussed in section 2 of this chapter. This section will discuss the changes and effects of furnish composition on quality.

In some regions, more than 50% of DIP furnish comes from old magazines. Up to 50% softwood kraft is used as reinforcement fiber in wood-containing magazine grades. In addition, wood-free printed products and some office waste enter the same recovered paper stream. This means that the chemical pulp content gradually grows in all products where deinked pulp is used. The accumulation of filler material in RCF furnishes occurs in the same way as the recovery rate grows.

In recycling, mechanical fibers are not prone to hornification. To some extent, they even gain in flexibility probably due to internal fibrillation caused by repeated drying and calendering. This is evident in the flattening of fibers with progressive recycling. Since chemical fibers are also accumulating in the recovered paper recovery chain, they will further enhance the densification of DIP-containing sheets because the lumens of chemical fibers remain collapsed. Figures 51 and 52 show a typical difference for virgin fiber sheet for cross sections of commercial newsprint made of 100% TMP and 100% DIP. Filler contents were 8.5% and 12%, respectively, and the DIP sheet contained about 35% chemical fibers.

Although calendering of the DIP-based newsprint of Fig. 52 has been softer to give it a higher roughness value, that sheet is considerably thinner than the TMP-based newsprint of Fig. 51. The tube-like summerwood fibers of virgin TMP can effectively resist calendering work and are not flattened.

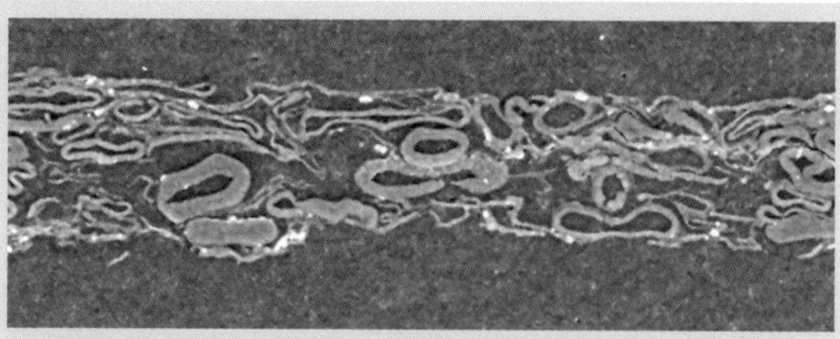

Figure 51. Cross section of 100% TMP-containing 45 g/m^2 newsprint.

Mathematically[107], Eq. 16 shows how the apparent density and relative bonded area of a fiber network in a sheet grow with a dimensionless flexibility number, F, that is affected by fiber character:

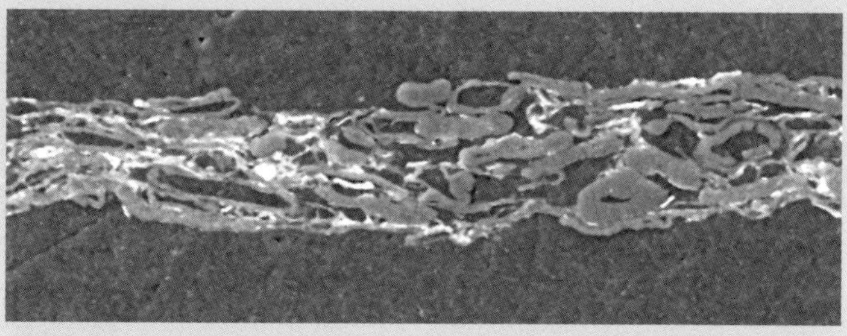

Figure 52. Cross section of 100% DIP-containing 45 g/m^2 newsprint.

$$F = (wf/tf)(C \cdot wf \cdot WFF)^{1/4} \qquad (16)$$

where WFF is the fiber flexibility in the wet state
wf the width of fiber
tf the thickness of fiber
C a constant that depends on the fiber material.

Although this formula ignores the importance of fines and fillers, it does show the effects of recycling. The apparent density and relative bonded area of a fiber network grow when the thickness of fibers decreases due to flattening. The increasing proportion of recycled chemical fibers affects the sheet density more strongly than any changes in mechanical fibers because they are flat and their flexibility value in the wet state is more than an order of magnitude higher. Despite hornification, the flexibility of chemical fibers will increase with repeated recycling by outer damage (twisting and kinking).

When DIP replaces mechanical pulp, the reduced thickness compared with the virgin fiber sheet will show declining rigidity. Mathematically, Eq. 17 shows sheet stiffness, S:

$$S = E/t^3 12 \qquad (17)$$

where E is the elasticity modulus
t the thickness of the sheet.

The increasing proportion of kraft pulp in a DIP sheet should increase the E-modulus and contribute to the sheet rigidity, but its effect is outweighed by the reduction of sheet thickness (third power in the formula). Theoretically, the dense sheet and low light scattering of chemical pulp should result in poor opacity that is a critical property in printing paper. Fine pigments coming from old magazines to DIP can partly compensate for this. Nevertheless, the limp sheet and show-through of print may become problematic in lighter basis weight grades.

4.10.1 Effects of increasing recycling rate on quality of newsprint

Newsprint is a printing paper segment in which the use of DIP is particularly well established. Newsprint with a DIP content of 100% is almost a standard in some regions. The progress in deinking and papermaking technology has eliminated the disadvantages in runnability and appearance of the sheet. For example, primary stickies and visible ink specks can be almost eliminated by hot dispersion. The formation of secondary stickies can be controlled by appropriate wet end chemistry (See Chapter 11).

As stated earlier, the proportion of chemical pulp in DIP-based newsprint will grow with a growing paper recovery rate. Whether this converts to improved sheet strength is unclear since the filler content will also become higher and the chemical pulp component is hornified. To maintain opacity and stiffness in newsprint, refining of DIP is not usual in practice. To some extent, fine pigments coming from old magazines can compensate for declining light scattering. The pigments can also contribute to brightness. The most noticeable difference between virgin fiber- and DIP-based newsprint is the denser DIP sheet of Figs. 51 and 52. It provides more meters in a reel, but the final newspaper will be more limp as noted.

The smoothness target measured by air leakage methods can be achieved by lighter calendering than with virgin TMP. The resistance against compression in a calender is small with chemical fibers and flat mechanical fibers as Fig. 52 shows. The micro-scale roughness measured by an optical tester can be higher[33]. A scanning electron microscope (SEM) picture can reveal an open surface for a DIP sheet[37]. A virgin TMP sheet is more closed. That difference is probably due to loss of fines during deinking.

Despite its high content of kraft fiber, the hydroexpansion of the DIP-containing web after water application on an offset press is not larger than that of a TMP web. This is because the chemical pulp component of DIP has also lost some swelling ability in the sheet through hornification[13]. Linting propensity on offset presses has been variable. Excessive filler content can cause some fibers on a paper surface to be more poorly bonded, but fewer loose fiber fines, e.g. parenchyma cells, exist in DIP because of fines losses in deinking. Less lipophilic extractives are present in a DIP sheet than in a TMP sheet due to the dissolving effect of alkaline deinking[25]. This is often evident in the higher friction coefficients and quicker wettability of DIP sheets.

Considering the paper requirements of the most common newspaper printing method, coldset offset, the micro-rough surface of a DIP sheet should provide a good base for anchoring inks. This is especially true for so-called low-viscosity inks. That kind of ink can cause problems with rub-off and set-off of print[108]. The more closed surface of virgin TMP newsprint might do a better job of preventing the excessive penetration of ink, i.e., so-called strike-through. The strike-through of newspaper ink can become significant with lightweight DIP-containing paper grades.

4.10.2 Deinked pulp in magazine paper grades

Traditionally, RCF have been considered less suitable than virgin fibers for use in magazine grades due to their high quality demands. Fear of unacceptably poor runnability has existed particularly for blade coaters due to defects caused by stickies. Market demand and the abundant supply of recovered paper have changed these attitudes. The most common processes in magazine printing, rotogravure and heatset offset, have different requirements on the substrate compared with the usual coldset offset process for newspapers.

By sequential peroxide-dithionite bleaching, a brightness level of only 71%–73% ISO can be guaranteed with normal household recovered paper. The use of such DIP in top-brightness paper grades is therefore questionable. The ash content in DIP is reduced to about 5% by washing for blade coating base paper when many fiber fines are also lost. Without refining, 10%–15% less fines exist in DIP than in magazine TMP. In base paper for film press coating and in SC paper, retaining more mineral filler and fines in the DIP is possible.

One way to improve the brightness potential of DIP is to use old magazines as furnish exclusively or at as high a level as possible. Another possibility is to fractionate the pulp after the first flotation in deinking and use the easily bleachable long fiber fraction for making magazine paper. Despite good brightness, the light scattering of the long fiber fraction remains low due to a high proportion of chemical fibers and lack of fines. The dark fines fraction of DIP alone seldom has any use in practical papermaking. IfP studied various deinking treatment combinations for long fiber and short fiber fractions on the laboratory scale[110]. Separate treatments and the subsequent combining of fractions did not give any significant advantages in optical properties compared with deinking and bleaching of the entire pulp.

DIP is a suitable furnish for coated heatset offset grades when substituting for mechanical pulps because of its sheet densifying character shown in Figs. 51 and 52. Fewer thick walled mechanical fibers exist in DIP than in TMP. In a series of pilot trials with regular DIP, the fiber rising (surface roughening) tendency in heatset offset printing was even less when mechanical magazine pulp was replaced by DIP[109]. This property can also be improved by high-consistency refining of DIP or its coarse fraction. As in newsprint, DIP provides in coating base stock an open, hydrophilic surface. Especially for film press coating, this is a beneficial property because of rapid dewatering and immobilization of the coating layer in the press nip.

The use of DIP in rotogravure grades is a more complex case. The critical quality characteristic in both LWC and SC rotogravure grades is printing smoothness. Insufficient smoothness leads to local noncontact between printing cylinder and the paper resulting in missing dots of the print. The thick-walled undeveloped mechanical fibers of DIP may cause missing dots despite the usual heavy calendering or electrostatic assisted ink transfer (ESA) in printing. Calendering does not level the peaks and particularly the valleys on the surface due to crossings of thick fibers.

Refining of the long fiber fraction or the entire DIP to a lower freeness can improve the flexibility of those coarse mechanical fibers and make them flatter. The long fiber fraction of DIP also contains chemical fibers that will receive the same treatment. As stated in connection with RCF refining, this causes secondary fines that show very high swelling and shrinking ability. When present in the sheet, these fines build bridges between fibers in the fiber network after wet pressing. Subsequently, these bonded bridges shrink greatly during drying and cause local contractions and deflections in the fiber network. The refining therefore not only results in densifying the sheet but also in increased micro-roughness caused by those local contractions. The high proportion of chemical fibers coming with DIP makes the sheet "hard". This is not advantageous for gravure printability. Accordingly, a clear need exists to develop a selective separation for the mechanical and chemical fibers of DIP and treat them separately before using deinked pulp in gravure grades.

4.11 Quality potential of RCF derived from office paper

Primary wood-free office paper is normally a blend of chemical hardwood and softwood fibers. The RCF derived from office paper therefore cannot reach the requirements set for reinforcement pulp that is normally pure softwood kraft. Wood-free recovered paper normally goes into production of tissue paper, office paper grades, or commercial market pulp that may have use in various ways. With increasing rates of office paper recovery, the purity of the "wood-free" quality will probably decline with the inclusion of some mechanical or brown chemical fiber components. Very strong bleaching stages are therefore sometimes included in the processes for office paper deinking to produce adequate brightness (See Chapter 9).

In wood-free office paper deinking processes, the most filler and fiber fines are normally removed due to quality requirements. New fines can be created again by refining to develop bonding potential in the stock. The ability of hornified wood-free DIP to resist refining is weak, i.e., its beating degree develops more rapidly than that of virgin pulp. Figure 53 shows typical refining curves in which wood-free recycled pulp derived from ledger paper grades is compared with virgin fibers using treatment in the Valley Beater[89].

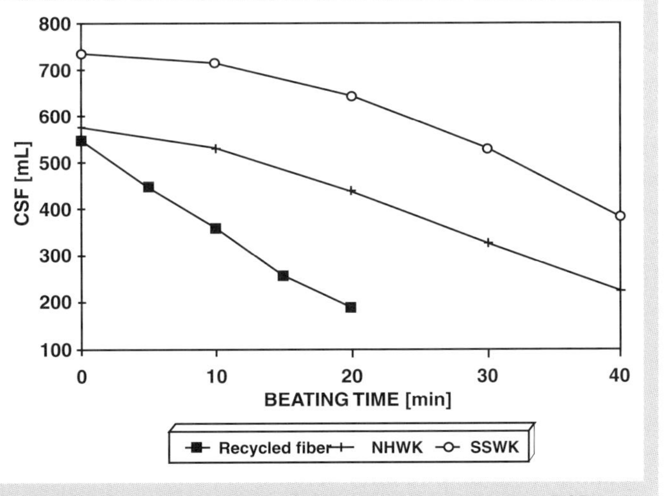

Figure 53. Freeness vs. Valley beating time for wood-free RCF and averages of four southern softwood (SSWK) and four northern hardwood (NHWK) kraft pulps[111].

The properties of Valley beaten RCF such as tensile strength in Fig. 54 and opacity develop at a similar rate as that of virgin hardwood kraft for which recycled wood-free pulp is usually substituting. Sometimes, recycled fibers can show slightly higher tear strength than virgin hardwood.

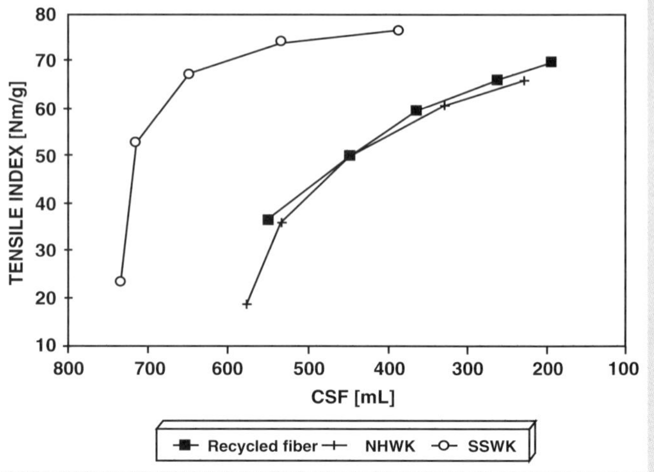

Figure 54. Tensile index vs. freeness for wood-free RCF and averages of four southern softwood (SSWK) and four northern hardwood (NHWK) kraft pulps[111].

In wood-free deinking processes, fibers have often been subjected to multiple hot disperger, kneading, or refining treatments to reach complete fragmentation of toner inks. As described in section 4.4., the swelling ability of chemical fibers suffers at elevated temperature. High-consistency treatments can cause permanent microcompressions and curling. Curl of fibers may be a benefit for softness of tissue paper but a disadvantage if high density and strength are necessary in the sheet.

Some evidence exists that drainage on the wire section and stiffness of final sheet can become worse when hardwood kraft pulp is replaced by wood-free RCF in office paper manufacture[112]. Probably, the wet sheet of curled, stiff fibers does not compress well on the suction boxes of the wire section and therefore responds poorly to their vacuum pulses that are squeezing water out. Naturally, the bonding ability of curled fibers decreases, and they cannot produce a high modulus of elasticity and stiffness in a sheet (See Eq. 17). The bonding ability of hornified or deswollen chemical fibers can be restored to some extent by gentle refining at low-consistency but at the expense of dewatering properties. Alkaline conditions or the use of enzymes for refining can possibly alleviate the problem as discussed in the following section.

4.12 Challenges for manufacturing of packaging grades from RCF furnish

4.12.1 Trends today and tomorrow

Most recycled chemical fibers are in the recovery chain of packaging papers and board in which the utilization rate has been higher than that of recycled printed products. In the future, an even greater use of recycled fiber in these grades may occur. International trade of used packaging material and especially old corrugated containers (OCC) and double lined kraft clippings (DLK) has always been high. The United States shipped about 6 million tons per year of OCC to supply packaging paper and board mills primarily in Asia and Europe in the middle 1990s. The material that was 70%–80% made of virgin fiber has been a welcome supplement to strengthen the worn, multi-recycled fiber in the containerboard of other continents. In the United States, all the new capacities of liner and fluting used recycled fiber during the 1990s. Most virgin fiber mills have started a partial replacement of their raw material with RCF. Consequently, the fresh recovered fiber supply from there may be dwindling.

Paper recovery will become even more intense because the disposal of organic waste into landfills is gradually becoming more difficult. This trend will cause quality changes in recovered paper supply that will have influence especially on packaging paper and board manufacturing. The conventional sources of OCC such as converters and the retail sector have already been well used in countries where paper consumption is high. The additional future sources will be in households and offices as mixed waste (MW) comprising variable proportions of mixed packages, office paper, old newspapers and magazines. If incineration is not possible for this material, very few uses exist for it other than as recycled fiber for packaging paper and board grades. This would lead to a gradual growth of mechanical fiber and mineral filler content in packaging grades.

The third trend is the reduction of fresh water consumption in papermaking by closing water circuits. Effluent reduction or even complete closure is easier to implement in packaging paper and board manufacturing than in a printing paper line including a deinking plant that always needs an outlet for ink. Besides control of dissolved and colloidal substances that was discussed earlier, the closure of a system can also affect runnability and quality if no countermeasures are taken by purification of the process water loop.

4.12.2 Optimization of refining for OCC pulps

A strong relationship exists between refining, drainage, strength, and drying in packaging paper and board production. When actions are taken to improve one feature, they also influence the other factors. Measures to improve drainage such as lightening of refining or increasing headbox consistency usually result in a loss of strength. The burst strength of commercial OCC liner generally stays at a lower level than that of virgin fiber liner although other properties such as compression strength are about the same[113].

Reduced bonding ability for OCC material is probably due to the hot disperger and other high-consistency treatments. As discussed in section 4.4, the high-consistency refining of chemical pulps produces a high degree of permanent fiber curl and microcompressions. These result in high tear strength, porosity, and bulk, but the burst strength develops less effectively than in low-consistency refining. High-consistency refining might therefore be of interest only for special packaging grades.

Increasing the recycling rate and the introduction of mixed recovered papers and closed water circuits have made packaging paper and board machines increasingly more drainage restricted. OCC lines have therefore become an appropriate field to adapt enzyme technology to increase the freeness of stock and machine speed. For example, a dosage of 0.4% cellulase enzyme increased freeness from 350 to 600 mL on a production scale[114]. Substantial improvements have occurred without essential loss in strength properties[114–116].

The mechanism of enzyme action is hydrolysis of fines and stripping of small fibrils or attached fines particles from the fiber surface and decomposing those particles into sugars. This is in contrast to conventional drainage aid polymers that flocculate fibers and bind fines by a mechanism that loses its effectiveness in systems highly loaded with fines or dissolved and colloidal substances. Enzymes also cut damaged fibers and swell cell walls at damaged spots[117]. In a study[17], the enzyme treatment of recycled kraft pulp reduced cuprethylenediamine (CED) viscosity and the strength properties related to bonding ability as Fig. 55 shows. CED viscosity is considered to be a measure for the degradation of cellulosic material and an indication of the general strength potential of chemical pulp.

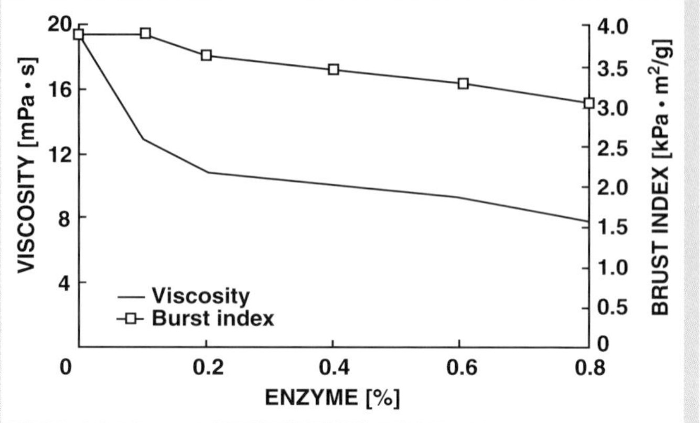

Figure 55. Effect of enzyme addition on burst strength and CED viscosity of dried, beaten kraft pulp.

Enzyme treatment hardly impairs the possibility of improving bonding potential by revealing new surfaces with refining. In the study quoted above, the original burst was achieved by subsequent beating during which the pulp showed a reduced refining resistance, i.e., it consumed less energy for the same drop of freeness[17]. Whether multiple enzyme treatment and refining speed the deterioration of recycled fibers in the long run is still an open question.

4.12.3 Challenges for strength properties of packaging grades

The increasing utilization rate and higher share of mixed recovered paper requires the use of more dry strength additives to avoid excessive refining when keeping strength specifications. Such additives probably do not increase the bonded area in sheets but the specific strength of bonds that results in bulk saving. Most mills use modified wet end starches. In some cases, the level is even 4%–5%. In tightly closed systems, the retention of cationic starches starts to decrease at 1.5%–2%, and strength gains achieve plateaus[118]. Most starch is adsorbed by fines, and the system can become overcationized. Adding anionic polymers can alleviate this condition[119]. In some cases, the additional use of polyacrylamide dry strength resins is effective.

For test liner, a size press often applies the additional starch. In highly humid conditions, the hygroscopic starch does not substitute for the lack of long fibers, and the compression strength or collapse time of boxes decreases. In the long run, the growing use of starch can also accumulate in the recovery chain. This causes microbiological problems in closed water systems as discussed earlier.

The introduction of mixed waste (MW) into the OCC chain will bring mechanical fibers and mineral fillers that reduce strength properties. Even additions of 2%–3% filler can be statistically seen in the loss of compression strength of liner. Figure 56 shows an experience from a linerboard machine where MW partly replaced OCC[118]. The compression strength loss began with 8%–12% substitution.

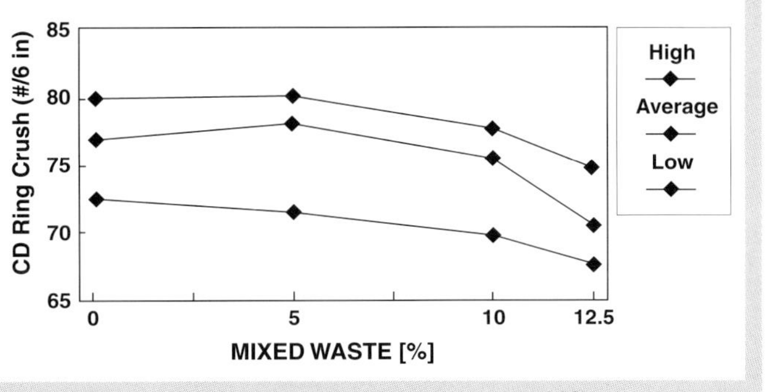

Figure 56. Substitution of OCC with mixed waste[118].

The probable accumulation of fillers and mechanical pulp fines may gradually become more and more problematic if that material cannot be removed in any way. Even in grades such as cardboard in which the use of mixed recovered paper as the furnish of a middle layer has been a tradition, the enrichment of filler and fines in the recycling chain has given a general loss in bending stiffness and bulk of the sheet. For example, a Central European cardboard mill was obliged to install a line to produce virgin refiner pulp (RMP) to improve the bending stiffness of board to an acceptable level[120]. The top-level recycling rate in that region had progressively increased the filler and mechanical fines contents in the components of mixed recovered paper. An increase of basis weight would also improve the compression strength and bending stiffness of board, but this is not desirable considering material use and cost considerations.

Mixed recovered paper also brings $CaCO_3$ pigment progressively into the system. This increases the content of dissolved calcium and requires running at higher pH than is customary. Neutral or alkaline board making requires modifications to a conventional acidic alum-rosin sizing system. The problems caused by dissolved ions can appear in many ways as discussed earlier. In one packaging board manufacturing plant for example, the retention of wet end starch significantly decreased as calcium hardness exceeded 0.01 mol/L (400 ppm) at the wet end[113].

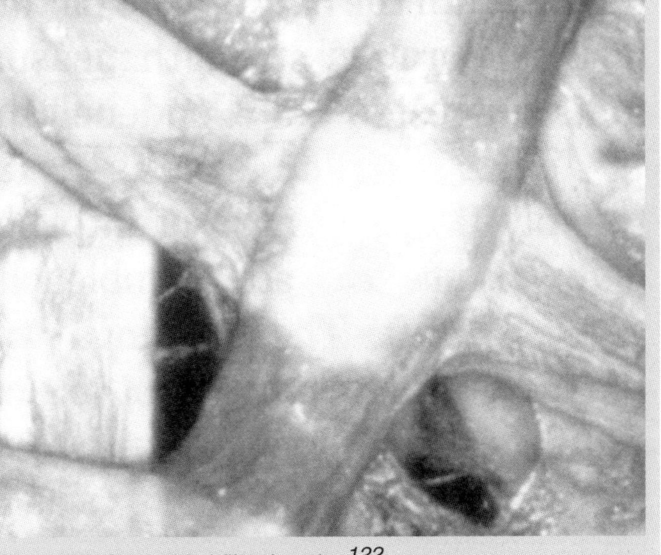

Figure 57. Press dried fiber junction[122].

Figure 58. Cylinder dried fiber junction[122].

Challenge and opportunity for tomorrow's papermaking technology are to use the unique properties of lignin-rich mechanical and other high-yield fibers in mixed recovered paper. The possible tools in that development might be impulse or press drying techniques. The latter has already been applied to the manufacture of liner, corrugating medium, and cartonboard[121]. In the process, a high pressure (up to 5 bar) and temperature are imposed on the web for a time during which lignin and probably also hemicelluloses soften and flow. This "welds" fibers together as confocal laser microscope pictures of fiber junctions in Figs. 57 and 58 show[122].

This "welding" contributes to all bonding related strength properties and especially to compression strength that can be used in the reduction of basis weight of the end product[121]. The flowing lignin also forms a protective layer on the fibers to give good resistance against moisture and an opportunity to omit rosin sizing. Some problems arising through the increasing use of mixed recovered paper can possibly be avoided by these new impulse or press drying techniques. In recycling trials, press dried fibers lost more swelling ability (WRV), but the strength properties of the pulp sheet were very similar to that made of recycled fibers from cylinder dried board[123].

References

1. Alanko, K., "Recyclability of thermomechanical fibers", M.Sc. thesis, Helsinki University of Technology, Department of Forest Products Technology, Espoo, Finland, 1993.
2. Maloney, T.C., Todorovic, A., and Paulapuro, H., Nordic Pulp Paper Res.J. 13(4):285(1998).
3. Korpela, A., personal communication.
4. Howard, R.C. and Bichard, W., J. Pulp Paper Sci. 18(4):J151(1992).
5. Stenius, P., Hynynen, R., and Laine, J., "The surface properties of virgin and recycled fiber", Interim Report of Nordpap Project: Properties of virgin and recycled fiber, Helsinki University of Technology, Helsinki, 1996, p. 23.
6. Weise, U., Paperi Puu 80(2):110(1998).
7. Jayme, G., Wochenbl. Papierfabr. 72(6):187(1944).
8. Scallan, A. and Tigerstrom, A.C., "Elasticity of fiber wall; effects of pulping and recycling", 1991 CPPA 1st Research Forum on Recycling, CPPA, Montreal, p.149.
9. Howard, R.C., Paper Technology 32(4):20(1991).
10. Carlson, G., Svensk Paperstidn. 87(15):R119(1984).
11. Bouchard, J. and Douek, M., "The effects of recycling on the chemical properties of pulps", 1993 CPPA 2nd Research Forum on Recycling, CPPA, Montreal, p.23.
12. Weise, U. and Paulapuro, H., Das Papier 50(6):328(1996).
13. Stürmer, L. and Göttsching, L., Wochenbl. Papierfabr. 107(3):69(1979).
14. Law, K.N., Valade, J.L., and Li, Z., Tappi J. 79(10):181(1996).
15. Brecht, W. and Klemm, K., Pulp Paper Mag. Can. 54(1):72(1953).
16. Lammi, T. and Heikkurinen, A., PSC Report No. 102, KCL, Espoo, Finland, 1997, p.18.
17. Rutledge-Cropsey, K.-R., Abubakr, S., and Klungness, J.H., "Performance of enzymatically deinked wastepaper on paper machine runnability", TAPPI 1995 Papermakers Conference Proceedings, TAPPI PRESS, Atlanta, p. 311.
18. Yamashki, Y. and Oye, R., Japan Tappi J. 35(9):33(1981).
19. Law, K.N., Progress in Paper Recycling 6(1):32(1996).
20. Gurnagul, N., Tappi J. 78(12):119(1995).

21. Klofta, J. and Miller, M.L., "Effects of deinking on the recycle potential of papermaking fibers", 1993 CPPA 2nd Research Forum on Recycling, CPPA, Montreal, p.207.

22. Koljonen, T. and Heikkurinen, A., "Characterization of refining effects on deinked pulp", 6th PTS-Deinking-Symposium, Papiertechnische Stiftung, Munich, 1994, p.43.

23. Tam Doo, P. and Kerekes, R., Pulp Paper Can. 83(2):T37(1982).

24. Jang, H.F., Howard, R.C., and Seth, R.S., Tappi J. 78(12):131(1995).

25. Hoppe, J. and Baumgarten, H.-L., Wochenbl. Papierfabr. 125(18):860(1997).

26. Gratton, M.F., "The recycling potential of calendered newsprint fibers", 1991 CPPA 1st Research Forum on Recycling, CPPA, Montreal, p.65.

27. Katz, S., Liebergott, N., and Scallan, A.M., Tappi J. 64(7):97(1981).

28. Laine, J. and Stenius, P., Paperi Puu 79(4):257(1997).

29. Laine, J.,"The effect of cooking and bleaching on the surface chemistry and charge properties of kraft pulp fibers", Ph.D. thesis, Helsinki University of Technology, Laboratory of Forest Products Chemistry, Espoo, Finland,1996.

30. Liebe, H., "Einfluss des Papierherstellungsprosesses auf das Festigkeitspotential von ECF- und TCF-Zellstoff", M.Sc. thesis, Darmstadt University of Technology, Darmstadt, Germany, 1995.

31. Mahagaonkar, M., Banham, P., and Stack, K., Appita J. 49(6):403(1996).

32. Nyström, M., Pykäläinen, J., and Lehto, J., Paperi Puu 75(6):419(1993).

33. Eriksson, I., Lunabba, P., and Pettersson, A., "Recycling potential of printed thermomechanical fibers for newsprint", 1995 CPPA 3rd Research Forum on Recycling, CPPA, Montreal, p.269.

34. Brandal, J. and Lindheim, A., Pulp Paper Can. 67(10):431(1966).

35. Villför, S., "Inverkan av extraktivämnen och polysackarider på egeskaper hos papper", M.Sc. thesis, Åbo Akademi University, Department of Forest Products Chemistry, Turku, Finland, 1996.

36. Okayama, T., Yoshinaga, N., and Take, Y., "Effect of recycling on wetting and liquid penetration of paper", TAPPI 1995 Coating Conference Proceedings, TAPPI PRESS, Atlanta, p.369.

37. Viitaharju, P., personal communication.

38. Chaterjee, A., Kortschof, M., and Roy, D.N., Tappi J. 76(7):109(1993).

39. Bayer, R., Allg. Papierrundschau 120(27):739(1996).

40. Lyne, M.B., Tappi J. 78(12):138(1995).

41. Lunabba, P., Petterson, A.M., and Eriksson, I.V.,"The effect of aging of waste paper on the deinking efficiency," 1996 SPCI Conference, SPCI, Stockholm, p. 179.

42. Tuovinen, J., "Welche Einflüsse haben Druckfarbenkomponenten auf die Deinking-Ergebnisse?" 5th PTS-Deinking-Symposium, Papiertechnische Stiftung, Munich, 1992, p.19.

43. Anon., "Altpapierliste der Deutschen Standardsorten und ihre Qualitäten", BDE/Köln, bvse/Bonn, VDP/Bonn, Germany, 1997, 2 pp.

44. Renner, K., Putz, H.-J., and Göttsching, L., Wochenbl. Papierfabr. 124(14/15):662(1996).

45. Ferguson, L.D. and Grant, R.L., "The state of the art in deinking technology in North America", 7th PTS-Deinking-Symposium, Papiertechnische Stiftung, Munich,1996.

46. Anon., Progress in Paper Recycling 2(3):80(1993).

47. Julien Saint Amand, F. and Celse, J.P., "Identification des freins au recyclage dans la fabrication des papiers et cartons", Journée technique CTP/REVIPAP, Grenoble, 1994.

48. Hornfeck, K., Liphard, M., and Schreck, B., Wochenbl. Papierfabr. 122(4):124(1994).

49. Geistbeck, M. and Wiese, H., Wochenbl. Papierfabr. 125(16):733(1997).

50. McCool, M.A. and Silveri, I., Tappi J. 70(11):75(1987).

51. Jordan, B.D. and Popson, S.J., J. Pulp Paper Sci. 20(6):J161(1994).

52. Yule, J.A.C. and Nielsen, W.J., 1951 TAGA Proceedings, p. 65.

53. Peel, J.D., Wochenbl. Papierfabr. 114(6):189(1986).

54. Selder, H., Das Papier 51(9):455(1997).

55. Roßkopf, J., "Optische Bewertung von Melierfasern in hellen Altpapierstoffen", M.Sc. thesis, Darmstadt University of Technology, Darmstadt, Germany, 1994.

56. Ackermann, C., Putz, H.-J., and Göttsching, L., Wochenbl. Papierfabr. 124(11/12):508(1996).

57. Linhart, F., Auhorn, W., Degen, H.J., et al., Wochenbl. Papierfabr. 116(9):329(1988).

58. Tremont, S.R., Pulp Paper Can. 96(12):T399(1995).

59. Hamm, U. and Putz, H.-J., "Untersuchungen zur möglichen Bildung aromatischer Amine bei der reduktiven Altpapierbleiche", INGEDE Report 4896 IfP, INGEDE, Munich, 1997, 10 pp.

60. Escabasse, J.Y., Vernac, Y., Lenon, G., et al., "Microbiological Decontamination of Recycled Papers and Boards", 4th Research Forum on Recycling, CPPA, Montral, 1997, 17 pp.

61. Hunold, M., "Experimentelle und theoretische Untersuchungen über quantitative und qualitative Auswirkungen steigender Altpapier-Einsatzquoten auf das

Recyclingsystem Papier – Altpapier", Ph.D. thesis, Darmstadt University of Technology, Darmstadt, Germany, 1997, 146 pp.

62. *Law, K. N., Hong, C. Z., and Valade, J. L., Progress in Paper Recycling 6(1):39(1996).*
63. *Korpela, A., personal communication.*
64. *Hanecker, E., "DIP-Charakterisierung", INGEDE Report 4796 PTS, Papiertechnische Stiftung, Munich, 1997, 30 pp.*
65. *Ajersch, M. and Pelton, R., "Mechanism of pulp loss in flotation deinking", 1995 CPPA 3rd Research Forum on Recycling, CPPA, Montreal, p. 27.*
66. *Turvey, R., J. Pulp Paper Sci. 19(2):J52(1993).*
67. *Le Ny, C., personal communication.*
68. *Mohlin, U.-B., J. Pulp Paper Sci. 23(1):J28(1997).*
69. *Lammi, T. and Heikkurinen, A., PSC Report No. 102, KCL, Espoo, Finland, 1997, p. 18.*
70. *Lumiainen, J., Paperi Puu 74(4):319(1992).*
71. *Lumiainen, J., Tappi J. 75(8):92(1992).*
72. *Selder, H. and Linck, E., World Paper 220(Jan.gFeb):26(1995).*
73. *Latomaa, A., personal communication.*
74. *Kibblewhite, R.P. and Bailey, D.G., Appita J. 41(4):297(1988).*
75. *Bawden, A.D. and Kibblewhite, R.P., J. Pulp Paper Sci, 23(7):J340(1997).*
76. *Houen, P. J. and Fjerdingen, H., "On the effect of recycling on cross-sectional shapes and dimensions of sulphate pulp fibers", TAPPI 1997 Recycling Symposium Proceedings, TAPPI PRESS, Atlanta, p. 347.*
77. *Waterhouse, J.F. and Omori, K., "Products of Papermaking", 1993 Fundamental Research Symposium, PIRA, Leatherhead, UK, p. 1261.*
78. *Poli, M., Progress in Paper Recycling 6(2):99(1997).*
79. *Maddern, R., "Experience of Krima dispersion unit in treating white waste for printing grades", Recycling Conference, PIRA, Leatherhead, UK, 1993, Paper No. 05.*
80. *Freeland, S.A. and Hrutfjord, B.F., Tappi J. 77(4):185(1994).*
81. *Drehmer, B. and Back, E., "Effect of dispersion variables on the papermaking properties of OCC", 1995 CPPA 3rd Research Forum on Recycling, CPPA, Montreal, p. 141.*
82. *McKinney, R.J.W., Pulp Paper International 39(1):37(1997).*
83. *Page, D.H., Barbe, M.C., and Seth, R.S., J. Pulp Paper Sci. 10(5):J74(1984).*

84. Lucas, R. and Williams, T., "Winding problems associated with recycled fibers and additives in today's papers", TAPPI 1997 Finishing & Converting Conference Proceedings, TAPPI PRESS, Atlanta, p.151.

85. Fernandez, C. and Garnier, G., J. Pulp Paper Sci. 23(4):J144(1997).

86. Larson, A. and Stenius, P., Svensk Papperstidn. 88(3):R2(1985).

87. Krook, K., "Stickies and wood resin in a deinking plant and on a paper machine". M.Sc. thesis, Åbo Akademi University, Faculty of Chem. Eng., Turku, Finland, 1997.

88. Haynes, D. and Marcoux, H., "Evaluation of fatty acid carryover in North American newsprint deinking mills", 1997 CPPA 4th Research Forum on Recycling, CPPA, Montreal, p. 163.

89. Holmbäck, Å., "Analys av fettsyratvålar och vedharts i en returpappersanläggning", M.Sc. thesis, Åbo Akademi University, Faculty of Chem. Eng., Turku, Finland, 1995.

90. Krüger, E., Göttsching, L., and Mönch, D., Wochenbl. Papierfabr. 125(20):986(1997).

91. Sjöström, L. and Ödberg, L., Das Papier 51(6A):V69(1997).

92. Grau, U., Schuhmacher, R., and Kleemann, S., Wochenbl. Papierfabr. 124(17):729(1996).

93. Ackermann, C. and Putz, H.-J., personal communication.

94. Terho, J., Jaakko Pöyry Client Magazine (1):16(1994).

95. Ventola, J., "Effect of dissolved and colloidal substances on bonding ability of TMP-fibers", M.Sc.thesis, Helsinki University of Technology, Paper Laboratories, Espoo, Finland, 1997.

96. Sundberg, K., "Effects of wood polysaccharides on colloidal wood resin in papermaking", Ph.D. thesis, Åbo Akademi University, Faculty of Chem. Eng., Turku, Finland, 1995.

97. Röring, A. and Wackeberg, E., Pulp Paper Can. 98(5):T136(1997).

98. Krogerus, B., PSC Report No. 71, KCL, Espoo, Finland, 1995, pp. 20-24.

99. Stetter, A. and Eisenschmid, K., Wochenbl. Papierfabr. 121(23/24):1018(1993).

100. Stetter, A. and Eisenschmid, K., Wochenbl. Papierfabr. 125(7):186(1997).

101. Upton, B.H., Krishnagopalan, G., and Abubakr, S., Tappi J. 80(2):155(1997).

102. Diedrich, K., Hamm, U., and Knelissen, J., Das Papier 51(6A):V153(1997).

103. Habets, L. and Knelisen, H., Pulp Paper Europe 2(6):8(1997).

104. Verkasalo, L., personal communication.

105. Kärenlampi, P. and Yu, Y., Alava, M., et al., "Performance of reinforcement fibers in paper", in Fundamentals of Papermaking Materials (C.F. Baker, Ed.), PIRA, Leatherhead, UK, 1997, p. 521.

106. "Fiber properties and paper fracture – fiber length and fiber strength", in Fundamentals of Papermaking Materials (C.F. Baker, Ed.), PIRA, Leatherhead, UK, 1997, p. 1270.

107. Niskanen, K., Nilsen N., and Hellen, E., "Simulation of the 3D structure of paper", in Fundamentals of Papermaking Materials (C.F. Baker, Ed.), PIRA, Leatherhead, UK, 1997.

108. Brinckman, M., "The printability of newsprint containing recycled fibers", IFRA Newspaper Techniques Bulletin, IFRA, Darmstadt, Germany, 1997, p. 24.

109. Kärnä, A., personal communication.

110. Ackermann, C., Putz, H.-J., and Göttsching, L., "Aufbereitung von Altpapier für höherwertige Papiere", 5th PTS-Deinking-Symposium, Papiertechnische Stiftung, Munich, 1992, p. 92.

111. Farrand, J.A., McCrory, S.J., and Forsman, D.L., "Secondary fiber plant operations does influence paper machine performance", TAPPI 1995 Papermakers Conference Proceedings, TAPPI PRESS, Atlanta, p. 327.

112. Korpela, A., Information Bulletin No. 190, KCL, Espoo, Finland, 1998, p.71.

113. Renders, R. J., "Brown grade paper and board – Requirements on RCF", The Future of RCF Technology, 11th Sunds-Defibrator Technical Seminar, Sunds-Defibrator, Stockholm, 1995, Paper No. 06.

114. Bhardwaj, N.K., Bajpaj, P., and Bajpaj, P.K., Appita J. 50(3):230(1997).

115. Sarkar, J., Appita J. 50(1):57(1997).

116. Jackson, L.S., Heitmann, J.A., and Joyce, T.W., Tappi J. 76(3):147(1993).

117. Gruber, E., Gelbrich, M., and Geistler, U., Das Papier 51(9):464(1997).

118. Latta, J. and Mangat, N., Paper Age 113(11):13(1997).

119. Scott, W. and Kumar, V., Paper Age 113(12):22(1997).

120. Mühlhauser, M., Stark, H., and Merckens, C., Das Papier 51(6A):V74(1997).

121. Ojala, T., Valmet Paper News 13(2):22(1997)

122. Kunnas, L., Lehtinen, J., and Paulapuro, H.,Tappi J. 76(4):95(1993).

123. Retulainen, E., Paperi Puu 80(2):104(1998).

124. Höötmann, U., Tappi J. 82(1):77(1999).

125. Danielsson, K., "Analysis of detrimental substances in two newsprint mills", M.Sc. thesis, Åbo Akademi University, Faculty of Chem. Eng., Turku, Finland, 1998.

CHAPTER 11

Stickies in recycled fiber pulp

CHAPTER 11

Hans-Joachim Putz

Stickies in recycled fiber pulp

1 What are stickies?

The term "stickies" does not have a precise, generally applicable definition. Despite this, papermakers understand it to mean those tacky substances contained in the pulp and process water system that deposit or attempt to deposit on paper machine clothing, cylinders, or rolls at certain stages of papermaking. The particles in the produced paper identified as tacky contaminants are also stickies. Stickies are only one of many different contaminants contained in recovered paper. Scott[1] gives a good general definition of contaminants:

"A contaminant is any fiber or nonfibrous component that is not intentionally added to the system to construct the paper web, to enhance paper properties, or to facilitate the operation or promote the cleanliness of the paper manufacturing system".

This definition does not regard as contaminants functional and process chemicals used for the first time as additives in the papermaking process. The same chemicals are often contaminants if they enter the recycling process with recovered paper.

Another pragmatic definition of contaminants distinguishes between external and internal contaminants (See Fig. 15 in Chapter 10). External contaminants are the materials not physically or chemically bound to paper and board and their fibers but contained in recovered paper such as sand, glass particles, wood pieces, screws, or cans. Considerable removal of these materials should be possible at the time of sorting recovered paper, or they can usually be easily removed from the pulp slurry during recovered paper processing[2].

Internal contaminants are bound physically or chemically to the paper or board surface or even embedded in the fiber structure. Some components that can later act as contaminants enter as additives during the papermaking process to control certain paper characteristics. Examples of these are the application of starch to enhance the strength of testliner and corrugating medium or the use of wet strength additives for kitchen towels. Other internal contaminants are added during paper converting or by private users. These include plastic coatings of catalog covers or liquid packaging; hot melt materials for binding book or magazine spines; ink use for newspapers, magazines, or books; addition of starch for gluing corrugated board; use of wax coatings for corrugated board; application of pressure sensitive labels in marking the prices of consumable goods; or the use of pressure sensitive note pads in private homes.

These internal contaminants can be soluble or insoluble in water. Water-soluble internal contaminants such as starch, gluing agents, or redispersible adhesives usually transform into the aqueous stage in the pulp and water loop and are frequently included in the term detrimental substances. An appropriate process water and effluent cleaning is necessary to manage these dissolved detrimental substances. Theoretically, an increasingly closed process water system causes an increased concentration of these dissolved substances and favors the formation of deposits and growth of microorganisms or slime.

Insoluble contaminants can be thermoplastic or nonthermoplastic materials. The thermoplastic behavior of contaminants is a precondition for permitting the adsorption of a substance on another solid material. For example, the adhesive of pressure sensitive labels is normally in a permanent thermoplastic state at room temperature. This provides the basis for their durable tack. With hot melts, a certain temperature is necessary to ensure that they melt to develop their tacky properties. This thermoplasticity of hot melts occurs not only between 65°C and 120°C as often claimed but can be reached even at room temperature depending on their chemical composition. Plastics such as polyethylene and polypropylene are also thermoplastic and therefore develop tack at softening temperatures such as encountered in the drying section of paper machines.

Many insoluble contaminants, both thermoplastic and nonthermoplastic, consist of different chemicals. Some are soluble in an aqueous environment. The insoluble components can be tacky but may not necessarily deposit as stickies.

For this reason, the term "stickie-" is generally used in the paper industry for indefinite mixtures of organic compounds that enter the recycling loop with the recovered paper. They can be screened from the pulp suspension as tacky substances, precipitated from the aqueous medium, or deposit within the recycled pulp process on the paper machine, on the paper machine clothing, or in the final product.

1.1 Classification of stickies

The distinguish features of stickies are the following:

- Size
- Type
- Behavior.

Classification by size uses a laboratory screening procedure. The stickies remaining on a slotted screen plate are called macro stickies. The stickies passing the plate are micro stickies. Micro stickies can be finely dispersed (100 μm to 100 nm), colloidal (100 nm to 10 nm), or molecular dissolved (<10 nm)[3].

Depending on the type, they can be primary or secondary stickies. Primary stickies are present from the beginning in the disintegrated pulp and can consist of micro and macro stickies. Secondary stickies can develop only during the processing operation or in the paper machine from micro stickies, finely dispersed, molecular dissolved or colloidal substances.

Another classification for stickies uses their behavior. They can be permanently tacky (permanent stickies such as macro and micro stickies) or potentially tacky (potential stickies under certain circumstances such as secondary stickies).

1.1.1 Size of stickies

The most common classification of stickies uses their size. Macro stickies are those that remain as screening residue after laboratory screening with a slotted plate equipped with a slot width of 0.10 or 0.15 mm. Micro stickies are those that can pass the screening slots with the accept after screening with a slotted plate equipped with the same slot width.

The reason for this type of distinction is that macro stickies from recovered paper in industrial processing systems can be removed from the pulp slurry by screens and cleaners. Micro stickies are usually so small that they cannot be eliminated even by the most effective screening units. If the macro stickies content in the headbox of a paper machine or in the produced paper is excessive, the screening of the pulp processing system, the approach flow system, or both require improvement. Micro stickies that pass the screens possibly agglomerate and lead to deposits on the paper machine or its clothing or pass into the product as newly formed secondary macro stickies.

Doshi[4] prescribes a slot width of 0.15 mm for distinguishing micro and macro stickies. Ackermann *et al.*[5] use 0.10 mm as the distinguishing criterion for deinked pulp. Since the slot widths in industrial recovered paper processing are becoming increasingly smaller and the developing fiber mat leads to an effective reduction of the actual screen perforation, a minimum slot width should be specified for laboratory screening operations during which no fiber mat is formed. A slot width of 0.10 mm cannot be applied for all recycled fiber pulps. For example, highly contaminated recovered paper grades or samples taken at the beginning of a recovered paper processing system cannot always be screened with 0.10 mm slots due to the risk of plugging the slotted screens or encountering excessive reject rates. In laboratory screening for control of stickies, the use of slotted plates with 0.10 mm slot width is therefore recommended for deinked pulp and with 0.15 mm for pulp from mixed (brown) recovered paper grades.

1.1.2 Type of stickies

A distinction is made between primary and secondary stickies. Primary stickies are identical with the aforementioned macro and micro stickies and characterized as intact tacky particles of adhesives such as hot melts or pressure sensitive adhesives, inks, binders, waxes, plastics, or wet strength resins.

Shock-type, chemical-physical alterations of a pulp suspension are a potential cause for formation of secondary stickies. This includes temperature, pH, or charge changes that cause a colloidal destabilizing and agglomeration of dissolved, colloidal, or both substances. Secondary stickies can then also lead to deposits on the paper machine or its clothing.

Incidentally, comparable problems also occur in the processing of virgin pulp. Papermakers then speak of pitch in connection with tacky substances from wood extractives or white pitch in connection with the processing of coated broke. The interaction of certain process chemicals or resin acids of either natural (wood) or artificial origin plays a dominant role in the formation of secondary stickies and white pitch[6].

When stickies-related problems occur at a paper mill, primary or secondary stickies may be the culprit. Combating the problem is usually easier if one can determine whether a problem is due primarily to primary or secondary stickies. If no recovered paper is used for example, the cause of stickies-related problems is usually the formation of secondary stickies. This can be counteracted by changing the wet end chemistry or the coating binder system.

1.1.3 Behavior of stickies

Interactions occur that can increase the tack of stickies and the problems experienced with stickies. Such interactions can involve primary or secondary stickies. Figure 1 shows in schematic form some possible interactions between virgin fibers, chemical additives, and contents of the recycled fibers that can cause deposits. For the formation of secondary stickies of Fig. 2, a sudden change (shock) of temperature, pH value, charge, shearing forces, or concentration is the cause of the agglomeration of colloidal or dissolved substances.

Figures 1 and 2 show a highly simplified representation of the complexity of the interactions that can contribute to formation of stickies. Because chemical measures to control stickies usually influence only one described mechanism, an absence of comprehensive knowledge of the interrelations that lead to formation of stickies continues despite many years of research into stickies. Major problems therefore continue to exist in the daily practice of recovered paper processing.

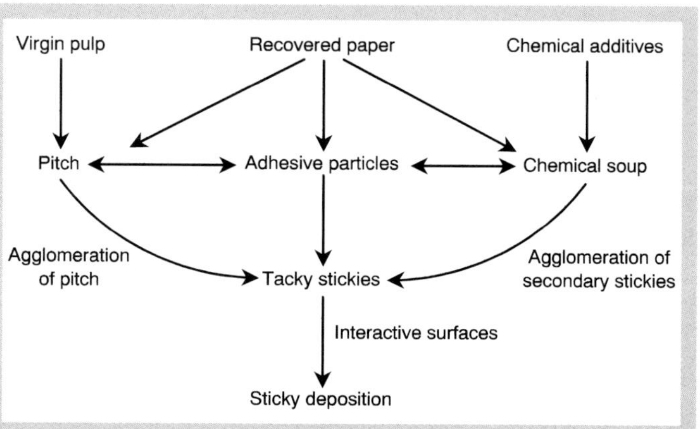

Figure 1. Interactions contributing to deposition of stickies[6].

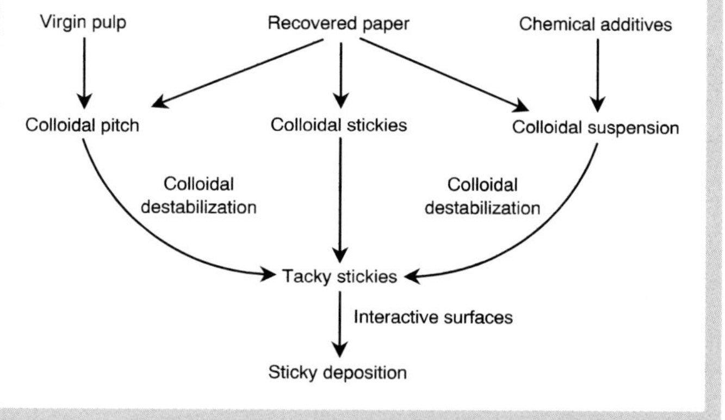

Figure 2. Interactions contributing to formation of secondary stickies[6].

1.2 Problems caused by stickies

The problems caused by tacky impurities are all too familiar. Deposits occur on stationary parts of a processing system such as pipes and chest walls and in the approach flow system. This includes the inner walls of the deculator in the area of the suspension level or in the headbox itself due to the adhesive character of these materials. Deposits primarily influence moveable parts such as wires, felts, or rolls. Deposits within the paper machine are especially critical since their release can directly cause a web break. A sufficiently intense contact of the running paper web with these tacky substances can also cause picking of fibers.

Disturbances caused in these ways have different effects. Weak spots in the paper web primarily as thinned parts or holes can cause web breaks in the paper machine and in the printing press. They result in lengthy nonproductive cleaning times and reduced levels of productivity. Even if breaks do not occur, holes or spots in the paper cause loss of quality that means the product has use only as second quality material or has to be processed as broke. All problems caused by stickies in the paper machine become more serious with lower basis weight or caliper, lower strength of the paper web, higher paper machine speed, and higher dynamic stress on the paper web[7].

Even more costly are the problems with stickies that occur after finishing of the paper or board sheet. Due to certain processing conditions of temperature or pressure, stickies bound into the inside of board or thicker paper can reach the outsides of paper or board sheets by melting. This causes unacceptable appearance after printing, varnishing, or laminating. The same applies to stickies on the surface of a paper or board web that have passed through the papermaking process including subsequent finishing without any problem but that cause problems during converting by their adhesion to paper or board blanks they contact. Sometimes, layers stick together during unwinding of a reel causing breaks or surface blemishes.

A particular problem occurs in corrugated board production by wax introduced by recovered paper. If this is insufficiently removed or not neutralized in recovered paper processing, it can result in a decreased coefficient of friction of the liner. Consequently, reels or rolls may telescope during transport. The finished corrugated board blanks could also slip. This would cause imperfect stacking and affect further processing.

As this brief description demonstrates, the problems caused by stickies are many and varied and can occur at practically every stage in the papermaking and paper conversion chain including printing.

1.3 Cost impact of stickies

Little information is available concerning the additional costs due to stickies in the paper production process. In 1984, Baumgarten estimated the costs due to tacky impurities contained in recovered paper as 5–25 US$/t of recycled fiber pulp but gave no information for this estimation[8]. The cost is higher as basis weight of the paper produced decreases. Included in the estimated figures are the costs incurred in the recovered paper processing system and at the paper machine.

Friberg gave a more recent and detailed estimation of the cost impact of stickies[9]. For the United States, he puts annual losses due to stickies at about 700 million US$. This figure covers recycled fiber pulp processing, paper production, and the collection and sorting of recovered paper. In relation to the recovered paper consumption in the United States of 28 million tons in 1994, this represents the average cost impact of stickies at 25 US$/t of recovered paper. Not considered here are the costs of the paper and board converters.

The costs estimated by Friberg in Fig. 3 do consider the additional disposal costs (2%) for stickies that have been removed and the losses for the recovered paper trade due to the lower sales prices for recovered paper grades having an unacceptably high content of stickies (22%). Also considered are the losses (76%) of the producers of different types of paper and board using recycled fiber pulp due to the following:

- Use of chemicals for control of stickies (5 US$/t)
- Nonproductive paper machine time caused by stickies (104–350 h/year)
- Costs per paper machine hour (10 000–30 000 US$/h).

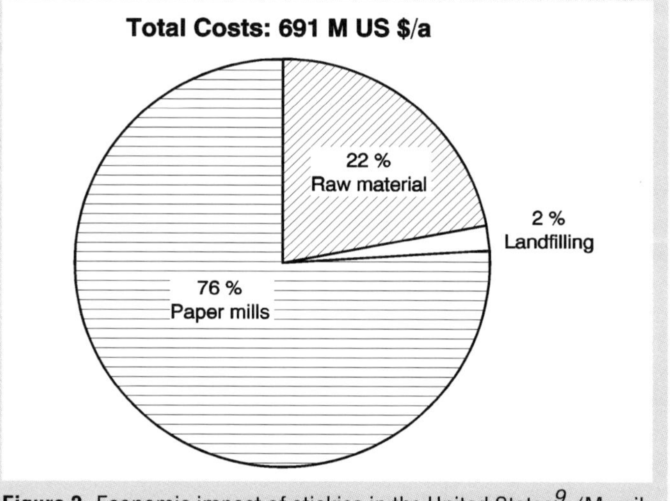

Figure 3. Economic impact of stickies in the United States[9]. (M: million)

These losses vary widely with the type of paper or board produced and the grade of recovered paper used. On this basis, the cost impact of stickies for the processing of recovered paper in major product groups incurs annual costs in the region of 130 million US$ in the German paper and board industry. The specific costs caused by stickies total an average of 15 US$/t of recovered paper processed. According to Fig. 3, these costs that are incurred directly at paper mills represent only 76% of the entire related costs of stickies. The total is therefore about 20 US$/t of recovered paper, and the annual cost to the economy is a minimum of 170 million US$ in Germany.

The paper and board production structure in the European Union with 50% graphic papers (including 12% newsprint), 40% packaging paper and board, and 10% other types of paper is comparable with the German structure. The related costs of stickies in the European Union may be calculated on the basis of these figures. In 1994 for a consumption of recovered paper of about 28 million tons, the costs were more than 525 million US$. This also shows that stickies have an enormous influence on the production costs of paper and board.

2 Origins of stickies

A frequently cited categorization of the chemical origin of tacky impurities comes from Moreland[10]. Figure 4 shows this work in a modified form. Starting with organic deposits, the figure follows first the division into synthetic and natural materials. The natural components originate as fatty acids or resin acids from the wood and cause the familiar pitch problems in virgin fiber processing. In accordance with this division, synthetic components originate from coating binders, printing inks, and adhesives. Two of the named groups are subdivided into product classes and listed as typical representatives of the chemicals used whose abbreviations are explained in the text.

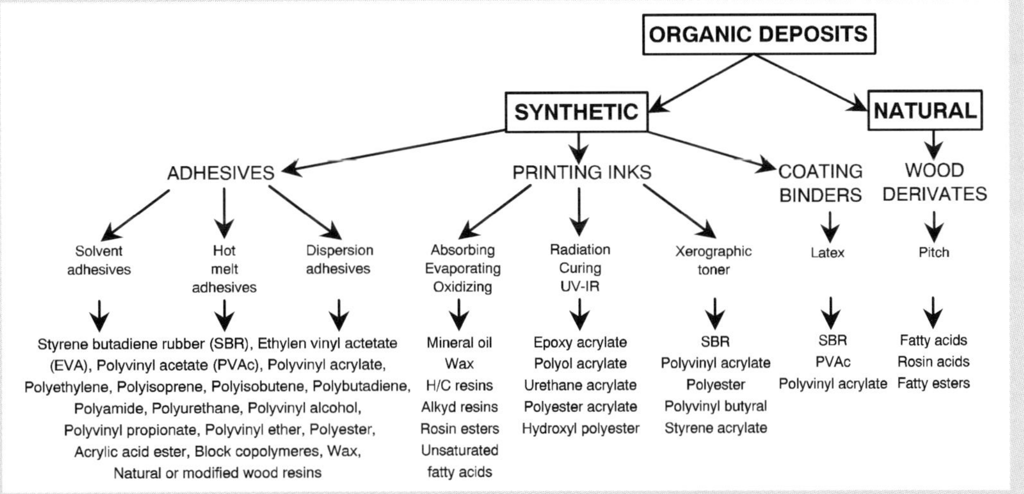

Figure 4. Contaminants frequently identified in paper machine deposits from recovered paper processing mills.

2.1 Adhesives

Adhesives bond two materials together by adhering to the surface of each. Figure 5 shows the use of glues and sealing agents in different industrial applications in Europe. With a total of 35%, the products used in the paper and board industries represent the second-largest market segment after the building industry. Typical applications in the paper converting sector are the binding of magazines and

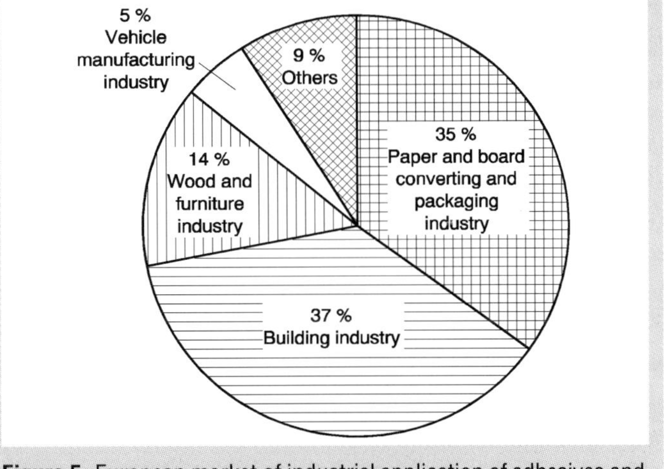

Figure 5. European market of industrial application of adhesives and sealing compounds (1996).

catalogs, coating of pressure sealing labels, and production of self-sealing envelopes, folding boxes, corrugated board containers, etc.

Adhesives must satisfy two major requirements, i.e., have sufficient adhesion and cohesion. At the time of transfer to the materials to be bonded, they should adhere well to the respective surfaces. The close contact of the adhesive molecules with the paper surface occurs by wetting and diffusion processes. To guarantee sufficient adhesive strength (adhesion) between the boundary layers of the adhesive film and the joined parts, the molecules of the adhesive must be in as close contact as possible with the surface of the joined parts. Application of the liquid adhesives accomplishes this technically.

Besides good adhesion, the adhesive requires sufficient internal strength (cohesion) to ensure that the connection between the bonded materials can survive mechanical stress. After application of a glue with good adhesion, the material must solidify into the solid phase to develop sufficient cohesive force. In paper converting and packaging production, the formation of the adhesive film occurs almost exclusively by physical bonding processes characterized by the evaporation of water (dispersion adhesives) or solvents (solvent adhesives) or the cooling of the adhesive film (hot melt adhesives).

Adhesive formulations usually contain several components to satisfy optimally all the gluing requirements. The processing method including its speed and the application systems also influences the composition of the adhesive. Common denominators do exist for the gluing systems used in paper and board converting:

- Aqueous systems have common use for the production of corrugated board. Starch products are dominant.
- Dispersion glues and hot melts have traditional use in bookbinding operations.
- Various glue systems find use in the paper converting industries. Dispersion glues have the greatest share[11].

The large number of adhesive formulations are due to the most important basic components. As Fig. 6 shows, dispersion glues at 40% presently account for the largest share of adhesives used in the paper converting industries in Europe. Natural colloidal systems such as starch, dextrin, and casein cover about 25% of the adhesive requirements. Another segment of 30% consists of hot melt adhesives including polyurethane hot melts processed in a liquid state at high

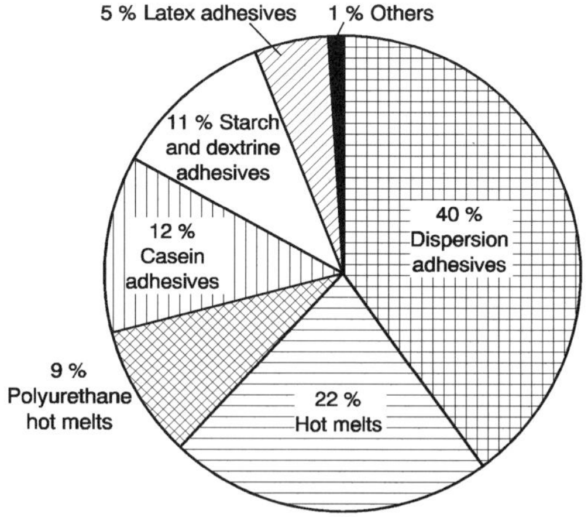

Figure 6. Estimated adhesive consumption in the paper and board converting industries in Europe (1995).

temperatures to form a solid film after cooling. Their market share is especially evident for gluing printed material such as the spines of mail order catalogs or telephone directories.

Conventional gluing uses the following sequential steps:

- Application of the liquid adhesive film
- Joining of the material surfaces to be bonded
- Hardening of the adhesive film by evaporation, cooling, or cross-linking.

With dispersion, solvent, and hot melt adhesives, pressure sensitive adhesives (PSA) are a major source of stickies, but they do not constitute an independent category of adhesive. Instead, all three types of adhesive can be PSA. The term pressure sensitive adhesive is application related and describes how bonding with a pressure sensitive material simply involves joining and exerting slight pressure. Examples are labels or adhesive tapes. Pressure sensitive adhesives simplify many work processes such as envelope sealing. Unfortunately, the lasting tack of the pressure sensitive film means that such adhesives automatically contribute to related problems with stickies. This is why a special section covers this adhesive application technique.

2.1.1 Hot melt adhesives

Hot melt adhesives (HMA) are applied in a liquid state and harden by cooling at room temperature. Their most important areas of application are binding magazines and books and gluing boxes. Ethylene vinyl acetate (EVA), styrene butadiene rubber (SBR), polyethylene, polyamide, polyurethane (PUR), and polyisoprene are common thermoplastic polymers or copolymers used as ingredients. Modified resins, unmodified resins, and synthetic resins can provide the necessary tack. Nonlinear or branched waxes and low molecular weight polyethylene are used to regulate the viscosity and melting point of hot melt adhesives. Small quantities of additives such as fillers, heat stabilizers, and antioxidants also have use[12].

Hot melts are solid at room temperature and usually become soft between 65°C and 120°C. Some hot melt formulations soften at room temperature. Hot melts are usually not redispersible in water under alkaline or acidic conditions but can be dissolved with several organic solvents such as dichloromethane or toluene. The density of hot melts is between 0.9 and 1.0 g/cm^3[13].

2.1.2 Dispersion adhesives

Dispersion adhesives are used for a wide range of different applications in paper converting. They contain a tacky polymer substance composed of different monomers that are present in the aqueous phase in a finely dispersed state.

The aqueous phase treats the dispersed medium as an external phase whose high polarity causes the dispersion adhesives to be especially suitable for gluing polar materials such as paper, wood, or textiles. Among adhesives, dispersions have the special advantage of achieving high levels of polymerization at a certain processing viscosity without requiring the solids content to be as low as is necessary when dissolving the adhesive in an organic solvent.

As an internal phase, the dispersed tacky substance always consists of several polymers formed from different monomers. The particles for dispersion require separation from each other by protective colloids or emulsifiers and distribution in the water to prevent reagglomeration. Polyvinyl alcohol is a common protective colloid or emulsifier. Among the monomers for use in making adhesive dispersions are vinyl acetate, ethylene, vinyl propionate, and acrylate. The characteristics of the adhesive dispersions vary within wide tolerances depending on the type of monomer and the different possibilities to form homopolymers, copolymers, or terpolymers[11].

Although particle size distribution influences the application characteristics of an aqueous polymer dispersion, the chemical nature of the dispersion largely determines the characteristics of the adhesive joint. Besides the chemical composition and the particle size distribution, the type and quantity of emulsifiers and protective colloids have a decisive influence on the characteristics. A larger inner surface of the polymer dispersion means a more finely dispersed system with a larger demand for these substances. Because these chemicals are integrated into the adhesive film, they have a major influence on its final characteristics such as water resistance of a paper sheet or paper product.

Adhesive films formed from polymer dispersions bond after removal of the aqueous phase. This occurs by absorption into the substrate and by evaporation. Removal of the water increases the particle concentration until the individual particles come into contact. In the intermediate spaces that become smaller, specific capillary forces develop that lead the particles of the internal phase to a homogeneous film due to their plastic behavior. To ensure that this process can occur, a characteristically minimum film forming temperature (threshold temperature) is necessary for each polymer dispersion so that a transparent adhesive film can develop.

Polymer dispersion adhesive films are thermoplastic materials that become soft and plastic at higher temperatures. Depending on the type of polymer dispersion involved, this process occurs between 60°C and 150°C. Within any one group of substances, the softening points are higher with a higher average degree of polymerization. Simultaneously, the higher polymerized products have better heat resistance of the glued joints[11].

2.1.3 Solvent adhesives

The classical areas of application for solvent adhesives are pressure sensitive adhesives and do-it-yourself applications. Originally, pressure sensitive adhesives consisted of rubber available only in solvent. With evaporation of the highly volatile solvent, the viscosity of the adhesive increased and its wetting capacity fell. With the increasing molecular weight of the adhesive components, the cohesion within the adhesive film also increased. Rising costs for organic solvents, the environmental legislation, and costs for drying and solvent recovery have resulted in an increasing substitution by solvent-free gluing techniques. The substitution of solvent adhesives in the paper sector occurred primarily by the development of acrylate dispersions that considerably broadened the technical limits on the use of dispersions and the area of application. Today, the share of solvent adhesives of the entire European adhesive market is less than 15% with a decreasing tendency. Such systems therefore have only minor importance in the paper sector.

2.1.4 Pressure sensitive adhesives

The term pressure sensitive adhesive (PSA) includes a range of different adhesives with a common application. PSA materials are extremely high-viscosity, elastic-viscous fluids that provide adhesion to nearly all surfaces in response to slight pressure due to their flow behavior and wetting characteristics. This "permanent adhesive" state distinguishes the management of PSA materials from other types of adhesives and means that they are generally tacky after use.

Products that typically use PSA materials are labels, adhesive tapes, and self-sealing envelopes. In these cases, permanent adhesion is necessary after the first application of the self-sealing material. In some cases, removable or repositionable characteristics are required (e.g., post it notes). Such applications should not leave any residue and attach again and again. As a result of these different requirements, many different chemical components are necessary to satisfy the product demands.

The production of self adhesive labels uses dispersion, solvent, or hot melt adhesives. Today, dispersion adhesives in Europe have 60%–70% share of the market for the production for self adhesive labels followed by hot melt and solvent adhesives that have approximately equal market shares. Due to a greater environmental awareness, expensive solvent recovery, and increased solvent prices, solvent adhesives of all types are increasingly undergoing replacement by other, solvent-free adhesives[7].

The most important polymers used for PSA materials of this type are styrene butadiene rubber (SBR), natural rubber, butyl rubber, polyisobutene, polyvinyl ether, and polyacrylate. Because most polymers are not tacky, they require mixing with natural or modified resins of which an almost limitless number exists to achieve the permanently adhesive state.

While solvent pressure sensitive adhesives primarily use SBR, dispersion PSA materials usually use a polyacrylate base. Typical recipes for PSA materials use the following[14]:

- Polymers: 30%–50%
- Resins: 20%–40%
- Softeners: 0%–40%
- Fillers: 0%–10%
- Stabilizers: 1%– 2%.

The following categories relate to specific features of processing and use:

- Type A: Adhesives with a natural rubber base

  Nonadhesive elastomer, "natural rubber", is made permanently adhesive with resins. Stabilizers are necessary, and plasticizer and fillers are optional. Processing uses organic solvents.

- Type B: Adhesives with a synthetic rubber base

  Nonadhesive elastomer, "block copolymer", is made permanently adhesive with resins. Stabilizers are necessary, and plasticizer and fillers optional. Processing uses organic solvents or melting.

- Type C: Adhesives with an acrylate solution base

  Elastomer, "polyacrylate", is resin-modified according to adhesion properties already present. Processing uses organic solvents.

- Type D: Adhesives with an acrylate dispersion base

  Elastomer, "polyacrylate", is resin-modified according to adhesion properties already present. Viscosity regulator and wetting agent are necessary. Processing uses water[15].

PSA materials attach under low pressure and room temperature. For certain applications, adhesion is also necessary under extreme conditions such as temperatures below 0°C for labeling of frozen food packaging or at high humidity in a subtropical climate. The density of pressure sensitive adhesives is between 0.9 and 1.1 g/cm^3. Conventional pressure sensitive adhesives are insoluble in water under neutral or mildly alkaline or acidic conditions but can dissolve quickly in several organic solvents.

The parameter that has a major influence on the adhesive strength of a PSA is one that simultaneously affects its recycling behavior. This is the thickness of the applied adhesive film. Typical layer thickness values are as follows[16]:

- Repositionable label: 5 μm
- Paper label: 25 μm
- Paper transfer tape: 25 μm
- Plastic label and decal: 25–40 μm
- Polyester packaging tape: 50–90 μm
- Polypropylene strapping base: 60–100 μm.

High adhesive layer thickness is primarily with tapes, and the lowest layer thickness is for repositionable labels.

2.2 Printing inks

Printing inks are presumably a source of stickies although to date no detailed studies have proven this. Remembering drying methods of the different printing processes, one can distinguish three major categories. Most printing inks belong to the classical printing processes that dry by absorption, evaporation, or oxidation. Figure 4 shows the most important binder components for these inks. For inks dried by infrared (IR) and ultraviolet (UV) radiation, various acrylates and hydroxyl polyester have frequent use[17]. For possible tacky impurities from printing inks, one must especially consider toner for copiers and laser printers since they "melt" onto the substrate paper during the printing process by heat and pressure and become fixed to it. Toners using styrene butadiene rubber and polyvinyl acrylate contain the same chemical components as adhesives. Chapter 8 provides more details.

2.3 Coating binders

Polymer dispersions and starch derivatives have use as coating binders. The synthetic latexes and especially those in the form of styrene butadiene rubber, polyvinyl acetate (PVAc), and polyvinyl acrylate are possible causes of stickies as Fig. 4 shows. Paper mills that process coated broke refer to the possible problem of tacky deposits as white pitch. By their selection of coating binders, paper mills that must process their own broke can influence the development of white pitch that leads to deposits primarily on places with considerable shearing force such as foil bodies and in the drying section. The binder particles released at the time of recycling are similar to resin. Both substances can also be easily recognized as colloidal particles under a microscope.

2.4 Wood derivatives

The problems experienced with tacky substances in the processing of virgin fibers especially in the use of mechanical pulps are considered under the term pitch. The tacky components originating from the resins and extractives of wood consist of a mixture of different components having similar physical characteristics. As low molecular weight substances, they are practically insoluble in water under alkaline or acidic conditions. Large proportions are saponifiable. Resin particles can result in deposits on the paper machine or its clothing with resultant impairment of paper quality. The quantity and chemical composition of resins and extractives in virgin pulps depend on the type and age of wood used and the pulping method applied.

2.5 Papermaking additives

Many chemical additives find use in the papermaking process to improve the paper production process or obtain certain paper characteristics. Because of the possibility that these chemicals may react with each other or with other substances of the pulp slurry, they require consideration as potential causes for the formation of secondary stickies. This is especially true because an accumulation of a wide range of different substances cannot be excluded as a result of increasingly closed process water loops. In the processing of wood-free recovered paper for hygienic papers for example, the use of a cationic wet strength agent with an anionic dye resulted in problems with stickies[18].

The collectors used in the flotation deinking process (soaps or saponified fatty acids) are largely removed with the flotation foam. In addition, the water loop of the pulp processing plant and the paper machine are usually separated by dewatering to high pulp consistencies followed by bleaching, pulp storage, or both to minimize carryover into downstream loops. A residual transfer of calcium soaps into the paper machine loop still occurs. Many paper machine deposits contain chemical components of the flotation agents used although they may not be the cause of the corresponding deposit. A deposit may also occur on top of an already existing deposit. According to Mattingley, this carryover of flotation chemicals can consist of particles that are solid but deformable. The application of calcium stearates as flotation chemicals is one reason why these particles do not develop a tacky behavior under pressure[19].

2.6 Chemical substances in potential stickies

The different categories of chemical substances listed for adhesives, printing inks, and coating binders demonstrate that common denominators for their properties exist especially for pressure sensitive adhesives, hot melts, xerographic printing inks, and synthetic coating binders. Figure 7 shows that some chemical compounds have use not only for the production of one type of adhesive, toner, or coating binder but also in two or several different product areas.

As an example, consider the two monomers of styrene and butadiene processed in an emulsion polymerization to a milky-white product with the aid of other chemicals as Fig. 8 shows. This styrene butadiene latex has use in papermaking in coating binders, in various laminates, or for increasing strength in the production of paper-based imitation leather. Its effect relies on the cross-linking of the polymers during water removal above the glass transition temperature.

The latex can also be processed into a polymer group by destabilization and water removal. This is dried styrene butadiene "rubber crumb" (SBR). This SBR can be mixed with silica or carbon black, cross-linked with sulfur

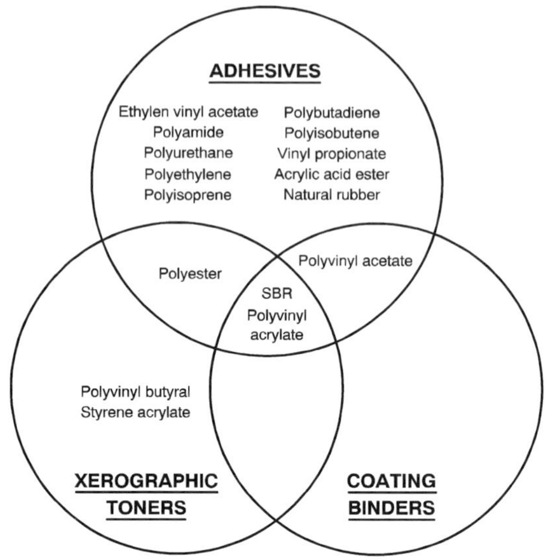

Figure 7. Major chemical components for paper coating and paper and board converting suspected to create problems with stickies in subsequent recovery processes.

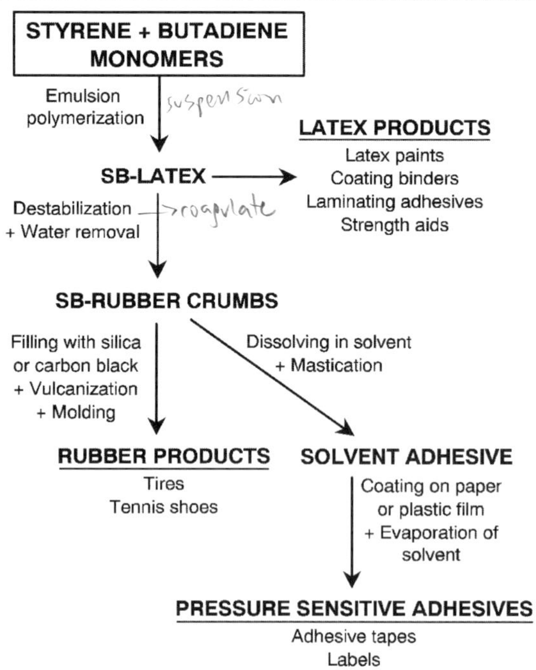

Figure 8. Styrene butadiene emulsion polymerization process[10].

(vulcanized), and molded into different rubber products such as automobile tires or athletic shoes. Alternatively, the dried rubber lumps can be treated with a solvent and homogenized for subsequent application to a specific substrate. After evaporation of the solvent, a pressure sensitive adhesive product is ready for use as address labels or for packaging purposes representing future sources of stickies in recovered paper processing[10].

The possibility to use the same chemical substances in completely different products makes interpretation of chemical analyses very difficult. Additives present in the recovered paper can also react with each other or with the chemicals used in the paper production process. These include wet strength agents, rosin size, aluminum sulfate, or retention agents. They cause secondary agglomerations of stickies[20].

3 Adsorption of stickies

The formation of deposits in a recovered paper processing system or on a paper machine is due to adsorption processes that occur exclusively on the surface of a solid phase. The term adsorbent describes the surface of the solid phase. The material to be adsorbed in its free form is the adsorptive as Fig. 9 shows. The term adsorbate covers the combined solid and deposited phase.

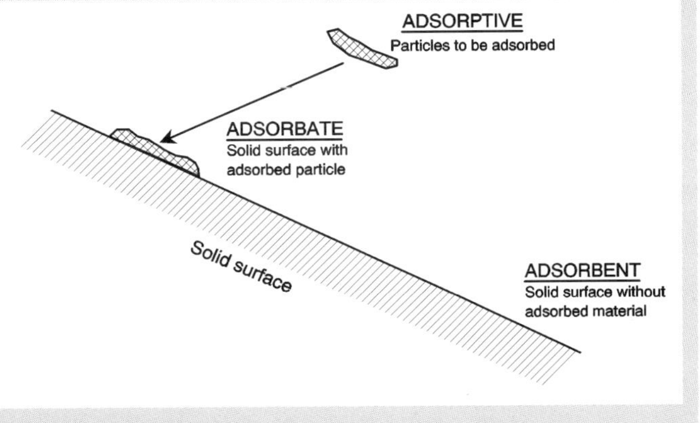

Figure 9. Definition of adsorption processes.

The bonding of the adsorptive to the adsorbent occurs by various types of bonding between atoms and molecules that can be physical or chemical bonding. The chemical bonds represent primary bonds that include the atomic bond (homopolar, nonpolar, and covalent bond), ionic bond (heteropolar, polar, and electrovalent bond), and the metallic bond. The physical bonds are secondary bonds whose bonding energies are considerably lower. These include van der Waals and hydrogen bonds. The forces responsible for the van der Waals bond are dipole forces (Keesom forces) with a range up to 5 Å (1 Å = 10^{-10} m), induction forces (Debye forces), and dispersion forces (London forces). The final two types have lower bonding energies than the dipole forces[21].

A basic distinction exists between physical adsorption (physisorption) and chemical adsorption (chemisorption). In addition, adsorption can occur due to electrostatic forces based on chemical and physical forces. This is often called chemisorption. The differences between physical and chemical adsorption are explained below.

Physisorption is characterized by physical bonding forces and especially van der Waals bonding forces. It is reversible, and the bonding energy is less than 60 kJ/mol. Under constant conditions, an adsorption balance develops between adsorption and desorption. With increasing temperature, the probability of desorption due to increasing molecular movements also increases.

In contrast, chemisorption represents a type of chemical bond. Due to the high bonding energy of 100–800 kJ/mol, the bonds developed by chemisorption are irreversible. The monomolecular layer of the adsorbate represents the maximum possible adsorbable quantity of the adsorptive. As the temperature increases, the adsorption speed increases due to thermal activation.

To permit deposition of finely dispersed tacky particles on solid surfaces, several conditions must be satisfied according to Kamutzki[22].

- The particles must be sufficiently close to the adsorbent to permit interaction between them.
- To enable particles to deposit, they must wet the adsorbent surface. Insufficient wetting can lead to release of the particles from the adsorbent.
- The tacky particles must be present in a viscous or plastic state to permit them to deposit via deformation.
- The finely dispersed or colloidal particles of the adsorptive must be destabilized, i.e., they may no longer have any active protective colloid, emulsifier cover layer, or both.

To satisfy the first demand, the adsorptive must approach nearer than the distance of the van der Waals interactive forces that are 2.6–5.0 Å when the adsorptive is being transported to the adsorbent. For the laminar flow of suspensions in pipe systems, this distance does not occur due to multiple layers of adsorbed water molecules. For direct contact of the adsorptive to the adsorbent, a turbulent flow is necessary whose kinetic energy allows penetration of the adsorbed water layer.

The surface energy of the fluid (γ_{LG}), the surface tension of the solid surface (γ_{SG}), and the boundary energy between the liquid and solid phase (γ_{SL}) determine the wetting of a solid surface by a fluid. The contact angle is an important thermodynamic parameter and an indicator of the spreading of a fluid on a solid surface. Figure 10 illustrates the two limiting cases: no wetting and total wetting. No wetting occurs when the surface tension of the solid surface (γ_{SG}) is much smaller than that of the fluid phase

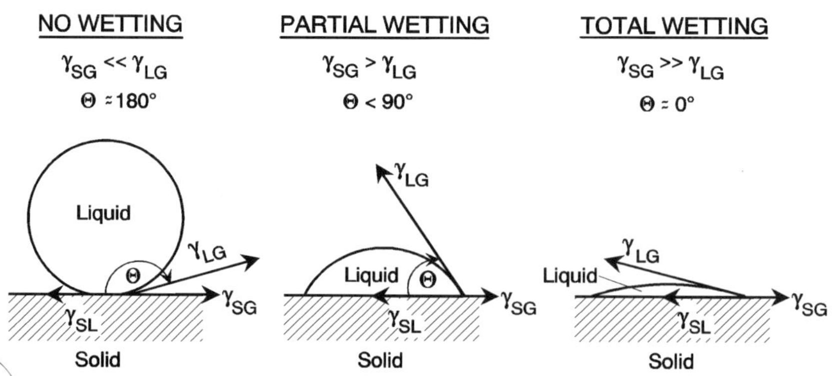

Figure 10. Surface forces involved in wetting.

(γ_{LG}). The contact angle θ then moves toward 180°. An example of such behavior is mercury on nearly all surfaces where liquid mercury drops freely roll around on the surface. At the other extreme, a spontaneous total wetting occurs if the contact angle θ moves toward zero when the surface tension of the solid surface (γ_{SG}) is much greater than that of the liquid phase (γ_{LG})[23].

Most fluid to solid surface systems reveal a wetting behavior that lies between both extremes (See middle representation in Fig. 10). For adhesives, the term critical surface tension describes this situation. Wetting occurs if the surface tension of the fluid is less than that of the solid surface[22].

Figure 11 shows that resin, pitch, and thermoplastic adhesives manufactured from petrochemical materials are nonpolar and have a low surface tension of 20–40 mJ/m^2. Because of these characteristics, the previously mentioned substances can wet nearly all hydrophobic solid surfaces whose surface tension is greater. An affinity with such surfaces therefore exists. An exception to this is polytetrafluoroethylene (Teflon) with a surface tension of only 18 mJ/m^2. Due to these reasons, deposits of stickies on polyester and polyamide surfaces such as usually used for paper machine clothing (wire and felt) can occur. In contrast, deposits on metallic surfaces such as copper or nickel despite their high surface tension cannot occur to the same degree because of their hydrophilic character. This is especially true because water has better qualifications for adhesion. It is present in much greater volumes, and its adhesion is strong due to polarity forces[22].

Practice shows that deposits can occur even on these hydrophilic surfaces. Weigl *et al.* explain this by the deposition of fatty acids or their esters on copper surfaces that then present hydrophobic parts of their molecules to the surrounding medium. Some work has also shown that changes in pH cause polarity changes so that copper surfaces in an alkaline medium have lower surface tension and polarity. This favors adhesion of tacky contaminants[24].

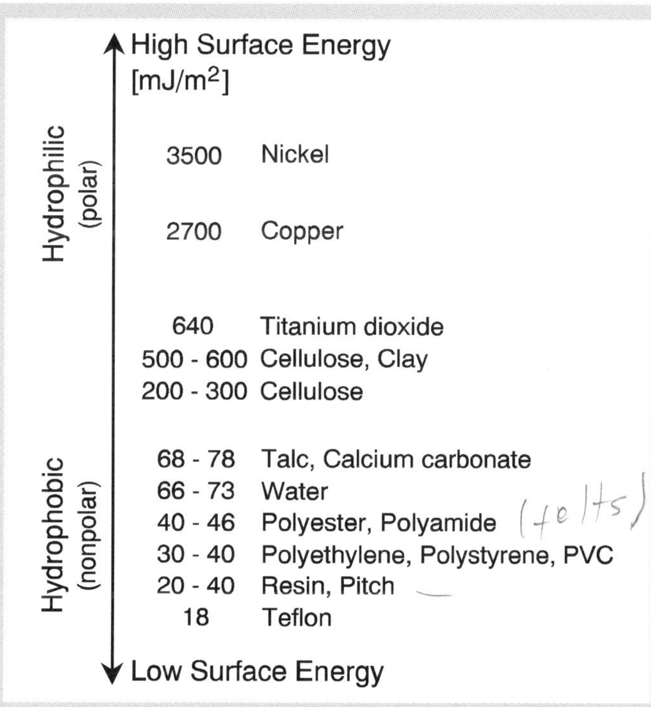

Figure 11. Surface energy of selected materials [12, 21, 22, 24].

These thermodynamic models for deposit-related problems are valid only for the surface conditions and tensions in each case. Surface active substances can dramatically alter surface characteristics. Remember that after the adsorption of a monomolecular layer of a substance on a solid surface the other surface characteristics are the result of the adsorbed substance. System parameters such as temperature, pH, or salt concentration are also important.

The different electrical charge of materials and therefore their electrostatic attraction may have secondary importance as the cause of adsorption processes on solid polar surfaces in aqueous systems. The explanation of agglomeration processes and the precipitation of negatively charged, potentially tacky contaminants must include the effects of the cationic additives on negatively charged fiber surfaces. In addition, electrostatic charges of wires, felts, or rolls in the paper machine can result in deposits of potentially tacky impurities that are not sufficiently fixed to the paper structure.

4 Characterization of stickies

To date, no standard test methods or standards exist at a national or international level for the characterization of stickies and their content in pulp slurries. This is undoubtedly a reason why individual companies in the paper and supply industries have developed very different test methods for stickies and might continue to do so in the future. Reasons for the absence of test methods are the varying requirements of the test methods including the different types of stickies for testing, the amount of time accepted for the analyses, and the type of conclusions desired. Test methods for stickies are urgently necessary for the following reasons:

- Judging the quality of recovered paper and recycled fiber pulp
- Checking the removal efficiency of stickies for pulp processing equipment
- Determining the effects of process parameters on the effectiveness of removal of stickies
- Verifying the effectiveness of chemicals used for control of stickies
- Judging the recyclability of paper and board products such as adhesive labels
- Establishing a quality control system that prevents the processing of recovered paper highly contaminated with stickies or permits the application of special countermeasures at the time of recycled fiber pulp processing and paper production.

The existing methods for quantifying stickies consider:

- The degree of contamination by stickies of recycled fiber pulp expressed as the count of stickies, the area of stickies, or the mass of stickies per mass unit of pulp or paper area observed
- The tack of the stickies that leads to attachment to different substrates
- The adhesive strength of stickies that has developed on different substrates[25].

Stickies that have formed deposits can be submitted for chemical analysis to determine their components. This analysis does not constitute a direct detection method for stickies nor does it supply any information concerning the content of stickies in recycled fiber pulp. Under certain conditions, it provides identification of specific related problems for stickies. The chemical analysis of deposits on paper machine components and paper machine clothing gives an insight into the wide range of chemical compounds that can be contained in the chemical "cocktail" of stickies.

4.1 Deposit analyses

Representative sampling is an important requirement of any analysis. The deposits to be examined consist of a large number of organic and inorganic components. To ensure that one is not merely determining the composition of a sample, the sample size must contain individual components of tacky impurities. This can provide insight into the causes of the deposits. For the chemical analyses, the following instrumental techniques are available:

- Energy dispersive spectroscopy (EDS) is a technique for analysis of inorganic components in deposits.
- Fourier transform infrared spectroscopy (FTIR) is the most common technique for analysis of stickies. It can analyze unknown substances even when only small amounts of deposits are available.
- Gas chromatography (GC) combined with mass spectrometry (MS) can identify unknown substances in raw materials or additives that may result in tacky deposits or both.
- Gel permeation chromatography (GPC) determines molecular weights of organic substances and their distribution.
- High performance liquid chromatography (HPLC) is used for the analysis of high molecular organic substances with boiling points above the optimum range for GC analyses.
- Thin layer chromatography (TLC) identifies oligomers and monomers.
- Inductively coupled plasma spectrometry (ICP) measures the amount of elements in inorganic and organic compounds and is used mainly for the analysis of heavy metals.
- Infrared spectroscopy (IR) is used for identification and quantitative analysis of organic compounds and functional groups of organic substances.
- Mass spectrometry (MS) is used with GC.
- Nuclear magnetic resonance spectroscopy (NMR) is used for the structural analysis of organic substances.
- Pyrolysis-gas chromatography (Py-GC) is gas chromatography with a preceding pyrolysis that results in a characteristic fingerprint for polymers.

- Scanning electron microscope with X-ray analysis (SEM-EDXA) detects inorganic elements with mass number greater than 23.
- Size exclusion chromatography (SEC) separates compounds into fractions having different mol masses.
- Thermogravimetric analysis (TGA) measures the behavior of inorganic and organic substances as a function of temperature and time. TGA is most suitable for the analysis of volatile inorganic and organic compounds.

Performing a chemical analysis must therefore consider the measuring possibilities, i.e. the available instruments and devices. Preparation of the sample and its separation into individual fractions for the selective determination of the chemical components presents major challenges. For certain deposits, the combinations of different analytical procedures described in the following have been successful[26–28].

In 1984, the Papiertechnische Stiftung (PTS) in Munich, Germany, established an analysis specification using IR, REM, and GC[26]. The scheme in Fig. 12 has had successful use for the chemical analysis of recovered paper suspected of causing deposits in paper machines and giving impurities in paper and board. The necessary extractions were carried out quantitatively with water and organic solvents including ethanol, acetic acid ethyl ester, and trichloroethylene in a given sequence in Soxhlet devices.

The advantages of a sequential extraction with organic solvents are that all components can be sufficiently dissolved, a preliminary separation is realized in accordance with the solubility and concentration, a small sample quantity is necessary, and the sample preparation is simple. The use of these three extraction agents represents a compromise between a high detection sensitivity and the amount of work necessary. Using a single extraction agent has the inherent risk that any statements concerning the organic components remain incomplete. Due to the potential health hazard of trichloroethylene, this solvent has no use today.

PAPRICAN introduced another analysis scheme for deposits from paper machines used to process recycled fiber pulp in 1995[27]. This uses FTIR analysis with GC, NMR, TGA, EDS, Py-GC, and ICP. Figure 13 shows this analysis using a sequential solvent extraction followed by analyses of the individual fractions by different techniques. Ethanol is the first solvent since most resin compounds of wood, hydrocarbon oils, and plasticizers have good solubility in hot ethanol. The ethanol extract is dissolved in hexane to separate resin compounds of the wood from other hydrocarbon compounds. The components that are insoluble in ethanol undergo an additional extraction with chloroform since a wide range of the synthetic polymers possibly present in deposits can dissolve in this solvent.

Detailed tests showed that the two-stage extraction with acetone and chloroform did not offer any advantage compared to direct chloroform extraction that is preferred. Using the analyses of deposits in press sections of different newsprint paper machines led to the conclusion that polymers with a wide molecular weight distribution are present in more than one fraction. The deposits also contain large proportions of the carryover of deinking chemicals and ink components that could also play an active part in the formation of deposits.

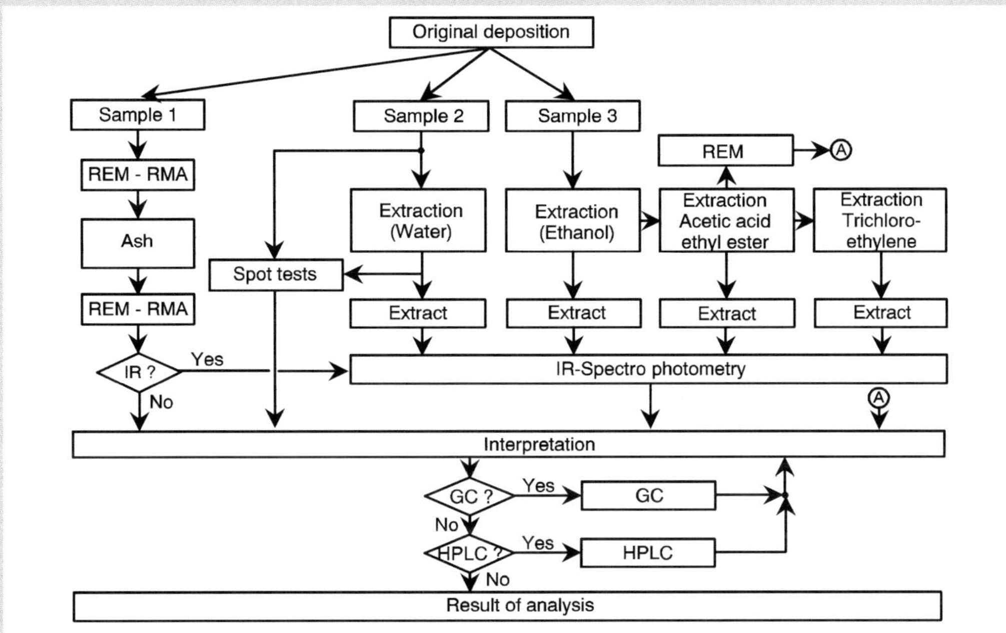

Figure 12. General scheme for deposit analyses according to PTS[26].

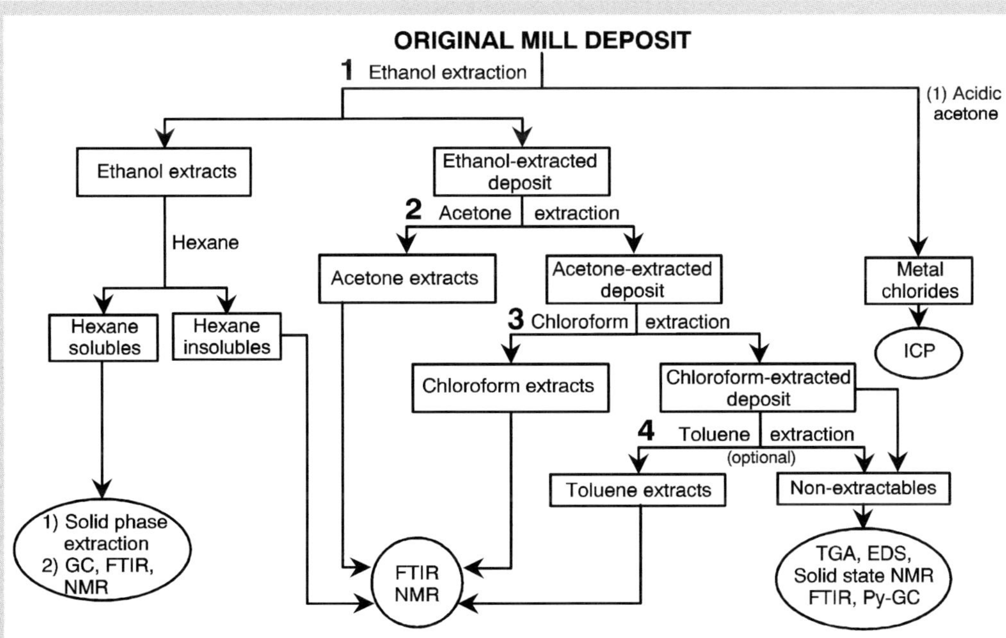

Figure 13. General scheme for deposit analyses according to PAPRICAN[27].

The Holmbom technique first uses microscopic examination[28]. Then the samples are soaked for 16 h in nonstabilized tetrahydrofurane (THF). To support the dissolution, an additional treatment in an ultrasonic bath occurs. The solution is evaporated under nitrogen atmosphere and then methylated with diazomethane in a mixture of diethyl ether and methanol. Finally, the solvent is exchanged for THF as Fig. 14 shows. Using SEC, the solution is separated into different polymers that are free of oligomers and can be analyzed by IR and Py-GC. Methylation is an efficient method for separating oligomer compounds and monomeric acids since methyl esters elute later than the corresponding free acids.

The described techniques allow determination of the mass flow of SBR, EVA, PVAc, and polyacrylates in a recycled fiber pulp processing system with double flotation. A level of 90%–95% of the EVA and PVAc present in recovered paper are removed by the process. Only about 5%–10% approaches the paper machine with the finished pulp. In contrast, the removal of SBR and polyacrylate at 60%–70% is clearly less efficient. Considerable quantities of these two components transfer to the paper machine with the processed pulp[28,29].

The disadvantage of chemical deposit analyses is that the detection of individual chemical compounds does not provide any information on the tack of the analyzed substances especially since deposits usually form over a certain period. This is because individual small particles agglomerate to larger particles. Deposits therefore consist primarily of inorganic fillers and pigments that are not naturally tacky and do not deposit. They only become tacky deposits by contact with other tacky substances. In addition, the time required for the described chemical analyses should not be underestimated.

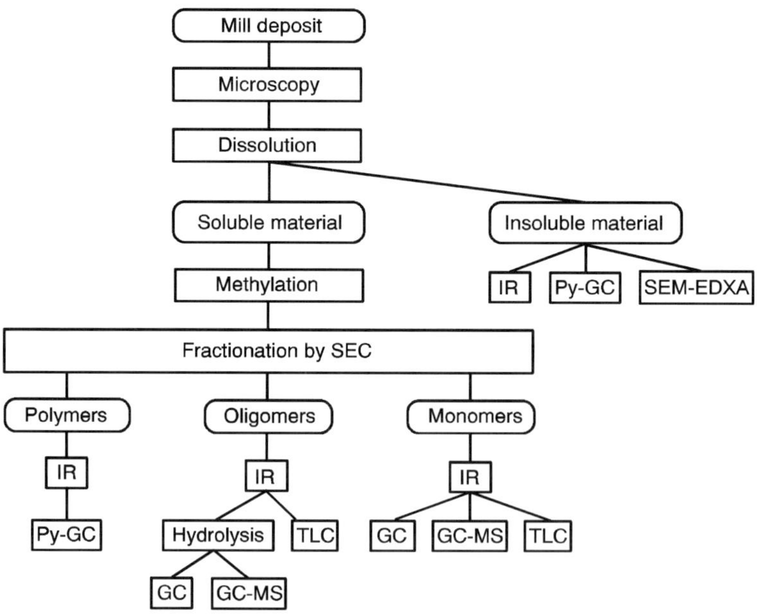

Figure 14. General scheme for deposit analyses according to Åbo Akademi[28].

The costs for the individual analysis, the capital costs of the various devices, and the costs of the highly qualified operators are also important factors.

4.2 Handsheet inspection methods

In contrast to chemical analyses, handsheet inspection methods for determining stickies in recycled fiber pulp can use much simpler techniques. The most important items of equipment necessary are a handsheet mold, a dryer, and possibly an image analysis system for evaluation of particles of stickies after their visualization.

In all cases, a handsheet is produced from a pulp sample. The number, area of detectable tacky impurities, or both are determined. Previously, contaminants were almost always visually counted by their number. A dirt estimation chart such as TAPPI T213 or T437 also had use to determine the size or area, respectively. This means the evaluation depends on subjective influences by the technician. In addition, this evaluation step is time consuming. This is especially true where numerous sheets of a pulp sample require analysis such as is necessary to obtain a representative result.

Recently, image analysis systems have found use to determine the number or the area of stickies. By using personal computer-supported image analysis systems, one can perform the determination of stickies very quickly. With sufficient mill experience on pulp samples, changes in the furnish from which a higher load of stickies results can be recognized quickly. The use of image analysis systems involves the risk that other, non-tacky particles such as shives or ink particles are detected with the stickies. This is because they have sufficient contrast to the background.

Most proposed test methods start with handsheet formation with 1.0–1.3 g o.d. recycled fiber pulp for basis weight 35–45 g/m^2. When using image analysis processes, dyeing the handsheets containing the impurities with black ink after drying is recommended. This results in the dyeing of the hydrophilic components of the handsheets that are primarily the fibers. The hydrophobic particles such as stickies retain their original color. White stickies offer a good contrast to the blackened background. With dark stickies the contrast is not sufficient for detection by image analysis.

Differences between the various methods exist especially in relation to the recommended drying processes using temperatures of up to 130°C for 20 min before dyeing. After dyeing, air-drying in drying rings or drying at 130°C are recommended. Specific drying conditions have enormous effects on the thermoplastic behavior and therefore on the deformation of the stickies. This influences the area of the stickies identified.

Camera systems and scanner systems have use for image analysis evaluation. The measured area of each handsheet should be as large as possible such as 314 cm^2 with the German method based on the Rapid-Köthen sheet mold. For different methods, 6–10 handsheets should be produced per pulp sample with evaluation of the top and bottom sides[30–32].

Modified processes measure the frequently present fluorescence of adhesives to determine visually the number of stickies. The use of such simplified methods that can facilitate the visual detection of stickies also means accepting the disadvantages that not all stickies fluoresce and not all fluorescing areas are necessarily stickies. This is because the pulp may contain optical brighteners from the preceding paper manufacture[25,33].

Compared with a previous screening procedure in the laboratory to concentrate stickies, handsheet forming has the advantage that only very small particles of stickies are removed during dewatering. This means a wide range of particle sizes of stickies and not merely macro stickies remain in the test sheet. Obtaining a statistically reliable number of detected stickies in recycled fiber pulp may be difficult especially where the stock concerned is efficiently processed such as deinked pulp for printing paper or market DIP. In these cases, previous screening as described in the following is necessary to concentrate the stickies to be analyzed. Otherwise, the number of handsheets to be produced for a representative result would be very large.

4.3 Screening methods

One reason why screening methods are the most common approach for determining the content of stickies of a pulp sample is that they simulate the removal possibilities of a recovered paper processing system. They therefore reveal whether the industrial process succeeded in removing at least the tacky contaminants from the pulp slurry that laboratory screening conditions can remove. Laboratory screening of recycled fiber pulp samples permits concentration of very small quantities of the stickies in an efficiently processed pulp. Simultaneously, the screening purifies the samples from fibers that could disturb the subsequent analysis or mask tacky particles. Note that when using a laboratory screening process the particles of stickies that are smaller than the perforation of the screening element can be removed as micro stickies with the fiber material as accept and cannot be detected in the reject material.

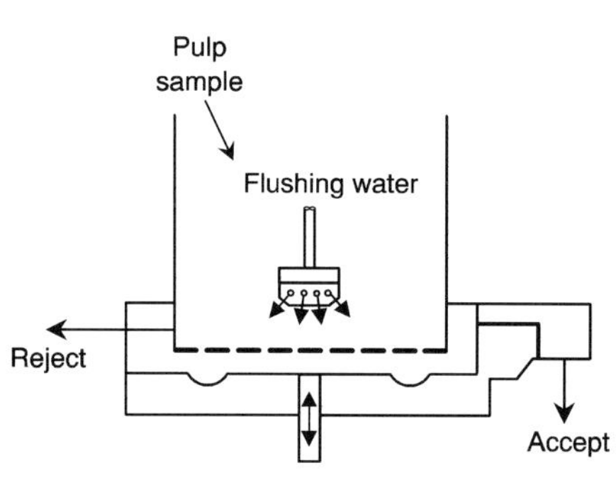

Conditions:
- Slot width: 0.1 mm
- Sample mass: 50 g o.d. DIP
- Time of sample input: 5 min
- Additional screening time: 5 min
- Stroke frequency: 480 double strokes/min
- Flushing water: 10 L/min

Figure 15. Screening with Haindl classifier.

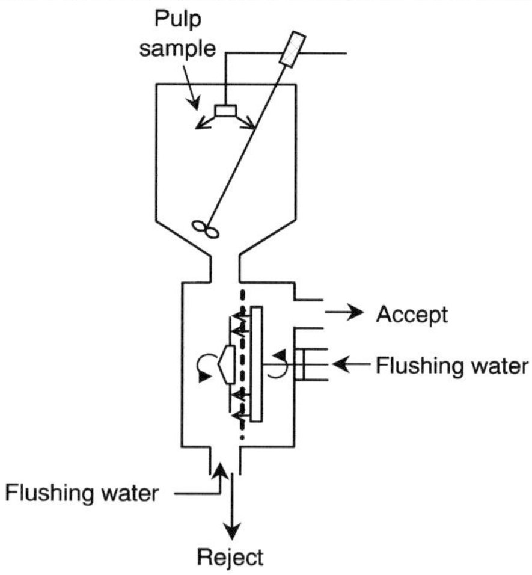

Conditions:
- Slot width: 0.1 mm
- Sample mass: 50 g o.d. DIP
- Automatic screening process

Figure 16. Screening with Pulmac MasterScreen.

Different screening devices have use for laboratory screening. The most common are the Haindl classifier of Fig. 15, the Pulmac MasterScreen of Fig. 16, the Somerville screen, and the Valley screen. All devices work with a flat slotted plate. Except the Pulmac MasterScreen in which a rotor produces the turbulence for the through-flow of the vertically arranged slotted plate, the other devices work with a horizontal slotted plate through which the pulp slurry is washed by flushing water. Although the Pulmac Master Screen was designed especially for screening stickies, the other devices are more common because of their use in determining the shives content of mechanical pulp.

For all four devices, screening is performed with slotted plates that differ in their external dimensions and the open area of the plate. Slotted plates of 100 μm and 75 μm slot width are presently available as a standard feature only for the Pulmac MasterScreen. For the other devices, the minimum slot widths are currently limited to 150 μm or 200 μm for the Valley screen. Test slotted plates that may differ from the original arrangement of the slots and the open area are now offered by other suppliers at least for the Haindl classifier and the Somerville screen.

Although the use of 150 μm slotted plates may be sufficient for evaluation of the content of stickies in brown recovered paper grades and old corrugated containers (OCC), the use of 100 μm slotted plates for laboratory screening is becoming increasingly necessary for characterizing deinked pulp including market DIP[34]. This is due to the greater cleanliness of the pulps that result from a more efficient industrial treatment, e.g., using centrifugal cleaners and pressure screens with 150 μm or even 100 μm slot width. In addition, the recovered paper grades used for these deinked pulps are contaminated by smaller quantities of stickies than brown recovered paper grades, e.g., supermarket waste or OCC.

When screening recycled fiber pulp for determination of stickies, always remember that the screening of larger quantities of pulp enhances the representative character of the sample analyzed. This is especially true for pulp samples with a low content of stickies. Devices having low mass throughputs require screening of several batches with increased testing time. In the Haindl classifier in a standard operation for example, 10 g o.d. virgin pulp screening takes 20 min. By modifying the device such as increasing the double stroke frequency of the membrane and enlarging the screening chamber, screening 50 g o.d. wood-containing or wood-free deinked pulp is possible in 10 min without any problem. The Pulmac MasterScreen or the Somerville screen permit screening of at least 50 g o.d. pulp in one run without modifying the regular operating conditions[5,34].

After feeding with the pulp sample, the Pulmac MasterScreen is the only device that can perform the entire screening process automatically. Filling the supply container with water, screening with flushing and washing cycles, and dewatering of the screening rejects can all be automatic. At the end of the screening process, the dewatered reject can therefore be removed from the device on a filter paper. The other screening devices require manual removal of the slotted plate after screening, rinsing of the reject, and dewatering onto a paper filter.

The different screening devices, quantities of pulp, openings of slotted plates, and applied operating conditions that various individual methods use produce significant differences in measured contents of stickies. Even greater differences result from the evaluation of the produced screening rejects that are usually first transferred from the slotted plate onto a filter paper. The processes used here are as follows:

- The quantitative determination of the screening reject using simple weighing: This process is unsuitable for an evaluation of the content of stickies of a pulp sample since residual fibers and other rejected, nontacky contaminants such as plastic film particles, ink particles, and shives can considerably influence the measured result[35].
- The visual control of the screening reject using a pair of tweezers, dissecting needle, and magnifying glass or microscope: In this case, the deformation potential and tack of each particle are examined in a highly time-consuming operation[36,37]. Due to the subjective estimation of stickies or particles that are not stickies, these methods offer poor reproducibility. The use of image analysis systems can improve the reproducibility although then one cannot directly distinguish between stickies and particles that are not stickies on the basis of tack or deformation potential.
- Image analysis systems using a sufficient contrast between the stickies to be detected and the image background (filter paper): The challenge here is to produce a sufficient contrast between stickies and image background for image analysis measurements. Care is necessary to ensure that it is only the stickies that provide the contrast to the background and not other, nontacky contaminants too.

One method uses white membrane filters of cellulose nitrate onto which the stickies are filtered. After covering with a black photographic board and drying under additional pressure, the stickies cause a tacky bond between the photographic board and the membrane filter. While the membrane filter can be removed at nontacky places, areas contaminated with stickies glue the filter to the photographic board. The stickies covered with the white membrane filter provide sufficient contrast to the black background. The work-intensive aspect of this method is the manual removal of the membrane filter to ensure removal of all nontacky components and retention of all tacky components completely on the black board[38].

A different method also uses the tack of the stickies by dusting a black silicon carbide onto the filter paper containing stickies and drying under pressure and heat. If the dimensions of the powder particles are correct, they will attach only to the tacky contaminants and have easy removal from the remainder of the filter paper. The stickies covered by the black powder offer sufficient contrast against the white filter background for measurement by image analysis. A disadvantage of this procedure is that other dark contaminants such as black ink particles and shives are also recorded as stickies when taking measurements[39]. This is especially disadvantageous when analyzing DIP.

The International Research Association for Deinking Technology (INGEDE) in Munich, commissioned development of an improved method. Dyeing of the filter specimen with a black ink including the filter and shives or fiber components lying on it is the first part of the test. With the method developed, the tack of the stickies gives adhesion of a white powder (special fused alumina). The stickies then appear white against a black background without black ink particles being falsely recorded as stickies[5,34]. This method is described in more detail in Book 17, *"Pulp and Paper Testing,"* of this series in chapter 4 on "Properties of recycled pulps".

Many other variations of measurements for stickies use the described screening processes. These include the use of chemicals for the agglomeration of stickies before the screening process[35] or the activation of the paper filters with "tackifiers"[40]. Publications are available concerning the addition of test screen rejects to virgin pulp from which handsheets are then produced. These are then judged visually by different methods[2,41]. At individual paper mills, dyed filter papers such as orange or plastic films onto which the stickies are transferred and then measured by image analysis are used for fast visualization in the production control of deinked pulp. This eliminates individual processing steps. The result cannot satisfy the general requirements concerning possible detection of all stickies. On the positive side, such individual "in-house methods" are especially fast and can suffice for certain applications.

Any application of the screening technique for determination of stickies involves tolerating the fact that the test result is influenced by the size and deformability of the stickies and by the physical characteristics during screening in the device used. This includes flow conditions, shearing forces, penetration velocity through the screening opening, slot width, open surface of the slotted plate, pulp consistency, and temperature. Small stickies can pass through the slots without detection, but they may agglomerate later and cause problems on the paper machine as secondary stickies.

The determination of the count of stickies and the measurement of the area of stickies depend on temperature. The applied drying temperatures and pressures vary greatly and influence the deformation behavior of thermoplastic, tacky contaminants. This influences the result of the measured area and the number of stickies since particles of stickies can only be detected above a certain size. For the naked eye, this is about 50 μm. If smaller thermoplastic stickies are not pressed or heated sufficiently, they may not be counted as stickies. To measure the effectiveness of additives used to reduce tack, one must appreciate that screening methods are suitable only if due consideration is given in the test to the actual tack of the screened stickies. For example, this includes adsorption of powder to the surface of the stickies.

Despite the various disadvantages, the screening technique is the most common method to assess the content of stickies for pulp samples. One reason is that many paper mills can adapt the methods easily and can use them for routine inspections. The method also has use to analyze the efficiency of commercial screening equipment regarding the removal of stickies and the effectiveness of chemical and mechanical dispersion on the development of particle size distribution of stickies.

For this reason, the screening technique also forms the basis for the first international draft standard developed by an ISO Task Force Group. In accordance with the draft standard, "Recycled pulps – estimation of stickies and plastics by a visual method" (ISO/DIS15360), an enriched screen reject sample of stickies is produced by laboratory screening and then inspected by magnifying glass or microscope and dissecting needle for the content of stickies. The objects identified as stickies are counted and their area estimated by visual comparison with an "estimation chart of contraries." Because this method is very time-consuming and susceptible to high coefficients of variation and poor reproducibility, it will probably not find routine use in paper mills. Consequently, image analysis techniques such as the INGEDE method described above are more promising.

4.4 Adsorption methods

Adsorption tests in which the stickies can attach to certain reference adsorbents use the tack of stickies in a pulp slurry. These tests cover tacky substances in a pulp slurry with no distinction for size as is the case with the screening methods. Nevertheless, it is frequently observed that especially large stickies do not reliably attach to the adsorbents, or they can be very simply sheared off by turbulence during the test or when cleaning the adsorbents.

With adsorption methods, the pulp suspension to be tested is brought into contact for a specified time and under defined conditions with a substrate on which the tacky contaminants should deposit. The following materials are used as adsorbents:

- Wire samples[42] or felt materials from paper machines[43]
- Polyethylene bottles[44]
- Low density polyethylene (LDPE) strips[35]
- Polypropylene foam[45]
- Microfoam[46]
- Other solid material surfaces made of copper, polyester resin, polyamide, polyethylene, or acrylonitrile butadiene styrene copolymer[47].

For the different methods, the pulp suspension remains in motion with an agitator, or the substrates are moved in the pulp slurry. After the programmed contact time with the pulp suspension, the adsorbent materials are removed and carefully freed from the attached fibers for a subsequent visual assessment of the deposits. Besides the microscopic and image analysis evaluations of the deposits, the increase in weight of the substrates caused by the stickies retained is usually determined gravimetrically. Because all test configurations work in batch operation with a finite quantity of pulp and the content of stickies of industrial pulp samples is very small (< 1% by weight), the content of stickies by weight is more accurately quantifiable when using lighter substrates[48].

With application of this method to industrial pulp samples, tacky printing ink binders and other contaminants can lead to additional deposits especially on paper machine wires and felts to influence the result. Counting the stickies visually is frequently necessary since sufficient contrast between the stickies and the adsorbents may not exist. The main advantage of the described procedure is that physico-chemical conditions at the paper machine such as pH, temperature, and water hardness can be simulated in the pulp suspension to be tested. For this reason, investigations of new materials are especially suitable under laboratory conditions. This includes studies of wires or felts that are believed to have a lower tendency to deposit stickies and tests to identify the effectiveness of specific additives intended to reduce the tack of stickies[49].

Moreland presented an objective test method to assess the effectiveness of chemicals intended to reduce tack[10]. In the test, an adhesive coated tape and a polyester strip are treated with a solution that should reduce tack. The two adhesive strips are then pressed together to determine subsequently the separating force between the two plastic strips. This process may well be categorized as objective for the evaluation of different additives with their effects on reducing tack. Where suspended fibers are present, such methods are not suitable for testing the efficiency of pulp processing because determining the separating force of particles of stickies without falsification of the result is not possible.

Despite their weaknesses, adsorption methods are the only test methods for stickies that actually consider the tack of the particles in a suspension. Accordingly, further improved adsorption methods are necessary. At the Department of Paper Science and Technology (IfP) in Darmstadt, Germany, a method was developed where adsorption occurs under additional pressure (forced adsorption contact) between two rolls to avoid the detachment of larger particles of stickies from a very thin adsorbent (basis weight is approximately 20 g/m^2). Compared with other test methods, higher temperatures of 80°C and higher consistencies can be used to ensure that increased proportions and amounts of stickies are adsorbed by the plastic substrate[50].

4.5 Extraction methods

Extraction methods use a solvent such as low molecular weight alcohols, ketones, ethylene, ethyl acetate, or chlorinated compounds such as methyl chloride or dichloromethane. The unscreened, recycled fiber pulp or rejects from laboratory screening are extracted by the solvent, the solvent evaporates, and the remaining residue is weighed.

These test methods are objective and have a low margin of analytical error. The results depend considerably on the solvent used and the applied extraction conditions. The long time necessary for analysis must be considered since different stickies have different dissolving behaviors in the solvents used. In addition, other substances besides stickies can also be extracted and falsify the result. The extraction results therefore do not provide any immediate indication of the content of stickies in a pulp sample. The results of extractions within a system permit the conclusion that higher measured values indicate a greater content of detrimental substances and therefore a greater potential for stickies.

Using this method for a dichloromethane (DCM) extraction, a Sankey diagram of the influence of detrimental substances in deinking systems is published[51]. The Sankey diagram for a recovered paper processing system with flotation of Fig. 17 clearly demonstrates that 174 kg DCM extract per hour are introduced by the recovered paper. This is then reduced to about one-third in the processed DIP. Important stages of the removal are screening and flotation. By dewatering the deinked pulp, a considerable DCM load enters the white water loop with the filtrate and comes again into repeated contact with the recycled fiber pulp.

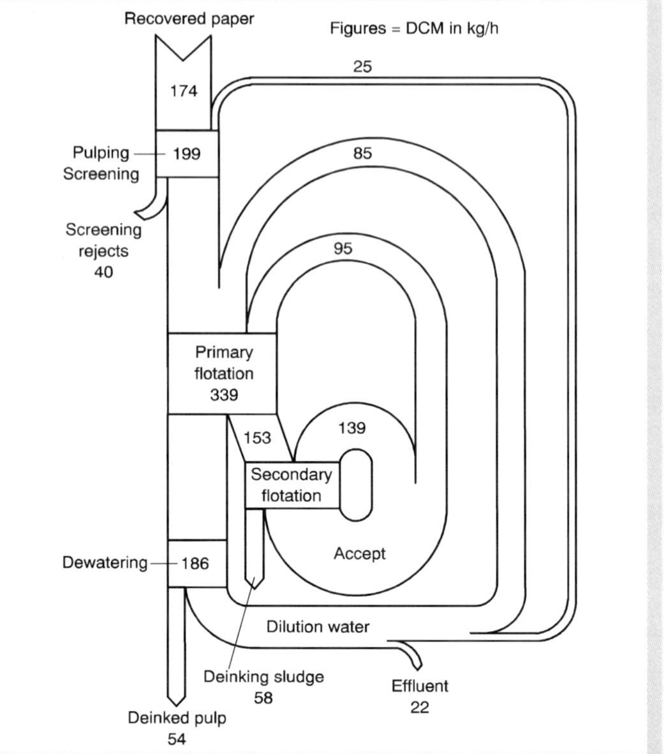

Figure 17. Flow of detrimental substances measured as DCM load in a deinking plant[51].

4.6 Precipitation methods

The methods described so far serve primarily to detect primary stickies in the form of macro stickies that can be screened or micro stickies whose hydrophobic character allows adsorption to solid surfaces. To date, no special methods are available to detect secondary stickies that result from the agglomeration of micro, molecular dissolved, or colloidal stickies. Although several of the aforementioned methods through modification of the test conditions can have use to assess the secondary formation of stickies, no relevant information is available.

Secondary stickies are created mainly by physico-chemical changes that occur following pulp processing. One such important change can be the addition of retention or drainage agents immediately before the headbox of the paper machine. This can partially or fully destabilize the pulp suspension. In this case, the secondary stickies result from the reaction of anionic colloidal substances that become detached from various paper components such as the coating layer, mechanical pulp, or pressure sensitive adhesive during pulping and precipitated by a cationic polyelectrolyte[52].

The process developed by the Centre Technique du Papier (CTP) in Grenoble, uses two paper machine wires made of polyester or polyamide that are introduced into an agitated pulp suspension under defined conditions of consistency, temperature, and time (Fig. 18). The test then uses controlled destabilization to change the charge of the pulp suspension. From the aqueous stage of the pulp suspension, the requirement for a cationic polyethyleneimine up to charge neutralization (= V_{th}) is determined. Then, different parts of the determined value (usually one-, two-, three-, and five-thirds related to V_{th}) are added to the pulp suspension samples not yet containing polyethyleneimine. Each volume of cationic polymer added has an equilibrium time of 10 min after which the wire is submerged in the suspension for 5 min under shearing action. The wire is removed and rinsed carefully with tap water and dried in an oven. The contamination of stickies is then measured by image analysis, i.e., the holes of the wire that are plugged by secondary stickies. Image analysis is performed by back-scattering illumination under a microscope with a black background (stickies look white). Results are expressed as percentage of wire hole area that is plugged as a function of destabilization. This method of testing shows that the deposition tendency can be increased considerably by charge changes if secondary stickies are present in a pulp suspension. In the case of an overdosage of the cationic polyethlyeneimine, the tendency toward deposition decreases again[53].

The precipitation method developed by the Papiertechnische Stiftung (PTS) examines finely dispersed and colloidal anionic substances in the pulp and water system. The dispersed coarse particles contained in the suspension are removed by a wire of the Schopper-Riegler drainage tester. A diluted precipitation agent (strongly cationic, long-chain polymer) is dosed into the filtrate until the resulting coacervate reaction has reached a balanced level. The filtrate with the coacervate is then passed through a suction filter. The reference for the precipitated substances is the evaporation residue or the filter residue without precipitation agent added. In principle, the precipitation may be achieved in several cycles.

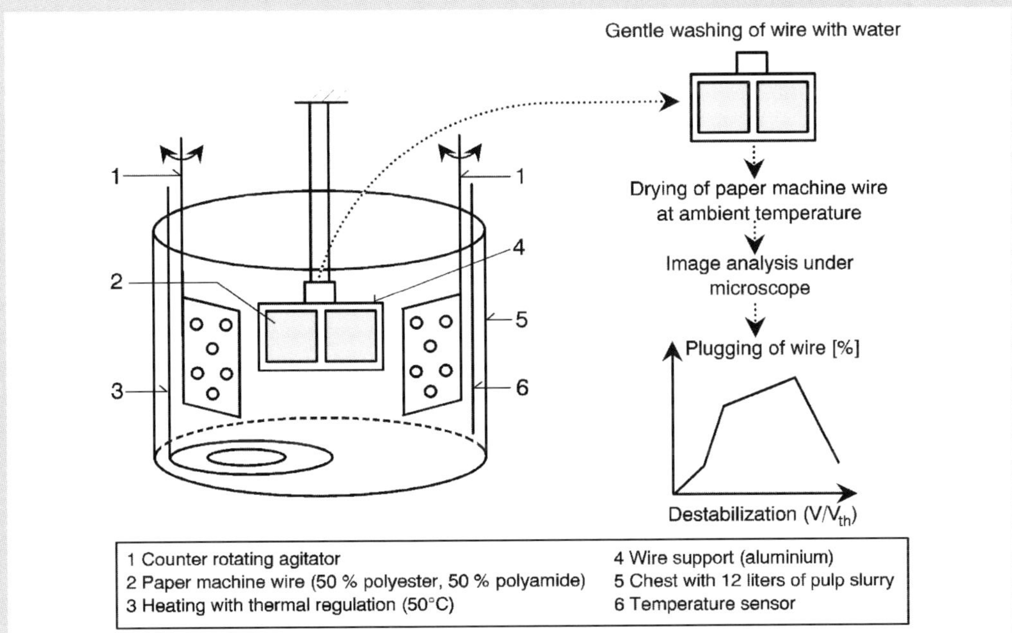

Figure 18. Measurement of secondary stickies by the paper machine wire test according to CTP[53].

Using the following formula, one can calculate with this method the minimum precipitation yield of finely dispersed and colloidal particles. The minimum precipitation yield expresses the amount of the coacervation tendency originating as coarsely dispersed stickies from finely dispersed and colloidal particles that are precipitated by adding a highly cationic, long-chain polymer. The expression considers the error variables such as fines, fillers, and even the precipitation agent.

$$P = \frac{m_{PP} - m_{PA} - m_{PF} - m_{FL}}{m_{ER} - m_{TF} - m_{FL}} 100[\%] \tag{1}$$

where P is the minimal precipitation yield
m_{PP} the mass of precipitated product
m_{PA} the mass of precipated additive
m_{PF} the mass of precipated fines
m_{FL} the mass of fillers calculated from ash content
m_{ER} the mass of evaporation residue
m_{TF} the total mass of fines calculated from evaporation residue.

Proving that different potential stickies have highly varying minimal precipitation yields is possible. For example, high coacervate masses develop from samples of redispersable adhesives since such adhesives largely redisperse at the time of pulping. Because the resulting coacervate is always tacky, it can be safely assumed that these deposition products cause problems in papermaking[54].

4.7 Other methods

A wide range of other, partly unpublished methods of determination of stickies are available. To separate contaminants having different densities for example, a laboratory cleaner with a 25 mm diameter was developed that operates with a low pulp consistency and high centrifugal forces. Instead of separating stickies by size using slotted plates, this provides a concentration in the reject by density. If only contaminants are supplied to the cleaner, realizing a separation into stickies having a low and high density is possible[55].

Another process describes the change that occurs when stickies obtained by laboratory screening are mixed into virgin pulp and dried under defined conditions. All those areas of stickies grow during drying and therefore indicate that they are thermoplastic stickies. In reaction to pressure and temperature, they begin to melt and flow. This method is recommended especially for testing market DIP[56].

Finally, a new process for identifying stickies uses an infrared technique. Using a thermographic camera in a wavelength range of 700–1 200 nm, the surface temperatures of handsheets are measured in relation to low- and high-temperature regions. Fibers and sticky materials such as polyester, polystyrene, or polypropylene differ significantly in thermal conductivity. By thermographic analysis, tacky surfaces in the fiber stock are supposed to be identified in model pulps[57].

4.8 Conclusions

Without appropriate and recognized test methods for stickies, progress in controlling problems related to stickies is difficult to achieve. Researchers setting out to develop devices or chemicals for control of stickies must direct their activities first to development of more reliable test methods by which the newly developed device or product can demonstrate its effectiveness. With such test methods, comparisons between different studies on problems related to stickies would be possible. This is impossible today due to the different conditions used for sample preparation and analysis. Nevertheless, justification exists for using different test methods for stickies since not all characteristics of the stickies such as number, size, mass, and tack can be measured comprehensively by one method.

5 Control of stickies

The increasing use of recovered paper inevitably means that growing numbers of paper producers encounter problems caused by stickies. Simultaneously, the intensified recycling of recovered paper contributes to an accumulation of tacky components. Stickies in paper grades containing recycled fibers re-enter the production process via recovered papers. No knowledge is available on how the tacky contaminants in the paper behave at the time of its reprocessing. One can assume that these substances contribute to problems related to stickies experienced during the following recycling loops with the result that an accumulation of stickies can occur with each loop.

Paper producers and the supply industry (machinery, chemistry, and clothing manufacturers) have had to be highly inventive to control stickies. Despite all endeavors, such efforts will undoubtedly continue to be necessary in the future because no quantitative solution is in sight to solve the problem of stickies. Lost production time due to stickies can significantly reduce the efficiency of paper machines and therefore the economy of paper production. Stickies can also be expected to cause a decrease in paper quality.

As Fig. 19 shows, the following control measures for stickies can occur in paper production:

- Alteration
  of the physical properties of the stickies by the manufacturers of the tacky substances so that any of the following methods work better

- Avoidance
  of stickies to ensure that potentially tacky substances do not enter the processing system with the recovered paper

- Removal
  of stickies by the separating steps within the recovered paper processing operation

- Dispersion
  of stickies to ensure that they remain in a very finely dispersed state and are therefore less harmful
- Passivation
  to reduce the tack of stickies by attachment of other nontacky materials to the surface of stickies
- Prevention
  to control the formation and deposition of stickies within the paper machine.

Included among these different strategies are several possible individual control measures for stickies that are described in greater detail in the following sections. They can be used separately or combined.

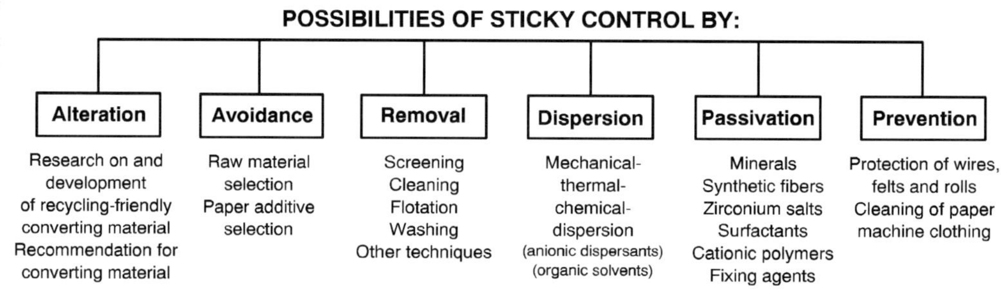

Figure 19. Control of stickies.

Only the first three measures, i.e., alteration, avoidance, and removal, offer the possibility of a long-term solution to the problems related to stickies. The other three measures relieve the problem only temporarily, i.e., directly at the time of paper production. Because of the usual practice of paper recycling, those three methods do not allow a long-term solution. Each industry segment must contribute to the listed control measures for stickies to improve paper recyclability.

Accordingly, every paper mill should closely examine its production process to avoid "homemade" problems with stickies. With a rigorous inspection of the recovered paper upon delivery, one can ensure that the supplied material has a certain quality.

5.1 Alteration methods

The most promising way to avoid problems of stickies in paper production is the development of more recycling-friendly paper and board products (including adhesives and binders) to satisfy the precondition of mastering the problem of stickies with complementary control measures. For example, adhesives with very strong cohesion offer the possibility for removal after disintegration almost completely. In contrast, weak cohesion of adhesives leads to a strong fragmentation during pulping that offers the opportunity that the adhesive particles pass subsequent screening and cleaning stages and cause problems with stickies on the paper machine. Research is therefore necessary to develop better screenable adhesives resulting in less problems with stickies for the paper manufacturers using recovered paper. In cooperation with the chemical industry, adhesive manufacturers, converters, paper industry, and research organization, one attempt was made to improve recycling-friendly adhesives by the United States Postal Service (USPS) as described in the following section.

USPS pressure sensitive adhesive program

Pressure sensitive adhesives are products that contribute considerably to problems with stickies due to their permanent tack. For this reason, the industry has demanded for some time that such adhesives during processing of recovered paper can be removed with screening systems with perforated and slotted screens. The development of adhesives that could be removed by flotation or separated by lightweight centrifugal cleaners would not suit the entire paper industry since not all processing plants have the necessary machines.

Especially praiseworthy is the USPS program. Because of the continually growing use of pressure sensitive adhesive stamps since 1990, USPS has launched a project with the United States Department of Agriculture (USDA) Forest Products Laboratory called "Research and Development of an Environmentally Benign Pressure Sensitive Adhesive for Use on Postage Stamps". In 1997, 82% of the USPS postage stamps were PSA issues. One estimate is that the USPS uses more than 160 km^2 of PSA postage stamps and labels annually. This is about 14% of the entire United States market.

As a result of a new USPS environmental strategic plan, the USPS has initiated industry-wide research to satisfy specifications for environmental friendly PSA postage stamps. In 1996, research began with a vertical team concept involving adhesive manufacturers, postage stamp printers, converters, face and liner paper suppliers, recovered paper processing mills, test laboratories, and machine manufacturers. The teams worked on a new type of solution to overcome the problem of the recyclability of PSA products that simultaneously had to satisfy the USPS specifications for use.

USPS wanted to develop the entire stamp laminate construction (face paper, adhesive, and release liner) to be recyclable. For this reason, the recycling test process included examination of two types of postage stamps: pre-consumer stamps with liner and post-consumer stamps without liner. The results of the joint consultations led to a three-stage plan with tests in the laboratory, pilot installations, and industrial scale.

Two philosophies of adhesive removal at the recovered paper processing stage emerged from the discussions held with the recycled paper processing industry. The first included the application of strict pulping conditions for maximum disintegration of the impurities. This used the theory of disintegrating unwanted adhesive particles so flotation removes them. A more comprehensive removal occurs at the water loop cleaning stage. The adhesives developed for this strategy preferably consisted of redispersible adhesives. Size of the dispersed adhesive particles after pulping received special attention. The redispersible adhesives also had negative effects including the possibility of reagglomeration during recovered paper processing and accumulation in closed water loop systems. In addition, these redispersible adhesives could not satisfy the specifications of the USPS high humidity aging performance test.

Consequently, the second philosophy of removing adhesives from recycled fiber slurries received greater attention. This used a minimum disintegration of adhesive particles by gentle pulping conditions at low temperatures[58]. In this case, the removal of the adhesive particles should occur at the screening stages of recycled fiber processing. The ideal adhesive for use with this method has sufficiently large particles after pulping that are relatively nontacky in the wet state. The members of the recovered paper processing industry favored adhesives that could be removed by screening from the fiber suspension as early in the recycling process as possible.

Laboratory tests established that pulping temperature has an influence on the particle size of standard adhesives[59]. Compared with a pulping temperature of 60°C for example, fewer particles with a larger area occur at a temperature of 40°C. The application of lower pulping temperatures therefore increases the removal efficiency of adhesive particles by screening. Under defined pulping conditions, smaller adhesive particles occurred in the pulp slurry with pre-consumer stamps with liner than with post-consumer stamps without liner.

In the pilot scale trials, control adhesives were used for benchmarking the linerless and the standard PSA stamp stock with and without liner. An experimental adhesive was also used. Figure 20 shows results of the contaminant level in the pulp determined by speck check analysis after pulping and pressurized screening[58,59]. As in the laboratory trials, the standard PSA stamp stock with the liner has a lower screening efficiency apparently due to some material in the liner such as clay or silicone causing some pacification and giving a higher adhesive speck area.

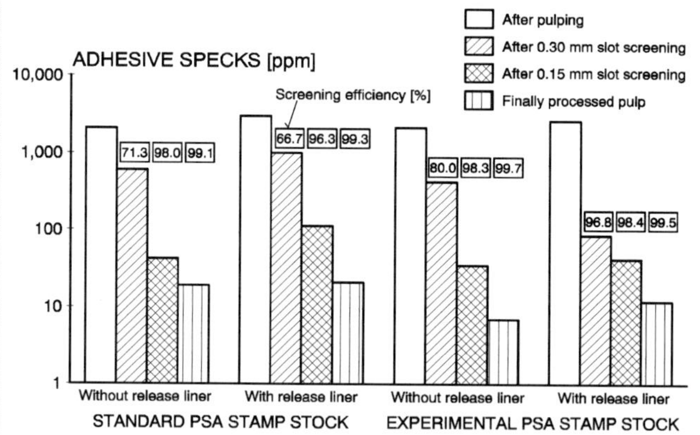

Figure 20. USPS pilot plant results of standard and experimental PSA stamp stock.

One experimental adhesive gave better screening. It should therefore cause less problems during industrial processing. Even with the release liner, contamination by the adhesive was the same as the control trial with the standard adhesive without liner. Early removal of most adhesive is important since the smaller quantity of adhesive remaining in the pulp has a higher probability of responding to the removal action of the subsequent unit operations. After additional forward cleaning, two-stage through-flow cleaning, flotation, side hill washing, and dewatering, the observed adhesive contamination was about 20 ppm for the standard PSA stamp stock and 7 and 12 ppm for the experimental PSA stamp stock with and without release liner, respectively[60,61].

The work performed in the USPS project is not complete. Additional progress in the development of recycling-friendly (screenable) adhesives resulting in a successful conclusion of the research project is still desirable. At the end of the project, PSA systems should be available that are technically suitable, economical, and screenable. To improve the recyclability of paper products, the developed screenable PSA should not only be used by USPS applications but also be used more widely in adhesive applications in the United States and Europe where recovered paper recycling has a much higher significance. Additionally, the knowledge obtained by the adhesive manufacturers should also be used for development of other recycling-friendly adhesive applications in the paper and board converting industry.

5.2 Avoidance methods

The avoidance of stickies is also a very effective way to cope with problems of stickies. Before this measure could be applied, knowing every source of the formation of stickies would be necessary. All parties participating in the paper chain would need to observe the proposals on avoidance. If all substances used at the preceding paper production and paper conversion that lead to problems related to stickies were substituted, significant reduction of the problems experienced with tacky contaminants during succeeding paper production based on recycled fibers would occur. The necessary preconditions for this will not occur in the foreseeable future. This is why it will not be sufficient to control the problem of stickies adequately by avoidance measures alone.

Because completely avoiding the entry of tacky substances into the paper cycle by recovered paper will never be possible and a total removal cannot be expected, a certain concentration of stickies will always remain in the processed recycled fiber pulp that is supplied to the paper machine. The paper quality and runnability in the paper machine are greatly influenced by the size of the residual stickies and their concentration. This will decide whether the concentration and size distribution of stickies offered by the recovered paper is acceptable. Because recovered paper processing plants offer only a certain efficiency for the removal of contaminants, increasing loads of stickies in the recovered paper result in a greater stickies contamination of the processed pulp. For this reason, care is necessary when using recovered paper so that it does not exceed a certain amount of tacky contaminants.

A prime difficulty is that seeing when recovered paper contains unacceptable quantities of stickies is almost impossible. Nevertheless, the papermaker knows the paper or board products that are critical for the stickies problem. For example, he is aware of the problems experienced in conventional recovered paper processing plants with excessive quantities of waxed corrugated containers, self-sealing envelopes, or pressure sensitive stickers as magazine inserts. "A conventional recovered paper processing plants" refers to a type of plant that can produce suitable pulp for a certain paper product from a regular recovered paper mix. The limitation to conventional processing plants must be made because special processing plants may exist in which certain problematical types of recovered paper can be processed. These include waxed corrugated containers or composite materials such as polyethylene coated liquid board or corresponding products such as used milk containers.

For purchased recovered paper, one should make the effort to assess its quality in relation to contamination with stickies and discuss any shortcomings with the recovered paper supplier. This allows demanding that waxed corrugated containers or self-sealing envelopes are removed from the supplied recovered paper to bring at least a reduction in these materials entering into the paper cycle by the recovered paper.

With the processing of over-issue magazines for example, a common stipulation when acquiring recovered paper is that pressure sensitive labels such as Easter rabbits and Santa Claus figures should be manually removed before supply to a paper mill. This removal of pressure sensitive stickers from the over-issues is only possible by using extensive labor. Another alternative would be storing these over-issues having a high content of pressure sensitive stickers separately and placing them in the recovered paper cycle in small volume to avoid putting an excessive load on the processing plant with these unwanted components.

Paper producers should also express their demands, wishes, or fears to other industry partners in the paper chain. Considering the political climate in Europe and elsewhere, all branches of industry must be receptive to the recycling concept and the extended closing of material loops. As a result, reports of problems experienced in the reprocessing of certain products (printed material or packaging) are taken seriously by the supplying industry, e.g., printing ink manufacturers and adhesive manufacturers, or the paper converters, e.g., label manufacturers, and give impetus for new developments of recycling-friendly materials.

Preconditions are that the problems have to be identified and adequately described. Included in this is a statement concerning the desired recycling process of the future. Naturally, this objective can be achieved more easily when more recovered paper processors cooperate and approach other industry partners with joint demands. An especially positive example in this connection is INGEDE. Since 1989, this organization with its 25 member companies has represented the interests of the deinking paper industry throughout Western Europe and can already claim success.

A paper producer must also ensure that he does not cause problems by producing secondary stickies in the production process. This requires careful selection of the retention and fixing agents used and avoidance of drastic pH changes or temperature shocks so that no destabilization of the pulp slurry occurs that can cause precipitation of colloidal or dissolved matter. Special attention is necessary to avoid charge transfer due to a change of retention agents such as cationic polymers since this may cause increased coverage by stickies of a paper machine wire[52]. The use of other substances such as wet strength agents, carboxymethyl cellulose (CMC), dyes, anti-foam agents, or anionic polymers can also influence the formation of stickies[18].

5.3 Removal methods

Chapter 5 of this book discusses methods and equipment to remove different types of contaminants in processing recovered paper. Aspects of stickies related to screening, cleaning, washing, and flotation are topics of the following sections.

5.3.1 Screening

General aspects

Screening of pulp slurries using screens with holes or slots is a basic process in stock preparation lines. Screening processes remove particles according to their size, shape, and deformation behavior. Particles that are larger than the size of openings (holes or slots) of the separating element in all three dimensions are retained provided the particles are not deformable and therefore enter the reject. If one dimension of a particle is smaller than the openings such as pulp fiber diameter compared with the hole diameter or plastic foil caliper compared to the slot width, flow conditions in the screening device and the flexibility of the particle will determine whether it enters the reject or accept of a screen.

The dimension of the perforations alone does not determine the separating efficiency of a screening operation and the resultant removal of stickies. Besides the combination of screening units in several stages, the design features of a pressure screen and the operating conditions also have important influence. The following criteria are examples of such design features:

- Flow inside the machine (centrifugal or centripetal)
- Type and geometry of the rotor (full rotors [bump, lobed, and stepped rotor] or open rotors [foil and blade rotor])
- Type and geometry of the screen (slotted or drilled screens, plain or contoured inlet surfaces of the screen, and shape and angle of the surface profile)
- Body shape (cylindrical or conical).

For example, a screen with 0.15 mm slot width can sometimes achieve a higher level of removal efficiency for stickies than a 0.10 mm slotted screen if a flatter, less angled surface profile is selected for the screen with the larger slot width[62–64].

For operating parameters, stock consistency, rotor speed, pulp throughput, and reject ratio have prime importance. The pulp characteristics such as freeness, fiber length distribution, ash content, contaminant content, and size distribution of contaminants also influence the screening result.

Model calculation of removal of stickies

The claimed levels of efficiency of removal of stickies for screening range from 40% to about 90%[7,63,65,66]. The percentages relate to the removal efficiency of individual pressure screens and a multi-stage screening system. Analyzing the numerical values requires extreme care. Basically, the removal efficiency depends largely on the determination method for the stickies. Important factors here include whether the content of stickies was determined gravimetrically, i.e., by mass, or optically, i.e., by number or area as discussed earlier in this chapter.

At this stage, a model calculation to illustrate these influences is used as Fig. 21 shows. A commercial pressure screen equipped with a screen of 150 µm slot width is fed with pulp with a total number of 1 500 stickies per unit mass of dry pulp. This figure comprises 1 000 particles with an idealized spherical diameter of 120 µm and 500 particles with a diameter of 200 µm. Where exact observance of the geometry of the screen perforations exists with nondeformable stickies, the screen removes all particles of stickes larger than 150 µm in the screening stage. For purposes of simplification, smaller particles are assumed to enter the accept flow. For our example here, this means that all 500 particles of 200 µm diameter are rejected, and all the particles with 120 µm diameter enter the accept. In calculating the number of stickies, this produces a removal efficiency of 33%. The calculation of the removal efficiency using the mass of stickies and a presumed constant specific density produces an efficiency of 70%.

Because the three-dimensional stickies in this example were considered spherical, an assessment via the removal by area of stickies can occur only if the area for the calculation has been agreed upon. If the area of stickies projected by a sphere is the basis, this removal efficiency would be 58%.

Real screening processes assume that many stickies have an almost two-dimensional expansion. For this reason, thin particles with a larger area than the slot width of the screen can enter the accept flow. Cubical stickies due to their deformability can also pass through perforations. In real screening processes as opposed to this theoretical observation, total removal (100% efficiency) of the particles that are only slightly larger than the perforations is not possible.

Quantitative analysis of the stickies in the feed and accept flow first uses a laboratory screening method with a 100 µm slotted plate as Fig. 22 shows. Presuming ideal laboratory screening conditions, all stickies in the feed and accept will be detected because all stickies in the model pulp are larger than the slot width of the screening plate of the laboratory classifier. An idealized screening implies that no particle agglomeration or fragmentation occurs during laboratory screening and that no fibers or other foreign matter enter the reject of the laboratory screening. The number and mass measured should then correctly reflect the stickies actually contained in the pulp flows. Assuming for purposes of the laboratory determination method for stickies that the numerically determined area corresponds to the projected stickies area, the calculated areas and removal efficiency in relation to number, area, and volume or mass in this example will agree with the conditions of a commercial screen.

If a laboratory screening procedure with 150 µm slotted plate is used for detection of stickies in the same commercial screening example of Fig. 21, all removal efficiencies will increase to 100% independent of whether the evaluation is realized with the number, area, or mass of stickies as Fig. 23 shows. These higher levels of removal efficiency result from the fact that only the 200 µm particles are intercepted with the laboratory screening in the inlet flow. In the commercial screen, these would ideally have been completely screened out. This is why with laboratory screening no more stickies can be detected in the accept with laboratory screening on 150 µm slots. Because some stickies with 120 µm diameter completely escape the laboratory screening in this example, the determined levels of removal efficiency do not reflect the actual conditions in the industrial screening.

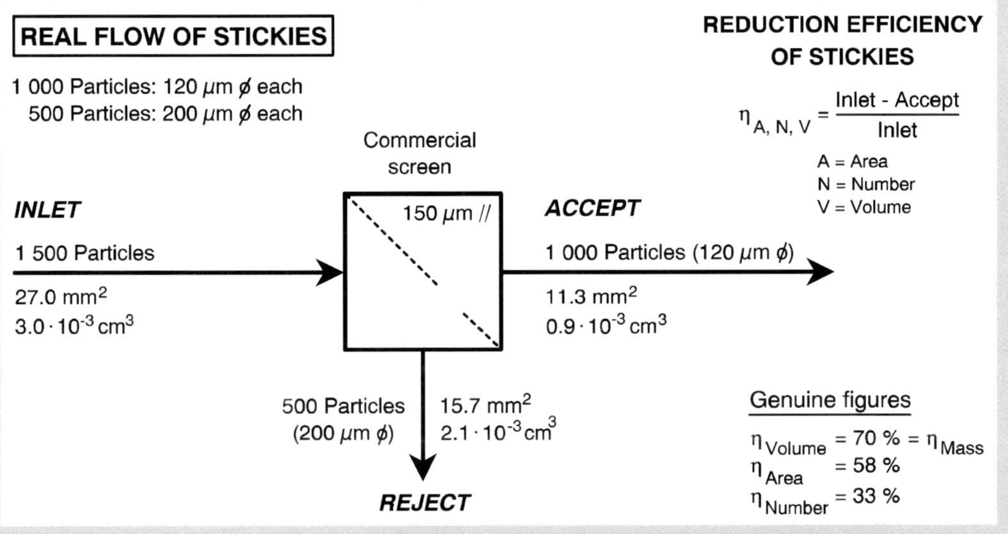

Figure 21. Theoretical flow of stickies in commercial screening equipment and reduction efficiency of stickies.

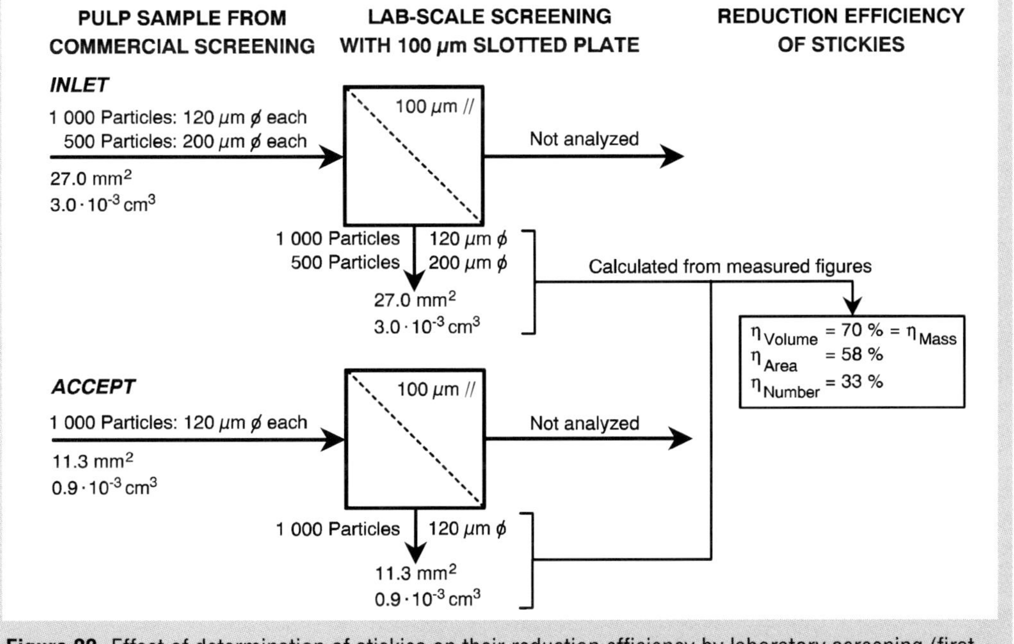

Figure 22. Effect of determination of stickies on their reduction efficiency by laboratory screening (first case).

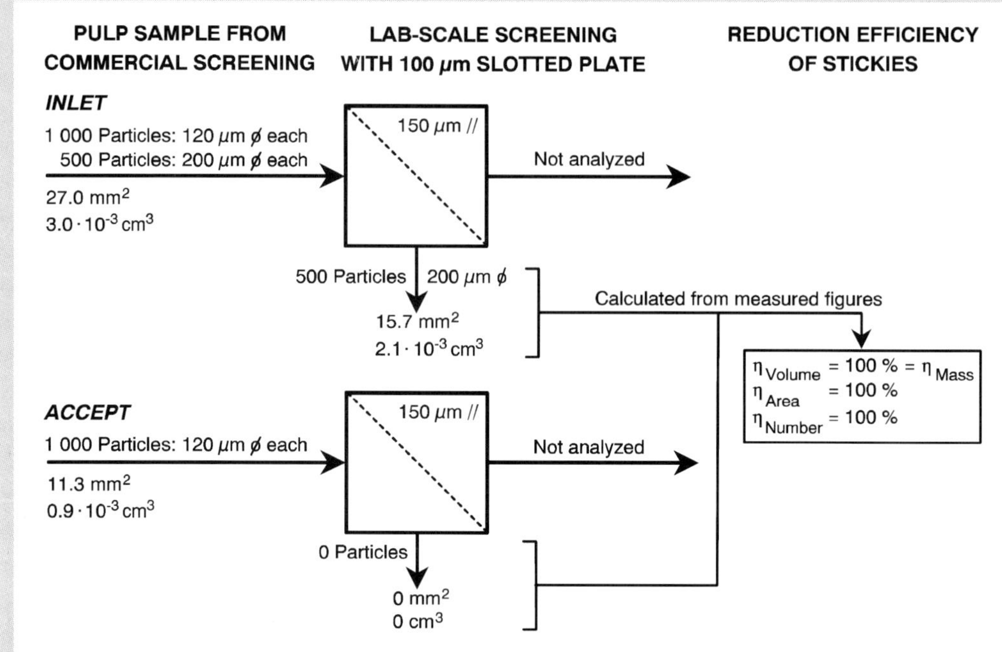

Figure 23. Effect of determination of stickies on their reduction efficiency by laboratory screening (second case).

One can assume that some particles contained in a pulp slurry always escape detection by a laboratory determination method for stickies because they are smaller than the laboratory slotted plate.

In another example, assume the previous 1 000 particles have a diameter of 80 μm instead of 120 μm. Using the 100 μm slotted plate for laboratory screening, the same levels of removal efficiency are calculated as in the first case study: 33% for number of stickies, 58% for area of stickies, or 70% for mass of stickies. Because the 80 μm particles cannot be rejected by laboratory or industrial screening, the actual removal efficiency of stickies for the considered industrial pressure screen decreases as follows:

- From 33% to 20% according to the number of stickies
- From 58% to 49% according to the area of stickies
- From 70% to 64% according to the volume or mass of stickies.

This example shows that the data relating to screening efficiency of stickies depend directly on the evaluation parameters such as number, area, or mass even when using identical methods of sample pre-treatment, i.e., laboratory screening. In addition, determinations using mass flows always produce more positive results in cases where disproportionately large amounts of larger size particles are included in the calculation. Recycled fiber pulp containing a high proportion of stickies with sizes ranging between the slot width of industrial screening and laboratory screening always produces a poor removal efficiency since the stickies belonging to this size category are not rejected by the industrial screen, whereas they are detected by the laboratory determination method.

Because no method of analysis for stickies currently exists that reflects the actual content in a pulp and because the measured values always depend on marginal conditions, any data relating to removal efficiency of stickies should be considered carefully depending on the slot width of the laboratory screen plates and the number, area, or mass of stickies identified.

5.3.2 Cleaning

The removal of contaminants by means of centrifugal forces is generally called cleaning and occurs in cleaners. In contrast to screening in which size, shape and deformation behavior of particles influence removal, cleaners permit a particle removal using differences of density and hydrodynamic characteristics. The advantage of using cleaners to clean pulp suspensions is the ability to remove very small and soft particles that due to their size and deformability would escape removal by screening since their density varies considerably from that of water and wetted fibers, respectively.

To minimize fiber loss, cleaner systems usually operate in three, four, or even five stages with different types of flow interconnections. An exception is the centrifugal cleaner with rotating housing. Several operate in parallel always in a single-stage process. Different sized devices depending on the volume or mass throughput rate are available[67–71].

According to Moreland, the following principles apply for operation of forward-, reverse-, and through-flow cleaners[72]:

- Efficiency of particle removal and throughput volume increase with increasing pressure difference
- Efficiency increases with higher temperature due to decreasing viscosity of the water and increased buoyancy of the light contaminant particles since these expand considerably more than water
- Efficiency decreases with increasing feed consistency due to the limited particle mobility
- Efficiency increases with lower cleaner diameter due to increasing centrifugal forces
- Efficiency increases and throughput volume decreases with smoother cleaner walls
- Efficiency increases with extended retention time with increasing cleaner length since the cone angle of the cleaner becomes smaller.

Although centrifugal cleaning is effective for particles that clearly deviate from the density of the water or wetted fibers, this process is ineffective for removal of particles of about 1.0 g/cm^3. Table 1 illustrates that the substances in adhesives have a density that is close to 1.0 g/cm^3.

The density of many components of tacky contaminants of about 1.0 g/cm^3 is the reason why the efficiency of removal of stickies by centrifugal cleaning is far from optimum. Figure 24 shows clearly that forward cleaners with reverse- or through-flow cleaners can be used very effectively for removal of particles that are denser or lighter than water. This example shows a cleanliness efficiency of nearly 100% for a density less than 0.95 g/cm^3 and efficiencies of more than 95% above a density of 1.05 g/cm^3. It also clearly shows that the removal efficiency drops dramatically in a density range of 0.98–1.03 g/cm^3 (73).

Barrett reaches a similar conclusion for the lower range of the density spectrum[74]. In model tests with specially marked hot melts of different densities from 0.89–1.02 g/cm^3, substances having a density below 0.95 g/cm^3 can be removed more efficiently independent of the chemical composition of the hot melts.

As stated in the section on screening, numerical values relating to the actual

Table 1. Density of contaminants in recycled fiber pulp[70].

Contaminants	Density [g/cm^3]
Metal	6.00–9.00
Sand	2.00–2.20
Clay	1.80–2.60
Printing ink	1.20–1.60
Asphalt	1.10–1.50
Polystyrene	1.04–1.10
Hot melt	0.95–1.10
Polyethylene	0.91–0.97
Latex, rubber	0.90–1.10
PSA	0.90–1.10
Wax	0.90–1.00
Styrofoam	0.30–0.50

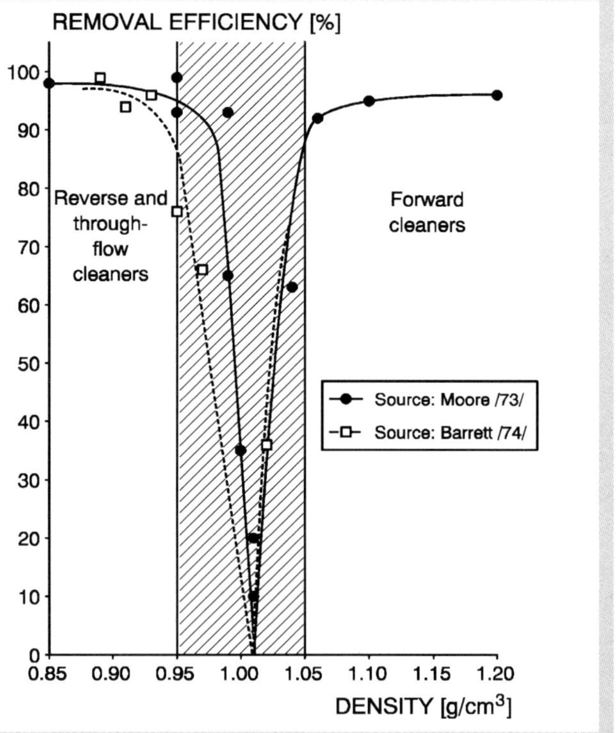

Figure 24. Removal efficiency by cleaning vs. density.

removal of stickies via cleaners in recovered paper processing are influenced by the applied method of determination of stickies. Independently of this, a determination test for stickies using different cleaners under identical operating conditions can provide valuable information. In comparative tests using optimum conditions with forward-, reverse-, and through-flow and rotating cleaners in pilot plant and industrial systems, mass-related reject efficiencies of stickies of about 40% were calculated for each individual cleaner except the through-flow cleaner. While the centrifugal cleaner with rotating housing could achieve this separation effect with a total reject rate of 2%, the reject of forward and reverse cleaners obtained with this reject efficiency for stickies was 20%–30%. This means that a multistage cleaning is absolutely necessary. Independent of the combination of the cleaners applied, the cleaning must result in an additional decrease of the total reject efficiency of stickies. This is not necessary with the centrifugal cleaner with rotating housing. Operating the through-flow cleaner with an overall reject rate of only 2% is possible but with only half the reject efficiency of stickies at approximately 20%[68].

Another comparison concludes that the rotation and through-flow cleaners installed at a paper mill for processing deinked pulp have approximately the same levels of efficiency for removal of stickies. The system under consideration had a two-stage through-flow cleaner with cascade arrangement and a one-stage cleaner with rotating housing used in parallel operation. Both cleaner systems were supplied from the same pulp chest and achieved a reject efficiency level for stickies of less than 20% by mass of stickies and less than 15% by volume of stickies. This efficiency for removal of stickies with light-weight cleaners is much too low. Based on its lower energy requirement, fiber loss, and water consumption, the cleaner with rotating housing had lower operating costs[75].

A new reverse cleaner with high g forces introduced in North America offers efficient removal of particles with density 0.95–0.99 g/cm^3[76,77]. In that range, conventional cleaner efficiency is very low. Due to its favorable operating conditions, the cleaner accepts streams can be better. The new reverse cleaning system also features a different through-flow cleaner to concentrate the light reject particles into a small volumetric stream. Chapter 5 of this book has more detail.

5.3.3 Flotation

The flotation process should be very suitable for removal of stickies for the following reasons:

- The usually hydrophobic nature of the stickies satisfies the precondition for ensuring the attachment of the air bubbles and guarantees the buoyancy of the particles
- A large proportion of the stickies to be removed is in a size range favorable for flotation because a pre-screening is usually undertaken before flotation so that large stickies should already have been separated.

Despite this, information on the removal of stickies by flotation is seldom available in the literature. A distinction is necessary between the flotation process within stock preparation and the dissolved air flotation (DAF) type treatment of the process water or effluent. The flotation in pulp processing accomplishes ink removal and should be considered a selective flotation of ink particles although fillers, fibers, and fiber fines are also partially floated.

Flotation in the process water and effluent purification is DAF that ideally addresses flotation of solid matter suspended in the water. Using suitable flocculation agents and precipitation reactions in the DAF system, one can possibly remove at least partly tacky solid particles and dissolved or colloidally dispersed stickies. Quantification of the efficiency of this removal method of stickies is not possible due to the incompleteness of the analytical methods for determining solid, dissolved, and colloidal stickies in the inlet and accept of a DAF system. The sludge of DAF systems is usually very tacky, i.e., it contains stickies. When such systems are switched off, problems related to stickies are experienced more often on the paper machine[78].

Besides these preconditions, Stratton notes that a certain air bubble to particle attachment force (F_a) of stickies has special importance for the removal of stickies. This attachment force can be calculated by the following formula[79,80]:

$$F_a \approx \pi \cdot D \cdot \gamma_{LV} \cdot (1 - \cos^2\theta) \qquad (2)$$

where D is the air bubble diameter
γ_{LV} the surface tension of the deinking liquor
θ the contact angle between the deinking liquor and the particles of stickies.

The contact angle depends on the surface tensions of the deinking liquor and the solid particles. Series of tests using four different hot melts having different surface tensions and different deinking chemical additions showed that the results of the mathematical model correlate with the hot melt removal in flotation steps. Figure 25 shows that the hot melt removal improves significantly when the attachment forces are greater than 1 dyne. A consistent removal of approximately 90% starts above about 1.5 dyne.

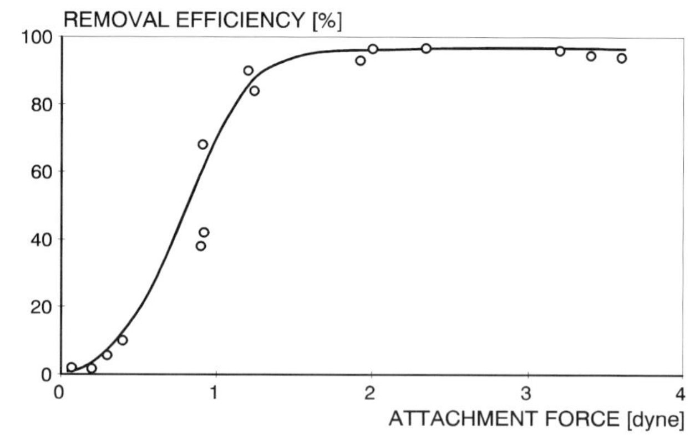

Figure 25. Flotation efficiency of hot melts vs. calculated attachment forces[79].

For a 90% removal of hot melts, Fig. 26 shows that the interfacial tension of the deinking liquor must increase from about 30 to 40 mN/m when the surface energy of the stickies increases from 30 and 60 mJ/m^2 during the tests. Note that the relationships are valid only with the flotation regime, e.g., size of the generated air bubbles or size range of stickies. The relationships highlight the fact that a high surface tension of the deinking liquor, a low surface tension of the stickies, or both improve their flotation. A disadvantage that emerged during the tests was that the surface tension of contaminants (stickies) is time-dependent. After contact with the deinking liquor, the surface tension of the stickies increases to a constant level within a period corresponding approximately to the pulping time after the addition of the deinking chemicals. This causes the stickies to become more highly polarized or hydrophilic and therefore hinder their removal behavior[80].

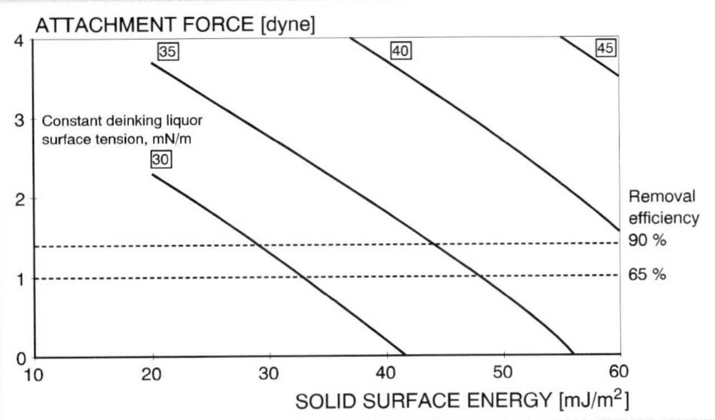

Figure 26. Attachment force between air bubble and stickies vs. surface energy of stickies for various levels of deinking liquor surface tension[79].

Figure 27 considers an industrial deinking plant for processing deinked pulp for newsprint. In the flotation stage including the existing reject (secondary) flotation, 32% of

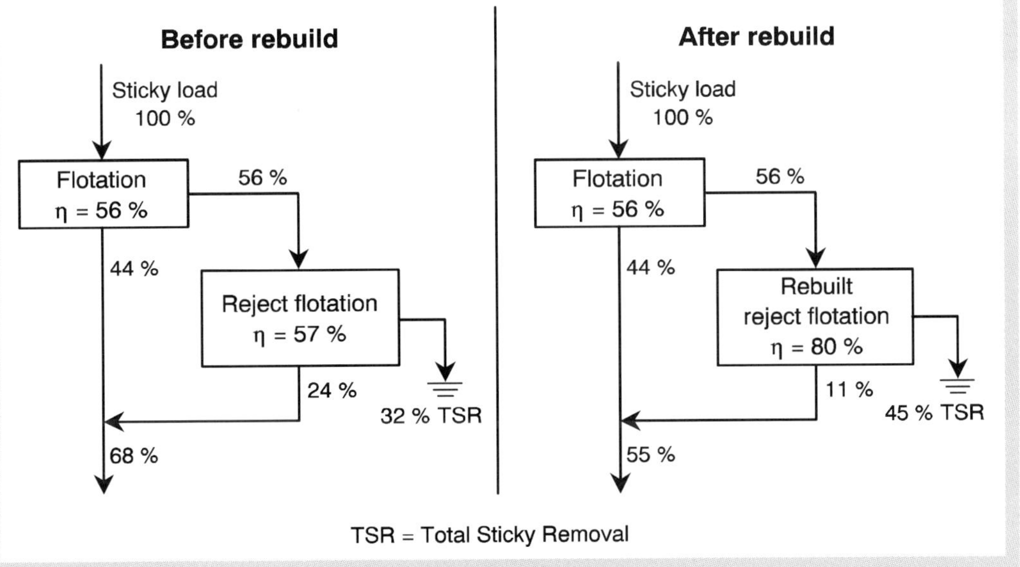

Figure 27. Removal of macro stickies by flotation before and after rebuild of the aeration elements in secondary flotation cells.

the macro stickies present were removed. Laboratory flotation tests with the feed of the reject flotation stage showed that a removal efficiency for stickies of 80%–90% instead of 57% was possible. For this reason, comparative tests were conducted in situ with a new industrial flotation cell in parallel that used a new type of aeration element to improve removal of stickies. After rebuilding the entire reject flotation stage, a removal for stickies of 80% was achieved to raise the total separation of stickies from 32% to 45%.

Simultaneously, increasing brightness by 5% ISO and reducing energy consumption by 33% was possible in the secondary stage. Using these results, flotation operated under appropriate conditions is an important process for elimination of macro stickies in the size range 150–500 μm. Although smaller stickies less than 150 μm were not analyzed, stickies in the size range 5–150 μm could be removed even more effectively. By improving the aeration elements in flotation cells, a removal rate of macro stickies with an efficiency similar to that of screening was possible[81].

For the production of graphic papers based on recovered paper processing with deinking when the screening and cleaning stages are carefully organized and operated and double flotation is integrated into the system, the residual content of stickies of processed pulp is still too high in many cases. This is surprising compared with the processing of recovered paper for packaging paper and board because considerably more intensive process stages are used. The graphic paper is in a clearly lower basis weight range resulting in reduced strength and a greater risk of breaks. This situation is exacerbated by the fact that paper machines for graphic papers are run at higher speeds compared with the packaging paper and board sector.

Despite the fact that flotation occurs in the production of graphic paper, removal of stickies is still not optimum although at least hydrophobic stickies should be floatable. The following factors are presumably the reason for insufficient removal of stickies[82–84]:

- Stickies of a size that cannot be optimally removed by flotation, e.g., too small
- Optimally sized stickies are insufficiently removed because they have a disk-shaped structure that often allows them to escape flotation in the case of toner particles
- By the adsorption of surfactants on the surface of stickies, the hydrophobic character and therefore the contact angle can be reduced to have a negative influence on the removal of stickies
- Surfactants used to stabilize the foam in flotation also influence the surface tension of the pulp to be deinked with possibly negative consequences.

Flotation units are intended primarily to ensure ink removal and are therefore optimized with this in mind. Any optimization of flotation for best removal of stickies in the production of DIP for graphic paper is possible only when undertaken without adversely affecting ink removal.

5.3.4 Washing

Besides flotation, washing is the second ink removal process used on an industrial scale. The washing process is especially efficient for removing very finely dispersed particles such as ink particles or fillers and dissolved substances from pulp suspensions. To ensure an efficient removal of solid matter by washing, the particles should not exceed about 30 μm in diameter. The extent of particle removal then depends largely on the washing device used and the operating conditions.

In principle, the same machines can be used for washing and thickening. The results depend to a large extent on how the filtrates from these machines are treated. Cleaning of the washing filtrates is therefore necessary to ensure the final removal of ink particles or fillers. The cleaned process water is then re-used in the course of the process. In contrast, thickening is usually applied only to increase the pulp consistency between two process steps in which different consistencies are necessary for optimum operating conditions, e.g., between cleaning and dispersion. Removal of the suspended solid matter and dissolved substances contained in the filtrate of the thickening machines does not occur. They recirculate with the process water.

Finely dispersed stickies should behave similarly to tiny ink particles or fillers in a washing process and be removable from the pulp suspension with the filtrate. No publications are available concerning the efficiency of removal of stickies by washing. One reason is that the available determination methods for stickies are insufficient especially for very small particles (micro stickies). Although determining the amount of stickies in the washing filtrate by precipitation is possible, this cannot be done on the inlet stock. This makes determination of removal efficiency of stickies impossible. To analyze the concentration of finely dispersed stickies in a pulp slurry, the solids would first have to be removed. This would cause a washing effect at the sample preparation stage. For this reason, only indirect conclusions based on the removal of ink particles by washing are possible.

In practice, washing results are usually assessed from the measurement of filler removal since the quantitative determination of the ash or filler content unlike the quantitative determination of the ink or even content of stickies can be carried out very simply. Table 2 shows the ranges of usual operating conditions for industrial washers.

Table 2. The operational range of industrial washers[85].

Washer	Inlet consistency [%]	Outlet consistency [%]	Filtrate consistency [%]	Ash removal (single-stage) [%]
Sidehill screen	0.6–1.0	3.0–4.0	0.15–0.35	55–60
Gravity decker	0.6–1.0	4.0–6.0	0.04–0.18	50–55
High speed belt	0.6–3.0	8.0–15.0	0.18–0.38	65–85
Screw extractor	2.0–4.0	8.0–12.0	0.25–0.40	40–45
Screw press	2.0–4.5	18.0–28.0	0.15–0.28	35–40

In older industrial plants with inefficient washers, several washing units are sometimes arranged in a series. The wash water is routed in counter-flow fashion to obtain efficient washing. Cleaned filtrate or even fresh water is used for diluting the stock before the final washing stage. The contaminated filtrate is routed countercurrently to the stock flow through earlier stages. At the first washer, the most heavily contaminated wash filtrate is directed to effluent or process water purification. Contained in this filtrate are heavy accumulations of finely dispersed contaminants in the form of ink particles, fillers, and stickies. To guarantee the definitive removal of these contaminants, ensuring the efficient operation of the process water purification system is important.

5.3.5 Other techniques

The processes described above are essentially the only ones that exist for removal of stickies although a wide range of different machine variations are offered on the market for screening, cleaning, flotation, and washing. To assure their most effective performance, the entire stock process system must be optimally sequenced and operated as Chapter 6 discussed. One important variable in this respect is pulper performance. The pulper must operate gently at high consistency and low temperature. This will ensure that contaminants and therefore also stickies are not made unnecessarily smaller since larger particles can be better eliminated by screening or cleaning.

The spinning disc developed by the Forest Products Laboratories in Madison, Wisconsin, offers an alternative to contaminant removal. With this separation technique, a pulp suspension is fed centrally to a fast-rotating disc as Fig. 28 shows. The disc sends the suspension into rotating movement that flows outside as a film. At the disc shoulder, most suspension follows the bend of the disc and is delivered as an accept flow at the lip edge. Two forces at the edge of the disc shoulder act on the particles in the liquid film that are responsible for their removal:

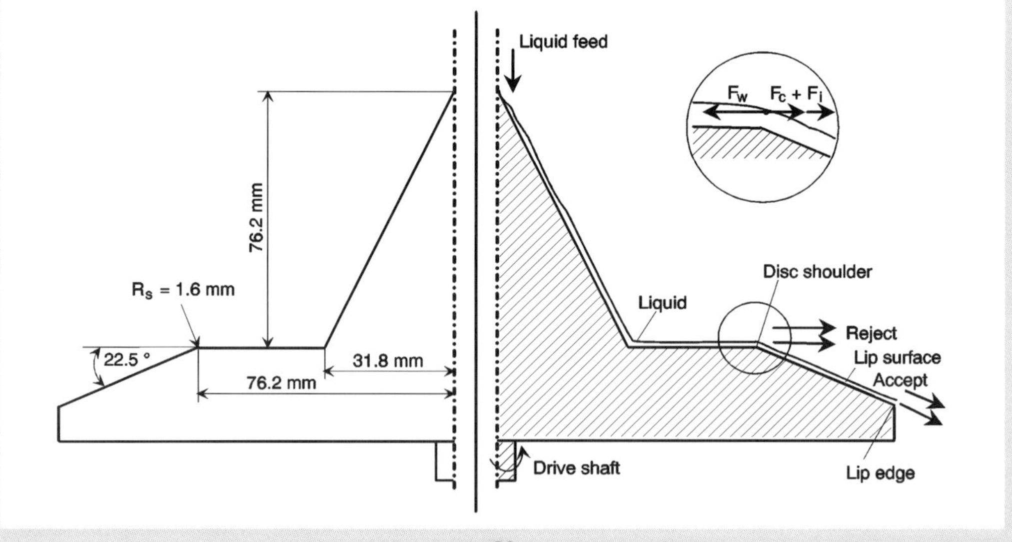

Figure 28. Forces and dimensions of disc separator[86].

- Inertial force (F_i) that results from the directional change
- Centrifugal force (F_c) that results from the rotational movement of the liquid film.

When the components of the surface tension of the liquid film (F_w) that act upon a particle are smaller than the sum of the two inversely acting removal forces $(F_i + F_c)$, the particles will separate from the liquid film and be removed as reject. The surface tension F_w is inversely proportional to the wetting angle. Because the inertial force, F_i, and the centrifugal force, F_c, are determined by the fiber length, the square of the fiber diameter, and the density, a separation of stickies by differences in the wetting angle should be possible[86].

Despite the positive screening results for hot melts and pressure sensitive adhesives for which removal efficiencies of 90% and 87%, respectively, were recorded, no industrial application worth mentioning has been realized. This is possibly due to scale-up problems. The pilot machines used to date with a disc diameter of 152 mm, a rotational speed of 4 000–9 000 min^{-1}, feed flows of 1–3 m^3/h, and consistencies of about 1% permitted a maximum throughput of only 25 kg/h[87].

5.4 Dispersion methods

Dispersion machines do not usually have use in recycled fiber pulp processing primarily for treating stickies. Instead, their use is directed toward the fragmentation of residual ink particles and the additional detachment of ink from the fiber surface in deinked pulp. They also provide a total improvement of the optical homogeneity of fiber stock. This applies for mixed (brown) recovered paper grades for the production of test liner or board and for the treatment of deinked pulps. The stickies also undergo a mechanical treatment although this does not result in their removal but causes a change in their size. Two different types of machines are available for dispersion (See Chapter 5):

- Kneaders have a slow-running rotor with protruding rhombohedrically shaped paddles on the rotor and protruding round or rectangular bars on the stator. Clearance between rotor and stator is large so the stock is treated by intense fiber-to-fiber rubbing action (inter-fiber friction).
- Dispergers consist of a high-speed rotating disk and a stationary disk both equipped with tooth-type designs that operate at much smaller clearances compared with kneaders. In the disperger, the stock undergoes acceleration and deceleration forces with internal pulp friction.

While the two dispersion machines operate at similar consistency, temperature, or specific power demand, the main difference between both machines is the circumferential speed. The kneader runs at a maximum circumferential speed of 15 m/s, but dispergers reach a circumferential speed of 50-100 m/s.

How does the mechanical treatment of the fiber stock carried out at a high consistency in the disperger change the stickies? In the high-speed disperger, the stickies are probably squashed and fragmented by impact against the rotor-stator elements. In the kneader, fiber-to-fiber friction forces crush the stickies or tear them apart. The possibility also exists that some stickies alter their shape in these machines from flat to round particles and even agglomerate. This facilitates their removal in a subsequent flotation or washing stage. The different behavior is probably due to the nature of the stickies. Strong ones roll into balls, and weak ones disintegrate[12,88,89].

One supplier has long included both types of machines in its production program and can therefore make a neutral evaluation using this experience. Comparative tests with both types of machines for office paper have shown that the disperger provides a larger reduction of the average size of the stickies. The range of size reduction of stickies is considerably greater in the case of the kneader. This is partly due to the opposing effects of fragmentation of stickies and agglomeration. Although the operating conditions in the described example are not necessarily identical, similar trends are also true for other types of recovered paper. For this reason, high-speed dispergers are recommended for the dispersion of stickies[88].

Besides the type of equipment used, temperature, consistency, and specific power demand especially influence the effects that occur during dispersion. Higher temperatures, consistencies, circumferential speeds and power demand give an increasing reduction of the size of stickies. One must always remember that even in cases where the size reduction of stickies would be very significant this does not correspond to a removal of stickies. The total area of stickies increases due to the fragmentation of the stickies and their increasing specific surface area. For the commonly used screening measurement methods for stickies, the size-fragmented stickies cannot be detected especially by a laboratory screening method using a 150 μm or even 100 μm slotted plate if the generated stickies are smaller. This results in a lower total area of stickies and a decreased average diameter of stickies. Hopefully, all stickies fed to the disperger are reduced in size such that the finely dispersed stickies enter the white water loop and are partly removed by process water treatment.

For these reasons, paper mills usually recommend much greater efforts to remove stickies from the systems by smaller slots in screens, highest efficiency reverse and forward cleaners, flotation, etc., rather than relying on dispersion[90].

5.5 Passivation methods

The final measure to control or take preventive measures against stickies is the use of additives. The additives used in a pulp suspension to reduce the negative effects of stickies can be organic or inorganic. They can also be used to a certain degree directly at the paper machine where they help prevent problems with deposits.

5.5.1 Minerals

A commonly applied method of using additives to control stickies is the addition of minerals to the recycled pulp slurry. The adsorption of the minerals to the surface of stickies reduces tack. Talc has most frequent use for this purpose in addition levels of 5–20 kg/t pulp. Talc is a soft, plate-like mineral that consists of a layered structure in which a magnesium sheet is embedded sandwich-like between two silica sheets. The individual layers are held together by van der Waals forces. This explains the simple delamination of

Table 3. Chemical and physical characteristics of the different minerals available for control of stickies[91].

Mineral	Particle shape	Color	Brightness [%]	B.E.T. surface area [m^2/g]	Charge	Particle size range [μm]	Remarks
Asbestos (Chrysotile)	Silica and magnesium sheets forming long, fibrous fibers	White, gray, green	n/a	4–50	Positive	Variable	1. Health hazard 2. Replaced by talc for pitch control
Bentonite	Aluminum octahedral sheet between two silica sheets	Gray to beige to white	70–90	130	Positive	20–80	1. Reduces sodium silicates 2. Works by entrapment 3. Improves efficiency of polyelectrolytes for retention 4. Replaced by talc for pitch control
Calcium carbonate	Ultrafine grind-irregular, precipitated-scalenoheron (rosette cluster)	White	96	5–11	Negative	0.1–8	1. Used in neutral and alkaline papermaking as a filler 2. Causes foaming in acid systems
Clay	Silica and aluminia thin hexagonal plates	White	83–93	13–22	Negative	0.2–30	1. Used in paper filling and coating for lightscattering characteristics, printability, brightness and TiO_2 extension 2. Comes in water-washed, air-floated and calcined forms
Diatomaceous earth	Siliceous skeleton of the diatomine (a unicellular microscopic algae)	Gray to white	75–90	26–28	n/a	0.2–73	1. Works by entrapment 2. Very abrasive 3. Replaced by talc for pitch control
Talc	Magnesium sheet between two silica sheets	White	80–90	6–17	Negative	0.2–20	1. Edge hydrophilic allowing dispersion in water, flat surface hydrophobic and organophilic

talc during the beating process. The special feature of talc is that its structure has a hy drophobic and an organophilic surface although the edges are hydrophilic. This permits its dispersion in water to provide the largest organophilic surface of all known materials. The hydrophobic characteristic is also the reason why talc has often replaced other minerals such as diatomite or clay for control of stickies.

Table 3 shows the physical characteristics of talc compared with several other minerals. Talc offers high specific surface area, low particle size, high brightness, and low abrasion. Due to its charge, talc adsorbs small stickies on the organophilic surface and removes them with the produced paper. Several talc particles can also adsorb on the surface of larger stickies and cause them to be less tacky. The ability of talc to be effective simultaneously on both small and large stickies provides an advantage over other additives. If stickies are no longer tacky due to the influence of talc, an agglomeration of stickies will also be prevented. This permits the removal of stickies with the paper and ensures that they no longer present any potential problems in further processing because the tack has been neutralized[12,91,92].

The correct use of talc can often reduce the deposition of stickies in the wet end and in the press section of the paper machine and contribute to considerably better runnability. A reported disadvantage to the use of talc is the poor shearing stability of the talc particles on the stickies, e.g., during pumping or refining processes. This can then be the cause of large agglomerations of stickies[93].

5.5.2 Synthetic fibers

Instead of minerals, synthetic fibers have also been proposed for control of stickies. Synthetic fibers with their large specific surface area and extensive affinity to the hydrophobic surface of stickies can almost entirely envelop stickies to deprive them of their tack. In practice, polypropylene fibers are especially suitable. They provide an optimum effectiveness for reducing tack of stickies and deposition of stickies when used at about 0.1–0.3% (1–3 kg/t pulp)[94].

The recommended synthetic fiber material has a fiber length of 0.8–1.0 mm and a brightness of 95%. Its specific surface area with 5–10 m^2/g is much larger than that of chemical pulp fibers. The synthetic fibers suspended in water (consistency of 1.0%–1.25%) should be added to the recycled fiber pulp in the beginning of its processing to allow contact with the stickies as early as possible while they are still predominantly hydrophobic. This causes small stickies to lose their tack before they can agglomerate to larger particles. For deinking systems, synthetic fibers should be used only after the ink removal process. Bleaching processes should be avoided because these process stages alter the surface characteristics of the synthetic fibers and can therefore reduce their effectiveness[95].

The following advantages can result with the continuous addition of synthetic fibers to a system in the processing of recycled fiber pulp[96]:

- Continuous cleaning of the system of small stickies due to removal via the produced paper

References

1. Scott, W.E., "A survey of the various contaminants present in recycled wastepaper white water systems", TAPPI 1989 Seminar: Contaminant Problems and Strategies in Wastepaper Recycling, TAPPI PRESS, Atlanta, p. 69.
2. Anon., Progress in Paper Recycling 2(3):80(1993).
3. Gruber, E., Großmann, K., and Pätzold, R., Wochenbl. Papierfabr. 126(13):645(1998).
4. Doshi, M.R., Progress in Paper Recycling 4(2):103(1995).
5. Ackermann, C., Putz, H.-J., and Göttsching, L., Progress in Paper Recycling 7(2):22(1998).
6. McKinney, R.W.J., "A review of stickie control methods including the role of surface phenomena control", TAPPI 1989 Seminar: Contaminant Problems and Strategies in Wastepaper Recycling, TAPPI PRESS, Atlanta, p. 101.
7. Ackermann, C., Putz, H.-J., and Göttsching, L., Wochenbl. Papierfabr. 124(11/12):508(1996).
8. Baumgarten, H.-L., Das Papier 38(10A):V121(1984).
9. Friberg, T., Progress in Paper Recycling 6(1):70(1996).
10. Moreland, R.D., "Stickies control by detackification", TAPPI 1986 Pulping Conference Proceedings, TAPPI PRESS, Atlanta, p. 193.
11. Onusseit, H., Allg. Papierrundschau 118(20):518(1994).
12. Galland, G., "Stickies: origins and solutions", 2nd Advanced Training Course on Deinking, CTP, Grenoble, 1995, Session 13, 36 pp.
13. Doshi, M.R., Progress in Paper Recycling 1(1):54(1991).
14. Müller, W.J., "Haftklebstoffe", Special Issue TI/ED 1194d, BASF, Ludwigshafen, Germany, 1993, 18 pp.
15. Kirchner, C., Labels & Labelling 14(2):48(1992).
16. Scholz, W., Progress in Paper Recycling 4(1):71(1994).
17. Bassemir, R.W., "The chemical nature of modern printing inks", TAPPI 1982 Annual Meeting Proceedings, TAPPI PRESS, Atlanta, p. 99.
18. de Jong, R.L., Wochenbl. Papierfabr. 118(17):763(1990).
19. Mattingley, J.T., Tappi J. 67(6):74(1984).

20. Miller, P.C., "Chemical treatment programs for stickies control", in Recycling Paper: From Fiber to Finished Product (M. J. Coleman, Ed.), TAPPI PRESS, Atlanta, 1990, Vol. 2, p. 471.

21. Klein, R. and Großmann, H., "Use of surface and rheological properties in stickies removal and control", PTS Report, Papiertechnische Stiftung, Munich, 1993, 61 pp.

22. Kamutzki, W., Wochenbl. Papierfabr. 111(20):753(1983).

23. Ling, T.F., Pulp Paper Can. 94(10):T273(1993).

24. Weigl, J. and Wilken, R., "Formation mechanisms of stickies in waste paper pulp used in papermaking", in Pulp, Paper and Board (I. F. Hendry and W. J. H. Hanssens, Eds.), Elsevier Applied Science, London, 1987, p. 60.

25. McKinney, R.W.J., "Test methods for assessing stickie contamination – a review", TAPPI 1987 Pulping Conference Proceedings, TAPPI PRESS, Atlanta, p. 725.

26. Wilken, R. and Strauss, J., Allg. Papierrundschau 108(11/12):292(1984).

27. Guo, X.-Y. and Douek, M., "A new scheme to analyze deposits/stickies from mills using recycled fiber", 1995 CPPA 3rd Research Forum on Recycling, CPPA, Montreal, p. 91.

28. Sjöström, J. and Holmbom, B., Paperi Puu 70(2):151(1988).

29. Asp, F., Engstrand, P., Holmbom, B., et al., "Evaluation of contaminant flow in a deinking process and estimations of the value of the raw material", 1994 PIRA International Paper Recycling Conference, PIRA, Leatherhead, UK, 25 pp.

30. Jean, M. and Nguyen, X.T., "Quantification and removal of stickies from OCC waste papers", 1993 CPPA 2nd Research Forum on Recycling, CPPA, Montreal, p. 61.

31. Klungness, J.H., Fernandez, L.E., and Plantinga, P. L., "Image analysis for measuring adhesive contaminants in pulp", in Recycling Paper: From Fiber to Finished Product (M. J. Coleman, Ed.), TAPPI PRESS, Atlanta, 1990, Vol. 1, p. 455.

32. Hacker, M.P., Tappi J. 75(7):63(1992).

33. Doshi, M.R., Dyer, J.M., and Krueger, W.C., Pulp Paper 58(3):113(1984).

34. Ackermann, C., Putz, H.-J., and Göttsching, L., Das Papier 51(6):271(1997).

35. Elsby, L.E., "Experiences from tissue and board production using stickies additives", TAPPI 1986 Pulping Conference Proceedings, TAPPI PRESS, Atlanta, p. 187.

36. Bittner, M.J., "Stickies collection and quantification within a production environment", TAPPI 1994 Pulping Conference Proceedings, TAPPI PRESS, Atlanta, p. 445.

37. Gordon, A., Paper Technology & Industry 29(8):399(1988).

38. Matzke, W. and Selder, H., "Additional benefits from pressureless high consistency disperging of secondary fiber stock", TAPPI 1986 Pulping Conference Proceedings, TAPPI PRESS, Atlanta, p. 597.

39. Biza, P. and Lippert, G.V., Wochenbl. Papierfabr. 120(11/12):439(1992).

40. Burrows, P.R.M. and Small, J.D., "Removal of sticky contaminants from waste-based papermaking furnishes" 1982 European Seminar on Technological Progress in Recycling of Paper and Board, European Communities, Brussels, Report No. 14, 20 pp.

41. Forester, W., "Methods for the detection of stickies in pulp", TAPPI 1993 Deinking Short Course Notes, TAPPI PRESS, Atlanta, Session 4-2, 14 pp.

42. Cathie, K., Haydock, R., and Dias, I., Pulp Paper Can. 93(12):T392(1992).

43. Fukui, T. and Okagawa, A., Tappi J. 69(9):134(1986).

44. Dykstra, G.M., Hoekstra, P.M., and Suzuki, T., "A new method for measuring depositable pitch and stickies and evaluating control agents", TAPPI 1988 Papermakers Conference Proceedings, TAPPI PRESS, Atlanta, p. 327.

45. Doshi, M.R., "Additives to combat sticky contaminants in secondary fibers", TAPPI 1989 Seminar: Contaminant Problems and Strategies in Wastepaper Recycling, TAPPI PRESS, Atlanta, p. 81.

46. Ling, T.F., Hall, J.D., and Walker, M.M., Pulp Paper Can. 94(12):T420(1993).

47. Klein, R. and Großmann, H., Wochenbl. Papierfabr. 123(19):871(1995).

48. Krauthauf, T., Ackermann, C., and Putz, H.-J., Wochenbl. Papierfabr. 126(3):81(1998).

49. Doshi, M.R., Progress in Paper Recycling 2(1):45(1992).

50. Krauthauf, T., Putz, H.-J., and Göttsching, L., "Forced adsorption contact – A new way of stickie evaluation in recycled pulps", TAPPI 1999 Recycling Symposium Proceedings, TAPPI PRESS, Atlanta, p. 661.

51. Kolb, W. and Heusser, U., Wochenbl. Papierfabr. 114(1):1(1986).

52. Carré, B., Brun, J., and Galland, G., "The incidence of the destabilization of the pulp suspension on the deposition of secondary stickies", 1995 CPPA 3rd Research Forum on Recycling, CPPA, Montreal, p. 187.

53. Carré, B., Fabry, B., and Brun, J., Progress in Paper Recycling 5(1):68(1995).

54. Klein, R. and Großmann, H., Paper Technology 38(6):45(1997).

55. Doshi, M.R., Progress in Paper Recycling 4(4):77(1995).

56. Darlington, W. B. and Blair, W.R., Progress in Paper Recycling 4(3):68(1995).

57. Lobbes, T.J. and Forester, W.K., Progress in Paper Recycling 5(2):78(1996).

58. Abubakr, S. and Bormett, D., "Pilot testing and recycling evaluation of newly developed environmentally benign pressure sensitive adhesives", TAPPI 1998 Recycling Symposium Proceedings, TAPPI PRESS, Atlanta, p. 339.

59. Abubakr, S. and Peng, J., "Development of environmentally benign pressure sensitive adhesives", 8th PTS-CTP-Deinking-Symposium Proceedings, Munich, 1998, No. 12E. 25 pp.

60. Peng, J.Y., "Environmentally benign pressure sensitive adhesive R&D program", TAPPI 1998 Recycling Symposium Proceedings, TAPPI PRESS, Atlanta, p. 327.

61. Donermeyer, D., "Laboratory recycling of benign pressure sensitive adhesive stamp material", TAPPI 1998 Recycling Symposium Proceedings, TAPPI PRESS, Atlanta, p. 417.

62. Brettschneider, W., Wochenbl. Papierfabr. 121(13):531(1993).

63. Rienecker, R., Wochenbl. Papierfabr. 125(17):787(1997).

64. Schabel, S. and Respondek, P., Wochenbl. Papierfabr. 125(16):736(1997).

65. Doshi, M.R., Progress in Paper Recycling 3(4):69(1994).

66. Vitori, C.M., "Stock velocity and stickies removal efficiency in slotted pressure screens", 1991 CPPA 1st Research Forum on Recycling, CPPA, Montreal, p. 133.

67. Fallows, J.D., "Forward cleaners", TAPPI 1993 Deinking Short Course Notes, TAPPI PRESS, Atlanta, Session 2-5, 6 pp.

68. Berger, J., Wochenbl. Papierfabr. 113(21):810(1985).

69. Serres, A. and Fejoz, R., Wochenbl. Papierfabr. 117(7):259(1989).

70. Ferguson, L.D., "Through-flow cleaners", TAPPI 1993 Deinking Short Course Notes, TAPPI PRESS, Atlanta, Session 3-1, 9 pp.

71. McBride, D., "High density cleaners", TAPPI 1993 Deinking Short Course Notes, TAPPI PRESS, Atlanta, Session 2-2, 9 pp.

72. Moreland, L.B., Progress in Paper Recycling 1(1):78(1991).

73. Pearson, N. and Moore, G.K., "Tackling stickies problems at source", 1995 EUCEPA Recycling Symposium, EUCEPA, Paris, p. 67.

74. Barrett, A.E., Progress in Paper Recycling 3(4):37(1994).

75. Rivals, P., "Stickies removal by hydrocloning – Gyroclean and uniflow comparison at a newsprint mill scale", 1995 PIRA International Wastepaper Conference, PIRA, Leatherhead, UK, Paper No. 14, 7 pp.

76. Maze, E.L., Progress in Paper Recycling 6(3):40(1997).

77. Shaw, M., Pulp & Paper 71(12):51(1997).

78. Eisenschmid, K. and Steinbeis, B., Wochenbl. Papierfabr. 118(16):709(1990).

79. Stratton, R.A., Progress in Paper Recycling 1(4):31(1992).

80. Stratton, R.A., J. Pulp Paper Sci. 18(6):J221(1992).

81. Geistbeck, M. and Wiese, H., Wochenbl. Papierfabr. 125(16):733(1997).

82. Hodgson, K.T., "Surface science and process factors affecting removal of sticky contaminants", in Paper Recycling Challenge (M.R. Doshi and J.M. Dyer, Eds.), Vol. 1, Doshi & Associates Inc., Appleton, 1997, p. 71.

83. Schmidt, D.C. and Berg, J.C., Progress in Paper Recycling 5(2):67(1996).

84. Ling, T.F., Pulp Paper Can. 95(12):T501(1994).

85. Horacek, R.G., "Deinking washers", TAPPI 1993 Deinking Short Course Notes, TAPPI PRESS, Atlanta, Session 3-2, 10 pp.

86. Klungness, J.H., Tappi J. 70(7):125(1987).

87. Klungness, J.H., Tappi J. 70(3):131(1987).

88. Mannes, W., Wochenbl. Papierfabr. 125(19):938(1997).

89. McBride, D.H., "Dispersion", TAPPI 1993 Deinking Short Course Notes, TAPPI PRESS, Atlanta, Session 3-4, 21 pp.

90. Pimley, I.I., Paper Asia 14(15):22(1998).

91. Williams, G.R., "Physical chemistry of the absorption of talc, clay and other additives on the surface of sticky contaminants", TAPPI 1987 Pulping Conference Proceedings, TAPPI PRESS, Atlanta, p. 563.

92. Doshi, M.R., "Additives to combat sticky contaminants in secondary fibers", TAPPI 1989 Seminar: Contaminant Problems and Strategies in Wastepaper Recycling, TAPPI PRESS, Atlanta, p. 81.

93. Fogarty, T.J., "Cost effective, common sense approach to stickies control", TAPPI 1992 Pulping Conference Proceedings, TAPPI PRESS, Atlanta, p. 429.

94. McKinney, R.W.J. and Currie, P.C.G., "Sticky pacification – new additive shows considerable promise", TAPPI 1986 Papermakers Conference Proceedings, TAPPI PRESS, Atlanta, p. 161.

95. Wade, D.E., Pulp Paper Can. 90(1):154(1989).

96. Wade, D.E., "Sticky pacification with synthetic pulps", in Recycling Paper: From Fiber to Finished Product (M. J. Coleman, Ed.), Vol. 2, TAPPI PRESS, Atlanta, 1990, p. 536.

97. Mann, S., "Sticky moments", 1993 PIRA Conference: Developments in Wastepaper Technology, Leatherhead, UK, Paper No. 10, 13 pp.

98. Goldberg, J.Q., "Use of zirconium chemical in sticky contaminants control", TAPPI 1987 Pulping Conference Proceedings, TAPPI PRESS, Atlanta, p. 585.

99. Hoekstra, P.M. and May, O.W., "Developments in the control of stickies", TAPPI 1987 Pulping Conference Proceedings, TAPPI PRESS, Atlanta, p. 573.

100. Gould, I., "Chemical control of stickies", 1993 PIRA Conference: Developments in Wastepaper Technology, PIRA, Leatherhead, UK, Paper No. 11, 11 pp.

101. Gould, I., "Contaminant control – stickies", 1995 PIRA International Wastepaper Technology Conference, PIRA, Leatherhead, UK, Paper No. 15, 14 pp.

102. Dykstra, P.M. and May, O.W., "Controlling stickies with water-soluble polymers", TAPPI 1989 Seminar: Contaminant Problems and Strategies in Wastepaper Recycling, TAPPI PRESS, Atlanta, p. 97.

103. Hooimeijer, A. and Blaak, R., "Chemicals to control secondary stickies – making money or a waste of money?" 1995 PIRA International Wastepaper Technology Conference, PIRA, Leatherhead, UK, Paper No. 16, 12 pp.

104. Ormerod, D.L. and Hipolit, K.J., "Aluminum control prevents stickies problems", in Recycling Paper: From Fiber to Finished Product (M. J. Coleman, Ed.), Vol. 2, TAPPI PRESS, Atlanta, 1990, p. 489.

105. Kenney, R.M. and Engstrom, G.G., "New technology for stickies/pitch control with increased production/profitability", TAPPI 1988 Pulping Conference Proceedings, TAPPI PRESS, Atlanta, p. 339.

106. Duffy, R.J. and Aston, D.A., Pulp Paper Can. 90(7):89(1989).

CHAPTER 12

Final fate of waste from recovered paper processing and non-recycled paper products

CHAPTER 12

CHAPTER 12

Udo Hamm

Final fate of waste from recovered paper processing and non-recycled paper products

1 Categories of paper mill waste

Besides graphic papers in books or archived material in libraries, paper and board products such as printed paper and packaging material have a short life span. After their use, they become waste or secondary raw material within relatively short periods. Recycling of recovered paper helps reduce the volume of waste. New waste is generated during the processing of recovered paper. Disposal of this waste is becoming an important issue for paper mills.

Figure 1 shows the solid wastes encountered in recovered paper processing mills. Besides material removed from the pulper or ejected from a drum pulper, solids separated by screening and cleaning are all rejects. Among the sludges encountered are deinking sludges, solids removed during the mechanical clarification of fresh water, process water and wastewater, and sludges from biological effluent treatment plants. Classified as incineration residues are ashes and slags from internal power plants and waste incineration plants and the residues from flue gas cleaning including fly ashes. In the category of other waste are chemical residues, used oil, wires and felts, and hazard-

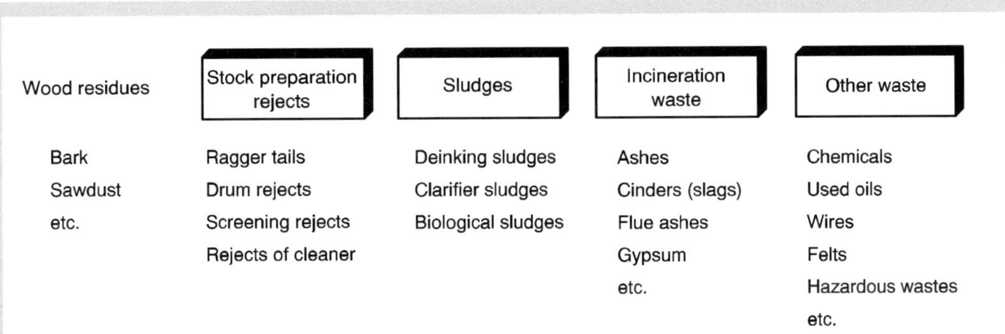

Figure 1. Solid wastes of recovered paper processing mills[1].

ous waste such as certain laboratory chemicals, batteries, or transformer fluids. Wood residues such as bark and sawdust occur in cases where mechanical pulp is also produced at the paper mills concerned.

Figure 2 shows the current possibilities for material and energy recovery and final disposal of these different types of waste. Landfill is obviously important for all types of waste. The dominant practice in the past for disposal by landfill will no longer be permissible in several European countries in a few years. Some European and non-European countries will continue to practice landfilling on a wide scale. Even in those countries, the shortage of landfill sites and cost-intensive specifications concerning the landfill operation with liner systems, leakage water collection, and treatment and landfill gas control and gas use will discourage this disposal method. For this reason, energy recovery and the material use of waste in other branches of industry will gain increasingly importance.

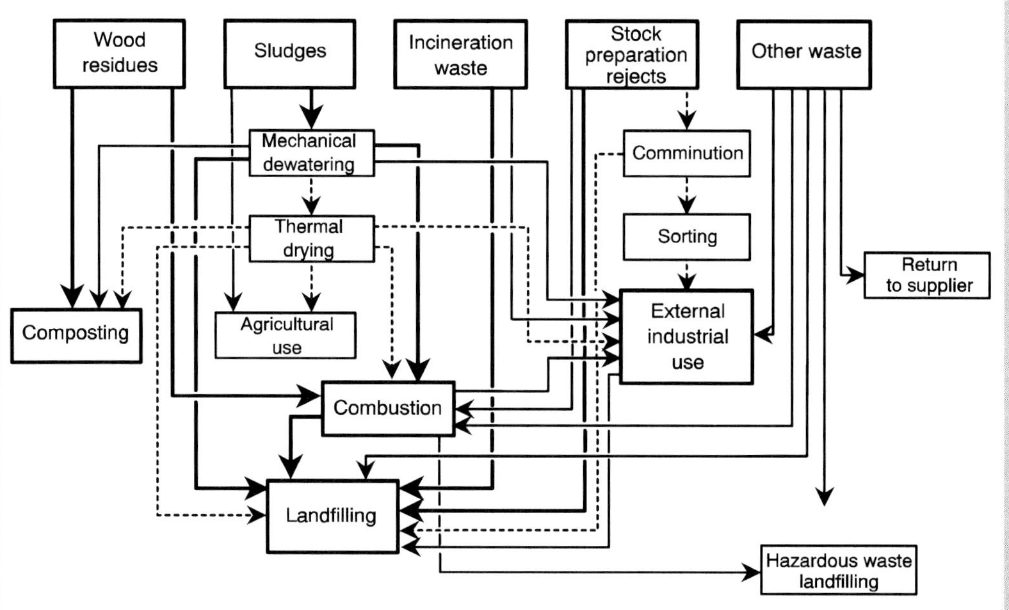

Figure 2. Use and disposal of solid wastes of recovered paper processing[1].

Depending on the recovery rate of used paper and paper products, a certain volume of used paper and board products remain that do not enter the recycling loop in paper mills. For some paper grades such as certain hygienic and specialty papers, no possibility exists for recycling. Recovered paper not processed in paper mills is usually a part of household waste, but it can also be recovered paper used for other purposes such as combustion for energy recovery, composting, or production of insulation material.

The disposal of household waste is subject to the same pressure as the disposal of waste from recovered paper processing. Only waste containing a small share of organic matter may be dumped in a landfill site. This makes thermal treatment of household waste including nonrecovered paper and paper products necessary. In some countries, organic components of household waste are already collected separately. Included in the components of this biological waste are paper products such as kitchen towels, handkerchiefs, napkins, soiled wrapping paper, and coffee filters. The biological household waste is composted or subjected to anaerobic digestion.

2 Use and final disposal of solid waste from recovered paper processing

The aim of recycling recovered paper in the paper industry is to produce recycled fiber pulp suitable for use in the manufacture of different paper and board grades. This presupposes that all substances likely to disturb the processing of recovered paper or the quality of the final product are removed as far as possible from the disintegrated recycled stock. Depending on the contamination of the recovered paper with nonpaper components and the paper grade produced, larger or smaller volumes of waste result (See chapter 1).

All solid substances removed during paper production from recycled fiber pulp by the screening, cleaning, and deinking processes as rejects and sludges and not returned into the production loop are classified as waste. This waste is waste for use and waste for final disposal. Removed materials destined for re-entry into the production process are not waste as Fig. 3 shows.

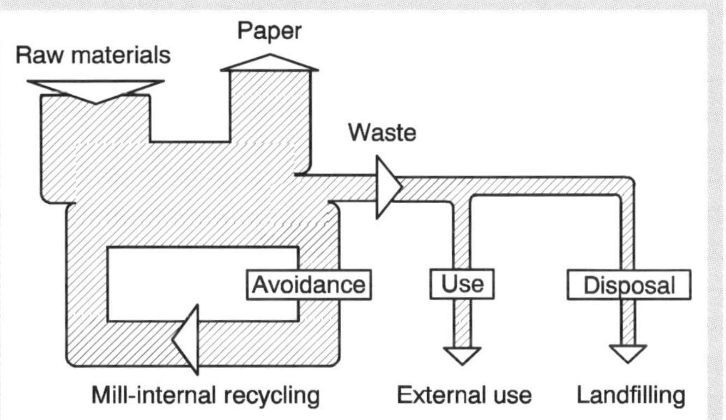

Figure 3. Simplified scheme of management of paper mill wastes.

The type and volume of waste that occurs in recovered paper processing depend on the recovered paper grades and the amount of effort invested in separation at the processing stages to guarantee the desired product quality and paper machine runnability. The waste from recovered paper processing can be heavy and coarse rejects, light and fine rejects, and sludges. A further distinction is type and origin between sludges from deinking flotation and sludges from the microflotation of process water or washer filtrates. Table 1 shows the percentages of waste amounts related to the recovered paper used dependent on the paper and board grades produced.

Table 1. Amounts of rejects and sludges dependent on recovered paper grade and the paper produced[2].

Produced paper	Recovered paper grade	Amount of total waste	Amount of waste [% by dry weight]			
			Rejects		Sludges	
		[% by dry weight]	Heavy-weight & coarse	Light-weight & fine	Flotation deinking	White water clarification
Graphic paper	News, magazines	15–20	1–2	3–5	8–13	2–5
	Superior grades	10–25	< 1	≤ 3	7–16	1–5
Hygienic paper	Files, office paper, ordinary, medium grades	28–40	1–2	3–5	8–13	15–25
Market DIP	Office paper	32–40	< 1	4–5	12–15	15–25
Liner, fluting	Sorted mixed recov. paper, supermarket waste	4–9	1–2	3–6	–	0–(1)
Board	Sorted mixed recov. paper, supermarket waste	4–9	1–2	3–6	–	0–(1)

Table 2. Solid waste from different stages in recovered paper processing[2].

Source of waste	Composition of waste	Characteristics
Pulper, drum pulper	Larger objects such as plastic bags, bookbinders, textiles, bottles, shoes, strings, tools, toys, wires, wood pieces, wet strength paper, ...	nonhazardous
High-density cleaning	Glass, nails, paper clips, textiles, pins, staples, ...	nonhazardous
Pre-screening	Long, thin and wide contaminants, plastic, polystyrene, stickies, ...	nonhazardous
Flotation deinking	Fillers, fibers, fines, printing ink, stickies	nonhazardous
Forward cleaning	Small compact particles with high density such as sand, shives, hard particles from UV-colors, coating colors, varnish, ...	nonhazardous
Fine screening	Plastic fragments, light-weight contaminants, hot melts, stickies	nonhazardous
Process water clarification	Colloidal material, fillers, fibers, fines, ink particles	nonhazardous

2.1 Solid waste composition and characteristics

Table 2 shows the composition of waste that occurs at different stages of recovered paper processing. The waste from disintegration, cleaning, and screening is reject material or rejects. The materials that occur during flotation deinking and the cleaning of process water from wash deinking are sludges. Because today's processing methods are selective only to a limited degree, both rejects and deinking sludges continue to contain a certain proportion of fibers and fiber fines. Sludges also result from clarification and biological treatment of wastewater.

2.1.1 Rejects

The amounts of rejects produced and their composition depend largely on the recovered paper grades used in paper production. Chryssos and Murr[3] examined the composition of rejects of different recovered paper processing mills and established clear differences related to the proportion of plastic. Figure 4 shows the composition of a reject sample from the production of newsprint. The paper mill concerned worked exclusively with deinked pulp (60% old newspapers and 40% old magazines) as base material. Slushing of the recovered paper used a drum pulper. The sample came from the reject of the slushing drum. The proportion of this reject fraction as dry solids was 0.7%–1% related to air-dried recovered paper. The dry content of the reject sample was about 70%. The plastic share was 52%. The combined share of flakes and fibers (27%) was primarily due to wet strength papers. In this mill during the subsequent screening and cleaning stages, an additional 3%–4% dry basis was generated. The weight of baling wires was irrelevant because of loose recovered paper deliveries.

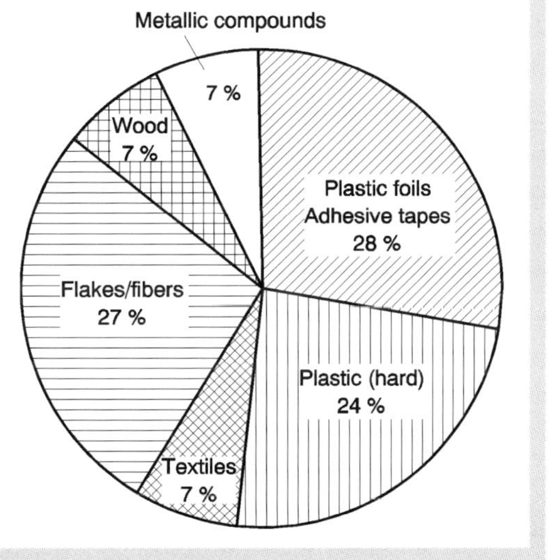

Figure 4. Composition of a reject sample from the production of newsprint based on recycled fibers[3].

Another test at a manufacturer of packaging paper revealed the reject composition of Fig. 5. At this mill, sorted mixed recovered paper, supermarket waste, and old corrugated containers (OCC) were the base material. The slushing of the recovered paper mixture used a conventional pulper with a ragger and a subsequent screening drum from which the reject sample was removed. The proportion of plastic, foil, and self-adhesive

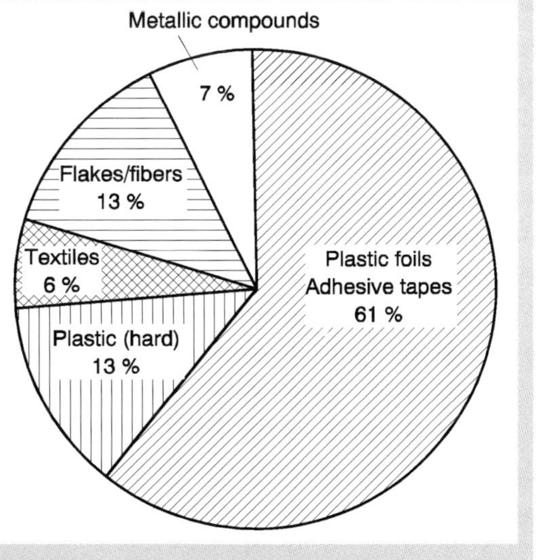

Figure 5. Composition of a reject sample from the production of packaging paper based on recycled fibers[3].

tapes in this reject at 60% is significantly higher than that in the reject from newsprint production at nearly 30%. The proportion of flakes and fibers is low at 13%. The weight of baling wires is part of the 7% metallic waste.

Fiber content of the rejects increases in the cleaning and screening stages following recovered paper disintegration. Figure 6 shows the composition of light and fine rejects from the production of corrugating medium and test liner. The data for the material and chemical composition represent averages from five paper mills in Germany[4]. The fiber portion related to dry substance now exceeds 35%. Due to the high proportion of combustible plastic materials, the heating value of more than 20 GJ/t of dry substance is considerably higher than that of lignite.

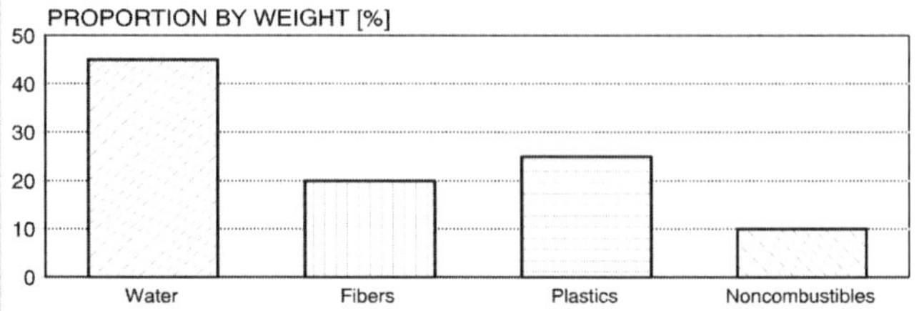

Chemical composition
(proportions by weight based on dry solids)
Averages of five recovered paper processing paper mills in Germany

Ash	9.4 %
Carbon	51.7 %
Hydrogen	8.3 %
Chlorine	1.1 %
Sulfur	0.2 %
Nitrogen	0.8 %
Phosphorus	0.1 %

Figure 6. Composition of screening rejects of corrugating medium and test liner production based on recycled fibers[4].

In the chemical composition of the rejects, the high total chlorine content as the sum of organic and inorganic chlorine is especially evident compared with sludges from recovered paper processing. It can amount to 5% by weight of rejects on a dry basis. Polyvinyl chloride (PVC)-containing materials such as the following are the main sources of this chlorine content:

- Self-adhesive tapes
- Carrier handles from solid board and corrugated board containers
- PVC foil laminates from the packaging sector
- PVC products thrown in error into recovered paper containers.

The high chlorine content of the rejects can limit their use as an energy source such as secondary fuels for cement production.

2.1.2 Deinking sludges

Deinking sludges consist of fillers and coating pigments, fibers, fiber fines and printing inks (black and colored pigments), and adhesive components. Figure 7 shows that the average contents of a deinking sludge from the production of wood-containing graphic papers with more than 55% of the solids removed by flotation are inorganic components. They are primarily fillers and coating pigments such as clay and calcium carbonate. The proportion of fibers at 7% is low. Materials extractable with methylene chloride are an average proportion of 8%. They contain wood components from the fibers such as resins, fats, and resin acids, soluble printing ink and adhesive components, and flotation deinking chemicals. The remaining 29% comprises fiber fines, insoluble ink components (mainly carbon black and colored pigments), and insoluble adhesive components.

The wash deinking sludges produced during the production of hygienic papers differ from flotation deinking sludges due to the more extensive cleaning measures involved and the nearly complete filler and pigment removal shown in Fig. 8. Because of the higher fiber and fines content at 11%, the filler and pigment proportion is generally less than 50%. Fines and insoluble ink and adhesive components account for 40%. At about 3%, the extractable material is low. This can be due to the higher proportion of wood-free recovered paper in the base material.

Table 3 provides a summary of the ash contents, heating values, elemental contents, and levels of different contaminants of deinking sludges. It shows minimum, maximum, and averages for sludges from different paper mills[6]. A characteristic for deinking sludges is their high ash content that is 40%–70%. The heating value depends on the ash content and is 4.7–8.6 GJ/t of dry substance. The sulfur, fluorine, chlorine, bromine, and iodine contents are low. For this reason, no costly flue gas purification systems are necessary when incinerating deinking sludge. Compared with sludges from biological effluent treatment plants, the nitrogen and phosphorus contents are very low. This is something that requires consideration when using deinking sludges for composting and agricultural and land application purposes[7].

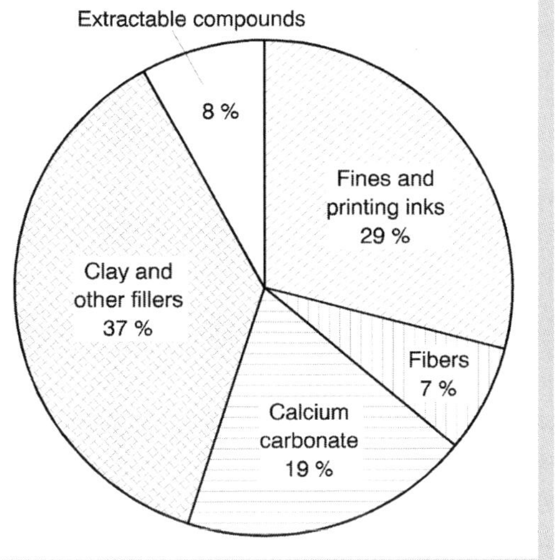

Figure 7. Composition of deinking sludges from the production of wood-containing graphic papers related to dry substances[5].

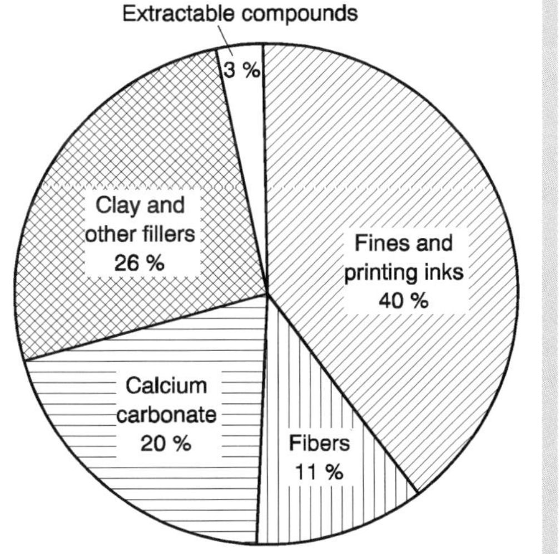

Figure 8. Composition of deinking sludges from the production of hygienic paper related to dry substances[5].

The level of heavy metals for sludges of recovered paper processing is generally low[8]. Figure 9 compares the concentrations of heavy metals in deinking sludge with the contents of heavy metals in biological sludge of wastewater treatment plants of paper mills and municipal sewage sludge. This data shows that sludges of deinking plants have less contamination than those of municipal wastewater treatment. The concentration of cadmium and mercury is especially insignificant and sometimes even below the detection limit of the test method applied (atomic absorption spectrometry). Only the concentration of copper has the same order of magnitude as that of municipal sewage sludge. The copper content of deinking sludge is primarily due to blue pigments of printing inks using phthalocyano compounds.

Table 3. Chemical characteristics of deinking sludges with data related to dry substance[6].

Parameter	Unit	n	Minimum	Maximum	Average
Dry content	Weight-%	11	38.1	62.1	51.2
Ash content	Weight-%	14	36.4	67.3	54.8
Heating value	kJ/kg (wet)	12	1 240	4 320	2 310
Heating value	kJ/kg (dry)	13	4 750	8 600	6 940
pH		13	6.8	8.3	7.5
Sulfur	Weight-%	3	0.06	0.12	0.08
Fluorine	Weight-%	14	n/a	< 0.1	n/a
Chlorine	Weight-%	14	0.01	0.1	0.05
Carbon	Weight-%	14	19.1	35.8	26.0
Hydrogen	Weight-%	14	2.6	4.2	3.3
Nitrogen	Weight-%	14	0.20	0.50	0.35
Phosphorus	mg/kg	14	300	560	464
Potassium	Weight-%	14	0.08	0.30	0.19
Calcium	Weight-%	14	1.5	14.9	9.3
Magnesium	Weight-%	14	0.16	0.79	0.37
Aluminium	Weight-%	6	2.8	6.5	6.5
Iron	Weight-%	6	< 0.1	0.5	0.5
Silicon	Weight-%	6	3.3	7.6	7.6
Lead	mg/kg	14	9.5	79.4	44.6
Cadmium	mg/kg	14	0.02	1.54	0.34
Chromium	mg/kg	14	4.8	96.6	39.1
Copper	mg/kg	14	64.2	345.0	186.2
Mercury	mg/kg	14	0.10	0.89	0.26
Nickel	mg/kg	14	< 10	31.3	n/a
Zinc	mg/kg	14	34.2	1 320	269
PCB	mg/kg	5	n/a	< 0.3	n/a
AOX	mg/kg	14	160	1 200	516
Dioxins/furans	ng I-TE/kg	3	26.5	58.3	37.9

n: Quantity of analyzed deinking sludge samples; n/a: not available
I-TE = International Toxicity Equivalent according to NATO/CCMS

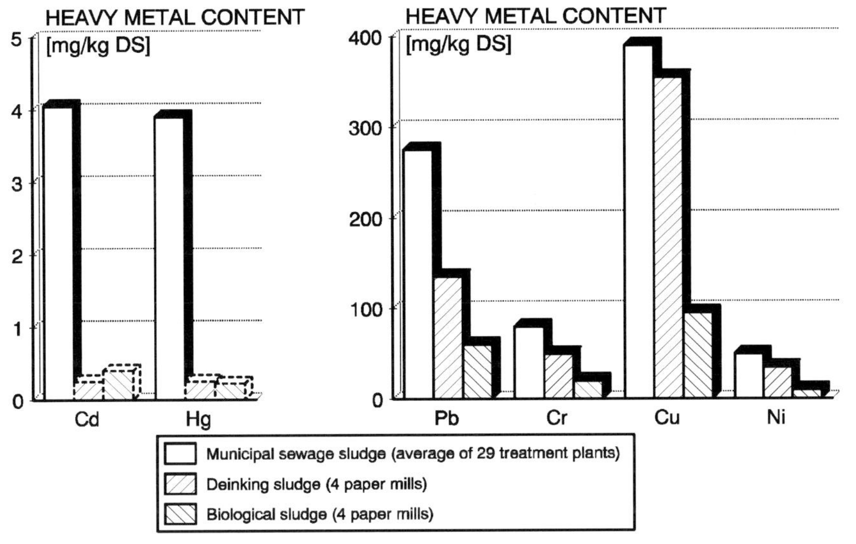

Figure 9. Heavy metal contents of deinking sludge, biological sludge, and municipal sewage sludge (Germany)[9].

Traces of halogenated organic compounds such as polychlorinated biphenyls (PCB), polychlorinated dibenzodioxins (PCDD), and polychlorinated dibenzofurans (PCDF) also require consideration. PCB compounds had use in manufacturing carbonless paper until the early 1970s. The PCB level of deinking sludge has decreased significantly since then. Recent data obtained from several deinking plants confirm that PCB concentration is below 0.3 mg/kg dry solids (0.3 ppm) using the most relevant seven congeners (See Chapter 13).

PCDD/PCDF concentrations of deinking sludges show a similar pattern of decline. Due to the continuing change from elemental chlorine bleaching of chemical pulp to chlorine dioxide and to oxygen bleaching, the PCDD/PCDF contents of deinking sludges of German paper mills has been decreasing significantly. Today, PCDD/PCDF concentrations of deinking sludge are 25–60 ng I-TE/kg dry solids (I-TE = International Toxicity Equivalent). These figures are not significantly higher than the average contents of PCDD/PCDF in municipal sewage sludge. As a result of modifications of the bleaching sequences in chemical pulping, dioxin formation does not occur in most pulp producing countries. Consequently, dioxin discharges from recovered paper processing mills that are already low will decrease further.

The parameter AOX (adsorbable organic halogen-containing compounds) plays an essential role in environmental regulations. In Germany for example, the direct application of sewage sludge on agricultural soil is regulated for heavy metal concentration, PCB and dioxin concentrations, and AOX level. In many cases, the AOX of deinking sludge is often above the acceptable limit of 500 mg/kg dry solids[10]. Investigations at the Department of Paper Science and Technology in Darmstadt showed that up to 80% of AOX of deinking sludge is due to chlorinated yellow pigments that are components of printing inks[11]. These pigments are water insoluble and nonbiodegradable.

2.2 In-mill pre-handling

The topics of the following sections are the dewatering and drying of rejects and sludges. Dewatering is an essential stage of waste handling regardless of the recycling and disposal method. All commonly used methods for material and energy recovery benefit from a high dry solids content of rejects and sludges.

2.2.1 Dewatering

Nearly all waste from recovered paper processing with the exception of the dewiring unit, ragger, and drum pulper rejects has an original dry content of only 1%–6%. For this reason, the first step in waste treatment is dewatering. Today, the ragger and the rejects from the pulper disposal system are usually not subjected to any special dewatering. Due to their material composition, "draining off" gives dry contents of 60%–80%. Screen spiral conveyors, vibrating screens, and screw and rake classifiers are useful for dewatering heavy and coarse rejects. Dewatering of light and fine rejects with screens, endless wires, or vibrating screens is usually followed by additional dewatering using a reject screw press[12] (See Chapter 6).

Two-stage processes are often used for sludge dewatering. An initial step for dewatering using gravity tables or drums and disk thickeners occurs before a high-consistency dewatering with belt filter presses or screw presses (See Chapter 5). Batch-agitated chamber filter presses have only minor importance due to their costly sludge conditioning with inorganic precipitation agents (metallic salts and chalk) or organic polymers and their high maintenance costs. Centrifuges are universally applicable and generally managed without steps before dewatering. With pure biological sludges, systems using these devices can achieve the highest dry contents.

Figure 10 shows the basic function of a belt filter press. The sludge is transported between two wire belts where it is subject to a gradually increasing pressure. Direct pressure forces and shearing forces squeeze water from the sludge. Conditioning with polyelectrolytes is necessary for most types of sludge[13]. Figure 11 shows the basic function of a screw press. The sludge is transported by a slowly rotating screw working against discharge restriction. The water squeezed out is

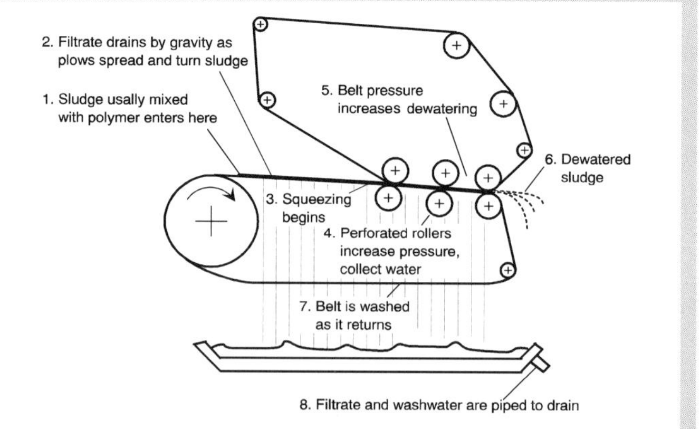

Figure 10. Basic function of a belt filter press.

released through openings in the cylindrical housing surrounding the screw. Some screw presses have the optional possibility of heating the sludge by injecting steam through the screw shaft. Conditioning of the sludge with polyelectrolytes is common.

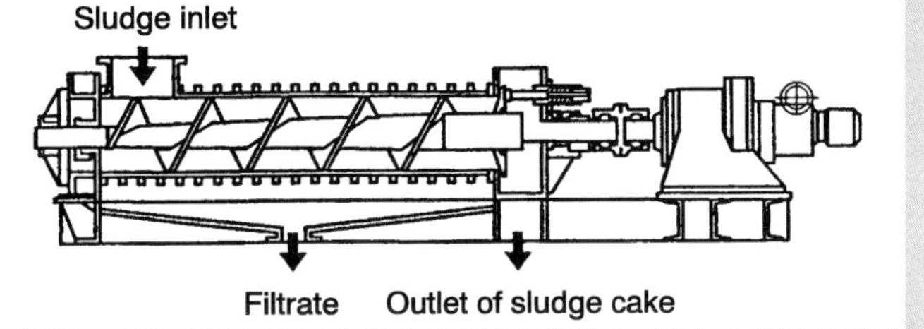

Figure 11. Basic function of a screw press.

Table 4 summarizes the most common dewatering systems for different types of waste. Indicated also are the achievable dry solids contents. The differences in the dry solids contents are due to different pressure levels and times. For example, screw presses operate with pressures up to 8 bar and times up to 10 min. Belt filter presses work with pressures only up to 0.5 bar and short pressing times of only 1–2 min. With increasing proportions of biological sludges from wastewater treatment plants, the dry solids contents decrease for all techniques. Independent of the technique applied, dry contents of only 15%–40% are possible with purely biological sludges.

Table 4. Composition and dewatering of rejects and deinking sludges[2].

	Heavy-weight & coarse rejects	Light-weight & fine rejects	Deinking sludges
Composition	Glass, nails, sand, stones, paper clips, pins, staples, textiles, wood pieces, wet strength paper	Sand, textiles, fibers, coating colors, plastic fragments, hot melts, stickies	Fillers, pigments, fibers, fines, printing ink, stickies
Dewatering facilities	Screens Vibrating screens Screw classifiers Rake classifiers	Screens Disk thickeners Dewatering drums Gravity tables Screw presses	Dewatering drums Gravity tables Belt filter presses Screw presses Chamber filter presses Centrifuges
Achievable dry solids contents	60%–80%	50%–65%	one-stage operation: <15% two-stage operation: <65%

2.2.2 Drying

Thermal drying processes are used for more comprehensive removal of water from mechanically dewatered sludges. Sludge dryers find use primarily when available or potential disposal processes are impossible or uneconomical, e.g., high landfill costs per ton of moist sludge due to insufficient dryness. Besides giving mass reduction, drying of the sludge has other advantages that broaden the waste management options[14]:

- Sludges at about 95% dry substance are odorless. The bacteria-free conditions mean that sludges can be stored in the open or in simple silos

- Dried sludges can be transported without any problem in the same way as bulk goods in regular transport systems
- Dosing and mixing with other waste and combustibles is simpler
- Combustion is easier and produces more heat per unit of mass of dry material.

Direct and indirect drying systems are in operation. Direct systems are convection dryers, and indirect systems are contact dryers as Fig. 12 shows. In convection drying, the drying media of flue gas, superheated steam, or vapor have direct contact with the sludge and absorb water evaporating from the sludge. In contact drying, the heat energy is applied indirectly to the sludge. The sludge and the heat medium are separated by a wall.

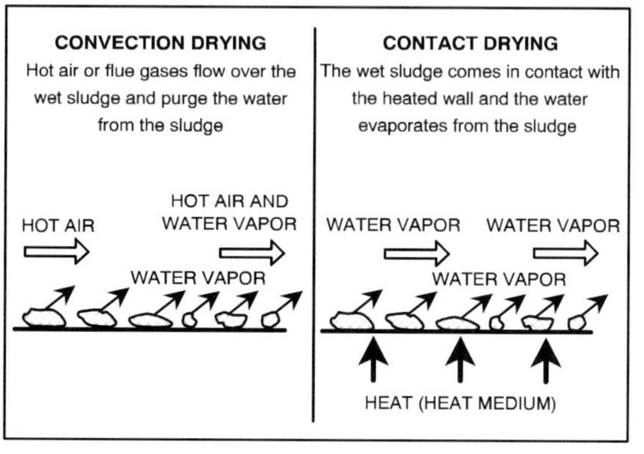

Figure 12. Thermal drying methods for sludges[15].

Contact or indirect dryers feed the energy required for water evaporation to the moist sludge usually by steam-heated surfaces. Because the temperature of the heating surfaces is higher than the required evaporation temperature, the sludge becomes overheated on the contact surfaces with a risk of crust formation. The condensed exhaust air of such systems shows high contamination, but the volume of exhaust air is low.

With convection dryers, heat transfer uses direct contact between hot air and the sludge. The hot air also provides pneumatic transport through the dryer. Convection dryers can be heated directly or indirectly. Compared with the contact dryers, the contamination of the resulting condensed exhaust air is low, but the volume of exhaust air is considerably higher.

Among the contact drying processes in the paper industry are steam centrifugal dryers and disk dryers. The types of convection dryers used are drum dryers, belt dryers, or rack dryers[16].

In all drying processes, flue gas and dry goods control have primary importance due to the danger of combustion and explosion. In this respect, indirect drying processes are less problematical. Depending on its structure in the moist state, the dried sludge takes on the form of fine granules or pellets. For direct and indirect drying, the heat requirement is 4.0–4.5 GJ/t of evaporated water.

Figure 13 shows a diagram of a drum drying system for drying a deinking sludge. In this case, the dried sludge goes to an electrical power plant located immediately beside the paper mill. It is used there with lignite and hard coal as a secondary fuel[17].

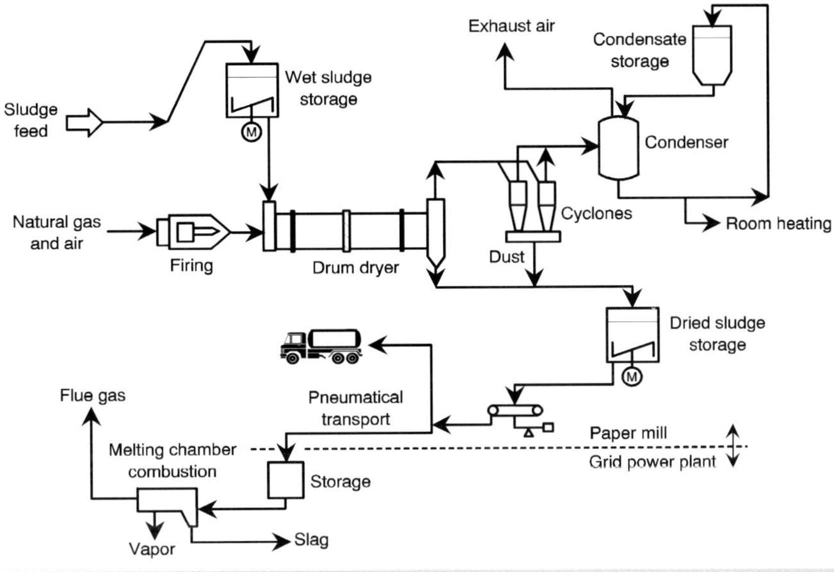

Figure 13. Deinking sludge drying with a drum dryer[17].

2.3 Energy recovery in internal and external combustion facilities

The external use of waste from recovered paper processing mills has encountered acceptance difficulties among some potential users. In addition, the risk exists of a monopolistic dependence on external use operations that fix the purchasing fees themselves. For these reasons, internal use of waste for energy recovery will become very important[18–22].

The main objectives of energetic waste treatment are as follows:

- Maximum volume and mass reduction of the waste
- Achieving inertness of the organic components
- Immobilization of harmful substances by integration into ashes and slags
- Use of the generated energy.

A long tradition of waste incineration exists in the paper industry. For example, consider the combustion of bark and wood residues. Recent years have seen a growing interest in the use of other types of waste such as sludges and rejects for energy production purposes. The reasons are as follows:

- Increased costs of fossil fuels and purchased power
- Reduced landfill capacities and increased landfill costs
- More stringent environmental regulations governing the use of this waste
- Development of new combustion technologies with highly effective flue gas cleaning technologies.

Due to their heating values and low content of harmful substances, most types of waste from recovered paper processing mills are suitable for energy recovery. Table 5 shows a comparison of heating values, ash contents, and water contents of different solid fuels with those of the types of waste concerned. Clearly evident is the high heating value of rejects due to the large share of plastic contained in this waste. The relationship between the combustible, i.e., organic part, ash content, and water content can be described by the so-called fuel triangle of Fig. 14. Marked especially in this triangle is the range of self-supporting combustion. This means that no additional fuels in the form of fossil fuels are necessary for the combustion of the materials listed in this range. Indicated with close parallel lines are the areas concerning deinking sludges and rejects from recovered paper processing to the degree that they were largely dewatered by mechanical processes. Self-supporting combustion is obviously a practical proposition not only for the rejects but also for the deinking sludges having high ash contents.

Table 5. Heating value, ash content, and water content of solid fuels[23,24].

Solid fuel	Heating value [MJ/kg]	Ash content [%]	Water content [%]
Wood (air-dried)	14–17	< 1	10–20
Peat (air-dried)	12–15	15	15–30
Lignite	8–12	2–8	15–35
Hard coal	28–34	3–12	0–10
Deinking sludges (wet)	7–10	40–60	40–60
Rejects (wet)	16–23	8–12	20–50
Bio sludges (wet)	7–9	20–40	70–85

Presently, different processes and system concepts are available on the market for waste combustion. The application possibilities depend on the total operational conditions and type of waste concerned. The important factors for all combustion processes are as follows:

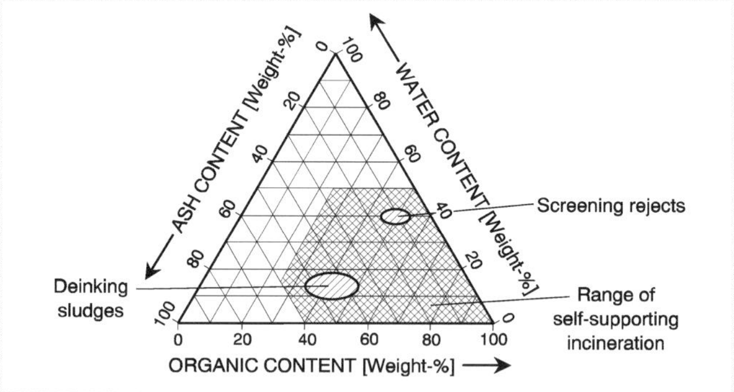

Figure 14. Fuel triangle of waste of recovered paper processing mills[4].

- Retention time in combustion zone
- Combustion temperature
- Air or oxygen supply
- Fuel density or fuel distribution in the combustion area.

In recovered paper processing mills, the following combustion technologies are primarily used for reject and sludge combustion:

- Grate combustion
- Fluidized and circulating fluidized bed combustion
- Multiple hearth combustion.

2.3.1 Grate combustion

In grate combustion, the fuel is fed mechanically onto the grate. Combustion air is fed from below through the gaps in the grate. This also helps to cool the grate rods as Fig. 15 shows. The free grate area formed by the gaps between the rods and the shape of the rods influence the quality of combustion. They are matched to the type of fuel and grain size of the fuel. The gap width and grain size of the fuel also determine the proportion of noncombusted material in the ash.

A basic distinction exists between grate combustion with a stationary or moving grate. Stationary grates are used primarily in furnaces up to 10 MW combustion heat generation. They reach grate heat loads or heat flow per grate area unit of about 1 MW/m^2. With combustion using a moving grate, grate heat loads of 1–2 MW/m^2 are achieved. When using large grate areas, combustion heat generation of a maximum of 100 MW is possible.

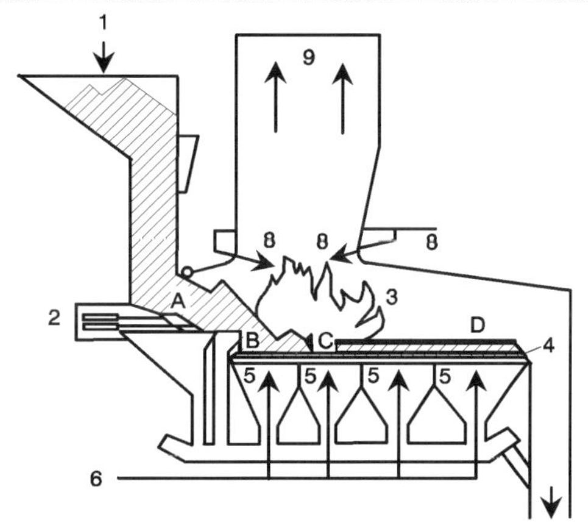

Figure 15. Grate combustion with a stationary grate[15].

In combustion with stationary grates, the fuel layer is distributed across the grate by inclining the grate and mechanically assisted movement using an inclined, stepped, stoker agitated grate, or both. The fuel passes through the drying zone, ignition zone, combustion zone, and the burn-out zone. The air supply is adjusted to suit the different combustion stages.

Among the grate combustion processes with moving grate, traveling grate combustion is the most common combustion system. The traveling grate consists of a continuous grate belt driven by two shafts. The fuel is supplied to the grate belt through the combustion zone at speeds of 100–300 mm/min. The combustion residues are ejected at the end of the grate. With larger traveling grates, the combustion air is fed through zoned air boxes positioned between the top and bottom grate belt. The number of air boxes depends on the grate load and the desired degree of burn-out. The traveling grate is automatically freed from baked-on combustion residues on the bottom of the returning grate usually with some form of mechanical assistance.

By controlled adjustment of the combustion conditions to the fuel characteristics, optimizing the combustion air feed, and using a suitable design of the combustion chamber, grate combustion systems can operate with low emissions and a high burn-out efficiency. With sufficient retention time of the fuel on the grate and return of ash with a high coke content, only small amounts of noncombusted material remain in the combustion residues. The burn-out of the combustion gases can improve significantly with the addition of secondary air immediately over the grate. This causes a reduction in the carbon monoxide and hydrocarbon emissions.

Grate combustion systems are especially suitable for lumpy fuels but not for moist, paste-like, and finely grained fuels. With moist sludges for example, the retention times on the grate or in the combustion area can be insufficient for complete combustion. Nevertheless, grate combustion systems are used in the paper industry primarily for co-firing of sludges with coal, bark, or wood residues. Long reaction times for control processes and the difficult combustion control on the grate by the addition of air presuppose that the moisture content and ash content of the fuels vary only within narrow tolerances. A range of 15–20 GJ/t of dry substance is the maximum limit for the heating value of the fuels.

2.3.2 Fluidized bed combustion

In fluidized bed combustion, the fuel is fired in a bed of inert solids such as silica, sand, or ash fluidized by air. The air is blown from below into the combustion area using a perforated bottom. The air serves as the medium for combustion. At velocities of 1–2.5 m/s, it fluidizes the inert solids layer.

The fuel is added above the fluidized bed. Due to its high heat capacity and dynamic fluid characteristics, the inert bed material provides heat transfer and temperature averaging. The fuel is therefore quickly ignited as it is fed. Due to intensive mixing of fuel and combustion air in the fluidized area, very favorable material and heat exchange conditions occur. For this reason, good combustion results even with fuels of low quality.

The combustion temperatures of fluidized bed combustion systems are 800°C–900°C. The combustion gases emitted from the fluidized bed flow into the free space above which an after-burning zone is created by the addition of secondary air. In this zone, the exhaust gases react at temperatures of 800°C–950°C. Retention times of 2–4 s guarantee complete combustion. The after-burn chamber can be separate outside the fluidized bed combustion chamber and optionally has an after-burner.

Depending on the dimension of the fluidized bed and the level of the operating pressure, the following techniques are possible:

- Stationary fluidized bed
- Circulating fluidized bed
- Pressurized fluidized bed.

Stationary fluidized bed combustion

The stationary fluidized bed has a slightly expanded fluidized bed and a defined bed surface as Fig. 16 shows. Fluidizing speed and solid grain size are matched to a tolerable loss of small amounts of particles from the fluidized bed. The heat generation related to the perforated bottom is 1–2 MW/m^2. A stationary fluidized bed has primary use in industrial facilities with heat outputs less than 50 MW. For furnaces with greater output capacities, unfavorable combustion conditions exist due to the increasing fluidized bed dimensions and resulting mixing problems.

Stationary fluidized bed furnaces offer very good operating conditions for low emissions of air-borne pollutants. The emissions of nitrogen oxides are especially low compared with grate combustion. At comparatively low combustion temperatures, thermal development of nitrogen oxides from the combustion air is largely eliminated. Engineering measures such as a stepped air supply can effectively reduce the nitrogen oxides resulting from the fuel nitrogen. Direct desulfurization is possible by the addition of basic sorbents such as limestone or dolomite into the body of the furnace.

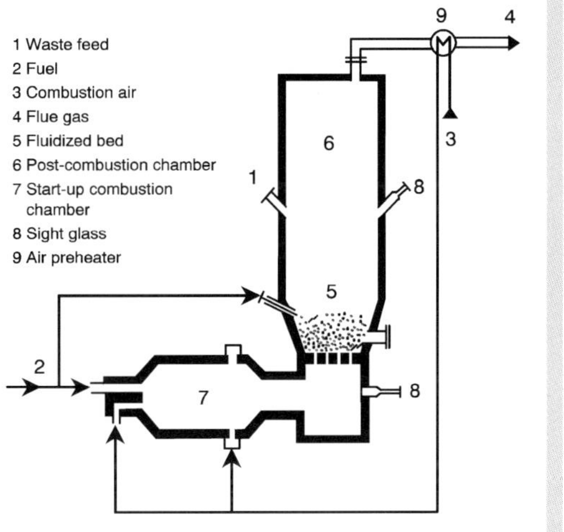

Figure 16. Stationary fluidized bed combustion[25].

Circulating fluidized bed combustion

Better results for the quality of combustion and emission reduction are possible with circulating fluidized bed combustion. These plants operate with higher gas speeds up to 8 m/s and considerably smaller grain sizes of the bed material. The high gas speed produces a strongly expanded fluidized bed and removes a large share of the solids from the furnace. The ash is removed from the exhaust gas in a cyclone and returned to the furnace as Fig. 17 shows. Depending on the heat requirement, a direct return or a return via fluidized bed heat exchangers (fluidized bed coolers) is possible. The continuous return of the finely grained solid material guarantees almost complete combustion and keeps the fluidized bed in the furnace in balance. Using multiple stepped air feed to the

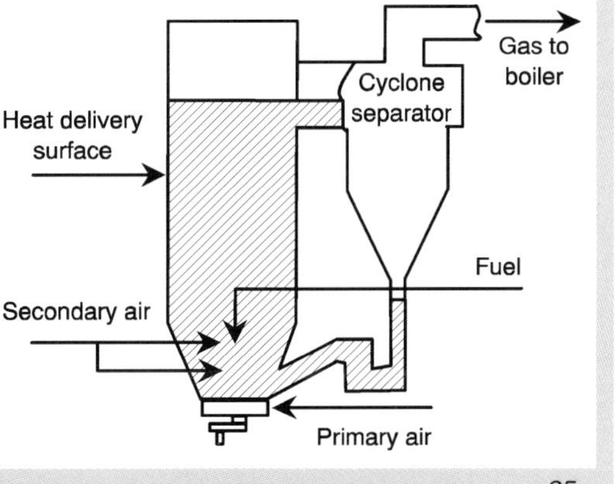

Figure 17. Circulating fluidized bed combustion[25].

furnace provides very low nitrogen oxides emission below 100 mg/m^3. The high air and solids throughput permits average thermal cross-sectional loads up to 12 MW/m^2. A heat output of 500 MW is the ceiling limit.

Pressurized fluidized bed combustion

Even higher heat outputs up to 1 000 MW are possible with pressurized fluidized bed furnaces. They can operate in stationary and circulating modes. Fluidized bed combustion under pressure improves the reaction processes in the furnace. The emissions decrease as pressure increases. A major advantage of pressurized fluidized bed combustion plants is their compact construction. The unit volume is only one-third of the unit volume of a conventional fluidized bed furnace of identical performance.

To reduce particle emissions from fluidized bed combustion plants, electrostatic precipitators, fabric filters, or both are used in flue gas purification. With the addition of limestone, filters help to reduce inorganic halogen compounds such as HCl and HF. Comparatively high compared to other combustion technologies are the emissions of nitrous oxides (NO_x) due to the lower combustion temperatures.

2.3.3 Multiple hearth combustion

Multiple hearth combustion plants are especially suitable for the combustion of moist and paste-like waste. They have had use for decades in the paper industry for energy production from primary and biological sludges often with bark. The fuel is not fed directly to the combustion zone of the oven but instead via trays in the tiered arrangement of Fig. 18. The fuel feed occurs in the top hearth where drying first occurs by upwardly directed hot flue gases. Good heat transfer happens from the direct contact of the flue gas stream with the fuel. The exchange surface necessary for the evaporation of the water content in the fuel occurs by constant circulation and spreading of the fuel on the hearth surfaces using agitators. Transport of the fuel occurs alternately to each hearth in the direction of the outside of the furnace or toward the center. At the center, it falls to the next hearth. In this way, the fuel passes through all furnace zones from top to bottom.

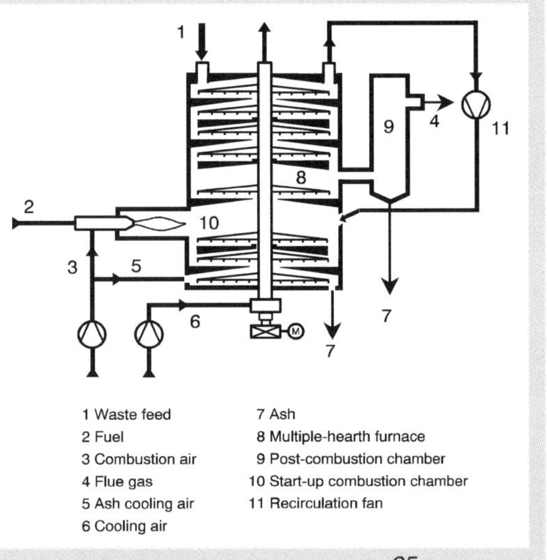

Figure 18. Multiple hearth combustion[25].

The temperature level at the top hearth, the drying zone, depends on the combustion temperature and is usually 250°C–400°C. Combustion occurs at 800°C–1 000°C. To achieve these temperatures, use of additional fuels may be necessary. Liquid or gaseous support fuels such as natural gas or fuel oil are fed via a starting combustion chamber.

Solid fuels such as coal and wood can be supplied directly to the hearths of the combustion zone. At the bottom hearth before removal, the ash is cooled to 200°C–400°C.

Two types of multiple hearth furnaces exist. In the classical version, the furnace has an external after-burning chamber for post-combustion of the flue gas. In a newer version, after-burning occurs by returning the flue gas via circulating blowers into the combustion zone.

The flue gas purification plant downstream from the multiple hearth furnace usually consists of wet scrubbers to remove dust and sulfur compounds. Use of lime is necessary for removal of the sulfur compounds. The flue gas finally enters the stack via a mist eliminator.

2.3.4 Which combustion technology for which type of waste?

The choice of combustion technology depends on the operating conditions and the nature of the waste. According to knowledge today, fluidized bed combustion is the most suitable for thermal treatment of sludges and rejects[26–30]. Combustion in the fluidized bed occurs quickly and completely. Neither fluctuations of heating value of the fuel nor changing proportions of noncombustible components such as sand, metals, or fillers have a negative influence on the combustion efficiency. For fuels with a low heating value such as deinking sludges, the stationary fluidizing is probably a more suitable option[31]. Fuels with a higher heating value such as wood and bark are incinerated more effectively in a circulating fluidized bed.

A disadvantage of fluidized bed combustion plants is the higher investment costs compared to grate and multiple hearth furnaces. Fluidized bed furnaces are also offered only in larger-scale plant units making them suitable only to a limited degree for smaller mills with minor waste volumes. Offsetting the additional costs by savings in flue gas purification is possible because of direct sulfur and halogen removal in the furnace. Due to mechanical flow requirements of the fuel, producing a consistent grain size distribution by mechanical processing such as extraction of metals and grinding may be necessary. This requires consideration especially when co-burning rejects with sludges in fluidized bed furnaces.

2.3.5 Emissions

General aspects

Atmospheric emissions of incineration plants contain several potential pollutants. The most important are solid particles or dust, sulfur dioxide (SO_2), nitrogen oxides (NO_x), hydrogen chloride (HCl), hydrogen fluoride (HF), various heavy metals, and trace organic compounds such as dioxins and furans.

The key emission problems discussed in previous years refer to dioxins and furans due to their notoriety with respect to waste incineration and their high toxicity. Emissions of carbon dioxide and its effect as a greenhouse gas have had wide discussion but have minor importance when burning sludge of recovered paper processing mills. Since the organic proportion of sludge is essentially biomass, its combustion does not contribute to the greenhouse effect. Trees that were the original source of such waste will reabsorb the released carbon dioxide (See Chapter 13).

The literature provides little information concerning emissions that occur from burning recovered paper processing waste. Any comparison of emissions from different combustion plants becomes difficult or even impossible when using different fuel mixtures such as different sludge and reject levels. Emission data from laboratory tests cannot always be applied in practice since the combustion conditions in industrial plants cannot be simulated with sufficient precision.

Emitted air-borne substances have the following classifications:

- Fuel-dependent substances
- Process-dependent substances
- Fuel- and process-dependent substances
- Solid particles (dust).

In the first group of fuel-dependent, air-polluting substances are sulfur dioxide, hydrogen fluoride, and hydrogen chloride that form from organic and inorganic sulfur, fluorine, and chlorine compounds. Also included here are the heavy metal compounds bound to dust particles in the flue gases or present in gaseous form. Even in an ideally complete combustion, emissions of this category of air-polluting substances cannot be avoided. The result is that purification of the flue gases is necessary.

Among the group of process-dependent substances are all substances that occur as products of incomplete combustion. Examples of these are carbon monoxide and different hydrocarbon compounds including methane, ethylene, benzene derivatives, and polycyclic aromatic compounds. The concentrations of these compounds in the flue gas depend strongly on the pre-treatment of the fuel and the conditions of furnace operation.

Categorized as belonging to the third group are those compounds whose occurrence is both fuel- and process-dependent. These include nitrogen oxides such as NO and NO_2 called NO_x compounds. They come from two reactions. One is the reaction of nitrogen in the fuel with oxygen from combustion air. The other reaction is that of nitrogen and oxygen as components of the combustion air. The latter reaction occurs only at high temperatures.

The formation of polychlorinated dioxins and furans is also fuel- and process-dependent. The occurrence of these compounds in exhaust gases from combustion plants may have several origins:

- Dioxins and furans are present in the fuel and are not decomposed or insufficiently decomposed at the time of combustion due to short retention times and low combustion temperatures in the furnace.
- Dioxins and furans are produced in the furnace from chlorinated organic compounds. These are the so-called precursors such as chlorophenol and polychlorinated biphenyls contained in the fuel, combustion gases, or both.
- Dioxins and furans are produced from carbon compounds and inorganic chlorine compounds in the presence of catalytic heavy metals (primarily copper) in a so-called DeNovo synthesis. This occurs after combustion in the cooling phase of the exhaust gases at 250°C–400°C.

The oxygen and CO contents in the combustion chamber and the retention time of the flue gas in the cooling system heavily influence the dioxin and furan contents of flue gases from combustion plants. Suppressing CO peaks and keeping the oxygen content in the combustion chamber as constant as possible can considerably reduce dioxin and furan emissions. Short retention times of the flue gases in the waste heat recovery boiler and a very fast lowering of the flue gas temperature below 250°C are additional measures to minimize dioxin and furan emissions. Added to these are steps to influence the input of a combustion plant:

- Continuous feed of fuel that is as homogeneous as possible by an appropriate feed system
- Reduction of the availability of halogens in the combustion chamber by the exclusion of fuels with a high halogen content such as PVC.

The fourth group of air-polluting substances consists of solid particles. Flue gases contain particles of different origins such as incompletely burned fuel particles and some fluidized bed particles, incombustible fuel contaminants, or carbon black as a decomposition product of the fuel.

The exhaust gas components, oxygen, nitrogen, water vapor, and carbon dioxide are components of the atmosphere and do not belong to the air-polluting substances. For global warming and its effects on world climate, CO_2 emissions are increasingly coming under critical scrutiny. An important consideration is whether CO_2 emissions are due to the combustion of fossil fuels or the combustion of renewable energy carriers (See Chapter 13).

Practical experiences

Table 6 summarizes the emission data of waste incinerating plants at three German recovered paper processing mills. Plant A is an inclined rotary grate furnace in which a mixture of fiber residues and biological sludge is fuel. The fuels of plants B and C include deinking sludges with a small amount of biological sludges. Plant C also uses rejects from a recovered paper processing plant. Plants B and C are stationary fluidized bed boilers.

The three plants have different flue gas purification systems. The flue gas purification system of plant A is a fabric filter for particle removal with wet scrubbers in two stages. In plant B, particle removal uses an electrostatic precipitator. Acid gases and sulfur dioxide are retained by wet scrubbers. Plant C applies fabric filters for particle removal and the injection of lime into the flue gas stream to neutralize acid compounds.

The flue gas purification equipment of the three plants includes no additional measures such as activated carbon injection in the flue gas or activated carbon filters to control dioxin and furan emissions. Additional control of dioxins and furans is unnecessary. If sludge incineration occurs properly with sufficient retention time, uniform and complete combustion, and efficient solid particle removal, flue gases with very low dioxin and furan concentrations will result.

Table 6. Emissions of in-mill incineration plants in Germany burning waste of recovered paper processing compared with legal standards.

Parameter		Unit	Plant A	Plant B	Plant C	Legal standards
Particles		mg/m^3	8.2	6.0	6.4	10
Carbon monoxide	CO	mg/m^3	< 5.0	8.0	8.5	50
Total carbon	C	mg/m^3	< 1.0	0.3	0.9	10
Hydrogen chloride	HCl	mg/m^3	1.7	0.5	2.8	10
Hydrogen fluoride	HF	mg/m^3	0.03	0.1	< 0.1	1.0
Sulfur dioxide	SO_2	mg/m^3	8.0	15	6.1	50
Nitrogen oxides	NO_x	mg/m^3	190	185	190	200
Cadmium + Thallium	Cd + Tl	mg/m^3	< 0.0002	0.002	0.01	0.05
Mercury	Hg		< 0.010	0.02	0.02	0.05
Other heavy metals	As, Co, Ni, Sb, Pb, Cr, Cu, Mn, V, Sn	mg/m^3	0.022	0.06	0.01	0.5
Dioxins/furans	PCDD/PCDF	ng I-TE/m^3	0.002	0.03	0.01	0.1

Table 6 compares the flue gas emissions of the incineration plants discussed with the standards of the 17th Regulation of the German Federal Emission Control Act that applies to plants that incinerate waste and similar matter. For sludge incineration in recovered paper processing mills, the following conclusions are possible:

- Emissions of solid particles, sulfur dioxide, hydrogen chloride, hydrogen fluoride, and carbon monoxide are low compared with the standards. Due to the high ash content of deinking sludge that is rich in calcium carbonate, this sludge should not release all sulfur as sulfur dioxide. State-of-the-art flue gas purification guarantees that no significant environmental impact arises.
- The emission of heavy metals is very low compared with the standards especially for emission of cadmium and mercury.
- The emission standard of nitrogen oxides can usually be met by improved combustion control (temperature and oxygen level). Sometimes, the application of specific measures to remove NO_x could be necessary.
- The stringent emission standard of dioxins and furans (0.1 ng I-TE/m^3) can be met without any additional measures in flue gas purification.

The latter reference has special interest for incineration plant B. In its fluidized bed, shredded rejects of recovered paper processing were used as additional fuel. The high plastic content of this material appears to increase dioxin and furan emissions but not above the specified limits.

2.3.6 Ash composition and ash use

Due to the different filler, coating pigment, residual chemical, and contaminant contents, ashes produced by the combustion of deinking sludges and rejects have no consistent composition. The ash properties vary from paper mill to paper mill depending on the fuel mixtures. Figure 19 shows the contents of different oxides present in the ashes that have special interest for the construction material industry. The oxide composition indicates that the combustion ashes are often suitable for use in cement production. Other areas of use for the ashes are road construction and production of sand-lime bricks and concrete. These are also the processes applied to use the ashes from the aforementioned combustion plants. The grate ash of plant A finds use in road construction. The ashes of plants B and C are used in cement production.

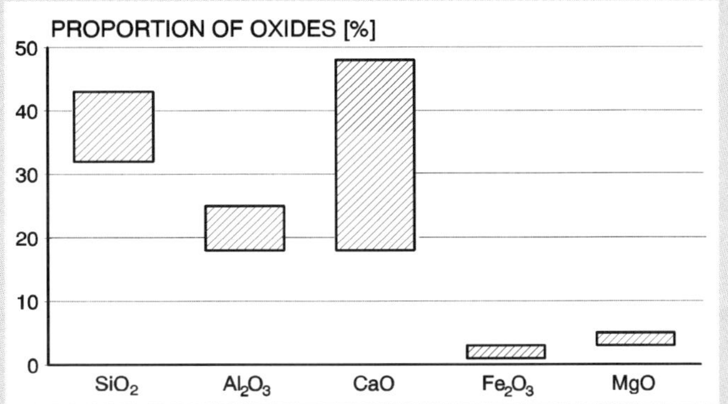

Figure 19. Oxides in ashes of deinking sludge combustion[28].

Table 7 compares the contents of different substances in the ashes of plants B and C with the contents of these substances in ashes from municipal waste incineration plants. The comparison shows that the ashes from the deinking sludge combustion are considerably less harmful than ashes from household waste incineration and are therefore suitable for use as secondary raw material.

Table 7. Composition of ashes from deinking sludge and municipal waste incineration.

Parameter	Unit	Ashes from deinking sludge incineration		Ashes from municipal waste incineration[32]	
		Plant B	Plant C	Ranges before 1990	Average after 1990
Aluminium oxide	%	20–25	20–27	18–36	19
Calcium oxide	%	15–25	20–30	4–20	14
Silicon dioxide	%	43–47	40–45	30–70	50
Chlorine	%	0.1–0.3	0.1–0.3	0.03–0.6	0.3
Sulfur	%	0.2–0.4	0.3–0.5	0.2–0.4	0.3
Lead	mg/kg	60–100	120–270	600–5 200	1 600
Cadmium	mg/kg	1.8–4.9	1.9–9.4	0.1–82	10
Chromium	mg/kg	90–100	75–130	100–9 600	500
Copper	mg/kg	360–390	360–490	200–7 000	2 200
Nickel	mg/kg	50–70	40–100	40–760	150
Mercury	mg/kg	0.05–0.4	0.1–0.4	0.1–20	0.6
Zinc	mg/kg	400–600	470–890	500–21 000	4 800
Dioxins/furans	ng I-TE/kg	n.a.	1.0–4.3	0.1	0.6

n.a.: not analyzed

2.3.7 Co-firing of sludges and rejects

The following plants may be considered for co-firing sludges and rejects from recovered paper processing mills:

- Hard coal power plants
- Lignite power plants
- Cement rotary kilns
- Blast furnaces
- Incineration plants for municipal sewage sludge
- Municipal waste incineration plants.

When using external incineration facilities, implementing the characteristics of the waste resembling those of the fuels incinerated becomes technically easier. Rejects from the pulping, screening, and cleaning stages of recovered paper processing are less suitable than deinking sludge and biological sludge due to their nonhomogeneity and different material composition, but their heating value is considerably higher.

For co-firing sludge in power plants operated with solid fuels, mixing mechanically dewatered sludges into the fuels is usually sufficient. The mixing can use a spreading device or the existing fuel feed systems. For coal dust fired power plants, the sludge must be pre-dried to about 90% dry content and added to the coal before grinding.

With co-firing deinking sludges and biological sludges in hard coal and lignite furnaces, the altered emissions in Table 8 generally result. Accordingly, the contents of sulfur dioxide, nitrogen oxides, hydrogen chloride, and hydrogen fluoride in the untreated exhaust gases of the incineration plants are lower. Especially for co-firing deinking sludges, the high alkaline content effectively suppresses the SO_2, HCl, and HF emissions. The cadmium and mercury contents of the untreated exhaust gas are also significantly lower since pollution of the deinking sludges with these heavy metals is low compared with hard coal and lignite.

Table 8. Expected changes of flue gas emissions when co-firing paper mill sludges in hard coal and lignite combustion plants[4].

Emission parameter		Expected changes
Particles		⇔
Sulfur dioxide	SO_2	⇓
Nitrogen oxides	NO_x	⇔
Hydrogen chloride	HCl	⇔ ⇓
Hydrogen fluoride	HF	⇔ ⇓
Carbon monoxide	CO	⇔
Total carbon	C	⇔
Cadmium	Cd	⇓
Mercury	Hg	⇓
Other heavy metals		⇔

⇔: no change; ⇓: reduction

In Germany, the annual capacities of coal-fired power plants are about 45 million tons for hard coal and about 95 million tons for lignite. Using the entire annual sludge output of Germany's deinking mills, totaling about 1 million tons as received, in these external power plants by mixing it with the fuel would be possible.

To date in Germany, only a limited amount of rejects and sludges is co-fired in the cement industry. With an annual production of about 33 million tons cement clinker at 65 locations and a total of about 100 cement rotary kilns, this also offers a considerably greater potential for the use of waste from recovered paper processing mills.

Until now, municipal waste incineration plants (52 units) in Germany have fired rejects from recovered paper processing but not sludges from deinking and clarification. A disadvantage of this is that waste incineration plants are less energy-efficient than in-mill waste incineration plants if co-generation is not practiced.

2.4 Composting and agricultural use

Composting is the oldest and most natural form of waste use. Only in the last ten years has it gained importance as a way to use organic waste in commercial composting plants. During this period as a consequence of an intensified use of separately collected organic household wastes and yard and park wastes, the composting technique has made major advances. With modern processes, annual throughputs of 30 000–70 000 t are possible at composting plants. The maximum throughputs attained by some composting processes even reach 150 000 t per year[33].

During composting, organic matter changes into a relatively stable decomposition product, the compost, by action of various microorganisms in an aerobic environment as Fig. 20 shows. Some substrates are mineralized to generate carbon dioxide that often results in a mass reduction. The remaining carbon is transformed into new cell material and humus-rich compost. The microbiological processes occur under the influence of biologically self-generated heat. This self-heating leads to evaporation of water and therefore to an additional mass reduction of the raw material. The average mass reduction in composting related to the organic proportion of the raw material is about 50%.

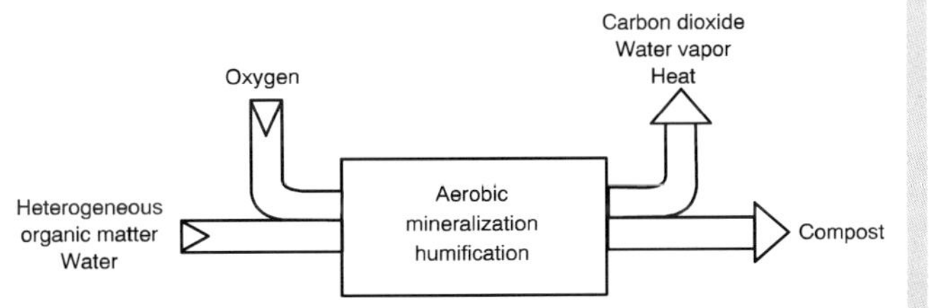

Figure 20. Simplified scheme of biological degradation during composting[11].

2.4.1 Composting conditions

Composting is a complex biological process involving many microorganism populations. Temperatures at which they develop their optimal microbial activity are the following:

- Mesophilic microorganisms (25°C–45°C)
- Thermophilic microorganisms (45°C–70°C).

Recording the temperature development can establish which microorganism groups in the biological reaction chain dominate at a particular stage in the composting process. Due to microbial activity, temperatures of 70°C–75°C are possible. A longer exposure to temperatures of 55°C–65°C is necessary to eliminate pathogens, parasites, vermin, and weed seeds.

To achieve optimum composting, various requirements related to the composting raw materials and processing are necessary:

- Ratio of carbon to nitrogen (C/N ratio)
- Microorganisms primarily need carbon (C) and nitrogen (N) as nutrients
- The optimal C/N ratio is 20:1–35:1. If less carbon is present, ammonia is released. This causes odor problems and nitrogen releases. At higher C/N ratios, the nitrogen content has a limiting effect because the decomposition of organic substances is slower.

Water content

Microorganisms can take up nutrients only from an aqueous solution. This means that sufficient water must always be present during the composting process. A water content of 40%–70% of the compost feedstock is optimum. The composting process will stop with a water content less than 30%. An excess of water displaces air from the intermediate spaces and pores of the raw compost materials and leads to an insufficient supply of oxygen or anaerobic conditions. The consequences are anaerobic decomposition processes that cause odor problems.

Structure of the compost feedstock

Dense structures and low porosity of the substrates used for composting prevent sufficient aeration. Highly structured raw compost materials such as wood residues, bark, or recovered paper are useful to improve the structure.

pH

For optimum composting, the pH must be 7–9. The pH decreases in the initial stage of the decomposition process due to formation of fatty acids, the production of CO_2, and nitrification. It increases again after bacterial populations change.

Temperature

The temperature during composting should not exceed 65°C. Above this temperature, thermophilic microorganisms are eliminated. The final temperature depends on the composting process.

Aeration

Aeration provides the oxygen supply for the microorganisms and removes carbon dioxide, vapor, and heat. At high temperatures, aeration also facilitates cooling. For this reason, about ten times more air is used than is necessary for oxygen supply.

Harmful substance content

The content of harmful substances in the raw compost materials must be low to ensure that the composting process is not hindered and that a nontoxic, marketable compost can be produced.

2.4.2 Composting processes

In commercial composting plants, the process described in Fig. 21 occurs. The three most important work phases are as follows:

- Pre-processing and mixing of compost feedstock
- Decomposition of the processed raw materials to fresh compost (initial biodegradation)
- Decomposition of fresh compost to finished compost (maturation).

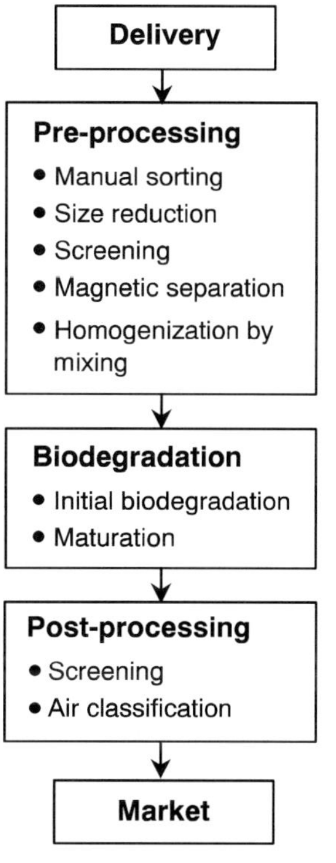

Figure 21. Process steps in commercial composting plants.

The differences among the types of present day composting processes relate primarily to the initial biodegradation systems. Initial biodegradation constitutes the first phase of decomposition of the organic substances. To prevent exhaust air emission problems and achieve a complete hygienic conditioning of the compost feedstock, different degrading systems are possible. These differ primarily in the initial biodegradation aggregate, the aeration system, and the transfer technique. In all processes, maturation occurs in a windrow that is mechanically turned or force-aerated to ensure the necessary oxygen supply.

The initial biodegradation techniques fall into the six groups of Fig. 22:

- Tower composting: The compost feedstock moves through the tower from top to bottom and is usually force-aerated. The retention time in the tower is 6–9 weeks.
- Drum composting: The compost feedstock enters force-aerated drums and exits at the other end. Constant rotation of the drum ensures good mixing and size reduction. Depending on the process concerned, the treatment in the drum lasts for 1–7 days. Maturation is done exclusively in windrows.

- Box/in-vessel composting: The compost feedstock enters a closed, force-aerated container with complete exhaust air and leakage water or condensate recording. The initial biodegradation takes 1–2 weeks. Maturation is carried out in windrows.
- Capsulated windrow composting: The compost feedstock enters windrows in an enclosed building and is force-aerated. Turning and loosening of the compost feedstock uses tractors or specialized windrow turning machines. Maturation is usually performed also in the enclosed building. The entire composting process lasts about 10–12 weeks.
- Tunnel composting: The compost feedstock enters an open row or a closed tunnel separated by fixed walls. Each row or tunnel is aerated and turned separately. Depending on the desired degree of degradation, composting lasts between 4–12 weeks.
- Briquette composting: The compost feedstock is pressed to form briquettes stored on top of each other on pallets in a compost hall. The pressing process displaces water in the briquettes from the inside to the outside. Simultaneously, the aerobic biodegradation begins. The entire composting process lasts about 3–4 weeks. The briquettes are ground in a grinder, and the resulting product is processed to finished compost on windrows.

In Germany, strict environmental regulations apply to composting plants. The laws primarily concern leakage and malodorous emissions. The COD concentration in leakage water can be 50 000–100 000 mg/L. Covering the windrows and installing spe-

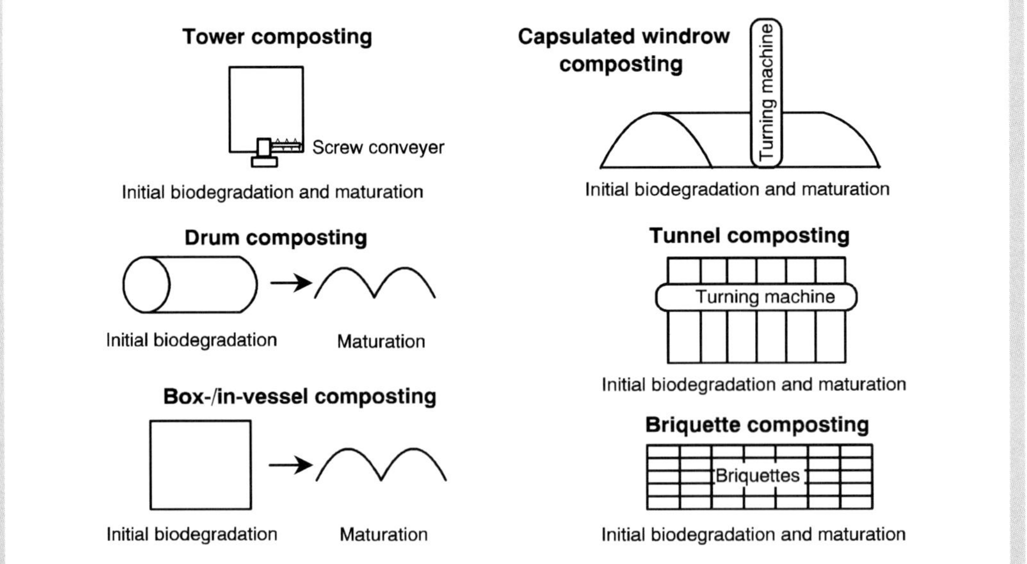

Figure 22. Simplified schemes of different composting techniques[33].

cial leakage loops or a separate collecting and treatment are suitable methods of preventing contamination of ground water. Malodorous emissions from anaerobic degradation processes can be prevented by suitable construction of the windrows, frequent turning, and more intensive aeration. Composting in closed vessel systems offers the advantage that the exhaust air is drawn off and fed to a deodorizing process. Exhaust air purification in compost filters (biofilters) has proven its value in practice.

2.4.3 Compost applications

The main areas of application of compost in agriculture and plant growing are the following:

- Farming
- Fruit growing
- Vine growing
- Vegetable gardening
- Ornamental plant growing
- Tree nurseries
- Forestry nurseries
- Mushroom cultivation
- Gardening, cemetery, and landscape cultivation.

Special applications such as the following also exist:

- Recultivation
- Revegetation of landfills
- Noise prevention walls
- Material for biofilters.

The fertilizing effect of compost is small. For this reason, compost has use primarily as a soil conditioner. Heavy, loamy soils are loosened to ensure better aeration. Sandy soils achieve a higher water retaining ability and have better protection against surface erosion. As humus, composts help limit the use of peat and therefore contribute to the protection of marshy biotopes. An especially important use of compost is for landscaping such as recultivating waste land resulting from lignite and potassium salt mining. Included among such recultivation measures is the revegetation of waste dumps.

A condition for the marketability of composts is a consistently high composting quality. Important criteria for the quality of compost are soil conditioning properties, guaranteed hygienic safety, and a low content of contaminants. Many countries have specifications concerning the heavy metal content of composts. A comparison of the guidelines and limit values is unsatisfactory because no uniform reference value exists for the allowable maximum contents of lead, cadmium, chromium, copper, nickel, mercury, and zinc. In some countries, the values relate to the dry substance content of the compost. Other countries relate the heavy metal content to a 30% organic substance content. The enrichment of heavy metals during composting caused by the mass reduction is theoretically compensated by this standardization. This provides the possibility of a direct comparison of the heavy metal content of compost raw materials and the composts they produce.

Besides heavy metals, some regulations also limit the content of harmful organic substances such as PCB, dioxin, furan, and lindane. Other stipulations concern the content of organic substance, water content, C/N ratio, pH, nutrient content, decomposition grade, guaranteed hygiene, compatibility with plants, and freedom from germinating seeds, stones, and foreign matter.

2.4.4 Composting of rejects and sludges

Nearly all types of paper industry waste are suitable for composting. These include fiber-containing sludges, deinking sludges, bark, wood residues, and biological sludges from effluent treatment plants. Rejects from recovered paper pulping and screening operations are not suitable or suitable only to a limited degree. With this type of waste, the content of plastic and other nonpaper components such as glass or stones has a detrimental influence.

In all cases, composting of sludges from recovered paper processing requires using additives. With the exception of biological sludges, the sludges have an unfavorable C/N ratio for microbial decomposition. They also have a dense structure that is unfavorable for composting. This requires addition of structure-improving materials that are ideally also nitrogen carriers. Biological waste from households, garden waste, cuttings from trees and plants, straw, bark, and waste from animal husbandry are suitable as such components.

The composting of fiber sludges and biological sludges usually with bark has been a practice on an industrial scale for a long time. In the 1980s, Mick *et al.*[34], Campbell *et al.*[35], Carter[36], and Smyser [37] published articles on the composting of fibrous sludge and biological sludge with bark and slaughterhouse waste. Campbell *et al.* investigated the possibility of adding urea to fiber sludges.

Systematic tests on a laboratory scale showed that compost corresponding to the quality specifications currently in force can also be produced with deinking sludges and the biowaste from households and yard and park wastes[38]. The addition of deinking sludges generally has a positive influence on the organic substance content, the salt content, and water retaining ability of the compost. The leakage during composting is less. Limiting factors relating to the composition of the compost feedstock could possibly be the copper and zinc contents of the deinking sludges. Printing inks using blue pigment based on copper phthalocyano compounds are primarily responsible for the higher copper contents. Higher zinc contents are due to several different sources such as metallic inks, coating components, galvanized wire stitching, or office staples. An important point is the previously mentioned item that the heavy metal content of deinking sludges is less than the heavy metal content of municipal sewage sludges. This is true also for copper and zinc.

When composted with biowaste from households, the proportion of deinking sludge can be up to 30% without exceeding permissible limits for heavy metal content of the finished compost. For cadmium, chromium, nickel, and mercury, the addition of deinking sludges to biowaste even contributes to a reduction of the content of these heavy metals in the finished compost.

The limitations to which composting of deinking sludges with biowaste from households would be subject can be shown by two case studies that use an annual occurrence of 10 000 and 100 000 t deinking sludge with a 50% dry content. The proportion of deinking sludge in the raw compost mix in relation to the dry substance is assumed to be 30% by weight. This means annual requirements for biowaste of about 24 000 or 240 000 t of fresh substance, respectively. With an average annual biowaste and garden waste amount of about 120 kg fresh substance per inhabitant, the composting plants would require separately collected biowaste of about 200 000 and 2 000 000 inhabitants, respectively. This theoretical example shows that composting of deinking sludges with biowaste from households makes sense only for smaller to medium-sized paper mills.

2.4.5 Agricultural use

The paper industry has a long tradition of using sludges in agriculture. This is true especially for virgin fiber sludges and for biological sludges since the first biological effluent treatment plants started operation. In Germany, the use of paper industry sludges comes under the scientific supervision of agricultural research organizations. Using the evaluation of all available research results, these organizations consider the agricultural use of these sludges to be positive.

With the exception of sludges from biological effluent treatment plants, sludges from the paper industry have a high carbon/nitrogen (C/N) ratio and therefore contain only a small proportion of nitrogen. For this reason, they have only a limited fertilizing effect. The advantage of their use in agriculture considers their soil enhancing properties. They not only contribute to covering its requirements of humus forming organic substance but also improve the aeration and cultivation of the soil, increase the water retention capacity, and prevent erosion. The two latter criteria have significance especially for use of fiber sludges from the paper industry for vineyards located at steep slopes. Other positive effects are the contribution to the supply of trace nutrients, the increased availability to plants of poorly accessible phosphates, and the enhanced activity of living organisms in the soil.

Use of deinking sludges in agriculture is still controversial. In North America, no objections exist concerning their use as a soil improving material provided the material does not exceed defined contaminant concentrations or harmful substance loads. The situation in Germany is different. In general, the contaminant concentrations of deinking sludges are well below the valid threshold limits for municipal sewage sludge defined by legislation governing sewage sludge. The German Environmental Protection Agency still considers the agricultural use as unjustifiable for reasons of soil protection including the irreversibility of numerous soil contaminants. The agency claims the ecological risk potential of deinking sludges is not sufficiently known. Pollution of the soil with persistent organic substances for which no analytical detection processes may exist is feared.

Due to this restrictive attitude of the German agency, a general prohibition on the application of deinking sludges to soil used for cultivating plants has been issued in several German federal states. In the Biowaste Ordinance issued in 1998, the use of deinking sludges is also not allowable on soil intended for agricultural, forestry, or gardening purposes.

Experts at the German agency were not persuaded by the positive aspects of the direct agricultural use of deinking sludges described primarily in North American publications[39–41]. Diehn[42] established an improvement in the soil characteristics and increased grain yield following the use of deinking sludges. Pridham and Yau[43] and Bailey *et al.*[44] reported positive influences on harvests following the use of mixtures of deinking sludges and biological sludges. The authors ruled out any risk to the ground water. Pridham and Yau testified to the suitability of deinking sludges for replanting waste land and former industrial sites with soils having no organic substance. Trepanier *et al.*[45] confirmed that additional nitrogen is necessary to avoid nitrogen immobilization associated with deinking sludge application on silty loam soil.

2.5 Use in other industries

Rejects, deinking sludges, and combustion ashes of recovered paper processing mills are used in many branches of industry for material and energy production purposes. The material use possibilities for deinking sludges and combustion ashes depend primarily on the composition of the inorganic components. The inorganic part of the deinking sludges consists mainly of calcium carbonate and clay. In the combustion ashes of deinking sludges, primarily calcium oxide and sintered clay are present. The aforementioned inorganic components also have use as raw materials in the construction industry. The primary use possibilities are therefore the following:

- Cement production
- Brick manufacturing
- Concrete production
- Mortar and sand lime brick production
- Road construction.

2.5.1 Cement production

Cement is a binding agent consisting essentially of compounds of calcium oxide with silicon dioxide, aluminum oxide, and iron oxide that form from components in the raw materials by burning at high temperatures. The production of cement includes pre-processing and burning of the raw material mixture to cement clinker, the processing of additives, and the blending of the cooled clinker with the additives as Fig. 23 shows.

A basic distinction occurs in cement production between raw materials, fuels, and additives for fine-tuning. The raw materials are either primary or secondary raw materials. The type and quantity of primary raw materials result from the geology and chemical composition of the deposit concerned and therefore vary among plants. The raw material mixture or raw meal is analytically controlled. The raw meal composition follows

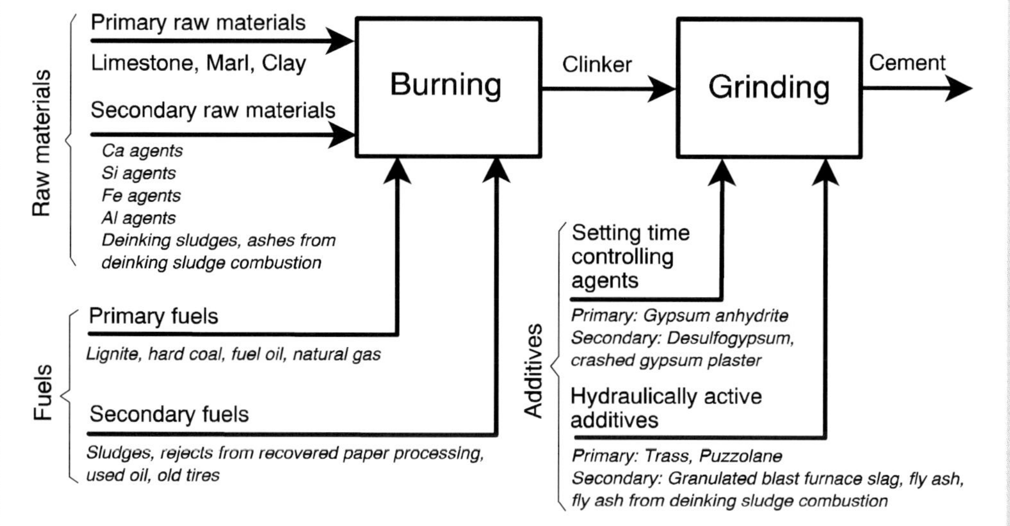

Figure 23. Mass flow in cement production[46].

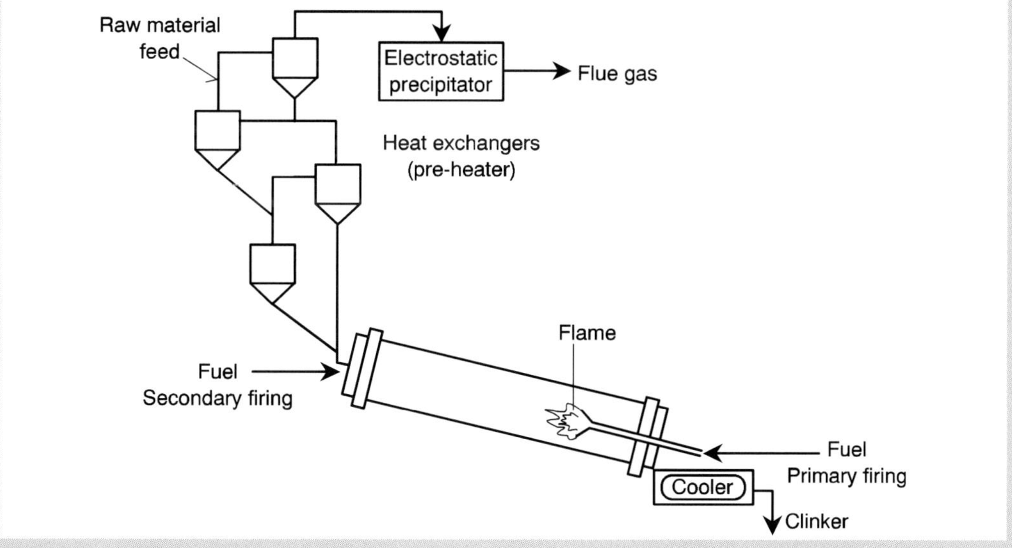

Figure 24. Cement kiln dry process.

given optimized target values. For this reason, Ca, Si, Fe, and Al agents must be added as secondary raw materials. Deinking sludges and ashes from energy recovery are also secondary raw materials.

With fuels, a distinction is made between primary fuels such as lignite, hard coal, fuel oil, and natural gas and secondary fuels. Besides old tires, used oils, halogen-free organic solvents, and petroleum coke, rejects and dewatered sludges from recovered paper processing mills are also used as secondary fuels. The secondary fuels are fed at the primary firing stage (feeding of liquid, gaseous, or solid fuels via firing pipes) or as is usually the case with rejects of recovered paper processing at the secondary firing stage at the entrance of the rotary kiln as Fig. 24 shows.

Fine-tuning materials set the properties of the clinker to the desired cement grade. In the production of Portland cement, up to 20% by volume of granulated blast furnace slag, trash, and fly ash are comminuted with the cement clinker as hydraulical additives. The fly ash from the fluidized bed combustion of deinking sludges is also suitable for this purpose. Figure 25 shows the ternary diagram Al_2O_3, SiO_2, and CaO for different cement clinker and ash types. For these oxide components, the composition of deinking sludge ash very closely resembles that of Portland cement clinker. Considering this, the small use of dewatered deinking sludges possibly mixed with biological sludges without preceding combustion is a type of energy and material use. This uses not only the heating value from the organic content but also the mineral components by integrating them into the finished cement product [47,48]. Deinking sludge alone or mixed with biological sludge does not have general use for industrial cement kilns.

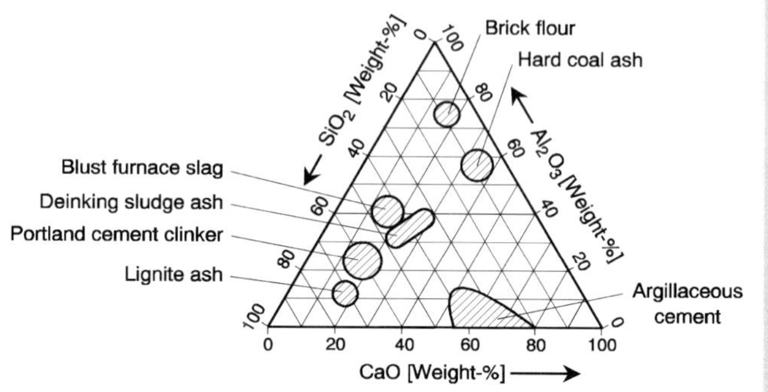

Figure 25. Ternary diagram Al_2O_3, SiO_2 and CaO for different cement clinker and ash types[4].

The use of rejects from recovered paper processing as secondary fuel presupposes treatment of this waste. This consists primarily of pre-comminution, screening, and post-comminution to produce a large surface of the reject particles. The screening can be combined with air separation and magnetic separation of ferrous metals to eliminate impurities that could damage the post-comminution aggregates.

Conditions in the rotary kiln during the clinker burning process with gas temperatures of more than 2 000°C and material temperatures of about 1 500°C guarantee that even very stable organic substances that are harmful such as dioxins, furans, and polychlorinated biphenyls are effectively destroyed to the extent that they enter with the fuels[49]. The temperature conditions in the heat exchanger ensure that a resynthesis of dioxins and furans does not occur. Chlorine from organic compounds reacts in the rotary kiln with the components of the raw meal to give alkali chlorides. These are highly volatile so their integration into the cement clinker is negligible.

The alkali chlorides enter the heat exchanger with the exhaust gas of the rotary kiln where they condense and return with the raw material flow into the rotary kiln. To prevent an enrichment of chlorides in the clinker, a partial flow of the kiln exhaust air can be separated via a bypass and the chlorides are removed with the dust in filters.

2.5.2 Brick production

In brick works, fiber-containing waste from the paper industry promotes porosity. The porosity of the bricks baked from clay and loam improves their heat insulating properties. When burning the bricks, the porous causing materials also burn and leave behind a large number of small pores that provide good heat insulation. Besides sludges from the processing of virgin fibers, increased use has been made in recent years of deinking sludges. As alternative porosity agents, sawdust and polystyrene pellets are possibilities.

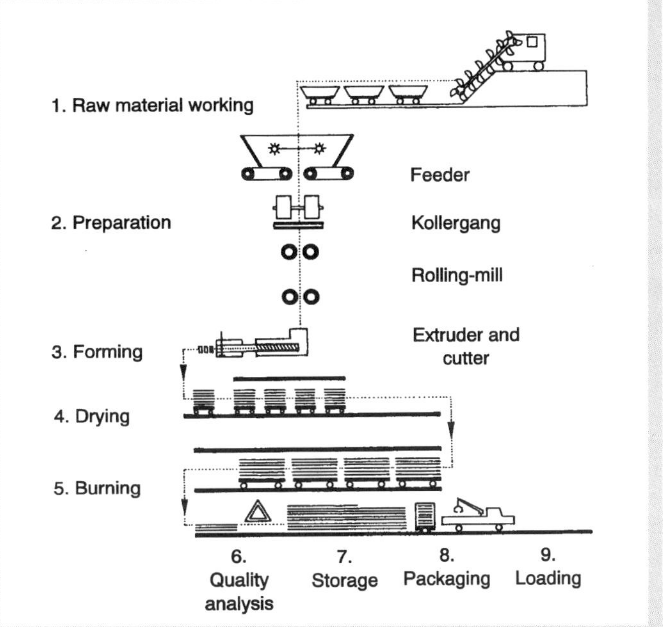

Figure 26. Brick production process.

Figure 26 shows brick production schematically. After mixing and crushing clay and loam with the porosity causing materials, the bricks are formed in molding machines. Upon emerging from the machine, the molded product is cut into the desired brick shape. The bricks then have a water content of about 50% and are called raw bricks.

The moist raw bricks are usually slowly dried in drying chambers at temperatures up to 100°C to a moisture content of 2%. The fibers contained in the paper industry sludges have a reinforcing effect and significantly lower the risk of tearing during the drying process. The finely grained clay and calcium carbonate contained in deinking sludges also improve the rheology of the raw brick material. This has a positive effect at the time of extrusion of the raw bricks. To this extent, deinking sludges provide a porosity causing material and influence positively the technical brick production process. The calcium carbonate share of the deinking sludges is especially advantageous with use of low lime or fatty clays. Calcium carbonate binds sulfur and fluorine from the clay to reduce the sulfur and fluorine emissions during burning of bricks. A proportion of 30% by volume of calcium carbonate in the raw brick material is the maximum permissible limit.

Burning of bricks uses a continuous tunnel furnace heated with oil, gas, or coal at about 900°C. In these burning furnaces, the bricks are transported in a counter stream to the air supply and pass through a pre-heating zone, a burning zone, and a cooling zone. During the pre-heating process to 150°C– 450°C, the organic components of the porosity causing materials smould. In the burning zone, the bricks sinter and develop their ceramic properties, compressive strength, and resistance to atmospheric influence. The heating value of the porosity causing materials lowers the primary energy requirement for the tunnel furnace operation. Where large amounts of porosity causing materials are present, an excess supply of energy occurs that is difficult to control. This causes lower quality. In principle, the mineral content of sludges from the paper industry has a strengthening influence. The minerals at least partially compensate for the loss of strength that occurs with increasing porosity. The bricks leave the tunnel furnace after the cooling zone. The recovered heat serves to pre-heat the air for drying.

The low temperature carbonization gases that form during brick warming are either directed into the burning zone of the tunnel furnace for after-burning or are almost completely incinerated in separate after-burning plants with heat exchangers. In all brick works, the remaining residue gases are removed via high chimney stacks. Relevant harmful substances that are emitted are as follows:

- Sulfur and fluorine compounds from clay and loam
- Benzene, acetic aldehyde, and formaldehyde from the porosity causing materials.

Of special importance is the benzene emission since this is a carcinogen and occurs primarily when polystyrene is a porosity causing material. The presence of benzene in the exhaust air has required emission-reducing measures at all brick works.

While producing similar compressive strengths of bricks, up to 20% by volume of sludges from the paper industry, 10% by volume polystyrene, and 5% by volume sawdust are used as porosity causing material. When using sludges from the paper industry, the sludge must not be biologically decomposed during temporary storage at the brick works. The resulting malodorous emissions would not be acceptable by personnel at the brick works or by people living nearby especially during summer. These problems can be significantly reduced by setting a higher pH by the addition of lime or avoiding temporary storage. As the most effective measure, a strict separation at the paper mills of fibrous sludge and sludge from biological effluent treatment plants should occur.

2.6 Landfilling

Landfilling is the most commonly applied method worldwide for the final disposal of wastes. This situation will not change in many more remote regions of the world in the medium term[49]. In contrast, landfill capacities have already become very limited in densely populated countries. It is becoming increasingly difficult to obtain authorization to open new landfill sites. Further development of landfilling techniques has dramatically increased costs for construction and operation of landfills. With the current technical state-of-the-art, operating a landfill site such that any annoyance to the population living nearby or environmental risk due to leakage, odors, fire and explosion risk can be largely eliminated[50].

Modern landfills have sealing systems with leakage and landfill gas collection systems. Any leakage is treated in physical, chemical, or biological effluent treatment plants. The landfill gas is either fired or after purification used in combustion facilities or natural gas engines for energy production.

Even after a landfill has closed, a potential environmental risk from harmful substances in the leakage or gases that are emitted exists for decades. Constant monitoring is necessary to ensure that even after long periods the behavior of a landfill is not negatively influenced by damage to the basic liner due to settling of the landfill.

2.6.1 Types of landfills

A distinction can be made between the following types of landfills:

- Compaction landfill
- Compost landfill
- Bale fill landfill
- Inert and monofill landfill
- Hazardous waste landfill.

Compaction landfills are usually municipal waste dumps in which the waste is piled in layers up to 2 m before being compacted. The objective is to achieve the highest level of compaction with minimal use of equipment. With compost and bale fill operation, pre-processing of the waste is necessary. In a compost operation, the waste is first crushed and possibly mixed with sludge from sewage treatment plants. The processed waste is piled in loose layers up to 2 m and subjected to a decomposing process lasting several weeks. The decomposed material is then taken with the standard compaction methods to the landfill site or after further composting used for the intermediate cover or replanting of the landfill.

Bale fill landfills operate in several countries such as Japan. The municipal waste after separating recyclable material such as paper, glass, and metals is compressed to bales. At the landfill, these bales are stacked by fork lifts. The bale fill landfills are dust free and can be operated with less time, equipment, and labor investment than is necessary for the disposal of unbaled waste. Considerable compaction problems result due to the irregular bale dimensions. These can be managed only by use of large amounts of covering material.

A single type of waste is stored in monofill landfills. This is usually industrial waste produced in large amounts and for which no economical re-use such as building waste, plastics, or sewage sludge and industrial sludges exists. The type, harmful substance content, and reaction behavior of the waste must be similar and compatible.

Waste that is largely mineralized is stored in inert landfills. This can be excavated earth or ashes from incineration of waste. The working and washing processes in such landfills are considerably smaller.

Hazardous waste landfills are characterized by the toxicity of the materials (e.g. halogenated organic matters, lacquer and paint sludges) to be disposed. Special requirements are made on the landfill design and operating parameters of these landfills.

2.6.2 Landfill gas generation

Waste stored in an environmentally safe landfill is protected from light and oxygen. After the existing oxygen in the air contained in the waste has been used, metabolic processes occur under anaerobic conditions. The waste is first metabolized to acids, alcohols, and other organic compounds by various bacterial populations. The decomposition products of this first step are then converted by another group of bacteria, the acidogenic bacteria, to acetic acids, carbon dioxide, and water. These substances are then biologically converted to methane. The landfill gas generation rates and the composition of the gases are determined in each case by the prevailing conditions in the gas leakage.

Figure 27 uses the example of the decomposition of household waste to show the development of landfill gas composition. The initially high CO_2 production resulting from the aerobic processes declines in favor of the production of methane. For a short time in a transitional phase, hydrogen occurs as a decomposition product. After this unstable initial phase, the methane (CH_4) and CO_2 concentrations in the landfill gas remain constant for years. The CH_4:CO_2 ratio then is 1.2–1.5. Using current knowledge, methane production starts intensively within 0.5–1 year after disposal begins and continues over a long period. In the case of older landfills, gas emissions were recorded after 75 years.

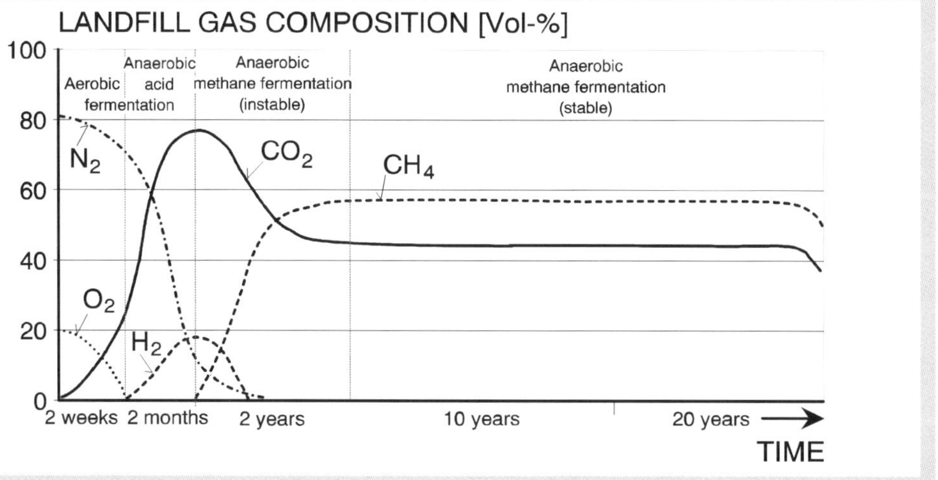

Figure 27. Composition of landfill gas during degradation of municipal waste[15].

Methane emissions from landfills have become increasingly important considering the current discussion of the contribution of different atmospheric trace gases to the greenhouse effect. Methane has a potential greenhouse effect greater than that of CO_2 by a factor of 22. Despite a low methane concentration of the atmosphere of about 2 ppm, its share in the greenhouse effect is 13%. Methane emerges from marshy areas, wet rice cultivation, cattle stomachs, the mining and transport of natural gas and coal, and to a considerable degree from landfills.

For this reason, all future efforts toward reduction and limitation of carbon dioxide and other greenhouse gas emissions should not fail to consider landfills. Environmentally unsafe landfilling as a simple and low-cost disposal of waste such as is still practiced primarily in developing countries must give way to a waste management that will include landfills but ideally only for inert materials with a low organic substance content.

2.6.3 Leakage generation

During its passage through the landfill, precipitation water picks up water soluble and water miscible waste components and products of the chemical and biological decomposition processes. The quality of leakage and leakage composition depend on a variety of factors:

- Physical and chemical properties and water content of the stored waste
- Landfill conditions
- Climate conditions and the seasons
- Biochemical decomposition processes in the landfilling site
- Age of the landfill
- Amount of water penetrating the landfill from below.

Some factors depend to a large degree on local conditions such as climate, topography, nature of soil, and hydrogeology.

Leakage can be polluted to different degrees depending on the landfill age. At the acid fermentation phase, leakage has especially a high COD of above 20 000 mg/L. These fall to less then 5 000 mg/L in the phase of methane fermentation.

Analyses of leakage components reveal a large range of organic and inorganic materials. A more accurate prediction of the nature of the leakage in a landfill is not possible. Frequently, long-term analyses are done on different types of artificial landfills to provide information on the distribution of precipitation water in the waste, storage density, leakage composition, and biochemical and hygienic parameters.

Reduction of the amount of leakage can occur by various means but especially by surface cover on the landfill. During operation of a landfill, the use of the smallest possible exposed surface area, a permanent cover or daily intermediate covering of the point of insertion, and the reduction of the moisture content of the supplied waste by appropriate pre-treatment processes may be suitable measures.

2.6.4 Landfill sealing systems

To prevent leakage penetrating the ground below the landfill, a base liner sealing with leakage collection and treatment is essential for an environmentally safe landfill. Such a step is absolutely vital when a risk of ground water contamination exists and when the ground water is used as a drinking water supply. The basic liner systems should simultaneously guarantee rapid leakage removal from the storage area to a leakage treatment plant.

The requirements of base liner systems include the following:

- Impermeability against precipitation and runoff
- Ability to withstand a temperature up to at least 70°C
- Impermeability to landfill gas migration
- Stability in case of additional top loads
- Minor sensitivity to settling
- Resistance to microorganisms, rodents, and the effects of root action of plants
- Easy installation.

So-called composite liners are used in landfills for household and hazardous waste. They comprise layers of different types of sealing materials in direct contact with each other. Figure 28 shows a schematic representation of such a composite liner. A three-layered lining usually of clay is applied to the geological barrier. The clay layer is covered with a synthetic liner having a minimum thickness of 2.5 mm. High density polyethylene (HDPE) is very suitable for this purpose. The synthetic liner is covered for protection with a layer of sand or similar fine-grained material. Added to this is a drainage layer of gravel or crushed stones in which the drainage pipes for leakage collection are installed.

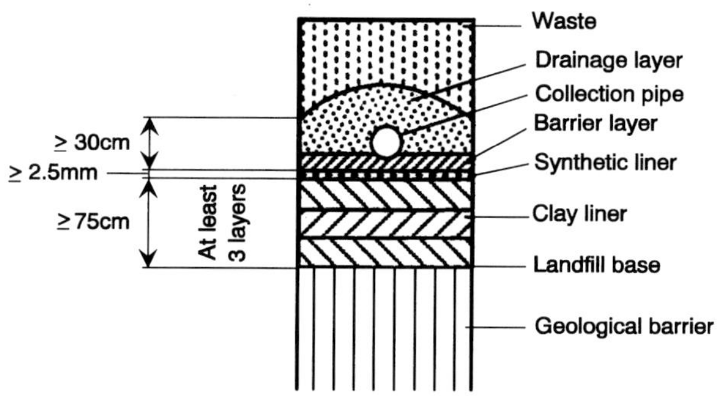

Figure 28. Scheme of a composite liner of landfill sites for municipal waste[15].

After the landfill has been filled, a surface cover is also necessary. Composite liners also have use for this purpose. Then, a final vegetative cover at least 1 m thick is added at the end. The use of deinking sludges as soil cover material is a common measure primarily in North America and in some European countries. The cultivation of this layer occurs in such a way that the synthetic liner remains protected from root and frost damage. Wind and water erosion is prevented, and penetration of precipitation water is minimal.

2.6.5 Landfilling of waste from recovered paper processing mills

Theoretically, the following possibilities exist for landfilling waste from recovered paper processing mills:

- Landfilling in monofill dumps for certain types of waste such as ash, sludge, and bark

- Works-owned landfills where all in-plant waste is stored
- Landfilling in public dumps where waste from other sources is also stored.

Depending on local circumstances and the types of waste concerned, pulp and paper mills sometimes use all three disposal methods.

Because the greatest share of the waste from recovered paper processing mills is not harmful, using municipal landfills is possible. This also applies for deinking sludges although they sometimes have a greater wet extraction potential for heavy metals. To minimize leakage and maintain landfill stability, an extensive dewatering of sludges and rejects is necessary. Dry solid contents should be above 40%. The landfill of ashes from reject and sludge incineration does not require any measures beyond those governing the dumping of household waste.

2.7 New developments

Besides extension of processes to use waste material described so far, further possibilities exist for the use of waste from recovered paper processing. These mainly concern fibrous sludges and deinking sludges. Only a small number of paper mills use these possibilities, or they have been limited so far to laboratory scale tests and pilot installations. New developments in the area of sludge use include:

- Wet oxidation processes
- Fermentation processes
- Hydrolysis processes
- Pyrolysis processes
- Production of cat litter and absorbent materials.

For ash use, filler recovery processes are under development that should permit the re-use of recovered fillers in the paper industry or other branches of industry.

2.7.1 Wet oxidation processes

Due to the thermal transformation processes of the mineral sludge components, the ashes resulting from use of sludges for energy production are not suitable for re-use as filler or pigment in paper production. For example, Table 9 shows the physical and chemical processes that occur in calcium carbonate and clay at the time of combustion of deinking sludges. Kaolinite loses its hydroxyl groups at 400°C–600°C. Metakaolinite is developed that is stable only up to 925°C. From 1 150°C, mullite and cristobalite develop via an intermediary spinel-type phase. Calcium carbonate decomposes at 700°C–920°C. When a mixture of calcium carbonate and clay is present, calcium silicate develops as a reaction of silicon dioxide and calcium oxide. These are therefore reactions that occur in part in the same temperature ranges. They can consequently run simultaneously in filler mixtures.

Table 9. Physical and chemical processes during thermal treatment of clay and calcium carbonate[55].

Temperature [°C]	Reaction		
400–600	$Al_2(OH)_4(Si_2O_5)$ Kaolinite	→	$Al_2O_3 \bullet 2\,SiO_2 + 2\,H_2O$ Metakaolinite
925	$Al_2O_3 \bullet 2\,SiO_2$	→	$g\text{-}Al_2O_3 \bullet 2\,SiO_2$ Spinel
1 100–1 150	$3\,Al_2O_3 \bullet 6\,SiO_2$	→	$3\,Al_2O_3 \bullet 2\,SiO_2 + 4\,SiO_2$ Mullite Cristobalite
700–920	$CaCO_3$ Calcium carbonate	→	$CaO + CO_2$ Calcium oxide
800–1 000	$CaO + SiO_2$	→	$CaSiO_3$ Calcium silicate

These transformation processes lead to an increased abrasion capacity of the combustion ashes. This eliminates their use as filler in paper production. Also unsatisfactory are the optical properties of the ashes. The ashes do have use as secondary raw material or additives in cement and concrete production, as construction material, or in road building.

Besides combustion, an extensive oxidation of organic material is also possible with wet oxidation processes. Generally speaking, wet oxidation is the flameless combustion (cold combustion) of organic compounds in an aqueous solution or suspension using atmospheric oxygen, pure oxygen, or other oxidation agents such as hydrogen peroxide, nitric acid, sodium hypochlorite, sodium peroxidisulfate, and ozone. If oxidation occurs under mild temperature conditions and normal pressure, it is a wet oxidation process. If higher temperatures of 120°C–330°C and pressures of about 3–300 bar are applied, these are thermal wet oxidation processes as Fig. 29 shows.

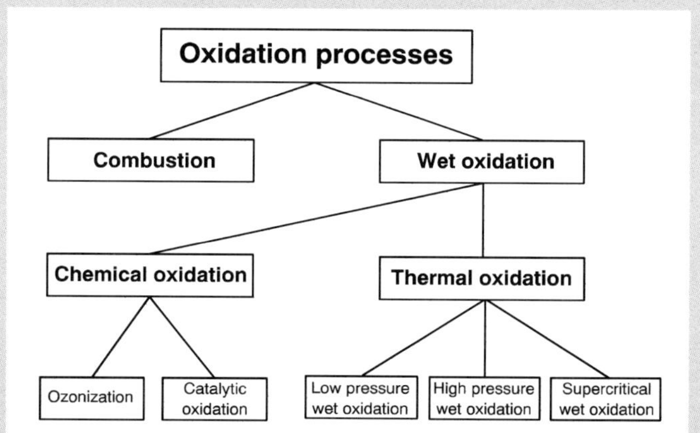

Figure 29. Oxidation processes[55].

Wet oxidation processes can be further subdivided into low pressure and high pressure processes and the process of supercritical wet oxidation. Due to the relatively "mild" reaction conditions, mineral wet oxidation residues occur with a structure identical to those of the fillers and pigments in the sludges. Table 10 shows that the aforementioned processes are used primarily for cleaning industrial and municipal wastewater, treating sewage sludge, and waste treatment. The basic principle of wet oxidation has been available since the beginning of the century. Its technical realization occurred after World War II initially in the pulp industry. The Swedish company, Stora Koppabergs AB in Falun, and the American Sterling Drug Company in New York patented the process of chemical recovery by wet oxidation. Starting in 1960, the American company in cooperation with the inventor Zimmermann built many wet oxidation plants for municipal sewage sludge.

Table 10. Classification and application of thermal wet oxidation processes[55].

	Wet oxidation processes		
	Low pressure	**High pressure**	**Supercritical**
Pressure [bar]	3–60	60–300	220–1 000
Temperature [°C]	50–250	250–370	370–600
Application	Regeneration of activated carbon	Treatment of process waters	Decontamination of soils
	Conditioning of sewage sludge	Oxidation of sewage sludge	Treatment of electronic waste (flame protection agents)
	Improvement of biological degradation of wastewater	Recovery of inorganic chemicals	Treatment of waste contamined by military activities
	Decolorization of wastewater		Treatment of halogene-containing wastewater

A high pressure wet oxidation plant working with the Zimmermann process (ZIMPRO) operated at the Biberist paper mill in Switzerland from 1978 to 1997[51]. The inorganic part of the sludge from a mechanical and biological effluent treatment plant was recovered and re-used as filler in the production of wood-free printing and stationery papers. The sludge entered the wet oxidation plant with 7%–10% dry substance and 70%–75% ash that was mainly clay. In the ZIMPRO reactor (a pipe with about 1 m diameter and 25 m length covered with titanium), the organic substances were extensively oxidized at about 300°C and 160 bar. The recovered clay had a brightness of 75%–80% ISO and could be returned to the production process. In 1997, the plant was shut for economical reasons.

The findings of Mertz and Jayme[52], who reported on experiences with a ZIMPRO plant operating at an American paper mill in 1982, confirmed the Biberist results. Gosling *et al.*[53] described the planning data for a wet oxidation plant at an American recovered paper processing mill. Filler as a mixture of clay and calcium carbonate recovered from deinking sludges is suitable for re-use in paper production. No information is available concerning the quality aspects of the recovered minerals.

To date, wet oxidation processes have not had use for recovering filler from deinking sludges on an industrial scale. Johnston and Wiseman[54] reported laboratory wet oxidation tests at 240°C and pressures of 40–50 bar. Wet oxidation residues with a brightness of 76% ISO resulted starting with a brightness of the deinking sludge of 17% ISO. No change in the filler particle structure occurred in the tests. Unfortunately, the authors provide little information on the practical range of test conditions.

Hamm and Göttsching[55] investigated the wet oxidation of deinking sludges on a laboratory scale. The wet oxidations at low pressure, high pressure, and in the supercritical range led to an almost 100% oxidation of the organic components of the deinking sludges. The wet oxidation residues in the low pressure and high pressure area did have a low brightness that came from incomplete oxidation of carbon black from black printing inks. The brightness in the residues obtained under supercritical conditions was higher, but the color of the wet oxidation residues shifted into the brown to orange area under the selected reaction conditions. The abrasion properties of the wet oxidation residues were comparable to those of fresh clay and calcium carbonate. More than 99% halogenated organic compounds measured as AOX were destroyed.

The supercritical wet oxidation process works under extreme conditions. It is usually called the supercritical water oxidation (SCWO) process. This is an oxidation process that occurs when water is under supercritical conditions. Figure 30 shows the phase diagram of water. The steam pressure curve of water ends at the critical point of 374°C and 221 bar. Above the critical point, a compression of water without condensation, i.e., without phase transition, is possible.

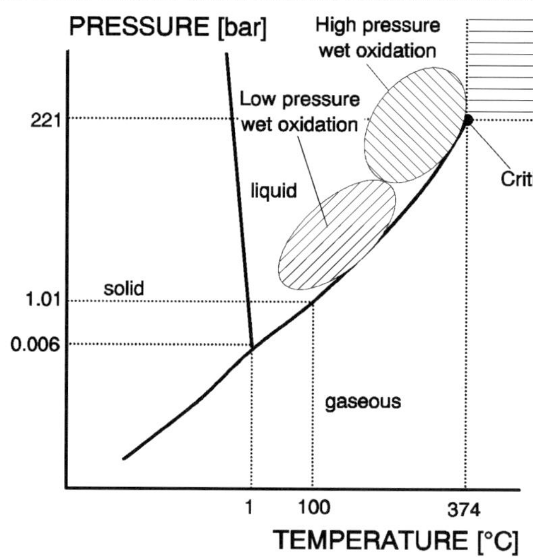

T_c	p_c	ρ_c
374 °C	221 bar	0.32 gcm^{-3}

Figure 30. Phase diagram of water[55].

The properties of supercritical water differ from those of liquid state water. Its density corresponds to approximately one-third of the density of "normal" water. In the critical temperature area, the solubility of organic compounds rises dramatically. In contrast to the low and high pressure oxidation processes, a total capability to mix with air or oxygen exists. The SCWO process works at 375°C–600°C and pressures up to 1 000 bar. Under these conditions, organic substances oxidize rapidly to carbon dioxide and water. Organic compounds containing sulfur, phosphorus, and chlorine transform into sulfuric acid, phosphoric acid, and hydrochloric acid, respectively. Elementary nitrogen and ammonia develop from organic nitrogen. Nitrogen oxides (NO and NO_2) occur only at temperatures above 600°C.

The degradation rates for organic substances possible with the SCWO process are 99.0%–99.99% related to total organic carbon (TOC) degradation. To operate the SCWO process independent of the energy supply, a 10% by weight proportion of organic substance in the wastewater or in an aqueous residue suspension suffices. The reductive heat released by oxidation heats the substrate via a heat exchanger. Fossil energy is necessary only for starting from the cold state.

The SCWO process was developed originally for the destruction of rocket fuels. Since then, the applicability of this process for treating wastewater containing toxic, biologically nondegradable substances and treating highly polluted sludges from the chemical and pharmaceutical industries has been tested in various pilot plants. Modell *et al.*[56] reported the SCWO treatment of sludges from the paper industry in a semi-industrial, continually working plant. The emphasis in those tests was not on filler recovery but on reducing the content of chloro-organic compounds (dioxins, furans) in the sludges. Cooper *et al.*[57] confirm the almost completely destruction of these compounds by SCWO application.

2.7.2 Filler recovery from combustion ashes

Two processes presently in the laboratory or technical testing stage involve the use of combustion ashes from deinking sludges used for energy production and for filler recovery. These ashes primarily comprise calcium-aluminium silicates and calcium silicates formed from the transformation of clay and calcium carbonate. In one process, the calcium silicates are treated with mineral acids. After separation from the undissolved components, calcium carbonate is precipitated upon introduction of CO_2.

Another process uses the ashes from deinking sludge combustion as a base material for calcium carbonate precipitation. An aqueous solution from slaked lime sparged with CO_2 gas is applied so that calcium carbonate crystallizes on the surface of the combustion ashes to form a precipitated calcium carbonate (PCC) coating[58]. Scanning electron microscope images confirm that the resulting fillers are not a mixture of coexisting separate particles from combustion ashes and PCC. Besides a slight influence on the brightness, color, and abrasion, the addition of combustion ashes to the precipitation process has no negative effects on the crystallizing behavior. The crystal morphology and particle size distribution are identical for PCC and ash precipitated PCC. Depending on the desired brightness, the ash contribution to the precipitated calcium carbonate can be 50% by weight.

2.7.3 Fermentation processes

In contrast to anaerobic wastewater treatment and sewage sludge fermentation, the development of anaerobic fermentation of organic waste from recovered paper processing is only beginning. Compared with composting in which organic material is primarily transformed under aerobic conditions into biomass in the form of compost, anaerobic fermentation essentially produces biogas of 50%–70% CH_4 and 30%–50% CO_2 and smaller quantities of biomass.

Anaerobic fermentation is especially suitable for waste with a high water content and a low proportion of structuring components. For this reason, it has long proven its value for sewage sludge treatment and for agricultural liquid manure use. This process is at the semi-industrial test stage of development for the use of waste from the pulp and paper industry[59,60]. The fermentation of different sludges varies considerably with regard to biogas production. Biological sludges from wastewater treatment plants or mixed sludges with a high proportion of biological sludges are most suitable. The specific biogas production rate of these types of waste is 400–600 m^3/t of organic substances. The methane content in the biogas is about 60% and therefore slightly less than the methane content in the biogas achievable in the fermentation of municipal sewage sludge. The potential biogas yield of fibrous sludges and deinking sludges at about 150–300 m^3/t of organic substance is considerably lower due among other reasons to the resistance of lignin and cellulose to anaerobic degradation. The degree to which the fermentation of these sludges can increase by pre-treatment such as steam, pressure, or chemical treatment is currently unknown.

As with composting in the anaerobic fermentation processes, the economic aspect is important. One must consider whether the obtained fermentation residues can be marketed or whether an environmentally safe disposal such as in landfills is possible.

2.7.4 Hydrolysis processes

As a result of the discussions concerning the increased use of renewable raw materials, the conversion of cellulose-containing material to glucose has gained importance again[61]. Besides the technically resolved hydrolysis, attention also focuses on enzymatic hydrolysis. Although this requires considerably less energy, its disadvantages are the slow reaction and high enzyme requirement. The products of hydrolysis are used as substrates for further fermentation processing. Ethanol can be obtained by fermentation to provide access to ethylene chemistry and ethanol fuels. Fermentation processes also permit the production of butanol, acetone, lactic acid, and gluconic acid.

Besides wood residues, annual plants, and recovered paper, various forms of fibrous waste from the paper industry also have consideration as raw materials for hydrolysis processes. Corresponding tests have been limited to date to the laboratory and pilot plant scale. They have also only used sludges from virgin fiber paper production. From the economic point of view, the aforementioned hydrolysis processes do not represent practical disposal methods as long as petroleum chemistry offers less expensive chemical raw materials. In the future, the prospects of success for these alternative waste treatment processes could grow due to ecological considerations and a shortage of fossil fuels.

2.7.5 Pyrolysis processes

Pyrolysis processes are used only in a few individual cases for producing energy from the waste of recovered paper processing. Thermal decomposition of organic components under the exclusion of air leads to formation of a pyrolysis gas with a high heating value, a coke-type residue, aqueous pyrolysis products, and pyrolysis oils and tars. Pyrolysis oils and tars are wastes that theoretically may be classified as recoverable such as activated charcoal or for petrochemical use but whose use causes practical problems. This disadvantage can be overcome by a combination of pyrolysis and combustion processes. At present, such combined processes are growing in importance. They seem to be more favorable than pure combustion technologies not only from the environmental point of view (burn-out, inert materials, and reduced emission) but also from the total energy efficiency. Three combined processes currently have use on a commercial scale for the thermal treatment of household waste:

- Thermoselect technology
- Siemens thermal waste recycling technology
- Noell conversion technology.

Some plants for all three technologies are under construction globally with capacities and processing technology designed for municipal waste. No experience with the treatment of paper industry waste is currently available. The aforementioned combined processes could prove to be suitable especially for the use of rejects from recovered paper processing for energy production purposes.

2.7.6 Production of cat litter and absorber materials

To save resources, cat litter that was produced almost exclusively from mineral components such as bentonite is now being made increasingly from organic waste. Besides sawdust and straw, recovered paper and fibrous sludges from the paper industry are suitable for this purpose. Development work is currently in progress with the objective of pelletizing and drying deinking sludges. Pelletizing is necessary to increase the liquid absorption and to reduce significantly the dust formation when using cat litter. The inorganic components of deinking sludges including clay and calcium carbonate guarantee good pellet stability. Especially suitable for pelletizing are sludges with a certain range of fiber length distribution. For this reason, some refining of sludges may occur before pelletizing.

Due to the high absorption capacity for liquids, pellets produced in this way are also suitable as absorption materials that are used for elimination of environmentally hazardous organic substances such as used oils and solvents in the event of accidents and leaks. The pellets may also find use for sealing intermediate layers and the tops of landfills to minimize leakage.

3 Use and final disposal of nonrecycled paper products

Figure 31 shows a schematic representation of the use and disposal of used paper products. In many countries, most paper collected separately is recycled to produce new paper. Other possibilities also exist such as the use of recovered paper as biofuel for energy production and its use as a bulking agent in composting or converting household biowaste to methane.

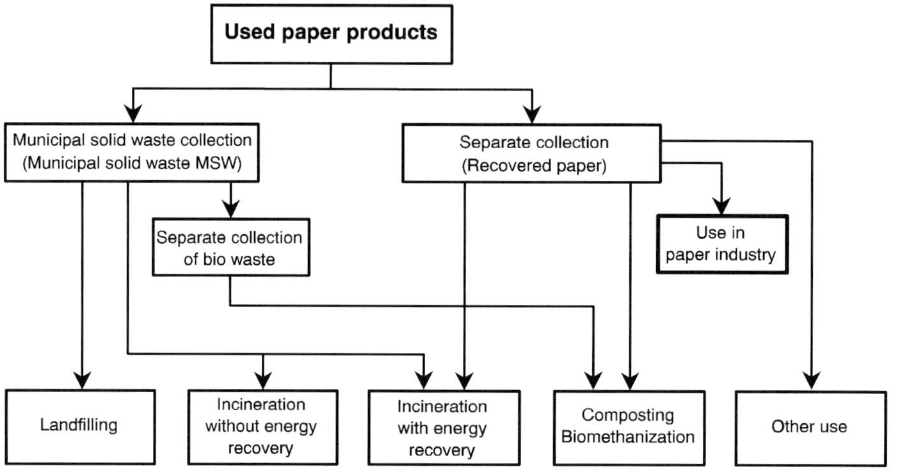

Figure 31. Use and final disposal of used paper products.

Used paper products can occasionally be a large proportion of municipal solid waste (MSW). This is especially true in developed countries with high paper consumption. In developing countries with lower consumption, the share is clearly lower as Fig. 32 shows. In Europe in 1993, the average share of paper in MSW was 27% by weight as Table 11 shows. Due to a high recovery rate of paper, Germany achieved a share in 1985 of only 18% by weight. After the introduction of separate collection of biowaste from households, this decreased in MSW to about 10%–15% by weight. The proportion of paper and board in separately collected biowaste is approximately 10% by weight. This is primarily in the form of kitchen towels, handkerchiefs, napkins, bags, and newspapers and magazines used to wrap household waste.

The most common method of MSW disposal worldwide is landfilling. As Figure 33 shows, some countries such as Japan and Switzerland favor incineration. In those locations, about 70%–80% of MSW undergoes incineration. The amount of municipal solid waste in Europe in 1993 totaled about 140 million tons. Of this, 28% or 40 million tons underwent incineration as Table 12 shows.

Table 11. Composition and energy data of European municipal solid waste (MSW)[62].

Material	Amount	Share	Heating value as received	Energy		
	[MT/a[1)]]	[%]	[MJ/kg]	[PJ/a[2)]]	[Mtoe/a[3)]]	[%]
MSW	141	100	10.0	1 410	35	100
out of which:						
- Putrescibles	56.3	40	5.6	315	7.7	22
- Paper	38.0	27	12.5	475	11.7	34
- Plastics	9.9	7	33.4	330	8.1	23
- Glass	11.3	8	–	–	–	–
- Metals	8.5	6	–	–	–	–
- Miscellaneous	18.0	13	16.0	290	7.1	20

1) *MT/a: million tons/a*

2) *PJ/a: 10^{15} J/a*

3) *Mtoe: Million tons of oil equivalents*

Table 12. European municipal solid waste data 1993[62].

Country	Total MSW [MT/a]	Combustion capacity [MT/a]	% of total capacity with energy recovery	MSW density [t/km^2]
Austria	2.8	0.3	100	33
Belgium	3.5	2.2	62	113
Denmark	2.6	2.1	100	60
Finland	2.5	0.1	100	7
France	20.0	10.3	75	37
Germany	25.0	11.2	100	101
Greece	3.2	–	–	24
Ireland	1.1	–	–	16
Italy	17.5	1.5	72	58
Luxembourg	0.2	0.2	100	69
Norway	2.0	0.4	100	6
Portugal	2.6	–	–	29
Spain	13.3	0.7	79	26
Switzerland	3.7	2.9	90	90
Sweden	3.2	1.8	100	7
The Netherlands	7.7	2.8	97	188
UK	30.0	3.6	29	123
Total	**141**	**40.1**	**Weighted mean: 83**	**58**

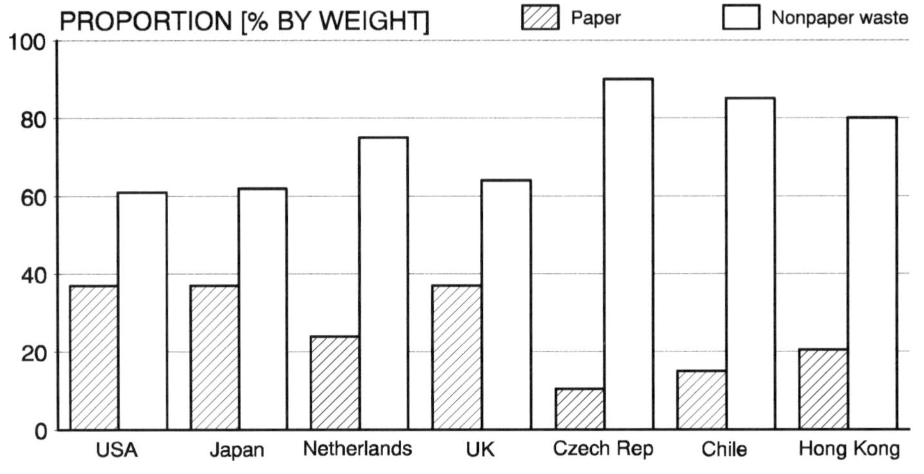

Figure 32. Proportion of nonrecovered paper in municipal solid waste in selected countries[63].

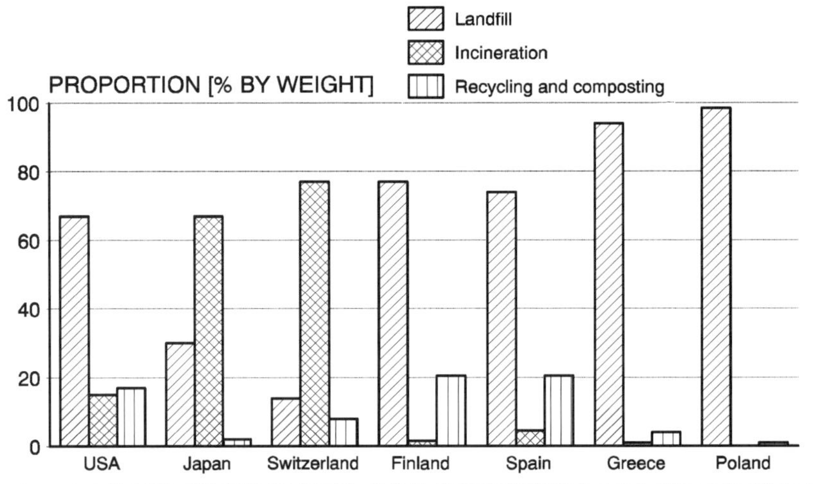

Figure 33. Proportion of various municipal solid waste disposal options in selected countries[63].

The separate collection of biowaste-containing used paper products and its composting or conversion to methane is a new development. In Germany, the per capita volume of separately collectable biowaste is 40–170 kg depending on the social structure of the collection areas and seasonal changes. In 1997, about 7 million tons of household waste underwent biological treatment for conversion to 3.5 million tons of compost. Assuming a 10% share of recovered paper in biowaste, this represents a use of 700 000 t of nonrecovered paper. This corresponds to about 30% of the total unrecovered paper that is discarded.

3.1 Landfilling of nonrecycled paper as a component of MSW

Landfilling has been the only waste management option for many years throughout the world. Currently, landfill is still the predominant form of disposal of MSW. In the previous few decades in developed countries, landfilling has advanced from uncontrolled dumping to an engineered waste disposal technology that increasingly has environmental protection measures.The standards of landfill operations even in industrial areas such as the United States, Japan, and Europe remain uneven, and uncontrolled dumping can still cause problems.

Due to its widespread use and importance as the final disposal method for waste, landfilling has been the subject of extensive research. Little research has been conducted on the degradation of paper and board in landfill sites. A key environmental concern associated with landfilling recovered paper is the generation of landfill gas. Pacey and De Gier[64] stated that methane production per unit volume is high for waste containing a high percentage of decomposable organic materials such as food residues, garden waste, and paper products. They assumed that 60% or more of the methane gas produced comes from biodegradation of paper. This estimate may be too high. Examples exist of certain types of paper and board products that have been disposed by landfilling remaining intact after many years or even decades.

Table 13 shows carbon dioxide and methane emissions from municipal waste dumps[64]. On a molecule for molecule basis, methane has an estimated global warming potential 21 times greater than carbon dioxide when considered for a 100 years[63]. Considering this higher global warming potential, a mean total equivalent carbon dioxide emission of 2 t per ton of MSW is possible.

Table 13. Emissions of methane and carbon dioxide from municipal solid waste (MSW)[64].

Emission	Unit	Minimum	Average	Maximum
Carbon dioxide	t/t MSW	0.116	0.193	0.231
Methane	t/t MSW	0.051	0.085	0.102
Equivalent carbon dioxide	t/t MSW	1.07	1.79	2.14
Total carbon dioxide	t/t MSW	1.18	1.98	2.37

In 1993, methane emissions generated worldwide from nonrecovered paper in landfills were about 12 million tons. In global warming equivalent terms, these methane emissions are almost as great as the heating effect of carbon dioxide released from fossil fuels combusted to produce pulp and paper[63].

These figures are uncertain. Estimates of global emissions of methane from waste disposal are 20–70 million tons per year. Much of the uncertainty in calculating methane generation from landfilling relates to the particular landfill conditions that can diminish the methane generation potential. Given the wide variation in landfill sizes and operation conditions, supplying exact methane emissions for any country is almost impossible.

For example, Fig. 34 shows the German situation regarding landfill gas collection and use and landfill base sealing. In 1995, half the 450 landfills had gas collecting systems. About 55% of the collected landfill gas was used for energy production purposes; the remaining 45% was burned off. To the extent that landfill gas is used for energy production, this is synonymous with the energetic use of the nonrecovered paper contained in MSW. One must remember that even sophisticated landfill gas collection systems do not guarantee complete collection of gas emissions. New landfill sites equipped with optimum collection systems can collect about 80% of the emitted landfill gas. Gas collection rates of 40%–60% are realistic for landfill sites in operation today.

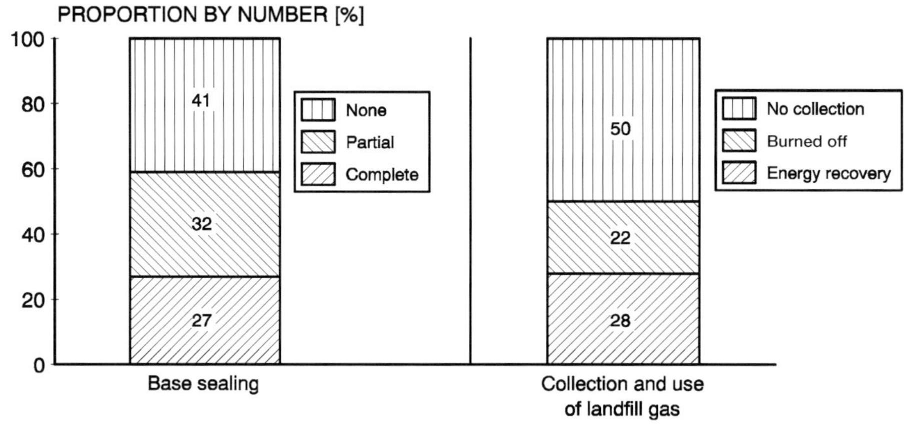

Figure 34. Sealing of landfill sites and collection of landfill gas in Germany in 1995[65].

One study[63] assumes that future methane emissions from landfill sites in industrialized countries will decrease faster than the predicted increase in paper consumption. A trend exists in many industrialized countries toward installing methane recovery technology at larger landfill sites, adopting mandatory recycling laws, and making increased use of incineration and composting alternatives. All these options will reduce methane emissions from nonrecovered paper disposal and other organic solid wastes.

3.2 Composting of nonrecycled paper as a component of biowaste from households

Used paper and paper products as components of biowaste from households, gardens, and parks intended for composting have several advantages at the collection and treatment stage. The presence of paper in biowaste can do the following:

- Reduce odors and humidity
- Reduce leakage
- Help achieve an optimum C/N ratio in the compost feedstock
- Reduce the quantity of other bulking material
- Improve the compost quality.

The leakage that occurs during the first three weeks of intensive degradation of paper in windrow composting decreases in accordance with the proportion of paper contained in the biowaste. A paper proportion of 20% by weight completely suppresses leakage. Investigations by Boelens *et al.*[66] and Wünschmann *et al.*[38] show that an addition of paper to biowaste leads to better humidity control, better aeration, and lower ammonia emissions as Fig. 35 shows. The composts produced with paper have a superior water retention capacity, lower salt content, and lower heavy metal content than the corresponding purely biowaste composts.

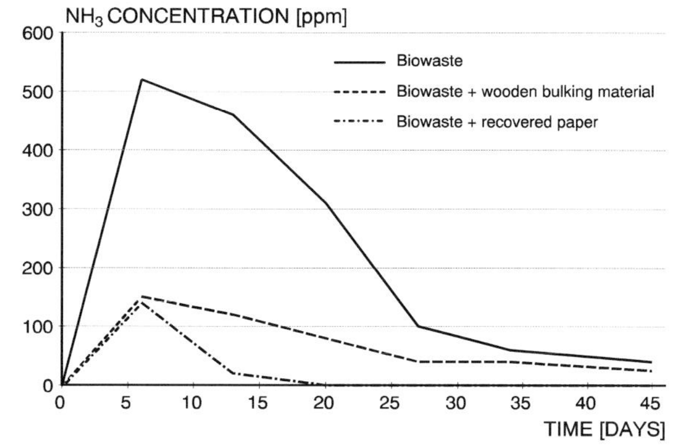

Figure 35. NH_3 emissions during biowaste composting tests with wooden bulking material and paper as co-substrates[66].

The use of paper in biowaste composting cannot increase indefinitely. Due to its low C/N ratio, the quantity of paper added to biowaste is limited to 30% by weight maximum.

3.3 Incineration of nonrecovered paper as a component of MSW

Municipal solid waste has been incinerated for many decades since the beginning of the last century in simple, small plants where the waste was treated without pollution control. The aim was simply to reduce the volume or mass of the waste. In the 1960s, the first large-scale plants for MSW incineration were built in the industrialized countries. Initially, only some areas such as Germany, Denmark, and Northern Europe attached priority to energy recovery by using the energy content of the flue gas. Hot water or steam was produced at low pressure and fed into a municipal heating network or the steam network of an industrial operation. In the late 1980s, MSW incineration was recognized as a source of energy. Since that time, all newly installed MSW incineration plants have operated with energy recovery. Due to the continued operation of many plants originally built in the 1970s and 1980s without energy recovery, this is still the predominant type in operation in some countries.

The key environmental issues associated with municipal solid waste incineration are the following:

- Atmospheric emissions from the stack
- Ash management.

The key emissions discussed and evaluated in the literature are carbon dioxide due to its effect as a greenhouse gas and dioxins and furans due to their toxicity. Until the beginning of the 1980s, waste incineration plants from today's ecological point of view were totally inadequately equipped. A dust separator that worked with questionable efficiency was often the only flue gas purifying element. The advances achieved since then in fuel engineering and flue gas treatment have lowered the emission of some toxic substances by a factor of 100 compared with the past and ensure that no significant environmental impact arises. Table 14 compares the annual emissions of 21 Swedish MSW incineration plants for the years 1985 and 1991[62]. Retrofitting existing plants with efficient flue gas purification technologies considerably lowered emissions of toxic substances.

Table 14. Annual emission of air pollutants from MSW incineration plants in Sweden in 1985 and 1991[62].

Substance	Unit	1985	1991	Reduction [%]
Particulates	t	420	45	90
Hydrogen chloride (HCl)	t	8 400	410	95
Sulfur oxides (as SO_2)	t	3 400	700	80
NO_x (as NO_2)	t	3 400	3 200	6
Cadmium (Cd)	kg	400	35	90
Lead (Pb)	kg	25 000	720	97
Mercury (Hg)	kg	3 300	170	95
Zinc (Zn)	kg	54 000	2 800	95
Dioxins (TCDD equivalents)	g	90	8	90

Paper as a component of MSW is unlikely to be a significant source for concerned substances in the context of atmospheric emissions. An analysis of household waste originally made in 1982 revealed that the paper and board in household waste had a considerably lower heavy metal content than the other components such as plastics, fine waste, and waste of a vegetable nature. Especially remarkable were the low cadmium, mercury, and lead concentrations in the paper compared with that in the plastic fraction. Svedberg[67] came to a similar conclusion in a Swedish project in 1992. According to the findings of that project, paper contributed only a very small percentage to the heavy metals content of municipal solid waste. The contributions were 5% cadmium, 12% chromium, 3% lead, and 1%–2% mercury of the total contents. According to the Swedish study, the chlorine, sulfur, and fluorine pollution of the household waste resulting from paper is negligible. The low sulfur content of household waste compared with fossil fuels is due to the paper contained in the household waste.

Another Swedish study[68] compared the content of heavy metals in packaging waste with industrial biofuels. The heavy metal content in packaging was in the same range as in wood and forestry residues.

MSW combustion normally uses grate-fired boilers. No pre-treatment of the waste is necessary. Another development for waste incineration is the fluidized bed system. This technique has had use for municipal waste combustion since the 1970s and has found extensive use in Japan and Sweden where energy recovery from waste has wide application. In addition, new combustion systems are under development that offer improved plant efficiency through such techniques as gasification, pyrolysis, and pressurized combustion systems.

3.4 Incineration of nonrecycled paper

Increased CO_2 concentration in the atmosphere and the feared consequences for the global climate are especially important factors to use energy resources efficiently in the future. Today's primary energy requirements are 80% from nonrenewable fossil fuels including 35% mineral oil, 25% coal, and 20% natural gas. Nuclear power contributes 7%, and renewable energy sources such as water and wind power, solar energy, and biomass account for 13%[69]. Estimates for how long the reserves of fossil fuels will last depend on the assumed future development of energy consumption. Certainly, the reserves of fossil fuels will be exhausted within the next few centuries.

In contrast to fossil fuels, biomass that is continually produced by photosynthesis now only has secondary importance as a source of energy. This applies especially for industrialized countries. The situation is somewhat different in developing countries. For example, about two-thirds of the world's population today continue to cover their domestic energy requirements with wood. This is humankind's oldest form of heat production. About 2 billion m^3 of wood are burned annually[70].

Due to its material composition, paper belongs to the category of biofuels. In the last few years as a result of the changing conditions in waste management (future reduced landfill capacities and limited marketing possibility for compost), a growing demand has existed to use nonrecycled paper for energy production to replace fossil fuels. For several reasons, Göttsching[71,72] considers the paper industry itself to be predestined to become involved in the energetic use of surplus quantities of used paper and paper products:

- The paper industry produces the electrical and thermal energy it requires with a high efficiency of 80%–90% by co-generation. Power generation plants in which nonrecycled paper could be incinerated achieve a thermal efficiency of only 38%–45%.
- The paper industry can classify used paper according to its fiber value. In this way, used paper can be separated into a superior quality for material recycling and a lower quality for thermal use.
- Surplus quantities of used paper can be incinerated with production residues to improve the compression of paper by pelletizing.

The CO_2 released by incinerating the nonrecycled paper fraction does not contribute to global warming since it is consumed again by photosynthesis of renewable biomass including wood. The substitution of fossil fuels simultaneously gives a reduction of global warming active CO_2 emission.

In Germany, the energetic use of nonrecycled paper in the paper industry could reduce the amount of fossil energy required by about one-third. Simultaneously, the emission of global warming active CO_2 would decrease by the same percentage[73].

3.4.1 Nonrecycled paper as a biofuel

Theoretically, all modern incineration technologies that permit the use of solid fuels are suitable for incinerating nonrecycled paper. Making a distinction between whether paper is the sole fuel or used with other solid fuels is important. Currently, experience is available only for mixtures in power plants. For this reason, we will concentrate in the following discussion on the most important material properties for incineration of nonrecycled paper by itself.

The heating value of paper depends essentially on its filler and pigment contents. Figure 36 shows the direct connection between ash content and heating value. The heating value of air-dried paper extrapolated to zero ash content at 16 MJ/kg corresponds to the heating value of wood with a moisture content of about 10%. Higher heating values for paper are sometimes given in the technical literature, but this can be attributed to the presence of heating value enhancing components in the paper mix such as adhesives, printing inks, varnishes, and laminating films.

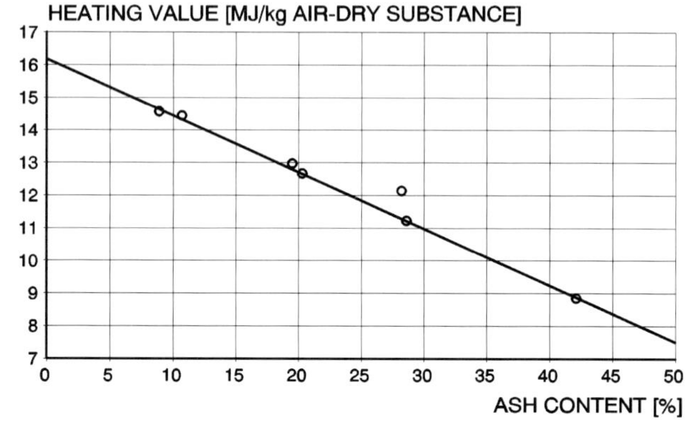

Figure 36. Heating value of air-dried paper vs. ash content analyzed at 575°C.

Compared with other solid fuels, paper has a low density. For this reason, the volume-related heating value is low. Considerable fluctuations in the heating value can also occur with low quality recovered papers of varying compositions. Both aspects require pre-treatment of the paper to obtain an easily manageable, homogeneous fuel.

With the incineration of loose, noncompressed recovered paper, the volume-related heating value corresponds to about 3%–4% of the volume-related heating value of hard coal. For identical heating performance in a combustion chamber, this means that the furnace admission rises dramatically with negative consequences for the thermal efficiency. Additional reasons exist for opposing incineration of loose paper by different incinerating techniques:

- Multiple hearth incineration: A risk exists for locally nonuniform incineration conditions within the furnace that can lead to incomplete incineration and increased CO emissions.
- Fluidized bed incineration: Carrying forward incompletely incinerated paper can give rise to sources of fire in the following flue gas purification plant and in the ash disposal system.
- Grate firing: A risk exists for reignition of the flame front in the paper feeding chute.

- Dust firing: The low density of paper, its high specific surface area in a ground state, and the high ignition capacity should constitute good prerequisites for burning ground paper in dust firing plants. Problems have occurred with grinding of paper from sizes as fine as dust to grain sizes of 25 mm.

3.4.2 Pre-treatment of recovered paper as a biofuel

Figure 37 shows the processes that have proven valuable for production of biofuel from recovered paper. They fall into classes of pre-treatment, densification, and after-treatment of the densified recovered paper.

When the recovered paper is delivered in bales, the pre-treatment consists of dewiring of the bales and loosening the recovered paper. Loosened paper and recovered paper delivered loose can then be manually pre-sorted to remove larger refuse. After pre-grinding by hammer mills and impact rippers, magnetic separators remove ferrous metals. In the subsequent after-grinding operation, the recovered paper is ground to a grain size of 40–50 mm. This size is suitable for densification. A subsequent air separation permits removal of smaller abrasive components that could influence the operating life of the densifying plant.

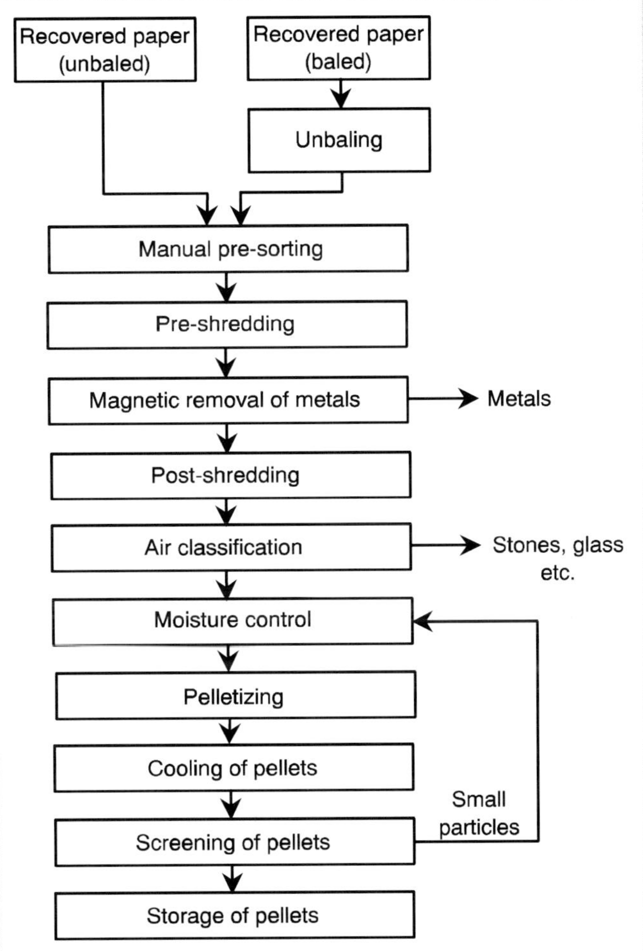

Figure 37. Production of fuel pellets using recovered paper.

Pelletizing processes are used to densify the pre-treated recovered paper. Successful pelletizing systems work according to the compression and agglomeration method or extrusion. The pre-ground paper is pressed through a shaping matrix using grinders, screws, or pistons. At the matrix delivery, the rod-like extruded material is cut to a given length with a knife. Binding agent additives are not necessary for pelletizing because plastic components in the recovered paper cause fusion during the extrusion process and guarantee the pellet stability.

Moisture content of the recovered paper has a decisive influence on the quality of the pellets. Moisture content of 15%–20% is optimum. This is possible by adding water or a water-containing material to the recovered paper. Use of sludges from a recovered paper process or from municipal wastewater treatment plants is also suitable. During pelletizing, the moisture content falls by about 5%. After pelletizing, cooling is necessary to enable the pellets to harden. During this phase, an additional loss of moisture occurs accompanied by a biological stabilization of the pellets. The cooled pellets are directed via a filter into a storage silo. The dust removed during this filtering process is returned to the pellet mill.

Recovered paper pellets are produced with diameters of about 5–30 mm and lengths of 20–50 mm. The bulk density of recovered paper pellets at 580–700 kg/m^3 is in the bulk density range of solid lignite at 500–800 kg/m^3 and close to the bulk density of solid hard coal at 700–850 kg/m^3.

3.4.3 Emissions

The air emissions resulting from the incineration of recovered paper in separate plants or the combustion with other fuels in power plants can presently only be forecast since no power plant uses exclusively or primarily recovered paper as fuel. One publication compares the air emissions resulting from incineration of recovered paper pellets and those of other solid fuels[74].

A study of the University of Wisconsin concludes that the SO_2 emissions of a lignite-fired power plant decrease by about 20% when replacing 50% of the lignite by recovered paper pellets. With the combined incineration of lignite and recovered paper pellets, the emissions of the heavy metals arsenic, chromium, lead, mercury, and selenium fall to the same extent as the proportion of recovered paper pellets is increased. The heavy metal content in the incineration ashes resulting from the incineration of recovered paper pellets is lower than with lignite incineration. According to the study of the University of Wisconsin compared to the SO_2 emissions resulting from the incineration of municipal solid waste, the incineration of recovered paper pellets can produce about 50% lower emissions.

Kara reports on the energy recovery of used paper in a lignite fluidized bed incineration at a paper mill in Germany[75]. 10% of the fuel heating value comes from recovered paper. The SO_2 and NO_x emissions are 20% and 10% lower, respectively, than with the exclusive burning of lignite.

Emissions published to date for burning recovered paper with other fuels correspond to the concentrations of emission-relevant contents of the fuels concerned. Figure 38 shows the ranges for the contents of chloride, sulfur, and nitrogen in recovered paper, lignite, hard coal, and wood. Shown also are the contents of sludges of the paper industry and of municipal solid waste. To facilitate an energetic comparison, the element contents relate to the same heating value in megajoules. The following conclusions arise from this comparison:

- Compared with lignite and hard coal, lower SO_2 and NO_x emissions are possible from the incineration of recovered paper. The HCl emissions will not change significantly.
- Compared with the incineration of municipal solid waste, considerably lower HCl, SO_2, and NO_x emissions occur with the incineration of recovered paper.
- The mixing of sludges from the paper industry with recovered paper lowers the heating value. This influences the heating value-related elemental contents of chlorine, sulfur, and nitrogen only slightly provided the proportion of sludges from biological wastewater purification is not too high.

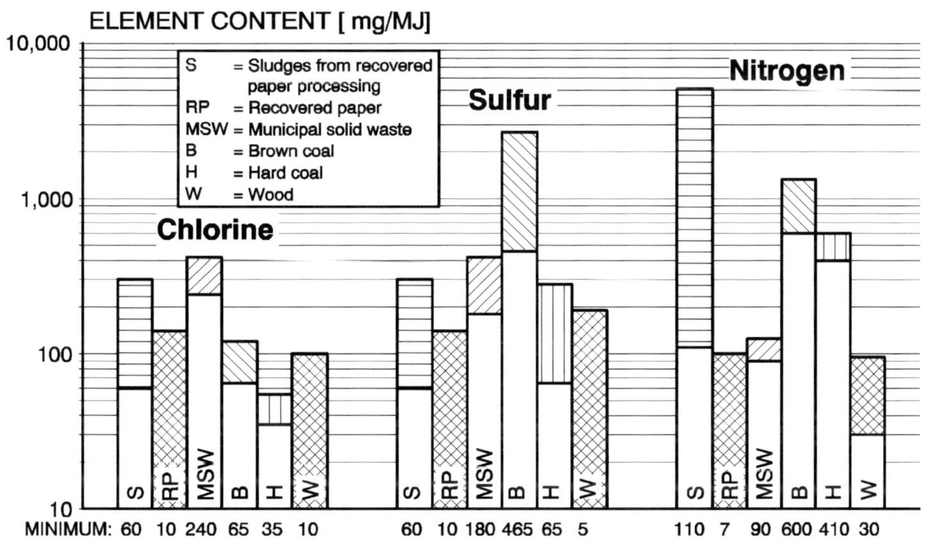

Figure 38. Element content of solid fuels related to heating value[73].

Figures 39 and 40 summarize the heating value-related heavy metal contents of solid fuels. Considerably lower contents of this group of toxic substances exist in the recovered paper than in the municipal solid waste. One especially positive aspect is that the cadmium and mercury load of the recovered paper is lower than the corresponding loads of lignite and hard coal. The contents of the other heavy metals are approximately at the same level as in coal fuels. Only the copper content may possibly be greater in recovered paper than in hard coal and lignite.

To summarize, no higher air emissions will result from the energetic use of recovered paper than from burning of lignite and hard coal. With some emission parameters such as SO_2, cadmium, and mercury, considerably lower amounts are emitted. The observance of emission thresholds of the exhaust gas can be guaranteed by a justifiable investment in flue gas purification.

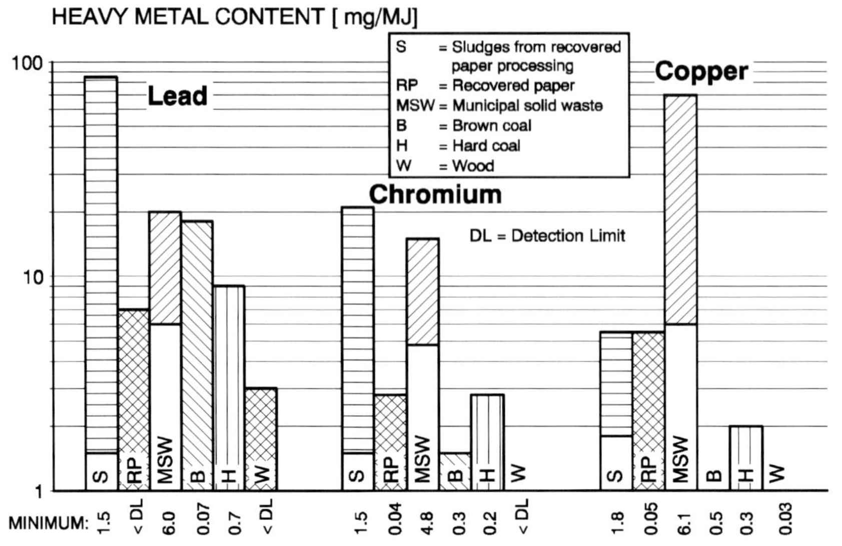

Figure 39. Heavy metal content of solid fuels related to heating value[73].

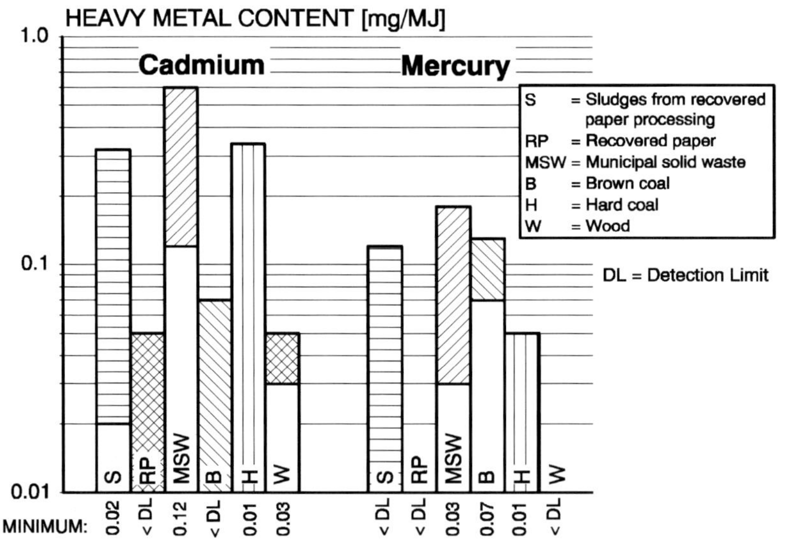

Figure 40. Heavy metal content of solid fuels related to heating value[73].

4 Wastewater from recovered paper processing and wastewater treatment

The water loops in processing of recovered paper generally minimize fresh water consumption. Process water is re-used several times as Fig. 41 shows. The main principle applied is the backward process water flow in the systems countercurrent to the fiber flow. Similar to virgin fiber based paper mills, recovered paper processing mills use untreated, fiber enriched white water from the paper machine for stock dilution in the mixing pump ahead of the paper machine (short circulation or primary circuit) and in the stock preparation line (long circulation or secondary circuit).

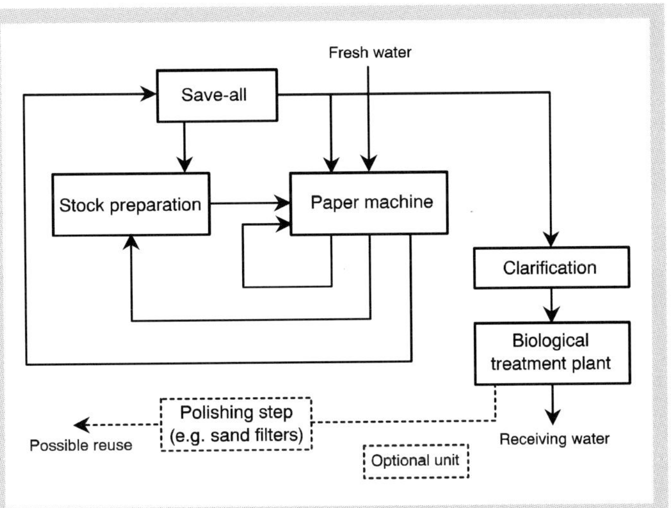

Figure 41. Water loops in recovered paper processing.

Some white water is clarified in save-alls applying filtration (polydisk filters or drum filters), flotation [dissolved air flotation (DAF)], or sedimentation (sedimentation cones or lamella separators) to re-use the clarified water for the replacement of fresh water at the showers used for cleaning machine clothing (wires and felts). Excess clarified process water goes to the wastewater treatment plant. In some cases, purified wastewater is partly re-used as process water after an additional polishing step.

4.1 Characterization of untreated wastewater

Recycled fiber processing has two main categories:

- Processes with exclusively mechanical upgrading, i.e., without deinking. This includes products such as test liner, corrugating medium, board, and carton board.
- Processes with mechanical and chemical unit processes, i.e., with deinking and bleaching. This includes products such as newsprint, hygienic paper, printing and copy paper, magazine papers (supercalendered or lightweight coated), or market deinked pulp (DIP).

Wastewater loads from processes of the first category are slightly higher than those from processes that must provide additional fiber processing such as deinking. The wastewater contains dissolved and undissolved substances of an organic and inorganic nature. Some solids are so finely dispersed that they behave almost like dissolved ones by forming a stable dispersion or colloids. These are considered under dissolved substances and require major attention in process and wastewater treatment. Among the undissolved substances are fibers, fines, fillers, and coating pigments. Removing these from the wastewater causes no problems using today's cleaning technology.

The dissolved wastewater substances originate from different sources. Small amounts enter with the fresh water into the production process. Various substances are released from the virgin pulp. These are carbohydrates, lignin, inorganic salts, and extractives. Of much greater importance are those coming from the recovered paper and the chemical additives used in paper production. The influence of chemical additives on the wastewater load is very complex. They form an extremely heterogeneous group and occur in different concentrations in a dissolved or dispersed state in the wastewater.

The composition of the untreated wastewater is generally described by the usual wastewater analysis parameters. These include chemical oxygen demand (COD), biochemical oxygen demand (BOD_5 or BOD_7), total organic carbon (TOC), adsorbable organic halogen-containing compounds (AOX), and suspended solids (SS). The concentrations of these parameters depend on the specific fresh water consumption that is in turn determined by the program of grades produced in a paper mill. More indicative are the specific loads, i.e., the loads in relation to the unit of weight of the produced paper given in g/t or kg/t. The specific emissions of recovered paper processing mills vary over a wide range. This has proven useful to distinguish between mills with and without deinking.

Packaging paper and board manufacturers working with recovered paper do not usually deink the recovered paper. Although no deinking chemicals are used, the untreated wastewater sometimes reveals a higher load of dissolved organic compounds than the wastewater of deinking mills. This is due to the greater share of strength-enhancing additives and especially starch in the recovered paper and in the production process and to steam-volatile organic acids. This is primarily acetic acid that is formed in tightly closed water circuits by microbial transformations. This can be an important part of the organic content of wastewater.

Table 15 shows the ranges of specific wastewater discharge, the COD and BOD_5 concentrations, and the COD and BOD_5 loads as they occur in the production of board and corrugating paper (test liner and corrugating medium) from 100% recycled fibers. Any comparison is more difficult due to the fact that the inventories at the mills have taken more than several years. The specific COD loads from the packaging paper mills using recovered paper vary over the wide range of 3–31 kg/t. Considering that very similar water loop systems are involved here, the variations can be due primarily to differences in stock and processing methods including the use of starch.

For board mills the AOX concentration in the untreated wastewater is 0.1–2.0 mg/L[76]. This corresponds to specific AOX loads of 0.001–0.008 kg/t. The solids contents total 100–1 000 mg/L. The BOD/COD ratio as the measure for the biological degradability of wastewater is 0.35–0.62 with an average of 0.44 for board and 0.52 for corrugating medium and test liner. This means that the wastewater has good biological degradability.

Table 15. Characterization of wastewater of RCF mills without deinking before biological treatment.

Paper grade	Number of mills	Specific wastewater volume [m³/t]			BOD_5 concentration [mg/L]			BOD_5 load [kg/t]			COD concentration [mg/L]			COD load [kg/t]		
		Min	Max	Average	Min	Max	Average	Min	Max	Average	Min	Max	Average	Min	Max	Average
Board	n.a.[78]	0.4	6.6	3.6	285*	4 500*	1 900*	1.6*	1.9*	1.8*	570	9 000	3 800	3.2	3.8	3.5
Board	5[76]	6	16	10	330*	1 750*	890*	3.5*	6.0*	5.0*	660	3 500	1 780	7.0	12.0	10.0
Board	12[77]	0	15	–	530	3 000	–	–	–	4.7	1 140	5 500	–	–	–	8.4
Test liner, fluting	13[76]	0	5.5	3.4	1 400*	3 500*	2 690*	5.5*	15.5*	9.7*	2 800	7 000	4 380	11.0	31.0	19.4
Test liner, fluting	29[77]	0	10	–	1 280	2 840	–	–	–	6.1	2 190	5 680	–	–	–	12.3

The BOD_5 values are estimated assuming that COD = 2 x BOD_5; n.a. = not announced

Table 16. Characterization of wastewater of RCF mills with deinking before biological treatment.

Paper grade	Number of mills	Specific wastewater volume [m³/t]			BOD_5 concentration [mg/L]			BOD_5 load [kg/t]			COD concentration [mg/L]			COD load [kg/t]		
		Min	Max	Average	Min	Max	Average	Min	Max	Average	Min	Max	Average	Min	Max	Average
Newsprint, tissue	n.a.[78]	9	39	15	220	995	550	3.5	20.0	10.0	440	1 900	1 100	7.0	40.0	20
Newsprint	7[76]	8	23	15	450*	1 150*	–	6.0*	13.0*	–	900	2 300	–	12.0	26.0	18
Newsprint	4[77]	7	20	–	460	1 270	–	4.0	11.8	–	960	2 400	–	9.6	22.3	–
Newsprint	4[79]	7.7	15.2	–							1 515	2 465	–	15.9	23.0	–
Recycled graphic paper	3[77]	10	20	–	250	400	–	2.5	8.0	–	540	790	–	5.4	15.8	–
Newsprint, tissue	n.a.[80]	15	38	27	–	–	–	18	36	27	–	–	–	–	–	--
Tissue	5[76]	8	10	–	1 000*	2 500*	–	8.0*	23.5*	–	2 000	5 000	–	16.0	47.0	–

** The BOD_5 values are estimated assuming that COD = 2 x BOD_5; n.a. = not announced*

Table 16 summarizes the wastewater ratings of deinking paper mills. No information on the process steps of deinking and recycled fiber pulp bleaching is available in the quoted literature. Note that the data originate from different years. Pfitzner[79] supplies current process water ratings from four German newsprint mills at which the proportion of recovered paper in the pulp input is 80%–100%. The specific COD load is 10–40 kg/t. The BOD/COD ratio of about 0.5 indicates good biological degradability. At 900–2 000 mg/L, the COD concentrations are in a range offering scant prospect for the successful application of anaerobic wastewater cleaning technologies with systems developed to date.

4.2 Wastewater treatment

The objective of wastewater treatment at recovered paper processing mills is to reduce pollution of the receiving waters. Oxygen consumption in the receiving water becomes lower and therefore gives considerable improvement of the water quality obtained.

A basic distinction exists between three cleaning stages in the treatment of process water in wastewater purification plants:

- Mechanical treatment
- Biological treatment
- Advanced treatment.

For mechanical treatment, suspended undissolved substances such as fibers, fines, fillers, and coating pigments are removed from the wastewater in the initial step. Addition of chemicals may enhance mechanical separation.

Biological treatment uses degradation of dissolved organic components by microorganisms. It usually occurs after a mechanical treatment. Aerobic processes mostly as an activated sludge stage, anaerobic processes, or a combination of both are applied. At recovered paper processing mills, these processes guarantee an almost 100% removal of wastewater components that have good biological degradability.

The rubble and sand filter, ultra- and nano-filtration, or ozone treatment processes that are already in existence at several mills belong to the additional or extended wastewater cleaning category. They achieve additional reduction of the solid load, decolorization of the wastewater, and an extensive COD reduction following a biological treatment.

4.2.1 Suspended solids removal

Mechanical cleaning of wastewater at paper mills almost exclusively uses sedimentation tanks. Preceding screening and sand-trapping systems ensure that bulky contaminants such as pieces of wood, plastics, foils, and metal and large amounts of sand do not enter the purification process. Particles of solid matter that have a higher density than water settle on the bottom of the sedimentation basin. The settled sludge is removed from the basin using sweeping or suction devices, thickened, dewatered, and finally re-used or discarded.

Sedimentation basins generally have a horizontal flow-through pattern. They are rectangular or circular units. Rectangular basins have flow-through in a longitudinal direction. Special devices such as scum boards, intake grids, or impact plates ensure uniform distribution of the entering material. A frequently applied form of sludge sweeping is the so-called conveyor rake. This consists of sweeping bars drawn across the bottom of the basin on a rotating chain that push the sludge into a collecting funnel from where it is pumped out as Fig. 42 shows.

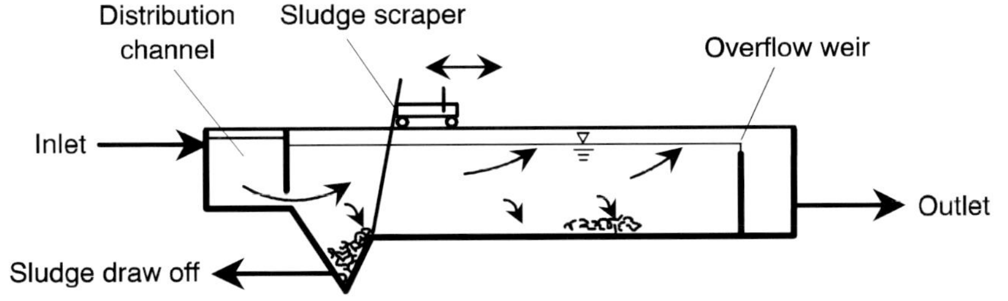

Figure 42. Rectangular clarifier.

With circular basins, the wastewater is introduced in the center, flows radially through the basin, and then over the edge. At the bottom of the basin, a revolving sweeping bar pushes the sludge into the center from where it is pumped out as Fig. 43 shows.

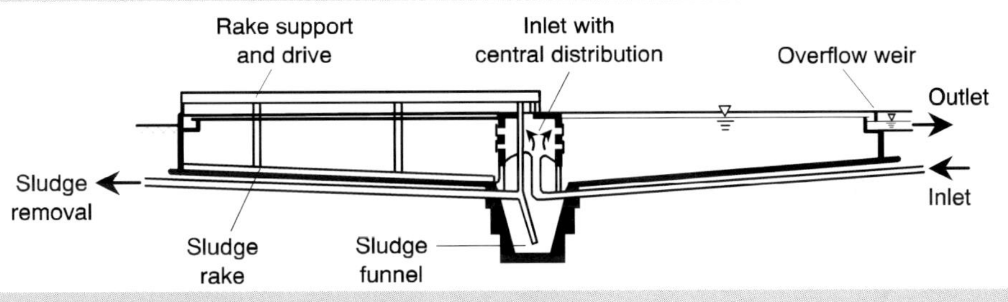

Figure 43. Circular clarifier.

The lamella clarifier is a modification of the conventional sedimentation basin. This type of clarifier produces a laminar flow that makes settling of sludge more efficient as Fig. 44 shows. The advantage of this settling process is the more compact construction and the reduced space requirement. A slime accumulation on the lamellae may occur and require maintenance.

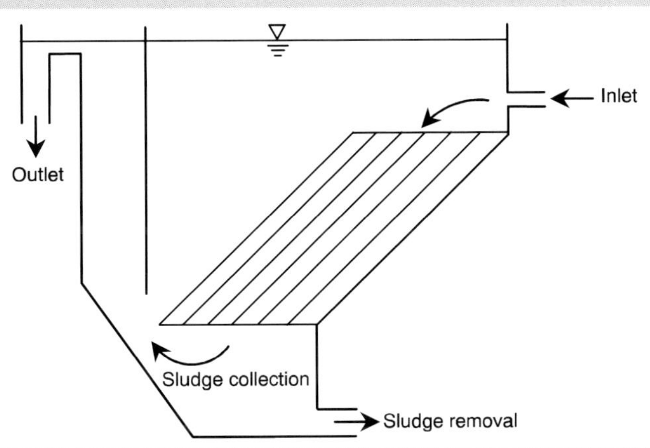

Figure 44. Lamella clarifier.

In the last few years, flotation clarifiers such as used for fiber recovery at paper machines have also proven their value in wastewater cleaning for separating solids that settle very slowly. This technique is also suitable for removal of suspended solids after biological treatment. The advantage of flotation clarifiers is their low space requirement. Disadvantages are a higher energy consumption and higher operating costs.

With all the named processes, the settling of suspended solids can improve significantly with the use of flocculation agents. The addition of chemicals including aluminum salts such as $Al_2(SO_4)_3$ and $Al_n(OH)_mCl_{3n-m}$, iron III salts such as $FeCl_3$ and $Fe_2(SO_4)_3$, lime, or certain polymers causes coagulation of originally very small solid particles to larger flocs that settle faster. To achieve optimum flocculation results, pH control is often necessary.

Clarifiers designed to remove suspended solids from paper and board mill effluents produce good results. Solids removal rates can be as high as 90%–98%. In many recovered paper processing mills, the separated sludge is re-introduced into production. This is done mainly in multi-ply board, test liner, and corrugating medium plants.

4.2.2 Biological treatment

After mechanical treatment, the material almost exclusively consists of only dissolved and colloidally dissolved substances that remain in the water. These are subjected to a degradation process in the biological clarification stage. Degradation uses bacteria and other microorganisms. The bacteria whose enzyme systems produce the biochemical degradation reactions grow and multiply. An excess bacterial mass develops that is referred to as activated sludge. With all biological treatment processes, the unbalanced composition of the mill wastewater requires metered addition of nutrients (N and P compounds) and in rare cases micronutrients such as salts of cobalt, selenium, and molybdenum.

Aerobic and anaerobic biological processes are used at recovered paper processing mills. Anaerobic technologies have grown in importance in the last twenty years. They find use mostly with aerobic process technologies because an exclusively anaerobic clarification alone does not reach the specified emission limits. Anaerobic technologies presuppose a COD over 1 500 mg/L in the wastewater to be treated. For this reason, anaerobic technologies have been applied almost exclusively for treatment of wastewater of the production of corrugating medium and test liner from recovered paper. More recent developments are oriented toward the treatment of wastewater from tissue production and for partial streams from deinking systems.

Aerobic processes

Preferred processes for the aerobic treatment of mill wastewater are as follows:

- Activated sludge processes
- Trickling filter processes
- Fixed bed processes
- Aerated lagoons.

Activated sludge processes

In activated sludge treatment, the biomass, i.e., activated sludge, and the wastewater are mixed to give a homogeneous suspension in the tank. The sludge-water suspension is removed from the aeration tank and directed to a secondary clarifier where the sludge settles. Some sludge is pumped back into the aeration tank. The excess is removed and recycled or disposed. Manufacturers of board, corrugating medium, and test liner frequently reintroduce this excess sludge with the sludge from primary clarification into production.

A precondition for good degradation is sufficient residence time of the wastewater in the aeration tank and a sufficient amount of activated sludge. For this reason, the size of the aeration tank and the separating efficiency of the secondary clarifier have special importance. Important parameters for activated sludge treatment processes include the following:

- Volume load (kg $O_2/m^3 \cdot d$) usually referring to the BOD_5 or COD in the inlet
- Sludge load (kg $O_2/kg \cdot d$) also referring to the BOD_5 or COD in the inlet
- Sludge volume index, SVI (mL/g of o.d. solids). (This parameter reflects the settling properties of activated sludge in the aeration tank. The target value should be less than 300 mL/g. If SVI exceeds 300 mL/g, bulking sludge with a very poor settlement behavior can occur.)
- Oxygen content (mg O_2/L) in the aeration tank. The content should be above a minimum level to ensure efficient functioning of the process.

In addition to one-stage activated sludge treatment processes, two-stage processes have frequent use in the paper industry. The first stage involves a low oxygen content in the aeration tank with a high sludge loading. The second stage is then a conventional activated sludge process designed according to the type of wastewater being treated. Two-stage systems are less sensitive to load fluctuations and bacterial-toxic wastewater components. The reduced sludge load of the second stage permits easier removal of poorly degradable substances. The COD reduction efficiency of two-stage plants is generally much superior to that of one-stage treatment. Möbius and Cordes-Tolle[77] gave conditions for one- and two-stage activated sludge treatment plants at paper mills as Table 17 shows.

Table 17. Sludge load, oxygen content, and sludge volume index in activated sludge treatment.

Type/stage	S_L [kg BOD_5/kg · d]	O_2 [mg/L]	SVI [mL/g]
1-stage aeration plant	0.25–0.30	2.0–3.5	< 300
2-stage aeration plant stage 1	1.40–2.00	0.4–0.8	< 100
2-stage aeration plant stage 2	0.12–0.22	2.5–3.5	< 200

SVI = Sludge Volume Index; S_L = Sludge load

The oxygen supply necessary for the biomass is usually provided by mechanical air input or with pure oxygen in special cases. In most plants, the input air is used simultaneously for the mechanical mixing of the aeration tank. Besides controllability, the decisive factor for the choice of aeration systems for recovered paper processing mills is the operating reliability. This primarily concerns the blocking of pressure-type ventilators and injectors caused by precipitation of calcium carbonate. Heavy-duty, pressure-type aerators are especially effective. The specific energy consumption related to 1 kg BOD_5 eliminated is 0.6–3 kWh/kg. Especially efficient designs of the aeration systems are necessary to reduce this to less than 1 kWh/kg.

Trickling filter

Trickling filter processes belong to the fixed bed process category. Many examples are available to show their successful application in recovered paper processing mills. The development of special plastic support media considerably reduced the risk of blocking and allowed operation at high volume load. Degradation occurs through an aerobic biofilm that forms on the support media sprinkled with the wastewater. Compared with activated sludge processes, trickling filters have a lower energy consumption and usually a higher operating reliability. The temperature of the treated wastewater is usually lower. This is a desirable side feature. Trickling filters are not efficient when used alone for purification of highly concentrated wastewater. For this reason, they operate primarily as the first stage of a two-stage biological process in which the trickling filter is followed by activated sludge treatment usually without intermediate clarification.

Figure 45 shows the scheme of a wastewater treatment plant for a German newsprint mill using 100% recycled fibers. Two highly loaded trickling filters are integrated into this plant. The flotation plant that follows the biological clarification (trickling filter, activated sludge treatment, and secondary clarification) acts as a "police filter" to ensure observance of the strict legislative stipulations in the biologically purified wastewater effluent.

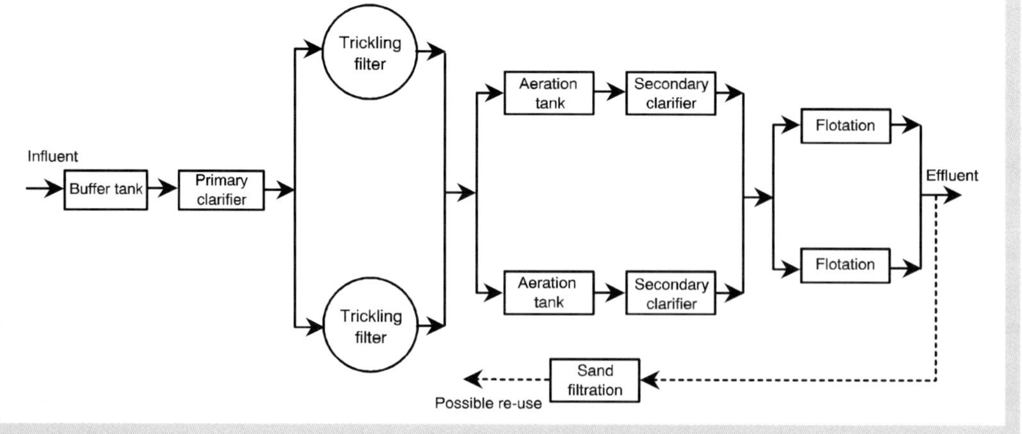

Figure 45. Wastewater treatment plant of a German newsprint mill using 100% recycled fibers.

Odor development can be a disadvantage of trickling filters. Such odors can be due to the anaerobic decomposition of wastewater contents or the evaporation of odorous substances contained in the wastewater. In both cases, a total enclosure of the trickling filter may be necessary.

Biofilter

Included in the fixed bed processes are biofilter processes. These separate suspended solids and provide biochemical degradation of dissolved organic substances as Fig. 46 shows. The plastic or porous ceramic filter bed remains aerobic by atmospheric oxygen input. The bed height is 2.5–4 m. The decisive factors for controlled operation of biofilters are the rinsing and rinsing sequence. Care is necessary to ensure that the entire filter bed is completely aerated. Besides being used as the sole biological stage for low-concentrated mill wastewater, biofilters are used as polishing treatment processes positioned downstream from other biological purification processes. The main advantages of biofiltration over activated sludge treatment are that the process is immune from bulking problems and that its energy demand is only one-third of that of an activated sludge plant of the same capacity.

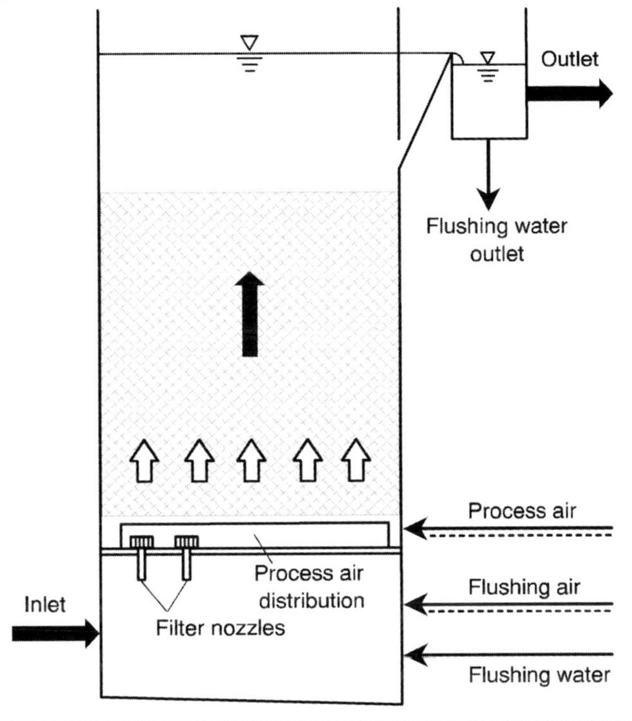

Figure 46. Scheme of a biofilter[81].

Biorotors

Special forms of aerobic fixed bed reactors are biorotors. These reactors are normally used for small plants. They consist of a fixed medium in cylindrical or disk form rotating on a horizontal shaft in a tank. The biofilm on the medium is alternatively in contact with wastewater and then with air. Usually, 3–5 rotors are connected in series.

Aerated lagoons

Aerated lagoons are less suitable for treating the wastewater of recovered paper processing mills. This system is primarily used when the BOD_5 concentration in the wastewater to be clarified is less than 100 mg/L.

Anaerobic processes

Anaerobic processes have had use in the pulp and paper industry for about 20 years. They offer several advantages compared with aerobic process technologies:

- No energy consumption for sludge aeration
- Energy comes in the form of biogas
- Compared with the sludge formation of aerobic processes, the amount of biological excess sludge is lower by about 90%–95%
- Need for nutrient salts such as nitrogen and phosphorus compounds is lower.

As a result of the increased costs of sludge disposal since the beginning of the 1980s, anaerobic processes have become established in recovered paper processing mills after previous development in other areas such as the sugar, beverage, and starch industries. The special process technology-related problem of anaerobic wastewater cleaning stems from the fact that the biomass required for the degradation has very small growth rates. For this reason, effective retention and enrichment of the biomass in the anaerobic reactor must be guaranteed or the degradation process quickly stops. Different process technologies have been developed to address the problem with varying results:

- The use of contact sludge reactors and the external separation of flushed biomass in settling basins, lamella separators, or inert gas-operated flotation systems and return to the reactor (contact sludge reactor as Fig. 47 shows)
- The use of granulated or pelletized biomass with a high settling rate (UASB reactor = Upflow Anaerobic Sludge Blanket reactor as Fig. 48 shows)
- Immobilization of the biomass by use of fixed beds of inert materials such as mineral fillers or plastic elements (fixed bed reactor as Fig. 49 shows)
- Fixing of the biomass by growing on particles of glass, sand, or activated carbon (fluidized bed reactors as Fig. 50 shows).

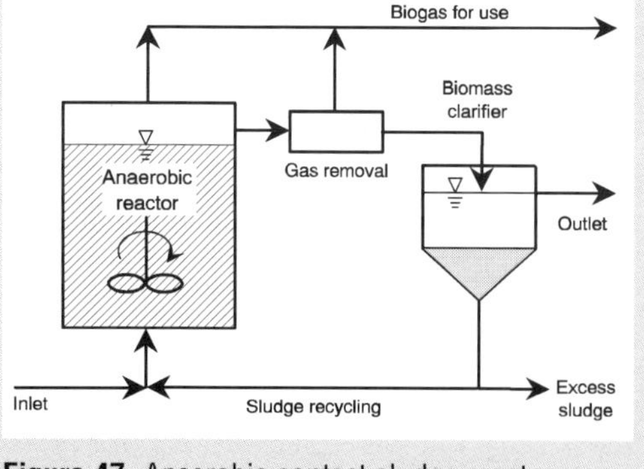

Figure 47. Anaerobic contact sludge reactor.

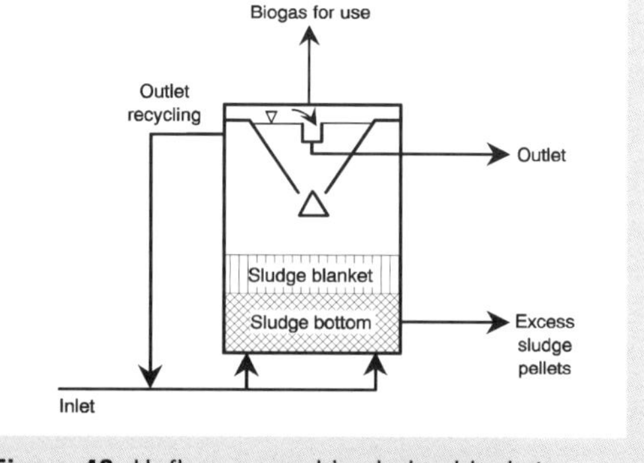

Figure 48. Upflow anaerobic sludge blanket (UASB) reactor.

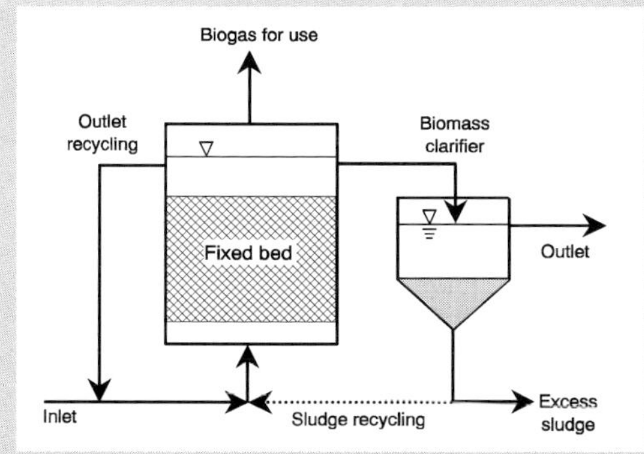

Figure 49. Anaerobic fixed bed reactor.

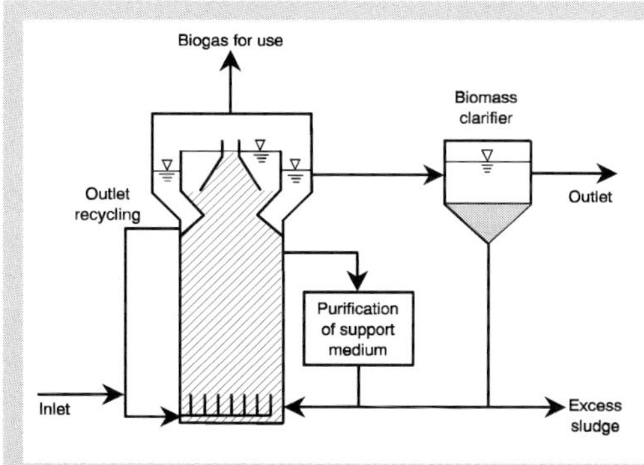

Figure 50. Anaerobic fluidized bed ractor.

Anaerobic degradation occurs in four steps in which different bacterial groups participate as Fig. 51 shows:

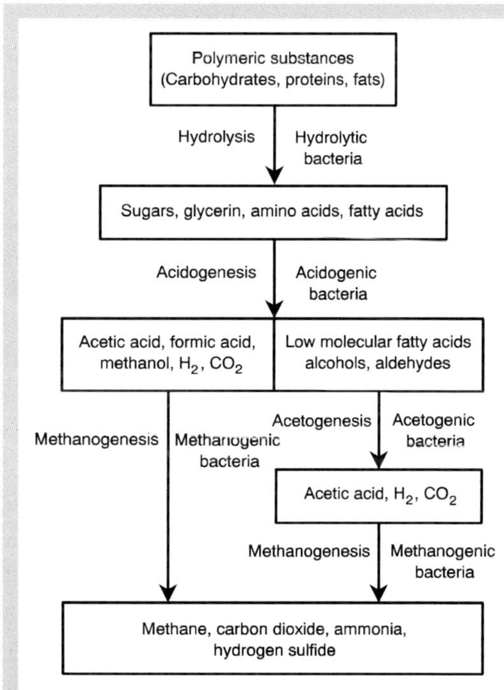

Figure 51. Four steps of anaerobic degradation.

- The first step is a hydrolytic degradation of polymer compounds. Macromolecules are split into low molecular weight substances by extra-cellular enzymes of the bacteria participating in this degradation process.
- In the next step, the hydrolysis products are fermented by acidogenic bacteria to short chain fatty acids. These are primarily acetic acid, propionic acid, butyric acid, and lactic acid. Other materials are simple alcohols, hydrogen, and carbon dioxide.
- In the third step, the intermediate products of the acidification process are degraded by acetogenic bacteria to acetic acid, hydrogen, and carbon dioxide.
- The final degradation step is the transformation of the products from the acetogenic phase by methanogenic microorganisms to methane and carbon dioxide, the main components of the biogas. The bacteria participating in this step have the lowest growth rates and are the most dependent on the environmental conditions.

The acetogenic and methanogenic bacteria must live in close symbiosis for the degradation processes to work efficiently. This imposes different types of conditions on the process technology. For example, high shearing forces caused by mechanical circulation should be largely avoided.

The optimum temperature for the growth of mesophilic anaerobic microorganisms is 35°C–38°C. For the anaerobic purification of the wastewater of recovered paper processing mills with tightly closed process water circuits, this means that heat exchangers must cool the wastewater to this temperature range. The optimum pH of 6.8–7.2 is usually present in process waters of recovered paper processing mills. The possibility for controlled addition of acids and alkaline solutions is necessary to prevent pH fluctuations that would disturb the activity of the bacteria in the wastewater treatment plant.

It has been agreed at an international level to base the process technology rating of anaerobic reactors on the COD volume load. This rating criterion depends largely on the different processes. In contact sludge reactors, the COD volume load is 2–5 kg $COD/m^3 \cdot d$. Fixed bed reactors operate with COD volume loads of 5–20 kg $COD/m^3 \cdot d$. Considerably higher volume loads are possible for the UASB reactor. Usually a value of 5–15 kg $COD/m^3 \cdot d$ is chosen. Newly developed pellet reactors can be loaded with 30 kg $COD/m^3 \cdot d$. With all reactor types, COD reductions of 70%–80% are possible. The reductions of COD_5 are 80%–95%. As part of this process, about 0.4 m^3 biogas is produced per kg COD eliminated.

The produced biogas comprises 70%–80% methane, 20%–30% CO_2, less than 2% hydrogen sulfide (H_2S), and traces of other gases. The heating value of the biogas is 22–30 MJ/m^3. In the anaerobic decomposition process, hydrogen sulfide forms from sulfur-containing organic compounds and sulfates. Caused by sulfate-reducing bacteria, the reaction is unavoidable since it is thermodynamically favored over methane formation. Any dissolved H_2S that is present has a toxic effect on the entire bacteria population and can cause a serious disturbance of the methane formation process. Because sulfate is usually present in the wastewater of recovered paper processing mills, problems in the anaerobic treatment are often due to high sulfate concentrations.

Biogas must usually be cleaned before using for energy purposes. The most important consideration is elimination of H_2S because H_2S and SO_2 produced by incineration cause corrosion in a combustion plant. Caustic soda washers are primarily used for desulfurization before burning the biogas. An interesting development is a bioreactor in which sulfide contained in the wash water is oxidized to elemental sulfur by microorganisms. The sulfur is separated and can be re-used elsewhere.

Problems can occur in anaerobic plants if the calcium concentration in the wastewater to be treated is more than 400 mg/L[82]. By reaction with carbon dioxide at a neutral pH, higher calcium concentrations lead to calcium carbonate precipitation. With processes using biomass carriers, precipitation can cause blocking of the growth surfaces. Especially with fixed bed reactors, an additional risk exists for blocking the reactor. To minimize calcium carbonate precipitation, running the anaerobic reactor in a lower acid pH range and accepting lower COD reduction levels is best. This is especially suitable where an aerobic treatment stage follows the anaerobic reactor.

After a long learning and understanding phase, experiences of working with anaerobic wastewater treatment plants at recovered paper processing mills are now primarily positive. An example of a successful technical application is the UASB reactor installed at the Roermond paper mill in the Netherlands[83]. Since being placed in operation at the end of 1983, the reactor has worked largely trouble-free. It reaches average reductions of 70% COD and 80% BOD_5.

The following preconditions are essential for the smooth operation of anaerobic reactor systems:

- Avoidance of pH and temperature fluctuations
- Avoidance of hydraulic impact loads by installing buffer tanks
- Use of chemical additives for paper production (These are primarily biocides, cleaning agents, and defoamers) that are compatible with the anaerobic microorganisms
- Effective retention of suspended solids.

The debate about efficiency of the anaerobic degradation process in a one-stage or two-stage operation continues. A two-stage process allows setting the optimum conditions of temperature and pH in a hydrolysis or acidification reactor, i.e., for the first two steps of anaerobic degradation. Acetification and methane formation are then done in a second reactor. This is the methane reactor of Fig. 52. A two-stage process has greater operating stability and reliability. Hydrolysis and acidification of the wastewater components occur in tightly closed water circuits at recovered paper processing mills. Because degradation products such as acetic acid, propionic acid, lactic acid, and butyric acid are formed that are easily transformed by acetogenic and methanogenic microorganisms, a one-stage treatment may also be suitable.

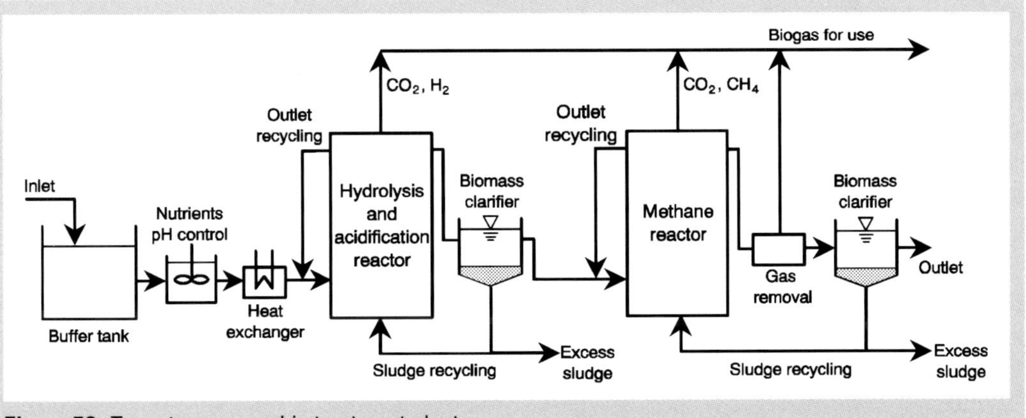

Figure 52. Two-stage anaerobic treatment plant.

4.3 Characterization of treated wastewater

Biological processes can eliminate nearly 100% of biologically degradable organic substances in the wastewater of recovered paper processing mills. Considering the sum of organic substances determined as COD, the level of elimination is 80%–95%. One must distinguish between a genuine biological degradation and a physico-chemical elimination, e.g., adsorption to activated sludge. Table 18 shows COD and BOD concentrations or loads in biologically treated wastewater from nondeinking paper mills. Table 19 summarizes the corresponding data for deinking mills.

Table 18. Characterization of wastewater of RCF mills without deinking after biological treatment.

Paper grade	Number of mills	BOD_5 concentration [mg/L]			BOD_5 load [kg/t]			COD concentration [mg/L]			COD load [kg/t]		
		Min	Max	Average	Min	Max	Average	Min	Max	Average	Min	Max	Average
Board	n.a.[78]	3	28	10	0.01	0.13	0.06	60	270	150	0.29	1.12	0.77
Board	4[76]	7	20	13	–	–	–	100	180	140	0.9	1.8	1.2
Test liner, fluting	7[76]	6	12	9	–	–	–	110	200	170	0.5	1.4	0.7
Test liner, fluting	n.a.[90]	–	–	–	0	0.2	0.1	–	–	–	0	2.2	1.0

n.a. = not announced

Table 19. Characterization of wastewater of RCF mills with deinking after biological treatment.

Paper grade	Number of mills	BOD_5 concentration [mg/L]			BOD_5 load [kg/t]			COD concentration [mg/L]			COD load [kg/t]		
		Min	Max	Average	Min	Max	Average	Min	Max	Average	Min	Max	Average
Newsprint	7[76]	10	25	16	–	–	–	110	420	260	2.1	4.2	3.1
Tissue	2[76]	3	21	–	–	–	–	100	225	–	1.0	2.3	–
Newsprint, tissue	n.a.[78]	–	–	9	–	–	0.1	–	–	292	–	–	3.1

n.a. = not announced

Treated wastewater from mills processing recovered paper like the wastewater from other mills is not toxic for water-borne organisms as evidenced by the results of fish and other biological tests[77]. The nitrogen and phosphorus concentration is low provided no excess dosing of these components as nutrients in biological wastewater treatment plants occurs. Only small risk of extensive growth of water plants in surface waters therefore exists.

The treated wastewater of recovered paper processing mills has a very low heavy metal content. To the extent that heavy metals are introduced into the recovered paper processing via printing inks such as copper from blue pigments, they are largely removed with the deinking sludge. A smaller amount is adsorbed to the biomass of biological wastewater treatment plants. In the treated wastewater, the heavy metal concentrations are considerably below 100 μg/L and are sometimes lower than the concentrations in drinking water.

The additives used in paper production generally contribute to the load of the treated wastewater only slightly. For example, some starches, soaps, or surfactants are easily biologically degradable. Others and especially higher molecular weight additives such as retention agents, wet strength agents, or coating binders adsorb strongly to the activated sludge.

Detailed research has been done on this subject. At the end of the 1970s, Göttsching and Lüttgen[84] demonstrated which additives are critical in this regard. More recent research shows that the manufacturers of additives today offer more environmentally friendly products. When used corrrectly, these products have neither a negative influence on the biological wastewater treatment plant nor on the receiving water[85]. This primarily applies to defoamers, cleaning agents, surfactants, and slimicides.

Due to their remobilizing effect on already separated heavy metals, chelating agents that are considered potentially dangerous in surface water particularly with regard to drinking water production are usually poorly degradable. This is true especially for diaminetriethylenepentaacetic acid (DTPA) and ethylenediaminetetraacetic acid (EDTA) that are sometimes used when bleaching recycled fiber pulp with peroxide or dithionite. Because no adsorption to the sludge of the wastewater treatment plants occurs, the chelating agents reach the effluent flow almost intact. Extensive research in the paper industry in Central Europe showed that the addition of chelating agents at deinking operations can largely be avoided. This is why no widespread use is made of chelating agents today (See Chapter 9). The manufacturers of packaging papers and board based on recycled fibers do not use any complexing agents.

4.4 Closed water circuits

In Germany at the beginning of the 1980s, about a dozen recovered paper processing mills had closed the process water circuit totally. Most were small paper mills producing corrugating medium and test liner[86]. The fresh water consumption in these mills was 1.0–1.5 m^3/t of paper produced. This corresponds to the volume of the water evaporated in the dryer section of the paper machine.

Total closure of the process water systems leads to considerable loading of the process water with colloidal and dissolved organic and inorganic compounds. This causes serious problems in the production process if no control measures are undertaken to avoid the following possible disadvantages:

- Significant decrease of oxygen content of the process water approaching anaerobic conditions with a microbiologically induced reduction of sulfate to hydrogen sulfide and the formation of odorous low molecular weight fatty acids
- Intensified growth of microorganisms
- Aggressive corrosion caused by high temperature and high contents of chlorides, sulfates, and organic acids

- Significant emission of odorous organic compounds from the dryer section of the paper machine to the surroundings of the paper mill
- Impaired quality of the paper produced influenced by odorous compounds
- Higher demand of slimicides.

To control these serious problems, some mills have re-opened their closed water circuit and are now producing with a small volume of wastewater in the range of 2.5–5.0 m^3/t of paper produced.

Because of local environmental requirements, two European paper mills applied new methods in the middle of the 1990s to control the demanding conditions of their closed effluent-free process water loops. They installed in-line treatment plants to reduce the organic load of the process water. A partial stream of the total process water is treated physically and biologically followed by a final physical treatment. The process water treatment plants therefore are "kidneys" in the totally closed water circuits.

Diedrich *et. al.*[87] describe the process water treatment plant at the Zülpich paper mill in Zülpich, Germany. The annual production of this paper mill is 360 000 t of test liner and corrugating medium produced on two machines with trims of about 5 m at maximum production speeds of 650 m/min and 1 200 m/min, respectively.

For economical and other reasons, the arrangement in Fig. 53 was chosen as a system of kidneys. A partial stream of the heavily loaded process water is treated physically and biologically followed by a final physical treatment. The process water volume

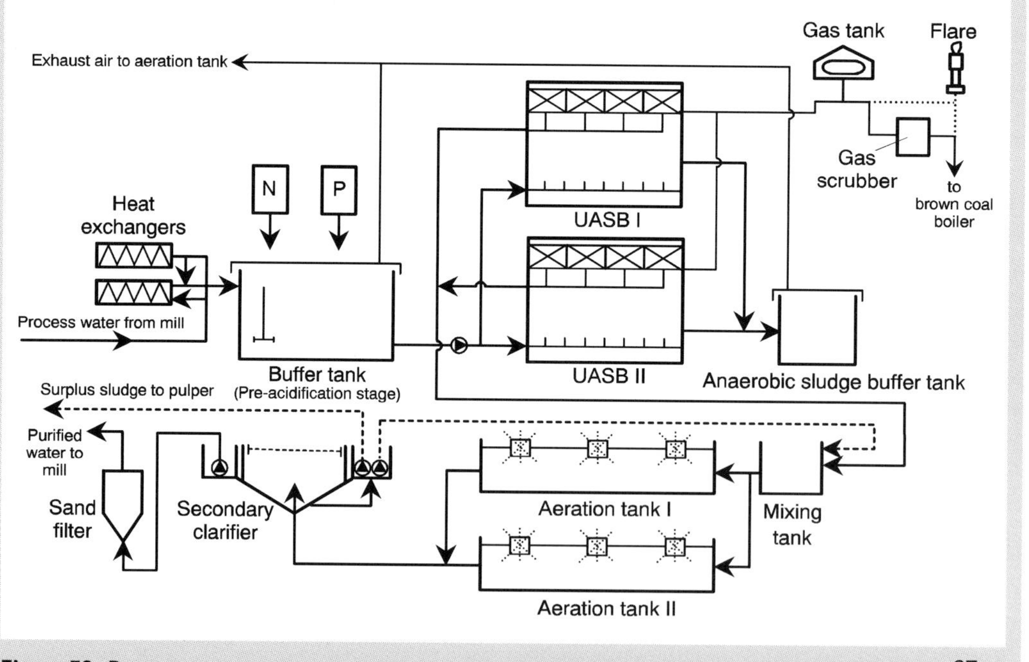

Figure 53. Process water treatment plant in the closed water system of Zülpich paper mill, Germany[87].

for treatment corresponds to about 4 m^3/t of paper produced. In the first step, a clarification is performed by disk filter and flotation in sequence. Fig. 53 does not show these clarification units. The next step reduces the temperature of the clarified process water from about 50°C to less than 40°C by heat exchangers fed with ground water. Before entering the bioreactors that operate in parallel, the cooled and clarified process water is pre-acidified. This occurs in a separate tank where the nitrogen and phosphorus nutrients in the form of urea and phosphoric acid are added. The residence time in the pre-acidification tank can be 2–6 h.

The biological treatment uses two UASB reactors with a total volume of 2 000 m^3. Both bioreactors run with recirculation of some reactor effluent not indicated in the figure to control flow in the bioreactors. Increasing the upstream velocity of the flow improves mixing in the bioreactors and contributes to better separation of the biogas produced. In addition, recirculation levels any COD and pH variations of the partial stream of the incoming process water. Such variations would influence the degradation efficiency of the bioreactors. On average, the recycled flow is about 50% larger than the output of the bioreactors. The excess pellets are separated from the effluent by a conventional screen and partly fed back to the UASB reactors.

The effluent of the UASB reactors then enters two activated sludge tanks in series. Both tanks have a volume of 900 m^3 and are aerated with three mammoth rotors. The aerobically treated process water is clarified in the following sedimentation basin and finally purified by sand filtration consisting of four units of which three are permanently operating. The fourth unit is a spare filter required for the washing intervals when the three operating filters are cleaned countercurrently.

The biogas produced in both anaerobic reactors contains a high concentration of hydrogen sulfide (H_2S). The biogas must therefore be desulfurized using a gas washer before the purified gas is burned for energy recovery. The sulfide-containing wash water is treated in a separate bioreactor using the principle of biological oxidation of sulfide. This results in the formation of sulfur whose sedimented flocs are then dewatered by a centrifuge approaching a dry substance of 60%. The efficiency of the desulfurization is surprisingly high at approximately 99.9%. This gives a final hydrogen sulfide concentration of the biogas of less than 15 ppm. The recovered sulfur is used for production of sulfuric acid.

The amount of sludge produced in the anaerobic and aerobic stages is not more than 1 000 kg dry substance per day. This corresponds to approximately 0.1% of daily paper production. Such a small volume is used as a raw material for the paper production without affecting runnability on the paper machine and the paper characteristics. Another option would be incineration of these sludges for energy generation after their dewatering. In this paper mill, only the sludge of the aerobic stage is re-used. The excess sludge pellets of the anaerobic stage go to other paper mills.

Before installation of the treatment plant, the COD concentration of the process water was as high as 34 000 mg/L. After the start-up of the treatment plant, the COD of the process water decreased to 8 500 mg/L after about eight weeks as Fig. 54 shows. The COD of the effluent of the treatment plant that is recycled into the process water circuit is about 800 mg/L. This corresponds to a total COD reduction rate of the treatment plant of 90%. This efficiency fulfills the high requirements of the process.

The BOD_5 of the original, untreated process water was about 20 000 mg/L resulting in a BOD_5/COD ratio of about 0.6. This is common for process water and untreated effluent of recovered paper processing. The ratio indicates excellent biodegradability. Thanks to the multistage treatment, the purified partial stream of the process water reached a BOD_5 concentration of less than 50 mg/L in the effluent of the post-clarification basin as Fig. 55 shows. The efficiency of the multistage treatment plant in terms of BOD reduction was as high as 99%. The process water today has BOD_5 of about 5 000 mg/L. This is very acceptable.

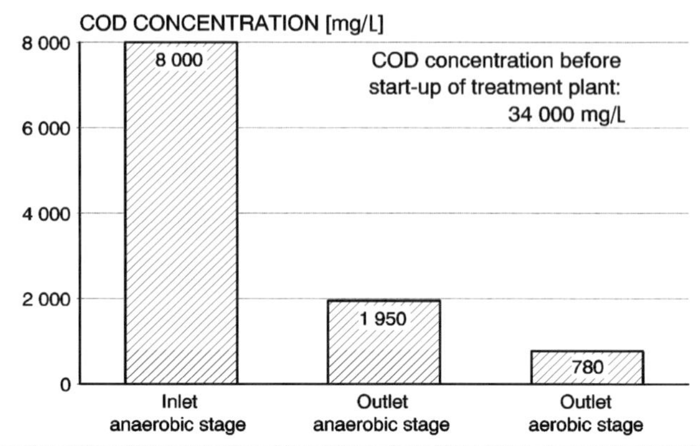

Figure 54. COD concentration of process water at different stages of the process water treatment plant (Zülpich paper mill).

Table 20 shows the concentration of organic acids in the process water before the installation of the kidneys. The total concentration of the above-mentioned organic acids was 10 000–13 000 mg/L. The COD proportion caused by the organic acids was 32%–43%. Figure 56 shows concentration of the organic acids of the process water in the different stages of the treatment plant and illustrates the high efficiency of the degradation of organic acids.

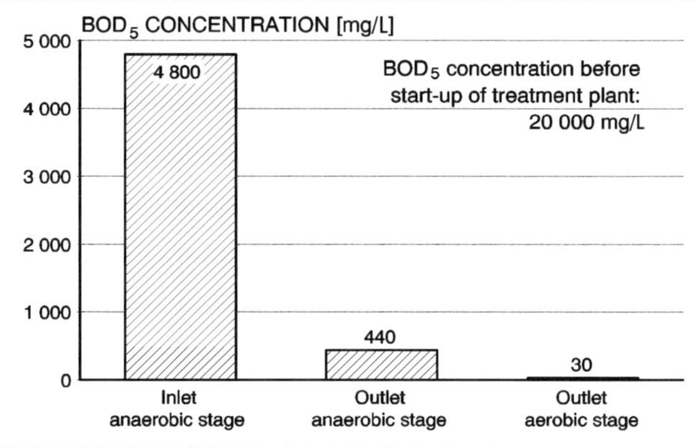

Figure 55. BOD_5 concentration of process water at different stages of the process water treatment plant (Zülpich paper mill).

Table 20. Concentration of organic acids in the process water before start-up of the process water treatment plant (Zülpich paper mill).

Organic acid	Concentration [mg/L]	COD proportion caused by organic acids [%]
Acetic acid	5 000–5 400	15–17
Propionic acid	800–900	3–4
Butyric acid	400–800	2–4
Lactic acid	3 800–5 700	12–18
Total	**10 000–12 800**	**32–43**

Papermakers have frequent concern that biologically purified effluent reused in the paper mill contributes to an intensified formation of microorganisms in the white water system. This relates to the formation of slime lumps and an increased frequency of web breaks on the paper machine. No negative effects on web breaks occurred after the start-up of the process water treatment plant. Due to purification of a partial stream of the process water, a positive effect occurred because the amount of biocides used in the paper mill could be reduced by 30% after the start-up of the process water treatment plant.

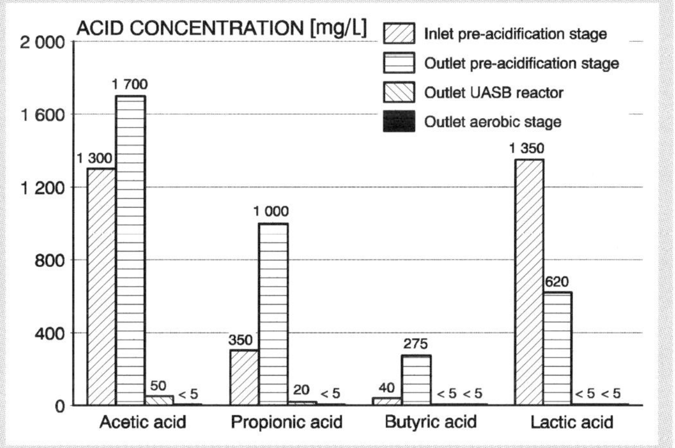

Figure 56. Concentration of organic acids in the process water at different stages of the process water treatment plant (Zülpich paper mill).

For paper characteristics, the most important properties of corrugating medium and test liner such as Concora medium test (CMT), short span compression, and burst have not been affected. Most significant was the effect on the content of organic acids of the paper produced. According to Table 21, the concentrations of low molecular fatty acids decreased significantly. The bad odor of the previous paper produced that had caused occasional complaints by customers disappeared. Sensory tests comparing the intensity of odor with reference samples produced by competitors using only a partially closed water system verified this success.

Table 21. Concentrations and reduction rates of organic acids in a closed process water system (Zülpich paper mill).

Organic acid	Concentration [g/kg paper]		Reduction rate [%]
	Before	After	
	start-up of treatment plant		
Formic acid	1.2	0.6	50
Acetic acid	5.9	0.6	90
Propionic acid	0.4	< 0.2	> 50
Butyric acid	0.6	< 0.2	> 65
Lactic acid	6.6	0.7	90

Before the start-up of the process water treatment plant, the area around the paper mill and neighboring villages occasionally complained of a bad odor. After the start-up of the treatment plant, the contaminants of the exhaust air of the paper machine and in the surrounding air of the paper mill were measured. Table 22 shows the emissions of organic acids released by the paper machine exhaust air before and after the start-up of the process water treatment plant. The substances that cause bad smells were decreased by as much as 97%. The total volume of odor-creating components and the total carbon emission by the paper machine exhaust air were reduced by 93% and 76%, respectively. The inhabitants of the neighboring villages no longer complain about bad odor, and the immediate vicinity of the paper company is also free from bad odor.

Table 22. Effect of process water treatment on the load of the exhaust air of paper machines (Zülpich paper mill).

Exhaust air compound	Unit	Before	After	Reduction rate
		start-up of treatment plant		[%]
Total C	kg/h	7.1	1.7	76
Odor mass stream	Odor units/h	443.0	31.0	93
Formic acid	kg/h	96.6	2.8	97
Acetic acid	kg/h	56.1	2.8	95
Propionic acid	kg/h	17.1	0.6	96
Butyric acid	kg/h	n.a.	0.6	–
Lactic acid	kg/h	n.a.	0.6	–
COD	kg/h	220.6	10.9	95

n.a.: not analyzed

Another in-plant process water project with the objective of total closure of the water circuit is the French Assi Domain Packaging Lecoursonnois mill. With an annual production of 75 000 t of corrugating medium, this is a medium-sized paper mill whose receiving water is used as drinking water for the city of Paris. The plant shown in Fig. 57 reveals several differences compared with the plant at Zülpich Papier. The anaerobic treatment in a fluidized bed reactor is not followed by an activated sludge stage. Instead, a re-aeration stage was incorporated in which dissolved sulfide is oxidized to sulfate. The odorous air of this stage is purified using a compost filter. The treated wastewater is returned 100% to the production process as is the excess sludge generated in the anaerobic fluidized bed reactor. The produced biogas is used in a boiler or burned.

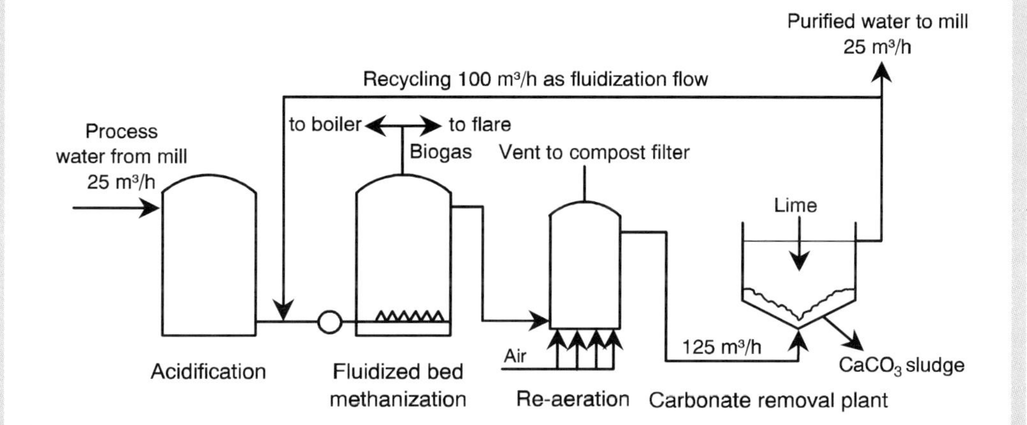

Figure 57. Process water treatment plant in the closed water system of Lecoursonnois paper mill, France[88].

The biological plant started in 1994. At a partial closure of 50%, a decrease in anaerobic reactor efficiency occurred. This was probably due to scaling of calcium carbonate. The re-aeration tank also accumulated scale after several weeks. After complete closure, the calcium concentration in the process water increased from 500 mg/L to 1 400 mg/L. Calcium comes from the fillers and pigments in the recovered paper furnish. No possibility to reduce the calcium input therefore is possible. To protect the methanization reactor against scaling, a calcium carbonate precipitation stage was added using lime as precipitating agent. The resulting calcium carbonate sludge is then spread on land. Modified in this way, the process achieved good working levels in 1997. Since that time, the water system of the paper mill has been completely closed. According to the findings of Nivelon *et. al.*[88], the runnability of the paper machine is not disturbed and the quality of the produced paper is at a constant high level.

In tests at other paper mills concerning the return of parts of biologically treated wastewater to the production process, the precipitation of calcium carbonate has also been a limiting factor[89]. Research is currently concerned with establishing the degree of effectively preventing calcium scaling of showers and rolls by using $CaCO_3$ precipitation processes, nano-filtration, inhibitors, and magnetic process water treatment.

References

1. Hamm, U. and Göttsching, L., Das Papier 45(10A):V146(1991).
2. Borschke, D., Gehr, V., and Mönnigmann, R., Das Papier 41(6A):V146(1997).
3. Chryssos, G. and Murr, J., "Einsparung von Deponieraum durch mechanische Fraktionierung und weitergehende Verwertung von Sortierrückständen aus der Altpapieraufbereitung", PTS-FB 30/96, Papiertechnische Stiftung, Munich, 1996, 25 pp.
4. Hamm, U., Wochenbl. Papierfabr. 124(9):409(1996).
5. Hanecker, E., "Alternative Verwertung von Reststoffen – außer thermische Nutzung – getrennt nach Fasern und Füllstoffen", INGEDE Report 2892 PTS, Papiertechnische Stiftung, Munich, 1994, 25 pp.
6. Bienert, C., "Untersuchung der Verwertungsmöglichkeiten von Schlämmen aus der Papierindustrie", PTS-FB 12/94, Papiertechnische Stiftung, Munich, 1994.
7. Raitio, L., Paperi Puu 74(2):132(1992).
8. Grünewald, H., Hogrebe, B., and von der Geest, C., Wochenbl. Papierfabr. 117(7):263(1988).
9. Hamm, U. and Göttsching, L., Das Papier 43(10A):V39(1989).
10. Welker, A. and Schmidt, T.G., Water Res. 31(4):805(1994).
11. Wünschmann, G., Hamm, U., and Göttsching, L., Wochenbl. Papierfabr. 123(9):395(1995).
12. Krieger, U., Paper Technology 39(4):45(1998).
13. Wenzel, L., Wochenbl. Papierfabr. 124(19):866(1996).
14. Azarniouch, M. K., Tappi J. 78(9):139(1995).
15. Bilitewski, B., Härdtle, G., Marek, K., et al., "Waste Management", Springer-Verlag, Berlin, Germany, 1994, 699 pp.
16. Kappel, J., Fluch, H. -W., and Brunnmair, E., Das Papier 51(6A):V182(1997).
17. Bertolotti, H. and Ploss, H., Wochenbl. Papierfabr. 123(10):456(1995).
18. Webb, L., Pulp Paper International 35(10):18(1994).
19. Kraft, D.L. and Orender, H.C., Tappi J. 76(3):175(1993).
20. Anthony, E.J., Herb, B.E., and Lewnard, J.J., Pulp Paper Can. 96(3):T109(1995).
21. Frederick, W.M. J., Lisa, K., Lundy, J.R., et al., Tappi J. 79(6):123(1996).

22. Bienert, Ch., Hanecker, E., and Murr, J., Das Papier 51(6A):V177(1997).
23. Latva-Somppi, J., Tran, H.N., Barham, D., et al., Pulp Paper Can. 95(10):T381(1994).
24. Gahr, A. and Metz, A. -M., Das Papier 47(10A):V126(1993).
25. Dörre, R., Paper Technology 36(5):26(1995).
26. Douglas, M.A., Latva-Somppi, J., and Razbin, V.V., Tappi J. 77(5):109(1994).
27. Lewnard, J.J., Herb, B.E., Siebert, K.J., et al., Progress in Paper Recycling 2(4):45(1993).
28. Chryssos, G. and Murr, J., Das Papier 50(4):178(1996).
29. Louhimo, J., Burelle, R., and Suoniemi, T., Pulp Paper Can. 96(4):T143(1995).
30. Gutjahr, A., Das Papier 51(6A):V208(1997).
31. Demharter, W., Allg. Papierrundschau 119(33):818(1995).
32. Reimann, D.O., Entsorgungspraxis (10):63(1995).
33. Wünschmann, G. and Hamm, U., Das Papier 47(10A):V119(1993).
34. Mick, A., Ross, D. and Flemming, J.D., "Processing primary clarifier sludge into compost", TAPPI 1982 Environmental Conference Proceedings, TAPPI PRESS, Atlanta, p. 93.
35. Campbell, A.G., Engebretsond, R.R., and Tripepi, R.R., Tappi J. 74(9):183(1989).
36. Carter, N., Pulp & Paper 57(3):102(1983).
37. Smyser, S., Biocycle 23(3):25(1982).
38. Wünschmann, G., Hamm, U., and Göttsching, L., Wochenbl. Papierfabr. 123(16):707(1995).
39. Einspahr, D. and Fisus, M., "Paper mill sludge as a soil amendment", TAPPI 1984 Environmental Conference Proceedings, TAPPI PRESS, Atlanta, p. 253.
40. Shimek, S., Charles, T., Nessman, M., et al., "Paper sludge land application studies for three Wisconsins mills", TAPPI 1988 Environmental Conference Proceedings, TAPPI PRESS, Atlanta, p. 413.
41. Hatch, C. J. and Pepin, R. G., Tappi J. 68(10):70(1985).
42. Diehn, K., "Recycling a deinking mill's waste through land spreading", TAPPI 1991 Environmental Conference Proceedings, TAPPI PRESS, Atlanta, p. 739.
43. Pridham, N.F. and Yan, A.Y., Das Papier 45(3):103(1991).
44. Bailey, S.L., O'Neill, D.C., and Smith, R.E., "Assessing the impact of winter land application of secondary sludge on water quality", TAPPI 1995 Environmental Conference Proceedings, TAPPI PRESS, Atlanta, p. 735.
45. Trepanier, L., Caron, J., Yelle, S., et al., "Impact of deinking sludge amendment on agricultural soil quality", TAPPI 1996 Environmental Conference Proceedings, TAPPI PRESS, Atlanta, p. 529.

46. Liebe, P. and Gerger, W., Zement-Kalk-Gips 46(12):E327(1993).

47. Ernstbrunner, L., Das Papier 51(6):284(1997).

48. Buchinger, H. and Ernstbrunner, L., Wochenbl. Papierfabr. 119(21):845(1991).

49. Kreft, W., AbfallwirtschaftsJournal 1(11):1(1989).

50. Russel, C. and Odendahl, S., Pulp Paper Can. 97(1):T17(1996).

51. Fenchel, U., Das Papier 32(10A):V54(1978).

52. Mertz, H.A. and Jayme, T.G., "Startup and operating experience with the ZIMPRO high pressure wet oxidation system for sludge treatment and clay reclamation", TAPPI 1984 Environmental Conference Proceedings, TAPPI PRESS, Atlanta, p. 75

53. Gosling, C.D., Lamparter, R.A., and Barna, B.A., Tappi 62(2):75(1981).

54. Johnston, J. H. and Wiseman, N., Appita J. 49(6):397(1996).

55. Hamm, U. and Göttsching, L., Wochenbl. Papierfabr. 126(1):15(1998).

56. Modell, M., Larson, J., and Sobzynski, S., "Supercritical water oxidation of pulp mill sludges", TAPPI 1991 Engineering Conference Proceedings, TAPPI PRESS, Atlanta, p. 393.

57. Cooper, S.P., Folster, H.G., Gairns, S.A., et al., Pulp Paper Can. 98(10):T365(1997).

58. Sohara, J.A., Internationale Papierwirtschaft (1):F46(1997).

59. Wünschmann, G., "Untersuchungen zur Kompostierung von Reststoffen der Papierindustrie und Altpapier unter besonderer Berücksichtigung von Schadstoffbilanzierungen", Ph.D. thesis, Darmstadt University of Technology, Darmstadt, Germany, 1997, 187 pp.

60. Welker, A., "Die Belastung von Reststoffen aus der Papierindustrie mit halogenorganischen Verbindungen – Aufklärung der chemischen Zusammensetzung und Folgerungen für die Reststoffentsorgung", Ph.D. thesis, University Kaiserslautern, Kaiserslautern,Germany, 1995, 170 pp.

61. Schurz, J., Cellulose Chemistry Technology 18(3):257(1984).

62. Anon., "Energy recovery from used packaging", SIS-Swedish Standards Institution, Stockholm, 1994, 39 pp.

63. Anon., "Towards a Sustainable Paper Cycle", International Institute for Environment and Development (IIED), London, 1996, 258 pp.

64. Pacey, J.G. and De Gier, J.P., „The factors influencing landfill gas production", Conference on Energy from Landfill Gas, London, 1992, 26 pp.

65. Anon., "Daten zur Umwelt, Der Zustand der Umwelt in Deutschland" (Umweltbundesamt, Ed.), Erich Schmidt Verlag, Berlin, 1997, 570 pp.

66. Boelans, J., De Wilde, B., and De Baere, L., Compost Science and Utilization 4(1):60(1996).

67. Svedberg, G., "Waste incineration for energy recovery", Royal Institute of Technology (KTH), Stockholm, Report 1, 1992.

68. Ekwall, K., Avfall-92, NUTEK Report R 1992:31, NUTEK, Stockholm, 1992,
69. Anon., Bild der Wissenschaft Plus (5):36(1998).
70. Vahrenholt, F., "Globale Marktpotentiale für erneuerbare Energien" (Deutsche Shell AG, Ed.), Reihe: Analysen und Vorträge 1/1998, Hamburg, Germany, 1998, 15 pp.
71. Göttsching, L., Papier aus Österreich (9):11(1993).
72. Göttsching, L., Paperi Puu 76(8):479(1994).
73. Göttsching, L., Hamm, U., and Putz, H.-J., Das Papier 48(10A):V93(1994).
74. Anon., "Putting new energy into paper recycling", Internal Report, Georgia-Pacific Corp., Atlanta, 1992.
75. Kara, M., Paperi Puu 76(1/2):44(1994).
76. Zippel, F., "Wasserhaushalt von Papierfabriken", Deutscher Fachverlag GmbH, Frankfurt, Germany, 1999, 302 pp.
77. Möbius, C.H. and Cordes-Tolle, M., Das Papier 47(10A):V53(1993).
78. Luttmer, W.J., "Dutch Notes on Best Available Techniques for Paper and Board Production from Recycled Fibers", Document 96, Institute for Inland Water Management and Wastewater Treatment RIZA, Lelystad, The Netherlands, 1996.
79. Pfitzner, T., Wochenbl. Papierfabr. 127(4):223(1999).
80. Badar, T.A., Progress in Paper Recycling 2(5):42(1993).
81. Möbius, C.H. , Das Papier 45(10A):V49(1991).
82. Hamm, U., Bobek, B., and Göttsching, L., Das Papier 45(10A):V55(1991).
83. Habets, L.H.A., Knelissen, J.H., and Hack, P.J.F.M., Wochenbl. Papierfabr. 112(20):731(1984).
84. Göttsching, L. and Lüttgen, W., Das Papier 32(10A):V46(1978).
85. Hamm, U., Bobek, B., and Göttsching, L., Wochenbl. Papierfabr. 127(9):599(1999).
86. Geller, A. and Göttsching, L., Tappi J. 65(9):97(1982).
87. Diedrich, K., Hamm, U., and Knelissen, J.H., Das Papier 51(6A):V153(1997).
88. Nivelon, S., Pichon, M., and Piollet, A., Pulp Paper Can. 99(11):T371(1999).
89. Hamm, U., Wochenbl. Papierfabr. 127(5):303(1999).
90. Göttsching, L., Hamm, U., and Putz, H.-J., "Report on Best Available Techniques (BAT) for the production of case making materials (Test liner and Wellenstoff)", Institute of Paper Science and Technology, Darmstadt University of Technology, Darmstadt, Germany, 1998, 57 pp.

CHAPTER 13

Environmental aspects

CHAPTER 13

Udo Hamm

Environmental aspects

1 Paper recycling and ecological labeling

Since the first introduction of an ecological label, the Blue Angel, in Germany in 1977, a major increase has occurred globally in environmentally related labels for different types of consumable goods including paper and paper products. Awarding ecological labels for environmentally friendly products should stimulate industry to give due consideration as early as the product planning stage of the consequences for the environment of a product. For the consumer, the ecological label is a purchasing aid in the sense of an "ecological guide."

Ecological labeling systems are now in operation in many countries and regions. In the wake of the success of the Blue Angel in Germany, Canada and Japan started their ecological labels at the end of the 1980s followed by other European states such as Austria and the Netherlands. The Nordic countries of Finland, Norway, and Sweden defined criteria for ecological labels. Table 1 shows some worldwide ecological label programs with their starting dates.

Table 1. Worldwide ecological label programs[1].

Eco label program	Start-up date
German Blue Angel	1977
Swedish Society for Nature Conservation	1987
Canadian Environmental Choice	1989
Japanese Ecomark	1989
Nordic White Swan	1989
Australian/New Zealand Green	1990
Brand Paper (UK) Eco Check	1992
Taiwan eco label	1992
French "Norme Française Environement"	1992
Dutch eco label	1992
EU eco label	1993
Austrian eco label	1994

All ecological labels lay claim to overall consideration of various types of potential environmental pollution that could occur throughout the lifetime of a product. This includes its production, distribution, use, recycling, and waste disposal. Evaluation criteria differ greatly in the individual countries. This results in different findings for environmental compatibility. For example, the Blue Angel in Germany and the Ecomark in Japan that are the most long-running labels for paper products place considerable emphasis on the content of recycled fiber. The Nordic Swan focuses primarily on the emissions from pulping, bleaching, and papermaking.

1.1 The German Blue Angel

Typical characteristics of the German ecological label are the nongovernmental decision process and the inclusion of all concerned parties in the establishment of specification criteria. The German Environment Protection Agency, the German Institute for Quality Assurance and Labeling (RAL), and the Eco Label Jury, an independent body comprising experts from the scientific, industrial, and environmental protection organizations, all participate. The ecological label is awarded only in market segments with a direct competition between products with greatly differing environmental impacts.

In 1997, the ecological label program consisted of 88 different product groups. In the paper sector, the ecological label is awarded for the product categories listed in Table 2. About 15% of those awarded the ecological label are non-German products. In all product categories, the awarding of the ecological label depends on the use of recovered paper in their manufacture. Hygienic papers must therefore be totally manufactured from recovered paper. For crêpe toilet papers and paper towels, only low-grade, medium-grade, and kraft-containing recovered paper may be used (See Chapter 4). With other hygienic papers (toilet paper, kitchen towels, handkerchiefs, etc.), the proportion of low-grade, medium-grade, and kraft-containing recovered paper must be at least 20%. The remainder can be better grades such as pre-consumer recovered paper.

Table 2. Paper and paper products in the ecological label Blue Angel (1997)[2].

Product category	RAL-UZ number	Number of label-appliers	Number of products
Recycled hygienic paper	5	43	182
Recycled graphic paper	14	119	300
Wallpaper and woodchip paper	35	58	109
Recycled cardboard	56	59	99
Newsprint made mainly from recycled fibers	72	24	30

RAL =Reichs-Ausschuß für Lieferbedingungen (German Institute for Quality Assurance and Labeling)
UZ =Umweltzeichen (Eco Label)

Recycled graphic papers and recycled board must also use 100% recovered paper. At least 51% of the input of recovered paper must be low-grade and medium-grade recovered paper (post-consumer recovered paper). Newsprint that is entitled to bear the Blue Angel must have at least 80% recovered paper. Of this level, a minimum of 88% must be low-grade recovered paper. In addition, neither halogen-containing bleaching chemicals (chlorine, hypochlorite, or chlorine dioxide) nor the chelating agent ethylenediaminetetraacetic acid (EDTA) and optical brighteners may be used in the processing of recycled fibers.

tion are considered, but those from the use of renewable resources for energy production are not considered.

Figure 1 shows the matrix for toilet paper and kitchen paper rolls. Indicated are the ranges for individual parameters and the resulting scores of loading points. For toilet paper, the total of 7.5 loading points may not be exceeded. Likewise, the values designated as "hurdles" for the parameters CO_2, SO_2, COD, AOX, and waste may also not be exceeded. With kitchen paper rolls, the total score that may not be exceeded is 6.5 points.

The criteria for awarding the EU ecological label for copying paper used a different basis with the following objectives:

- Reduction of the release of toxic and other polluting substances into water
- Reduction of the environmental pollution and risks of energy consumption (global warming and acid rain) by reducing energy consumption
- Voluntary undertaking related to responsible management to protect forest resources.

To reduce water pollution, the chemical oxygen demand (COD) of the release from pulp and paper production may not exceed 30 kg/t of paper. The quantity of adsorbable organic halogen compounds (AOX) from pulp and paper production must be less than 0.30 kg/t of paper. The sulfur dioxide discharges into the air may not exceed 1.5 kg S/t of paper. Discharges from electric power generation at the production site must not be considered. The total energy consumption for the entire pulp and paper production process may not exceed 30 GJ/t of pulp or paper. For energy drawn from outside sources, a limit of 18 GJ/t of pulp or paper has been defined.

Companies applying for ecological labeling must present confirmation that the principles of sustainable management are observed in the forests from which the wood comes. If the paper is produced at nonintegrated mills, corresponding statements must be presented by all pulp suppliers.

Parts of the pulp and paper industry have heavily criticized the EU ecological labels developed to date. Industry concerns about the EU schemes vary depending on the country or region concerned. Some common themes are critical[7]:

- Lack of objective or scientific validity in the choice of criteria and definition of loading points
- Use of AOX as a toxicity parameter
- Location-specific nature of environmental concern
- Preference given to recycled fibers over virgin fibers although the latter come from sustainable managed sources
- Criteria and procedures for certifying sustainable forest management.

Against this background, the Confederation of European Paper Industries (CEPI) in Brussels has withdrawn from any further involvement in the development of criteria for other paper grades.

2 Paper recycling and CO_2 balance

2.1 Causes and consequences of the greenhouse effect

Most scientists in the field believe today that heating of the Earth's atmosphere represents a threat to the world climate. Humankind is contributing to global warming by accelerating the natural greenhouse effect. This effect is a result of the fact that the atmosphere retains heat over the Earth's surface. With other trace gases and water vapor in the form of clouds, CO_2 especially ensures that the radiant energy received by the Earth from the sun is not entirely reflected back into space but partially absorbed in the atmosphere as Fig. 2 shows. Water vapor, CO_2, and other trace gases allow visible light to pass largely unhindered but absorb reflection from the Earth's surface toward space in the long-wave infrared region. This energy is then almost entirely returned to Earth as heat radiation. This acts in a similar way to the glass in a greenhouse. Without this effect, the Earth would have an average surface temperature of minus 18°C instead of the actual 15°C and would therefore be uninhabitable.

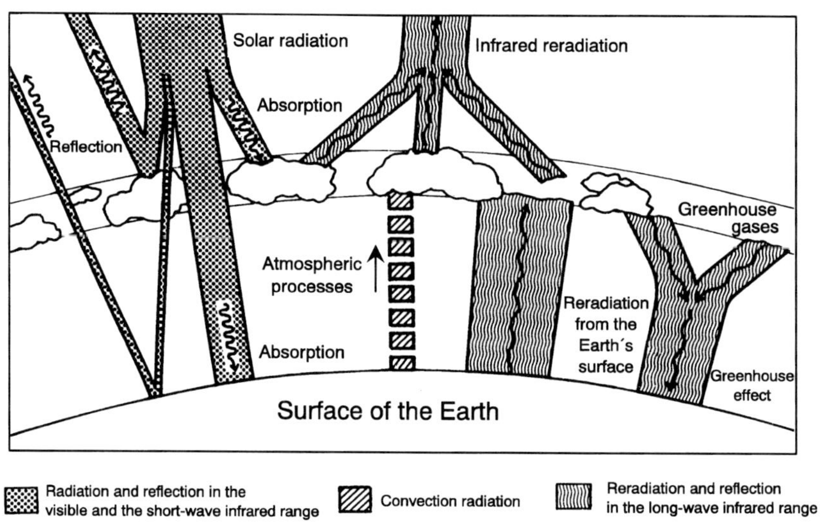

Figure 2. Global energy balance and greenhouse effect[8,9].

In the last 100 years, the global air temperature calculated from the mean value recorded over land and sea regions has increased by 0.3°C–0.6°C. A comparison of the observed global temperature change with model simulations leads to the conclusion that the observed temperature rise is not explainable by natural fluctuation in climatic conditions. Instead, the increase is the anthropogenic intensification of the greenhouse effect. This is the external factor causing the increase. The Intergovernmental Panel on Climate Change (IPCC) has concluded that "the balance of evidence suggests a discernible human influence on global climate" [10].

Since the beginning of the industrial age, the concentration of natural trace gases has increased considerably because of human activities such as combustion of fossil fuels, harvesting and destruction of tropical rain forests, or industrial animal husbandry.

Emissions of halogenated hydrocarbon compounds (fluorochloro-hydrocarbons) have also enriched anthropogenic trace gases in the atmosphere. The contribution of individual trace gases to the anthropogenic greenhouse effect is determined by the quantities in which they are present in the atmosphere and the greenhouse potential of the gases. This greenhouse potential depends on the infrared absorption capacity and retention time of the trace gas concerned in the atmosphere.

Due to its overall prevalence, carbon dioxide contributes 50% to the anthropogenic greenhouse effect. Another 22% is contributed by fluorochloro-hydrocarbons that are also responsible for the hole in the ozone layer although they constitute only 0.4 ppb by volume of the atmosphere. The third most important greenhouse gas is methane from natural and anthropogenic sources. It escapes from rice paddies, the stomachs of cattle, rubbish dumps, and extraction of natural gas. Other greenhouse gases are ozone, nitrous oxide, and stratospheric water vapor as Fig. 3 shows.

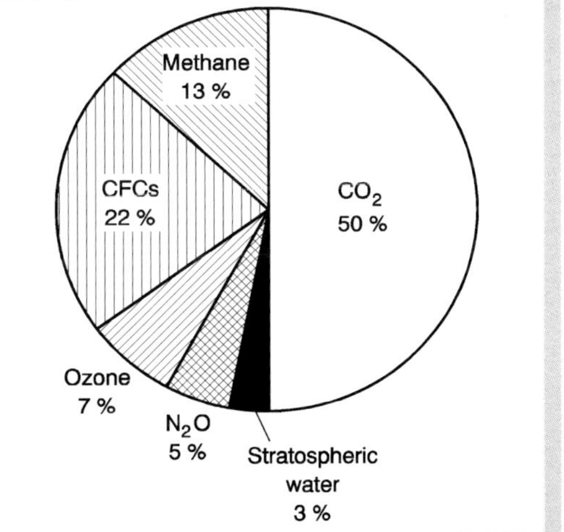

Figure 3. Contribution of various gases to the anthropogenic greenhouse effect[8].

Since the start of industrialization, the level of CO_2 in the atmosphere has risen by one quarter to 358 ppm. Initially, the annual rate of increase was 0.2 ppm. Figure 4 shows that it is now eight times this level. CO_2 emissions couple closely with energy production of which 85%–90% worldwide uses the combustion of fossil fuels such as coal, mineral oil, and natural gas. Figure 5 shows the share of primary energy sources in the global anthropogenic CO_2 emissions. About 40% is from mineral oil, 37% from hard coal and lignite, and 14% from natural gas. The renewable fuel wood has an average share of nearly 10% although this is significantly higher in Third World countries.

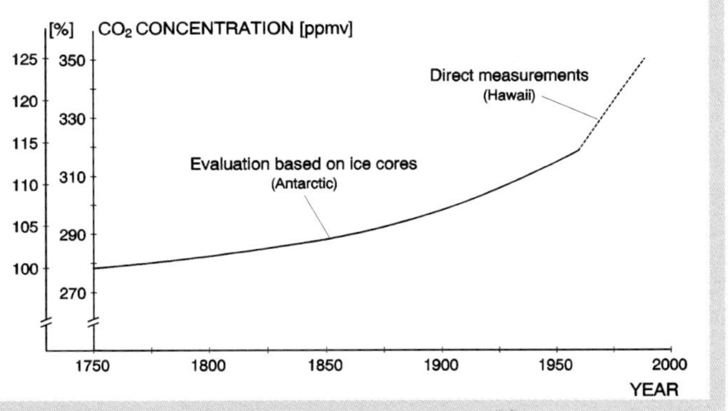

Figure 4. CO_2 concentration in the atmosphere[10].

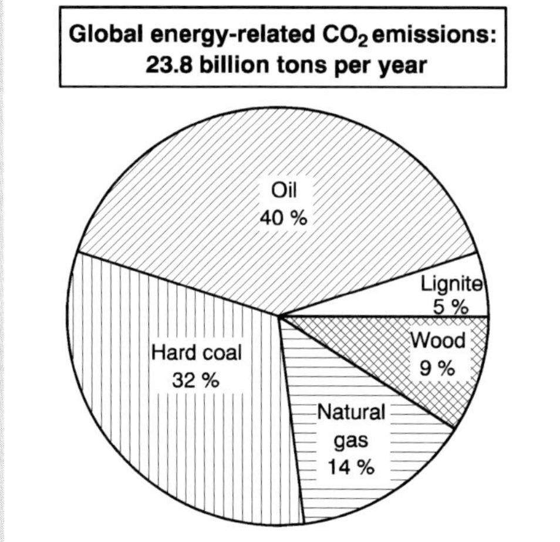

Figure 5. Contribution of primary energy sources to global anthropogenic CO_2 emissions in 1996[11].

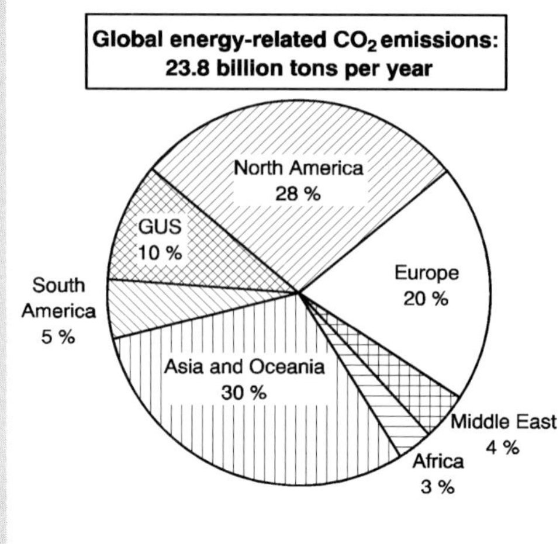

Figure 6. Contribution of different regions to energy-related CO_2 emissions in 1996[11].

The western industrialized countries consume nearly 50% of fossil fuels as Fig. 6 shows. Their share in the worldwide CO_2 emissions is also correspondingly high. In 1996 from a total of 23.8 billion tons, 28% was contributed by North America and 20% by Europe. Asia, and Oceania together had a share of 30%. China at 13% and Japan at 6% accounted for the highest CO_2 emissions.

2.2 Future strategies for counteracting the greenhouse effect

To limit the anthropogenic greenhouse effect, reducing or limiting the emission of greenhouse gases is vitally important. In 1992, the Conference of the United Nations on the Environment and Development in Rio de Janeiro created a basis for global climate protection under international law. 150 states signed a Framework Convention on Climate Change that came into force in March 1994. The objective of this Convention according to Article 2 is "to achieve a stabilization of the greenhouse gas concentrations in the atmosphere at a level where a dangerous anthropogenic disturbance of the climate system is prevented. Such a level should be achieved within a period that allows the ecological systems to naturally adapt to the climatic changes, ensures that food production is not endangered, and provides that economic development can be continued in a sustainable manner." The Framework Convention on Climate Change obliges the industrialized countries to reduce their emissions of CO_2 and other greenhouse gases not regulated by the Montreal Protocol to the level of 1990 by 2000.

After two conferences of the signatory states to the Framework Convention on Climate Change held in Berlin in 1995 and Geneva in 1996, the measures implemented then in a number of industrialized countries would clearly not suffice to meet their obligations under the Framework Convention. The results of the third conference of the signatory states held in Kyoto in 1997 were therefore all the more surprising. Instead of reducing emissions from 2008–2012 by 20%–25% as many scientists and individual countries had demanded, an average value of 5.2% was agreed. The positions of the industrialized countries at the beginning of the conference were too far apart to define a more ambitious reduction objective. The EU countries wanted a 15% reduction, Japan wanted 5%, and the United States wanted no reduction. The European Union and Japan were not prepared for the proposal put forward by the United States to introduce emission certificates.

According to the protocol, the United States will reduce emissions by 7% by 2012 and the EU and Japan have agreed on 8% and 6%, respectively. Russia, the Ukraine, and New Zealand need not make any reductions. In contrast, Australia won the right to increase its climatic gas emissions during the period by 8%.

In June 1998, the Ministers for the Environment of the European Union agreed on the internal allocation of the tasks for the 8% reduction of greenhouse gases. Accordingly, highly industrialized EU states such as Germany and the United Kingdom must reduce their emission by 21% and 12.5%, respectively, as Fig. 7 shows. Due to its high share of CO_2-free nuclear energy, France was exempt from any obligation to implement reductions. Less highly developed members can even increase their emissions, e.g., Greece by 25% and Portugal by 27%.

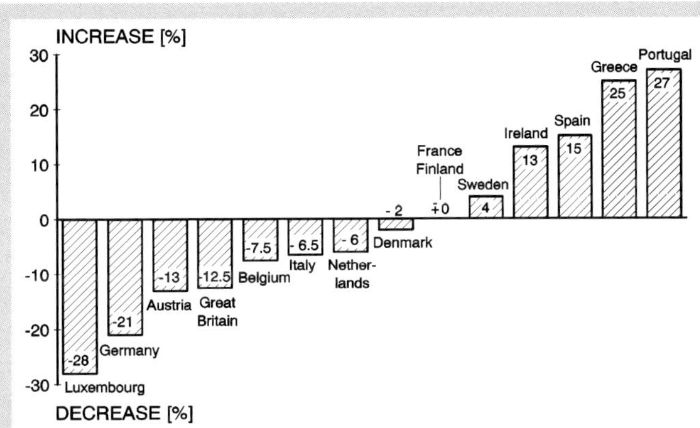

Figure 7. Changes of greenhouse gas emissions in the EU states until 2012 related to the emissions in 1990[12].

In Kyoto, the most important points especially for the United States were postponed until the following conference in 1998 in Buenos Aires. These included the following:

- Joint implementation: This is a decree allowing countries to achieve their reduction objectives by joint projects in the knowledge that reducing the globally effective greenhouse gases may be more economically efficient where this can be done with minimum cost. Compared with the industrialized countries where the level of energy efficiency is high, this would be the case in developing and threshold countries. The necessary principles, models, rules, and guidelines still require negotiation.
- Clean development mechanism: The solidarity principle intends to permit the realization of environmentally friendly measures in developing countries. For this purpose, the industrialized countries shall contribute into a fund.
- Trade in emission certificates: The trade in emission certificates should be allowed only between industrialized nations. Regulations, procedures, limitation rules, etc., for this market that as yet does not exist still must be formulated. Such a rule would allow the United States or the EU states to acquire credits from Russia to reach the agreed objectives in this way.

The results of the Kyoto conference are the subject of heated discussions at environment protection organizations and in industry. Industry thinks the reduction objectives for many countries are too ambitious. In addition, industry considers the target time of 2008–2012 too soon. No major changes in human behavior and technologies will probably occur during this brief period. In addition, the matter of financing requires resolution.

Table 5 compares the energy-related CO_2 emissions of different regions in the year 1996 to those of the year of reference, 1990. The table indicates how difficult reaching the reduction targets by 2012 will be. During the period under consideration, energy-related CO_2 emissions rose additionally in most regions. The only significant reductions occurred in the successor states of the former Soviet Union. The slight reduction of 2.5% in the member states of the European Union is a result of the temporary slackening of economic growth, reorganization of the energy supply and of the economy in the "new" federal German states, closing of coal mines in the United Kingdom, and switching of power plants and heating plants to natural gas.

Table 5. Global energy-related CO_2 emissions[11].

Region	1990 [MT]	1996 [MT]	Change [%]
Europe	4 801	4 679	-2.5
North America	6 214	6 762	+8.8
South America	712	886	+24.4
Asia, Oceania	5 650	7 303	+29.3
CIS (USSR)	3 907	2 430	-37.8
Middle East	728	922	+26.6
Africa	675	795	+17.8
World	22 687	23 777	4.8

In their evaluation of the Kyoto protocol, environment protection organizations primarily oppose the trade in emission certificates in which they see a modern sale of indulgences. If operated skillfully, a counter argument is that such a trade could contribute to a genuine north to south transfer and simultaneously protect the natural environment on which economic and social development in the Third World will also depend.

The Kyoto protocol calls for "palpable progress" by the year 2005. Article 2 lists policies and measures for use to fulfill the obligations. The most important of these are the following:

- Increase energy efficiency in suitable sectors
- Promote a strong agricultural sector
- Promote research into and use of renewable energies and progressive, innovative, and environmentally friendly technologies
- Protect and develop sinks and reservoirs for greenhouse gases.

The recognition and assessment of sinks and stores of these gases such as by additional forestation to absorb CO_2 are matters negotiated at the fourth conference in Buenos Aires in 1998.

This conference concluded by adopting a two-year Plan of Action to reduce the risk of global climate change. The plan established deadlines for finalizing the outstanding details of the Kyoto Protocol so that the agreement will be fully operational when it enters into force sometime after 2000. Besides the mechanism of the protocol, the Action Plan addresses work on compliance issues and policies and measures. The Plan of Action boosts work on transferring climate-friendly technologies to developing countries and addresses the special needs and concerns of countries affected by global warming and the economic implications of response measures.

No decision was made regarding direct human-induced activities of forestation, reforestation, and deforestation resulting in a net sink or a net source of CO_2. Due to many questions raised by the inclusion of sources and sinks, the Intergovernmental Panel on Climate Change (IPCC) was requested to prepare a special report on land-issue and land-use change and forestry. After completion of this report expected in 2000, a decision will be made on the inclusion of sources and sinks in joint implementation and clean development mechanism.

Figure 8 showing global carbon flux demonstrates the importance of forests considering the threat of climate changes caused by CO_2 emissions. This is determined by two cycles:

- A biological cycle characterized by CO_2 absorption with the photosynthesis of plants and CO_2 formation due to the decay of biomass
- A cycle determined by the absorption and desorption of CO_2 in the world's oceans.

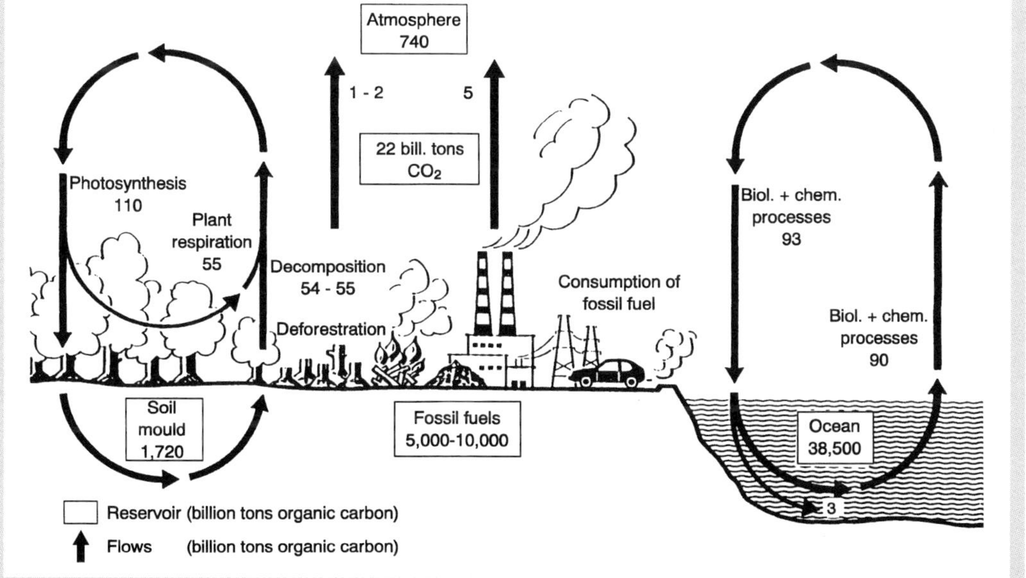

Figure 8. Global flow and reservoirs of organic carbon with figures in billion tons of organic carbon where 1 t of C = 3.7 t of CO_2[8].

Only about 4% of the annual CO_2 emissions have anthropogenic origin. This is primarily from the combustion of fossil fuels and deforestation. These 22 billion tons of CO_2 are responsible for disturbance of the dynamic equilibrium between the carbon stores that are the atmosphere, biosphere, and hydrosphere. The carbon bound into the biosphere includes the living and dead biomass. This is more than in the atmosphere and the close-to-the-surface, exchange-intensive water layers of the oceans combined. Forests are especially important in this respect since they contain 90% of the living and a very high proportion of the dead biomass. The biomass stored in the tropical forests contains approximately as much carbon as the total carbon content of the forests in the moderate and boreal zones.

2.3 CO_2 balance of the pulp and paper industry

The products of the pulp and paper industry consist primarily of wood as a renewable raw material. The CO_2 released by burning, composting, and partially by landfilling of used paper products is reabsorbed via photosynthesis by sustainable managed forests. A closed, CO_2-balanced paper cycle forms. If the life cycle of paper is examined in a cradle to grave analysis, significant additional quantities of process and transport energy are necessary. Except chemical pulp production, these largely come from fossil energy sources as Fig. 9 shows. The CO_2 emissions resulting from the combustion of fossil fuels contribute to the anthropogenic greenhouse effect.

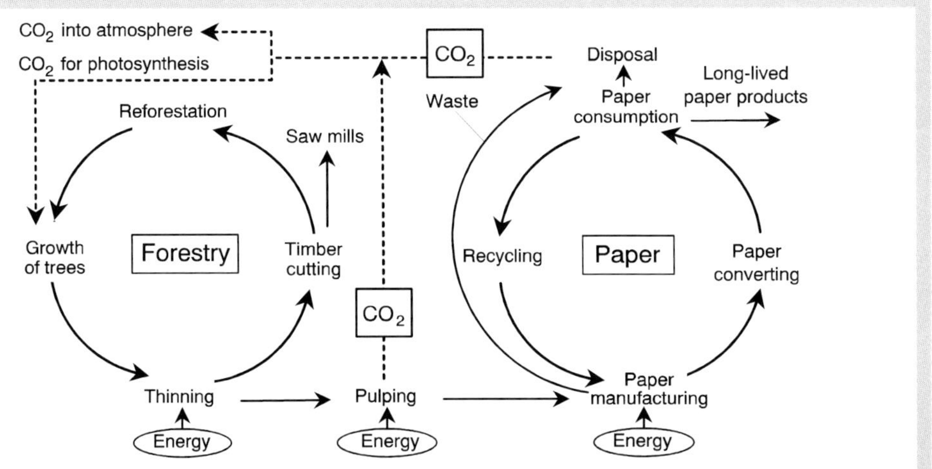

Figure 9. Cycle of organic carbon from wood to paper including its disposal and re-use [8].

An estimate exists for the greenhouse gas emissions of the pulp and paper industry globally[7]. The basis for the estimate includes consumptions of renewable and fossil fuels according to the data of the Organization for Economic Cooperation and Development (OECD) and the International Energy Agency (IEA) for 1991. The CO_2 emissions of the pulp and paper industry resulting from the consumption of fossil fuels were about

290–315 million tons for 1991. This corresponded to a share of 1.3%–1.4% of the anthropogenic CO_2 emissions. Although carbon dioxide is emitted when burning wood and other regenerative fuels, these emissions are not part of the above figures because this is temporary.

Hamm and Göttsching[8,9] calculated the proportion of greenhouse CO_2 emissions for the German paper cycle on the total emissions in Germany at 1.7% for 1990. According to the authors, the same amount of CO_2 in relation to the daily per capita paper consumption is emitted when a 100 W light bulb burns for nine hours or when a person drives 2.5 km in an automobile per day.

The Center for Social and Economic Research on the Global Environment (CSERGE) at the University of East Anglia and University College in London did an assessment of the sources and sinks of greenhouse gases in the global paper cycle[7]. Besides carbon dioxide, the greenhouse gas methane was included in the assessment to the extent that it is emitted during extraction of fossil fuels and landfill storage of used paper products.

In a cradle to grave analysis, the stations in the life cycle of paper shown in Fig. 10 were examined. Starting with the fiber supply and continuing with pulp and paper production, paper converting including printing, recovered paper recycling and re-use in the paper industry, transportation between stages, recovered paper disposal by burning, composting, and landfilling.

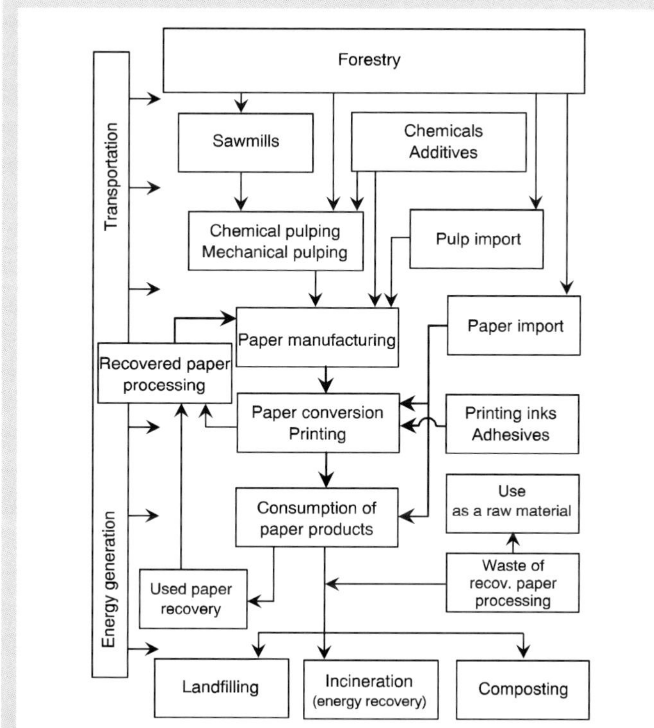

Figure 10. Paper chain from cradle to grave[14].

For carbon balance, virgin fiber supply had three categories:

- Plantations: Harvests from tree plantations established on land that was previously in agriculture or nonforest use
- Regenerating secondary forests: Harvests from natural forests on land that has been forested through much of this century and that could regenerate to its original biomass level

- Old-growth forest removals: Forests converted from high biomass original forest to other forms of vegetative cover such as secondary forests, plantations, or other uses such as agriculture with lower biomass production.

Table 6 shows the results of the CSERGE assessment of the net emissions and storage of greenhouse gases that occur during the paper cycle. Indicated are the annual emissions of CO_2 and methane, the CO_2 equivalence calculated from these values, and the annual carbon emissions considering only the carbon mass in carbon dioxide. This global analysis attributes a significant emission of greenhouse gases to the pulp and paper industry. Production, consumption, and disposal of paper products contributes to an emission of about 700 million tons in carbon dioxide equivalent units each year.

Table 6. Annual emission of greenhouse gases from the global paper cycle in million tons as of 1995[7].

	Greenhouse gas	Annual gas emissions	CO_2 emissions	C emissions	Certainty
Sources					
Energy use	CO_2	287	287	78	high
Energy extraction	CH_4	1	32	9	medium
Energy use recycling	CO_2	4	4	1	medium-low
Transport	CO_2	13	13	4	medium-low
Landfilling	CH_4	12	284	78	medium
Old-growth forest removal	CO_2	74	74	20	low-medium
Natural forest management	CO_2	0	0	0	low
Total sources		–	**694**	**190**	
Sinks					
Waste energy recovery	CO_2	-4	-4	-1	medium
Plantation stable harvest	CO_2	-249	-249	-68	medium-high
Plantation new establishment	CO_2	-21	-21	-6	low-medium
Total sinks		**-274**	**-274**	**-75**	
Net emissions flux			**420**	**115**	

The highest CO_2 emissions result from the extraction and burning of fossil fuels at 319 million tons CO_2 equivalent units. The second most important source in terms of CO_2 equivalents is the landfilling of recovered paper. For the methane emissions of 12 million tons per year, a heat-trapping capability 24 times greater than carbon dioxide was assumed.

The CO_2 emissions given for the recycling of recovered paper relate solely to the additional energy consumption in paper sorting operations at 22–27 kWh/t of recovered paper and to the transport to and from collection banks, conversion facilities, and users.

Excluded is the energy use after the pulp from recycled fibers joins new paper in the repulping, conversion, and distribution phases that is captured in the main energy statistics for the pulp and paper industry.

Managed forests are considered to be CO_2-neutral in a carbon balance. These forests are only carbon storages if the timber is converted into durable wood products such as construction material or furniture. For short-term forest products such as pulp and paper, the potential for net carbon storage is minor. This author thinks the CO_2 released by the burning of wood, production residues, and paper products at different stages in the paper cycle exceeds the cyclical uptake of the regenerating forest.

Old-growth forest removal is treated in the CSERGE study as a CO_2 source. Harvests from old-growth forests bring about many new types of forest including conversion into land used for agricultural purposes or the transformation into plantations or lower-density secondary forests. Annual CO_2 emissions due to forest conversion are 74 million tons. The assumption here is that the average carbon storage capacity after harvesting for pulp fibers accounts for about half the original level.

In the CSERGE study, plantations are CO_2 sinks to the extent that the plantations are in areas that were previously forested. Recent estimates are that this is the case for about 85% of plantations harvested to pulp. The carbon stored consists of the fibers used for pulp and paper production and the biomass and detritus that is not harvestable or recorded in the commercial harvest. Because only a minority of fibers are provided from such plantations (about 30%), only some fiber production may be credited as net uptake in the carbon dioxide emission balance. Nevertheless, plantations appear to be a significant carbon sink in the paper cycle by storing about 270 million tons of carbon dioxide each year.

CSERGE considers that the paper cycle contributes to the risk of global warming. The CO_2 imbalance is offset only by the storage of carbon dioxide in plantations and by the substitution of fossil fuels by wood residue products and nonrecyclable used paper products for energy production. These sinks do not compensate for the greenhouse gas emissions during manufacturing, transport, recycling, disposal in landfills, and deforestation. The annual carbon emissions of the global paper cycle of 115 million tons in 1995 account for about 1.7% of the global anthropogenic carbon emissions including the methane emissions on a carbon dioxide equivalent basis.

Landfills were almost as great a source of greenhouse gas emissions as energy use in manufacturing. For this reason, the final fate of paper has special importance. Alternative waste disposal practices such as material recycling, composting, and burning reduce the emissions from landfills. For greenhouse gas emissions, the burning of nonrecyclable used paper products offers the greatest reduction potential. Only negligible amounts of methane form, and fossil fuels can be saved. A reduction of methane emissions from landfills is also possible by an increased level of paper recycling.

Attempts have been made in life cycle analyses (LCA) to compare the environmental impact of recycling of recovered paper to the impact caused by the incineration (energy recovery) of paper. Several studies conclude that under certain conditions the energetic use of recovered paper produces lower CO_2 emissions than paper recycling.

In contrast, other studies claim ecological advantages for recovered paper recycling. Plätzer and Putz[13] take the example of newsprint produced from thermomechanical pulp or deinked pulp to demonstrate that the following factors have a major influence on CO_2 emissions:

- National or local energy carrier mix for purchased power production
- Share of self-generated power and purchased power for paper production
- Fuels used at paper mills for production of steam and electrical power.

In countries with a high proportion of nuclear power, hydropower, and biomass in their energy carrier mix for purchased power production, the resulting CO_2 emissions are lower than in countries that use primarily coal and mineral oil. Paper mills that use mostly renewable energy sources such as wood residues, bark, or spent liquors emit less greenhouse-active CO_2 than mills that make greater use of fossil energy sources. The considerably lower energy requirements for recovered paper processing compared with mechanical pulping have no effect in relation to greenhouse-active CO_2 discharges if the power used to produce the mechanical pulp comes from biogenious and CO_2-free energy sources, or both.

Considering this, a generally valid answer is not possible for the question whether the use of recovered paper for energy recovery represents the better alternative to paper recycling from an ecological point of view. Critical factors are the nature of the pulp and papermaking processes, the level of technology at all stages of the life cycles, and the energy structure of the countries. Even if the use of recovered paper for energy recovery compared with paper recycling causes a reduction of greenhouse-active CO_2 discharges, other environmentally critical factors require consideration. Plätzer and Putz[13] consider a scenario in which Germany's newsprint needs are covered completely by newsprint manufactured from virgin fibers. The newspapers and magazines collected from the consumers are used for energy recovery in power plants. The authors conclude that the resulting reduction of greenhouse-active CO_2 discharges must be considered against an increase in other air- and water-borne discharges such as CO, dust, VOC, quantity of wastewater, COD, and BOD_5. Even today, an assessment of the different types of environmental pollution requires subjective weighting of the damages caused by the impacts.

In addition, the problem of allocation is largely excluded from the international discussion about LCA methods. In LCA for paper and paper products, this means the allocation of environmental impacts from virgin pulp production to papers containing recycled fibers or to virgin fiber paper grades from recovered paper sorting, recovered paper processing, production waste disposal and the disposal of recovered paper that does not re-enter the loop system due to non-comprehensive collection.

The choice of allocation criteria has a major influence on the result of a LCA. Most LCA calculations to date for paper and paper products consider the primary life cycle and the secondary life cycle separately, i.e., environmental pollution loads are not allocated. Plätzer *et al.*[14] investigated the differences in the environmental impacts between paper grades manufactured from virgin and recycled fibers by applying different allocation criteria. The results related to greenhouse-active CO_2 discharges indicate that establishing the optimum use of recovered paper is not possible as long as a uniform allocation method is not available such as defined in an EN or ISO standard.

3 Heavy metal content of recovered paper

The increasing awareness by the general public about environmental and consumer protection caused a situation in the middle 1980s in which the paper industry was increasingly confronted with the problem of contamination of its products and production waste with harmful substances. Against the background of consumer protection, water protection, waste recycling and disposal, questions concerned the following possibilities:

- Does the content of harmful substances in paper represent a hazard to consumers?
- Do the chemical additives used in paper production result in increased concentrations of harmful substances?
- Does the recycling of recovered paper causes an accumulation of harmful substances in paper?
- Does the recycling of recovered paper and disposal of used paper products represent a hazard to the environmental media of water, soil, and air?

These and other questions were and continue to be asked repeatedly especially to paper mills that process recovered paper. Contamination with harmful substances contained in recovered paper can be due to chemical additives introduced at the preceding paper production and also by substances that come into contact with the paper during paper conversion including printing, paper consumption, and collection of used paper products.

The pulp and paper industry has invested considerable financial resources in the past years to obtain more detailed knowledge about the origin and fate of different potentially harmful substances in the paper production process and in the paper recycling loop. The initial research projects focused on the heavy metal concentrations in paper and paper products after reliable analytical processes had been established in the middle 1970s that the use of printing ink pigments containing lead, chromium, and cadmium and the use of mercury-containing anti-slime agents caused a high concentration of these heavy metals in paper products containing recycled fibers[15–18].

3.1 Origin of heavy metals

An extensive research project concerning the heavy metal content of wood, chemical and mechanical pulps, fillers, coating pigments, and recovered paper was done at the Department of Paper Science and Technology (IfP) in Germany in 1985[19]. Figure 11 shows some results of analyses of recovered paper. The heavy metal contents were determined after pressurized nitric acid decomposition and therefore represent the total content of the corresponding heavy metals in recovered paper. Shown in each case are the maximum and minimum measured values provided they do not lie below the detection threshold. The horizontal links in the bar represent the mean values. Shown also are the detection thresholds for each of the heavy metals concerned as horizontal lines. The figures below these lines indicate by what percentage the measured values lie below the detection threshold. The heavy metal contents use a logarithmic scale. The types of recovered paper examined were old newspapers, magazines, mixed recovered paper, supermarket waste, and OCC II (lower grade of old corrugated containers compared with American OCC).

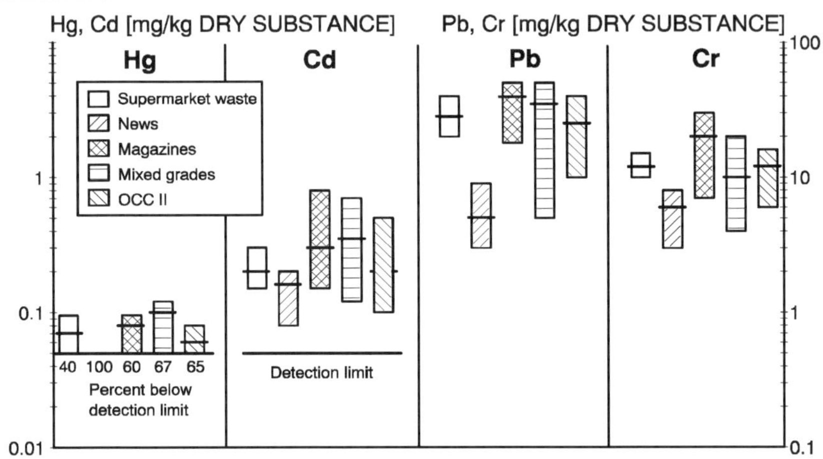

Figure 11. Heavy metal content of different recovered paper grades for 1985[19].

Compared with the situation ten years earlier, the results of the analyses revealed that a considerable reduction in the heavy metal concentration had occurred. The mercury content had fallen significantly. In most samples, the mercury content was below the detection threshold of 50 µg/kg of dry substance. The cadmium contents of the different grades of recovered paper were likewise at a very low level. They were only slightly higher than in the base material wood or the virgin chemical and mechanical pulps.

Attributing the lead and chromium concentrations of the recovered paper to printing ink pigments was not possible. Instead, the results of the analyses indicated that contamination with these heavy metals was due primarily to fillers and pigments used in the preceding paper production. Analyses of fresh fillers and coating pigments confirmed this. Depending on the geographical origin of fillers and pigments, high lead and chromium contents were sometimes measured. The heavy metal contents of these minerals (clay and calcium carbonate) have geological origins and vary from one deposit to another.

Table 7 shows that clays from Germany have a higher lead content than other clays. The chromium content of North American clays is higher than of clays from Europe. Because of the heavy metal content of fillers and coating pigments, one cannot categorically state that paper and paper products manufactured from virgin fibers naturally have a lower lead and chromium content than paper and paper products manufactured from recycled fibers. As Fig. 12 shows, virgin fiber paper (0% RCF) with a high filler content (SC paper) can have greater concentrations of heavy metals than paper manufactured from recycled fibers with a lower filler content (newsprint, 50% RCF).

Table 7. Heavy metal content of clays from different geographical origin[20].

Clay from	Ba [ppm]	Cr [ppm]	Cu [ppm]	Mn [ppm]	Pb [ppm]	Rb [ppm]	Sr [ppm]	Zn [ppm]
Hirschau Germany	700–2 500	13–29	91–226	8–111	250–1 060	25–268	170–363	30–91
Tirschenreuth Germany	270–380	15–20	5–29	7–100	50–80	133–543	34–44	54–82
Zettlitz Czech Republic	220–390	8–12	2–8	18–22	67–198	172–262	81–99	43–66
Schwertberg Austria	220–360	8–14	0–4	–	52–75	115–199	32–45	38–58
Cornwall United Kingdom	180–480	1–9	29–40	16–63	20–40	276–498	139–184	32–118
Bretagne France	150–380	0–37	8–19	3–5	112–135	256–260	26–56	18–46
Burela Spain	55–280	12–23	2–19	29–37	12–32	248–290	0–4	28–119
Georgia USA	110–184	79–159	12–17	0–18	22–61	3	30–53	16–264
South Carolina USA	280–300	78–116	10–14	–	29–46	9–19	92–123	41–70

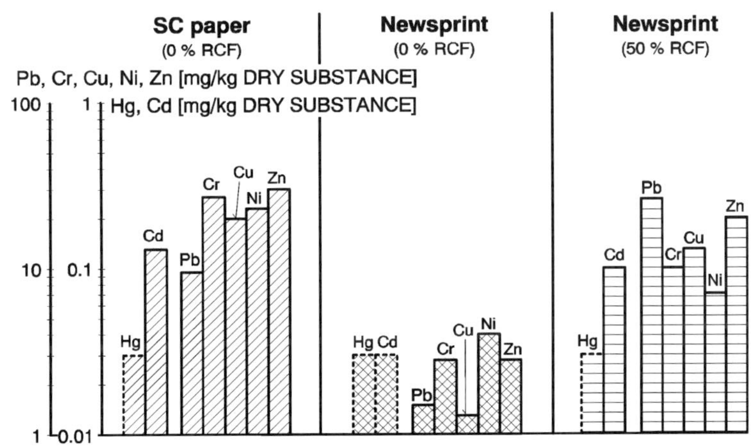

Figure 12. Heavy metal content of different printing papers[19].

Heavy metals in fillers and coating pigments are firmly fixed in a crystal lattice and are therefore almost insoluble in water. The heavy metal concentrations of aqueous paper extracts are in the lower ppb range. The limited solubility has another positive effect. White water contamination by heavy metals is insignificant. Even in recovered paper processing mills, the heavy metal concentrations of effluents discharged are far below the concentration limits set by the German Waste Water Act. They are even below the limits set by the German regulations for drinking water.

Another comprehensive study on heavy metal concentrations of different recovered paper grades was performed by the Papiertechnische Stiftung (PTS) in Munich in 1993[21]. Table 8 shows the composition of the analyzed paper samples. The determination of heavy metal concentrations used the following:

- Cold water extracts (CW extracts)
- Hot water extracts (HW extracts)
- Hydrochloric acid extracts of paper ashes (HCl ash)
- After pressurized nitric acid decomposition (HNO_3 pressure decomposition).

Table 8. Composition of recovered paper samples[21].

Sample	Composition of recovered paper samples
RP-1	News (DIP + virgin pulps)
RP-2	Packaging paper, paper sacks, OCC II
RP-3	Writing and printing papers, office paper
RP-4	Recycled graphic paper, grey board
RP-5	Composite materials (incl. liquid board)
RP-6	Pamphlets, magazines, catalogs

Figures 13–16 give the results of chromium, copper, lead, and nickel analyses as mg/kg oven-dried recovered paper. For all recovered paper samples investigated, the total concentrations of cadmium and mercury obtained by HNO_3 pressure decomposition are very low. The highest total concentrations of chromium, copper, lead, and nickel were analyzed in pamphlets, magazines, and catalogs. Heavy metals in these

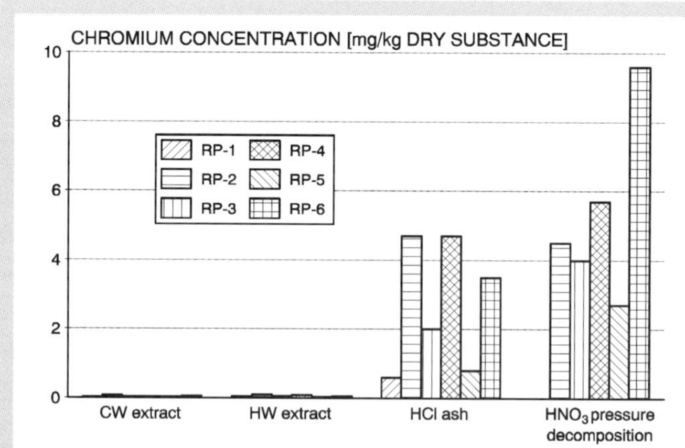

Figure 13. Chromium concentration of different recovered paper grades depending on sample preparation[21].

papers obviously originate primarily from filler and coating pigments that are present to a higher extent than in other paper grades.

According to the findings of Wünschmann *et al.*[22], the copper concentration in these papers is due primarily to blue copper phthalocyanine-based ink pigments. To date, the printing ink industry has offered no copper-free pigments as alternatives that guarantee equivalent degrees of quality and efficiency.

In most cases, the zinc contents of paper containing recycled fibers are higher than in paper produced from virgin fibers. The sources of higher zinc concentrations are not yet completely identified. Possible sources include printing inks (bronze inks), zinc resinates as binder in printing inks, coating recipes from former years, erosion of galvanized piping, and galvanized office and wire staples.

The heavy metal concentration measured in the water extracts of recovered paper are consistently much lower than the analyzed total contents. Only insignificant differences exist between hot and cold water extracts. The low contents of the water extracts are due to the fact that the total contents of the recovered paper or of paper produced from recy-

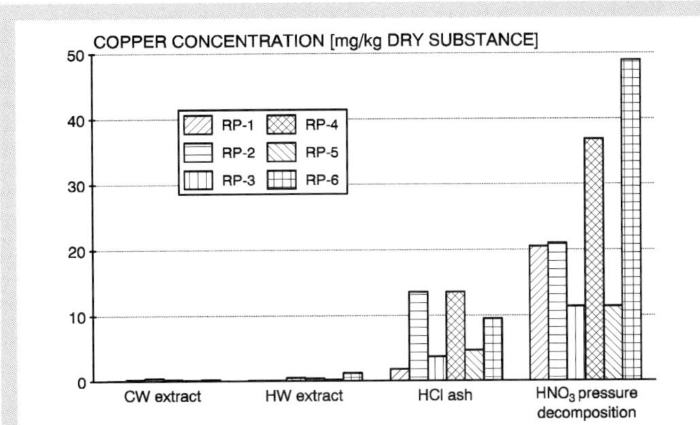

Figure 14. Copper concentration of different recovered paper grades depending on sample preparation[21].

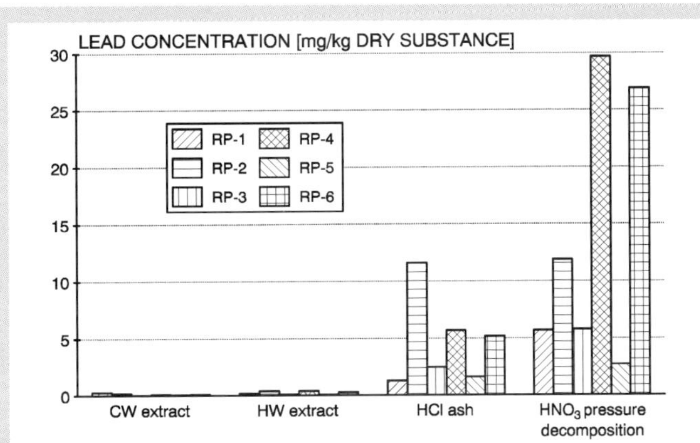

Figure 15. Lead concentration of different recovered paper grades depending on sample preparation[21].

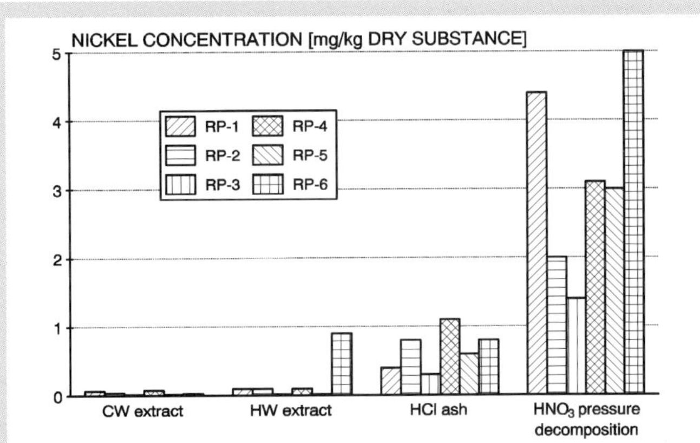

Figure 16. Nickel concentration of different recovered paper grades depending on sample preparation[21].

cled fibers are already at a very low level like the case of mercury and cadmium. Heavy metals such as lead, chromium, and nickel that are introduced primarily by fillers can only be detached under drastic conditions. One consequence of this is that the proportions of heavy metals from used paper products that enter the effluent of rubbish dumps or composting plants constitute only a small share of the total contents. In accordance with these results and against the background of consumer protection, the use of paper and paper products manufactured from recycled fibers does not represent a hazard to consumer health.

Many papers by different authors confirm the results of the aforementioned investigations of heavy metal concentrations of paper products and recovered paper[23–27]. Any comparison of the determined heavy metal contents must consider that different techniques have use for decomposing paper samples. In some cases, these decomposing techniques are not stated. Publications in which the heavy metal contents of water extracts are compared with the total contents are absolutely misleading.

Finally, one can state that the discrimination of paper containing recycled fibers on the grounds of excessive heavy metal contents as is occasionally practiced in ecological marketing is not justified. Instead, modification of chemical additives and printing inks has allowed considerable reduction of the heavy metal concentrations of all paper grades and paper products during the last two decades[28]. The concentrations of toxicologically relevant heavy metals all lie below physiologically critical threshold values.

3.2 Sinks of heavy metals in processing of recovered paper

To obtain knowledge of the amounts of heavy metals introduced via raw materials and additives and to establish with which waste streams heavy metals are removed from the production process, balances in the form of input and output analyses were done at several mills that process recovered paper in Germany. Flotation and wash deinking gave an effective lowering of the (low) heavy metal concentration. In all investigated paper mills, the input of the heavy metals cadmium, chromium, copper, lead, mercury, nickel, and zinc was due almost 100% to the recovered paper used. The heavy metal contents of chemical additives including deinking chemicals and fresh water had almost no importance for the purposes of the balance. For all considered additives, the proportion of the heavy metal input was well below 1%.

In the production of recycled graphic paper, flotation deinking considerably reduces the heavy metal content of the recycled fiber stock[29]. This is true especially for copper where the content is lower by about 95%. The reductions of chromium and lead are about 55% and 10%–15% for cadmium and nickel. This does not mean that the deinked sludges have a high heavy metal concentration. They still have a lower heavy metal content than municipal sewage sludge for example (See Chapter 12).

Another extensive balance was done at a German paper mill where recovered paper is used exclusively as raw material for the production of hygienic paper with annual production of 60 000 tons. At this mill, the recycled fiber pulp is deinked by a two-stage washing process as Fig. 17 shows[30]. Between the two wash deinking stages, the recycled fiber stock is dispersed and treated in a flotation deinking system.

Figure 18 illustrates the total mass balance of that paper mill in terms of oven-dried substance. In total, 30% of the recovered paper input becomes rejects and sludges. The largest volume of sludge comes from a flotation clarifier that is fed with the process water from the two washing stages.

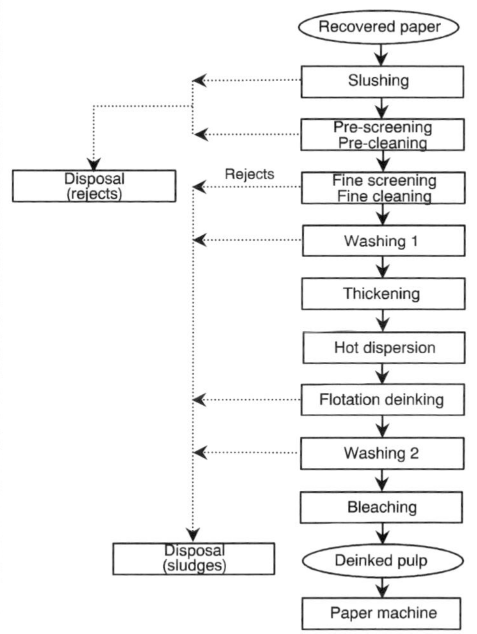

Figure 17. Scheme of the stock preparation of a paper mill producing hygienic paper totally using recycled fibers[30].

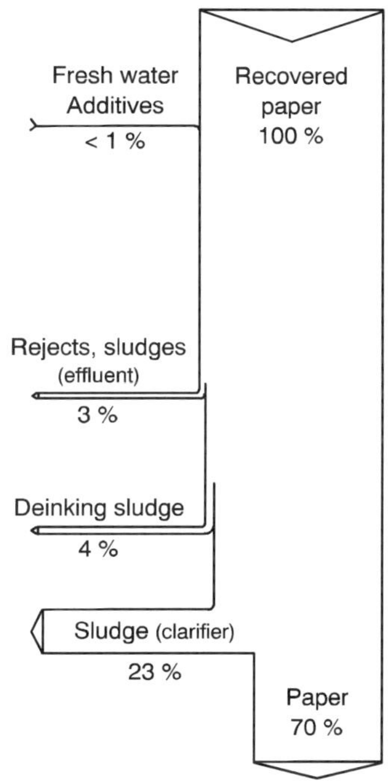

Figure 18. Mass balance of a paper mill producing hygienic paper totally using recycled fibers[30].

Figure 19 shows that 96% of the lead contained in recovered paper is removed from the production process. Nearly 80% is contained in the flotation sludge of the process water clarifier, and only 15% is in the flotation sludge of the flotation deinking system. The removal levels of copper are similarly high as Fig. 20 shows. The total copper content falls by 93%. The proportion of cadmium, chromium, nickel, and zinc removed via the residuals is 70%–80% or about 50% in the case of mercury. The heavy metal removed via the untreated effluent accounts for less than 1% of the input. Another effective reduction of the heavy metal contents of the effluent occurs by adsorption at the biomass of aerobic, anaerobic, or both effluent treatment plants.

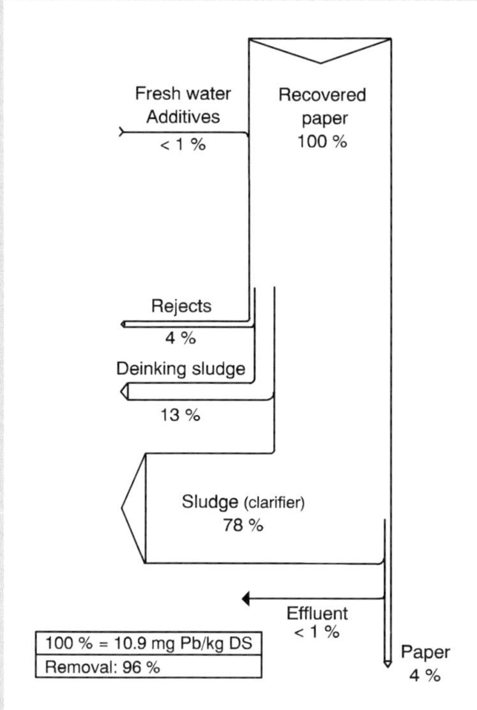

Figure 19. Sankey diagram of total lead load with stock preparation from Fig. 17[30].

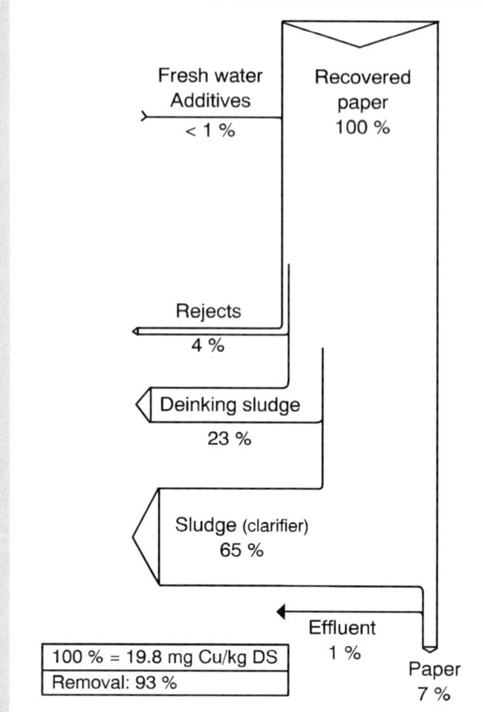

Figure 20. Sankey diagram of total copper load with stock preparation from Fig. 17[30].

Strictly speaking, the described results of the contents' balances apply only to the paper mills investigated with their specific technical factors within recovered paper processing. Using the results of numerous analyses of recovered paper, paper products, and waste from recovered paper processing, the following conclusions generally apply:

- Recovered paper and its filler content are the primary source of the heavy metals, cadmium, chromium, copper, lead, nickel, mercury, and zinc in the production of paper from recovered paper. Fresh water and additives are very minor sources.
- Because recovered paper undergoes extensive processing including deinking, the heavy metal content of the processed recycled fiber stock and the produced paper are considerably lower than of the used recovered paper. Flotation, wash deinking, or both processes applied for removing ink, fiber fines, and fillers for hygienic paper represent extremely effective ways of removing heavy metals.
- The produced deinking sludge is not hazardous waste for its heavy metal content.
- The sinks for heavy metals provided in the deinking processes counteract the accumulation of heavy metal concentrations in the recovered paper loop.

4 Content of chloro-organics in recovered paper

Besides heavy metals, chloro-organic compounds are the most commonly investigated components when assessing environmental hazards resulting from production processes and products. In past years due to the potential human and ecological toxicological hazards of these compounds and a frequently insufficient biological degradability, environmental protection organizations and others called for their minimum release into the environment. This has had consequences for the pulp and paper industry especially in the middle1980s. Polychlorinated dibenzodioxins and dibenzofurans that have a high toxicological potential were then found to occur in waste and effluent from pulp mills and in the pulp and paper produced.

Chloro-organic compounds are present in nearly all paper products. They can originate in the following:

- Chemical pulp bleached with elemental chlorine and chlorine compounds such as chlorine dioxide and hypochlorite
- Chemical additives used for paper production and paper conversion especially epichlorohydrin-based wet strength agents
- Printing ink pigments
- Chlorination of fresh water.

Especially in Northern and continental Europe but also in other parts of the world, the entry of chloro-organic compounds into the paper recycling loop resulting from pulp bleaching with chlorine and chlorine compounds has decreased considerably in the 1990s. It will decrease even more in the future as use of ECF and TCF pulps in paper production occurs. The change of the concentration of chloro-organic compounds in paper and paper products over the years provides impressive evidence that the content of a harmful substance is not a static but a dynamic factor. It is the result of a wide range of different technical influences such as the substitution of chlorine and chlorine compounds in pulp bleaching, integration of the deinking process into recovered paper processing, or the modifying of paper additives containing chloro-organic compounds.

4.1 Origin of chloro-organic compounds

4.1.1 AOX compounds

From an analytical point of view, the presence of chloro-organic compounds in paper and paper products can be quantified simply by the AOX (adsorbable organic halogens) parameter. Bromo- and iodo-organic compounds have little importance in the pulp and paper industry. The measured AOX are therefore due almost exclusively to chloro-organic compounds. AOX encompasses a wide range of groups of chloro-organic compounds such as dioxins, furans, polychlorinated biphenyls, chlorophenols, and chlorinated homologues of the aliphatic carboxylic acids, alkanes, alkenes, catechols, cymenes, guaiacols, naphthalenes, syringols, vanillins, and azo dyes and azo pigments of printing inks. AOX is not a sufficient indicator for the presence of toxicologically relevant substances in paper. As a total parameter, it covers an excessively wide range of chloro-organic compounds varying from harmless, easily degradable to toxic, persistent substances. Since its introduction, AOX has nevertheless proven itself as a "screening parameter" for controlling and preventing chloro-organic substances in effluents.

The analytical methods for determining AOX content depend on the sample matrix. In an aqueous environment and in effluents, the adsorbable portion of chloro-organic compounds is that adsorbed on activated carbon from the aqueous phase. After combustion of the activated carbon, AOX is determined. Only the soluble components are therefore analyzed. Solid samples such as papers or sludges are burned directly with the addition of activated carbon. The total content of chloro-organic compounds (OX content) is analyzed. The detection threshold of this method is 30 mg/kg of dry substance[31].

To date, no legal threshold values in relation to the OX content of paper and paper products exist. In Germany for the agricultural use of municipal sewage sludges, a threshold value of 500 mg/kg dry substance has been stipulated. Compost produced from municipal sewage sludges may not exceed an OX content of 250 mg/kg dry substance.

OX contents measured in papers and paper products depend largely on the fiber stock used for the paper production. In mechanical pulp and totally chlorine-free (TCF) chemical pulp, the OX content is usually below the detection threshold of the analytical method. Elemental chlorine-free (ECF) pulps have OX contents of 100–200 mg/kg dry substance. Chlorine bleached pulps occasionally have OX contents of 1 000 mg/kg dry substance. The range of OX contents of papers using virgin chemical fibers is correspondingly wide. It is from below 30 mg/kg dry substance to about 500 mg/kg dry substance. Note that the trend in the last years is toward a clear reduction in the maximum levels.

Contributing to this situation is the development of low AOX epichlorohydrin-based wet strength agents as Fig. 21 shows[32]. In 1986, the AOX content of polyamidoamine-epichlorohydrin resins was still 0.8% by weight related to commercial product. In 1988, low AOX wet strength agents had become available whose AOX contents were decreased by about 50% by changing the production process. This development continued during the following years. In 1994, an almost "AOX-free" product with an AOX content below 0.01% was available for the first time.

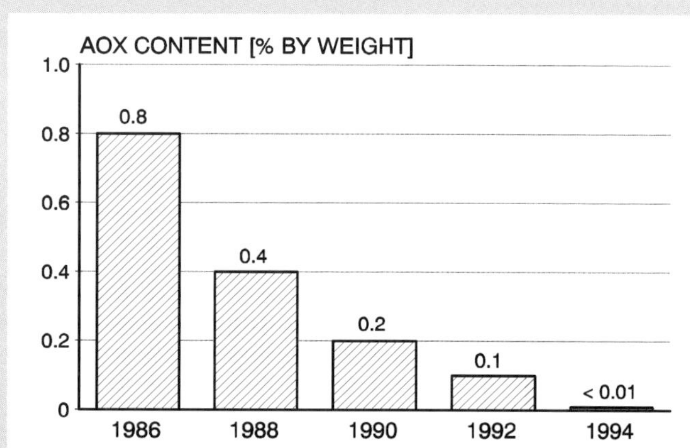

Figure 21. AOX content of polyamidoamine-epichlorohydrin resins calculated as commercial product containing 12.5% solid substance[32].

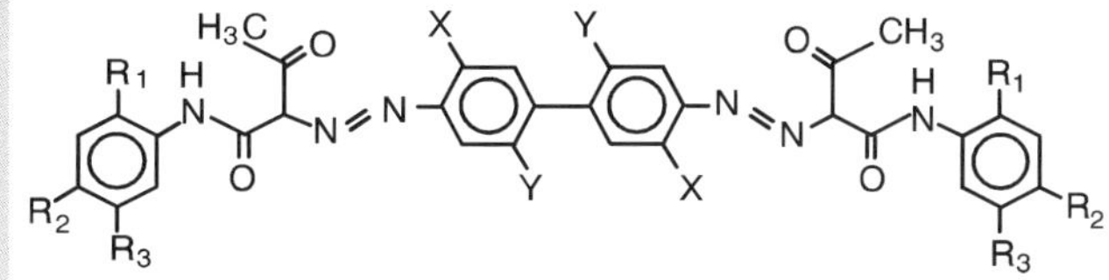

	X	Y	R_1	R_2	R_3
C.I. Pigment yellow 12	Cl	H	H	H	H
C.I. Pigment yellow 13	Cl	H	CH_3	CH_3	H

Figure 22. Structural formula of diaryl yellow pigments[22].

The OX contents of recovered paper are occasionally at a higher level than for paper before its conversion and printing. For paper products such as news, magazines, and catalogs, Wünschmann *et al.*[22] proved that the higher levels were due to the inks with chlorine-containing yellow pigments used in web offset and rotogravure printing. The chlorine-containing pigments concerned are primarily the diaryl yellow pigments PY 12 and PY 13 of Fig. 22 with a chlorine content of about 10%. This corresponds to an OX content of 100 g/kg dry substance. According to ink manufacturers and printers, replacing these diaryl yellow pigments has not been possible for quality and cost reasons. These yellow pigments and the corresponding yellow printing ink are resistent against any aerobic and anaerobic degradation when landfilling or composting. No threat to the soil or ground water therefore exists.

Higher OX contents in packaging material such as corrugated board containers, folding boxes, etc., are due to PVC or PVDC coatings, or both. Because the use of polyvinyl chloride and polyvinylidene chloride in foodstuff packaging is decreasing, lower OX contents in these recovered papers will result.

4.1.2 Dioxins and furans

Dioxin is the frequent name in popular science publications for polychlorinated dibenzodioxins (PCDD). A total of 75 PCDD congeners can contain between one and eight chlorine atoms at the positions in the benzene rings. Closely related chemically with the category of polychlorinated dioxins are polychlorinated dibenzofurans (PCDF) of which 135 isomers exist. They are simply known as furans.

Dioxins and furans are extremely toxic and reveal clear carcinogenic effects in tests using rats, mice, and hamsters. Especially important from the toxicological point of view are dibenzodioxins and dibenzofurans that are chlorine-substituted in positions 2, 3, 7, and 8. With the aid of toxicity equivalency factors (TEF), the higher-chlorinated dibenzodioxins and dibenzofurans are converted to the toxicity of the 2, 3, 7, 8 chlorinated dibenzodioxins and dibenzofurans. The total toxicity equivalency values (TE values) are calculated by multiplying the individual PCDD and PCDF concentrations with the TEF and adding the resulting values. The I-TEF (International Toxicity Equivalency Factors) proposed by NATO/CCMS* today have common international use.

*CCMS: Committee of the Challenge of Modern Society

In 1957, Sandermann showed the possibility of formation of polychlorinated dibenzodioxins and polychlorinated dibenzofurans when bleaching sulfite and sulfate pulp[33]. In the middle 1980s, the first measured results of the dioxin content in chemical pulps, fiber sludges, and papers products were published in the United States and Canada. Since then, numerous tests concerning individual aspects of the pulp and paper industry as sources of dioxin have been carried out[34–38]. Besides analyses of individual finished products, the emphasis in these projects involved investigation of dioxin formation during pulp bleaching.

In Germany, the first measurements were carried out by the then Federal German Health Agency (BGA) in 1987[39,40]. The emphasis in those analyses was on hygienic products and foodstuff packaging papers, coffee filters, and milk cartons. In 1990, BGA published dioxin analyses of newsprint, supercalendered (SC) and lightweight coated (LWC) papers and recycled graphic and hygienic papers with a high recycled fiber content. The dioxin contents of SC and LWC papers produced from virgin fibers were 30–35 ng I-TE/kg paper. The contents in hygienic papers containing recycled fibers were 6–17 ng I-TE/kg paper. In (wood-containing) recycled graphic papers, the values were 6–8ng I-TE/kg paper. The PCDD/F pattern familiar from the chlorine bleaching of pulp dominated in virgin fiber papers and recycled fiber papers.

By 1992, the dioxin content of papers had changed fundamentally. In an investigation on the occurrence and retention of toxic halogen-organic compounds in the pulp and paper industry conducted by the Fraunhofer Institute for Foodstuff Technology and Packaging in Munich[41–43], the positive consequences of the global changes in the pulp bleaching processes that had occurred became evident. For all tested LWC, SC, hygienic papers and newsprint and packaging papers produced with virgin fibers, the dioxin contents were less than 0.5 ng I-TE/kg. The PCDD/F concentrations measured in paper containing recycled fibers were 0.8–11.5 ng I-TE/kg. The lowest contents of 0.8–3.2 ng I-TE/kg were determined in papers produced with deinked pulp such as hygienic papers, newsprint, and recycled graphic paper. The higher PCDD/F concentrations were in corrugated board, test liner, grey board, and white lined chipboard.

The pattern of the tetrachlorinated dibenzodioxins and dibenzofurans considered typical for chlorine bleaching played only a minor role in all papers containing recycled fibers. In another extensive test conducted by the Fraunhofer Institute in 1995, dioxin was detected in recovered paper used to produce paper for corrugated board. Within the framework of this project, the dioxin content compared with 1992 had fallen on average by a factor of three. The average content was 4.3 ng I-TE/kg. Most investigated paper additives including starch-based adhesives, lignin-based additives, cationic starch, dyes, and anti-foaming agents have almost no traces of dioxin. The only relevant dioxin came from use of rosin sizes. Its value averaged 0.16 ng I-TE/kg of paper produced.

4.1.3 Pentachlorophenol

Its biocidal properties make pentachlorophenol (PCP) an extremely effective substance for preventing slime in paper production. PCP has serious toxicological and ecological effects that have motivated several European countries including Austria, The Netherlands, Denmark, and Sweden to prohibit its use in industrial production processes.

In 1989, the German Government banned the production, use, and distribution of PCP and its salts. This ban also covers distribution of products treated with PCP if the PCP concentration is greater than 5 mg/kg. In 1971, i.e., 18 years before the PCP prohibition, the German paper industry announced a unilateral abandonment of the use of PCP as a slimicide. To this extent, one is initially surprised to learn that still today PCP concentrations up to 0.2 mg/kg are detectable in papers produced with recovered paper. In European and non-European countries where the use of PCP is still permitted, PCP evidently continues to be used as an anti-slime agent in paper production. As a consequence of the international trade in paper and recovered paper, a global PCP contamination of recovered paper used for paper production occurs.

The PCP concentrations of paper containing recycled fibers reveal broad fluctuations. Higher contents occur in packaging papers containing recycled fibers. This is undoubtedly also a consequence of the fact that the process water systems are very effectively closed in this product segment. This results in a wider use of anti-slime agents. The PCP contents of papers produced from deinked pulp are considerably lower. With these papers including recycled graphic paper, newsprint, and hygienic paper, the flotation or wash deinking or both processes act as a PCP sink (see Chapter 6). Despite this, the PCP concentrations of the papers concerned are occasionally above the analytical detection threshold.

The possibility of a PCP contamination of paper by PCP impurities in chemical additives used in paper production is only slight. The manufacturers of packaging paper and board from recycled fibers usually work with the same chemical additives as used to produce the corresponding papers from virgin fibers. In the latter papers, the PCP contents are usually all below the analytical detection threshold. A contamination of paper by the use of PCP in forestry is therefore also unlikely. If this source of contamination was relevant, higher PCP contents should also be detectable in virgin fiber papers.

The PCP problem in paper containing recycled fibers could concern at least the European paper industry for some time to come. To date, the toxicologists on the Committee of Experts on Materials coming into Contact with Food of the European Council (Strasbourg, France) have taken the position that no PCP should be detectable in paper intended to come into contact with foodstuff in the interest of preventive consumer protection. Consequently, this makes the methods of detection and their detection threshold especially important. So far no method of analysis has been agreed upon.

Investigations to date on the dioxin contamination of paper containing recycled fibers have clearly shown that a correlation exists between PCP content and dioxin content. A further reduction in dioxin contamination is only possible if PCP can be kept from the paper recycling loop. The example of PCP shows that national prohibitions are insufficient to influence the contamination of secondary base materials when these materials are traded internationally.

4.1.4 Polychlorinated biphenyls

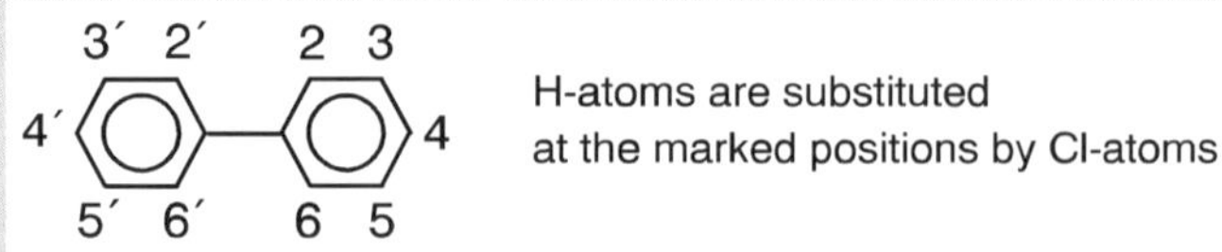

Figure 23. Structural formula of polychlorinated biphenyls.

For polychlorinated biphenyls (PCB), a distinction is made between 209 isomer compounds that result due to substitutions of hydrogen atoms by chlorine atoms in the structural formula shown in Fig. 23. These are toxic and carcinogenic substances that are soluble in fat and have a high bioaccumulation tendency since they are persistent and poorly biologically degradable.

Until the middle 1970s, PCBs with a chlorination level of about 40% were used as solvents for coloring agents in the capsules of carbonless paper. These papers with a retention time of several years entered the recovered paper loop. Until only a few years ago, PCBs were detectable in all papers containing recycled fibers.

In a test in 1983 in Germany by the Fraunhofer Institute for Foodstuff Technology and Packaging in Munich, an average PCB contamination of papers of 4–5 mg/kg was determined from a large number of individual analyses[44]. The test showed unequivocally that PCB contaminations relate to use of recovered paper. No PCB isomers were detectable in paper produced from virgin fibers. Figure 24 shows the results of follow-up tests in 1991, 1995, and 1996. These tests indicate that the PCB contamination of paper and board manufactured from 100% recycled fibers has decreased considerably.

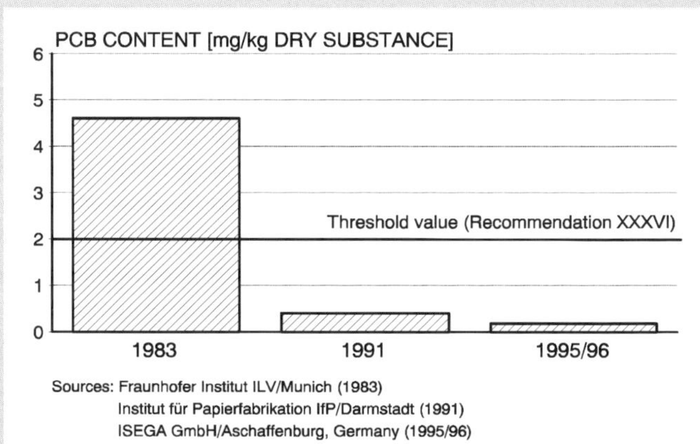

Figure 24. Average PCB content of paper and board made of 100% recycled fibers in Germany.

Today, typical contents of PCB are less than 0.005 to 0.4 mg/kg paper. This is therefore clearly below the threshold of 2 mg/kg stipulated by the German Health Agency in its Recommendation XXXVI. According to the current discussion in the Committee of Experts of the European Council, a threshold is irrelevant for paper intended to come into contact with foodstuff. Because no additional sources of PCB input have been determined to date, i.e., no PCBs can be detected in virgin fiber papers, an additional reduction of the PCB contamination of paper containing recycled fibers is probable. This is due to the fact that PCB has had no use as a solvent in carbonless paper production globally for about 20 years.

4.2 Sinks of chloro-organic compounds in processing recovered paper

In contrast to the input and output balances of heavy metals in paper production using recycled fiber pulp that were established for different product groups at several paper mills, the balances of chloro-organic compounds to date are limited to a small number of paper mills. The examined aspects were the input and output paths of chloro-organic compounds measured as AOX and dioxins and furans in the production of hygienic papers and recycled graphic papers from deinked pulp. These balances revealed results similar to those of the heavy metal balances. Flotation and wash deinking can also provide a significant sink for chloro-organic compounds as explained below. Most important is the fact that chlorine and chlorinated compounds are significantly lower in pulp bleaching and as additives in papermaking.

4.2.1 OX balance

Figure 25 shows OX balances at the tissue paper mill for which Fig. 18 shows the mass balance of recovered paper processing. The results indicate the following:

- Nearly 100% of the chloro-organic compounds entered the production process via the recovered paper used.
- The proportion of the OX input due to the fresh water and the chemical additives used in production including deinking chemicals is considerably below 1%. High OX contaminations were found in some chemicals such as wet strength agents, anti-foaming agents, and anti-slime agents, but the OX content caused by these chemicals was very low due to the small quantities used.
- Almost 70% of the OX input into the production process was removed by deinking processes (both flotation and wash deinking). Accordingly, only 30% of the OX quantities entering with the recovered paper was present in the produced paper.
- The proportion of OX input removed via the effluent was approximately 1%.

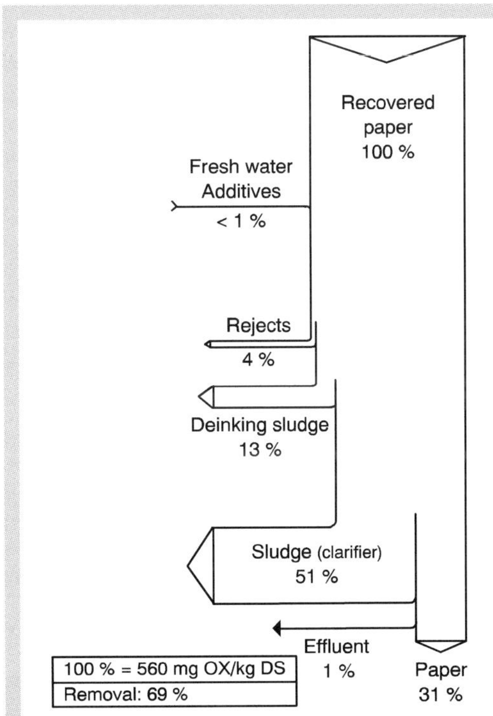

Figure 25. Sankey diagram of chloro-organic load[30].

- A high concentration of the OX compounds occurred in the deinking sludges. This has been confirmed by more recent tests according to which about 80% of the OX contamination of deinking sludges is due to chloro-organic yellow pigments in printing inks.

4.2.2 Dioxin and furan balance

The Fraunhofer Institute for Foodstuff Technology and Packaging in Munich did dioxin and furan balances at several paper mills that process recovered paper[43]. The following results are examples from the production of recycled graphic paper using deinked pulp:

- The primary input (nearly 100%) of dioxins and furans was the recovered paper used.
- Deinking chemicals and other additives did not represent a source of dioxins and furans. At the investigated paper mills, only one colophony-based sizing agent had contamination with dioxins and furans.
- About 75% of the dioxins and furans introduced into the process were removed with the flotation deinking sludge. The dioxin and furan content measured in the deinking sludge was 25 ng I-TE/kg dry substances. The dioxin and furan content in the produced paper was 1.1 ng I-TE/kg dry substance.
- No evidence occurred for an evaporation of dioxins and furans in the dryer section of the paper machine.
- No additional dioxin and furan formation occurred in the entire production process.

References

1. Webb, L., Pulp Paper International 36(11):39(1994).
2. Anon., "Umweltzeichen. Produktanforderungen, Zeichenanwender und Produkte" (RAL Deutsches Institut für Gütesicherung und Kennzeichnung e.V., Ed.), Sankt Augustin, Germany, 1997, 191 pp.
3. Anon., Norwegian Foundation for Environmental Labeling, "Nordic Eco-Labelling, Ecolabelling of Printing Paper", Criteria document Version 1.0, Oslo, 1996, 12 pp.
4. Anon., Commission of the European Communities, "Establishing the Ecological Criteria for the Award of the Community Eco-Label to Toilet Paper", CEC 94/924, Brussels, 1994.
5. Anon., Commission of the European Communities, "Establishing the Ecological Criteria for the Award of the Community Eco-Label to Kitchen Paper Rolls," CEC 94/925, Brussels, 1994.
6. Anon., Commission of the European Communities, "Establishing the Ecological Criteria for the Award of the Community Eco-Label to Copying Paper", CEC 96/467, Brussels, 1996.
7. Anon., "Towards a Sustainable Paper Cycle", International Institute for Environment and Development (IIED), London, 1996, 258 pp.
8. Hamm, U. and Göttsching, L., Das Papier 46(10 A):V53(1992).
9. Hamm, U. and Göttsching, L, Appita J. 46(3):179(1992).
10. Anon., "Daten zur Umwelt, Der Zustand der Umwelt in Deutschland" (Umweltbundesamt, Ed.), Erich Schmidt Verlag, Berlin, 1997, 570 pp.
11. Anon., "Energie Daten '97/'98, Nationale und internationale Entwicklung" (Bundesministerium für Wirtschaft, Ed.), Bonn, 1998, 72 pp.
12. Anon., Stromthemen, Information zu Energie und Umwelt (8):1 (1998) (Informationszentrale der Elektrizitätswirtschaft e.V., Ed.), Frankfurt/Main, Germany, 8 pp.
13. Plätzer, E. and Putz, H.-J., Das Papier 51(6 A):V215(1997).
14. Plätzer, E., Hamm, U., and Göttsching, L., Das Papier 50(10 A):V63(1996).
15. Watkins, D., Corbyons, T., and Bradshaw, J., Analytica Chem. Acta 85(2):403(1976).
16. Heichel, G. H., Hankin, L., and Botsford, R. A., Milk Food Technology (USA) 37(10):499(1976).
17. Croce, A. and Tonini, C., Ind. carta 14(6):259(1976).

18. *Simon, P.J., Giessen, B.C., and Copeland, T.R., Analytica Chem. 49(14):2285(1977).*

19. *Hamm, U., Geller, A., and Göttsching, L., Das Papier 40(10A):V37(1986).*

20. *Grenz, R. and Weigl, J., "Untersuchungen verschiedener Aufschlußverfahren für die Schwermetall-Analytik in der Papierindustrie", PTS-FB 04/89, Papiertechnische Stiftung, Munich, 1989, 70 pp.*

21. *Erhard, K. and Alber, W., "Charakterisierung von minderen Altpapierqualitäten im Hinblick auf ihre stoffliche Verwertung außerhalb der Papierindustrie", PTS-FB 18/94, Papiertechnische Stiftung, Munich, 1994, 109 pp.*

22. *Wünschmann, G., Hamm, U., and Göttsching, L., Wochenbl. Papierfabr. 123(9):395(1995).*

23. *Derra, R., Wochenbl. Papierfabr. 124(5):195(1996).*

24. *Jokinen, K., Siren, K., and Osmonen, R. -M., World Paper (11/12):31(1995).*

25. *Jokinen, K. and Siren, K., Paperi Puu 77(3):106(1995).*

26. *Muvundamina, M., Tappi J. 80(9):129(1997).*

27. *Mjöberg, J., Staffner, S., and Ullmann, P., Paper Technology 34(7/8):26(1993).*

28. *Bremer, H., Das Papier 40(10A):V46(1986).*

29. *Hamm, U. and Göttsching, L., Das Papier 43(10A):V39(1989).*

30. *Hamm, U., "Schadstoffbilanzierung bei der Herstellung von Tissue-Hygienepapier aus Altpapier", in UBA-Report 39/94, Aufbringung von Deinkingschlämmen auf landbaulich genutzte Flächen (Umweltbundesamt, Ed.), Berlin, 1994.*

31. *PTS-Method 012/90, Pulp testing: Determination of the total halogenated organics (OX content), Papiertechnische Stiftung, Munich, 1990.*

32. *Hamm, U., Das Papier 51(6A):V118(1997).*

33. *Sandermann, W., Naturwissenschaften 61:207(1974).*

34. *Axegard, P. and Remberg, L., Chemosphere 19:661(1989).*

35. *Kringstadt, K.P., Johansson, L., Kolar, M.-C., et al., Tappi J. 70(6):163(1989).*

36. *De Sousa, F., Kolar, M.-C., Kringstadt, K.P., et al., Tappi J. 70(4):147(1989).*

37. *Rappe, C., Swanson, S.E., Glas, B., et al., Pulp Paper Can. 90(8):T273(1989).*

38. *Swanson, S.E., Rappe, C., Malmström, J., et al., Chemosphere 17:681(1988).*

39. *Beck, H., Eckart, K., Mathar, W., et al., Chemosphere 17:51(1988).*

40. *Beck, H., Eckart, K., Mathar, W., et al., Chemosphere 16:1997(1987).*

41. *Santl, H., Gruber. L., and Stöhrer, E., Chemosphere 29:1987(1994).*

42. *Santl, H., Gruber, L., and Stöhrer, E., Chemosphere 29:1995(1994).*

43. *Santl, H., Gruber, L., and Stöhrer, E., Chemosphere 28:1633(1994).*

44. *Robinson-Maud, L. and Piringer, O., Das Papier 38(1):1(1984).*

Annex

EUROPEAN LIST OF
STANDARD GRADES OF RECOVERED PAPER AND BOARD

February 1999

This document has been developed jointly by the Confederation of European Paper Industries (CEPI) and the Bureau of International Recycling (BIR) as a revision of the CEPI list dated January 1995.

PREFACE

This list of European standard grades of recovered paper and board gives a general description of the standard grades by defining what they do and do not contain. The list is for use by industry professionals, organizations and individuals with an interest in the recovered paper sector to assist in the buying and selling of this raw material intended for recycling by the paper and board industry. The list also provides help and support for Customs and Excise Officers who are required to classify these raw materials from waste in the context of supranational legislation on the control of waste movement.

It is not the purpose of this list to specify all the qualities of recovered paper and board that exist in the different markets, but rather to define those qualities most commonly traded in Europe. The description of the grades is brief, and for this reason it is recognized that specific deals between buyer and supplier for grades with special specifications will still be necessary to meet individual requirements and will not be excluded by any implied regulations associated with the publication of this list.

Paper and board mills may ask for a declaration from the supplier about the origin of the material, in relation to national regulations or standard requirements. Recovered paper from refuse sorting stations is not suitable for use in the paper industry.

Recovered paper and board originating from multi material collection systems, containing only material of a valuable, recyclable nature, has to be specifically marked. It is not permissible to mix it unmarked with other recovered paper and board.

The list contains a group of recovered paper grades (Group 5 "Special Grades") that, in most cases, can only be recycled using specific processes, or can cause some particular constraints to recycling. Their inclusion in the list is justified by the existence of a significant European market. Actual recycling of the specific grades can only be done by a limited number of mills located in a few countries only.

DEFINITIONS

Unusable materials

Materials which are unusable in the production of paper and board consist of "non-paper components", and "paper and board detrimental to production". Recovered paper and board should in principle be supplied free of unusable materials, but where for specific grades a certain proportion of unusable materials is agreed between purchaser and supplier, it shall refer solely to the element described as "paper and board detrimental to production".

Non-paper components

These consist of any foreign matter in the recovered paper and board which, during processing, may cause damage to machines or interruptions to production or may reduce the value of the finished product, such as:

- Metal
- Plastic
- Glass
- Textiles
- Wood
- Sand and building materials
- Synthetic materials
- "Synthetic papers".

Paper and board detrimental to production

These are grades of paper and board which have been recovered or treated in such a way that they are, for a basic or standard level of equipment, unsuitable as raw material for the manufacture of paper and board, or are actually damaging, or whose presence makes the whole consignment of paper unusable.

A growing number of mills have, however, adapted treatment plants to handle such grades and the range of papers and boards capable of being recycled is increasing all the time as technology develops. The criteria for defining the percentage of "unusable materials" for these grades will be subject to individual mills' specifications.

Moisture content

Recovered paper and board will, in principle, be supplied with moisture of not more than the naturally occurring level. Where the moisture content is higher than 10% (of air dried weight), the additional weight in excess of 10% may be claimed back – with the method of testing and sampling to be agreed between buyer and seller.

To the numbering system

Recovered paper grades have been numbered in this list according to a numerical code system as follows: x.yy.ww

where: x: group
y: grade
w: subgrade

Group 1 – Ordinary Grades
Group 2 – Medium Grades
Group 3 – High Grades
Group 4 – Kraft Grades
Group 5 – Special Grades

Group 1 – Ordinary Grades

1.01 Mixed paper and board, unsorted, but unusable materials removed

A mixture of various grades of paper and board, without restriction on short fiber content.

1.02 Mixed papers and board (sorted)

A mixture of various qualities of paper and board, containing a maximum of 40% of newspapers and magazines.

1.03 Grey board

Printed and unprinted white lined and unlined grey board or mixed board, free from corrugated material.

1.04 Supermarket corrugated paper and board

Used paper and board packaging, containing a minimum of 70% of corrugated board, the rest being solid board and wrapping papers.

1.05 Old corrugated containers

Used boxes and sheets of corrugated board of various qualities.

1.06 Unsold magazines

Unsold magazines, with or without glue.

1.06.01 Unsold magazines without glue

Unsold magazines without glue.

1.07 Telephone books

New and used telephone books, with unlimited content of pages colored in the mass, with and without glue. Shavings allowed.

1.08 Mixed newspapers and magazines I

A mixture of newspapers and magazines, containing a minimum of 50% of newspapers, with or without glue.

1.09 Mixed newspapers and magazines II

A mixture of newspapers and magazines, containing a minimum of 60% of newspapers, with or without glue.

1.10 Mixed magazines and newspapers

A mixture of magazines and newspapers, containing a minimum of 60% of magazines, with or without glue.

1.11 Sorted graphic paper for deinking

Sorted graphic paper from households, newspapers and magazines, each at a minimum of 40%. The percentage of non-deinkable paper and board should be reduced over time to a maximum level of 1.5%. The actual percentage is to be negotiated between buyer and seller.

Group 2 – Medium Grades

2.01 Newspapers

Newspapers, containing a maximum of 5% of newspapers or advertisements colored in the mass.

2.02 Unsold newspapers

Unsold daily newspapers, free from additional inserts or illustrated material colored in the mass.

2.02.01 Unsold newspapers, no flexographic printing allowed

Unsold daily newspapers, free from additional inserts or illustrated material colored in the mass, strings allowed. No flexographic printed material allowed.

2.03 Lightly printed white shavings

Lightly printed white shavings, mainly mechanical pulp based paper.

2.03.01Lightly printed white shavings without glue

Lightly printed white shavings, mainly mechanical pulp based paper, without glue.

2.04 Heavily printed white shavings

Heavily printed white shavings, mainly mechanical pulp based paper.

2.04.01 Heavily printed white shavings without glue

Heavily printed white shavings, mainly mechanical pulp based paper, without glue.

2.05 Sorted office paper

Sorted office paper.

2.06 Colored letters

Correspondence, in mixed papers colored in the mass, with or without print, of printing or writing paper. Free from carbon paper and hard covers.

2.07 White wood-free books

Books, including misprints of books, without hard covers, mainly of wood-free white paper, black printed only. Containing a maximum of 10% of coated paper.

2.08 Colored wood-free magazines

Coated or uncoated magazines, white or colored in the mass, free from non-flexible covers, bindings, non-dispersible inks and adhesives, poster papers, labels or label trim. May include heavily printed circulars and colored in the mass shavings. Containing a maximum of 10% mechanical pulp based papers.

2.09 Carbonless copy paper

Carbonless copy paper.

2.10 Bleached wood-free PE-coated board

Bleached wood-free PE-coated board from board manufacturers and converters.

2.11 Other PE-coated board

Other PE-coated board. May contain unbleached board and paper from board manufacturers and converters.

2.12 Mechanical pulp based computer print-out

Continuous computer print-out, mechanical pulp based, sorted by colors, may include recycled fibers.

Group 3 – High Grades

3.01 Mixed lightly colored printers shavings

Mixed shavings of printing and writing papers, lightly colored in the mass, containing a minimum of 50% of wood-free paper.

3.02 Mixed lightly colored wood-free printer shavings

Mixed shavings of printing and writing papers lightly colored in the mass, containing a minimum of 90% of wood-free paper.

3.03 Wood-free binders

White wood-free lightly printed shavings with glue, free from paper colored in the mass. May contain a maximum of 10% of mechanical pulp based paper.

3.04 Tear white shavings

White wood-free lightly printed shavings without glue, free from wet strength paper and paper colored in the mass.

3.05 White wood-free letters

Sorted white wood-free writing papers, originating from office records, free from cash books, carbon paper and non-water soluble adhesives.

3.06 White business forms

White wood-free printed business forms.

3.07 White wood-free computer print-out

White wood-free computer print-out, free from carbonless paper and glue.

3.08 Printed bleached sulphate board

Heavily printed sheets of bleached sulphate board, without glue, polycoated or waxed material.

3.09 Lightly printed bleached sulphate board

Lightly printed sheets of bleached sulphate board, without glue, polycoated or waxed material.

3.10 Multi printing

Wood-free, coated, lightly printed, free from wet strength paper or paper colored in the mass.

3.11 White heavily printed multiply board

New cuttings of heavily printed white multiply board, containing wood-free, mechanical or thermo-mechanical pulp plies, but without grey plies.

3.12 White lightly printed multiply board

New cuttings of lightly printed white multiply board, containing wood-free, mechanical or thermo-mechanical pulp plies, but without grey plies.

3.13 White unprinted multiply board

New cuttings of unprinted white multiply board, containing wood-free, mechanical or thermo-mechanical pulp plies, but without grey plies.

3.14 White newsprint

Shavings and sheets of white unprinted newsprint, free from magazine paper.

3.15 White mechanical pulp based coated and uncoated paper

Shavings and sheets of white unprinted coated and uncoated mechanical pulp based paper.

3.15.01 White mechanical pulp based paper containing coated paper

Shavings and sheets of white unprinted mechanical pulp based coated paper.

Group 3 – High Grades

3.16 White wood-free coated paper, without glue

Shavings and sheets of white unprinted wood-free coated paper, without glue.

3.17 White shavings

Shavings and sheets of white unprinted paper, free from newsprint and magazine paper containing a minimum of 60% of wood-free paper. May contain a maximum of 10% of coated paper. Without glue.

3.18 White wood-free shavings

Shavings and sheets of white unprinted wood-free paper. May contain a maximum of 5% of coated paper. Without glue.

3.18.01 White wood-free uncoated shavings

Shavings and sheets of white unprinted wood-free paper, free from coated paper. Without glue.

3.19 Unprinted bleached sulphate board

Unprinted sheets of bleached sulphate board, without glue, polycoated or waxed materials.

Group 4 – Kraft Grades

4.01 New shavings of corrugated board

Shavings of corrugated board, with liners of kraft or testliner.

4.01.01 Unused corrugated kraft

Unused boxes, sheets and shavings of corrugated board, with kraft liners only, the fluting made from chemical or thermo-chemical pulp.

4.01.02 Unused corrugating material

Unused boxes, sheets and shavings of corrugated board, with liners or kraft or testliner.

4.02 Used corrugated kraft I

Used boxes of corrugated board, with kraft liners only, the fluting made from chemical or thermo-chemical pulp.

4.03 Used corrugated kraft II

Used boxes of corrugated board, with liners of kraft or testliners but having at least one liner made of kraft.

4.04 Used kraft sacks

Clean used kraft sacks. Wet strength and non wet strength.

4.04.01 Used kraft sacks with polycoated papers

Clean used kraft sacks. Wet strength and non wet strength. May include polycoated papers.

4.05 Unused kraft sacks

Unused kraft sacks. Wet strength and non wet strength.

4.05.01 Unused kraft sacks with polycoated papers

Unused kraft sacks. Wet strength and non wet strength. May include polycoated papers.

4.06 Used kraft

Used kraft paper and board of a natural or white shade.

4.07 New kraft

Shavings and other new kraft paper and board of a natural shade.

4.08 New carrier kraft

New carrier kraft, may include wet strength paper.

Group 5 – Special Grades

5.01 Mixed recovered paper and board

Unsorted paper and board, separated at source.

5.02 Mixed packaging

A mixture of various qualities of used paper and board packaging, free from newspapers and magazines.

5.03 Liquid board packaging

Used liquid packaging board including used PE-coated liquid packaging board (with or without aluminium content), containing a minimum of 50% by weight of fibers, and the balance being aluminium or coatings.

5.04 Wrapper kraft

Poly-lined, sprayed, or laminated used kraft. Must not contain bitumen or wax coatings.

5.05 Wet labels

Used wet labels from wet strength papers, containing a maximum of 1% glass content, and a maximum of 50% moisture, without other unusable materials.

5.06 Unprinted white wet strength wood-free papers

Unprinted white wet strength wood-free papers.

5.07 Printed white wet strength wood-free papers

Printed white wet strength wood-free papers.

Glossary

Age distribution — The single fibers of any paper made from recovered paper have traveled the recycling loop one or more times. The number of times is the age of the recycled fibers considering their generation. In a specific recovered paper grade consisting of a mixture of fibers of different ages, a mass-weighted age distribution can be statistically calculated. This also applies to paper furnishes that include virgin fibers whose age and generation are zero.

Anionic trash — A collective term for different dissolved and colloidally dispersed (less than 1 μm in size) anionic charged particles or polymers. Titration with cationic polyelectrolyte (cationic demand) can estimate the amount present. Anionic trash can disturb papermaking because of its interaction with cationic process chemicals such as retention aids, but some anionic substances can be beneficial in papermaking.

AOX — Abbreviation for Adsorbable Organic Halogens (X) in aqueous solutions or aqueous extracts from solid matter. The halogens can be Cl, Br, and I with concentration expressed in mg/L and kg/t as appropriate. In the pulp and paper industry, organic chlorine compounds are primarily the contributors to the total AOX. These compounds originate in the bleaching of chemical pulp with chlorine and chlorine dioxide, in the chlorination of fresh water, or the use of some chemical additives for papermaking and converting.

Determination of AOX uses activated carbon as the adsorbent. The activated carbon is burned with the adsorbed organic halogens in a furnace at temperatures above 950°C. The resulting hydrogen halide acids (HCl, HBr, and HI) are detected by coulometry, ion chromatography, or other methods (See also OX).

Biodegradation — Attack by living microorganisms causes degradation of organic material. Biodegradation occurs in aqueous systems (process water and waste water), particularly in effluent treatment systems, in soils, or during composting of organic matter. Depending on the local environment and the species of microorganisms, a distinction occurs between anaerobic and aerobic degradation.

Brown recovered paper grades — Brown recovered paper grades such as mixed paper grades, supermarket waste, and old corrugated containers (OCC) contain large proportions of kraft pulp based packaging paper and board or consist entirely of them. They are not used in deinking processes. Depending on the recovered paper grade, they might also contain an unknown amount of printed products such as old newspapers and magazines. In most cases, brown recovered paper grades are post-consumer grades. Brown recovered paper grades are used in the production of packaging paper and board such as test liner, corrugating medium, or chipboard.

Catalase During some stages of life, the metabolism of aerobic microorganisms produces hydrogen peroxide. This peroxide is toxic to the cell. To avoid spoiling the cell structure, it must be destroyed. A cellular enzyme called catalase is therefore produced during metabolism of aerobic living systems. This enzyme catalyses the breakdown of hydrogen peroxide produced during metabolic reactions to oxygen and water. If catalase is present in recycled pulps and in process water of recovered paper processing, it has a detrimental effect on the decomposition of valuable peroxide applied as a bleaching agent and causes low brightness gains.

Chloro-organics A collective term for organic substances in which the carbon-bonded hydrogen atoms are replaced partially or totally by chlorine atoms. Some are very toxic. Formation of chloro-organics such as chlorinated dibenzodioxins and dibenzofurans was the reason for changing conditions in bleaching of chemical pulp.

Curbside collection Curbside collections of recovered paper are those performed by charitable organizations or other groups voluntarily. By selling the recovered paper to mills, they earn money. The collecting organizations normally collect printed products such as old newspapers and magazines at specific times announced to the public.

Deinkability Deinkability is the extent to which ink is removed from pulp slurries of white recovered paper grades such as old newspapers and magazines in a standard or defined deinking procedure. It is characterized by optical properties such as the brightness of the deinked pulp, the brightness gained by deinking, the (visible) speck content, or the reduction of speck content. A (theoretical) removal of 100% printing ink corresponds to perfect deinkability. Printing paper grade, printing process, printing inks applied, aging of printed products, and the deinking technology in terms of the chemical and hydrodynamic regime influence the deinkability.

Deinked pulp (DIP) Deinked pulp is recycled fiber (RCF) treated in a deinking process with removal of a large proportion of inks and some other contaminants (specks). Recycled fibers can be mechanical (wood-containing DIP) or chemical fibers (wood-free DIP) depending on the recovered paper grade and its composition by paper grades.

Detrimental substances These are dissolved and colloidal substances in a paper stock that interfere with papermaking. Most harmful are the particles that aggregate into larger particles or which deposit themselves on surfaces within the process, e.g., secondary stickies. Sometimes substances that interfere with functioning of process chemicals are also called detrimental substances, e.g., anionic trash.

Dissolved and colloidal (DC) substances These substances are polymers or small particles less than 1 μm in paper stock or papermaking white waters that are dissolved or colloidally dispersed. They can be biological, organic, or inorganic in nature. A water sample containing mainly DC substances can be obtained from stock or white water by centrifugation or other suitable means to remove the suspended solids. In deinked pulps for example, DC substances can include wood components such as hemicelluloses, lignins, lignans, wood resin, colloidally sized fiber fragments, coating ingredients, adhesive particles, and various salts.

External substances

External substances in recovered paper are impurities not bound chemically or physically to the paper or board components. They include sand, glass particles, wood pieces, plastic bottles, screws, or cans. During recovered paper processing, external substances should be removed almost completely by screening and cleaning.

Hornification

An irreversible reduction of the fiber swelling capability in chemical pulps when they are dried and later reslushed. It is the decrease in water retention value (WRV) as a percentage of the original value. Hornification is due to structural changes in the cell wall. The process does not normally occur in mechanical pulp fibers during corresponding treatment.

Internal substances

Internal substances in recovered paper are impurities bound chemically or physically to the surface of the paper or board components or imbedded in the fiber structure of a sheet. They also include chemicals used during the previous papermaking process such as wax or wet strength additives. Most internal substances originate from paper converting and include plastic coatings, plastic laminates, adhesives, or printing inks. Internal substances can be soluble in water and influence the white water loops and the effluent treatment, or they can be insoluble in water. Thermoplastic insoluble internal substances can become tacky during paper manufacturing to create problems as stickies.

Macro stickies

Tacky materials present in recycled pulp slurries with a particle size of more than 100 or 150 μm quantified by laboratory screening with a slotted plate (slot width 100 or 150 μm) are macro stickies. They are removed by slotted screens in recovered paper processing to a high degree. In a broader sense, macro stickies can also be present in furnishes of virgin pulps as pitch from wood resins and white pitch caused by coating binders of papermaking broke.

Micro stickies

Tacky materials present in recycled pulp slurries with a particle size less than 100 or 150 μm are micro stickies. They pass slotted screens in recovered paper processing and therefore enter the paper machine where they might deposit on clothing, rolls, or cylinders. They also remain tacky in the paper web produced if no countermeasures are taken such as some form of passivation.

Nonpaper components

Nonpaper components consist of any foreign matter in the recovered paper that can cause damage during processing to machines, interruptions to production or reduction in the value of the finished product. They include metal, string, glass, textiles, wood, sand, building materials, and synthetic materials.

OX

Abbreviation for Organic Halogens (X) in pulp, paper, and solid waste. In the pulp and paper industry, organic chlorine compounds primarily contribute to the OX level. Contrary to the method used for the determination of the total AOX, the entire pretreated solid sample is burned directly in a furnace for OX determination. Activated carbon is added to promote the burning with oxygen. Analytical methods for detection of the resulting hydrogen halide acids are identical with AOX determination.

Paper and board detrimental to production

Paper and board detrimental to production are such grades in recovered paper treated such that they are unsuitable as raw material for the manufacture of paper and board for a basic or standard level of equipment, are actually damaging, or whose presence makes the entire consignment of recovered paper unusable. The European standard EN 643 gives no specific description of these paper and board grades. In other recovered paper grade lists, a description of paper and board detrimental to production is given. This includes bituminized board, carbon paper, parchment and grease proof papers, wet strength paper, or waxed paper. Especially in the (white) recovered paper grades intended for deinking, (brown) packaging materials and colored papers are unusable recovered paper components.

Paper chain

This new term refers to the total life cycle of paper from "cradle to grave" including pulping, papermaking, printing, and paper converting. It begins with the production of raw materials such as pulpwood harvesting, mineral exploitation, and chemical additive production and finishes with collection and recycling of used paper products and the final disposal of nonrecoverable waste paper by incineration, land filling, or composting.

Primary stickies

Primary stickies are variously sized, intact tacky particles of adhesives such as hot melts or pressure sensitive adhesives, inks, binders, waxes, plastics, wet strength resins, etc., present in the incoming deinked or nondeinked pulp. They are tacky at certain temperatures under pressure. The particle size of primary stickies is reduced in recovered paper processing by slushing, deflaking, dispersing, kneading, or refining. Primary stickies can be removed to a high degree by screening in recovered paper processing if the particle size exceeds approximately 100 µm.

Recovered paper

This term includes all types of used paper and board products recovered and used for the manufacture of paper or board or for other industrial purposes. Formerly known as waste paper, recovered paper can be preconsumer recovered paper obtained during converting operations such as broke or cuttings and post-consumer recovered paper collected from consumers such as households, offices, or supermarkets. When processed into pulp, recovered paper becomes recycled fiber (RCF) or recycled pulp.

Recovered paper utilization rate

This is the ratio of the amount of recovered paper used as furnish for paper and board manufacturing to the total paper and board production in a certain country or region. The utilization rate can also alternatively relate to specific paper or board grades produced such as newsprint in a certain country or region such as the EU.

Recovery rate of paper

This is the ratio of the amount of collected paper used for industrial purposes to the total paper and board consumption in a certain country or region such as the EU. Beside the recovered paper used for paper production this also would include papers which are used for, e.g., insulation fluff, filler in construction material, or energy recovery.

Recyclability

Recyclability applies to paper and board products and their suitability for use in material recycling such as papermaking. Paper and board products suitable for recycling should be designed in a way that disintegration and fiber recovery can be performed in typical plants for processing recovered paper without affecting negatively the white water, effluent, and chemical nature and amount of rejects, paper production and productivity, and paper converting and quality. No international standards exist for testing the recyclability of single paper and board products, and no common definitions exist for minimum requirements on recyclability. As a general rule, all nonpaper components should remain in a solid form and be as large as possible so they can be removed by screening and cleaning during recovered paper processing.

Recycled fiber (RCF)

This is paper or board furnish derived from recovered paper by multistage processing in the paper industry. Besides various fiber categories, it contains other materials used in the previous papermaking or converting operations such as fillers, coating pigments, residual ink, and adhesives (stickies). Recycled fiber can be treated by a deinking process in which case it is called deinked pulp (DIP).

Recycled paper

In general, recycled paper is made from 100% recycled fibers. In Europe the term recycled paper is commonly used for graphic paper grades made from 100% recycled fibers originating from white or nearly white recovered paper grades. Recycled paper can be made from chemical or mechanical recycled fibers. Some recycled paper therefore comprises only graphic or office paper such as copy paper, writing, or printing paper. Other recycled grades such as newsprint, supercalendered paper, or lightweight coated paper use old news or magazines. The most important operations in processing recycled pulp fibers are flotation deinking and partial washing with partial flotation deinking. (The term recycled paper is not identical with recovered paper, q.v.).

Secondary stickies

Secondary stickies are tacky agglomerates of dissolved, colloidally sized, or small suspended particles of adhesives, deinking soaps, waxes, latexes, wood resins, papermaking chemicals, salts, and mineral pigments present in recycled or virgin pulp slurries and white water systems of paper mills. Sometimes, fibers or fiber fragments are included. Formation of secondary stickies is due to sudden changes in the surrounding temperature, chemistry, turbulence, or pressure that can break the colloidal stability or solubility of dissolved and colloidal substances. Depending on the size of the sticky agglomerates, this type of stickie cannot be controlled by screening or other mechanical measures in recovered paper processing (See primary stickies).

Soaps

In deinking technology, sodium, potassium, and ammonium salts of fatty acids are water soluble soaps. Insoluble fatty acid salts (soaps) are formed by reactions with calcium, magnesium, and aluminum ions. The latter soaps play an important role in the performance of flotation deinking but can also be an important component in the formation of secondary stickies.

Stickies

This term covers all tacky material particles present in a recycled pulp slurry. Stickies originate in various ways including different types of adhesives used in paper converting such as book binding, self-adhesive labels, or self-adhesive envelopes. The content of these materials is highest in recovered paper and is stepwise reduced in recovered paper processing by screening, cleaning, and flotation deinking. The size of particles of stickies is diminished by deflaking, refining, dispersing, or kneading.

The presence of stickies in recycled fiber furnishes is a serious problem in papermaking affecting the runnability of the paper web in the paper machine especially in the manufacture of paper of a low basis weight on fast running machines such as newsprint or improved printing paper grades.

Stickies can cause deposition problems in the papermaking process or in paper converting. They cause paper defects because of their tacky nature. The material may be present as entire particles (primary stickies, q.v.). Dissolved or colloidally sized particles generate larger agglomerates: secondary stickies, q.v.

Unusable material

Unusable materials in recovered paper are defined according to the European standard EN 643 as materials unusable in paper and board production. They consist of nonpaper components and paper and board detrimental to production. Recovered paper should be supplied free of these unusable materials. For some specific recovered paper grades, a certain permitted content is specified between paper mills and suppliers.

Waste paper

Waste paper was a term previously used simply to describe used paper and paper products (post-consumer waste) and waste from converting operations (pre-consumer waste). Now, such paper re-used for material recycling is "recovered paper," q.v. Nonrecycled, used papers, and paper products can be a component of garbage disposed of by incineration or landfilling. As a component of biowaste from households, used papers and paper products can also be composted primarily in industrial composting plants.

White recovered paper grades

These grades are printed products and office papers made from graphic paper grades such as newsprint, supercalendered paper, and lightweight coated paper that are wood-containing paper grades and copy papers and other wood-free paper grades. Most grades contain post-consumer recovered paper and some pre-consumer recovered paper.
The grades are predominantly processed by deinking. Packaging paper and board are unsuitable components in white recovered paper grades due to their unbleached fiber content.

Wood-containing DIP

This is deinked pulp (DIP) that contains a large proportion of fibers manufactured by mechanical pulping methods. The paper used for its manufacture, the so-called deinking grade of recovered paper, is normally defined so that it guarantees a quality characteristic such as a given proportion of old news and magazines.

Wood-free DIP

This is deinked pulp (DIP) that consists of bleached fibers made by chemical pulping methods besides traces of mechanical fibers. The source of its furnish, e.g., recovered paper collected from offices, normally guarantees the desired quality characteristic. Sometimes, certain limits are set for the composition of the recovered paper such as at least 90% wood-free paper and chemical fibers, respectively.

Conversion factors

To convert numerical values found in this book in the RECOMMENDED FORM, divide by the indicated number to obtain the values in CUSTOMARY UNITS. This table is an excerpt from TIS 0800-01 "Units of measurement and conversion factors." The complete document containing additional conversion factors and references to appropriate TAPPI Test Methods is available at no charge from TAPPI, Technology Park/Atlanta, P. O. Box 105113, Atlanta GA 30348-5113 (Telephone: +1 770 209-7303, 1-800-332-8686 in the United States, or 1-800-446-9431 in Canada).

Property	To convert values expressed in RECOMMENDED FORM	Divide by	To obtain values expressed In CUSTOMARY UNITS
Area	square centimeters [cm^2] square meters [m^2] square meters [m^2]	6.4516 0.0929030 0.8361274	square inches [in^2] square feet [ft^2] square yards [yd^2]
Burst index	kilopascal sq. meters per gram [kPa • m^2/g]	0.0980665	grams-force per square centimeter per (gram per square meter) [(gf/cm^2)(g/m^2)]
Density	kilograms per cubic meter [kg/m^3] kilograms per cubic meter [kg/m^3]	16.01846 1 000	pounds per cubic foot [lb/ft^3] grams per cubic centimeter [g/cm^3]
Energy	joules [J] joules [J] millijoules [mJ] kilojoules [kJ] megajoules [MJ] megajoules [MJ] kilojoules [kJ] joules [J]	1.35582 9.80665 0.0980665 1.05506 2.68452 3.600 4.1868 1	foot pounds-force [ft • lbf] meter kilogams-force [m • kgf] centimeter grams-force [cm • gf] British thermal units, Int. [Btu] horsepower hours [hp • h] kilowatt hours [kW • h or kWh] kilocalories, Int. Table [kcal] meter newtons [m • N]
Force	newtons [N] newtons [N] newtons [N] millinewtons [mN]	4.44822 0.278014 9.80665 0.01	pounds-force [lbf] ounces-force [ozf] kilograms-force [kgf] dynes [dynes]
Force per unit length	newtons per meter [N/m] kilonewtons per meter [kN/m]	9.80665 0.1751268	grams-force per millimeter [gf/mm] pounds-force per inch [lbf/in]
Length	nanometers [nm] micrometers [μm] millimeters [mm] millimeters [mm] meters [m] kilometers [km]	0.1 1 0.0254 25.4 0.3048 1.609	angstroms [Å] microns mils [mil or 0.001 in] inches [in] feet [ft] miles [mi]
Mass	grams [g] kilograms [kg] metric tons (tonne) [t] (= 1000 kg)	28.3495 0.453592 0.907185	ounces [oz] pounds [lb] tons (= 2000 lb)

Property	To convert values expressed in RECOMMENDED FORM	Divide by	To obtain values expressed In CUSTOMARY UNITS
Mass per unit area	grams per square meter [g/m^2] grams per square meter [g/m^2] grams per square meter [g/m^2] grams per square meter [g/m^2] grams per square meter [g/m^2] grams per square meter [g/m^2]	3.7597 1.4801 1.4061 4.8824 1.6275 1.6275	pounds per ream, 17 x 22 - 500 pounds per ream, 25 x 38 - 500 pounds per ream, 25 x 40 - 500 pounds per 1000 square feet [lb/1 000 ft^2] pounds per 3000 square feet [lb/3 000 ft^2] pounds per ream, 24 x 36 - 500
Mass per unit volume	grams per liter [g/L] kilograms per liter [kg/L] kilograms per cubic meter [kg/m^3] megagrams per cubic meter [Mg/m^3] kilograms per cubic meter	7.48915 0.119826 1 27.6799 16.0184	ounces per gallon [oz/gal] pounds per gallon [lb/gal] grams per liter [g/L] pounds per cubic inch [lb/in^3] pounds per cubic foot [lb/ft^3]
Power	watts [W] watts [W] kilowatts [kW] watts [W]	1.35582 745.700 0.74570 735.499	foot pounds-force per second [ft Clbf/s] horsepower [hp] = 550 foot pounds-force per second horsepower [hp] metric horsepower
Pressure, stress, force per unit area	kilopascals [kPa] Pascals [Pa] kilopascals [kPa] kilopascals [kPa] kilopascals [kPa] kilopascals [kPa] kilopascals [kPa] megapascals [Mpa] Pascals [Pa] Pascals [Pa] kilopascals [kPa]	6.89477 47.8803 2.98898 0.24884 3.38638 3.37685 0.133322 0.101325 98.0665 1 100	pounds-force per square inch [lbf/in^2 or psi] pounds-force per square foot [lbf/ft^2] feet of water (39.2°F) [ft H_2O] inches of water (60°F) [in H_2O] inches of mercury (32°F) [in Hg] inches of mercury (60°F) [in Hg] millimeters of mercury (0°C) [mm Hg] atmospheres [atm] grams-force per square centimeter [gf/cm^2] newtons per square meter [N/m^2] bars [bar]
Speed	meters per second [m/s] millimeters per second [mm/s]	0.30480 5.080	feet per second [ft/s] feet per minute [ft/min or fpm]
Surface tension	millinewtons per meter [mN/m]	1	dynes per centimeter [dyn/cm]
Tear index	millinewton sq. meters per gram [mN • m^2/g]	0.0980665	100 grams-force per (gram per square meter) [100 gf/(g/m^2)]
Tensile index	newton meters per gram [N • m/g]	9.80665	kilometers breaking length [km]
Tensile strength	kilonewtons per meter [kN/m] kilonewtons per meter [kN/m] newtons per meter [N/m] kilonewtons per meter [kN/m] kilonewtons per meter [kN/m] kilonewtons per meter [kN/m] newtons per meter [N/m] newtons per meter [N/m]	0.175127 0.29655 10.945 0.65378 0.39227 0.980665 9.80665 66.6667	pounds-force per inch [lbf/in] pounds-force per 15 millimeter width [lbf/15mm] ounce-force per inch [ozf/in] kilograms-force per 15 millimeter width [kgf/15mm] kilograms-force per 25 millimeter width [kgf/25 mm] kilograms-force per centimeter [kgf/cm] grams-force per millimeter [gf/mm] newtons per 15 millimeter width [N/15 mm]
Thickness or caliper	micrometers [μm] millimeters [mm] millimeters [mm]	25.4 0.0254 25.4	mils [mil] (or points or thousandths of an inch) mils [mil] (or 0.001 in.] inches [in]
Viscosity, dynamic	Pascal seconds [Pa • s] millipascal seconds [mPa • s]	0.1 1	poise [P] centipoise [cP]

Property	To convert values expressed in RECOMMENDED FORM	Divide by	To obtain values expressed In CUSTOMARY UNITS
Volume, fluid	milliliters [mL] liters [L]	29.5735 3.785412	ounces [oz] gallons [gal]
Volume, solid	cubic centimeters [cm^3] cubic meters [m^3] cubic meters [m^3] cubic millimeters [mm^3] cubic centimeters [cm^3] cubic decimeters [dm^3] cubic meters [m^3]	16.38706 0.0283169 0.764555 1 1 1 0.001	cubic inches [in^3] cubic feet [ft^3] cubic yards [yd^3] microliters [µL] milliliters [mL] liters [L] liters [L]

Index